Spectroscopic Properties of Inorganic and Organometallic Compounds

Volume 18

A Specialist Periodical Report

Spectroscopic Properties of Inorganic and Organometallic Compounds

Volume 18

A Review of the Recent Literature Published up to Late 1984

Senior Reporters
G. Davidson, *Department of Chemistry, University of Nottingham*
E. A. V. Ebsworth, F.R.S.E., *Department of Chemistry, University of Edinburgh*

Reporters
S. J. Clark, *City University, London*
S. Cradock, *University of Edinburgh*
K. B. Dillon, *University of Durham*
J. D. Donaldson, *City University, London*
S. M. Grimes, *City University, London*
B. E. Mann, *University of Sheffield*
D. W. H. Rankin, *University of Edinburgh*
H. E. Robertson, *University of Edinburgh*

The Royal Society of Chemistry
Burlington House, London W1V 0BN

ISBN 0-85186-163-6
ISSN 0584-8555

Printed in Great Britain by Henry Ling Ltd., at the Dorset Press, Dorchester, Dorset

Foreword

This is the first of this series of SPR volumes to be produced from camera-ready copy. As a result, there have been some minor changes in the format, but I very much hope that these will not detract from the value of the book to research workers. The coverage is the same as in previous volumes, except that once again the proportion of relevant published work that has been reported is reduced because of the need to keep length and cost down. The future of this series of volumes is very much in the balance; anyone who finds the books useful could help to sustain the series by buying this volume, encouraging other people to do the same, and letting me know his or her opinion. Once more I must thank the reporters who have written so effectively and come to terms with the demands of camera-ready copy; I am most grateful to them for their excellent work.

October 1985 E. A. V. Ebsworth

Foreword

Contents

Conversion Factors

$1\ kJ\ mol^{-1}$	
2.3901×10^{-1}	$kcal\ mol^{-1}$
1.0364×10^{-2}	$eV\ atom^{-1}$
8.3593×10	cm^{-1}
2.5061×10^{6}	MHz

$1\ kcal\ mol^{-1}$	
4.1840	$kJ\ mol^{-1}$
4.3364×10^{-2}	$eV\ atom^{-1}$
3.4976×10^{2}	cm^{-1}
1.0486×10^{7}	MHz

$1\ cm^{-1}$	
1.1963×10^{-2}	$kJ\ mol^{-1}$
2.8592×10^{-3}	$kcal\ mol^{-1}$
1.2399×10^{-4}	$eV\ atom^{-1}$
2.9979×10^{4}	MHz

1 MHz	
3.9903×10^{-7}	$kJ\ mol^{-1}$
9.5370×10^{-8}	$kcal\ mol^{-1}$
4.1357×10^{-9}	$eV\ atom^{-1}$
3.3356×10^{-5}	cm^{-1}

$1\ eV\ atom^{-1}$	
9.6485×10	$kJ\ mol^{-1}$
2.3060×10	$kcal\ mol^{-1}$
8.0655×10^{3}	cm^{-1}
2.4180×10^{8}	MHz

Mössbauer spectra: $E_v(^{57}Fe) = 14.413$ keV

$1\ mm\ s^{-1}$	
4.639×10^{-6}	$kJ\ mol^{-1}$
1.109×10^{-6}	$kcal\ mol^{-1}$
4.808×10^{-8}	$eV\ atom^{-1}$
3.878×10^{-4}	cm^{-1}
1.162×10	MHz

For other Mössbauer nuclides, multiply the above conversion factors by E_v(keV)/14.413

1

Nuclear Magnetic Resonance

BY B. E. MANN

1 Introduction

Following the criteria established in earlier volumes, only books and reviews directly relevant to this chapter are included, and the reader who requires a complete list is referred to the Specialist Periodical Reports 'Nuclear Magnetic Resonance',[1] where a complete list of books and reviews is given. Reviews which are of direct relevance to a section of this Report are included in the beginning of that section rather than here. Papers where only ^{1}H n.m.r. spectroscopy is used are only included when the ^{1}H n.m.r. spectra make a non-routine contribution, but complete coverage of relevant papers is still attempted where nuclei other than protonsare involved. In view of the greater restrictions on space, and the ever growing numbers of publications, many more papers in marginal areas have been omitted. This is especially the case in the sections on solid-state n.m.r. spectroscopy, silicon and phosphorus.

Two books have appeared: 'N.m.r. spectroscopy of organometallic compounds: testing the use of n.m.r. in the chemistry of organic compounds of heavy nontransition metals',[2] and a two-volume book 'N.m.r. of newly accessible nuclei' with volume 1 containing 'Chemical and biochemical applications',[3] while volume 2 contains 'Chemically and biochemically important elements'.[4] These two volumes contain chapters on 'N.M.R. of less common nuclei'[5] and numerous specific chapters which are referenced in the relevant section of this Report.

A number of reviews have appeared including 'Physical methods and techniques. N.m.r. spectroscopy',[6] 'N.m.r. in the eighties: new nuclei, new pulses',[7] 'N.m.r of other nuclei',[8] 'Multinuclear n.m.r studies of naturally occurring nuclei',[9] '^{25}Mg, ^{43}Ca, and ^{113}Cd n.m.r. studies of regulatory calcium binding proteins',[10] 'N.m.r. of the first transition series nuclei (scandium to zinc)',[11] 'Reaction of heteropolymolybdic acid with metal ions',[12] 'Synthetic and spectroscopic studies of haems and haem proteins',[13] 'New potentialities of n.m.r spectroscopy in the study of catalysis',[14] 'N.m.r. spectroscopy of molten salts',[15] 'Experimental access to the molecular and electronic structures of organo-*f*-element complexes by n.m.r. spectroscopy',[16] 'Biological aspects of ^{23}Na imaging',[17] and 'N.m.r. studies of intracellular metal ions in intact cells and tissues'.[18]

A number of papers have been published which are too broadly based to fit into a later section and are included here. Isotopic effects on X-nuclei screening and X-H(D) scalar coupling in XH_4 and XO_4 have been determined for ^{14}N, ^{15}N, ^{27}Al, ^{51}V, ^{53}Cr, ^{69}Ga, ^{71}Ga, and ^{99}Tc.[19] An n.m.r. relaxometer for measuring T_1 and T_2 for ^{1}H, ^{7}Li, and ^{23}Na has been described.[20] The effect of ro-vibration on the chemical shifts of metals has been examined.[21] For d^6 transition-metal ions, ^{59}Co, ^{99}Ru, ^{103}Rh, and ^{195}Pt n.m.r. chemical shifts have been parameterised.[22] Rules have been discussed for predicting arrangements of point charges that give a zero electric field gradient at a point and applied to quadrupolar relaxation.[23] CNDO calculations have been performed on $Cr(CO)_6$, $Fe(CO)_5$, and $Ni(CO)_4$ to determine charge densities, which are consistent with ^{13}C and ^{17}O chemical shifts.[24] ^{15}N Chemical shifts and coupling constants have been determined for terminal N_2 complexes of Mo, W, Re, Fe, Ru, Os, and Rh. The ^{15}N shielding showed periodic variation with the central metal, increasing with increasing atomic number.[25] The n.m.r. chemical shifts of some metalloporphyrins have been calculated.[26] ^{13}C n.m.r spectroscopy has been used to demonstrate intramolecular co-ordination of quinone to some capped metalloporphyrins.[27] $^{1}J(M^{13}C)$ values, taken from the literature for compounds of ^{199}Hg, ^{207}Pb, and ^{119}Sn, increase with increasing hybridisation of the carbon atom attached to the metal. A method of obtaining the valence and non-valence components of the coupling constants was discussed.[28]

2 Stereochemistry

This section is subdivided into ten parts which contain n.m.r. information about Groups IA and IIA and transition-metal complexes presented by Groups according to the Periodic Table. Within each Group, classification is by ligand type.

Complexes of Groups IA and IIA.——Two reviews have appeared: 'Alkali metals'[29] and 'Magnesium-25 and calcium-43'.[30]

The essentially identical $^{1}J(^{13}C,^{1}H)$ for the central carbon atom in the contact and solvent-separated ion pairs of some allyl alkali-metal compounds indicates that this coupling constant is not dependent on out-of-plane distortion at C-2.[31] The solution structure of PhLi has been determined from $\underline{T}_1$ measurements using ^{7}Li/^{6}Li and ^{1}H/^{2}H isotopic substitution methods.[32] For some ^{6}Li and ^{13}C sulphur- and selenium-substituted organolithium derivatives, $^{1}\underline{J}(^{13}C,^{6}Li)$ is observed. $^{1}\underline{J}(^{13}C,^{1}H)$ and $^{1}\underline{J}(^{13}C,^{13}C)$ decrease and $^{1}\underline{J}(^{77}Se,^{13}C)$ increases upon lithiation.[33] The ^{7}Li n.m.r. spectrum of $[Li(\mu\text{-}PR_2)]_2$ is a triplet, while a doublet is observed for $Li(PR_2)$(tmed). The ^{31}P n.m.r. spectra were also reported.[34] ^{7}Li n.m.r. spectroscopy has been used to detect different Li species in human serum.[35] N.m.r. data have also been reported for M^- (M = Na, Rb; ^{23}Na, ^{87}Rb),[36] $(THF)_3Li(\mu\text{-}H)_3BC(SiMe_2Ph)_3$ (^{7}Li, ^{11}B),[37] (tmed)$LiCH_2PMe_2Li$(tmed)CH_2PMe_2 (^{13}C, ^{31}P),[38] PhS(O)^{13}CH^{6}LiF (^{13}C),[39] Li(NHAr)(tmed) (^{7}Li),[40] $[KMgPh(OCH_2CH_2OEt)_2]_{\underline{n}}$ (^{13}C).[41]

The ^{25}Mg n.m.r spectra of Grignard reagents depend on concentration and temperature.[42] The ^{1}H and ^{13}C n.m.r. spectra of Mg(anthracene) indicate maximum interaction with the 9 and 10 positions.[43] ^{1}H n.O.e. and ^{13}C n.m.r measurements have been used to determine the structure of 9-{(THF)$_{\underline{n}}$Mg}-10-(AlR^{1}R^{2}H)-anthracene.[44] Using 500 MHz ^{1}H n.m.r spectroscopy, the signals of chlorophyll $\underline{a}$ have been assigned.[45] ^{1}H $\underline{T}_1$ measurements and ring-current shifts have been used to investigate dimer structure of chlorophyll.[46] The complexation of Mg^{2+} and Cd^{2+} by guanosine nucleotides has been investigated by ^{31}P n.m.r. spectroscopy.[47] Intracellular calcium measurement has been described using ^{19}F n.m.r. signals of fluorine labelled chelators.[48] Prothrombin fragment 1-membrane interactions have been studied using ^{43}Ca n.m.r. spectroscopy.[49] The n.m.r. quadrupole-coupling constants for ^{43}Ca bound to either $\underline{a}$-lactalbumin or trypsin is significantly smaller than for EF-hand Ca^{2+} binding proteins. The ^{113}Cd chemical shifts indicate that only O ligands are involved in metal-ion binding.[50] The conformation of the barium perchlorate complex of valinomycin has been investigated using ^{13}C n.m.r. spectroscopy.[51] N.m.r. data have also been reported for $(Me_3Ge)_2Mg.2DME$ (^{13}C),[52] and $MgCl_2.TiCl_4.4MeCO_2Et$ (^{13}C).[53]

Complexes of Groups IIIA and IVA, the Lanthanides, and Actinides. All coupling constants have been determined for $(\eta^5\text{-}C_5H_5)_2Sc(BH_4)$ using ^{11}B and ^{45}Sc n.m.r. spectroscopy. The exchange process in $(\eta^5\text{-}C_5D_5)_2Zr(BH_4)_2$ has been reinvestigated using ^{2}H n.m.r. spectroscopy.[54] $\{(\eta^5\text{-}C_5H_5)Fe(\eta^5\text{-}C_5H_4)\}_2Yb$ shows $^1\underline{J}(^{171}Yb,^1H)$ of 20 and 36 Hz.[55] The ^{31}P n.m.r. spectra of $M\{(Ph_2PO)_3C\}_3$ (M = Sc, Y, Lu, Pr) have been measured and $^1\underline{J}(^{89}Y,^{31}P)$ determined.[56] The binding of $[UO_2]^{2+}$ to L-histidine, imidazole, and N-methylimidazole in aqueous and MeOH solutions has been studied by ^1H and ^{13}C n.m.r. spectroscopy.[57] Similarly the interaction with pyridoxal has been studied.[58, 59] ^{17}O n.m.r spectroscopy shows three peaks for $[^{16}O{=}U{=}^{17}O]^{2+}$, $[^{17}O{=}U{=}^{17}O]^{2+}$, and $[^{18}O{=}U{=}^{17}O]^{2+}$.[60] The ^{31}P n.m.r. spectrum of $UO_2\{S_2P(OPr^i)_2\}_2$ shows $^4\underline{J}(^{31}P,^{31}P)$ of 9 to 19 Hz, and the mechanism of coupling was discussed.[61] The first observation of ^{235}U n.m.r. spectroscopy has been reported for UF_6: $\underline{g}(^{235}U)$ = 492.6 ± 0.2 rad s^{-1} G^{-1}.[62] N.m.r. data have also been reported for $Sc(RCOCHCOR')_3$ (^{13}C, ^{19}F),[63] $[La\{C_5H_3N(CH{=}NCH_2CH_2N{=}CH)_2C_5H_3N\}]^{3+}$ (^{13}C),[64] $(\eta^5\text{-}C_5Me_5)_2M(\mu\text{-}CH_3)_2Li(tmed)$ (M = La, Lu; ^{13}C),[65] $(\eta^5\text{-}C_5Me_5)Lu\{(CH_2)_2PMe_2\}_2$ (^{13}C, ^{31}P),[66] $(\eta^5\text{-}C_5H_5)Lu(CH_2)_2PBu^t_2$ (^{13}C, ^{31}P),[67] $Lu(\underline{o}\text{-}C_6H_4CH_2NMe_2)_3$ (^{13}C),[68] $(\eta^5\text{-}C_5H_5)_2\overline{LuC_6H_4PPh_2CH_2}$ (^{13}C, ^{31}P),[69] $(\eta^5\text{-}C_5Me_5)_2Yb(NH_3)(THF)$ (^{13}C),[70] $Yb\{N(SiMe_3)_2\}_3$ (^{13}C),[71] $[Ln\{C_5H_3N(CH{=}NCH_2CH_2N{=}CH)_2C_5H_3N\}]^{3+}$ (^{13}C),[72] $[UO_2(NIPA)_2\text{-}EtOH]^{2+}$ (^{15}N, ^{31}P),[73] $[UO_2\{3,5\text{-}Me_2\text{-}1\text{-}(2'\text{-}pyridyl)pyrazole\}]^{2+}$ (^{13}C),[74] $UF_4O.SbF_5.2MeCN$ (^{19}F),[75] $(\eta^5\text{-}C_5Me_5)Th(CH_2)_2CMe_2$ (^{13}C),[76] and $M(CH_2Ph)_4(dmpe)$ (M = Th, U; ^{31}P).[77]

^{13}C and ^{31}P n.m.r. spectra show that $ZrMe_4(dmpe)_2$ is rigid at -60°C.[78] In $(\eta^5\text{-}C_5H_5)_2Zr(CH_2PPh_2)_2Rh(CO)Cl$, $^2\underline{J}(^{31}P,^{31}P)$ = 317 Hz.[79] The ^{47}Ti and ^{49}Ti n.m.r. chemical shifts of a series of titanocene derivatives have an inverse relation to the $Ti(2\underline{p}_{3/2})$ binding energies.[80] The ^{47}Ti and ^{49}Ti n.m.r. spectra of some mono- and bis-cyclopentadienyl compounds have been reported.[81] The ^{91}Zr chemical shifts of organozirconium compounds with π-ligands such as $\eta^8\text{-}C_8H_8$, $\eta^5\text{-}C_5H_5$, $\eta^3\text{-}C_3H_5$, and $\eta^4\text{-}C_4H_6$ involve distinctly different ranges for 16- and 18-electron environments at zirconium.[82] ^1H n.O.e. and ^{13}C n.m.r. spectroscopy have been used to differentiate between isomers of $(\eta^5\text{-}C_5H_5)Zr(\eta^3\text{-}C_3H_5)(\eta^4\text{-}C_4H_6)$.[83] The ^{13}C n.m.r. spectrum of $(\eta^5\text{-}C_5H_5)Zr(\eta^4\text{-}C_4H_6)$ shows a mixture of cisoid and transoid.[84] Ti, Zr, and Hf can be simultaneously determined in solution as heteropoly complexes with molybdophosphate by ^{31}P n.m.r. spectroscopy.[85] N.m.r. data have also been reported for $[Zr\{Mo(CO)_3\text{-}(\eta^5\text{-}C_5H_5)\}_2(NEt_2)_2(NHEt_2)]_2$ (^{13}C),[86] $\overline{ZrOC\{=M(CO)_2\}CH_2CHCHCH_2}\text{-}\eta^3\text{-}(\eta^5\text{-}C_5H_5)_2$ (M = Cr, Mo, W; ^{13}C),[87] $(\eta^5\text{-}C_5H_5)_2Ti_2(\eta^5\text{-}C_5H_4AlHR_2)_2$ (^{13}C),[88] $\{(\eta^5\text{-}C_5Me_5)_2ZrH\}_2O$ (^{17}O),[89] $(\eta^5\text{-}C_5H_5)_2Zr(BH_4)_2$ (^{11}B,

^{13}C),[90] $(\eta^5\text{-}C_5H_5)_2Zr(PMe_3)\text{=}CHOZrH(\eta^5\text{-}C_5Me_5)_2$ (^{13}C),[91] $(\eta^5\text{-}C_5H_5)_2$-$Ti(CH_2)_2AlMe_2$ (^{13}C),[92] $(\eta^5\text{-}C_5H_5)_2Ti(CH_2)_2MMe_2$ (M = Si, Ge, Sn; ^{13}C),[93, 94] $(\eta^5\text{-}C_5H_5)_2TiCl(\mu\text{-}OCCH_2)Ti(\eta^5\text{-}C_5Me_5)Cl_2$ (^{13}C),[95] $(\eta^5\text{-}C_5H_5)M(CHSiMe_3)_2C_6H_4$ (M = Ti, Zr, Hf; ^{13}C),[96] $(\eta^5\text{-}C_5H_5)_2ZrMe$-$(\mu\text{-}OCCH_2)Pt(PMe_3)_2Me$ (^{13}C, ^{31}P),[97] $Bu^t{}_4Zr$ (^{13}C, ^{91}Zr),[98] $(\eta^5\text{-}C_5H_5)_2ZrCH_2PMe_2C(O)CH_2PMe_2$ (^{13}C, ^{31}P),[99] $ZrOCH_2CH_2SiMe_2CH_2$-$(\eta^5\text{-}C_5H_5)_2$ (^{13}C),[100] $Rh_2(\mu\text{-}SBu^t)_2\{\mu_2\text{-}(Ph_2PCH_2)_2Zr(\eta^5\text{-}C_5H_5)_2\}(CO)_2$ (^{13}C, ^{31}P),[101] $(\eta^5\text{-}C_5Me_5)_2ZrOCMe_2CH_2CR\text{=}CHCH_2$ (^{13}C),[102] $(\eta^5\text{-}C_5H_5)_4$-$Zr_2Cl_2\{\mu\text{-}OC(CH_2PPh_2)_2\}$ (^{31}P),[103] $(\eta^5\text{-}C_5H_5)_4Zr_2Cl_2(\mu\text{-}OC\text{=}CHPPh_2)$ (^{31}P),[104] $(\eta^5\text{-}C_5H_5)_4Zr_2(\mu\text{-}Me)(\mu\text{-}OC\text{=}CHR)_2AlMe_2$ (^{13}C),[105] $(\eta^5\text{-}C_5H_5)_2$-$Zr\{Ru(\eta^5\text{-}C_5H_5)(CO)_2\}_2$ (^{13}C),[106] $[C_4Ph_4TiCl_3]^-$ (^{13}C),[107] $(\eta^5\text{-}C_5H_5)_2$-Ti(dmpe) (^{31}P),[108] $(\eta^5\text{-}C_5H_5)_2Ti(OCH\text{=}CMe_2)_2$ (^{13}C),[109] $(\eta^5\text{-}C_5H_5)Ti$-$S_5C_5H_5$ (^{13}C),[110] $(RC_{11}H_8N_2O_2)_4Ti_2(\mu,\eta^5\text{-}C_5H_4)_2$ (^{13}C),[111] $(\eta^5\text{-}C_5H_5)$-$ZrX\{OsH_3(PMe_2Ph)_3\}$ (^{31}P),[112] $(\eta^5\text{-}C_5H_5)Rh(CO)_2Zr(\eta^5\text{-}C_5Me_5)_2$ (^{13}C),[113] $\{\eta^5\text{-}C_5H_3(SiMe_3)_2\}_2M(CO)_2$ (M = Zr, Hf; ^{13}C),[114] $(\eta^5\text{-}C_5H_5)_2Zr(ER)_2$ (E = S, Se; ^{13}C, ^{77}Se),[115] $(\eta^5\text{-}C_5H_4R^1)_2Zr$-$Se_2C_6H_3R^2$ (^{77}Se),[116] $M(\underline{O}\text{-}2,4,6\text{-}Bu^t{}_3C_6H_2)_2(NMe_2)_2$ (M = Ti, Zr; ^{13}C),[117] $Ti(dppe)Cl_2(SBu^t)_2$ (^{31}P),[118] Zr{1-(2-hydroxyphenylazo)-2-naphtholate}$_2(OH)_2$ (^{13}C),[119] and $Ti\{OC(CF_3)_3\}_4$ (^{19}F).[120]

Complexes of V, Nb, and Ta.—^{51}V n.m.r. spectroscopy is a useful probe for studying the hydridation of $[V(CO)_5]^{3-}$. ^{93}Nb n.m.r. spectra were also recorded for $[HNb(CO)_5]^{2-}$.[121] Even at 120°C, $(\eta^5\text{-}C_5H_5)_2Ta(\eta^2\text{-}C_2H_4)H$ is not fluxional according to ^1H and ^{13}C n.m.r. spectra.[122] For $(\eta^5\text{-}C_5H_5)_2Nb(CO)SnR_nCl_{3-n}$, $\underline{J}(Sn,{}^1H)$ correlates with the electronegativity of the substituents.[123] Shielding of ^{51}V decreases in the order Sn > Ge > Pb for $[(\eta^5\text{-}C_5H_5)V(MPh_3)$-$(CO)_3]^-$ (M = Ge, Sn, Pb) and related compounds.[124] ^{31}P chemical shift data for the complexes $(\eta^5\text{-}C_5H_5)M(CO)_3PR^1R^2{}_2$ have been interpreted by varying $\underline{P}(3\underline{p}_\sigma)$ and $\underline{P}(3\underline{d}_\pi)$ contributions to the bond-order density matrix in the paramagnetic term.[125] N.m.r. data have also been reported for $[HV(CO)_5]^{2-}$ (^{13}C, ^{51}V),[126] $(\eta^5\text{-}C_5Me_5)TaH\{Zn$-$(\eta^5\text{-}C_5H_5)\}_2$ (^{13}C),[127] $(\eta^5\text{-}C_5H_5)_2NbBu(\eta^2\text{-}SCHSR)$ (^{13}C),[128] TaMe-$(\text{=}CH_2)(OAr)_2$ (^{13}C),[129] $(\eta^5\text{-}C_5Me_5)Nb(\eta^2\text{-hexyne})Cl$ (^{13}C),[130] $(\eta^5\text{-}C_5H_4CHR_2)V(CO)_2(dppe)$ (^{51}V),[131] and $(\eta^5\text{-}C_5H_4Me)_2VS_3(N_3C_2O_2Ph)$-$Pt(PPh_3)_2$ (^{13}C, ^{31}P).[132]

The presence of X in the first coordination sphere of $VX(NO)_2L_3$ (X = Cl, Br, I) was determined by the normal halogen dependence of the ^{51}V shielding.[133] $V(NR)Cl_3$ and its adducts display unusually high-frequency ^{51}V chemical shifts. ^{51}V line widths in this series are typically > 500 Hz.[134] The feasibility of n.m.r. of various nuclei during the study of iso- and hetero-poly compounds of V(V), Mo(VI), and W(VI) has been studied.[135] N.m.r. data have also been reported for oxovanadium(V) meso-tetraphenylporphyrin (^{51}V),[136] $[MCl_2(NBu^t)(NHBu^t)(NH_2Bu^t)]_n$ (M = Ta, Nb; ^{13}C),[137] $Nb_2Cl_6(PCy_3)_3$ (^{31}P),[138] oxosulphatovanadates (^{51}V),[139] $[V_2O_3]^{4+}$ (^{51}V),[140] $[(\eta^5-C_5Me_5)Rh(cis-Nb_2W_4O_{19})]^{2-}$ (^{13}C, ^{17}O, ^{183}W),[141] $[(\eta^5-C_5Me_5)-RhSiW_9Nb_3O_{40}]^{5-}$ (^{183}W),[142] and $F_5Ta.OPMe_2(C_6H_4R)$ (^{19}F).[143]

Complexes of Cr, Mo, and W.—^{13}C and ^{31}P n.m.r. spectra show that $MoH_2(PMe_3)_5$ is fluxional, even at 190K.[144] The ^1H n.m.r spectrum of $WHD(CO)_3(PPr^i_3)_2$ shows $^2J(^2H,^1H)$ = 33.5 Hz compared with 43.2 Hz in HD. The ^{31}P n.m.r. spectrum was also recorded.[145] ^{13}C n.m.r. data and CO stretching force constants of 23 anionic and neutral $W(CO)_5L$ have been given. Changes in $\delta(^{13}CO)$ arise mainly from the Q_{AA} and Q_{AB} terms. On the basis of $^1J(^{183}W,^{13}C)$ of the CO group, the trans-influence series is Ph_2C>CO>olefins>$[CN]^-$, $P(OMe)_3$, PR_3 > H^-, $[CH_3]^-$, $[NCS]^-$, $[O_2CR]^-$, py > RNH_2, $[S_2CR]^-$, AsR_3 > SbR_3, $[SH]^-$ > Cl^- > Br^- > I^-.[146] N.m.r. data have also been reported for $[DM(CO)_4L]^-$ (M = Cr, Mo, W; ^2H),[147] $[M_2(\mu-H)(CO)_{10}]^-$ (^{13}C),[148] $(OC)_4M(\mu-PPh_2)(\mu-H)Pt(PR_3)_2$ (M = Cr, Mo, W; ^{13}C, ^{31}P),[149] $MH_6(PR_3)_3$ (M = Mo, W; ^{31}P),[150] $[MoH_2(dmpe)_3]^{2+}$ (^{31}P),[151] $[(\eta^5-C_5H_5)HMo(CO)_2-P(OCMe_2CH_2)_2NH]^+$ (^{31}P),[152] $[MoH(N_2)_2(depe)_2]^+$ (^{15}N, ^{31}P),[153] $[MoH(CN)_7]^{4-}$ (^{13}C),[154] $(\eta^5-C_5H_5)_2Mo_2(CO)_4(\mu-H)(\mu-PhHPC_6H_4PPh)$ (^{31}P),[155] $(\eta^5-C_5H_5)Mo(CO)_2(\mu-H)\{\mu-P(p-tol)_2\}Mn(CO)_4$ (^{13}C, ^{31}P),[156] $(\eta^5-C_5H_5)WH_5(PMe_3)$ (^{31}P),[157] $WH_2(\eta^2-CH_2O)(PMe_3)_4$ (^{13}C, ^{31}P),[158] $[WH_2F(PMe_3)_4(OH_2)]^+$ (^{19}F, ^{31}P),[159] $[WH_2S_3(PMe_2Ph)_3]^{2+}$ (^{31}P),[160] $[(\mu-H)FeW(CO)_9]^-$ (^{13}C),[161] $[Os_3W(\mu-H)(CO)_{14}]^-$ (^{13}C),[162] $WIrH(\mu-PPh_2)_2\{C(OMe)Ph\}(CO)_4(PPh_3)$ (^{13}C, ^{31}P),[163] and $(\eta^5-C_5H_5)(CO)_2W(\mu-H)\{\mu-CH(p-tol)\}Pt(PR_3)_2$ (^{13}C, ^{31}P, ^{195}Pt).[164]

^1H n.O.e. measurements have been used to determine the isomer of $W(\eta^5-C_5H_4CH_2CH_2O_2C)(CO)_2\{P(p-tol)_3\}$. The ^{31}P n.m.r. spectrum was also determined.[165] The ^{13}C and ^{31}P n.m.r. spectra of $[AuW(\mu-CR^1)-(CO)_2(PR^2_3)(\eta^5-C_5H_5)]^+$ shows that the cation dissociates in solution.[166] N.m.r. data have also been reported for $Cr(CH_2ER_2CH_2ER_2)-(CO)_4$ (E = P, As; ^{31}P),[167, 168] $Cr(CH_2PPh_2NRPPh_2)(CO)_4$ (^{31}P),[169] $M(CO)_3(\mu-dppm)_2(\mu-C_2R)Pt(C_2R)$ (M = Cr, Mo, W; ^{31}P),[170] $(\eta^5-C_5H_5)-$

$Cr(NO)(CO)C(OSiMe_3)=CHPMe_3$ (^{31}P),[171] (η^5-C_5H_5)M{C(O)N(SNHEBu$^t_2$)-EBu$^t_2$}$(CO)_2$ (M = Cr, Mo, W; E = P, As; ^{13}C, ^{31}P),[172] M{=CH(p-C_6H_4R)}$(CO)_5$ (M = Cr, W; ^{13}C),[173] Cr{=C(OMe)C_6H_4OMe}$(CO)_5$ (^{13}C),[174] (η^5-C_5H_5)$_2Cr_2(CO)_4(C_4Ph_2)$ (^{13}C),[175] $[Cr(CNR_2)(CO)_5]^+$ (^{13}C),[176] (η^5-C_5H_5)MoMe(CO)dppm (^{31}P),[177] {(η^5-C_5H_5)$M^1(CO)_3CH_2C_6H_4NC$}M^2Cl_2-(PPh_3) (M^1 = Mo, W; M^2 = Pd, Pt; ^{13}C, ^{31}P),[178] (η^5-C_5H_5)Mo$(CO)_2$-(PPh_3)CO_2H (^{13}C),[179] Mo(CH)(OBu^t)$_3$(quinolinato) (^{13}C),[180] Mo(CPh)-$(CO)_2${HB(3,5-$Me_2C_3HN_2$)$_3$} (^{13}C),[181] (η^5-C_5H_5)M$(CO)_3$M(η^5-C_5H_5)-$(CO)_2$=$\overline{CCH_2CH_2CH_2O}$ (M = Mo, W; ^{13}C),[182] (η^5-C_5H_5)$_2Mo_2(CO)_3$(μ-CO)-(μ-OCHMe) (^{13}C),[183] $[W(NPh)Me_3]_2$(μ-η^1,η^1-NH_2NH_2)(μ-η^2,η^2-NHNH) (^{13}C, ^{15}N),[184] (η^5-C_5H_5)$WOs_3(CO)_6$(μ-O){μ_3-CCH_2(p-tol)} (^{13}C),[185] C_6Cl_5HgM(η^5-C_5H_5)$(CO)_2(PPh_3)$ (M = Mo, W; ^{31}P),[186] W{$(CH_2)_2C_6H_4$}$_3$ (^{13}C),[187] W(C_3Me_2Bu^t)$(py)_2Cl_3$ (^{13}C),[188] $W_2O_3(CH_2Bu^t)_6$ (^{13}C),[189] $W_2(OBu^t)_6$(CHCHCPhN) (^{13}C),[190] $\overline{W\{C(CF_3)=C(CF_3)C(O)}$SMe}$(CO)_2$($\eta^5$-$C_5H_5$) ($^{19}F$),[191] ($\eta^5$-$C_5H_5$)W(CH=CHCO$_2$Me)(NO)X ($^{13}C$),[192] (1) ($^{13}C$),[193]

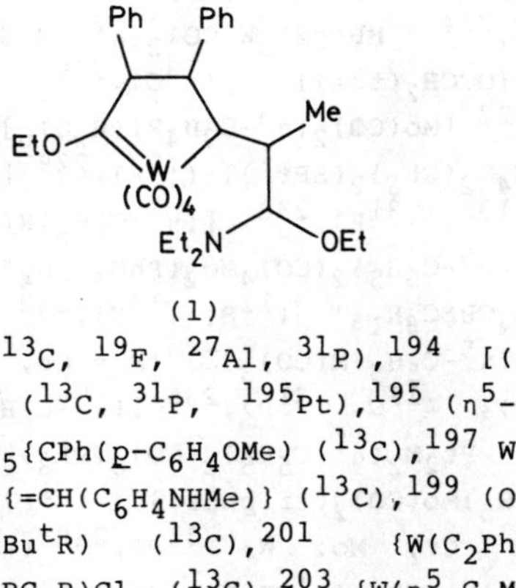

(1)

W(CHAlMe_3)$(PMe_3)_3$Cl (^{13}C, ^{19}F, ^{27}Al, ^{31}P),[194] [(η^5-C_5H_5)(CO)W{CH-(p-tol)}(CO)Pt$(PR_3)_2]^+$ (^{13}C, ^{31}P, ^{195}Pt),[195] (η^5-C_5H_5)W$(CO)_2$Cl(CH-NEt_2) (^{13}C),[196] W$(CO)_5${CPh(p-C_6H_4OMe) (^{13}C),[197] W$(CO)_4$(CNtol)(CPh-OMe) (^{13}C),[198] W$(CO)_5${=CH(C_6H_4NHMe)} (^{13}C),[199] $(OC)_5$W{=C(OR)NCPh$_2$} (^{13}C),[200] $(OC)_4$W(CCHBu^{t}R) (^{13}C),[201] {W(C_2Ph_2)(OBu^t)$_2$}$_2$(μ-N_2) (^{13}C),[202] W(CBu^t)(η^2-RC_2R)Cl_3 (^{13}C),[203] {W(η^5-C_5Me_4Bu^t)(CBu^t)Cl}$_2$-(μ-N_2H_2) (^{13}C),[204] (η^5-C_5H_5)W(CO)(PMe_3){(PMe_3)CHCHCOMe} (^{13}C, ^{31}P),[205] RhW{μ-C(tol)CPhCPh}$(CO)_2$(η^5-C_5H_5)(indenyl) (^{13}C),[206] CoW(μ-CR)$(CO)_3$(η^5-C_5H_5)(η^5-C_5Me_5) (^{13}C, ^{19}F),[207] [(η^5-C_5H_5)(η^2-$R^1C_2R^1$)W(μ-CO){μ-CH(p-tol)}Pt$(PR^2{}_3)_2]^+$ (^{13}C, ^{31}P, ^{195}Pt),[208] $[AuW_2(\mu$-$CR_2)_2(CO)_4$(η^5-C_5H_5)$_2]^+$ (^{13}C),[209] ReW_2(μ-Br)(μ-L){μ-C-(p-tol)}{μ_3-C(p-tol)}$(CO)_3$(η^5-C_5H_5)$_2$ (L = O, CO; ^{13}C),[210] and HB(pz)$_3(OC)_2$W(μ_3-CR){Rh(η^5-indenyl)}$_2$(μ-CO) (^{13}C).[211]

^1H and ^2H n.m.r. spectroscopy has been used to determine the structure of the product from D^- attack on endo- and exo-[(η^5-C_5H_5)Mo(NO)(CO)(η^3-C_8H_{13})]$^+$.[212] In (η^5-C_5H_5)W(η^3-$C_3Bu^t{}_2$Me)(PMe_3)-Cl_2 the two Bu^t groups are inequivalent in both the ^1H and ^{13}C

n.m.r. spectrum.[213] The positions of conformational equilibria in disubstituted arene chromium tricarbonyl complexes have been determined by ^{1}H n.m.r. spectroscopy.[214] The ^{1}H n.m.r. spectrum of (η^{6}-[10]paracyclophane)$_2$Cr has been measured to determine ring current quenching on co-ordination.[215] The ^{13}C chemical shifts of some $Cr(CO)_3$ complexes of arylmethyl and dibenzotropenyl carbanions correlate with rates of α-metallation.[216] ^{1}H n.m.r. studies have indicated that (η^{6}-4-chromanone)$Cr(CO)_3$ has its ring in a sofa conformation.[217] The ^{13}C, ^{14}N, and ^{95}Mo n.m.r. spectra have been determined for the pairs of nitrosyl/thionitrosyl complexes $LM(CO)_2$-(NE) [L = (η^{5}-C_5H_5), $HB(pz)_3$; M = Cr, Mo, W; E = O, S]. The ^{14}N and ^{95}Mo chemical shifts change markedly, while the ^{13}C chemical shifts are insensitive to the replacement of O by S.[218] N.m.r. data have also been reported for $Cr(CO)_5$(η^{2}-cyclooctene) (^{13}C),[219] Ph_3P=CH-(CN)$M(CO)_5$ (M = Cr, Mo, W; ^{31}P),[220] [(η^{6}-C_6H_6)Mo(dmpe)(η^{2}-olefin)]$^{+}$ (^{13}C, ^{31}P),[221] Mo(η^{2}-$COCH_2SiMe_3$)Cl(CO)(PMe_3) (^{13}C, ^{31}P),[222] [(η^{5}-C_5H_5)M(CO)(CN)(RCCO)]$^{+}$ (M = Mo, W; ^{13}C),[223] (η^{5}-C_5H_5)W(NO)L(η^{2}-CHXCHCOMe) (^{13}C),[224] $HB(pz)_3W(CO)_2${η^{2}-CH(SMe)}CF_3SO_3 (^{13}C),[225] (η^{5}-C_5H_5)WCl_2(CO){$OCCH_2$(tol)} (^{13}C),[226] (η^{5}-C_5H_5)$_2W_2Ir_2(CO)_8$-($C_2R^{1}R^{2}$) (^{13}C),[227] [$Mo(CO)_2$(η^{3}-C_3H_4R)(H_2O)$_{\underline{n}}$]$^{+}$ (^{13}C),[228] (η^{5}-C_5H_5)-W(CO)($CNBu^{t}$){η^{3}-$C_4O_2(CF_3)_2(SPr^{i})$} (^{19}F),[229] [(η^{5}-C_5H_5)W(η^{3}-$R^{1}{}_2PCR^{2}$-CO)(CO)(PMe_3)]$^{+}$ (^{13}C, ^{31}P),[230] [(η^{5}-C_5H_5)M(η^{4}-C_4MeR_3)$(CO)_2$]$^{+}$ (M = Mo, W; ^{13}C),[231] (η^{5}-C_5H_5)$_2(CO)_4Mo_2(PhC_4C_6H_4F)$ (^{13}C, ^{19}F),[232] (η^{5}-C_5H_5)$W(CO)_2(MeC_6H_4CHBC_8H_{14})$ (^{11}B, ^{13}C),[233] W(SPr^{i})(η^{4}-C_4F_6)$_2$(η^{5}-C_5H_5) (^{19}F),[234] [(η^{5}-C_5H_5)$M(CO)_3$]$^{-}$ (M = Cr, Mo, W; ^{13}C),[235] P_3N_3-Cl_5Cr(η^{5}-C_5H_5)$(CO)_3$ (^{13}C, ^{31}P),[236] [(η^{5}-C_5H_4CHO)$M(CO)_3$]$^{-}$ (M = Cr, Mo, W; ^{13}C),[237] Pt_2M_2(η^{5}-C_5H_5)$_2(CO)_6(PR_3)_2$ (M = Cr, Mo, W; ^{13}C, ^{31}P),[238] [(η^{5}-C_5H_5)$Mo(CO)_2$(triphos)]$^{+}$ (^{31}P),[239] (η^{5}-C_5H_5)$M(CO)_2$-{=P(OMe_2)$_2$} (M = Cr, Mo, W; ^{31}P),[240] (η^{5}-C_5H_5)$Mo(CO)_2P_3$ (^{13}C, ^{31}P),[241] (η^{5}-C_5H_5)$Mo(CO)_2${η^{1}-P=C($SiMe_3$)$_2$} (^{31}P),[242] (η^{5}-C_5Me_5)-$M(CO)_3PCl_2BH_3$ (M = Mo, W; ^{11}B, ^{31}P),[243] (η^{5}-C_5H_5)$Mo(CO)_2${$Ph_2P(S)$-C(S)NR} (^{31}P),[244] (η^{5}-C_5H_5)$Mo(CO)_2POCH_2CH_2NR$ (^{13}C, ^{31}P),[245] (η^{5}-C_5H_5)$_2(CO)_4Mo_2P_2${$Cr(CO)_5$}$_2$ (^{13}C, ^{31}P),[246] (η^{5}-C_5H_5)M{$P(OR)_3$}$(CO)_2$-($AsMe_2S$) (M = Mo, W; ^{31}P),[247] (η^{5}-C_5H_5)$Mo(CO)_2Se_2CR$ (^{13}C, ^{31}P, ^{77}Se),[248] {(η^{5}-C_5H_5)$Mo(CO)_2$}$_2$(B-CF_3N=C) (^{13}C, ^{19}F),[249] (η^{5}-C_5H_5)$_2$-$Mo_2(CO)_4$(μ-$Bu^{t}CP$)$M_3(CO)_{11}$ (M = Ru, Os; ^{31}P),[250] (η^{5}-C_5H_5)M(μ_3-CO)-{Rh(CO)(PPh_3)}$_3$ (M = Mo, W; ^{13}C, ^{31}P),[251] [(η^{5}-C_5H_4Me)$Mo(CO)_2$-(μ-dppm)Pt(dppm)]$_2$$^{2+}$ (^{31}P),[252] (μ-CH_2S){(η^{5}-C_5H_5)$W(CO)_2$}$_2$ (^{13}C),[253] (η^{5}-C_5H_5)$W(CO)_2${Cu(η^{5}-C_5Me_5)}$_2$C(p-tol) (^{13}C, ^{31}P),[254] [(η^{5}-C_5H_5)W-(CO)(PMe_3)(Ph_2PCMe)]$^{+}$ (^{13}C, ^{31}P),[255] (η^{5}-C_5H_5)W(CO)(PMe_3)($ER^{1}{}_2$-

$CR^2{=}C{=}O)X$ (E = P, As; ^{13}C, ^{31}P),[256] $\{(\eta^5-C_5H_5)W\}_2S_3(S_2C_2H_4)$ (^{13}C),[257] $Cr(\eta^6$-arene)$(PPh_3)(CO)_2$ (^{13}C, ^{31}P),[258] $Cr(\eta^6$-9,10-dihydroanthracene) (^{13}C),[259] $Cr(\eta^6-2,6-Bu^t{}_2-4-PhC_5H_2N)(CO)_3$ (^{13}C),[260] $Mo\{(\eta^6-Ph)PMePh\}L(PMe_2Ph)_2$ (L = N_2; ^{31}P),[261] {L = $P(OMe)_3$, $CNBu^t$; ^{31}P),[262] and $Cr(CO)_2(C_{16}H_{16}R^1R^2)$ (^{13}C).[263]

Linear correlations between υ(CO), $\delta(^{13}CO)$, and XPS C1s(CO) and O1s(CO) binding energy have been found for $Ph_nPX_{3-n}Cr(CO)_5$ and were discussed in terms of π-backbonding.[264] For $\{Cr(CO)_5\}_2PBu^t$, the ^{31}P chemical shift is 1362.[265] ^{95}Mo chemical shifts of $Mo(CO)_5L$ and $Mo(CO)_4L_2$ are in the range -1380 to -1626.[266] Excellent correlations have been found between ^{13}C, ^{17}O, ^{31}P, and ^{95}Mo chemical shifts of $Mo(CO)_5(PPh_2OC_6H_4R$-p) and the Taft $\sigma_{R°}$ and σ_I parameters.[267] For $Mo(CO)_5(PPh_2ER)$, the chemical-shift correlations between $\delta(^{13}CO)$, $\delta(^{17}O)$, $\delta(^{31}P)$, $\delta(^{95}Mo)$, and $^1J(^{31}P,^{13}C)$ depend upon the similarity in size of the XR groups.[268] Trans-RhCl(CO){cis-Mo-$(CO)_4(Ph_2P)_3CH\}_2$ has an unusual, virtually coupled ^{31}P n.m.r. spectrum.[269] For fac-$Mo(CO)_3(PR_3)_3$, a satisfactory linear correlation was found between $\delta(^{13}C)$ and E°. A similar correlation was not found for $\delta(^{31}P)$ or $\delta(^{95}Mo)$.[270] The ^{19}F and ^{31}P chemical shifts of $W(CO)_5PR_3$ have been interpreted as the W-P bond being a combination of σ-donor and π-acceptor.[271] The ^{13}C, ^{31}P, and ^{183}W n.m.r. spectra, including the sign of $^2J(^{31}P,^{31}P)$ in $W(CO)_4\{Ph_2P$-$CH_2CH_2CH(CH_2R)PPh_2\}$ have been determined.[272] Low-temperature ^{31}P n.m.r. spectra show that $Ru_5WC(CO)_{17}(AuPR_3)_2$ exists in solution as an isomeric mixture.[273] N.m.r. data have also been reported for $M(CO)_5(CNC_6H_4CH_2)PtI(PPh_3)_2$ (M = Cr, W; ^{13}C, ^{31}P),[274] $(CF_3NC)M(CO)_5$ (M = Cr, W; ^{13}C, ^{19}F),[275] $(Pr^i{}_2N)_2PhM(CO)_5$ (M = Cr, Mo, W; ^{31}P),[276] $Cy_2NPH_2M(CO)_5$ (M = Cr, W, ^{31}P),[277] $Ph_{3-n}PX_nCr(CO)_5$ (^{13}C),[278] $(OC)_5CrPX_2PX_2Cr(CO)_5$ (X = Cl, Br; ^{31}P),[279] $MeC(CH_2P)_3Cr(CO)_5$ (^{31}P),[280] $(OC)_5Cr(\mu-PC_4H_3Me_2)_2Cr(CO)_5$ (^{13}C, ^{31}P),[281] $(OC)_5Cr(\mu$-$Bu^tP)Rh(cod)$ (^{31}P),[282] $RC(O)CH(PPh_2)_2M(CO)_4$ (M = Cr, Mo, W; ^{31}P),[283] $(R_3P)PtCl_2PPh_2CH(PPh_2)_2M(CO)_4$ (M = Cr, Mo, W; ^{31}P),[284] $M(CO)_4\{(CF_3)_2PEMe\}_2$ (E = S, Se; ^{19}F, ^{31}P),[285] $Cr(CO)_4(PPh_3)_2$ (^{31}P),[286] $Cr(CO)_5\{S{=}C(SCH{=}CH_2)SEt\}$ (^{13}C),[287] $(\eta^5-C_5H_5)Rh(PMe_3)$-$CH_2SeCr(CO)_5$ (^{13}C),[288] $M(CO)_n\{P(C_2Ph)_mPh_{3-m}\}_{6-n}$ (^{31}P),[289] $M(CO)_3$-$(dmpe)_2$ (M = Cr, W; ^{31}P),[290] $[M(CO)_3(dppe)(SO_2SiMe_3)]^+$ (M = Cr, Mo, W; ^{29}Si, ^{31}P),[291] $[(\mu-N_3)_3M_2(CO)_6]^{3-}$ (M = Cr, Mo, W; ^{13}C),[292] $[Cr_4(CO)_{12}(\mu_3-OMe)_4]^-$ (^{13}C),[293] $[Fe_3CrC(CO)_{13}]^-$ (^{13}C),[294] $Mo(CO)_5PPh_2OPPh_2Cr(CO)_5$ (^{31}P),[295] $Mo(PhPC_4H_2Me_2)(CO)_5$ (^{13}C, ^{31}P),[296] $M(2,2'-pyPC_5H_2Me_2)(CO)_4$ (M = Mo, W; ^{13}C, ^{31}P),[297]

$[(OC)_4M(\mu-PPh_2)M(PPh_2H)(CO)_4]^-$ (M = Mo, W; ^{31}P),[298] $(OC)_3Mo(\mu-dppm)_2RhCl(CO)$ (^{31}P),[299] $M\{CH_2(CH_2SCH_2CH_2SCH_2)_2CH_2\}(CO)_3$ (M = Mo, W; ^{13}C),[300] $Mo(CO)_3(S_2PPh_2)_2$ (^{13}C, ^{31}P),[301] $[Mo(CNEt)_3(bipy)_2]^{2+}$ (^{13}C),[302] $Mo(CO)_2(CNR)(PMe_3)_3$ (^{13}C),[303] $Mo(CO_2)_2(PMe_3)_4$ (^{13}C, ^{31}P),[304] $MX_2(CO)_2(PMe_3)_2$ (M = Mo, W; ^{31}P),[305] $[Mo(CO)_2\{Ph_2PC(S)-NMe\}\{\mu-Ph_2PC(S)NMe\}]_2$ (^{31}P),[306] $\{MeN(PF_2)_2\}_2Mo_2(CO)_3(PF_2NHMe)-(PF_2)Cl$ (^{13}C, ^{19}F, ^{31}P),[307] $Mo_2(OR)_4(acac)_2(CNBu^t)$ (^{13}C),[308] $(XPH_2)W(CO)_5$ (^{31}P),[309] $(OC)_5WPHPhCHPr^iR$ (^{13}C, ^{31}P),[310] $(OC)_5W-PR^1C_2H_2R^2R^3$ (^{13}C, ^{31}P),[311] $W\{PPhF(CH_2CH=CH_2)\}(CO)_5$ (^{31}P),[312] $(C_5H_5)PX_2W(CO)_5$ (^{13}C, ^{31}P),[313] $Pd\{(PhP=PPh)[W(CO)_5]_2\}(dppe)$ (^{19}F, ^{31}P),[314] $(OC)_5WSCHCPh=CPh(OEt)$ (^{13}C),[315] $(PhPC_2Ph_2)W(CO)_4I_2$ (^{13}C, ^{31}P),[316] $(OC)_4W(\mu-PPh_2)_2Os(CO)_3L$ (^{31}P),[317] $(OC)W(PPh_2Et)(\mu-PPh_2)-Rh(PPh_3)(CO)$ (^{31}P),[318] $(OC)_4W(\mu-PPh_2)_2M(PPh_3)$ (M = Ni, Pd, Pt; ^{31}P),[319] and $[(phen)W(RC_2O)(CN)_2(CO)]^-$ (^{13}C).[320]

^{15}N chemical shifts have been compared for three types of N_2 bridge complexes between Mo, W, Re, or Os and acceptors such as $AlMe_3$, $TiCl_4$, $ZrCl_4$, $HfCl_4$, $NbCl_5$, or $TaCl_5$.[321] ^{15}N chemical shifts have been reported for trans-$M(^{15}N_2)(NCR)(dppe)$ (M = Mo, W). Linear correlations were observed for $\delta(^{15}N)$, oxidation potentials, and Hammett σ_p parameters for X substitution of XC_6H_4CN. The shielding of both nitrogens in the N_2 ligand decreases with increasing electron-releasing properties of the trans ligand.[322] ^{15}N chemical shifts and coupling constants have been determined for hydrazido, imido, and nitrido complexes of Mo and W, e.g. $MoI_2-(^{15}N^{15}NH_2)(PMe_2Ph)_3$.[323] The ^{95}Mo chemical shifts of a wide variety of complexes containing the cis-$[Mo(NO)_2]^{2+}$ unit range from +200 to −1000 p.p.m. and do not correlate with the NO stretching frequency or the energy of the first electronic transition.[324] N.m.r. data have also been reported for trans-$Mo(N_2)(PMePh_2)_2(PPh_2CH_2-CH_2SMe)$ (^{31}P),[325] mer-$Mo(N_2)_3(PPr^n_2Ph)_3$ (^{31}P),[326] $[Mo(\mu-\eta^1,\eta^6-PPh_3)(N_2)(PPh_3)]_2$ (^{13}C, ^{31}P),[327] $MoO_2\{HB(pz)_3\}X$ (^{95}Mo),[328] $Mo_2(OCH_2Bu^t)_6(Ph_2CNN)_2(HNMe_2)$ (^{13}C),[329] $[Mo(NC_6H_4Me)\{S_2P-(OEt)_2\}]_2(\mu-S)(\mu-SH)(\mu-O_2CCF_3)$ (^{31}P),[330] cis-$Mo(NO)_2(PF_6)_2$ (^{19}F, ^{31}P),[331] $(PMe_3)_3MoCl_3$ (^{31}P),[332] $MoWCl_4(PMePh_2)_4$ (^{31}P),[333] $Mo_2\{Ph_2PC(S)NR\}_4$ (^{31}P),[334] $W(NO)_2Cl_2(dppm)_2$ (^{31}P),[335] $WOCl_4-\{PN(CH_2CR_2O)_2\}_2$ (^{31}P),[336] and $\{W(SEt)O(PMe_2Ph)\}_2(\mu-S)(\mu-SEt)_2$ (^{31}P).[337]

At high field, ^{95}Mo n.m.r. spectroscopy can be used to study polyoxomolybdates.[338] An initial survey of ^{95}Mo chemical shifts of a range of polyoxomolybdates has been presented and covers a range

$\delta(^{95}Mo)$ = -18 to +140. ^{17}O n.m.r. spectroscopy was also used to assist in the structure determination of β-$[Mo_8O_{26}]^{2-}$.[339] For $H_4SiM_{12}O_{40}$ (M = Mo, W) ^{29}Si chemical shifts are determined by the decrease in the π-component of the Si bond.[340] A ^{31}P n.m.r. study of solutions of mixed heteropolymolybdo(tungsto)phosphates has been reported.[341] One and 2D ^{183}W n.m.r. techniques have been applied to $[PVW_{11}O_{40}]^{4-}$, $[SiVW_{11}O_{40}]^{5-}$, and $[BVW_{11}O_{40}]^{6-}$ using 2D INADEQUATE to establish the α-Keggin structure. $^{2}\underline{J}(^{183}W,^{51}V)$ was estimated from line broadening. $^{51}V,^{51}V$ COSY was used to get V-V connectivity.[342] In the ^{95}Mo and ^{183}W n.m.r. spectra of $[MO_{4-\underline{n}}E_{\underline{n}}]^{2-}$ (M = Mo, W, E = O, S, Se) a systematic decrease in shielding occurs with successive substitution of O by S or Se.[343] According to ^{19}F n.m.r. data, WOF_4 exists in solution as a tetramer.[344] N.m.r. data have also been reported for $MoO(3\text{-}NS_2)(OPPh_3)$ (^{31}P),[345] Mo_2O_5(1-Pr-2-α-hydroxybenzylbenzimidazolate)$_2$ (^{95}Mo),[346] $MOF_4(OPPh_3)$ (M = Mo, W; ^{19}F),[347] $MoO_2(OR)_2$ (^{17}O),[348] $[PVMo_{11}O_{40}]^{2-}$ (^{31}P),[349] $[E_2Mo_{12}O_{40}]^{6-}$ (E = P, As; ^{17}O),[350] $[Mo(S_2CNEt_2)_4][FeCl_3(PhCH_2SSCH_2Ph)]$ (^{13}C),[351] complexes of W(VI) with thiomalic acid (^{13}C),[352] $[(XCu)_{\underline{n}}MoS_4]^{2-}$ complexes (^{95}Mo),[353] $W_2(O_2CCF_3)_4(PR_3)$ (^{19}F, ^{31}P),[354] $[(Hg_2)_2WO(H_2O)(AsW_9O_{33})_2]^{10-}$ (^{183}W),[355] $[ThW_{10}O_{38}]^{8-}$ (^{17}O, ^{183}W),[356] $[P_2W_{12}Mo_6O_{62}]^{6-}$ (^{183}W),[357] and $[WOF_6]^{2-}$ (^{19}F).[358]

Complexes of Mn, Tc, and Re.—The ^{55}Mn nuclear magnetic screening constants of $Mn(CO)_5X$ (X = H, CN, Me, Cl) have been studied theoretically, and are determined predominantly by the 3$\underline{d}$ contribution of the paramagnetic term.[359] $^{1}\underline{J}(^{187}Os,^{1}H)$ and the ^{13}C n.m.r spectrum have been determined for $HOs_3M(CO)_{14}(C_2Ph)_2$ (M = Mn, Re).[360] N.m.r. data have also been reported for $(\eta^5\text{-}C_5H_5)LMnH(SiHPh_2)$ (^{29}Si),[361] $(Pr^i{}_2N)_2PMn_2(CO)_8H$ (^{31}P),[362] $ReH_6(AlMe_2)(PMePh_2)_2$ (^{31}P),[363] $ReH_5(PMePh_2)_3$ (^{31}P),[364] $[Re_2(\mu\text{-}H)_3H_2(PPh_3)_4(CNBu^t)_2]^+$ (^{31}P),[365] $[Re_2(\mu\text{-}H)_4H_3(PPh_3)_4(NCR)]^+$ (^{31}P),[366] $\{(\eta^6\text{-}C_6H_6)Re\}_2$-$(\mu\text{-}H)_2(\mu\text{-}CR^1R^2)$ (^{13}C),[367] $(\eta^6\text{-}C_6H_6)Re(\mu\text{-}H)_2Re(\eta^4\text{-}C_6H_8)_2$ (^{13}C),[368] $Re_2(CO)_8H(COR)$ (^{13}C),[369] and $[Re_6H_2C(CO)_{18}]^{2-}$ (^{13}C).[370]

^{13}C n.m.r. spectra indicate the presence of a hydrogen bond to the acyl in (2).[371] Anomalous ratios of the direct dipole-coupling constants have been found for $(\eta^5\text{-}C_5H_5)M(CO)_3$ (M = Mn, Re) dissolved in two different nematic liquid-crystal solvents.[372] The ^{1}H n.m.r. spectrum of 3-Me-phosphacymantrene in a nematic solvent has been interpreted.[373] An inverse temperature relationship of the ^{31}P n.m.r. spectrum has been observed for $[(\eta^5\text{-}C_5H_4CR_2)Mn(CO)L_2$.[374] N.m.r. data have also been reported for $(OC)_4Mn(CH_2)_{\underline{n}}OPPh_2$ (^{13}C,

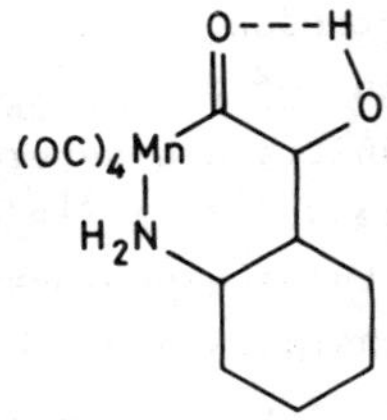

(2)

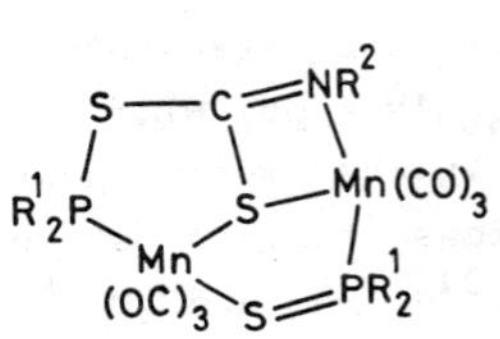

(3)

^{31}P),[375] $(OC)_4MnC_6H_4PC_5H_2Me_2Ph$ (^{13}C, ^{31}P),[376] $(\eta^5-C_5MeH_4)Mn(CO)_2-C_2PMe_3$ (^{13}C, ^{31}P),[377] $(OC)_5MnC(O)CHROSiMe_3$ (^{13}C),[378] $(\eta^5-C_5H_4Me)-(CO)_2Mn=C=C=CR_2$ (^{13}C, ^{31}P),[379] $X(OC)_4M=COCH_2CH_2Y$ (M = Mn, Re, Y = NH, O; ^{13}C),[380] $Mn_2\{\mu_2-C(O)CPh=CHPh\}(\mu_2-PPh_2)(CO)_6(CNBu^t)_2$ (^{13}C),[381] $FeMn(\mu-PPh_2)(\mu-COMe)(CO)_5(\eta^5-C_5H_5)$ (^{13}C, ^{31}P),[382] $Mn_2\{\mu-Au(PEt_3)\}(\mu-PPh_2)(CO)_8$ (^{31}P),[383] $Mn_2(\mu-HgCl)(\mu-PPh_2)(CO)_8$ (^{13}C, ^{31}P),[384] $(\eta^5-C_5H_5)Re(PMe_3)_2Ph$ (^{13}C, ^{31}P),[385] $(OC)_4ReC(O)NPhCMe=NPh$ (^{13}C),[386] $Ru(CO)(PPh_3)_2(CCl_2)Cl_2$ (^{13}C), $[Mn(CO)_6]^+$ (^{55}Mn),[387] $[(\eta^5-C_5H_5)Re(NO)(PPh_3)(\mu_2-CH_2=S)]^+$ (^{13}C, ^{31}P),[388] $(\eta^5-C_5H_5)ReX(NO)(PPh_3)$ (^{13}C, ^{31}P),[389] $[(OC)_3Re(\mu-CHR)(dppm)W(CO)_4]^-$ (^{13}C, ^{31}P),[390] $Re(O)-I(\eta^2-MeC_2Me)_2$ (^{13}C),[391] $(\eta^3-C_5H_7)Mn(CO)_3PR_3$ (^{13}C),[392] $[L_nM\{\eta^3-CH_2-COCO(R)BF_2\}]^-$ {L_n = $(\eta^5-C_5H_5)Fe(CO)_2$, $M(CO)_4$, M = Mn, Re; ^{13}C},[393] $Mn_2(CO)_8C_3H_4$ (^{13}C),[394] $(C_6H_8Ph)Mn(CO)_3$ (^{13}C),[395] $(\eta^5-C_5R_5)Mn(CO)_2L$ (^{13}C, ^{31}P, ^{55}Mn),[396] $\{(\eta-C_5H_4Me)Mn(CO)_2\}_2\{4-pyridyl(CH_2)_2-4-pyridyl\}$ (^{13}C),[397] $Mn(\eta^5-thiophene-H)(CO)_3$ (^{13}C),[398] $(\eta^5-C_5Me_5)_2Mn_2-(\mu-CO)_3$ (^{13}C),[399] $(\eta^5-C_5Me_5)_2Mn_2(CO)_4(\mu-AsCHR^1R^2)$ (^{13}C),[400] $(\eta^5-C_5-Me_5)ReO_3$ (^{13}C, ^{17}O),[401, 402] $[(\eta^5-C_5H_5)(C_5H_4)Re]_2$ (^{13}C),[403] and $[Mn(\eta^6-C_{10}H_8)(CO)_3]^+$ (^{13}C).[404]

$^1\underline{J}(^{55}Mn,^{13}C)$ has been observed in the ^{13}C, but not the ^{55}Mn n.m.r. spectrum of $[(RNC)_6Mn]^+$ and is between 111 and 121 Hz.[405] N.m.r data have also been reported for $(C_6H_4O_2)_2PMn(CO)_5$ (^{13}C, ^{31}P),[406, 407] $Mn(CO)_5(OTeF_5)$ (^{19}F),[408] $(OC)_5Mn(\mu-MeP=S)Mn(CO)_4$ (^{31}P),[409] $(OC)Mn\{(PPh_2O)_2H\}$ (^{31}P),[410] $[M(CO)_4PO(OEt)_2]_2$ (M = Mn, Re; ^{31}P),[411] $(OC)_4MnO_2CCH_2CH_2PPh_2$ (^{13}C, ^{31}P),[412] $(OC)_4Mn(PPh_2CR-OSiMe_3)$ (^{13}C, ^{31}P),[413] $M_2Pt(CO)_{12}$ (M = Mn, Re; ^{13}C),[414] $(OC)_3Mn-(NR)_2PR(NR_2)$ (^{31}P),[415] $(OC)_3Mn\{Ph_2C(S)NMeR\}X$ (^{13}C, ^{31}P, ^{55}Mn),[416] $\{(OC)_3BrMn\}_2\{\mu-(EtO)_2POP(OEt)_2\}_2$ (^{31}P),[417] (3) (^{13}C, ^{31}P),[418] $FeMn(CO)_6(Bu^tNCHCHNHBu^t)$ (^{13}C),[419] $XPd(\mu-dppm)_2Mn(CO)_3$ (^{13}C, ^{31}P),[420] $ReBr(CO)_3\{PPh_2(\underline{p}-tol)\}$ (^{13}C),[421] and $Re_2(CO)_6(\mu-Br)_2\{\mu-R^1P(NCR^2_3)_2PR^1\}$ (^{13}C, ^{31}P).[422]

The ^{31}P and ^{99}Tc n.m.r spectra of $[Tc\{P(OMe)_3\}_6]^+$ show

$^{1}J(^{99}Tc,^{31}P)$ = 910 Hz.[423, 424] N.m.r. data have also been reported for $[Re(NO)(CN)_5]^{3-}$ (^{13}C),[425] Mn{NMeP(O)(OMe)$_2$}(tetraphenylporphyrin) (^{31}P),[426] $MnI_2(PEt_3)_2$ (^{13}C),[427] $[Tc(dmpe)_3]^+$ (^{31}P, ^{99}Tc),[428] Re{N(SiMe$_3$)$_2$}(NSiMe$_3$)$_2$(OSiMe$_3$)$_2$ (^{13}C),[429] Re(SC$_6$H$_3$Pr$^i_2$)$_3$(MeCN)PPh$_3$ (^{31}P),[430] Re$_2$Cl$_5$(OR)(dppm)$_2$ (^{31}P),[431] Re$_2$Cl$_3$(Ph$_2$Ppy)$_2${Ph(C$_6$H$_4$)Ppy} (^{31}P),[432] [ReO$_2$(PMe$_3$)$_4$][ReO$_4$] (^{31}P),[433] and $[ReO_2(OEt)F_3]^-$ (^{19}F).[434]

Complexes of Fe, Ru, and Os.—Results of a ^{57}Fe n.m.r. investigation of characteristic relaxation times and chemical shifts of some substituted ferrocenes have been reported. Substituent effects and relaxation mechanisms were discussed. ^{15}N n.m.r spectra were also given.[435]

The ^{13}C n.m.r spectrum of (η^5-C$_5$Me$_5$)Fe(PMe$_3$)H(η^2-CH$_2$=CH$_2$) shows fast exchange between the hydride and one CH$_2$ of the ethylene. As a result, the ^{13}C signal is a quartet with $^{1}J(^{13}C,^{1}H)$ = 101Hz. The ^{31}P n.m.r. spectrum was also recorded.[436] ^{1}H and ^{13}C n.m.r. spectroscopy has been used to investigate the addition of H$^+$ to HFe$_4$-(CO)$_{12}$CMe to give [H$_2$Fe$_4$(CO)$_{12}$CMe]$^+$.[437] N.m.r. data have also been reported for FeH(HBH$_3$)(dmpe)$_2$ (^{11}B),[438] (η^5-C$_5$Me$_5$)FeH(CO)(dppe) (^{13}C, ^{31}P),[439] HM(SiEt$_3$)(CO)$_3$(PPh$_3$) (M = Fe, Ru; ^{13}C),[440] Fe$_3$(μ-H)-{μ-Pr^i(R)PCH$_2$PPr^i}(CO)$_9$ (^{13}C),[441] (μ-H)(μ-CO)Fe$_3$(CO)$_9$BH$_2$ (^{11}B),[442] (μ_2-H)Fe$_3$(CO)$_9$(μ_3-CMe) (^{13}C),[443] and (μ-H)$_2$Fe$_3$(μ_2-PPh$_2$)$_2$(CO)$_8$ (^{31}P).[444]

The ^{1}H and ^{31}P n.m.r spectra of Ru$_3$H$_6$(dppm)$_2$ have been reformulated.[445] For HRu$_3$(CO)$_9$(μ_3-R) (R = alkenyl, allyl, alkynyl) a linear relationship has been found between electrode potential and $\delta(^{13}C)$.[446] ^{2}H n.m.r. spectroscopy has been used to identify the products of the reaction of Os$_3$D$_2$(CO)$_{10}$ with C$_2$H$_2$.[447] N.m.r. data have also been reported for RuH$_4$(PR$_3$)$_3$ (^{31}P),[448] RuH$_2$(dppm)(PPh$_3$)$_2$ (^{31}P),[449] RuH$_2$(CO)(PMe$_2$Ph)$_3$ (^{31}P),[450] RuH(Et)(PMe$_3$)$_4$ (^{13}C, ^{31}P),[451] RuHCl{Ph$_2$P(CH$_2$)$_3$PMe$_2$} (^{31}P),[452] (η^5-C$_5$H$_5$)MH(η^2-CH$_2$=CHC$_6$H$_4$PPh$_2$) (M = Ru, Os; ^{13}C, ^{31}P),[453] [RuH(CO)(dppe)$_2$]$^+$ (^{13}C, ^{31}P),[454] RuH(CO)-{C$_6$H$_3$MeCO(tol)} (^{13}C),[455] RuHCl(η^6-C$_6$Me$_6$)L (^{13}C, ^{19}F, ^{31}P),[456] RuHPh(η^6-C$_6$Me$_6$)(PHCy$_2$) (^{31}P),[457] [Ru(η^6-arene)]$_2$H(O$_2$CCF$_3$)Cl$_2$ (^{19}F),[458] {RuH(pz)(η^4-cod)}$_2$pzH (^{13}C),[459] {RuH(PMe$_3$)$_3$}$_2$(μ-PMe$_2$CH$_2$)$_2$ (^{13}C, ^{31}P),[460] Ru$_3$RhH(PPh)(CO)$_{10}$(PPh$_3$) (^{13}C, ^{31}P),[461] [H(CO)$_3$Ru(μ-PPh$_2$)Co(CO)$_3$]$^-$ (^{13}C, ^{31}P),[462] (μ-H)Ru$_3$(CO)$_7$(μ_4-η^4-CH$_2$=CCPr^i=CH)-(μ-PPh$_2$) (^{13}C, ^{31}P),[463] HRu$_5$(CO)$_{10}$(PPh$_2$)(PPh)$_3$ (^{31}P),[464] [OsH$_5$-(PMe$_2$Ph)$_3$]$^+$ (^{31}P),[465] [(η^5-RC$_6$H$_6$)OsHI(PMe$_3$)$_2$]$^+$ (^{13}C, ^{31}P),[466] HOs$_3$(CO)$_{10}$N$_2$Ar (^{13}C, ^{15}N, ^{19}F),[467] Os$_3$(μ-H)(μ-CCH$_2$Pr^i)(CO)$_{10}$

(^{13}C),[468] $Os_3(CO)_8H(C_6H_4PPhCH_2PPh_2)$ (^{31}P),[469] $Os_3(\mu-H)(\mu_3-\eta^2-CF_3C{=}CHCF_3)(CO)_{10}$ (^{19}F),[470] $(\eta^5-C_5H_5)CoOs_3H_2(CO)_{10}$ (^{13}C),[471] and $Os_3Ni(\eta^5-C_5H_5)(CO)_8(PR_3)(\mu-H)_3$ (^{31}P).[472]

A simple empirical approach has been used to assign the ^1H n.m.r. spectra of $(MeSiRO)_{\underline{n}}$ [R = $(\eta^5-C_5H_5)Fe(CO)_2$, $Co(CO)_4$].[473] N.m.r. data have also been reported for $[(diars)Fe(CO)_2Me(4\text{-}Et\text{-}2,6,7\text{-}trioxa\text{-}1\text{-}bicyclo[2.2.2]octane)]^+$ (^{13}C),[474] $(\eta^5-C_5H_5)Fe(CO)_2R$ (^{13}C),[475] *cis*-$FeR(SiMe_3)(CO)_4$ (^{13}C),[476] $(OC)_3Fe(L)CH_2CH_2CH_2CH_2$ (^{13}C),[477] $(\eta^5-C_5H_5)Fe(CO)_2CH(OEt)$norbornane ($^{13}C$),[478] $(\eta^5-C_5H_5)-(OC)_2Fe{=}C(OMe)NBzCHMeCH_2$ (^{13}C),[479] $[(OC)_4Fe(CH_2CO_2Me)]^-$ (^{13}C),[480] $(\eta^5-C_5H_5)Fe(CO)_2CH(OMe)_2$ (^{13}C),[481] $(OC)_4FeCMe{=}CMeC(OEt)Ph$ (^{13}C),[482] $(C_3F_7)Fe(CO)_3CR{=}CRPEt_2{=}S$ (^{31}P),[483] 1,5-$\{(\eta-C_5H_5)Fe(CO)_2\}_2C_8F_6$ (^{19}F),[484] $(\eta^5-C_5H_5)Fe(CO)(CH{=}CMe_2)\{P(OCH_2CH_2)_2N\}$ (^{13}C),[485] $(\eta^5-C_5H_5)(OC)_2FeCH{=}CH{=}CR^1{=}CHR^2$ (^{13}C),[486] $(\eta^5-C_5H_5)Fe(CO)(PPh_3)-\{\eta^1-C(CH{=}CH_2){=}CMe_2\}$ (^{13}C),[487] $(\eta^5-C_5H_5)Fe(CO)SC(NEt_2)SCNC(O)Ph$ (^{13}C),[488] $(\eta^5-C_5H_5)FeAr(CO)L$ (^{13}C),[489] $(\eta^5-C_5H_5)Fe(CO)_2(COCH_2CH_2R)$ (^{13}C),[490] $(\eta^5-C_5H_5)Fe(CO)\{P(OPh)_3\}COCMe{=}CMePh$ (^{13}C),[491] $(\eta^5-C_5H_5)-Fe(CO)_2(CONHR)$ (^{13}C),[492] $Fe\{C{=}S(OMe)\}(CO)(PC_2HBu^tS)(PMe_2Ph)_2$ (^{31}P),[493] $Fe(CO_2)\{PMe_2(CH_2)_3PMe_2\}_2$ (^{31}P),[494] $[(\eta^5-C_5H_5)Fe({=}CH-CH{=}CMe_2)(CO)_2]^+$ (^{13}C),[495] $(\eta^5-C_5H_5)Fe(CO)(PPh_3)C(OMe){=}CMe_2$ (^{13}C),[496] $Fe({=}C{=}C{=}CBu^t)(CO)_4$ (^{13}C),[497] $(Ph_3P)(OC)_2Fe\{\mu-S_2C{=}C-(CMe_2CH_2)_2CH_2\}Fe\{{=}C{=}C(CMe_2CH_2)_2CH_2\}(CO)_2$ (^{13}C),[498] $[(\eta^5-C_5H_5)Fe-({=}CMeOMe)(PMe_3)_2][SO_3F]$ (^{13}C, ^{19}F, ^{31}P),[499] Fe(tetraphenylporphyrin)$C(ER^1)R^2$ (^{13}C),[500] $Ph_3PFe(CO)_3C(OEt){=}NPr^i_2$ (^{13}C, ^{31}P),[501] $Fe(HgCl)\{{=}CS_2C_2(CO_2Me)_2\}(CO)_2(PMe_2Ph)_2$ (^{31}P),[502] $[(\eta^5-C_5H_5)Fe(CO)_2-C(SR)SRe(CO)_5]^+$ (^{13}C),[503] $(\eta^5-C_5H_5)_2Fe_2(CO)_4(\mu-CO)(\mu-C_3H_4)$ (^{13}C),[504] $(\eta^5-C_5Me_5)_2Fe_2(\mu-CO)(CO)_2(\mu-CH_2)$ (^{13}C),[505] $Fe_2(CO)_6(\mu-COEt)(\mu-CHPh)$ (^{13}C),[506] $Fe_2(CO)_6(\mu-SMe)\{\mu-CS(OR)\}$ (^{13}C),[507] $H_3MRu_4(CO)_{12}PPh_3$, $MFe(\mu-COMe)(CO)_{10}PPh_3$ (M = Cu, Ag, Au; ^{13}C, ^{31}P),[508] $(\eta^5-C_5H_5)Fe(CO)_2HgX$ (^{199}Hg),[509] $(\eta^5-C_5H_5)Fe(SiH_3)(PMe_3)_2$ (^{31}P),[510] and $(\eta^5-C_5H_5)Fe(CO)_2Si(NMe_2)_3$ (^{13}C, ^{29}Si).[511]

^{119}Sn n.m.r. spectra of $[MCl_{\underline{n}}(SnCl_3)_{6-\underline{n}}]^{4-}$ (M = Ru, Os) and $[IrCl_2(SnCl_3)_4]^{3-}$ have been reported. The ^{119}Sn chemical shifts of the Ru compounds move to high frequency with increasing $[SnCl_3]^-$ ligands. $^1\underline{J}(^{119}Sn,^{117}Sn)$ varies between 11 250 and 40 759 Hz. $^1\underline{J}(^{119}Sn,^{99}Ru)$ is 756 Hz in $[Ru(SnCl_3)_6]^{4-}$ and $^1\underline{J}(^{187}Os,^{119}Sn)$ is 1123 Hz in $[Os(SnCl_3)_6]^{4-}$.[512] ^{13}C labelling has been used to show that, in the reaction of $Os_3(CO)_{11}(\mu\text{-}CH_2)$ with ^{13}CO to give $Os_3(CO)_{12}\{\eta^2\text{-}(C,C);\mu_2\text{-}CH_2CO\}$, the ketene CO derives from one of the original cluster carbonyls rather than added CO.[513] N.m.r. data have also been reported for cis-$\{RuMe(PMe_3)_4\}_2Hg$, $[Ru_2(\mu\text{-}Cl)_3(PMe_3)_6][BF_4]$ (^{13}C, ^{19}F, ^{31}P),[514] $(\eta^5\text{-}C_5H_5)RuR(PMe_3)_2$ (^{13}C, ^{31}P),[515] $Ru_2(\mu\text{-}O_2CMe)_2(CH_2Bu^t)_4$ (^{13}C),[516] $Ru\{C_6H_4OP(OPh)_2\}\text{-}\{P(OPh)_3\}_2$ (^{31}P),[517] $RuR(COMe)(CO)(CNBu^t)(PMe_2Ph)_2$ (^{13}C),[518] $Ru(COEt)Cl(CO)(PPh_3)_2$ (^{31}P),[519] $(\eta^5\text{-}C_5H_5)Ru(CO)(PPh_3)CO_2H$ (^{13}C),[520] $Ru_2(CO)_3(\eta^5\text{-}C_5H_5)_2(\mu\text{-}CHMe)$, $Os_3(CO)_{11}(\eta^2\text{-}C_2H_4)$ (^{13}C),[521] $(\eta^5\text{-}C_5H_5)\text{-}Ru_2(CO)_2(\mu\text{-}CO)(\mu\text{-}CR^1R^2)$ (^{13}C),[522] $RuCo(CO)_5(\mu\text{-}CO)(\mu\text{-}PPh_2CPhCPh)$ (^{31}P),[523] $(OC)_5OsM^1(CO)_3(M^2Cl_3)Br$ (M^1 = Ru, Os; M^2 = Si, Ge; ^{13}C),[524] $OsR_2(PMe_3)_4$ {R = Me, $(CH_2)_2SiMe_2$, $CH_2C_6H_4$; ^{13}C, ^{31}P),[525] $[NOsR_4]^-$ (R = Me, CH_2MMe_3; M = C, Si; ^{13}C),[526] $(\eta^6\text{-}C_6H_3Me_3)_2Os_2\text{-}(\mu\text{-}CHC_6H_3Me_2)$ (^{13}C),[527] $[Os_3(CO)_{11}CDO]^-$ (^2H, ^{13}C),[528] $[Os_3(CO)_{11}COMe]^-$ (^{13}C),[529] and $Os_3(CO)_{10}(\mu\text{-}RC{=}O)_2$ (^{13}C).[530]

The use of ^{17}O $\underline{T}_1$ measurements to determine correlation times and quadrupolar-coupling constants of metal carbonyls has been reported and applied to $Fe_2(CO)_6(Bu^tC_2H)_2$.[531] The rotational correlation time for 6 organometallic carbonyl complexes, e.g. $(\eta^4\text{-isoprene})Fe(CO)_3$, has been determined from ^{13}C $\underline{T}_1$ and $^{13}C\text{-}\{^1H\}$ n.O.e. measurements and yields ^{17}O electric quadrupole-coupling constants. The values show structural variation and so could be useful in the assignment of ^{17}O resonance.[532] The conformation of $(butatriene)Fe_2(CO)_6$ has been determined in three nematic phases from ^1H n.m.r. measurements.[533] N.m.r. data have also been reported for $Fe(\eta^2\text{-cyclooctene})_2(CO)_3$ (^{13}C),[534] $Fe_2Ru(CO)_9(alkyne)$ (^{13}C),[535] $[(\eta^5\text{-}C_5H_5)Ru(\eta^2\text{-}C_2H_4)_3]^+$ (^{13}C),[536] $Fe(\eta^3\text{-}C_5H_7)_2(PMe_3)_2$ (^{13}C),[537] $Ni_2M_2(\eta^5\text{-}C_5H_5)\{C_2HC({=}CH_2)Me\}$ (M = Fe, Ru; ^{13}C),[538] $[(\eta^3\text{-enyl})RuL_3]^+$ (^{13}C),[539] $Ru_4(CO)_5\{P(OMe)_3\}(\mu_3\text{-}C_5H_4)_2(\eta^5\text{-}C_5H_5)_2$ (^{13}C),[540] $Ru_2(CO)(\mu\text{-}CO)(\mu\text{-}CMeCHCH_2)(\eta^5\text{-}C_5H_5)_2$ (^{13}C),[541] $(C_7H_6\text{-}Me_2CO)Fe(CO)_3$ (^{13}C),[542] $(\eta^6\text{-}C_6Me_6)Fe(\eta^4\text{-diene})$ (^{13}C),[543] $(\eta^4\text{-}C_6H_8)\text{-}Fe(CO)_2L$ (^{13}C, ^{31}P),[544] $(\eta^4\text{-}C_4H_2Me_2O)Fe(CO)_3$ (^{13}C, ^{31}P),[545] $(\eta^4\text{-}C_6H_7CMe_2CN)Fe(CO)_2$ (^{13}C),[546] $(\eta^4\text{-}C_9H_6Ph_2O)Fe(CO)_3$ (^{13}C),[547] $(\eta^4\text{-tropone})Fe(CO)_3$ (^{13}C),[548] $\{\eta^4\text{-}C_{10}H_6(NR_2)_2\}Fe(CO)_3$ (^{13}C),[549] $(\eta^6\text{-}C_6Me_6)Fe\{\eta^4\text{-}C_6Me_4({=}CH_2)_2\}Fe$ (^{13}C),[550] $(\eta^4\text{-5-}\underline{exo}\text{-}C_6H_7X)Fe(CO)_3$ (^{13}C),[551] $Fe(CO)_3(\eta^4\text{-}C_7H_7CH{=}CHPh)$ (^{13}C),[552] $(\eta^4\text{-pulegone})Fe(CO)_3$ (^{13}C),[553] $(\eta^4\text{-}C_9H_8R^1R^2O)Fe(CO)_3$ (^{13}C),[554] $[Fe_2(CO)_6\{\mu\text{-}CPhCPhC(CF_3)\text{-}C(CF_3)H\}]^-$ (^{13}C, ^{19}F),[555] $[Fe_2(CO)_6[\eta^4{:}\eta'^4\text{-}C_{16}H_{16}(PPh_3)_2\}]^+$ (^{13}C, ^{31}P),[556] $Ru(CO)(\eta^4\text{-}C_6H_{10})_2$ (^{13}C),[557] $Ru(C_8H_8)(dppe)_2$ (^{31}P),[558] and $[Ru(C_{10}Me_8H)(C_6H_6)]^+$ (^{13}C).[559]

$\underline{T}_1$ values of ^{57}Fe have been determined for ferrocene and

$Fe(CO)_5$. Predominant relaxation mechanisms are chemical-shift anisotropy for the former and spin rotation for the latter.[560] The ^{31}P n.m.r. chemical shift of $(\eta^5\text{-}C_5Me_5)Fe(=PNPr^i_2)(CO)_2$ is 954. The ^{27}Al n.m.r. spectrum of $[AlCl_4]^-$ was also recorded.[561] $\delta(^{13}CO)$ for $(\eta^5\text{-}C_5H_5)Fe(\eta^5\text{-}C_5H_4)COC_6H_4X$ correlates with that for benzophenones.[562] In $(\eta^5\text{-}C_5H_5)Fe(\eta^5\text{-}C_5H_4)C_6H_4X$, $\delta(^1H)$ and $\delta(^{13}C)$ correlate with σ and σ^+ values of X.[563] In some protonated ferrocenophanes, coupling is observed between the hydride and ring protons.[564] INEPT and DEPT, as well as 2D n.m.r. spectroscopy, have been used to assign the ^{13}C n.m.r. spectra of some ferrocene and ruthenocene cryptands.[565] ^{19}F n.m.r. spectra have shown that, in $[(\eta^5\text{-}C_5H_5)M\text{-}(\eta^5\text{-}C_5H_4{=}CHC_6H_4F)]^+$, the ruthenocenyl group is more effective in charge delocalization than ferrocenyl.[566] N.m.r. data have also been reported for $[Fe(\eta^5\text{-dienyl})(CO)_2(CNR)]^+$ (^{13}C),[567] $[(\eta^5\text{-}C_5H_4\text{-}Me)Fe(CO)_3]^+$ (^{13}C),[568] $[(\eta^5\text{-}Me_3SiC_6H_3R^1R^2R^3)Fe(CO)_3]^+$ (^{13}C),[569] $[(\eta^5\text{-}C_5H_5)Fe(CO)_2(NC_2H_5)]^+$ (^{13}C),[570] $(\eta^5\text{-}C_5H_5)Fe(CO)_2PR_2S$ (^{13}C, ^{31}P),[571] $[(\eta^5\text{-}C_5H_5)Fe(CO)_2L]^+$ (^{13}C),[572] $FeRu(CO)_2(\mu\text{-}CO)_2(\eta^5\text{-}C_5H_5)_2$ (^{13}C),[573] $(\eta^5\text{-}C_5H_5)FeI\{P(OCH_2CF_3)_3\}_2$ (^{19}F, ^{31}P),[574] $[(\eta^5\text{-}C_5H_5)Fe\text{-}(PhPC_4H_2Me_2)_3]^+$ (^{13}C),[575] $(OC)_2Co(\mu\text{-}CO)[\mu\text{-}C(SMe)_2\}Fe(\eta^5\text{-}C_5H_5)(CO)$ (^{13}C),[576] $P_3N_3Cl_5N{=}PCl\{(\eta^5\text{-}C_5H_4)Fe(\eta^5\text{-}C_5H_5)\}_2$ (^{31}P),[577] $N_3P_3(OCH_2\text{-}CF_3)_5(\eta^5\text{-}C_5H_4)Fe(\eta^5\text{-}C_5H_5)$ (^{13}C, ^{19}F, ^{31}P),[578] $(\eta^5\text{-}C_5H_5)Fe(\eta^5\text{-}C_5H_4R)$ (^{13}C),[579] $(\eta^5\text{-}C_5H_5)Fe(\eta^5\text{-}C_5H_4NO_2)$ (^{13}C),[580] $Fe(\eta^5\text{-}C_5H_4NMe_2)_2$ (^{13}C),[581] some ferrocenophanes (^{13}C),[582-584] $[(Ph_3P)Ru(3,3',4,4'\text{-}Me_4\text{-}1,1'\text{-diphosphaferrocene})(\mu\text{-}Cl)Cl]_2$ (^{13}C, ^{31}P),[585] $[Ag\{\eta^5\text{-}C_5H_4\text{-}S(CH_2CH_2O)_{\underline{n}}CH_2CH_2SC_5H_{4\text{-}\underline{n}}\}Fe]^+$ (^{13}C),[586] $[(\eta^5\text{-}C_5H_5)Fe\{\eta^6\text{-}C_{13}H_8\text{-}(CH_2CH_2Y)_2\}]^+$ (^{13}C),[587] $(\eta^5\text{-}C_5H_4CO_2)(\eta^6\text{-}C_6Me_6)Fe$ (^{13}C),[588] $[(\eta^5\text{-}C_5Me_5)Fe(\eta^6\text{-cot})]^+$ (^{13}C, ^{31}P),[589] $[(\eta^5\text{-}C_5H_5)Fe(\eta^6\text{-thioxanthone})]^+$ (^{13}C),[590] $[(\eta^5\text{-}C_8H_9)Ru(PMe_2Ph)_3]^+$ {^{13}C, including $\underline{J}(^{13}C,^1H)$},[591] $(\eta^5\text{-}C_5Me_5)Ru(CO)_2(PBu^tCl)$ (^{31}P),[592] $(\eta^5\text{-}C_5H_5)Ru\text{-}(PPh_3)(CO)Cl$ (^{31}P),[593] $Ru(\eta^5\text{-}C_5H_4R)(CO)X(PPh_3)$ (^{13}C, ^{31}P),[594] $(\eta^5\text{-}C_5Me_5)Ru(PMe_3)_2I$ (^{31}P),[595] $[(\eta^5\text{-}C_5H_5)Ru(Ph_2PCH_2CHMePPh_2)\text{-}(NCMe)]^+$ (^{31}P),[596] $(\eta^5\text{-}C_5H_5)RuCl(PPh_3)(\eta^1\text{-dppm})$ (^{13}C, ^{31}P),[597] $(\eta^5\text{-}C_5H_5)RuCl(PPh_3)\{P(OPh)_3\}$ (^{13}C),[598] $(\eta^5\text{-}C_5H_5)Ru(PPh_3)_2CN$ (^{31}P),[599] $O(CH_2CH_2OCH_2CH_2S\text{-}\eta^5\text{-}C_5H_4)_2Ru$ (^{13}C),[600] $(\eta^5\text{-}C_5H_5)Ru\text{-}(CPhCPhCPhCHPh)$ (^{13}C),[601] $[(\eta^6\text{-}C_6Me_6)RuCl(PR_3)(\text{acetone})]^+$ (^{13}C, ^{31}P),[602] $[Ru_2(\eta^6\text{-}C_6H_6)_2(OMe)_3]^+$ (^{13}C),[603] $(\eta^6\text{-}\underline{p}\text{-cymene})RuCl_2(PPh_2X)$ (^{31}P),[604] and $(C_7H_7)FeRh(CO)_3(PR_3)_2$ (^{13}C, ^{31}P).[605]

The ^1H n.m.r. spectrum of $Fe(NH{=}NAr)(CO)\{P(OEt)_3\}_4$ shows $^1\underline{J}(^{15}N,^1H)$ = 67 Hz. The ^{31}P n.m.r. spectrum was also recorded.[606] The ^{31}P n.m.r. spectrum of $Fe_2(CO)_6(P_4Ph_4)$ is $[AX]_2$.[607] N.m.r. data have also been reported for $(Me_3Si)_2C{=}P\{Fe(CO)_4\}N(SiMe_3)_2$

(^{13}C),[608] $Bu^t{}_3C_6H_2P{=}P(C_6H_2Bu^t{}_3)Fe(CO)_4$ (^{13}C, ^{31}P),[609] [FeL(dppe)-$(CNR)_3]^{2+}$ (^{31}P),[610] $(OC)_3Fe(PhPCH_2C_6H_4CHMePPh)Fe(CO)_3$ (^{31}P),[611] $Cl_2Pt\{(Ph_2P)_2C_4Bu^t\}Fe(CO)_3$ (^{31}P),[612] $[(OC)_5FeRh(PPh_2)_3CH]^+$ $[Rh(CO)_2Cl_2]^-$ (^{31}P),[613] $Fe(CO)(PMe_3)_2\{CHSMeCOC(CO_2R{=}C)(CO_2R)S\}$ (^{13}C, ^{31}P),[614] (η^2-N,μ_2-N,η^2-C=N-glyoxal-bis-isopropylimin)Fe_2(^{13}C)$(CO)_5$ (^{13}C),[615] $Fe_2(CO)_6(RNS)$ (^{13}C),[616] $[Fe_2(\mu\text{-}CO)(\mu\text{-}PPh_2)(CO)_5(PPh_2)]^{2-}$ (^{13}C, ^{31}P),[617] $P_3N_3Cl_4Fe_2(CO)_8$ (^{31}P),[618] $Fe_2Pt(dppe)(CO)_6(Bu^tCP)$ (^{31}P),[619] $\{HC(S)OEt\}_2Fe_2(CO)_6$ (^{13}C),[620] $Fe_3(\mu\text{-}CNCF_3)(CO)_{11}$ (^{13}C),[621] $[BiFe_3(CO)_{10}]^-$ (^{13}C),[622] (μ_3-S)(μ_3-p-tolP)$Fe_3(CO)_9$ (^{31}P),[623] $FeRu_2(\mu\text{-}Cl)_2(CO)_8(dppe)$ (^{31}P),[624] $RuCl(CO)(PPh_3)_2$(oxine) (^{31}P),[625] $Ru(OC_6H_4NO)(OC_6H_4NO_2)(CO)(PPh_3)_2$ (^{13}C, ^{31}P),[626] $SRuCo_2$-$(CO)_8(PMe_2Ph)$ (^{31}P),[627] $Ru_5C(CO)_{14}(ArPR_3)_2$ (^{31}P),[628] [Os(phen)-$(PR_3)_2(CO)Cl]^+$ (^{13}C, ^{31}P),[629] 1,2-$Os_3(CO)_{10}$(diphosphine) (^{13}C, ^{31}P),[630] $PtOs_3(CO)_{10}(PMe_2Ph)_2(\mu_3\text{-}S)$ (^{31}P),[631] and $PtOs_3(CO)_9$-$(PPh_3)_2(\mu_3\text{-}S)_2$ (^{31}P).[632]

$^1J(^{13}C,^{13}C)$ in $[Fe(CN)_5NO]^{2-}$ and $^2J(^{15}N,^{15}N)$ in $[Fe_4E_3(NO)_7]^-$ (E = S or Se) have been determined.[633] The ^{31}P n.m.r. spectrum of $[Fe(dmpe)_2(acac)]^+$ has been analyzed as A_2B_2 rather than as $[AB]_2$.[634] Chiral Eu shift reagents have been shown to shift $[Ru(phen)_3]^{2+}$ and hence determine optical purity.[635] ^{15}N n.m.r. evidence has been given of the rapid interconversion of bent and linear NO ligands in $[RuCl(NO)_2(PPh_3)_2]^+$ from a $^{14,15}N$ equilibrium isotope effect.[636] N.m.r. data have also been reported for Fe(tetraphenylporphyrin)(CNCOR)(py) (^{13}C),[637] [Fe(2,2'-bipyrimid-ine)$_3]^{2+}$ (^{13}C),[638] $(NO)_2Fe(\mu\text{-}PPh_2)(\mu\text{-}NO)Fe(NO)(PPh_2H)$ (^{31}P),[639] $[Fe\{P(OR)_3\}_4(NO)]^+$ (^{31}P),[640] $Ru(NHC_6H_4PPh_2)(NH_2C_6H_4PPh_2)Cl_2$ (^{31}P),[641] $[M(R\text{-}bipy)_3]^{2+}$ (M = Ru, Os; ^{13}C),[642] $[Ru_2N(en)_5]^{5+}$ (^{13}C),[643] $Ru_2Cl_4\{(-)\text{-}2,2\text{-dimethyl-}4,5\text{-}(PPh_2CH_2)_2\text{-}1,3\text{-dioxolane}\}$ (^{31}P),[644] $(Me_3P)_3MS_6$ (M = Ru, Os; ^{31}P),[645] $(Me_3P)_3MS_7$ (M = Ru, Os; ^{31}P),[646] $Ru_3(\mu\text{-}dppm)_2(CO)_8(\mu\text{-}A)$ {A = AgO_2CCF_3, $Hg(O_2CCF_3)_2$, $Cu(NCMe)BF_4$, HO_2CCF_3; ^{31}P},[647] $Ru(O_2CCF_3)_2(dppm)_2$ (^{31}P),[648] $RuCl_2\{Ph_2P(CH_2)_nPPh_2\}(PPh_3)$ (^{31}P),[649] $RuCl_2\{P(tol)_2C_{12}H_7S\}$ (^{31}P),[650] and $Os_2(\mu\text{-}O)(\mu\text{-}dppm)_2Cl_6$ (^{31}P).[651]

Complexes of Co, Rh, and Ir.—Reviews have appeared on 'Cobalt-59'[652] and 'Rhodium-103'[653] n.m.r. spectroscopy. 'The use of Cr and Co complexes of nucleotides or pyrophosphate as probes for structural features of binding of natural metal substrate to enzymes acting on phosphorus containing compounds' has been reviewed.[654]

Some organometallic Rh(I) derivatives have been studied by ^{103}Rh

n.m.r. spectroscopy. The chemical-shift range extends from 609 to 2714.7 p.p.m. The ^{13}C and ^{31}P n.m.r. spectra were also recorded. For compounds containing cyclo-octa-1,5-diene, a linear correlation was observed between δ(^{103}Rh) and δ(^{13}C) (olefinic carbon atoms).[655]

The ^{1}H n.m.r spectrum of fac-$RhH_3(PMe_3)_3$ has been analyzed as $[AX]_3M$. The ^{13}C and ^{31}P n.m.r. spectra of $Ru(CH_2PMe_2)_2(PMe_2)_2$ were also analyzed.[656] A detailed analysis of the ^{1}H and ^{13}C n.m.r. spectra of $[Rh_4(\eta^5\text{-}C_5Me_5)_4H_4]^{2+}$ has been performed and $^{1}J(^{103}Rh,^{103}Rh)$ deduced.[657] N.m.r. data have also been reported for $CoH(BH_3CN)(PPh_3)_3$ (^{31}P),[658] $(\eta^5\text{-}C_5Me_5)RhH_2(SiEt_3)_2$ (^{13}C, ^{29}Si, ^{103}Rh),[659] $RhH_2Cl\{P(p\text{-tol})_3\}_2$ (^{31}P),[660] $(\eta^5\text{-}C_5Me_5)RhH(c\text{-Pr})(PMe_3)$ (^{13}C, ^{31}P),[661] $[RhH(CN)_5]^{3-}$ (^{13}C),[662] $[RhClH(SH)(PPh_3)_2]_2$ (^{31}P),[663] $[(Pr^i{}_2PCH_2CH_2PPr^i{}_2)Rh]_2(\mu\text{-H})(CH{=}CH_2)$ (^{13}C, ^{31}P),[664] $Rh_2H_2(CO)_2(dppm)_2$ (^{31}P),[665] $[(dppe)Rh(\mu\text{-H})(\mu\text{-Cl})IrH(PEt_3)_3]^+$ (^{31}P),[666] $\{IrH_3(PMe_2Ph)_3\}_2CuPF_6$ (^{31}P),[667] $(\eta^5\text{-}C_5Me_5)Ir(SiEt_3)_2H_2$ (^{13}C),[668] $[Ir(CO)H_2(PPh_3)_3][PF_2Se]$ (^{19}F, ^{31}P, ^{77}Se),[669] $IrH(Ar)(O_2CR)(CO)(PPh_3)_2$ (^{31}P),[670] $(\eta^5\text{-}C_5Me_5)(PMe_3)IrH(C_8H_{15})$ (^{13}C),[671] $IrHCl(PMe_2Ph)_3(R^1R^2PO)$ (^{31}P),[672] $IrHCl(CO)(CH{=}CMeCH_2PBu^t{}_2)\{PBu^t{}_2CH_2(c\text{-Pr})\}$ (^{13}C, ^{31}P),[673] and $[IrX(CO)(dppe)_2]^{2+}$ (X = Cl, Br, H; ^{31}P).[674]

The structures of heptamethyl α- and β-methylcob(III)yrinate have been investigated using n.O.e measurements.[675] For $(PPh_3)_2Rh\{=C(NR)CH_2CH_2NR\}$, $^{1}J(^{103}Rh,^{31}P)$ indicates that the carbene has a greater trans influence than PPh_3.[676] The ^{199}Hg n.m.r. spectrum of $RhCl_2(HgCl)(CO)(PEtPh_2)_2$ shows the first $^{1}J(^{199}Hg,^{103}Rh)$. The ^{31}P n.m.r. spectrum was also recorded.[677] Analysis of the ^{31}P n.m.r. spectrum of $(\eta^5\text{-}C_5Me_5)_2Rh_2(\mu\text{-}CH_2)_2X_2$ has given $^{1}J(^{103}Rh,^{103}Rh)$ = 12.2Hz. The ^{13}C n.m.r. spectrum was also recorded.[678] N.m.r. data have also been recorded for alkylcobalamins (^{13}C),[679] $MMe_3(PR_3)_3$ (M = Co, Rh, Ir; ^{31}P),[680] $[(\eta^5\text{-}C_5Me_5)CoEt(PR_3)]^+$ (^{13}C),[681] $CoMe_2(\eta^2\text{-}CH_2{=}PMe_2)(PMe_3)_2$ (^{31}P),[682] $Co(CMe_2C_2Me)(dmgH)_2py$ (^{13}C),[683] difluoromethylcobalamin (^{13}C, ^{19}F),[684] $(\eta^5\text{-}C_5H_5)Co_2(CO)_2(\mu\text{-}C{=}CH_2)$ (^{13}C),[685] $(\mu\text{-}R_2C{=}CS)_2Co_2(CO)_4$ (^{13}C),[686] $MeRhI_2(PPh_3)_2$ (^{13}C),[687] $(\eta^5\text{-}C_5H_5)Rh(PMe_3)_2(Al_2Me_4Cl_2)$ (^{13}C, ^{31}P),[688] $(\eta^5\text{-}C_5Me_5)Rh(CO)(CH_2PPh)$ (^{31}P),[689] $[Rh(nbd)\{CH_2(4\text{-}NO_2pz)_2\}]^+$ (^{13}C),[690] $Rh(p\text{-tol})\{Ph(CH_2CH_2CH_2PPh_2)_2\}(CO_2)$ (^{31}P),[691] $Rh(mesityl)\{P(OMe)_3\}_3$ (^{31}P),[692] $[M(C_6H_4py)_2Cl]$ (M = Rh, Ir; ^{13}C),[693] $Rh(C_4H_3N\text{-}2\text{-}py)Cl_2(PBu^n{}_3)_2$ (^{13}C),[694] $Rh(C_6F_4PPh_2)$(8-hydroxyquinolinato)$_2$ (^{19}F, ^{31}P),[695] $RhCl_2(COEt)(PPh_3)_2$ (^{31}P),[696] Rh{8-C(O)quinoline}$(CH_2Ph)(py)Cl$ (^{13}C),[697] $(\eta^5\text{-}C_5Me_5)Rh(CO)(CO_2R)_2$ (^{13}C),[698] $(\eta^5\text{-}C_5Me_5)Rh\{o\text{-}C(O)C_6Cl_4O\}(PPh_3)$

(^{31}P),[699] $\{\eta^2\text{-}(\eta^5\text{-}C_5H_5)Rh(PPr^i{}_3)\text{=}C\text{=}CHR\}RhCl(PPr^i{}_3)_2$ (^{13}C),[700] $(\eta^5\text{-}C_5Me_5)_2Rh_2(CO)_2(\mu\text{-}CR^1R^2)$ (^{13}C),[701] $(\eta^5\text{-}C_5Me_5)IrMe_4$ (^{13}C),[702] $[(\eta^5\text{-}C_5H_4Me)IrMe\{P(C_6H_4F)_3\}_2]^+$ (^{13}C),[703] $IrR(CO)\{PhP(CH_2CH_2CH_2\text{-}PPh_2)_2\}$ (^{31}P),[704] $Ir(PMe_3)_3Br\{CH_2C(tol)\text{=}Ctol\}$ (^{13}C),[705] $[Ir(CH_2O\text{-}CMe\text{=}NH)(PMe_3)_4]^{2+}$ (^{13}C),[706] $[Ir(C_5H_3NC_5H_4N)(bipy)_2]^{2+}$ (^{13}C),[707] and <u>trans</u>-$Ir(CO)Cl\{Zn[$<u>o</u>-(Ph_2P)benzoylpinacolonato$]_2\}$ (^{31}P).[708]

A drastic decrease in $^1\underline{J}(^{13}C,^{13}C)$ is observed on going from C_2H_2 to the complexes $Co_2(CO)_6(C_2H_2)$ and $Co_4(CO)_{10}(C_2H_2)$. The relationship between $^1\underline{J}(^{13}C,^{13}C)$ and the hybridization state of the carbon atoms was evaluated by CNDO calculations.[709] $Rh_2(O_2CCF_3)_4$ has been shown to be a shift reagent for olefins.[710] The ^{31}P n.m.r. spectrum of $[Ir_2(CO)_3(\mu\text{-}MeO_2CC\text{=}CCO_2Me)(dppm)_2]^+$ has been analyzed as $[AB]_2$.[711] N.m.r. data have also been reported for $Co_2(CO)_4(dppm)\text{-}(CF_3C_2CF_3)$ (^{19}F, ^{31}P),[712] $(\eta^2\text{-}C_2H_4)_2Rh(\mu\text{-}Cl)_2Rh\{(Ph_2PCH_2)_2NCHMe\text{-}CO_2Me\}$ (^{13}C, ^{31}P),[713] $Rh(\eta^2\text{-}C_2H_4)Cl\{ROP(OCH_2CH_2NMeCH_2)_2\}$ (^{13}C, ^{31}P),[714] $\{(\eta^2\text{-}CH_2CHCH_2)(\eta^2\text{-}MeCH\text{=}CH)SiMe_2\}Rh(\eta^5\text{-}$indenyl$)$ (^{13}C),[715] $Rh_2X_2(\mu\text{-}CF_3C_2CF_3)(dppm)_2$ (^{19}F, ^{31}P),[716] $Ir_2Cl_2(CO)_2(\mu\text{-}CF_3CCCF_3)\text{-}(\mu\text{-}dppm)_2$ (^{19}F, ^{31}P),[717] $[(R^1C_2CR^2R^3)Co_2(CO)_6]^+$ (^{13}C),[718] $[Rh(\eta^3\text{-}C_3H_5)(\eta^5\text{-}C_5H_5)(PPh_3)]^+$ (^{13}C, ^{31}P),[719] $(\eta^3\text{-}2\text{-}MeC_3H_4)Rh(PPr^i{}_2PCH_2CH_2\text{-}PPr^i{}_2)$ (^{31}P),[720] $(\eta^3\text{-}2\text{-}MeC_3H_4)Rh\{PH(c\text{-}hexyl)_2\}_2$ (^{31}P),[721] $[($triphos$)Rh(\mu\text{-}C_2S_4)Rh($triphos$)]^{2+}$ (^{31}P),[722] and $(\eta^5\text{-}C_5H_5)_2Rh_2\{\mu_2\text{-}[\eta^3\text{-}C(CF_3)C(CF_3)CONPh]\}$ (^{13}C, ^{19}F).[723]

A correlation between the ^{13}C and ^{59}Co n.m.r. spectra of substituted $(\eta^5\text{-}C_5H_5)Co$ complexes and their catalytic properties in the synthesis of pyridine derivatives has been examined.[724] Shifts induced by $M(fod)_3$ (M = Eu, Yb, or Pr) in some $(\eta^5\text{-}C_5H_5)Co$ complexes have been reported.[725] ^{1}H n.O.e. measurements have been used to investigate complexes $[Rh(\eta^4\text{-}cod)(NH_3)_2]^+$ and dibenzo-crown ethers.[726] The ^{1}H and ^{13}C n.m.r. spectra of 10 pyrazolate-bridged rhodium complexes with CO or 1,5-cod ligands have been recorded and the central N_4Rh_2 ring shown to have a boat conformation.[727] ^{13}C n.m.r. spectroscopy has been used to determine the hapticity of indenyl and naphthalene in some complexes of iridium.[728] N.m.r. data have also been reported for $\{\eta^4\text{-}C_4H(SiMe_3)(SiEt_3)(CHMePh)\}\text{-}(\eta^5\text{-}C_5H_5)Co$ (^{13}C),[729] $(\eta^4\text{-}cod)(\eta^5\text{-}PhC_5H_4)Co$ (^{13}C),[730] $(\eta^5\text{-}C_5H_5)Co\text{-}(PhHC\text{=}CPhCPh\text{=}C\text{=}O)$ (^{13}C),[731] Rh(hepta-2,5-diene)(acac) (^{13}C),[732] $(\eta^4\text{-}cod)RhCl\{PMe(SiMe_3)_2\}_2$ (^{31}P),[733] $[(\eta^4\text{-}cod)Rh\{Bu^tO_2CNC_4H_6\text{-}(PPh_2)(CH_2PPh_2)\}]^+$ (^{31}P),[734] $[(\eta^4\text{-}cod)Rh(SC_4H_8)(PPh_3)]^+$ (^{31}P),[735] $[(\eta^4\text{-}nbd)Rh\{Ph_2PCH_2CH_2OCH_2CH_2N[(CH_2CH_2)_2S]_2\}Zn]^{3+}$ (^{31}P),[736] $[(\eta^4\text{-}nbd)Rh(Ph_2PNC_4H_7CH_2NRPPh_2)]^+$ (^{31}P),[737] $[(\eta^4\text{-}nbd)RhCl\{PPh_2(CH_2)_2\text{-}$

NMe$_3$}]$^+$ (^{31}P),[738] [(η^5-C$_5$H$_5$)Rh(C$_7$H$_7$CHOR)]$^+$ (^{13}C),[739] Ir{η^4-C-(CH$_2$)$_3$}Cl(CO)(PPh$_3$) (^{13}C),[740] [(η^4-cod)Ir(μ-pz)]$_2$ (^{13}C, ^{31}P),[741] Ir(η^4-cod){(CN)$_2$C$_3$N$_2$}$_2$(PPh$_3$) (^{13}C, ^{31}P),[742] (η^4-nbd)$_2$Ir$_2$(μ-PPh$_2$)$_2$ (^{31}P),[743] [(η^4-cod)Ir(η^2-BrC$_6$H$_4$PPh$_2$)]$^+$ (^{31}P),[744] [(η^4-cod)Ir(CS$_2$)-(PR$_3$)$_2$]$^+$, [Rh(CO)(PPh$_3$)$_2$(S$_2$CPPh$_3$)]$^+$ (^{31}P),[745] (η^5-C$_5$H$_5$)Co(NHC$_6$H$_3$FN-C$_6$H$_3$F$_2$) (^{13}C, ^{19}F),[746] (η^5-C$_5$Me$_5$)Co(O$_2$C$_6$H$_4$) (^{13}C),[747] (η^5-C$_5$Me$_5$)$_2$-M$_2$(CO)$_2$(μ-SO$_2$) (M = Co, Rh; ^{13}C),[748] (η^5-C$_5$H$_5$)Rh(PhCHCHCHCH$_2$CH=CHR) (^{13}C),[749] [(η^5-C$_5$H$_5$)$_2$Rh$_2$(CO){μ-N(O)CRCR}]$^+$ (^{13}C),[750] [{η^5-C$_5$H$_4$C-(C$_6$H$_4$X-4)$_2$}RhL$_2$]$^+$ (^{13}C),[751] (η^5-C$_5$Me$_5$)$_2$Rh$_2$(μ-O$_2$CCR^{1}CR^2)(μ-CO) (^{13}C),[752] (η^5-C$_5$Me$_5$)RhLPMe$_2$PMe$_2$Rh(η^5-C$_5$Me$_5$)(μ-CO)$_2$Rh(η^5-C$_5$Me$_5$) (^{13}C, ^{31}P),[753] and [(η^5-C$_5$Me$_5$)Ir(C$_6$H$_7$)]$^+$ (^{13}C).[754]

$^4\underline{J}$(^{31}P,^{31}P) is large in [Co(CO)$_3$(Ph$_2$PNMeCHMePh)]$_2$ but small in [Mn(CO)$_4$(S)(Ph$_2$PNMeCHMePh)]$_2$.[755] N.m.r. data have also been reported for Rh$_4$(CO)$_9$(PR$_3$)$_3$ (^{31}P),[756] Co$_4$(CO)$_9$(Ph$_3$P)$_3$CH (^{13}C),[757] PdPtCo$_2$(CO)$_7$(dppm)$_2$ (^{31}P),[758] MeC{CH$_2$SRh(CO)[P(OMe)$_3$]}$_3$ (^{31}P),[759] Rh(CO)(2-amino-1,3,2-dioxaphosphorinane)(acac),[760] RhCl(CO)$_2$(PPh$_2$-C$_4$H$_3$S) (^{13}C, ^{31}P),[761] Rh(CO)(PPh$_3$)$_2$Cl (^{31}P),[762] Rh(CO)Cl{R^{1}P(NR^2)$_2$-PR^1}$_2$ (^{31}P),[763] Rh(CO)Cl{PhP(OCR$_2$CH$_2$)(OCH$_2$CHMe)NH} (^{31}P),[764] Rh(CO)Cl(PPh$_3$)$_2$(1,2-O$_2$C$_6$Br$_4$) (^{31}P),[765] [Rh(CO)(PPh$_3$)$_2$(6-mercapto-guanosine)]$^+$ (^{13}C, ^{31}P),[766] Rh(CO)(PCy$_3$)$_2$(O$_2$CH) (^{13}C, ^{31}P),[767] Rh(CO)X(Ph$_2$PCH$_2$CO$_2$H)$_2$ (^{31}P),[768] Rh$_2$Cl$_2$(CO){SCNC(S)N(CO$_2$Et)$_2$}(dppm)$_2$ (^{13}C, ^{31}P),[769] Rh$_2$(μ-dppm)$_2$(1,2-O$_2$C$_6$Cl$_4$)(CO) (^{31}P),[770] [Rh$_2$(μ-CN)-(μ-CO)(CO)$_2$(μ-dppm)$_2$]$^+$ (^{13}C, ^{31}P),[771] [Rh$_2$(CO)$_2$(SCNMe$_2$)(dppm)$_2$]$^+$ (^{13}C, ^{31}P),[772] [Rh$_2$(μ-dppm)$_2$(μ-S$_2$COEt)(μ-CO)(CO)$_6$]$^+$ (^{31}P),[773] {Rh(CO)(dppe)(MeCO)$_2$C}C$_6$H$_4$ (^{31}P),[774] [Rh$_2$(μ-Cl)(μ-Ph$_2$PNHPPh$_2$)$_2$-(CO)$_2$]$^+$ (^{31}P),[775] (OC)$_2$ClRh(μ-Cl)Rh(CO){HPhP(OCMe$_2$CH$_2$)$_2$N} (^{31}P),[776] Rh$_2$(CO)$_2$(PBu^t$_3$)$_2$Cl(SR) (^{31}P),[777] Cl(CO)Rh(μ-Ph$_2$AsCH$_2$PPh$_2$)$_2$ (^{31}P),[778] [Rh$_3$(μ_3-E)$_2$(CO)$_6$]$^-$ (E = S, Se; ^{13}C),[779] [Rh$_3${μ-(Ph$_2$P-CH$_2$)$_2$PPh}$_2$(CO)$_2$I$_2$]$^+$ (^{31}P),[780] Rh$_4$(CO)$_9$(SCH$_2$SCH$_2$SCH$_2$) (^{13}C),[781] Ir(PPh$_3$)(CO)(PS) (^{31}P),[782] Ir(CO)Cl$_2$(PEt$_3$)$_2$L {L = PCl$_2$BH$_3$, PH(Se)-SeH; ^{11}B, ^{13}C, ^{31}P, ^{77}Se},[783] Ir{3,5-(CF$_3$)$_2$pz}(CO)(PPh$_3$)$_2$ (^{19}F, ^{31}P),[784] IrCl$_2$(CO)(PEt$_3$)$_2$PF$_4$ (^{19}F, ^{31}P),[785] [Ir$_2$Cl(CO)$_3$(μ-CO)-(dppm)$_2$]$^+$ (^{31}P),[786] Ir$_2$X$_2$(CO)$_2$(dppp)$_2$ (X = Br, I; ^{31}P),[787] Ir$_4$(CO)$_{11}$(PF$_2$OR) (^{19}F, ^{31}P),[788] Ir$_4$(CO)$_9$(Ph$_2$P)$_3$CH (^{13}C),[789] Ir$_4$(CO)$_4$(PPh$_3$)$_4${[(NC)$_2$C$_3$N$_2$]$_2$}$_2$ (^{13}C, ^{31}P),[790] and Ir$_4$(CO)$_{12-\underline{n}}$L$_{\underline{n}}$ (L = olefin, PR$_3$; ^{13}C, ^{31}P).[791]

The ^{59}Co, ^{14}N, and ^{13}C $\underline{T}_1$ values have been reported for [Co(CN)$_6$]$^{3-}$ in aqueous solutions as functions of temperature and counter-ion. The ^{59}Co relaxation rate is dominated by the nuclear electric quadrupole relaxation mechanism, but measurements of the

^{13}C and ^{15}N relaxation rates show that modulation of the electric field gradient at the Co nucleus is not coupled to rotation of the symmetric complex ion.[792] The very large chemical-shift range of ^{59}Co has yielded well resolved lines for each of the 19 isotopic species of $[Co(NH_{\underline{n}}D_{3-\underline{n}})_6]^{3+}$ and each of the 13 isotopic species of $[Co(H_{\underline{n}}D_{2-\underline{n}}NCH_2CH_2NH_{\underline{m}}D_{2-\underline{m}})_3]^{3+}$.[793] ^{13}C n.m.r. assignments in cis-$[CoL_2(en)_2]^{\underline{n}+}$ [L_2 = $(CN)_2$, phen, $(NO_2)_2$, oxalato] have been made using ^{15}N-enriched ethylene diamine.[794] ^{15}N shifts have been reported for 5-co-ordinated Co(III) complexes with Schiff bases and bent apical nitrosyl ligands. As in C-nitroso groups, the ^{15}N is strongly deshielded by 500 to 800 p.p.m. relative to comparable linear NO ligands. The ^{15}NO and ^{59}Co shielding tends to decrease with a decrease in the MNO angle and in the energy of the longer wavelength electronic absorption.[795] N.m.r. data have also been reported for $[(NH_3)_5Co(NC)Fe(NO)(CN)_4]^+$ (^{13}C),[796] $Co(NH_3)_5(H_2P_3O_{13})$ (^{31}P),[797] $Co(NH_3)_5(PO_4)$ (^{31}P),[798] (triphosphate)$(NH_3)_4$Co-containing species (^{31}P),[799] $Co(NH_3)_4$ complexes of adenosine 5'-O-(1-thiodiphosphate) (^{31}P),[800] $Co(NH_3)_4\{O_6(S)P_2(OR)\}$ (^{31}P),[801] and $[Co(HN{=}CH{-}C_6H_4O)_2(NH_3)_2]^+$ (^{13}C).[802]

The different isomers of μ-peroxy$(en)_2(tren)_2Co_2$ have been identified by ^{59}Co n.m.r. spectroscopy. The ^{59}Co n.m.r. signals were assigned on the basis of a point - charge model for the chemical shifts and line widths.[803] N.m.r. data have also been reported for $[Co(pren)en]^+$ (^{13}C),[804] $[\{O_2NC(CH_2NC_6H_4NCH_2)_3NNO_2\}Co]^{3+}$ (^{13}C),[805] $[(2,3,2\text{-tet})\overline{Co(NH{=}CRCO_2})]^{2+}$ (^{13}C),[806] [Co(1,2-diamino-3-methylcyclohexane)$_3]^{3+}$ (^{13}C),[807] cyclodextrin-(μ-α,ω-diaminoalkane)-bis$[Cl(en)_2Co]$ (^{13}C),[808] $[Co(NH_2CH_2CH_2SO_2\text{-}\underline{N,O})(en)_2]^{2+}$ (^{13}C),[809] $[Co(en)_2(OH_2)(4\text{-Me-imidazoleH})]^{3+}$ (^{13}C),[810] $[Co(en)_2(maleato)]^+$ (^{13}C),[811] [Co(3,6-Me$_2$-3,6-diazaoctane-1,8-dithiolato)en]$^+$ (^{13}C),[812] [Co(2-Me$_2$-8-quinolinol)$_2$(en)]$^+$ (^{13}C),[813] [Co(en)(2,9-Me$_2$-4,7-diazadecane-2,9-dithiolate)]$^+$ complexes (^{13}C),[814] $[Co(dien)(bn)Cl]^{2+}$ (^{13}C),[815] and $Co(dien)(NO_2)_2Cl$ (^{59}Co).[816]

The ^1H n.m.r spectra of the diastereomeric complexes formed by the co-ordination of nitrogenous enantiomeric bases with Co(III) meso-tetraphenylporphyrin has allowed the differentiation of these complexes.[817] ^1H and 2D $^{13}C/^1H$ n.m.r. spectra have been used to determine the assignment of signals due to heptamethyldicyanocobyrinate.[818] ^{13}C n.m.r. spectra have been used to demonstrate the origin of hydrogen atoms at C-18 and C-19 in the biosynthesis of vitamin B_{12}.[819] 1,9-Diamino-4-methyl-3,7-diazanonane-3,7-diacetato(2-)cobalt(III) nitrate monohydrate exhibits large ^{13}C n.m.r. steric shifts.[820] The effect of equatorial ligands on the ^{19}F n.m.r. chemical shifts of the C_6H_4F group bonded to some cobalt chelates has been examined. The trend of increasing inductive donation is $Co(dmgBF_2)_2py$ < $Co(dpgH)_2py$ < $Co(dmgH)_2py$ < Co(SALEN) < Co(SALOPH) < Co(BAE).[821] The ^{13}C n.m.r spectra of the corresponding Rh and Co diastereomers of $[M\{(\pm)\text{-bn}\}_3]^{3+}$ are virtually identical. The possible contribution of magnetic anisotropy of the metal ion to ^{13}C n.m.r. chemical-shift differentiation of diastereotopic carbon atoms in these complexes was discussed.[822] The first direct observation of $^{1}\underline{J}(^{103}Rh,^{103}Rh)$ has been reported for $Rh_2(mhp)_4$ from direct ^{103}Rh n.m.r. measurements.[823] N.m.r. data have also been reported for (1-hydro-8H-HDP)Co(I) (^{13}C),[824] $[\{RC(CH_2NHCH_2CH_2NHCH_2)_3CCl\}Co]^{3+}$ (^{13}C),[825] [S-(4-amino-2-azabutyl-5-nitro-3,7-diaza-1,9-nonane-diamine)Co]$^{3+}$ (^{13}C),[826] [Co(2,9-diamino-3,3,8,8-tetramethyl-4,7-dithia-1,10-decanedioate)]$^{+}$ (^{13}C),[827] $[N(CH_2CH_2PPh_2)_3Co(CS)]^+$ (^{31}P),[828] $[Co(CO_3)(Me_2cyclen)]^+$ (^{13}C),[829] [S-(2-aminoethyl)-L-homocysteinato]Co(III) complexes with bidentate ligands (^{13}C),[830] Co(III) complexes with $[CN]^-$ and tetraamines (^{13}C),[831] O_2 complexes of $(\text{L-histidinato})_2Co(II)$ (^{13}C, ^{59}Co),[832] trans-(Cl,NO_2)-$[CoCl(NO_2)(ampr)_2]^+$ (^{13}C),[833] 5-nitrosalicylato-(quadridentateamine)Co(III) complexes (^{13}C),[834] $[\{Co(SALEN)\}_2O_2(H_2O)]_2$ (^{59}Co),[835] $[M(en)_3]^{3+}$ (M = Rh, Ir; ^{13}C),[836] Cl(partially hydrogenated tetraphenylporphyrinato)Rh(III) (^{13}C),[837] $[Rh(Ph_2PC_6H_4CH{=}NCH_2CH_2N{=}CHC_6H_4PPh_2)]^+$ (^{31}P),[838] Rh, Ir, Pd, and Pt complexes of oxazepam (^{13}C),[839] $[Ir(Me_3\text{-barralene})(PhNRPh)]^+$ (^{13}C),[840] and $[Ir(NO)\{3,6\text{-bis}(2'\text{-pyridyl})pyridazine\}(PPh_3)_2(PdCl_2)]^{2+}$ (^{31}P).[841, 842]

Linear n.m.r. correlations have been found for $\delta(^{31}P)$, $\delta(^{59}Co)$, and coupling constants for $[Co(PR_3)_6]^{3+}$.[843] For $Co(acac)_{\underline{n}}\{PhC(S)\text{-}CHC(O)Ph\}_{3-\underline{n}}$, there is an additivity effect on the ^{59}Co n.m.r. chemical shift of the stepwise replacement of O by S.[844] The products of bromination of $Co(acac)_3$ have been identified by ^{13}C n.m.r. spectroscopy.[845] ^{31}P and ^{103}Rh n.m.r. studies of $Rh_{55}(PBu^t{}_3)_{12}Cl_{20}$ show high mobility either of the ligands or of the rhodium atoms or of both.[846] The chemical shifts have been studied for ^{17}O and ^{103}Rh in chloroaquo complexes of Rh(III).[847] Concentrated solutions of Rh(III) chloride have been studied by ^{17}O, ^{35}Cl, ^{103}Rh, and ^{133}Cs n.m.r. spectroscopy. The compounds are polymeric, but are converted to monomers on standing.[848] N.m.r. data have also been reported for $[Co\{CMe(CH_2PPh_2)_3\}P_2E]^+$ (E = S,

Se; ^{31}P),[849] $[(triphos)Co(E_2S)Pt(PPh_3)_2]^+$ (E = P, As; ^{31}P),[850] $[Co(acac)_2(dppe)]^+$ (^{31}P),[851] $[M(Ph_2PC_6H_4CH{=}NC_6H_3MeC_6H_3MeN{=}CH{-}C_6H_4PPh_2)]^+$ (M = Rh, Ir; ^{31}P),[852] $[Rh(dppe)_2]^+$ (^{31}P),[853] $Rh_2(\mu{-}PPh_2)_2(PEt_3)_4$ (^{31}P),[854] $[MO_2(Ph_2PCH_2CH_2SR)_2]^+$ (M = Rh, Ir; ^{31}P),[855] $[RhO_2(Ph_2PCH_2CH_2PPhCH_2CH_2PPhCH_2CH_2PPh_2)]^+$ (^{31}P),[856] $[MeC(CH_2PPh_2)_2{-}Rh(\mu{-}C_2Se_4)Rh(Ph_2CH_2)_3CMe]^{2+}$ (^{31}P),[857] $M(PPr^i_3)_2(SO)Cl$ (M = Rh, Ir; ^{31}P),[858] $M(PPr^i_3)_2(SO)Cl$ (M = Rh, Ir; ^{31}P),[859] $Rh_2(O_2CCF_3)_4L$ (^{19}F, ^{31}P),[860] $Rh(H_2O)_4(ADP)$ (^{13}C, ^{31}P),[861] $[Rh_2(O_2CR)_2]^{2+}$ (^{13}C),[862] $[Ir(\eta^2{-}E_2O)(dppe)_2]^+$ (E = S, Se; ^{13}C, ^{31}P),[863] and $IrCl(PF_2NMe_2)_3$ (^{19}F, ^{31}P).[864]

Complexes of Ni, Pd, and Pt.—'N.m.r. techniques for studying platinum catalysts' has appeared.[865] ^{119}Sn and ^{195}Pt n.m.r. solution data for $[PdCl_2(SnCl_3)_2]^{2-}$, $[Pd(SnCl_3)_5]^{3-}$, $[PtCl(SnCl_3)_3]^{2-}$, $[Pt(SnCl_3)_4]^{2-}$, $[PtH(SnCl_3)_4]^{3-}$, and $[PtH(SnCl_3)_2(PEt_3)_2]^-$ have been reported and compared with those for known related complexes. An increasing number of $[SnCl_3]^-$ ligands per metal results in deshielding of the ^{119}Sn, but a shielding of the ^{195}Pt nuclei. The values of $^1\underline{J}(^{195}Pt,^{119}Sn)$ and $^2\underline{J}(^{119}Sn,^{117}Sn)$ are especially useful probes for molecular structure.[866] From the ^{31}P n.m.r. spectrum of $PtH(C_6H_4Y)(PEt_3)_2$, there is a linear correlation between $^1\underline{J}(^{195}Pt,{-}^{31}P)$ and (Pt-H).[867] N.m.r. data have also been reported for trans-$PdH(PBu^n_3)L_n$ (^{31}P),[868] $[\{2,1,3{-}(Ph_2PCH_2)C_6H_3Pt\}_2(\mu{-}H)]^+$ (M = Pd, Pt; ^{31}P),[869] $PtH_2Ph(SnMe_3)(PEt_3)_2$ (^{31}P),[870] $PtX(C_6Y_5)(PEt_3)_2$ (X = H, MeOH, Cl, Br, MeCN, HCO_2, $SnCl_3$; Y = Cl, F; ^{31}P),[871] $PtHCl\{PMe(OMe)_2\}_2$, $PtCl\{PMe(OMe)_2\}_2CHMeCO_2Me$ (^{31}P, ^{195}Pt),[872] $PtH(PPh_3)\{R_2PC(S)NPh\}$ (^{31}P, ^{195}Pt),[873] $Pt_2(\mu{-}H)(\mu{-}PBu^t_2)H(CH_2CMe_2{-}PBut_2)(PPh_3)$ (^{31}P),[874] and $[Pt_4H_2(PBu^t_3)_4]^{2+}$ (^{31}P).[875]

The ^1H n.m.r. assignments in $[Pd(C_7H_8OMe)Cl]_2$ have been revised.[876] For $[\{(p{-}tol)N{=}CHC_6H_3Y\}Pd(OAc)]_2$, there is a linear correlation between the ^{13}C shift of the palladated ring and the unsubstituted ring chemical shift.[877] For cis-$Pt(C_6H_4X)_2(PPh_3)_2$, the $\Delta\underline{G}^{\ddagger}$ of $(XC_6H_4)_2$ elimination is linearly related to $^1\underline{J}(^{195}Pt,^{31}P)$.[878] N.m.r. data have also been reported for $EtNiL_2(CN)$ (^{13}C, ^{31}P),[879] $(dppe)\overline{NiCH_2CH_2CO_2}$ (^{13}C, ^{31}P),[880] $L_2\overline{NiOC(O)CR{=}CR}$ (^{13}C),[881] $Ni(C_2Cl_3)_2(PEt_3)_2$ (^{13}C, ^{31}P),[882] $[(cdt)NiCOMe]^-$ (^{13}C),[883] (4) (M = Pd, Pt; ^{31}P, ^{195}Pt),[884] $PdR_2(PMe_3)_2$ (^{31}P),[885] $Pd(CH_2Bu^t)_2(dppe)$ (^{13}C),[886] $Pd(CH_2CN)(OCMeCHCO_2Et)(PPh_3)$ (^{13}C, ^{31}P),[887] $Pd(CHMe{-}CHMeNMe_2)Cl(NHMe_2)$ (^{13}C),[888] $\{C_4H_8O_2C(2{-}pyridyl)_2\}Pd\{CH(CO_2R)_2\}$ (^{13}C),[889] $[\overline{Pd\{CR^1R^2CR^3R^4ON}(O)\}Cl]_2$ (^{13}C),[890] $Pd(COPh)(CO)(PMePh_2)_2$

(^{13}C, ^{31}P),[891] $Pd(COCOR)Cl(PMePh_2)_2$ (^{13}C, ^{31}P),[892] $M(COCOPh)Cl(PPh_3)_2$ (M = Pd, Pt; ^{31}P),[893] Cl(N,N-dialkylbenzylamine-6,CN)-$(PPh_3)Pd$ complexes (^{13}C),[894] $(C_6H_2R^1R^2CH{=}NR^3)Pd(PPh_3)Br$ (^{13}C),[895] (5) (^{13}C),[896] $PdCl\{C(NHR){=}CHCOMe\}(PPh_3)_2$ (^{31}P),[897] $[M(C_{10}H_6CHMeNMe_2)\{(PPh_2)_2C_6H_4\}]^+$ (M = Pd, Pt; ^{13}C, ^{31}P),[898] $M(C_5H_3NCl)Cl(PPh_3)_2$ (M = Pd, Pt; ^{31}P),[899] $(R_3P)(C_6F_5)Pd(\mu\text{-}S_2CS)Pd(C_6F_5)(PR_3)_2$ (^{31}P),[900] $[M(C_6F_5)(PR^1R^2{}_2)_2(S_2CPR^3R^4{}_2)]^+$ (M = Pd, Pt; ^{31}P),[901] and $[(PPh_3)_2M\text{-}TlSn_8]_5$ (^{31}P, ^{119}Sn).[902]

The dependence of $^1J(^{195}Pt,^{31}P)$ on the electronic character of R in $(Ph_3P)_2MePtHgR$ has been studied.[903] The ^{31}P n.m.r. spectrum of trans-$Pt(CH_2CMe_2PBu^t{}_2)(SnCl_3)L$ shows $^2J(^{31}P,^{31}P)$ of 307 to 339Hz. The trans influence of L = $(p\text{-}XC_6H_4)_3P$ follows the order Cl < F < H < Me < MeO.[904] ^{13}C and ^{31}P n.m.r. spectroscopy and the Karplus equation have been applied to $Pt\{CH(py)Et\}X_2(PPh_3)$ and related compounds to determine conformation.[905] ^{19}F and ^{31}P n.m.r investigations of $(Ph_3P)_2PtR(O_2Bu^t)$ (R = CF_3, o-NCC_6H_4, Ph) have suggested a considerable covalent content in the Pt-O bond.[906] In the ^1H, ^{13}C, and ^{31}P n.m.r. spectra of $(\eta^4\text{-cod})Pt(C_6H_4R\text{-}p)_2$, the coupling constants show a clear correlation to the ligand configuration at Pt.[907] ^{195}Pt n.m.r. spectra have been measured for various ArPt(IV) derivatives. With increasing electron-withdrawing ability of R, the ^{195}Pt resonances shift towards low frequency. The ^{13}C chemical shifts were also given.[908] High-resolution ^{31}P n.m.r. spectra of lyotropic liquid crystals of metal poly-yne polymers, e.g. $[(Bu_3P)_2Pt(C_2)_2]_n$, have shown that magnetic anisotropy can be systematically controlled by including different transition metals in the polymers.[909] The ^{31}P n.m.r. spectra of trans-$(Ph_3P)_2ClPtHgR$ have been recorded. Values of ΔP, $^1J(^{195}Pt,^{31}P)$, and $^2J(^{199}Hg,^{31}P)$ varied slightly with changes in the electronic properties of the ligands and R. The Pt-Hg bonds are predominantly σ.[910] N.m.r. data have also been reported for $[Me_3Pt[(Me_2pz)_2CH_2]py]^+$ (^{13}C),[911] (2,2'-bipyrimidyl)$Pt(CF_3)_2$ (^{19}F),[912] $[PtMe_2(OR)(bipy)(OH_2)]^+$ (^2H, ^{13}C),[913] $PtMe(dppe)(CF_2CF_2OMe)$ (^{19}F, ^{31}P),[914] $PtMeCl(PEt_3)(AsPh_3)$ (^{31}P),[915] $[Pt_2Me_2(\mu\text{-dppm})_2Me_2(\mu\text{-}C_2Me)]^+$ (^{31}P),[916] $[Pt_2Me(py)(\mu\text{-dppm})_2]^+$ (^{31}P),[917] $[MePt_2(\mu\text{-dppm})_2(\eta^1\text{-dppm})]^+$ (^{31}P),[918] $Pt_2R^1{}_2(C_2R^2)(PEt_3)(\mu\text{-Li})_2$ (^7Li, ^{13}C, ^{31}P, ^{195}Pt),[919] $PtClR(PMe_3)_2$ (^{31}P),[920] $(CH_2)_4Pt(dppm)_2$ (^{31}P, ^{195}Pt),[921] $Pt_2(CH_2CMe_2PBu^t{}_2)_2(\mu\text{-}pz)_2$ (^{31}P),[922] $BrPt(PPh_3)_2(CH_2CH_2OPh)$ (^{13}C),[923] $[Pt(CH_2C_6H_4PPh_2)(PPh_3)_2]^+$ (^{13}C, ^{19}F, ^{31}P),[924] cis-$Pt(CH_2SOMe)_2(PEt_3)_2$ (^{31}P),[925] $Pt(CF_3)_2(NCPh)_2$ (^{19}F),[926] (6) {M = $Pd(\eta^3\text{-}2\text{-}MeC_3H_4)$, $Rh(\eta^4\text{-cod})$;

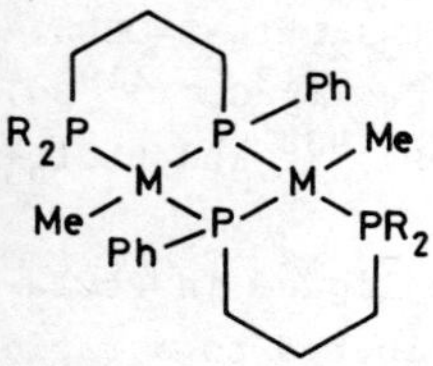

(4)

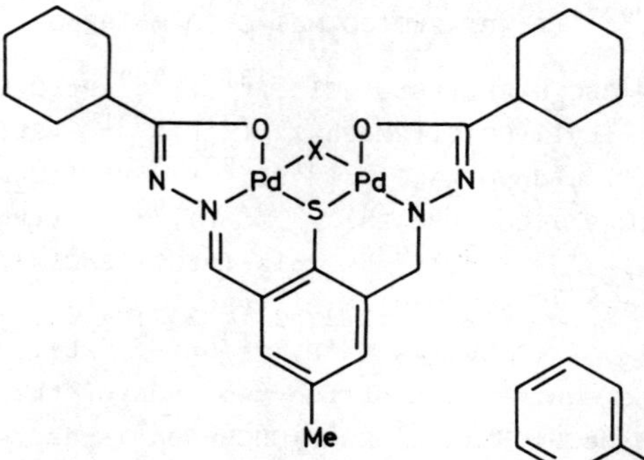

(5) X =

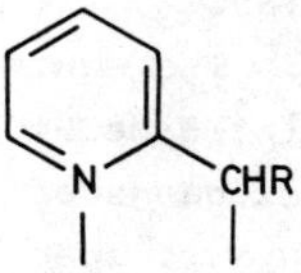

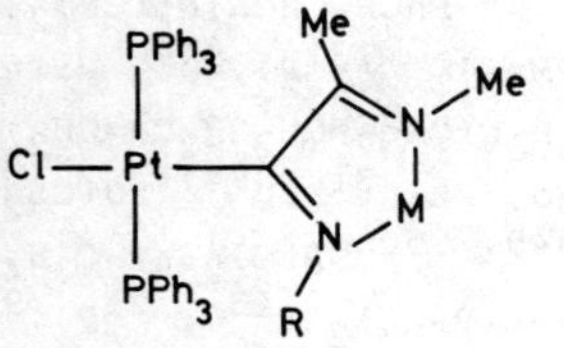

(6)

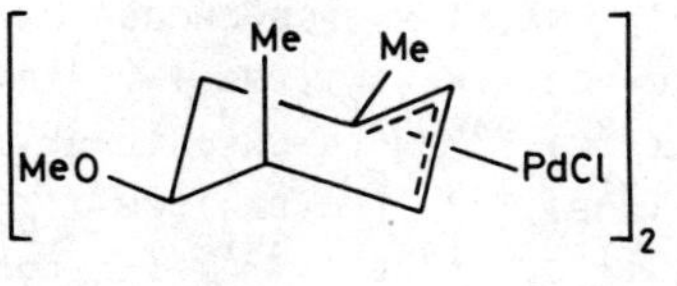

(7)

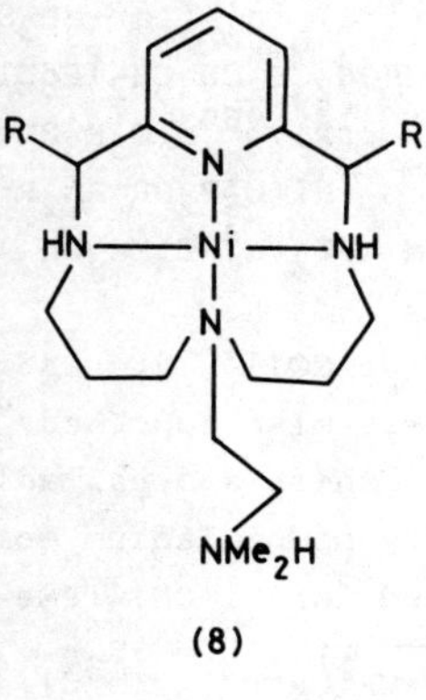

(8)

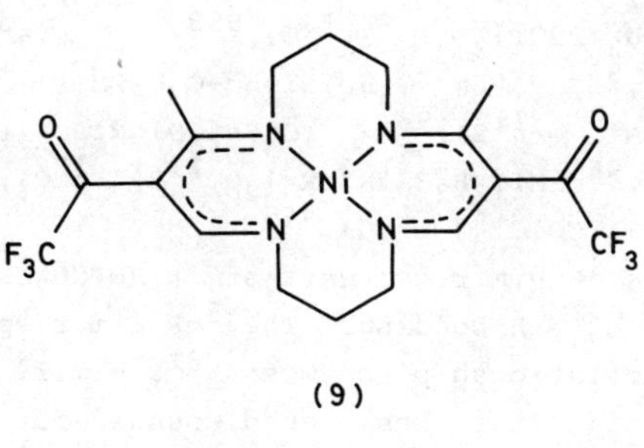

(9)

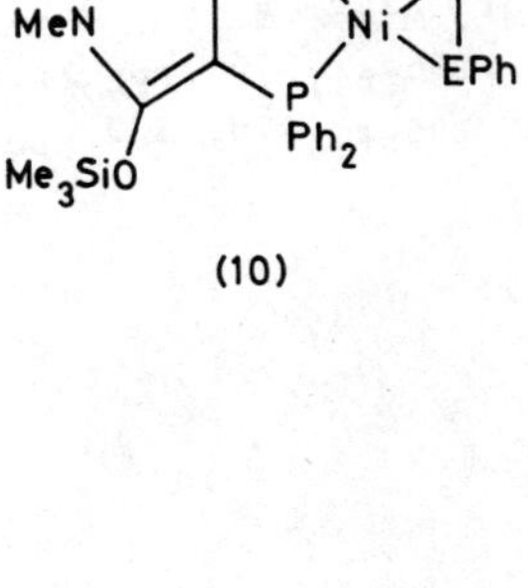

(10)

$^{31}P\}$,[927] $(Ph_3P)_2\overline{PtC(CO_2Me)=C(CO_2Me)NPhO}$ (^{31}P),[928] $PtCl(Et_3P)_2C_6H_3R$-$CH=NN=CHC_6H_3RPt(PEt_3)_2Cl$ (^{31}P),[929] $\overline{Pt(C_6H_4C_5H_3N)}_2$ (^{13}C, ^{195}Pt),[930] $Pt(mesityl)(CO)Cl(PMePh_2)$ (^{31}P),[931] $[Pt(C_6F_5)_3(dppm)]^-$ (^{31}P),[932] $RC_2PtCl(\mu\text{-}dppm)_2AgCl$ (^{31}P),[933] $PtCl(C_2R)(CO)L$ (^{13}C, ^{31}P),[934] $PtCl(C_2Ph)(CO)(PMe_2Ph)$ (^{31}P),[935] $[(PhC_2)Pt(PEt_3)_2Pt(\mu\text{-}C=CHPh)(PEt_3)_2]^+$ (^{13}C),[936] <u>cis</u>-$[PtCl_2(SnCl_3)(olefin)]^-$ (^{13}C, ^{119}Sn, ^{195}Pt),[937] $Pt_2(SnCl_3)_2Cl_2(PEt_3)_2(C_4F_6)_2$ (^{19}F, ^{31}P),[938] and $PtCl_{\underline{n}}(SnCl_3)_{2-\underline{n}}\{P(OR)_3\}_2$ (^{31}P, ^{119}Sn, ^{195}Pt).[939]

The equivalence of the two ends of the di-imine ligand in $PtCl_2(\eta^2\text{-}\underline{E}\text{-}MeCH=CHMe)(\eta^1\text{-}Bu^tN=CHCH=NBu^t)$ has been attributed to a rapid exchange of the two nitrogen atoms in the complex.[940] ^1H and ^{13}C n.m.r. spectroscopy has been used to determine the configuration of compounds such as (7).[941] N.m.r. data have also been reported for $(dmpe)(PR_3)Ni(\eta^2\text{-}C_2H_4)$ (^{13}C, ^{31}P),[942] $[(\eta^2\text{-}PhCH=CHCN)Ni(CN)_2]^{2-}$ (^{13}C),[943] $Ni(\eta^2\text{-}Me_3PCH=CHCOCH_2CMe_2Ph)Cl(PMe_3)$ (^{31}P),[944] $Ni\{\eta^2\text{-}(Me_3Si)_2C=PCH(SiMe_3)_2\}(PMe_3)_2$ (^{31}P),[945] $[C_5H_3N(CMe=NCH_2CH_2CH=CH_2)_2PdCl]^+$ (^{13}C),[946] $M^2(\mu\text{-}CS_2)_2(PBu^t_3)_2$ (M = Pd, Pt; ^{31}P),[947] $Pd(CS_2)(PR_3)_2$ (^{31}P),[948] $[Pd(PR_3)(CS_2)]_{\underline{n}}$ (^{31}P),[949, 950] $PtCl_2(\eta^2\text{-}C_2H_4)(imine)$ (^{13}C, ^{195}Pt),[951] $(\eta^2\text{-}norbornene)Pt(PPh_3)_2$ (^{13}C, ^{31}P),[952] $[(\eta^3\text{-}C_3H_5)_2\{(Bu^tC_6H_4O)_3P\}_2Ni_2(\mu\text{-}O_2PF_2)]^+$ (^{31}P),[953] $(\eta^5\text{-}C_5H_5)Ni(\eta^3\text{-}C_6H_9)$ (^{13}C),[954] $\overline{Ni(\eta^3\text{-}CHR^1CHCR^2CO_2})L$ (^{13}C),[955] $[(\eta^3\text{-}3\text{-}Me\text{-}butenyl)PdCl]_2$ (^{13}C),[956] $[Pd(\eta^3\text{-}CH_2CMeCMeCH_2OAc)Cl]_2$ (^{13}C),[957] $[(\eta^3\text{-}Me_3Si\text{-}CHCHCH_2)PdCl]_2$ (^{13}C),[958] $[(\eta^3\text{-}CH_2CClCHCH_2CH_2CH=CH_2)PdCl]_2$ (^{13}C),[959] $[(\eta^5\text{-}C_5H_5)Ni(\eta^4\text{-}C_7H_8PMe_3)]^+$ (^{13}C, ^{31}P),[960] $(C_3H_3BMeN\text{-}But)_2Ni$ (^{11}B),[961] $(C_4Ph_4CO)_2Ni$ (^{13}C),[962] $Ni(C_3B_2HR^1R^2{}_2R^3)_2$ (^{11}B),[963] $M(C_3B_2HR^1R^2{}_2R^3)_2$ (^{11}B, ^{13}C),[964] and $[Pt_2(\eta^4\text{-}C_4Me_4Ph_2)_2Cl_3][Pt(CO)Cl_3]$ (^{13}C).[965]

The ^1H n.m.r. spectra of $[\{HN(CH_2CH_2PPh_2)_3\}Ni(CO)]^+$ are indicative of N-H bonding. The ^{31}P n.m.r spectrum was also reported.[966] The relationship between ^{31}P n.m.r. chemical shift and palladium oxidation state has been discussed for a variety of palladium complexes.[967] N.m.r. data have also been reported for $Ni(CN)_2(PMe_3)_3$ (^{31}P),[968] $NiCl_2(CO)(PMe_3)_2$ (^{31}P),[969] $PdCl_2(\overline{N=CR^1C}R^2R^3)_2$ (^{13}C),[970] $Pd_2(O_2CCF_3)_2(\mu\text{-}CO)(\mu\text{-}dppm)_2$ (^{31}P),[971] $Pd_2Cl_2(\mu\text{-}CO)(\mu\text{-}dppm)$ (^{13}C, ^{31}P),[972] $[(RNC)MCl(\mu\text{-}dppm)_2AgCl]^+$ (M = Pd, Pt; ^{31}P, ^{195}Pt),[973] $M^1M^2Cl_2(\mu\text{-}CO)(PPh_3)_3$ (M^1, M^2 = Pd, Pt; ^{31}P),[974] $(diop)Pt(CO)_2$ (^{13}C, ^{31}P),[975] $Pt(CO)_2Cl_2$ (^{195}Pt),[976] $[Pt_2(P_2O_3H_2)_4CNBr]^{4-}$ (^{31}P),[977] $Pt_4(CO)_5(PEt_3)_4$ (^{31}P, ^{195}Pt),[978, 979] and $[Pt_3Au(\mu_2\text{-}CO)_3(PCy_3)_4]^+$

(^{31}P).[980]

^{13}C n.m.r. studies have shown that $[Ni\{MeN(CH_2CH_2CH_2NHCH_2CH_2)_2$-$NH\}]^{2+}$ is one isomer while the Zn^{2+} anologue is a 1:2 mixture of species.[981] The structure of a Ni derivative of chlorophyll c related petroporphyrin has been determined using selective decoupling and n.O.e. studies.[982] The ^{13}C chemical shifts of (5,10,15,20-tetramethylchlorinato)nickel(II) support the model of asymmetric charge distribution in this chlorine.[983] N.m.r. data have also been reported for $M(H_2NCSCHCSNH_2)_2$ (M = Ni, Pd, Pt; ^{13}C, ^{195}Pt),[984] $(Me_2Ga)_2\{M(dmg)_2\}$ (M = Ni, Pd, Pt; ^{13}C),[985] (8) (^{13}C),[986] (9) (^{13}C),[987] $[Ni(PR_3)_3NO]^+$ (^{31}P),[988] and (NC)Ni(μ-dppm)$_2$NiNCBH$_3$ (^{31}P).[989]

N.O.e. experiments and chemical-shift analysis have permitted the identification of all signals in some complexes between $[(1,1,4,7,7\text{-pentamethyldien})Pd]^{2+}$ and some nucleosides.[990] N.m.r. data have also been reported for M(meso-tetraphenylporphyrin) (M = Pd, Pt; ^{13}C),[991] Pd(amine)(carboxylato) (^{13}C),[992] Cl^2Pd(dipeptide) (^{13}C),[993] $[Pd(pro)Cl]_2$ (^{13}C),[994] Pd^{2+} and Pt^{2+} complexes with aminomonosaccharides (^{13}C),[995] [{Pd(8-methylquinolinato)}$_3$(μ^3-Ph_2PCHCO_2Et)(μ$_3$-OH)]$^+$ (^{31}P),[996] $[Pd(NCMe)_{\underline{n}}(PPh_3)_{4-\underline{n}}]^{2+}$ (^{31}P),[997, 998] and $Pd(NHRCSCSNHC_5H_{10})_2$ (^{13}C).[999]

Reactions of cis-$[Pt(^{15}NH_3)_2(H_2O)_2]^{2+}$ with a range of anions have been studied by ^{15}N, ^{31}P, and ^{195}Pt n.m.r. spectroscopy.[1000] Complexes between $[Pt(NH_3)_2(OH_2)_2]^{2+}$ and methylated guanines have been characterised using ^{195}Pt n.m.r. spectroscopy.[1001] The interaction of $PtCl_2(NH_3)_2$ with purine mononucleotides has been investigated using ^1H n.m.r. spectroscopy.[1002] ^1H n.m.r. spectroscopy has been used to investigate the co-ordination of 9-Me-adenine to cis-$[Pt(NH_3)_2]^{2+}$ and $[Pt(dien)_2]^{2+}$.[1003] The interaction of trans-$Pt(NH_3)_2Cl_2$ with thymidine, cytidine, guanosine, and 5'-adenosine phosphate has been studied by ^{13}C n.m.r. spectroscopy.[1004] A ^1H n.m.r. study of the reaction of $PtCl_2(NH_3)_2$ with 5-substituted uracils has been reported.[1005] The chelation of cis-$[Pt(NH_3)_2$-$(OH_2)_2]^{2+}$ by the guanosine of some oligonucleotides has been investigated.[1006] The nature of the cis-$PtCl_2(NH_3)_2$-induced distortion of a double-helix DNA fragment has been studied using ^1H n.O.e measurements.[1007] ^{195}Pt n.m.r. spectra have been used to identify some trans-dihydroxo-Pt(IV) anti-tumour complexes.[1008] The co-ordination of 9-methylhypoxanthine to cis-$[Pt(NH_3)_2]^{2+}$ and $[Pt(dien)_2]^{2+}$ has been studied by ^1H n.m.r. spectroscopy.[1009] The interaction between poly(I) and poly(I).poly(C) with $[Pt(dien)Cl]^+$ has been studied.[1010] ^{13}C n.m.r. spectra have been used to study the nature of succinamide platinum blues.[1011, 1012] N.m.r. data have also been reported for cis-$PtCl_2X_2(NH_3)_2$ (^{195}Pt),[1013] $Cl(PEt_3)$-$Pt\{P(O)(OMe)NHNMeCMe{=}NH\}$ (^{31}P, ^{195}Pt),[1014] cis-$PtCl_2(PBu^n{}_3)(H_2NCH_2$-$COX)$ (^{15}N, ^{31}P),[1015] $PtCl_2$(8-formylquinoline)(PEt_3) (^{13}C, ^{31}P, ^{195}Pt),[1016] $[PtCl_2$(2,4,6-Me_3-pyridine)$]_2$ (^{13}C),[1017] and cis-$PtCl_2$-(2,6-Me_2-pyrazine)$_2$ (^{13}C).[1018]

The ^{195}Pt n.m.r. chemical shifts of $Pt(PR_3)_n$ have been reported. The shifts follow the order: $n = 3 > n = 4 > n = 2$.[1019] For $(Ph_3P)_2PtEP(NBu^t)NBuSiMe_3$ (E = S, Se) temperature-dependent ^1H n.m.r. spectra give evidence of 5-co-ordinate silicon. The ^{13}C, ^{31}P, and ^{195}Pt n.m.r. spectra were also reported.[1020] A full analysis of the ^{31}P and ^{195}Pt n.m.r. spectra of $[Pt_3S_2(PMe_2Ph)_6]^{2+}$ has yielded $^1J(^{195}Pt,^{31}P)$ = 3202Hz, $^3J(^{195}Pt,^{31}P)$ = -25Hz, and $^2J(^{195}Pt,^{195}Pt)$ = 476Hz.[1021] In $PtCl_2\{P(C_6H_4X^1)(C_6H_4X^2)(C_6H_4X^3)\}_2$, the substituent constant σ is correlated with $^1J(^{195}Pt,^{31}P)$.[1022] N.m.r. data have also been reported for (10) (E = P, As; ^{31}P),[1023] $N(CH_2CH_2PPh_2)_3M(P_4E_3)$ (M = Ni, Pd, E = S, Se; ^{31}P),[1024] (11) (M =

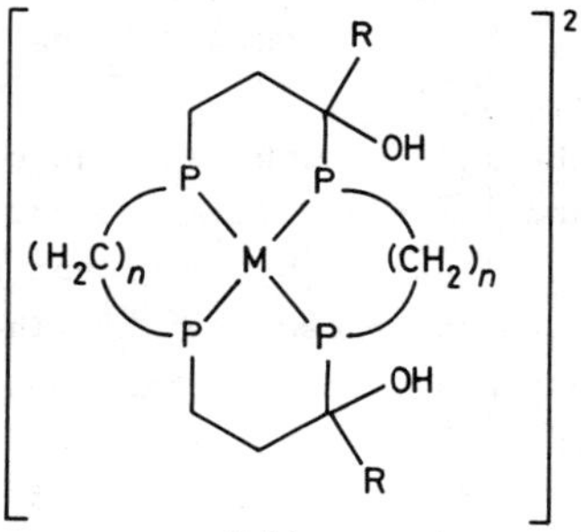

(11)

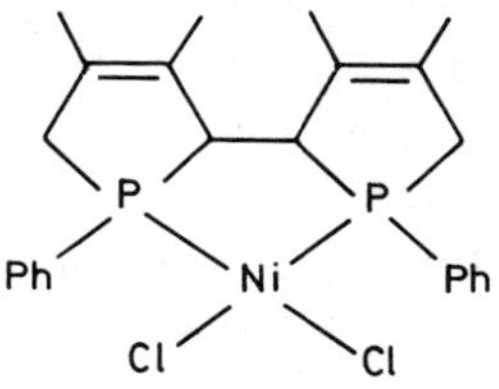

(12)

Ni, Pd, Pt, Zn; ^{31}P),[1025] $[N(CH_2CH_2PPh_2)_2(CH_2CH_2PPh_2CS_2Me)Ni]^+$ (^{31}P),[1026] $Ni_2X_2\{S[(CH_2)_3PPh(CH_2)_2PPh(CH_2)_3]_2S\}$ (^{31}P),[1027] $[\{E(CH_2$-$CH_2PPhCH_2CH_2PPhCH_2CH_2)_2E\}Ni]^{2+}$ (E = O, S; ^{31}P),[1028] $Ni(SPh)_2(PEt_3)_2$ (^{31}P),[1029] Ni(ß-phenylsulphonylacrylate)(dmpe) (^{13}C, ^{31}P),[1030] $[M\{1\text{-}(MePhAs)\text{-}2\text{-}(MePhP)C_6H_4\}_2]^{2+}$ (^{31}P),[1031] (12) (^{13}C, ^{31}P),[1032] $NiBr\{(HO_2CCH_2)_2PCH_2CH_2)_2PCH_2CH_2P(CH_2CO_2H)_2\}_2$ (^{31}P),[1033] $[Ni_2(SCH_2$-$CH_2PPh_2)_3]^+$ (^{31}P),[1034] $[Ni_9E_9(PEt_3)_6]^{2+}$ (E = S, Se; ^{31}P),[1035] $Pd(PEt_3)\{P(OPh)_3\}_3$ (^{31}P),[1036] $[Pd(PEt_3)_3X]^+$ (^{31}P),[1037] $[M(Ph_2PCH_2$-$CH_2SR)_2]^{2+}$ (M = Pd, Pt; ^{31}P),[1038] $(R^1{}_3P)_2M(S_2CNR^2)$ (M = Pd, Pt; ^{13}C, ^{31}P, ^{195}Pt),[1039] $[Pt(PR_3)_4]^{2+}$ (^{31}P),[1040] $[Pt(PEt_3)_4]^{2+}$ (^{31}P, ^{195}Pt),[1041] $[Ni\{S(CH_2CH_2PPhCH_2CH_2PPhCH_2CH_2)_2S\}]^{2+}$ (^{31}P),[1042] 2,11-(Ph_2PCH_2)benzo[c]phenanthrene complexes with Pt(0) (^{31}P),[1043]

Pt(dppe)(dppm) (^{31}P),[1044] $[PtCl\{R^1{}_2P(CH_2)_3PPh(CH_2)_{\underline{n}}PR^2{}_2\}]^+$ (^{31}P, ^{195}Pt),[1045] $Pt\{P(OH)(OR)_2\}_2\{P(O)(OR)_2\}_2$ (^{31}P),[1046] $[Pt_2(PPh_3)_2(\mu\text{-}dppm)_2]^{2+}$ (^{31}P),[1047] $Pt_2Br_2(\mu\text{-}dppm)_2$ (^{31}P),[1048] $Pt(O_2)(PPh_3)(PBu^t{}_3)$ (^{31}P),[1049] $[Pt_2(\mu\text{-}S)(\mu\text{-}SMe)(PPh_3)_4]^+$ (^{13}C, ^{31}P, ^{195}Pt),[1050] and $[Cl(Et_3P)Pt\{Au(PPh_3)\}_2]^+$ (^{31}P).[1051]

A ^{195}Pt n.m.r. spectral study of mixtures of Na_2PtCl_6 and $Na_2Pt(OH)_6$ in alkaline solution has indicated the formation of $[PtCl_{6-\underline{n}}(OH)_{\underline{n}}]^{2-}$.[1052] ^{195}Pt $\underline{T}_1$ of $Pt(acac)_2$ has been analyzed as a function of temperature and magnetic field. The relaxation is dominated by the chemical-shift anisotropy of ca. 10,000 p.p.m.[1053] For a wide range of Pt ß-diketonate complexes, a good correlation has been found between (M-O) and $\delta(^{195}Pt)$.[1054] A ^{105}Pd n.m.r. signal has been observed for H_2PdCl_6. The linewidth is 1200Hz.[1055] N.m.r data have also been reported for $[PtX_{\underline{n}}(OH_2)_{4-\underline{n}}]^{(2-\underline{n})+}$ (^{195}Pt),[1056] $Pt(S_2PR_2)_2$ (^{195}Pt),[1057] *cis*-$Pt\{MeSCH(CF_3)CH(CF_3)SMe\}Cl_2$ (^{19}F),[1058] and $Pt_2(S_2CPr^i)_4I_2$ (^{13}C).[1059]

Complexes of Cu, Ag, and Au.—'Silver-109' n.m.r. spectroscopy has been reviewed.[1060] The nuclear magnetic shielding constants of the metals in Cu, Zn, Ag, and Cu complexes have been studied theoretically and are mainly due to the paramagnetic term.[1061]

The ^{13}C n.m.r. spectrum observed upon mixing Me_2CuLi and $PhCH{=}CHCO_2Bu^t$ shows that the signals due to the olefinic carbon atoms move to low frequency and this was attributed to π-complexation.[1062] N.m.r. data have also been reported for $B_2H_4.2PMe_3.Cu(PPh_3)X$ (^{11}B),[1063] $\{(Me_2N)_3PCH_2\}_2MCl_{\underline{n}}$ (M = Cu, Ag, Au, Zn, Cd, Hg, n = 1 or 2; ^{13}C),[1064] $Ag_2Li_2(C_6H_4CH_2NMe_2\text{-}2)_4$ (^6Li, ^7Li),[1065] $[Ag\{C(SiMe_3)_3\}_2][Li(thf)_4]$ (^7Li, ^{109}Ag),[1066] $R_2Au(PPh_2)_2CHAuC_6F_5$ (^{31}P),[1067] (13) (^{31}P),[1068] $(C_6H_4)(AsMe_2)_2Au(C_6F_4C_6F_4)Au(C_6F_4C_6F_4)Au(C_6F_4C_6F_4)Au(Me_2As)_2C_6H_4]^{2+}$ (^{19}F),[1069] and $(RC_2H)Cu(O_3SCF_3)$ (^{13}C).[1070]

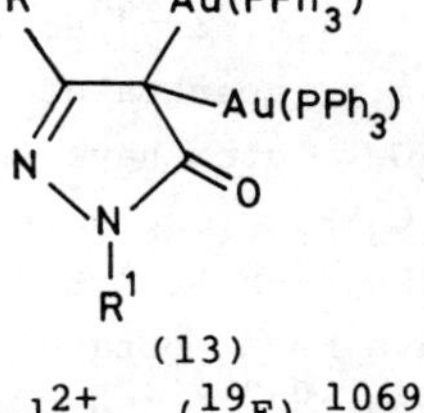

(13)

The structures of $[M_2\{\mu\text{-}1,2\text{-}(py\text{-}2\text{-}CH{=}N)_2Cy\}_2]^{2+}$ (M = Cu, Ag) and related complexes have been studied in solution by ^1H, ^{13}C, ^{15}N,

^{19}F, and ^{109}Ag n.m.r. spectroscopy. The ^{15}N and ^{109}Ag n.m.r. spectra were recorded using INEPT.[1071, 1072] Similarly $[M(6\text{-}R^1C_5H_4N\text{-}2\text{-}CH{=}NR^2)_2]^+$ (M = Cu, Ag) was studied using ^1H, ^{15}N, and ^{109}Ag n.m.r. spectroscopy.[1073] Analysis of the ^{31}P n.m.r. spectrum of $[Ar(Ag)P{=}P(Ag)Ar]^{2+}$ has yielded $^1\underline{J}(^{31}P,^{31}P)$ and $^1\underline{J}(^{109}Ag,^{31}P)$.[1074] For species of the type $[Ag(PR_3)_{\underline{n}}]^+$ ($\underline{n}$ = 1 to 4) linear correlations have been observed between $^1\underline{J}(^{109}Ag,^{31}P)$ and the electronegativity of R.[1075] N.m.r. data have also been reported for $[CuL_4]^+$ (L = RCN, RNC, py; ^{63}Cu),[1076] $[Ag(CNR)_4]^+$, $[Ag(CNR)(Ph_2PCH_2)_3CMe]^+$ (^{13}C, ^{31}P),[1077] $[Cu(NCS)(PPh_3)_2]_2$ (^{13}C, ^{31}P),[1078] $(Me_2PhP)_3Cu(NC)_2BH_2$ (^{11}B, ^{31}P),[1079] $[Ag(glutarimide)_2]^-$ (^{13}C),[1080] $[Ag(succinimide)_2]^-$ (^{13}C),[1081] Ag(I) complexes with cyclo(glycyl-L-histidyl) and cyclo(L-methionyl-L-histidyl) (^{13}C),[1082] $[(ArHNCNHAr)_2Au]^+$ (^{13}C),[1083] $M_2\{Ph_2PC_6H_4C(O)CHCBu^tO\}_2$ (M = Cu, Ag; ^{31}P),[1084] $[Ag\{P(NR_2\}_3]^+$ (^{13}C, ^{31}P),[1085] $HC\{P(S)Ph_2\}_3Ag(PBu^{\underline{n}}_3)$ (^{13}C, ^{31}P),[1086] $\{(R_2PCH_2)_2$benzo[c]phenanthrene$\}AgX$ (^{31}P),[1087] $(AuCl)_2(dppe)$ (^{31}P),[1088] $[Au(PR_3)_2]^+$ (^{31}P),[1089] $[Au_8(PPh_3)_6I]^+$ (^{31}P),[1090] $[Au_9(PPh_3)_8]^+$ (^{31}P),[1091] $[Au\{P(C_6H_4OMe)_3\}_8]^{3+}$ (^{31}P),[1092] and $\{NHC_2H_4NEtC(S)\}AuX$ (X = Cl, Br; ^{13}C).[1093]

Complexes of Zn, Cd, and Hg.—A review entitled 'Cadmium-113 n.m.r' spectroscopy has appeared.[1094] The mercury cations Hg^{2+}, $[Hg_2]^{2+}$, and $[Hg_4]^{2+}$ have been studied by ^{199}Hg n.m.r. spectroscopy. The ^{199}Hg n.m.r. spectrum of $[Hg_3]^{2+}$ shows second-order effects due to coupled ^{199}Hg atoms, with $^1\underline{J}(^{199}Hg,^{199}Hg)$ = 139 600 Hz, which is the largest nuclear spin-spin coupling constant reported to date.[1095]

The ^1H, ^{13}C, and ^{199}Hg n.m.r. spectra of Me_2Hg in the nematic phase have been reported.[1096] In general, changes in electron ionizations of various alkynyl mercury compounds agree with changes in the magnitude of $\delta(^{13}C)$, $\delta(^{199}Hg)$, $^1\underline{J}(^{199}Hg,^{13}C)$, and $^1J(^{13}C,^{13}C)$.[1097] ^{19}F n.m.r. spectra of $R_{\underline{n}}MC_2C_6H_4F\text{-}\underline{p}$ ($R_{\underline{n}}M$ = H, PhHg, Ph_3Sn, Ph_3Pb, Ph_3Ge, etc.) have shown that the electron-acceptor properties of the $C_2MR_{\underline{n}}$ group increase in the series CyHg < MeHg < PhHg ~ Ph_3C ~ Ph_3Pb < Ph_3Sn < H < Ph_3Ge.[1098] ^1H, ^{13}C, ^{14}N, and ^{199}Hg n.m.r. data have been obtained for RHgCNO, and even $^1\underline{J}(^{199}Hg,^{14}N)$ was observed.[1099] The interaction between $[MeHg]^+$ and xanthosine has been investigated using ^1H and ^{13}C n.m.r. spectroscopy. Correlations were drawn between $^2\underline{J}(^{199}Hg,^1H)$, $p\underline{K}_a$, and ^{13}C chemical shifts.[1100] ^{199}Hg n.m.r. spectroscopy has been shown to be a sensitive probe to the binding point of $[MeHg]^+$ to nucleic acid constituents.[1101] EtHgX (X = Cl, Br) has been investigated in the nematic phase using ^{1}H and ^{199}Hg n.m.r. spectroscopy.[1102] Linear relationships have been found between ^{1}H, ^{13}C, ^{119}Sn, and ^{199}Hg chemical shifts and coupling constants in $Hg(CH_2MMe_3)_2$, $Bu^iHgCH_2MMe_3$, $Sn(CH_2MMe_3)_4$, and $Me_3SnCH_2MMe_3$ (M = C, Si, Ge or Sn).[1103] ^{1}H, ^{29}Si, and ^{199}Hg spectroscopy has been used to show Hg-R co-ordination interaction in $(Me_{3-\underline{n}}X_{\underline{n}}SiCH_2)_2Hg$ (X = Cl, MeO).[1104] ^{1}H n.m.r. spectroscopy has shown that $(2\text{-}Me_2NCH_2C_6H_4)_2M$ (M = Zn, Cd) undergoes rapid intramolecular bond fission and formation of the M-N bond.[1105] In 1,5-$(CF_3CO_2Hg)_2$-2-(MeO)-3-$(O_2N)C_6H_2$, the ^{199}Hg n.m.r. spectrum shows $^{4}\underline{J}(^{199}Hg,^{199}Hg)$.[1106] N.m.r. data have also been reported for $[MeHg(8\text{-thioguanosine})]^+$ (^{13}C),[1107] $[MeHg(uridine)]^+$ (^{13}C, ^{199}Hg),[1108] $[(AcOHg)_{\underline{n}}CH_{2-\underline{n}}\{PPh_2(HgOAc)\}_2]^{2+}$ (^{31}P, ^{199}Hg),[1109] $\{Cl(CH_2)_3CH{=}CH\}_2Hg$ (^{13}C),[1110] $(1\text{-decenyl})_2Hg$ (^{13}C),[1111] $Bu^tMe_2SiOCHMe(CH_2)_3CH{=}CHHgCl$ (^{13}C),[1112] $(MeO_2C)(NC)C{=}C(CO_2Me)OHgPh$ (^{13}C),[1113] and $(F_4C_6C_6F_4)_3Hg_3$ (^{19}F).[1114]

A new method for the assignment of Zn-N co-ordination mode has been reported. ^{15}N n.m.r spectra are recorded for complexes enriched in ^{67}Zn and complexes at natural abundance. Resonances due to N atoms co-ordinated to the ^{67}Zn are broadened by scalar interaction while unco-ordinated N atoms show no effect. The method was tested for bis(dien) and bis(histidine) complexes.[1115] ^{1}H and ^{13}C n.m.r spectroscopy has been used to investigate tautomerism in Zn(II) complexes of the condensation products between (1$\underline{R}$)-3-hydroxymethylenebornan-2-one and $\underline{l}$-amino acids.[1116] N.m.r. data have also been reported for Zn(protoporphyrin di-methyl ester) (^{13}C),[1117] Zn(II), Cd(II), and Pb(II) complexes of 4 different 3-dentate azomethine ligands (^{13}C),[1118] Zn(II), Cd(II), and Hg(II) complexes of pyrimidine-2-thione (^{13}C),[1119] Zn(II), Cd(II), and Hg(II) complexes of 1,4,5,6-tetrahydropyrimidine-2-thione (^{13}C),[1120] Zn(II) complexes of thiazolidine-4-carboxylic acid (^{13}C),[1121] Zn(II) and Cd(II) complexes of $\underline{N}$-(2-pyridyl)thioacetamide (^{13}C),[1122] Zn(II), Cd(II), and Hg(II) complexes of creatinine (^{13}C),[1123] Zn(II) complexes of cysteine (^{15}N),[1124] and Zn(II)-Cd(II), and Zn(II)-Mg(II) complexes of alkaline phosphatase (^{15}N, ^{113}Cd).[1125]

Solution conditions have been reported which allow the easy observation of the ^{113}Cd n.m.r. spectra of amine complexes in D_2O and in DMSO-$\underline{d}_6$. For a series of $\underline{N}$-donor ligands in DMSO-$\underline{d}_6$ with 1M $Cd(NO_3)_2$, the following linear relationship was observed:

$$\delta(^{113}Cd) = 75\underline{A} + 51\underline{B} + 31\underline{C}$$

where A, B, and C are the numbers of primary, secondary, and tertiary amines in the co-ordination sphere, respectively.[1126] ^{113}Cd n.m.r studies of dicysteinyl-edta complexes in aqueous solution have yielded the first example of $J(^{113}Cd,^{113}Cd)$ (80 Hz) in a model system. The chemical shifts are comparable with those observed in metallothionein.[1127] The ^{13}C n.m.r. spectra of protonated thiamine and $[Cd(thiamine)Cl_3]^-$ have been compared to determine the effect of co-ordination on chemical shifts. The results were inconclusive.[1128] The binding of Cd(II) to ATP has been investigated by ^1H and ^{31}P n.m.r. spectroscopy. It was concluded that the binding is to the ß- and γ-phosphates and N^7 of the adenine ring.[1129] N.m.r. data have also been reported for $[M\{RN(CH_2CH_2CH_2NRCH_2CH_2CH_2)_2NR\}]^{2+}$ (M = Cd, Hg; ^{13}C),[1130] 1,5,9,13-tetraazacyclohexadecane complexes of Cd(II), Hg(II), and Pb(II) (^{13}C, ^{15}N),[1131] Cd(imidazoline-2-thione)$_2Cl_2$ (^{13}C),[1132] $Cd(NBu^t)(NSiMe_3)$ (^{29}Si, ^{113}Cd),[1133] arsenate complexation of cadmium(II) alkaline phosphatase (^{113}Cd),[1134] yeast inorganic pyrophosphatase (^{31}P, ^{113}Cd),[1135] chloride binding to alkaline phosphatase (^{35}Cl, ^{113}Cd),[1136] metallothioneins (^{113}Cd),[1137, 1138] calmodulin (^{113}Cd),[1139, 1140] Cd(II) substituted blue copper proteins (^{113}Cd),[1141] and Hg(II) complexes with NADPH (^{13}C).[1142]

^{31}P and ^{199}Hg n.m.r. spectra have been reported for $HgX_2(PCy_3)(PBu_3)$. It was found that $^2J(^{31}P,^{31}P)$ decreases with increasing co-ordination ability of X.[1143] For $[HgCl_2\{P(C_6H_4\text{-}4\text{-}X)_3\}]_2$ and $HgCl_2\{P(C_6H_4\text{-}4\text{-}X)_3\}_2$, the ^{31}P and ^{199}Hg n.m.r. spectra have been reported. $^1J(^{199}Hg,^{31}P)$ increases with increasing phosphine basicity, decreasing temperature, and decreasing sample concentration.[1144] According to ^{31}P n.m.r. data for Hg{2,11-$(Ph_2PCH_2)_2$benzo[c]phenanthrene}X_2, the values of $^1J(^{199}Hg,^{31}P)$ increase in the order X = CN < I < SCN < Br < Cl < NO_3.[1145] The ^{31}P and ^{199}Hg n.m.r. spectra of $[XHg(PR_2)]_4$ have shown the first $^2J(^{199}Hg,^{199}Hg)$ in the range 2435 to 2959 Hz.[1146] $^2J(^{113}Cd,^{111}Cd)$ = 46 Hz has been observed in $[Cd_4(SPh)_{10}]^{2-}$.[1147] N.m.r. data have also been reported for $Cd(O_2CCF_3)_2(PPh_3)_n$ (^{19}F, ^{31}P),[1148] $I_2Cd(\mu\text{-}I)_2Hg(PPr_3)_2$ (^{31}P, ^{199}Hg),[1149] $MX_2(PHCy_2)_2$ (M = Cd, Hg; ^{31}P, ^{113}Cd, ^{199}Hg),[1150] and $[Hg(PR_3)_n]^{2+}$ (^{31}P, ^{199}Hg).[1151]

3 Dynamic Systems

This section is in three main parts: (i) 'Fluxional Molecules', dealing with rate processes involving no molecular change, (ii)

'Equilibria', dealing with the use of n.m.r. spectroscopy to measure the position of equilibria and ligand-exchange reactions, including solvation, and (iii) 'Course of Reactions', dealing with the use of n.m.r. spectroscopy to monitor the course of reactions. Each section is ordered by the Periodic Table.

Three relevant reviews have appeared: 'Magnetic resonance studies of alkali metals in non-aqueous solvents', which contains ^{13}C, ^{23}Na, and $^{85, 87}Rb$ n.m.r. spectra,[1152] 'Characteristic diffusion phenomena of fluoride in molten $LiF-BeF_2$ and $NaF-BeF_2$ systems', which contains ^{7}Li and ^{19}F n.m.r. spectra,[1153] and 'The stereodynamics of sulphur-, selenium-, and tellurium-containing ligands'.[1154]

Fluxional Molecules.—Groups IA and IIA. ^{1}H n.m.r. spectroscopy has been used to investigate restricted rotation of the CH_2 group in 2-pyridylCH_2Li.[1155] The ^{1}H, ^{11}B, and ^{13}C n.m.r. spectra of $C_6H_4(CH_2NMe_2)BR_3LiOEt_2$ have shown dynamic behaviour due to N-Li interaction.[1156] The ^{13}C n.m.r. spectra of cryptands [2.1.1], [2,2,1], and [2.2.2] in the presence of Li^+, Na^+, K^+, Ca^{2+}, and Ba^{2+} have been analyzed. The ^{13}C $\underline{T}_1$ values show a surprising increase in magnitude for most alkali metal but not alkaline-earth metal complexes. This increase in relaxation times was interpreted in terms of ligand compression desolvation and solvent reorganization.[1157]

Yttrium, Cerium, and the Lanthanides. The ^{1}H n.m.r. spectrum of $[Ce(edta)]^-$ and $[Y(edta)]^-$ in solution exhibits a characteristic AB multiplet signal. The broadening of the signal with increasing H^+ concentration was attributed to proton inter-change. In the case of the Ce complex $\Delta\underline{G}^{\ddagger} = 53 \pm 5$ kJ mol^{-1}, closely equal to the proton-catalyzed dissociation of the Ce complex.[1158] The conformational behaviour of $[Eu(1,4,7,10$-tetraazacyclododecane-1,4,7,10-tetraacetato$)]^-$ has been investigated.[1159] The ^{1}H n.m.r. spectrum of $Yb_2Co_3(\eta^5-C_5Me_5)_4(\eta^5-C_5H_4SiMe_3)_2(CO)_4$ shows two C_5Me_5 signals which interchange with $\Delta\underline{G}^{\ddagger} = 41.0$ kJ mol^{-1}.[1160]

Uranium. The kinetics of the intramolecular exchange of methyl groups of the co-ordinated acac in $UO_2(acac)_2L$ have been studied, and correlate with the activation energy of intermolecular exchange of L.[1161]

Titanium, Zirconium, and Hafnium. ^{1}H n.m.r. spectroscopy has been used to determine the rate of conformer interchange in $(\eta^5-C_5H_5-SiMe_3)_2\overline{Hf\{C(O)CH_2C_6H_4CH_2CO}\}$. ^{13}C n.m.r. data were reported for $M(\eta^5-C_5H_4SiMe_3)_2\{o-(CH_2)_2C_6H_4\}$ (M = Ti, Zr, Hf).[1162] Comparison of the barriers for ME_n (M = Ti, Zr, Hf, E = S, Se) ring reversal in $(\eta^5-C_5Me_5)_2ME_n$ and related compounds has revealed that when n = 5 the barrier decreases in the order M = Ti > Zr > Hf.[1163] The inversion barrier for ring inversion has been determined for $(\eta^5-C_5H_5)_2TiS_4CR_2$.[1164] ^{1}H n.m.r. studies of $(\eta^5-C_5H_5)M(S_2CNMe_2)_3$ (M = Ti, Zr, Hf) exhibit evidence of three kinetic processes: (i) exchange of the S_2CNMe_2 groups within the equatorial ligands, (ii) exchange of the Me groups within the unique ligand, (iii) exchange of the equatorial unique ligands. Rates and activation energies were determined by total line-shape analysis.[1165] A rapid intramolecular rearrangement of $(\eta^5-C_5H_5)\overline{ZrR^1(S{=}CR^2N}CH_2Ph)\{(PhCH_2NC)(S)-R^2\}$ has been observed by ^{1}H n.m.r. spectroscopy.[1166]

Niobium and Tantalum. ^{1}H, ^{31}P, and ^{93}Nb n.m.r. spectra of $[NbH(CO)_3\{PhP(CH_2PPh_2)_2\}]^-$ show fluxional behaviour.[1167] Two stereochemically non-rigid isomers of $RB(pz)_3TaMe_3Cl$ have been shown using ^{11}B and ^{13}C n.m.r. spectroscopy.[1168]

Chromium, Molybdenum, and Tungsten. The ^{1}H and ^{13}C n.m.r. spectra of $WH(CO)_2(\eta^2\text{-olefin})(\eta^5-C_5R_5)$ show restricted rotation of the olefin.[1169] ^{13}C and ^{31}P n.m.r. spectra of $R^1{}_3PM^1(\mu-H)M^2(CO)_4PR^2{}_3$ (M^1 = Ag, Au, M^2 = Cr, Mo, W) show dynamic behaviour due to tertiary phosphine exchange.[1170] ^{1}H and ^{13}C n.m.r. spectroscopy has been used to investigate exchange of the p-tolyl methyl groups in $(\eta^5-C_5H_5)WOs_3(CO)_9(\mu^3\text{-Ctol})_2H$.[1171] The ^{13}C n.m.r. spectrum of $[Mo_2(\mu-C_8H_8)(\mu_{Mo,C}-H)(\eta^5-C_5H_5)_2]^+$ shows that μ-H undergoes site exchange. The ^{95}Mo n.m.r. spectrum of $Mo_2(\mu-C_8H_8)(\eta^5-C_5H_5)_2$ was also recorded.[1172]

Analysis of the ^{1}H and ^{13}C n.m.r. spectra of $(\eta^5-C_5H_5)Cr({=}CPh_2)(CO)(NO)$ and $(\eta^5-C_5H_5)Mn({=}CPh_2)(CO)_2$ has shown that the Cr-carbene conformation is locked, whereas the Mn-carbene rotation is fast.[1173] ^{1}H n.m.r. spectroscopy has shown reversible benzyl-alkoxy migration in $(Pr^iO)_3Mo\ Mo(CH_2Ph)_2(OPr^i)(PMe_3)$.[1174] The barrier to rotation of the M≡M bond in $M^1{}_2\{M^2(M^2Me_3)_3\}_2(NMe_2)_4$ (M^1 = Mo, W, M^2 = Si, Sn) has been determined and the ^{119}Sn n.m.r. spectrum recorded.[1175] On the ^{1}H and ^{13}C n.m.r. time-scales, the μ-CMeCMeCSiMe$_3$ ligand in $(Me_3SiCH_2)_4W_2(\mu-C_3Me_2SiMe_3)(\mu-CSiMe_3)$ undergoes facile site exchange of the 1 and 2 CMe groups. A facile interconversion between a 1,3-dimetalla-allyl and a μ-cyclopropenyl form of the bridging $C_3Me_2SiMe_3$ ligand was proposed.[1176] ^1H and ^{13}C n.m.r. spectra of (14) have revealed fluxional processes due to

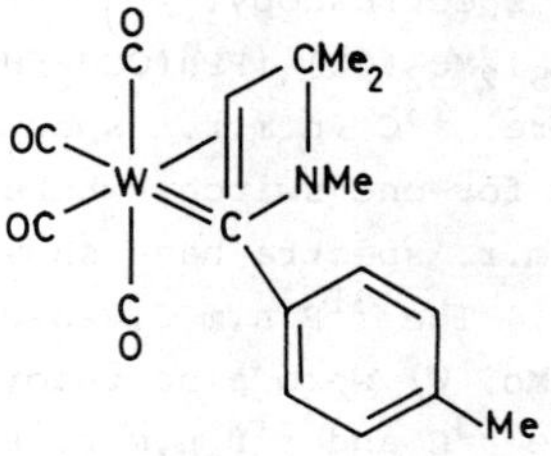

(14)

slow rotation of the p-tolyl group and to reversible decomplexation of the alkene ligand.[1177]

Intramolecular exchange processes in $[(\eta^5-C_5H_5)Mo(\eta^2-HC_2R)_2(CO)]^+$ have been studied by 2D n.m.r. exchange techniques.[1178] ^{13}C n.m.r. spectra show butyne rotation in $[(\eta-C_5H_5)Mo(PPh_2C_6H_4CH{=}CH_2)(\eta^2-MeC_2Me)]^+$. The ^{31}P n.m.r. spectrum was also recorded.[1179] The ^{13}C and ^{31}P n.m.r. spectrum of $(\eta^5-C_5H_5)Mo(\eta^2-MeC_2Me)(SePh)\{P(OMe)_3\}$ has been recorded and the activation energy for alkyne rotation determined.[1180] For $(\eta^5-C_5H_5)Mo(\eta^2-CH_2PPh_2)(CO)_2$, variable-temperature ^1H n.m.r. spectroscopy has been used to determine $\Delta G^\ddagger$ = 53.9 kJ mol^{-1} for $\eta^2-CH_2PPh_2$ rotation. The ^{13}C and ^{31}P n.m.r. spectrum was also reported.[1181] Dynamic ^{13}C n.m.r. studies of $[W(\eta^2-PhC_2Ph)_3SnPh_3]$ have shown that the alkyne ligand rotation is rapid. The ^{119}Sn n.m.r. spectrum was also recorded.[1182] In $W(\eta^2-C_2H_4)_2Cl_2(PMe_3)_2$, ethylene rotation occurs with $\Delta G^\ddagger$ = 17.9 kcal mol^{-1}. The report includes ^{13}C and ^{31}P n.m.r. data.[1183] The ^1H and ^{13}C n.m.r. spectra of $W_2(OR^1)_6(\mu-C_4R^2{}_4)(\eta^2-R^2C_2R^2)$ have shown it to be fluxional.[1184] $W_2(OR^1)_6(\mu-R^2C_2R^2)(py)_2$ is fluxional on the ^1H and ^{13}C n.m.r. time-scale, and the low-temperature limiting spectra are in agreement with the solid-state structure.[1185] The barrier to $\eta^4-C(CH_2)_3$ ligand rotation in $[(\eta^5-C_5H_5)Mo(CO)_2\{\eta^4-C(CH_2)_3\}]^+$ has been determined. The ^{13}C n.m.r. spectrum was also recorded.[1186]

Variable-temperature ^{13}C n.m.r. studies of $CrRh(\mu-CO)_2(CO)_2(\eta^5-indenyl)(\eta^6-1,3,5-Me_3C_6H_3)$ have shown dynamic behaviour with hindered rotation of the indenyl ligand with $\Delta G^\ddagger_{245}$ = 45.2 kJ mol^{-1}.[1187] ^1H n.m.r. studies of $(\eta^6-C_7H_8)_2Mo$ have shown it to be fluxional with twisting of the mutually staggered cycloheptatrienyl rings. It subsequently rearranges to $(\eta^7-C_7H_7)Mo(\eta^5-C_7H_9)$.[1188] The activation energy for the rotation of the $CoRh(\mu-CO)_2(\eta^5-C_5Me_5)_2$ fragment in $MoCoRh(\mu-CO)_2(CO)_5(\eta^5-C_5Me_5)_2$ has been determined using ^{13}C n.m.r. spectroscopy.[1189] The dynamic behaviour of ^{13}CO enriched $(\eta^5-C_5H_5)_2Mo_2(CO)_3\{PPh(OCH_2CH_2)_2N\}$ has been studied by variable-temperature ^{13}C n.m.r. spectroscopy. Two carbonyl groups underwent a one-for-one switch, whereas the third remained unchanged.[1190] ^{13}C n.m.r. spectra have shown that $CuW(CO)_3(PPh_3)_2(\eta^5-C_5H_5)$ is fluxional. The ^{31}P n.m.r. spectra of $MRh(CO)(\mu-CO)_2(PPh_3)_2(\eta^5-C_5H_5)$ (M = Mo, W) were also recorded.[1191]

The low-temperature ^{13}C and ^{31}P n.m.r. spectra of $Cr(\eta^6-C_6Et_6)(CO)_2(PMe_3)$ were explained by slowed ethyl-group rotation, resulting in multiple C_6Et_6 stereoisomers co-existing in equilibrium.[1192] A similar explanation was given for the variable-temperature ^{13}C n.m.r. spectrum of $(\eta^6-C_6Et_6)Cr(CO)_2(CS)$.[1193] The solid-state ^{13}C and solution ^1H and ^{13}C n.m.r. spectra of $(\eta^6-C_6Et_6)Cr(CO)_2(CS)$ show slow tripodal rotation.[1194] The low-temperature ^1H n.m.r. spectra of $\{\eta^6-1-(MeO)-2,6-Bu^tC_6H_3\}Cr(CO)_2(PPh_3)$ show dynamic behaviour which was interpreted in terms of restricted rotation about the arene-OMe bond.[1195]

The ^{31}P n.m.r. spectrum of $(OC)_4(Me_3P)Cr(\mu-PBu^t)NiCl_2(PMe_3)$ shows dynamic behaviour.[1196] Total band-shape fitting of variable temperature ^1H n.m.r. spectra of $M(CO)_5(MeSCH_2SCH_2SMe)$ (M = Cr, Mo, W) has shown inversions of the pyramidal sulphur atoms. At higher temperatures, 1,3 intramolecular shifts of the $M(CO)_5$ moieties between ligand sulphur atoms were observed.[1197] The ^{13}C n.m.r. spectrum of $[S\{Cr(CO)_4\}_3]^{2-}$ shows two separate dynamic processes.[1198] Variable-temperature ^{13}C n.m.r. spectra of $[Fe_3M(CO)_{14}]^{2-}$ (M = Mo, W) have been used to study dynamic processes.[1199]

Variable-temperature ^1H n.m.r. studies of meso-substituted Mo(II) porphyrin dimers provide evidence for a barrier to rotation about the metal-metal quadrupole bond of 10.1 ± 0.5 kcal mol^{-1}.[1200] ^1H n.m.r. studies have shown restricted rotation about the $Mo-NMe_2$ bond in $Mo_2(NMe_2)_2(OAr)_4$.[1201] ^1H, ^{13}C, and ^{95}Mo n.m.r. spectroscopy has been used to demonstrate the existence of diastereoisomers in solutions of $MoO_2\{\underline{O}-Me-(\underline{S})-penicillaminato\}_2$, and the activation energy for interconversion was determined. The mechanism of inversion at Mo is intramolecular, involving Mo-N bond cleavage, rotation of the intermediate, and recyclization.[1202] For a given Mo(VI) complex, ^{17}O and ^{1}H n.m.r. spectra are consistent with the adoption in solution of the same octahedral or skew-trapezoidal-bipyramidal structure that is found in the solid state. The skew-trapezoidal-bipyramidal complexes display temperature-dependent n.m.r. spectra which were interpreted in terms of configurational averaging probably caused by Mo-N bond cleavage.[1203] The ^{1}H n.m.r. spectra of $[WBr(dppe)_2\{NN{=}\overline{CNR(CH_2)_nNR}\}]^+$ are fluxional due to N=C rotation.[1204] The ^{1}H n.m.r. spectra of $M_2(OR^1)_4(R^2COCHCOCR^3)_2$ (M = Mo, W) have been reported. For M = Mo, R^2 = Me, R^3 = Ph, at 110°C, there is evidence for a rapid reversible chelate-monodentate behaviour for the ß-diketonate ligands.[1205]

Manganese. The ^{13}C n.m.r. spectrum of $(R^1C_2R^2)CoMn(CO)_7$ indicates that the molecule is static at room temperature, but becomes fluxional on heating.[1206]

Rhenium. The ^{1}H n.m.r. spectrum of $[Ag\{ReH_7(PPr^i{}_2Ph)_2\}_2]^+$ shows the molecule to be fluxional. The ^{31}P n.m.r. spectrum does not show $J(^{109}Ag,^{31}P)$.[1207] ^{1}H n.m.r. saturation transfer measurements have shown exchange between the hydride and ethylene in $ReH_3(\eta^2\text{-}C_2H_4)_2(PR_3)_2$. The ^{31}P n.m.r. spectrum was also recorded.[1208] The fluxionality of $ReH_3(\eta^4\text{-}1,3\text{-diene})(PPh_3)_2$ has been studied by n.m.r. spectroscopy. Ligand interchange, reversible migration of the hydrogen ligand to the diene ligand, and isomerization of the diene ligand were observed.[1209] ^{1}H n.m.r. spectroscopy has demonstrated dynamic behaviour by the organic group in $[Re_3(\mu\text{-}H)_3(CO)_{10}(\mu\text{-}\eta^2\text{-}OCPhNH)]^-$.[1210] ^{1}H and ^{13}C n.m.r. spectra have indicated that, in $Re_3(\mu\text{-}H)_2(CO)_{10}(\eta^5\text{-}C_7H_9)$, the cycloheptadienyl ligand is fluxional and the two carbonyl groups, bound to the same rhenium, scramble.[1211] The ^{1}H n.m.r. spectrum of (15) shows CH_2 exchange. The ^{13}C n.m.r. spectrum was also recorded, along with the ^{31}P n.m.r. spectrum of $Re_2(CO)_8(PMe_3)_2$.[1212] The acetylide in $Re_2(\mu\text{-}H)(\mu\text{-}C_2Ph)(CO)_6(PPh_3)_2$ is fluxional, as demonstrated by ^{13}C n.m.r. studies of CO exchange.[1213] $Re_2(\mu\text{-}H)(\mu\text{-}C_2H)(CO)_8$ shows hindered ligand movement of the ethynyl bridge, with $\Delta G^\ddagger$ = 48.2 kJ mol^{-1}.[1214]

Variable-temperature ^{1}H n.m.r. spectra of $(\eta^5\text{-}C_5H_5)Re(CH_2C_6H_2Me_3)(NO)(PPh_3)$ show restricted rotation about the CH_2-mesityl bond. The ^{13}C n.m.r. spectrum was also recorded.[1215] The ^{19}F n.m.r. spectrum of $Re(CO)_5CBr_2CF_3$ shows restricted rotation about the CBr_2-CF_3 bond to give an A_2B n.m.r. spectrum. The ^{11}B, ^{13}C, and ^{55}Mn n.m.r. spectra of this compound and related manganese compounds were also reported.[1216] Variable-temperature ^{13}C n.m.r. spectra of $Re_2(CO)_8(dppe)$ show CO exchange.[1217] The energies associated with the pyramidal inversion of the sulphur atoms and rotation of the chelated ligands in fac-$ReX(CO)_3(MeSCH_2SCH_2SMe)$ have been obtained.[1218] Similar studies have been performed on fac-$ReX(CO)_3$(o-$MeSC_6H_4SeMe$).[1219]

Iron, Ruthenium, and Osmium. ^{31}P n.m.r. spectra, including the use of the DANTE pulse sequence, have shown exchange between the two different phosphorus sites in (16), and it was suggested that the mechanism involved (17) as an intermediate. The ^{13}C n.m.r. spectrum was also recorded.[1220] The ^{1}H n.m.r. spectrum of $(\mu\text{-H})Fe_3(CO)_9(\mu_3\text{-}BH_4)$ at 60°C shows exchange of the Fe-H-B and Fe-H-Fe protons.[1221] Low-temperature ^{1}H and ^{13}C n.m.r. spectra of $[HFe_4(CO)_{13}]^-$ demonstrate the presence of two distinct hydride species which are in equilibrium. Both species begin to equilibrate at -30°C. The ^{17}O n.m.r. spectra were also recorded.[1222]

Fluxional processes in $[Ru(\eta^3\text{-}C_8H_{11})(PR_3)_3]^+$ involve the mutual exchange of the three phosphorus donor ligands, and the endo hydrogen atoms on the two carbon atoms adjacent to the allylic functionality. ^{13}C and ^{31}P n.m.r. spectra were also recorded.[1223] In (18), only the two phosphorus ligands A and B exchange according to ^{31}P n.m.r. data.[1224] For $Ru_3(\mu\text{-H})(\mu\text{-X})(CO)_{10}$ (X = Cl, Br, I) an intramolecular polytopal rearrangement has been observed for the three CO groups on each of the two Ru atoms along the doubly bridged edge by ^{13}C n.m.r. spectroscopy, with a coalescence temperature of -40°C. In contrast for $Ru_3(\mu\text{-H})(\mu_3\text{-I})(CO)_9$, it was not possible to slow the exchange, even at -80°C.[1225] Kinetic ^{2}H isotope effects have been observed on the axial-radial carbonyl exchange process at the hydride-bridged metal atoms in $Os_3(\mu\text{-H})H(CO)_{10}L$ by variable-temperature ^{1}H and ^{13}C n.m.r. spectroscopy. Kinetic ^{2}H isotope effects were also measured directly on the site exchange of the magnetically inequivalent hydrides in $[(\mu\text{-H})_2Ru_3(CO)_9(\mu_3\text{-}CH{=}CBu^t)]^{2+}$ by ^{1}H and ^{2}H n.m.r. spectroscopy.[1226] Variable-temperature ^{13}C and ^{31}P n.m.r. spectra have been used to study carbonyl scrambling in $Ru_3(\mu\text{-H})(\mu\text{-Cl})(CO)_9(PPh_3)$.[1227] $Au_3Ru_4(\mu_3\text{-H})(CO)_{12}(PPh_3)_3$ and $Au_3Ru_4(\mu_3\text{-H})(\mu\text{-dppm})(CO)_{12}(PPh_3)$ undergo intramolecular rearrangement of the metal core according to ^{1}H, ^{13}C, and ^{31}P n.m.r. measurements.[1228]

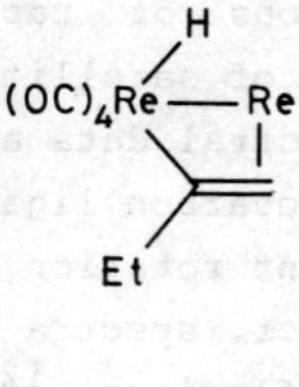

(15)

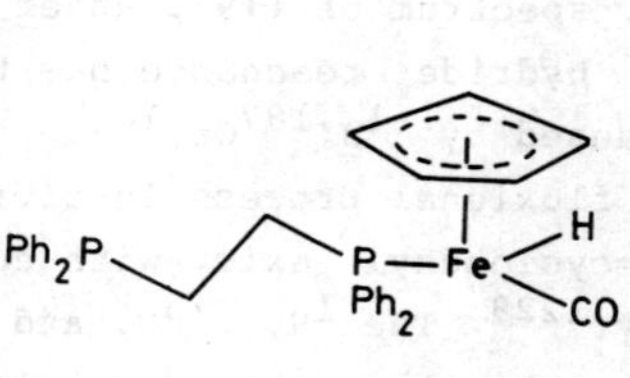

(16)

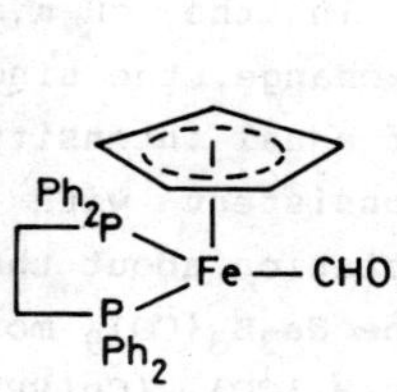

(17)

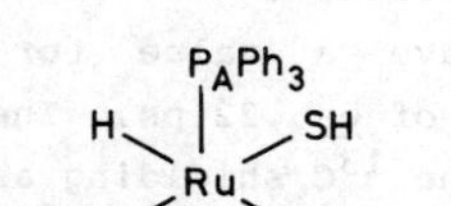

(18)

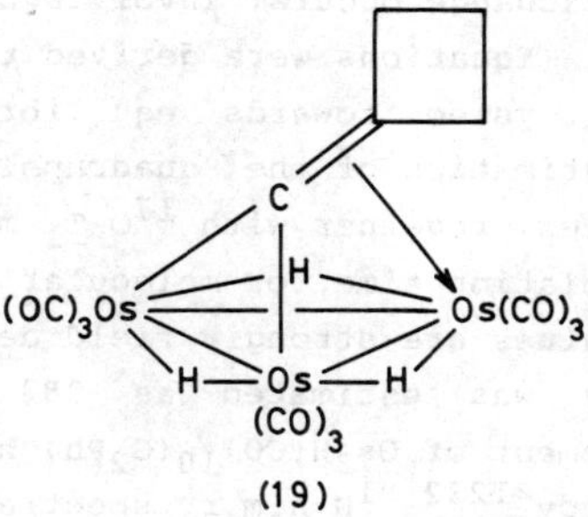

(19)

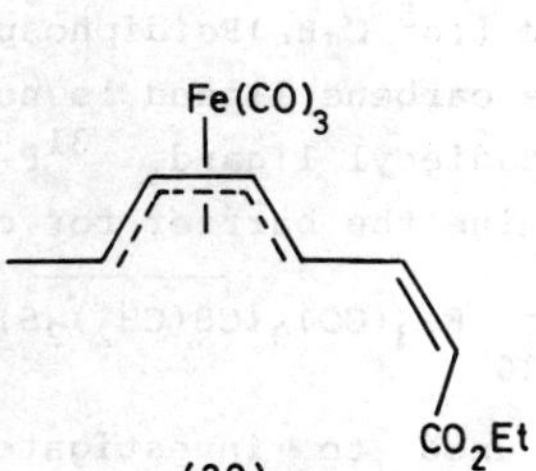

(20)

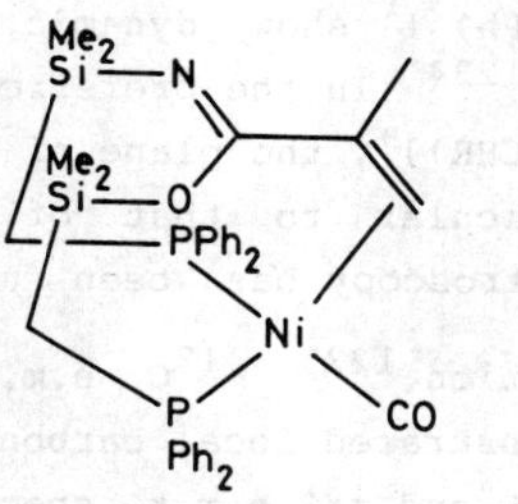

(21)

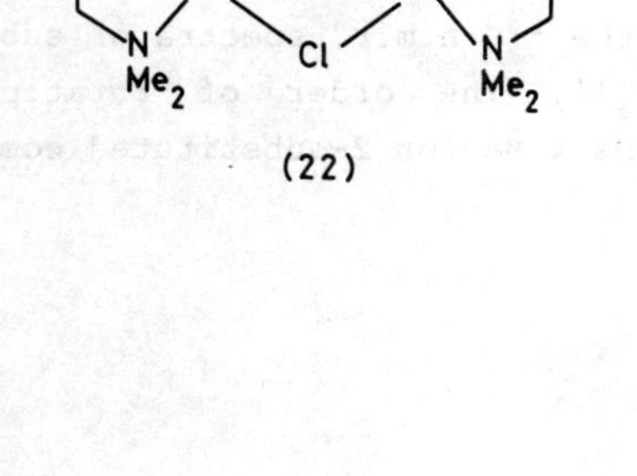

(22)

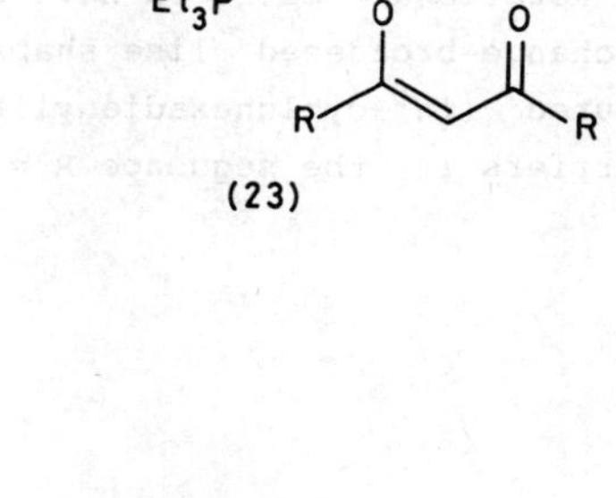

(23)

In the ^{1}H n.m.r. spectrum of (19), under conditions of rapid exchange, the single hydride resonance has two sets of satellites of equal intensity caused by $^{1}\underline{J}(^{187}Os,^{1}H)$. The spectral data are consistent with a fluxional process involving hydrocarbon ligand rotation about the C=cyclobutyl axis, with concomitant rotation of the $Os_3H_3(CO)_9$ moiety.[1229] The ^{1}H, ^{13}C, and ^{31}P n.m.r. spectra of $Os_3H_3(CO)_{10}\{Cu(PPh_3)\}$ show that the CO groups are fluxional.[1230] Application of the saturation transfer technique to the ^{13}C n.m.r. spectra of ^{13}C-enriched $Os_3(\mu\text{-}H)_2(CO)_{10}$ shows that only localized ^{13}C exchange occurs, involving 2 of the 4 types of carbonyl ligands. Equations were derived to describe the recovery of the AX_2 spin system towards equilibrium, after the initial perturbation. An estimation of the quadrupole-coupling constant for ^{17}O in the complex, together with ^{17}O $\underline{T}_1$ measurements, gave a value for the correlation time for molecular reorientation of ca. 22 ps. The ^{13}C $\underline{T}_1$ values are strongly field dependent and the ^{13}C shielding anisotropy was estimated as 383 p.p.m.[1231] The intramolecular rearrangement of $Os_3H(CO)_{10}(C_2Ph)$ has been studied by ^{13}C n.m.r. spectroscopy.[1232] ^{1}H n.m.r. spectra indicate restricted rotation of the CN bond in $Os(\mu\text{-}H)(\mu\text{-}O{=}CNR^1R^2)(CO)_{10}$.[1233]

The ^{1}H and ^{13}C n.m.r. spectra of $(\eta^5\text{-}C_5H_5)Fe(\eta^1\text{-}CMe{=}CMeL)(CO)\text{-}\{P(OPh)_3\}$ show dynamic behaviour due to Fe-C restricted rotation.[1234] In the preferred geometry for $[(\eta^5\text{-}C_5H_5)Fe(\text{diphosphine})\text{-}({=}C{=}CHR)]^+$, the plane of the alkylidene carbene ligand is not perpendicular to that of the cyclopentadienyl ligand. ^{31}P n.m.r. spectroscopy has been used to determine the barrier for carbene rotation.[1235] ^{13}C n.m.r. spectra of $Fe_3(CO)_9\{\overline{CS(CH_2)_2S}\}S$ have demonstrated local carbonyl exchange.[1236]

^{1}H and ^{13}C n.m.r. spectra have been used to investigate metal shifts in (20).[1237] The fluxional behaviour and equilibria of $(\eta^4\text{-enone})Fe(CO)_2L$ have been examined using ^{13}C and ^{31}P n.m.r. spectroscopy.[1238] ^{1}H n.m.r. spectroscopy has been used to demonstrate that $[(C_7H_7)Fe_2(CO)_6]^-$ is highly fluxional in solution.[1239] The dynamic behaviour of $Ru_4(CO)_{11}(\mu_4,\eta^2\text{-}MeC_2Ph)_2$ has been investigated using ^{13}C n.m.r. spectroscopy. At room temperature there is one ^{13}CO n.m.r. signal, and two at -80°C.[1240]

Rotational barriers have been determined from the analysis of exchange-broadened line shapes in the ^{13}C n.m.r. spectra of substituted $\{(\eta^5\text{-cyclohexadienyl})Fe(CO)_3]^+$. The order of rotational barriers in the sequence R = H < OMe < Me for 2-substituted complexes can be explained on steric grounds by the smaller bulk of the OMe group compared to that of the Me group.[1241] The activation energy for ring rotation and molecular tumbling in $(\eta^5\text{-}C_5H_5)(\eta^5\text{-}C_5H_4Bu^n)Fe$ and $(\eta^5\text{-}C_5H_5)(\eta^5\text{-}C_5H_4CMe_2Et)Fe$ has been deduced from variable-temperature ^{13}C $\underline{T}_1$ and n.O.e measurements.[1242] The stereodynamics of $Fe(\eta^5\text{-}C_5H_4SR)_2MX_2$ (M = Pd, Pt) have been examined by variable-temperature n.m.r. spectroscopy. $\Delta\underline{G}^\ddagger$ for S-inversion was determined.[1243] Rotation about the C-N bond in $(\eta^5\text{-}C_5H_4S_2CNR_2)_2Fe$ and $(\eta^5\text{-}C_5H_5)(\eta^5\text{-}C_5H_4S_2CNR_2)Fe$ has been studied by ^1H and ^{13}C n.m.r. spectoscopy.[1244] The ^{13}C and ^{31}P n.m.r. spectra of $(\eta^5\text{-}RC_6H_6)Os(PMe_3)_2I$ indicate restricted rotation around the metal-ring axis.[1245] ^{13}C n.m.r. spectroscopy and ^1H spin saturation transfer have been applied to the dynamic behaviour of $(1\text{-}3,5,6\text{-}\eta^5\text{-}C_8H_{11})_2Ru$.[1246] For $[(\eta^5\text{-}6\text{-}R^1{}_3PC_6H_6)Ru(PR^1{}_3)(PR^2{}_3)(PR^3{}_3)]^{2+}$, rotation around the metal-ring bond is so hindered below -30°C that rigid conformers are detected by ^{31}P n.m.r. spectroscopy.[1247]

From the ^{13}C n.m.r. spectrum of $FeRu(CO)_6(Pr^iDAB)$, carbonyl exchange is rapid on Fe and slow on Ru.[1248] Carbonyl scrambling in $[Ru_3\{\mu\text{-}O{=}C(NMe_2)\}(CO)_{10}]^-$ has been studied using ^{13}C n.m.r. spectroscopy. ^{31}P n.m.r. spectra were also recorded for related compounds.[1249] According to variable-temperature ^{13}C n.m.r spectroscopy, $[RuRh_4(CO)_{15}]^{2-}$ is static at -103°C.[1250] Variable-temperature ^{13}C n.m.r. spectroscopy has shown $[Os_3(CO)_{10}(NO)]^+$ to be fluxional.[1251]

The rotational motion of $[RuL_3]^{2+}$ and $[CoL_3]^{3+}$ (L = bipyridyl or phenanthroline) has been investigated using ^{13}C $\underline{T}_1$ data.[1252] Variable-temperature ^1H n.m.r. spectra of $RuCl_2(MeNCMeCMeNMe)_2$ shows intramolecular exchange.[1253]

<u>Cobalt</u>. According to ^1H and ^{11}B n.m.r. data, the barrier to exchange of bridging and terminal hydride ligands in $Co(terpyridyl)H_2BH_2$ is 11.1 kcal mol^{-1}.[1254] Spin saturation ^1H n.m.r. spectroscopy has been used to examine <u>syn</u>-<u>anti</u> exchange in $(\eta^3\text{-}CH_2CMeCHCMeCH_2)Co(PMe_3)_3$. ^{13}C and ^{31}P n.m.r. spectra were also recorded.[1255] Fluxionality of $(\eta^6\text{-}C_6H_5BPh_3)Co(PMe_3)_2$ has been studied using ^{13}C and ^{31}P n.m.r. spectroscopy.[1256]

<u>Rhodium and Iridium</u>. The ^1H and ^{31}P n.m.r. spectra of $Rh_2Cl_2(\mu\text{-}H)(\mu\text{-}CO)(dppm)X$ indicate that they are symmetric on the n.m.r. time scale at room temperature owing to a fluxional process that either results from anion dissociation and re-co-ordination or from rever-

sible loss of HX.[1257] Stopped-flow n.m.r. spectroscopy has been used to investigate the co-ordination of cyclo-octadiene and norbornadiene with $[IrH_2(PPh_3)_2(OCMe_2)_2]^+$ to give $[IrH_2(PPh_3)_2(\text{diene})]^+$. The ^{31}P n.m.r. spectrum was also recorded. Spin saturation transfer shows hydrogen exchange.[1258] Hydride exchange has been observed in $Ir_2(\mu\text{-H})_2H_2(PMe_2Ph)_4$. The ^{31}P n.m.r. spectrum was also recorded.[1259]

The ^1H n.m.r. spectrum of $[(\eta^5\text{-}C_5Me_5)_2Rh_2(\mu\text{-}CH_2)_2Me(NCMe)]^+$ shows methyl exchange between the two rhodium atoms. The ^{13}C n.m.r. spectrum was also recorded.[1260] The ^1H n.m.r. spectra of $(\eta^5\text{-}C_5Me_5)Rh(PR^1_3)R^2X$ have been used to investigate restricted rotation about Rh-P and Rh-R bonds and activation energies determined. The ^{31}P n.m.r. spectra were also determined.[1261]

The non-rigidity of $\{Me_2Ga(pz)_2\}RhL_2$ [L_2 = cod, $(CO)_2$, $(PPh_3)_2$, $(CO)(PPh_3)$] has been investigated by ^1H n.m.r. spectroscopy.[1262] Conformational equilibria in $[(\eta^4\text{-cod})M\{Ph_2(CH_2)_4PPh_2\}_2]^+$ (M = Rh, Ir) have been investigated using ^1H and ^{31}P n.m.r. spectroscopy.[1263] The ^{13}C n.m.r. spectrum of $Ir_3Pt_3(\mu\text{-CO})_3(CO)_3(\eta^5\text{-}C_5Me_5)_3$ shows dynamic CO scrambling.[1264]

Low-temperature ^{31}P n.m.r spectra of trans-$Rh(CO)Cl(PBu^t{}_2R)_2$ (M = Rh, Ir) are consistent with slow rotation about the metal-P bonds on the n.m.r. time-scale. ^{13}C n.m.r. data were also published.[1265] The temperature-dependent ^1H n.m.r. spectrum of $Rh_2(CNR)_8$ suggests harder intramolecular and easier intermolecular ligand exchange than those of the corresponding Co analogue.[1266] Internuclear site exchange of a bridging PPh_2 ligand is rapid at room temperature for $Rh_4(\mu\text{-CO})_2(CO)_4(\mu\text{-}PPh_2)_4$. A second higher-temperature process allows one PPh_2 ligand to migrate between six sites about three Rh-Rh edges. The fluxionality was investigated using 2D ^{31}P n.m.r. spectroscopy.[1267] The fluxionality of $[PtRh_5(CO)_{15}]^-$, $[PtRh_4(CO)_{14}]^{2-}$, and $[PtRh_4(CO)_{12}]^{2-}$ has been established by ^{13}C, $^{13}C\text{-}\{^{103}Rh\}$, ^{103}Rh, and ^{195}Pt variable-temperature n.m.r measurements.[1268]

The ^{31}P n.m.r. spectrum of $[Rh(PPh_3)_3]^+$ is a doublet at room temperature and A_2BX at -50°C. The ^{13}C n.m.r. spectrum shows no $\underline{J}(^{103}Rh,^{13}C)$ coupling.[1269] Variable-temperature ^{31}P n.m.r. spectroscopy has shown that $[\{RhCl(PPh_2)(PPh_2O)\}_2(\mu\text{-}Cl)_3]^+$ undergoes facile intramolecular scrambling at room temperature.[1270] A line shape analysis of the temperature-dependent ^1H n.m.r. spectrum of $Rh(S_2CNMePh)_3$ has yielded a value of $\Delta G^\ddagger = 61.5 \pm 4.2$ kJ mol^{-1} for C-N bond rotation. This barrier to rotation is large compared to similar complexes of Fe(II) and Fe(III), but similar to values for Fe(IV) and Co(III). Metal-centred optical inversion is slow over the temperature range studied.[1271]

Nickel. ^1H n.m.r. spin saturation transfer measurements have been used to demonstrate intramolecular exchange in $[NiH\{P(O\text{-}p\text{-}tol)_3\}_3]^+$.[1272] A theoretical study of the barrier to olefin rotation in $NiMe(\eta^2\text{-}C_2H_4)(PH_3)X$ has been reported and compared with n.m.r. results.[1273] Using variable-temperature ^1H and ^{29}Si n.m.r. spectroscopy, rapid silyl-group exchange between O and N centres has been demonstrated for (21) and related compounds. A comparison of $\Delta G^\ddagger$ values for the silyl exchange in $Ph_2PCH_2SiMe_2OCPh{=}NSiMe_2CH_2\text{-}PPh_2$ and its nickel derivative indicates a substantial enhancement for this migration when the ligand is bound to nickel.[1274] Olefin rotation has been studied in $(\eta^5\text{-}C_5H_5)Ni(\eta^2\text{-}alkene)R$ compounds using ^{13}C n.m.r. spectroscopy.[1275]

Palladium and Platinum. ^1H n.m.r. spectroscopy has shown synchronous inversion of sulphur or selenium in $(PtMe_3X)_2(Me_2\overline{CCH_2EECH_2})$ (E = S, Se), and $\Delta G^\ddagger$ was determined.[1276] The non-rigidity of $(PtMe_3Cl)_2(\overline{SCH_2SCH_2SCH_2})$ has been studied using ^1H n.m.r. spectroscopy. Both metal-pivot and PtMe rearrangements were found.[1277] The activation energy for inversion of the 8-membered rings in $Me_4Pt_2(\mu\text{-}dppm)_2$ and related compounds has been determined.[1278] The ^1H and ^{13}C n.m.r. spectra of $Pt_2(\mu\text{-}Cl)_2Me_2(\eta^2\text{-}C_2H_4)_2$ show restricted rotation of the olefin. ^{19}F n.m.r. spectra were also reported for $Me(CF_3)C{=}C(CF_3)Pt(py)_2Cl$.[1279] ^1H, ^{13}C, and ^{31}P n.m.r. spectra of $\overline{Pt(CHRCOCHR)}(PPh_3)_2$ have been interpreted in terms of inversion of the platinacyclobutan-3-one ring. The magnitude of ^{195}Pt coupling to the axial and equatorial ring hydrogen atoms of the platinacyclobutan-3-one ring is dependent upon the orientation of the CH bonds with respect to the square plane.[1280] A systematic study of (22) [M = $(\eta^5\text{-}C_5H_5)W(CO)_3$] and related molecules has revealed that the value of $\Delta G^\ddagger$ depends only on the nature of the π-bonded carbon atom.[1281] ^1H and ^{13}C n.m.r. studies have shown that the palladium undergoes an intramolecular sigmatropic [1,5] shift between the oxygen atoms in (23) (R = Me, CF_3).[1282]

^{13}C and ^{31}P n.m.r. spectra have been reported for $[Pt(\eta^2\text{-}alkene)(\eta^3\text{-}2\text{-}MeC_3H_4)(PPh_3)]^+$ and the barrier to olefin rotation determined.[1283] ^1H n.m.r. spectroscopy has been used to investigate olefin rotation in Pt(η^2-alkene)(p-toluidine)Cl_2.[1284] Intramolecular mobility of (η^3-allyl)PdCl complexes of different tertiary phosphines has been studied by 2D n.m.r. spectroscopy.[1285]

Variable-temperature ^{31}P n.m.r. spectra of $Pt_3(CO)_3\{P(p\text{-tol})_3\}_4$ show one ^{31}P n.m.r. signal at room temperature and two at low temperature. ^{31}P n.m.r. spectroscopy was also used to investigate the reaction of cis-$PtCl_2(PEt_3)_2$ with MeOH and N-Et-piperidine.[1286] The fluxional behaviour of some $[B(pz)_4]^-$ complexes of Pd(II) has been studied.[1287] ^1H, ^{13}C, ^{19}F, and ^{31}P n.m.r. studies have shown that $M(hfac)_2L$ (M = Pd, Pt) is fluxional above -60°C.[1288]

Copper and Gold. The ^1H and ^{13}C n.m.r. spectra of (η^1-C_5H_5)$CuPR_3$ and (η^1-C_5Me_5)$AuPR_3$ show fluxionality.[1289] For Au(SC_6H_4R-p), the ^1H and ^{13}C n.m.r. spectra show several sets of signals which coalesce to a single set at high temperature.[1290]

Zinc. ^1H and ^{13}C n.m.r. spectra of (η^1-C_5H_5)$_2$(η^5-C_5H_5)$\{\mu$-N$(SiMe_3)_2\}$ show exchange between the two types of cyclopentadienyl groups.[1291]

Cadmium. The kinetics of inversion of configuration of the nitrogen atoms in $[Cd(edta)]^{2-}$ have been studied by ^1H n.m.r. spectroscopy, including full line-shape analysis. Addition of dioxane to the aqueous solution increases the rate of nitrogen inversion.[1292] N.m.r. spectra, including 2D, of Cd$\{PhC(S)CPr^iCHNCy\}_2$ have shown that intermolecular exchange occurs, involving a 5-coordinate Cd in a dimer.[1293] Temperature-dependent ^{113}Cd n.m.r. spectroscopy has been used to investigate the flexible nature of metallothioneins.[1294]

Boron. The reversible cage rearrangement of $R_4C_4B_8H_8$ has been examined using ^1H, ^{11}B, and ^{13}C n.m.r. spectroscopy.[1295] The structure and fluxional character of derivatives of $[B_9H_{14}]^-$ have been examined by ^1H and ^{11}B n.m.r. spectroscopy.[1296] Variable temperature ^1H and ^{31}P n.m.r. spectra have been used to show B-H-B/Rh-H-B exchange in Rh(7,8-$C_2B_9H_{10}Me_2$)$(PPh_3)_2$.[1297] In EtBCEt($SnMe_3$)CMe=CMeCEtSnMe_3, ^1H, ^{11}B, ^{13}C, and ^{119}Sn n.m.r. spectra have been used to show $SnMe_3$ migration around the ring.[1298] ^{11}B and ^{13}C n.m.r. spectra have been used to show that (24) is fluxional.[1299] Rotational barriers about the B-NMe_2 bond in $(Me_3Si)_2NBX(NMe_2)$ have been determined by ^1H and ^{13}C n.m.r. spectroscopy and the results discussed in terms of steric interactions.[1300] The structurally inequivalent MeN and PhB groups in (25) interchange at room temperature.[1301] $\Delta\underline{G}^\ddagger$ for ring inversion in (26) has been determined from ^1H n.m.r. measurements to be 14.4 kcal mol^{-1}.[1302]

<u>Aluminium and Gallium</u>. A dynamic ^1H n.m.r. study of $M(tfac)_3$ (M = Al, Ga) has been reported. A bond-rupture mechanism for <u>cis</u>-<u>trans</u> isomerization is indicated by the positive entropy of activation.[1303]

<u>Silicon, Germanium, and Tin</u>. The ^1H n.m.r. spectrum of $C(SiMe_3)_2(SiMePh_2)(SiMe_2ONO_2)$ shows two signals for the $SiMe_3$ and $SiMe_2$ protons at room temperature and 8 at -90°C.[1304] Fluxionality of the trigonal-bipyramidal silicon in 1-R_3Si-2-$(Me_2NCH_2)C_6H_4$ has been shown by ^1H, ^{13}C, and ^{19}F n.m.r. spectroscopy.[1305] According to the ^1H, ^{13}C, and ^{31}P n.m.r. data, the activation barrier for [1,3] silyl exchange in $(Me_3Si)_2NPHR{=}NSiMe_3$ lies between 12 and 19 kcal mol^{-1}.[1306] Exchange of R^1 = $SiMe_3$ or $P(O)(OR^2)_2$ in (27) between the <u>exo</u>-cyclic nitrogen atoms has been investigated by n.m.r. spectroscopy and has an activation energy of between 23 and 27 kcal mol^{-1}.[1307] The $\Delta\underline{G}^\ddagger$ for ring inversion in some dibenzo[<u>d</u>,<u>g</u>][1,3,2]-dioxasilepin rings has been measured by ^1H and ^{13}C n.m.r. spectroscopy.[1308] The ^1H n.m.r. spectra of $(\underline{o}\text{-}Et_2NCH_2C_6H_4)SnMe_2X$ exhibit dynamic properties due to N dissociation.[1309] The phosphorus inversion barrier in $C_6H_4(PR^1)_2MR^2{}_2$ (M = C, Si, Ge, Sn) is low, <84 kJ mol^{-1}, compared with 1,3-diphospholanes. The ^{31}P n.m.r. spectra were also recorded.[1310] ^1H n.m.r. spectra have shown (28) to be fluxional.[1311]

Molecular motion in the liquid-crystal phase of polyethylsiloxane has been studied by n.m.r. spectroscopy.[1312] Ring inversion of <u>cis</u>-$\overline{SiF_2SiF_2CH_2CHXCHXCH_2}$ (X = H, Me, Br, Cl) has been investigated by ^{19}F n.m.r. spectroscopy and $\Delta\underline{H}^\ddagger$, $\Delta\underline{S}^\ddagger$, and $\Delta\underline{G}^\ddagger$ determined.[1313] The ^1H and ^{13}C n.m.r. spectra of Sn(η^1-indenyl)$_4$ and SnBu(η^1-indenyl)$_3$ are temperature dependent, resulting in the interconversion of all the isomers by sigmatropic [1,5] shifts.[1314] ^1H n.m.r. spectroscopy has been used to study molecular motion in $Ph_8Si_4O_4$.[1315] The activation energy for CF_3 exchange in $\overline{Si\{C_6H_4C(CF_3)_2O\}_2}$ has

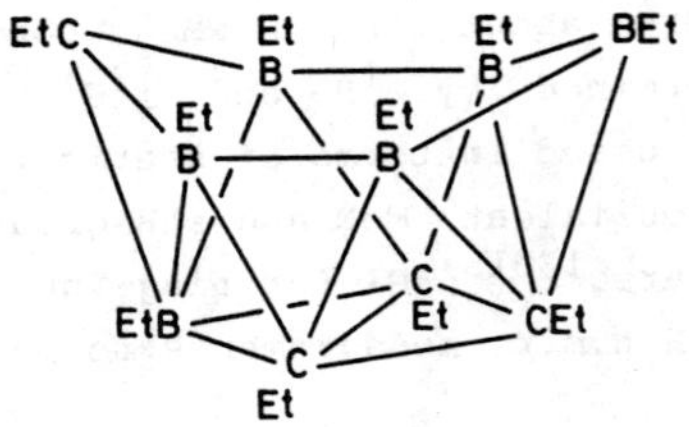

(24)

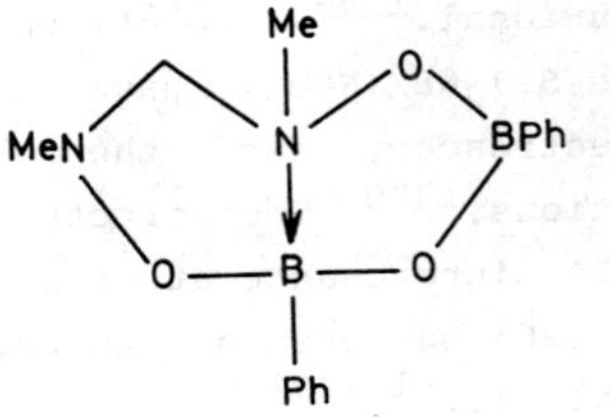

(25)

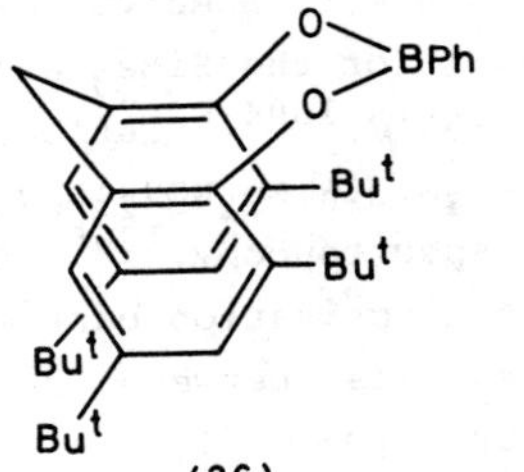

(26)

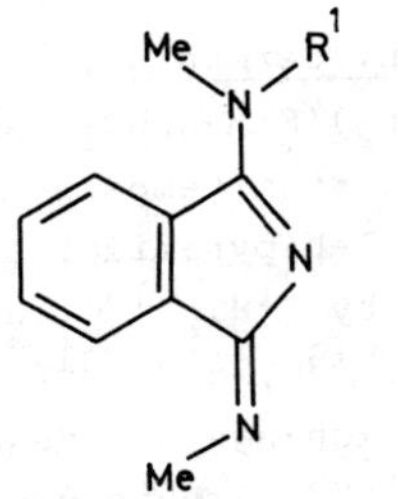

(27)

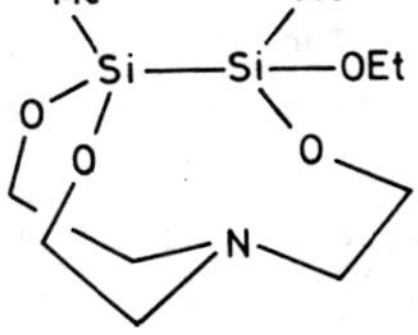

(28)

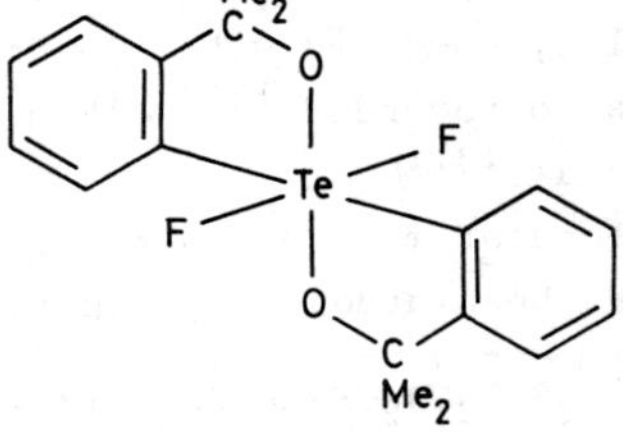

(29)

been determined using ^{19}F n.m.r. spectroscopy.[1316] ^{1}H and ^{19}F n.m.r. spectra have been used to investigate temperature-dependent fluorosilanes containing intramolecular Si-N coordination. A clear distinction was found between pseudo-rotation and ring-opening processes.[1317]

Phosphorus. [1,5] Sigmatropic rearrangements have been observed in $(\eta^1\text{-}C_5H_5)PR_2$ and $(\eta^1\text{-}C_5H_5)PRCl$. Substituents at phosphorus drastically influence the speed of these rearrangements.[1318] A low temperature ^{13}C n.m.r. study of $(R_2N)_2PPh$ has indicated the conformer with the nitrogen lone pair orthogonal to the phosphorus lone pair. P-N rotation barriers were determined as 7.2 and 8.4 kcal mol^{-1}.[1319] $\Delta G^{\ddagger}$ for rotation about the P-N bond in $\overline{MeNCH_2CH_2OP}NR_2$ has been determined. The ^{13}C and ^{15}N n.m.r. spectra were recorded.[1320] Permutational isomerism in $PhC(NMe_2)_2PF_4$ has been investigated by ^{31}P n.m.r. spectroscopy and the activation energy determined.[1321] Variable-temperature ^{19}F and ^{31}P n.m.r. spectra have been reported for four 1-piperidyl-trifluorophosphoranes, substituted with a methyl group in the piperidyl ring. Restricted rotation of the P-N bond was found.[1322] Variable-temperature ^{19}F spectra have shown a steric effect on the intramolecular ligand reorganization in $Ph_{\underline{n}}P\{OCH(CF_3)_2\}_{5-\underline{n}}$ and $PhSP\{OCH(CF_3)_2\}_4$, and were discussed in terms of pseudo-rotation.[1323] The hindered rotation of bulky alkyl groups on Me and Et groups in $(R^1O)_2P(S)SCH_2R^2$ and $(R^1O)_2P(S)OR^2$ has been studied by ^{1}H n.m.r. spectra.[1324]

Arsenic. ^{13}C $\underline{T}_1$ measurements have been performed on 2-Cl-1,3,2-dioxarsolane. Below 320K, the dipole-dipole relaxation mechanism is dominant, but at higher temperatures spin-rotation relaxation becomes significant. Activation energies for tumbling were determined.[1325]

Antimony. The barrier to internal rotation of $\underline{N},\underline{N}$-dimethylacetamide, when coordinated to $SbCl_3$, has been determined from total line shape analysis of the ^{13}C n.m.r. spectra. The rotational barrier in DMA decreases by a small amount on formation of this adduct in contrast to the increase normally observed for similar complexes.[1326]

Oxygen. The dynamics of pressurised and super-cooled water and aqueous solutions have been studied by ^{1}H, ^{2}H, and ^{17}O n.m.r. spectroscopy.[1327, 1328] ^{1}H, ^{2}H, and ^{17}O n.m.r. spectroscopy has been used to investigate water reorientation in a lyotropic lamellar phase.[1329]

Sulphur. The ^{33}S chemical shift and ^{33}S and ^{17}O T_1 data have been reported for liquid SO_2. The temperature dependence of the ^{33}S and ^{17}O T_1 data was described by an average activation energy of 6.3 ± 0.6 kJ mol^{-1}. The ^{17}O quadrupolar-coupling constant was calculated as 6.6 ± 0.7 MHz in SO_2.[1330]

Selenium. $PhSeSePr^{i}$ has been characterized by ^{13}C and ^{77}Se n.m.r. spectroscopy. ^{77}Se DANTE selective-transfer magnetization measurements show exchange.[1331]

Tellurium. ^{1}H, ^{19}F, and ^{125}Te n.m.r. spectroscopy has been used to determine the activation energy for fluxionality of (29).[1332]

Equilibria.—Solvation Studies of Ions. The principles of the structure of hydrate shells of ions according to n.m.r. relaxation data have been discussed.[1333] A complex method for studying aqueous electrolyte structure, based on diffusion and magnetic relaxation measurements, has been described.[1334] The motion of a complex in solution has been described as an oscillation of a non-linear harmonic oscillator, disturbed by thermal motion. From the experimental data on the temperature dependence of T_1, the lifetimes of molecules of water around the ions were found.[1335]

Group IA. The Knight shifts and T_1 of ^{7}Li in liquid NH_3 and $MeNH_2$ solutions have been investigated with particular attention to the transition from localized to itinerant electron states. A considerable enhancement in the Korringa relaxation rate was observed in Li-$MeNH_2$ solutions, while it is less pronounced in Li-NH_3 solutions.[1336] Molecular dynamic simulations have been performed to investigate the molecular origin of the nuclear quadrupole relaxation mechanism for $^{7}Li^{+}$, $^{23}Na^{+}$, and $^{35,37}Cl^{-}$ ions in dilute aqueous solution. The field gradients are almost completely due to the water molecules in the first shell.[1337] Structural-dynamic features of some aqueous solutions of LiCl and $CaCl_2$ have been based on ^{1}H nuclear magnetic relaxation data in the presence of $MnCl_2$.[1338] A n.m.r. relaxation study has been carried out to determine the nature of the structural change in concentrated LiCl solutions.

The anomalous nature of the physicochemical properties of concentrated LiCl solutions was explained in terms of the increase in H_2O reorientation energy when $[Li(OH_2)_4]^+$ becomes the primary kinetic unit instead of free water molecules.[1339] Solvation interactions and ion association in solutions of $LiClO_4$ in DMF have been studied using ^7Li, ^{13}C, ^{14}N, ^{17}O, and ^{35}Cl n.m.r. spectroscopy. Solvation is through O, and the equilibrium constant for the binding of $[ClO_4]^-$ was determined.[1340] The ^{13}C n.m.r. spectra of Me_2CO, EtCOMe, Me_2C=CHCOMe, cyclopentanone, and cyclohexanone have been recorded in the presence of Li^+, Na^+, and Mg^{2+}. The solvation numbers of Na^+ with Me_2CO and cyclohexanone were determined from ^{23}Na n.m.r. spectra.[1341]

A modified Poisson-Boltzmann equation has been applied to calculate the distribution of Na^+ counterions in the vicinity of micellar SDS aggregates. The model was tested using ^{23}Na n.m.r. line widths.[1342] The Knight shift and $\underline{T}_1$ of ^{23}Na in Na-NH_3 solutions have been measured. The electronic contribution to $\underline{T}_1$ suggests the existence of 2 different composition regions.[1343] The Knight shifts and $\underline{T}_1$ of ^1H and ^{14}N in Na^- and K-NH_3 solutions have been measured as a function of temperature and composition.[1344] The fractionation factor for Cl^- has been determined by studying the dependence of the ^1H chemical shifts of H_2O on the concentration of NaCl or KCl and the ^2H content.[1345, 1346] The pressure and temperature dependence of the ^2H $\underline{T}_1$ of aqueous electrolyte solutions of LiCl, NaCl, and $MgCl_2$ have been measured over a large concentration range.[1347] ^{23}Na $\underline{T}_1$ and $\underline{T}_2$ have been measured in pigs and are suitable for imaging.[1348] $\underline{T}_1$ of $^{23}Na^+$, $^{87}Rb^+$, and ^{14}N in MeCN-H_2O mixtures has been determined and is an excellent probe for the direct interaction with other molecules.[1349] $\underline{T}_1$ of $^{23}Na^+$ and $^{87}Rb^+$ in MeCN increases strongly when small amounts of water are added, due to the disturbance of symmetry of the solvation shell. The $^{23}Na^+$, $^{87}Rb^+$, and ^{14}N relaxation rates were measured as a function of temperature to determine activation-energy values.[1350] ^1H $\underline{T}_1$ values have been determined to investigate the NaI-CdI_2-MeOH system.[1351] ^{133}Cs and ^{127}I n.m.r. spectra have been used to investigate solutions of Cs-CsI-I.[1352]

Magnesium. Low-temperature ^1H n.m.r. spectra have been used to investigate the primary solvation sphere of Mg^{2+} in DMF-H_2O mixtures. The stepwise equilibrium constants for water replacement by DMF were determined.[1353]

The Lanthanides. ^1H, ^{13}C, and ^{17}O n.m.r. spectroscopy has been used to investigate DMF exchange and the equilibrium between $[Ln(DMF)_8]^{3+}$ and $[Ln(DMF)_9]^{3+}$.[1354]

Uranium. T_1 and T_2 have been determined for ^1H in aqueous solutions of $UO_2(NO_3)_2$, and the cation hydration number determined.[1355] The exchange rate of DMF on $UO_2(acac)_2(DMF)$ has been determined by ^1H n.m.r. spectroscopy.[1356]

Vanadium. The kinetics of DMSO exchange in $[VO(DMSO)_5]^{2+}$ have been studied by high-pressure n.m.r. spectroscopy. The volume of activation for DMSO exchange at the basal positions of $[VO(DMSO)_5]^{2+}$ is -5.3 ± 0.4 cm^3 mol^{-1}, consistent with an associative mechanism.[1357, 1358]

Manganese. The analysis of the enhancement of H_2O n.m.r. relaxation rates by paramagnetic metal ions bound to macromolecules has been re-examined. By measuring T_1 and T_2 for both ^1H and ^2H in the same solution and at high field, it is possible to obtain the hydration number, exchange lifetime, and correlation time of the bound water molecules, as well as the magnitude of the additional contribution from outer-sphere water molecules. The technique was applied to Mn(II) bovine carboxypeptidase A.[1359] ^1H n.m.r. studies of DMF exchange on $[M(Me_6tren)(DMF)]^{2+}$ (M = Mn, Co, Cu) have yielded the rate of exchange, $\Delta H^{\ddagger}$, $\Delta S^{\ddagger}$, and $\Delta V^{\ddagger}$.[1360]

Ruthenium. ^1H n.m.r. chemical shifts have been used to study the hydrophobic interaction between $[Ru(phen)_3]^{2+}$ in aqueous solution.[1361]

Cobalt and Nickel. The theory has been developed for the dependence of the rotating-frame relaxation rate of bulk solvent nuclei on the magnitude of the spin-locking field for dilute solutions of a paramagnetic ion which undergoes exchange between bulk and coordinated sites. This approach was applied to Co(II) in CH_3OD and Ni(II) in CH_3CN.[1362] The hydration equilibrium of $[Ni(1,4,8,11\text{-}Me_4\text{-}1,4,8,11\text{-tetraazacyclotetradecane})]^{2+}$ has been investigated by ^1H n.m.r. spectroscopy.[1363] Similar studies of DMF exchange on this complex using ^{17}O[1364] or ^{13}C n.m.r.[1365, 1366] spectroscopy have been reported. ^1H n.m.r. spectroscopy has been used to determine the DMF exchange rate on $[Ni\{N(CH_2CH_2NMe_2)_3\}(DMF)]^{2+}$.[1367]

Palladium. Water exchange on $[Pd(OH_2)_4]^{2+}$ has been studied as a function of temperature and pressure by measuring ^{17}O n.m.r. linewidths. The exchange rate for $[Pd(OH_2)_4]^{2+}$ is 1.4×10^6 times faster than for $[Pt(OH_2)_4]^{2+}$ at 298K.[1368]

Copper. The spin-spin relaxation time of H_2O protons in aqueous solutions of the Cu(II)-NH_3 system has been measured as a function of free NH_3 concentration. Rate constants were determined for consecutive stepwise complex formation reactions.[1369] ^{1}H and ^{17}O n.m.r. spectroscopy has been used to investigate the interaction of water with bovine copper zinc superoxide dismutase.[1370]

Silver. A ^{109}Ag n.m.r. study of the selective solvation of Ag^{+} in solvent mixtures of water and pyridine, MeCN, or DMSO has been reported.[1371]

Boron. Equilibrium constants and kinetic parameters for the hydration of boric acid have been determined.[1372]

Chlorine. The nuclear magnetic relaxation rates of ^{1}H, ^{2}H, ^{17}O, and ^{35}Cl in aqueous solutions have been reported. The self-diffusion rate of H^{+} was estimated.[1373] A method has been described for measuring the difference between a ^{1}H frequency and the corresponding ^{2}H frequency. This measurement is a sensitive way of measuring equilibrium isotope effects for ^{1}H-^{2}H exchange and was applied to the $[H_3O]^{+}$-$[D_3O]^{+}$ system.[1374] The structure of aqueous solutions of $HClO_4$ has been investigated at high field.[1375]

Ionic Equilibria. The interaction between tetraphenylporphyrin or its metal derivatives with 2,4,7-trinitrofluorenone has been studied by ^{1}H n.m.r. spectroscopy. The equilibrium constants decrease in the order Cd > Pd > Co > Cu > Ni > $2H^{+}$ > Zn.[1376] ^{13}C n.m.r. spectroscopy has been used to demonstrate that acetate complexes as a mono-dentate ligand to Mg(II), Mn(II), Co(II), and Zn(II), but is bidentate to Cd(II).[1377] ^{13}C chemical shifts and relaxation times have been used to study alkali-metal formate and acetate eutectic melts containing transition metals. Activation energies for viscous flow derived from T_1 measurements are 10 to 11 kcal mol^{-1}.[1378] For complexes of octamethylpyrophosphoramide with Mg(II), Ca(II), Zn(II), Cd(II), Al(III), Ga(III), and In(III), two kinetic processes have been measured by ^{2}H n.m.r. spectroscopy: intramolecular optical inversion at low temperatures, and intermolecular ligand exchange between free and bound ligands at higher temperatures.[1379]

Group IA. The stoichiometry of complexation of an amide macrocycle with Li^{+}, Na^{+}, K^{+}, Rb^{+}, and Cs^{+} has been investigated by ^{1}H n.m.r. spectroscopy.[1380] ^{23}Na n.m.r. spectra of 1,4,8,11-tetrakis(2-hydroxyethyl)-1,4,8,11-tetra-azacyclotetradecane solutions containing Na^{+} with Dy^{3+} as the shift reagent have indicated the formation of a 1:1 complex with Na^{+}.[1381] ^{7}Li n.m.r. spectroscopy has been used to investigate the complexation of Li^{+} by some crown ethers.[1382] A simple estimation of Na^{+} complex stabilities has been reported and applied to various polyether ligands.[1383] ^{7}Li, ^{23}Na, and ^{133}Cs n.m.r. spectroscopy has been used to study MCl and some macrocyclic compounds in $AlCl_3$-[Bu^npy]Cl room-temperature melts. Formation constants were determined.[1384] ^{31}P chemical shifts and $\underline{T}_1$ have been studied as a function of concentration for Li, Na, and Ba dibutylphosphates and Na monohexylphosphate. On micellization, the ^{31}P resonance moves to lower frequency.[1385] An experimental method has been described that uses ^{31}P chemical shifts to determine association constants. The method was applied to association between $[Me_2PO_4]^-$, [c-AMP]$^-$, $[EtPO_4]^{2-}$, and [AMP]$^{2-}$ with Li^+, Na^+, K^+, Mg^{2+}, and Ca^{2+}.[1386] The ordered forms of [GMP]$^{2-}$ in the presence of Na^+ have been investigated by ^{1}H and ^{31}P n.m.r. spectroscopy.[1387] The binding of Na^+ and K^+ to double-helix DNA has been investigated by ^{23}Na and ^{39}K n.m.r. spectroscopy.[1388] A ^{17}O, ^{23}Na, and ^{25}Mg n.m.r. study of the counterion binding and hydration of hyaluronate and chondroitin has been published.[1389] The binding of Na^+ and K^+ to hen-yolk phosvitin has been studied by ^{23}Na and ^{39}K n.m.r. spectroscopy.[1390] ^{23}Na n.m.r. spectra have been measured in blood. The intracellular and extracellular Na signals were separated by adding Dy^{3+} and tripolyphosphate to the plasma.[1391] Micelle formation of anionic and cationic surfactants has been studied using ^{1}H and ^{7}Li n.m.r. spectroscopy.[1392] ^{2}H n.m.r. spectroscopy has been used to determine the order parameters for K palmitate.[1393] ^{23}Na quadrupole splittings have been measured in the hexagonal phase of a zwitterionic surfactant. ^{2}H n.m.r. spectra were also recorded.[1394] ^{23}Na n.m.r. spectra have been reported for the various lyomesophases of disodium cromoglycate.[1395] The aggregation of double-tail sulphonate surfactants has been probed by ^{23}Na n.m.r. spectroscopy.[1396] The diffusion of F^- and cations in melts of Li_2BeF_4, $LiBeF_3$, and $NaBeF_3$ has been studied by ^{19}F n.m.r. spectroscopy.[1397]

Group IIA. ^{1}H n.m.r. spectroscopy has been used to study the equilibrium between $LiAlH_4$ and Me_2Mg.[1398] ^{1}H n.m.r. spectroscopy has been used to study Mg^{2+} and Ag^+ co-ordination to biotin derivatives.[1399] ^{13}C, ^{15}N, ^{17}O, and ^{31}P n.m.r. spectra, including ^{13}C $\underline{T}_1$ measurements, have been used to investigate Mg^{2+}, Mn^{2+}, and Cu^{2+} interaction with L-aspartic acid, O-phospho-L-serine, and 2-mercapto-L-succinic acid.[1400] The phase behaviour in Ca(Mg) bis(2-ethylhexyl)sulphosuccinate-H_2O systems has been studied by ^{2}H and ^{25}Mg n.m.r. spectroscopy.[1401] Mg^{2+}, Ca^{2+}, Mn^{2+}, Zn^{2+}, and Cu^{2+} interactions with c-AMP and its derivatives have been studied by ^{1}H

n.m.r. spectroscopy.[1402] Stability constants for the Mg^{2+} and Cd^{2+} complexes of ATP and other nucleotides have been determined by ^{31}P n.m.r. spectroscopy.[1403] The dissociation constant of Mg^{2+}-ATP has been measured at physiological nucleotide levels by ^{31}P n.m.r. spectroscopy.[1404, 1405] 2D ^{31}P n.m.r. studies have been performed on myocardial ATP and phosphocreatine.[1406] ^{31}P n.m.r. spectroscopy has been used to investigate the binding of ATP to Mg^{2+} and Ca^{2+}, and the exchange rates were determined.[1407] Ca^{2+} binding to pancreatic phospholipase has been studied using ^{43}Ca n.m.r. spectroscopy.[1408] ^{43}Ca n.m.r. spectroscopy has also been used to study trifluoroperazine binding to calmodulin.[1409]

The Lanthanides. The determination of binding constants, bound chemical shifts, and stoichiometry of lanthanide-substrate complexes has been examined.[1410] A ^{1}H n.m.r. study of the equilibrium between lanthanide shift reagents and crown ethers has been published.[1411] Dynamic properties of Gd^{3+} complexes with ATP, including the temperature dependence of the correlation times and ATP exchange, have been studied by field-dependent ^{1}H $\underline{T}_1$ and by temperature dependent ^{31}P $\underline{T}_1$ studies.[1412] ^{1}H n.m.r. spectroscopy has been used to study the complexation of Dy^{3+} with malic acid. Equilibrium constants were determined.[1413] A similar study was also performed using Gd^{3+}.[1414] ^{13}C n.m.r. spectroscopy has been used to study the mode of binding of Gd^{3+} to various glycopeptides.[1415] The binding of Gd^{3+} and Mn^{2+} to a tri-O-D-galactosylated hexapeptide has also been studied using ^{13}C n.m.r. spectroscopy.[1416] ^{1}H n.m.r. spectroscopy has been used to determine the stablity constant between monoacetylacetone and La^{3+}, Nd^{3+}, and Lu^{3+}.[1417]

Uranium. The kinetics of the exchange reaction of acac in $UO_2(acac)_2(DMSO)$ have been studied by ^{1}H n.m.r. spectroscopy. The exchange rate depends on the concentration of the enol isomer of acac in its low region and approaches the limiting value in its high region.[1418] A full pH range ^{1}H and ^{13}C n.m.r. study has been performed on the complexation of $[UO_2]^{2+}$ with D-tartaric acid. Evidence was found for at least 6 complexes.[1419]

Vanadium. The speciation and equilibrium constants in the H^{+}-$[HVO_4]^{2-}$ system have been determined from ^{51}V n.m.r. measurements. Formation constants and n.m.r. data for the complexes have been determined.[1420]

Niobium and Tantalum. ^{19}F n.m.r. spectroscopy has been used to show that Nb and Ta are extracted by bis(2-ethylhexyl)phosphate as $HNbF_6$ and $HTaF_6$.[1421]

Chromium. The electron-transfer reaction between $Cr(CO)_3(PR_3)_3$ and $[Cr(CO)_3(PR_3)_3]^+$ has been investigated by ^{13}C n.m.r. spectroscopy.[1422] The pH dependence of the structure of $[Cr(edta)]^-$ has been investigated by ^2H n.m.r. spectroscopy.[1423]

Molybdenum and Tungsten. Complex formation between molybdate ions and *R,R*-(+)-tartaric acid has been studied as a function of concentration and pH using ^1H and ^{13}C n.m.r. techniques.[1424] A similar study was performed using a mixture of molybdate and tungstate.[1425] ^{31}P and ^{51}V n.m.r. studies have yielded the equilibrium constant for[1426]

$$11[H_2PMo_{10}V_2O_{40}]^{3-} + zH^+ \rightleftharpoons 10[HPMo_{11}VO_{40}]^{3-} + 12[VO]^+ + H_3PO_4 + 12H_2O$$

The stages in the formation of $H_5PMo_{10}V_2O_{40}$, $H_6SiMo_{10}V_2O_{40}$, and $H_5SiMo_{11}VO_{40}$ in the H_3PO_4-Na_2MoO_4-$NaVO_3$ and Na_2SiO_3-H_2MoO_4-$NaVO_3$ systems have been studied by ^{31}P and ^{51}V n.m.r. spectroscopy.[1427] The protonation and complexing of $[MMo_{12}O_{42}]^{8-}$ (M = Ce, U) heteropoly anions in aqueous solution have been studied using ^{17}O n.m.r. spectroscopy.[1428]

Manganese. The rate of electron self-exchange between $[Mn(CNCy)_6]^+$ and $[Mn(CNCy)_6]^{2+}$ has been measured by ^{55}Mn n.m.r. line broadening.[1429] The ^{13}C T_1 and T_2 of a Mn^{2+}-glycine system have been measured as a function of temperature, magnetic-field strength, and glycine concentration. Two complexes were detected.[1430] The penetration of electrolytes into the hydroxyethyl layer of non-ionic surfactants has been investigated by n.m.r. in the presence of Cu^{2+} and Mn^{2+}.[1431]

Rhenium. ^1H n.m.r. spectra of $[(\eta^5-C_5H_5)Re(CO)_2(NHNMeC_6H_4Me-p)]^+$ show the presence of two rotamers. The activation energy for rotation about the N-N bond was determined.[1432] A variable-temperature ^{19}F n.m.r. investigation of $[Re(CO)_3(tmen)(OH_2)]^+[BF_4]^-$ has demonstrated an equilibrium with $Re(CO)_3(tmen)(FBF_3)$.[1433]

Iron and Ruthenium. According to ^1H and ^{13}C n.m.r. spectroscopy, *cis*- and *trans*-$[Ru_2(CO)_2(\mu-CO)(\mu-Me)(\eta^5-C_5H_5)_2]^+$ interconvert with $\Delta G^\ddagger$ = 55 kJ mol^{-1}.[1434] ^{13}C and ^{15}N n.m.r. spectroscopy has been used to determine the equilibrium between two isomers of $[FeRu_3N(CO)_{12}]^-$ as a function of temperature, and hence ΔH° and ΔS°.[1435] The binding of $[Co(CN)_6]^{3-}$, $[Fe(CN)_6]^{3-}$, and $[Ru(CN)_6]^{3-}$ to horse cytochrome *c* has been studied by ^{59}Co n.m.r. spectroscopy, and association constants determined.[1436] The self-association of urohaemin I has been investigated by ^1H n.m.r. spectroscopy.[1437] The equilibrium between $[M(SR)_4]^{2-}$, $[M_2(SR)_6]^{2-}$, and $[M_4(SR)_{10}]^{2-}$

(M = Fe, Co) has also been investigated by ^{1}H n.m.r. spectroscopy.[1438] The self-exchange rate for the $[Ru(OH_2)_6]^{2+}/[Ru(OH_2)_6]^{3+}$ couple has been determined using ^{17}O and ^{99}Ru n.m.r. spectroscopy.[1439]

Cobalt. The interaction of methylcobalamin with $[Pt(CN)_4]^{2-}$, $[Pt(SCN)_4]^{2-}$, and $[PtCl_4]^{2-}$ has been investigated by ^{1}H n.m.r. spectroscopy. The site of coordination was tentatively identified.[1440] Dynamic ^{1}H n.m.r. spectra have been used to estimate ligand – exchange rates for (3,5-lutidine)Co(saloph)CH_2CN. The exchange rate is 10^{10} times faster in the saloph complex than in the corresponding dmgH complex.[1441] Both ^{13}C and ^{31}P n.m.r. spectra have been used to demonstrate the equilibrium between the base-on and base-off forms of cyanocobalamin. pK_a values were determined for the phosphate in both forms.[1442] Nuclear magnetic relaxation rates have been reported for $[Co(NH_3)_6]^{3+}$ and $[Co(en)_3]^{3+}$. The changes in the rotational correlation times of the metal complexes caused by ion pairing were derived from ^{13}C relaxation rates.[1443] The redox properties of $Co(dmgH)_2$ in base have been studied by n.m.r. spin echo. Dimerisation and oxidation were examined.[1444] The ^{31}P n.m.r. spectrum of aquocobalamin show separate signals for the base-on and base-off forms, and hence can be used to investigate equilibria.[1445] The binding of 4,5-$Pr^n{}_2$-imidazole to Co(II) and Zn(II) have been investigated by n.m.r. spectroscopy.[1446] ^{14}N n.m.r. spectroscopy has been used to study the exchange of pyridine on $M(acac)_2(py)_2$ (M = Co, Ni).[1447] A method has been described and used to measure the stability constant of the complex formed between pyridine and $[Co(CD_3OD)_6]^{2+}$.[1448] ^{1}H n.m.r. studies have been performed on the inter- and intra-molecular ligand mobility in the Co(II)/18-crown-6 system.[1449] Nuclear magnetic relaxation has been used to establish the formation of 1:1, 1:2, 2:2, and 1:3 metal:ligand complexes between Co(II) or Ni(II) and citric acid.[1450] The degree of association of Co(II) and Zn(II) di(2-hexyl)phosphates in CCl_4, C_6H_6, and hexane has been investigated by ^{31}P n.m.r. spectroscopy.[1451]

Nickel. Magnetic relaxation studies have shown the formation of a wide range of complexes between Ni(II) and 1-hydroxyethane-1,1-diphosphonic acid.[1452] ^{14}N n.m.r. spectroscopy has been used to measure the rates of pyridine exchange of a series of bis-pyridine adducts of square-planar nickel complexes containing dithiocarbamate, acac, and similar ligands.[1453]

Palladium. ^{1}H and ^{13}C n.m.r. spectra have been used to investigate the equilibrium between (30) and (31).[1454]

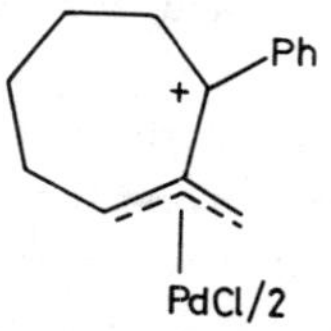

(30)

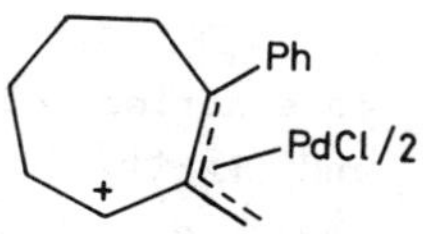

(31)

Platinum. ^{1}H and ^{13}C n.m.r. spectra have been used to determine the equilibrium between the Z- and E-olefin complex and $[Pt(\eta^{2}-CH_2=CHPh)(\eta^{3}-2-MeC_3H_4)(PPh_3)]^{+}$ for MeCH=CHR.[1455] ^{119}Sn and ^{195}Pt n.m.r. spectroscopy has been used to investigate the interaction of $[PtCl_6]^{2-}$ with $SnCl_2$ in acetone,[1456] and $Pr^{i}OH$.[1457] ^{19}F n.m.r. spectroscopy has been used to investigate the equilibrium between $H_2[Pt(SO_3F)_6]$ and HSO_3F.[1458]

Copper. Stacking in the Zn(II)/phenanthroline/benzyl alcohol system has been confirmed by ^{1}H n.m.r. measurements.[1459] ^{1}H n.m.r. spectroscopy has been used to determine the self-exchange rate for

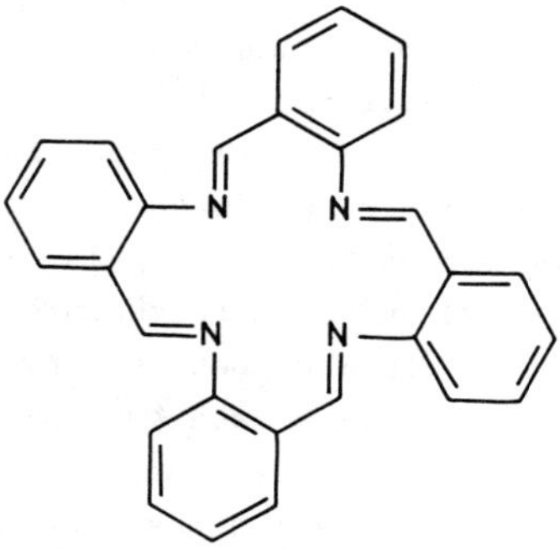

(32)

the Cu(II)/Cu(I) couple complexed to (32).[1460] ^{1}H and ^{13}C n.m.r. spectroscopy has been used to investigate the interaction between Cu(II) and ribavirin.[1461] The kinetics of the exchange of some amino acids in the ternary Cu(II) hydroxyethylidenediphosphonic acid complexes have been studied using nuclear magnetic relaxation.[1462] The polycondensation of aqua ions in mixed Al(III)/Cu(II) solutions has been studied using multinuclear n.m.r. spectroscopy.[1463]

Silver. The equilibrium between Ag^{+} and thiophosphonyl containing macrocyclic polyethers has been investigated by ^{1}H n.m.r. spectroscopy.[1464]

Zinc and Cadmium. ^{67}Zn linewidths have been determined in aqueous solutions of $ZnSO_4$ as a function of concentration of sulphate and citrate.[1465] ^{13}C and ^{113}Cd n.m.r. spectra of dibromo[5-(1,2,5-di-thiazepan-5-ylmethylene)-4-methyl-2-ethylimid-azole]Cd(II) have shown the use of two N donors and 1:1 complex formation. A stability constant of 60 M^{-1} was determined from ^{13}C n.m.r. measurements.[1466] ^{31}P n.m.r. spectroscopy has been used to investigate the equilibria in the system Zn(II)-pyridoxal-5'-phosphate-2-amino-3-phosphonopropionic acid.[1467] The stability and structure of monomeric Cd(II) and Zn(II) complexes of the 5'-triphosphate of adenosine and cytidine have been determined from ^{1}H n.m.r. chemical shift measurements.[1468] ^{113}Cd n.m.r. spectroscopy has been used to investigate the binding of zinc and cadmium to 5-aminovulinic acid dehydratase,[1469] and cadmium to skeletal troponin C.[1470]

Mercury. The kinetics of the exchange of $[MeHg]^{+}$ between thiol ligands in aqueous solution have been studied by ^{1}H n.m.r. spectroscopy over a range of pH values.[1471] The binding of Hg(II) to selenomethionine has been studied by ^{1}H n.m.r. spectroscopy.[1472]

Boron. The system Me_2CO-H_3BO_3-H_2O-MeOH has been investigated by ^{1}H n.m.r. spectroscopy.[1473] The behaviour of $NaBF_3(OH)$, $[SnF_5(OH)]^{2-}$, and $[Sn(OH)]^{2-}$ in H_2O_2 has been studied by ^{1}H n.m.r. spectroscopy.[1474]

Aluminium and Gallium. ^{27}Al n.m.r. spectroscopy has been used to identify the constituents of fresh and aged Al plating baths containing $AlCl_3$, $LiAlH_4$, and THF as $AlHCl_2.2THF$ and $[AlHCl_3]^{-}$.[1475] A method for the determination of the thermodynamic properties and n.m.r. parameters of weak intermolecular complexes in binary systems with fast equilibria reactions has been described. The system mesitylene/$AlBr_3$ was examined by ^{13}C n.m.r. spectroscopy, but the results were ambiguous until ^{27}Al data were also used.[1476] Mixtures of 1-methyl-3-ethylimidazolium chloride and $AlCl_3$ are liquid at room temperature, and ion interactions were investigated by ^{1}H n.m.r. spectroscopy.[1477] ^{27}Al $\underline{T}_1$ measurements have been made on molten 1-n-butylpyridinium chloride/$AlCl_3$ mixtures and various species identified.[1478] ^{19}F n.m.r. spectra indicate that $[M(HOR)_6]^{3+}$ (M = Al, Ga) form outer-sphere complexes with F^{-}, Cl^{-}, and Br^{-}.[1479] The solvation of 3-methylsydnone molecules to Al^{3+} in a water-3-methylsydnone mixture has been studied by ^{1}H and ^{13}C n.m.r. measurements.[1480] Complex formation between Al^{3+} and nitrilotris(methylenephosphonic acid) has been studied using ^{27}Al

n.m.r. spectroscopy.[1481] ^{1}H, ^{13}C, ^{27}Al, and ^{31}P n.m.r. spectroscopy has been used to study the kinetics and equilibria of Al^{3+}-ATP complexes.[1482]

Thallium. The thermodynamic parameters for the reaction of thallium perchlorate with 18-crown-6 in aqueous solution have been determined by n.m.r. spectroscopy.[1483] ^{205}Tl n.m.r. chemical shifts and line widths have been measured in molten $TlCl-Tl_2S$ mixtures as a function of composition.[1484] ^{1}H, ^{13}C, and ^{205}Tl n.m.r. spectroscopy has been used to investigate metal-binding sites of the transferrins.[1485]

Silicon. ^{29}Si n.m.r. spectroscopy has given evidence for chemical exchange between small silicate anions in alkaline aqueous solutions of Na and K silicates.[1486] A variable-temperature study using inversion recovery ^{29}Si n.m.r. spectroscopy has yielded the first definitive information on the exchange between silicate anions.[1487] Temperature-dependent broadening has been observed in the ^{29}Si n.m.r. spectrum of Na_2SiO_4 solution indicative of the dynamic exchange of $[SiO_4]^{4-}$ groups between different silicate anion species.[1488]

Tin. ^{19}F and ^{119}Sn n.m.r. spectroscopy has been used to study the species formed in the SnF_2-MF-H_2O system.[1489]

Phosphorus. ^{1}H, ^{13}C, and ^{31}P n.m.r. spectra of $MePO_3H_2$ and $H_2N(CH_2)_nPO_3H_2$ have been studied as a function of pD in D_2O solution.[1490] Intermolecular fluorine exchange in the $PhPF_4$-$[PhPF_5]^-$ system has been studied by ^{31}P n.m.r. spectroscopy. The exchange is first order in both $PhPF_4$ and $[PhPF_5]^-$ and is slower in donor solvents such as MeCN.[1491]

Equilibria Among Uncharged Species. The association constants and thermodynamic parameters for the 1:1 complexes of metallo tetraphenylporphyrins with 4,6-dinitrobenzofuroxan have been determined using ^{1}H n.m.r. spectroscopy. The donor abilities are in the order Pd(II) > Co(II) > Cu(II) > Ni(II) ~ VO(IV) ~ 2H > Zn(II).[1492]

Lithium. A ^{13}C and ^{6}Li n.m.r. study has shown that $Bu^{s}Li$ exists as a dimer, hexamer, and tetramer mixture that undergoes fast intra- and inter-aggregate exchange and slower inversion at C(2)-Li.[1493]

Magnesium. The Schlenk equilibrium of $Bu^{t}MgBr$ in THF has been investigated using ^{1}H and ^{13}C n.m.r. spectroscopy.[1494]

Thorium. ^{1}H n.m.r. spectroscopy has been used to demonstrate exchange between $(\eta^5-C_5Me_4Et)_2ThD(OCHBu^{t}_2)$ and $(\eta^5-C_5Me_4Et)_2Th(OCD)(OCHBu^{t}_2)$ in the presence of CO. The ^{13}C n.m.r. spectrum of

$(\eta^5-C_5Me_4Et)_2Th(\eta^2-{}^{13}COCH_2Bu^t)(OCH_2Bu^t)$ was also recorded.[1495] The kinetics of ligand exchange between $Th(acac)_4$ and free acetylacetone has been studied by ^1H n.m.r. spectroscopy.[1496]

Uranium. The temperature dependence of the ^1H n.m.r. spectrum of $(\eta^5-C_5H_4Me)_3U(NCBH_3)$ has been described in terms of an equilibrium between $(\eta^5-C_5H_4Me)_3U(\eta^3-H_3BCN)$ and $(\eta^5-C_5H_4Me)_3NCBH_3$.[1497] The dimer-monomer equilibrium in $[U\{\eta^5-C_5H_3(SiMe_3)_2\}_2(\mu-BF_4)(\mu-F)]_2$ has been monitored by ^{11}B and ^{19}F n.m.r. spectroscopy.[1498] ^{19}F n.m.r. spectroscopy has shown that UF_6 undergoes facile halide exchange with $SiMe_3Cl$ or UCl_6.[1499]

Titanium. ^{13}C and ^{31}P n.m.r. spectra have been used to show that (33) and (34) are in equilibrium when R = C_6H_4OEt-p.[1500]

Zirconium. ^1H n.m.r. spectroscopy has been used to investigate the

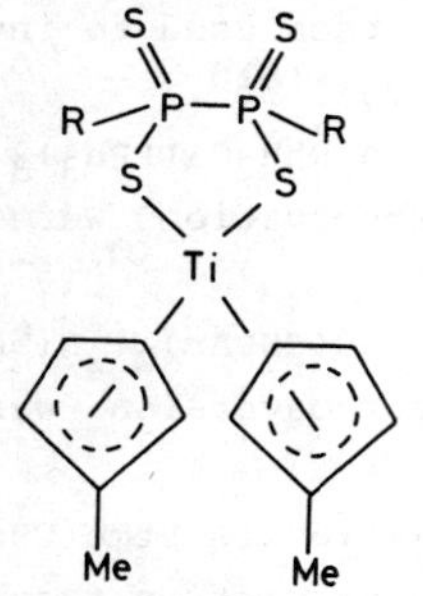

(33)

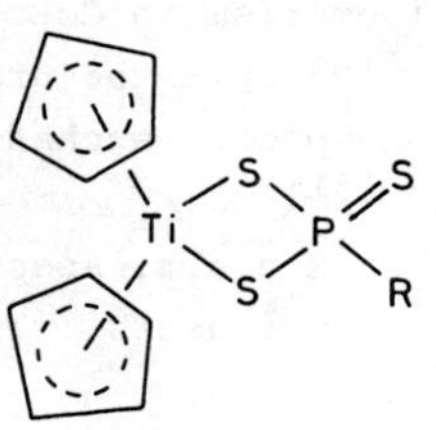

(34)

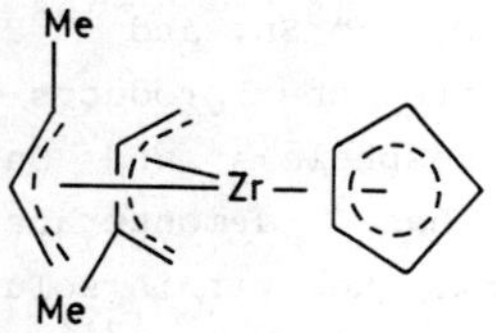

(35)

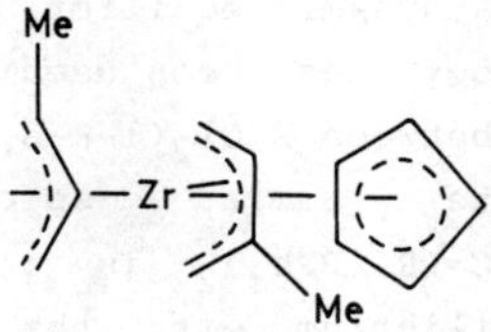

(36)

equilibrium between (35) and (36).[1501]

Chromium. ^1H and ^{13}C n.m.r. spectroscopy has been used to investigate the equilibrium between the η^5- and η^6-isomers of $[(fluorenyl)Cr(CO)_3]^-$.[1502] The n.m.r. spectra of $Cr(acac)_3$ in $CCl_2=CHCl$ have indicated the formation of a 1:1 complex.[1503] The ^1H T_1 of the CH_2 protons is lower than that of the CH_3 protons in $SnEt_2Cl_2$ in the presence of $Cr(acac)_3$, which supports a specific interaction between these compounds.[1504]

Molybdenum. ^1H, ^{13}C, and ^{95}Mo n.m.r. spectra have shown that

MoO_2(L-Cys-OMe)$_2$ exists as two isomers which exchange with $\Delta G^{\ddagger}$ = 64 kJ mol^{-1}.[1505]

Tungsten. ^1H n.m.r. spectroscopy has been used to investigate the temperature-dependent η^1-η^2 equilibrium in $(OC)_5W(Se{=}CHPh)$.[1506] A ^1H and ^{13}C n.m.r. study has shown that $(Me_3P)(OC)_4OsW(CO)_5$ exists in solution as a mixture of two isomers which are in dynamic equilibrium. Spin-saturation transfer was used.[1507]

Iron. Variable-temperature ^{11}B and ^{31}P n.m.r. spectra of $Fe(NO)_2$-$\{PhP(OCH_2CH_2)_2NHBPh_3\}_2$ indicate stepwise reversible dissociation of BPh_3 groups.[1508]

Ruthenium. ^{31}P n.m.r. spectroscopy has been used to investigate exchange between $Ru(CO)_2(^{13}CO)(PPh_3)(SiCl_3)_2$ and $Ru(CO)_3\{P(p\text{-}tol)_3\}(SiCl_3)_2$.[1509]

Cobalt. ^1H n.m.r. line-shape analysis has been used to investigate cis-trans isomerism in $CoEt(dmgH)_2\{C(OMe){=}NH\}$.[1510]

Rhodium. ^{31}P spin-saturation transfer on $RhH_2Cl(PPh_3)_3$ and PPh_3 has shown phosphorus exchange. Similar observations were made for $RhCl(PPh_3)_3$.[1511]

Nickel. The ^1H n.m.r. spectrum of $\{ArN{=}CHC(CHNAr)_2\}_2Ni$ shows the presence of two isomers. The rate of interconversion was determined.[1512]

Platinum. The rate of cis-trans isomerization in $PtMe(CH{=}NC_6H_4Cl\text{-}4)(dmpe)$ has been determined by ^1H spin-saturation transfer.[1513] According to ^{119}Sn and ^{195}Pt n.m.r. spectroscopic studies of the interaction of $(C_3H_6)PtCl_2$ with $SnCl_2$, three types of complexes exist in a dynamic equilibrium.[1514] ^{31}P, ^{119}Sn, and ^{195}Pt n.m.r. spectroscopy has been used to investigate the products of the reaction between $PtCl_2$(1-R-3,4-dimethylphosphole)$_2$ and $SnCl_2$.[1515] Solid-state n.m.r. spectroscopy has demonstrated that $Pt(\eta^1\text{-}Ph_2C{=}PR)(PPh_3)_2$ is in the η^1-form. However, in solution it is in equilibrium with the η^2-form according to ^{31}P and ^{195}Pt n.m.r. studies.[1516, 1517]

Zinc. The temperature dependence of the ^1H and ^{13}C n.m.r. spectra of some alkenylzinc compounds has shown that these η^1-allyl metal systems are fluxional. The rates of exchange depend upon the structure, solvent, and concentration, and at low temperatures are reduced sufficiently to be in the slow-exchange region. The mechanism is second order.[1518] The extra-planar ligand-exchange dynamics of (tetraphenylporphinato)Zn with pyridine and N-methylimidazole has been studied by ^1H n.m.r. spectroscopy.[1519]

Mercury. ^{19}F n.m.r. spectroscopy has been used to investigate exchange equilibria in $HC_2C_6H_4F$, HC_2R, and $CyHgC_2R$.[1520] ^{199}Hg n.m.r. spectra of mixtures of different $Hg(S_2CNR^1R^2)_2$ have shown that exchange processes in solution to form mixed-ligand complexes are extremely fast.[1521]

Boron. ^1H, ^{11}B, ^{13}C n.m.r. spectra of (37) and (38) show

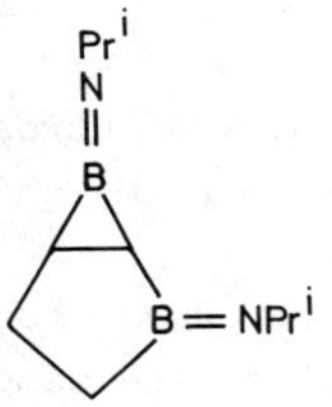

(37)

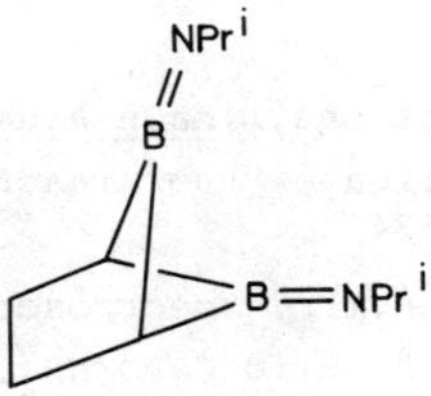

(38)

tautomeric skeletal rearrangement between them with $\Delta G^‡ = 45 \pm 3$ kJ mol^{-1}.[1522]

Aluminium. Low-temperature ^1H n.m.r. studies of Et_3Al-diglyme in the presence of an excess of Et_6Al_2 have resulted in an activation energy of 6.1 kcal mol^{-1} for the exchange of complexed and free Et_3Al.[1523] ^{13}C and ^{27}Al n.m.r. spectroscopy has been used to investigate the interaction of methylbenzenes with $AlBr_3$.[1524]

Tin. ^1H and ^{119}Sn n.m.r. studies of the interacion of $Co(acac)_2$ and $SnPh_2Cl_2$ with pyridine have revealed ligand exchange.[1525] Conventional ^1H n.m.r. lineshape analysis alone is not capable of distinguishing between the 3 possible exchange processes occurring for cis- and trans-$SnCl_4(SMe_2)_2$ in the presence of SMe_2. ^{119}Sn n.m.r. magnetization transfer has allowed the unequivocal assignment of the different processes and their relative importance. The fastest process is that of ligand exchange between cis-coordinated and free Me_2S.[1526]

Lead. ^{207}Pb n.m.r. spectroscopy has shown that $PbMe_3(OAc)$ and $PbMe_4$ exchange rapidly. The ^{13}C n.m.r. spectrum of $(\eta^5-C_5H_5)Mo-(CH=CHCN)_2$ was also reported.[1527]

Phosphorus and Arsenic. The aldo-enol tautomerism and cis-trans isomerism in $R^1R^2P(O)CR^3R^4CHO$ and $R^1R^2P(O)CR^3=CHOR^4$ have been studied by ^1H and ^{13}C n.m.r. spectroscopy.[1528] ^1H and ^{31}P n.m.r. spectra have been used to determine $\Delta G^‡$ for cis-trans isomerism in $Bu^t_3C_6H_2P=PC_6H_2Bu^t_3$.[1529] Exchange between P_2Ph_4, As_2Ph_4, and Ph_2PAsPh_2 has been studied by ^{31}P n.m.r. spectroscopy, and equilibrium constants determined.[1530] The kinetics of the equilibrium between $Et_2NP(OCH_2CH_2)_2CH_2$ and its polymer have been determined using ^1H, ^{13}C, and ^{31}P n.m.r. spectroscopy.[1531] ^{31}P n.m.r. spectroscopy has been used to show that at -80°C the reaction of $(CyN=)_2C$ with $(RO)_2PS_2H$ yields $CyN=C(NHCy)SP(OR)_2S$.[1532] The transesterification of $2\text{-}HOC_6H_4OP(OCH_2)_2CH_2$ and $C_6H_4O_2PO(CH_2)_3OH$ has been investigated by ^{13}C and ^{31}P n.m.r. spectroscopy. The mechanism is believed to involve $(C_6H_4O_2)\{CH_2(CH_2O)_2\}PH$ as an intermediate.[1533]

Course of Reactions.—Magnesium. ^{13}C n.m.r. spectroscopy has been used to investigate the reaction between Et_2Mg and ^{13}CO to give $EtCOCEt_2MgEt$.[1534]

Calcium. ^{31}P n.m.r. spectroscopy has been used to monitor the decomposition of solid $CaPO_3NH_2.H_2O$ by dissolving samples.[1535]

Titanium. The mechanism of activation of $(\eta^5\text{-}C_5H_5)_2TiEtCl$ as a Ziegler-type catalyst by Al_2Et_6 has been investigated by ^{27}Al n.m.r. spectroscopy.[1536] ^{13}C and ^{15}N n.m.r. spectroscopy has been used to investigate the reaction of $TiCl_4(thf)_2/Mg$ with N_2 to give $TiNMg_2Cl_2(thf)$ and with ^{13}CO to give $TiCONMg_2Cl_2(thf)$.[1537]

Hafnium. The hydration of hafnium oxide chloride in an aqueous solution has been studied by ^1H n.m.r. spectroscopy.[1538]

Vanadium. ^{31}P n.m.r. spectroscopy has been used to monitor the reaction of paramagnetic $V(CO)_5P(OMe)_3$ with $P(OEt)_3$ by observing the ligand resonances.[1539]

Chromium, Molybdenum, and Tungsten. The facile exchange of ^1H for ^2H in $[HM(CO)_4L]^-$ (M = Cr, W) has provided a synthesis of $[DM(CO)_4L]^-$. The reduction of organic halides by $[DM(CO)_4L]^-$ has been studied using ^2H n.m.r. spectroscopy.[1540] The reaction of $MoMe(\eta^5\text{-}C_5H_5)(CO)_3$ with $LiEt_3BH$ has been monitored by ^1H, ^2H, and ^{13}C n.m.r. spectroscopy. The reaction goes *via* $(\eta^5\text{-}C_5H_5)MoMe(CHO)(CO)_2$ and $(\eta^5\text{-}C_5H_5)MoH(COMe)(CO)_2$ to give $(\eta^5\text{-}C_5H_5)Mo(\eta^2\text{-}OCHMe)(CO)_2$.[1541] ^1H n.m.r. spectroscopy has been used to investigate the reaction of $W(C_3R_3)(\underline{O}\text{-}2,6\text{-}C_6H_3Pr^i{}_2)_3$[1542] and $W(CBu^t)\{OCH(CF_3)_2\}_3(py)_2$[1543] with hex-3-yne. The ^{13}C n.m.r. spectra of $W(CBu^t)(\underline{O}\text{-}2,6\text{-}C_6H_3Pr^i{}_2)_3$[1542] and $W(CBu^t)\{OCH(CF_3)_2\}_3(py)_2$[1543] were also reported. ^1H n.m.r. spectroscopy has been used to monitor the kinetics of the reaction of $(\eta^6\text{-}\underline{p}\text{-xylene})Cr(CO)_3$ exchange with benzene.[1544] The rate of *cis*-*trans* isomerism of $M(CO)_4L_2$ (M = Cr, Mo, W) has been determined by ^1H, ^{13}C, and ^{31}P n.m.r. spectroscopy.[1545] The *cis*-$M(CO)_4(^{13}CO)PR_3$ to *trans*-$M(CO)_4(^{13}CO)PR_3$ (M = Cr, W) isomerizations have been studied by ^{13}C n.m.r. spectroscopy. The mechanism is intramolecular.[1546] ^{17}O n.m.r. spectroscopy has been used to investigate exchange of $MoO_2(SCH_2CH_2)_2NCH_2CH_2NMe_2$ with

$H_2{}^{17}O$.[1547] The stepwise hydrolysis of $MoOF_4$ in MeCN has been studied by ^{19}F n.m.r. spectroscopy.[1548]

<u>Manganese</u>. The reaction of $Mn(\eta^5-C_{13}H_9)(CO)_3$ with an excess of $PBu^n{}_3$ has been monitored by ^{13}C n.m.r. spectroscopy.[1549]

<u>Technetium</u>. ^1H n.m.r. spectroscopy has been used to monitor deuteriation of the $O{=}CCH_2N$ group in (39). The ^{13}C n.m.r. spectrum was

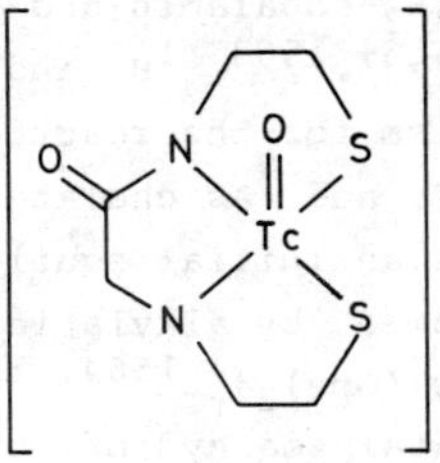

(39)

also recorded.[1550]

<u>Rhenium</u>. The reaction of $(\eta^5-C_5H_5)Re(COR)(NO)(PPh_3)$ with $[NPr^i{}_2]^-$ has been monitored by ^{31}P n.m.r. spectroscopy. Subsequent addition of MeI produces $(\eta^5-C_5H_4COR)ReMe(NO)(PPh_3)$.[1551] ^2H, ^{13}C, and ^{31}P n.m.r. spectroscopy has been used to investigate the deuteriation of $ReH_7(PCy_3)_2$ in the presence of D_2.[1552]

<u>Iron and Ruthenium</u>. The reaction between $HFe(CO)_4SiCl_3$ and several conjugated dienes produces a CIDNP effect which is consistent with a hydrogen-atom-transfer free-radical-type mechanism.[1553] The rate of the conversion of $HRu_3(CO)_{10}CH$ to $H_2Ru_3(CO)_9(CCO)$ has been determined by ^1H n.m.r. spectroscopy.[1554] The mechanism of CO insertion in $FeMe(CO)_2(PMe_3)_2X$ has been studied by ^{13}C n.m.r. spectroscopy. The results were explained in terms of a mechanism involving a trigonal-bipyramidal intermediate, which undergoes a reversible attack by the group entering from 3 different directions.[1555] ^2H n.m.r. spectroscopy has been used to determine the mode of reaction of $(\eta^5-C_5H_5)_2Fe_2(CO)_2(\mu-CD)(\mu-CO)$ with cyclohexane.[1556] Fragmentation studies of iron and ruthenium carbonyls have been made using ^{13}C n.m.r. spectroscopy. Thus the rate of CO exchange on $Fe_3(CO)_{12}$ has been determined, and when $Fe_3(CO)_{12}$ and $Ru_3(CO)_9\{P(OMe)_3\}_3$ are allowed to react, $Fe_2Ru(CO)_{11}\{P(OMe)_3\}$ and $FeRu_2(CO)_{10}\{P(OMe)_3\}_2$ are formed.[1557] ^{13}C n.m.r. spectroscopy has been used to examine the mechanism of base-catalysed ^{13}CO exchange on $Ru_3(CO)_{12}$.[1558] $^2H/^1H$ exchange in chloro Fe(III) etioporphyrin has been investigated using ^2H n.m.r. spectroscopy.[1559] The reaction between (tetramesitylporphyrin)Fe(II) and O_2 has been monitored by

^{1}H n.m.r. spectroscopy.[1560]

Cobalt. The reaction of $[(\eta^5-C_5H_5)CoMe(CO)]_2$ with $[(\eta^5-C_5H_5)Co(CH_2CF_3)(CO)]_2$ under CO has been monitored by ^{1}H and ^{19}F n.m.r. spectroscopy.[1561] The transfer of the methyl group from methylcobalamin to diaquocobinamide in aqueous solution has been demonstrated by ^{1}H, ^{13}C, and ^{31}P n.m.r. spectroscopy.[1562] ^{13}C n.m.r. spectra have been reported for $^{13}CH_3$-cobalamin and have been used to investigate its reaction with Fe^{3+}.[1563] ^{1}H and ^{31}P n.m.r. measurements support a two-step mechanism for the reaction between $[(\eta^5-C_5H_5)Co(LL)X]^-$ (LL = various N, P, and As chelate ligands, X = Cl, Br, I, CN) and $P(OMe)_3$. There is an initial equilibrium to give $[(\eta^5-C_5H_5)Co(LL)P(OMe)_3]^{2+}$ followed by alkylation of the liberated X^- to give $[(\eta^5-C_5H_5)Co(LL)P(O)(OMe)_2]^+$.[1564, 1565] Formate formation during the $Co_2(CO)_8/PR_3$-catalyzed hydroformylation has been investigated by ^{31}P n.m.r. spectroscopy.[1566] ^{59}Co n.m.r. spectroscopy has been used to determine the mechanism and kinetics for the deuteriation of $[Co(NH_3)_6]^{3+}$ and $[Co(en)_3]^{3+}$.[1567, 1568] The cis-trans product ratio from the hydrolysis of cis- and trans-$[Co(NH_3)_4(NH_2R)Cl]^{2+}$ has been determined from ^{1}H n.m.r. spectroscopy.[1569] The proton-exchange kinetics of $[Co(NH_3)_5(SO_3)]^+$ have been studied by ^{1}H n.m.r. spectroscopy and rates determined.[1570] Proton exchange in $[Co(cyclam)Cl_2]^+$ has been studied using ^{1}H and ^{2}H n.m.r. spectroscopy.[1571] The rapid cleavage of $[P_2O_7]^{4-}$ in $[Co(en)_2(P_2O_7)]^-$ by $[Co(tn)_2(OH)(OH_2)]^{2+}$ has been monitored by ^{31}P n.m.r. spectroscopy.[1572] ^{13}C and ^{31}P n.m.r. spectroscopy has been used to follow the hydrolysis of $[(en)_2Co\{(\mu-O_3P)OC_6H_4NO_2\}_2Co(en)_2]^{2+}$.[1573]

Rhodium. The reaction between $(\eta^5-C_5Me_5)RhHPh(PMe_3)$ and C_6D_6 to give C_6H_6 and $(\eta^5-C_5H_5)RhD(C_6D_5)(PMe_3)$ has been monitored by ^{1}H n.m.r. spectroscopy and the activation parameters determined. ^{31}P n.m.r. spectra were also recorded.[1574] ^{31}P n.m.r. spectroscopy has been used to investigate the reaction between trans-$Rh(CO)Br(PEt_3)_2$ and PF_2Br to give $Rh(CO)Br(PEt_3)_2(PF_2Br)$, which rearranges to $RhBr_2(CO)(PEt_3)_2(PF_2)$. ^{13}C n.m.r. spectra were also reported.[1575] ^{13}C n.m.r. spectroscopy has been used to monitor the rate of ^{13}CO incorporation into ArCOCl in the presence of $RhCl(PPh_3)_3$.[1576] Deuteriation of the o-hydrogen atoms of the $P(OPh)_3$ in Rh(acac)-$\{P(OPh)_3\}_2$ has been investigated using ^{1}H n.m.r. spectroscopy.[1577, 1578] ^{31}P n.m.r. spectroscopy has been used to investigate the selectivity and mechanism of $[1,1'\text{-bi-(naphthyl-2-PPh}_2)RhS_2]^+$ as a hydrogenation catalyst.[1579]

Iridium. ^{1}H n.m.r. spectroscopy has been used to demonstrate the regiospecific addition of $Ph_2PCH_2CH_2SiMe_2D$ to $IrH(CO)(PPh_3)_3$ to give $\overline{Ir(PPh_2CH_2CH_2SiMe_2})(CO)(PPh_3)(^1H)(^2H)$ followed by redistribution of the hydrides with $\Delta \underline{G}^{\ddagger}_{298}$ = 19.3 ± 3.2 kcal mol^{-1}.[1580, 1581] ^2H n.m.r. spectroscopy has been used to determine the mechanism of the reaction of $[IrH_2(OCMe_2)_2(PPh_3)_2]^+$ and $ArCH{=}CH_2$ to give $[Ir(\eta^6\text{-}ArEt)(PPh_3)_2]^+$. ^{13}C and ^{31}P n.m.r. spectra were also recorded.[1582] ^1H n.m.r. spectroscopy has been used to investigate the reaction of $[(\eta^4\text{-cod})Ir(Ph_2PC_4H_2OPPh_2)_2Ir(\eta^4\text{-cod})]^{2+}$ with H_2.[1583] ^{31}P n.m.r. spectroscopy has been used to demonstrate ^{18}O incorporation into $[(H_3N)_5Ir\{OP(OMe)_3\}]^{3+}$ on hydrolysis.[1584]

Palladium. The mechanisms of the intramolecular acetylene insertion reactions of $Pd\{O_2C(CH_2)_{\underline{n}}C_2Me\}Cl(PR_3)_2$ and olefin insertion reactions of $Pd(O_2CCH_2CH_2CH{=}CH_2)Cl(PPh_3)_2$ have been investigated using ^{31}P n.m.r. spectroscopy.[1585] ^{31}P n.m.r. spectroscopy has been used to investigate the reaction of $C_5H_3N(PPh_2)_2$ and $(\eta^4\text{-cod})MX_2$ (M = Pd, Pt). The palladium complex forms $Pd_3\{\mu\text{-}(Ph_2P)_2py\}_3Cl_6$ in high yield.[1586] The palladium-assisted amination of olefins has been studied by low-temperature ^1H n.m.r. spectroscopy.[1587] The rates of the reactions between $PdCl_2(\underline{o}\text{-}Ph_2PC_6H_4SMe)$ and $[Pd(\underline{o}\text{-}Ph_2PC_6H_4SMe)_2]^{2+}$ with benzylamine to give $[PdCl(\mu\text{-}SC_6H_4PPh_2\text{-}\underline{o})]_2$ and $Pd(\underline{o}\text{-}Ph_2PC_6H_4S)_2$, respectively, have been monitored by ^1H n.m.r. observation of the decay rate of the methyl signal. The rate is first order in Pd complex and benzylamine. ^{13}C n.m.r. spectra were also recorded.[1588] ^{31}P n.m.r. spectroscopy has been used to monitor the rate of trans-cis isomerism of $Pd(NO_2)_2(PPr^n{}_3)_2$ produced by irradiation.[1589] ^{17}O n.m.r. spectroscopy has been used to investigate the mechanism of Pd-catalysed rearrangement of $MeCH{=}CHCH(^{17}OCOMe)CH{=}CHMe$.[1590] The $[PdCl_4]^{2-}$-catalysed isomerization of allyl-1,1-$\underline{d}_2$ alcohol into an equilibrium mixture with allyl-3,3-$\underline{d}_2$ alcohol has been monitored by ^1H n.m.r. spectroscopy.[1591]

Platinum. ^{31}P n.m.r. spectroscopy has been used for kinetic studies of the reaction of cis-PtR_2L_2 with H_2.[1592] A preliminary kinetic study by ^1H and ^2H n.m.r. spectroscopy has shown that (40) rearranges to (41) reversibly with first-order kinetics at 125°C.[1593] Flow ^1H n.m.r. spectroscopy has been used to investigate the exchange of DMSO and the replacement by ligands on cis-$PtPh_2(DMSO)_2$.[1594, 1595] ^{31}P and ^{195}Pt n.m.r. spectroscopy has been used to investigate the reaction of $PhCOCH_2CH_2NMe_2$ with $PtCl_2(\eta^2\text{-}C_2H_4)(PPh_3)$ to give first $PhCOCH_2CH_2NMe_2CH_2CH_2PtCl_2(PPh_3)$ at -40°C

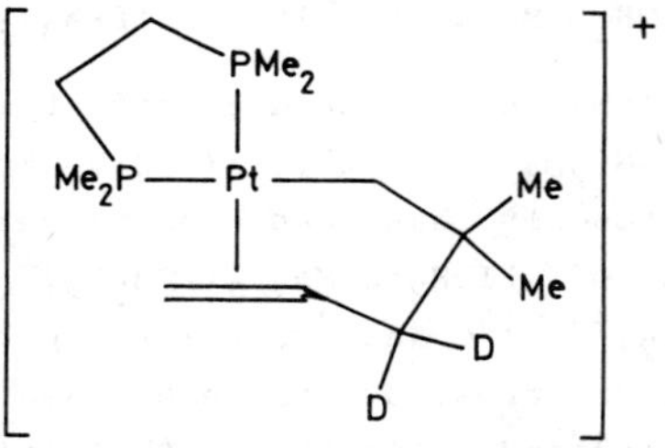

(40)

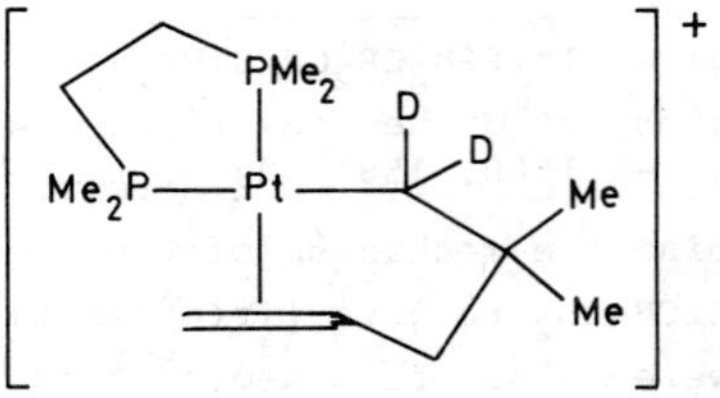

(41)

and then $\overline{Pt(CH_2CH_2NMe_2)}Cl(PPh_3)$ at room temperature.[1596] ^1H n.m.r. spectroscopy has been used to monitor olefin exchange on $(\eta^2$-R^1-$CH=CH_2)PtCl_2\{NC_5H_2Me_2(CH_2)_nCHR^2OH\}$.[1597] The reactions between $Pt(\eta^2$-$C_2H_4)$ and $Ph_3SnC(S)SR$ have been studied by ^{31}P, ^{119}Sn, and ^{195}Pt n.m.r. spectroscopy.[1598] ^{31}P n.m.r. spectroscopy has been used to monitor the formation of ortho-, pyro-, and tri-phosphate complexes of cis-$PtCl_2(NH_3)_2$.[1599] ^1H n.m.r. spectroscopy has been used to monitor the reaction of trans-$PtCl_2(SR_2)$(R,S-Val-OMe) with 1-phenylethylamine. The ^{13}C n.m.r. spectra were also given.[1600] ^{31}P n.m.r. spectroscopy has been used to characterize the kinetic intermediates in the Pt(II)-catalyzed hydrolysis of pyrophosphate and triphosphate.[1601]

Copper. ^{77}Se n.m.r. spectroscopy has been used to investigate the reaction of $R^1SeCR^2R^3Li$ with CuI and SMe_2 at low temperatures.[1602]

Gold. ^2H n.m.r. spectroscopy has been used to monitor ligand scrambling between $[Au(CH_2)_2PPh_2]_2MeBr$ and $[Au(CH_2)_2PPh_2]_2$-$(CD_3)I$.[1603] The reactions of albumin and SH-blocked albumin with $AuCl(PEt_3)$ have been monitored by ^{31}P n.m.r. spectroscopy.[1604]

Zinc. ^1H and ^{13}C n.m.r. spectroscopy has been used to investigate the reaction of CH_2Br_2 and metallic zinc. Evidence was found for $BrCH_2ZnBr$, Me_2Zn, and two other species.[1605]

Mercury. ^{13}C n.m.r. spectroscopy has been used to investigate methyl transfer from methylcobalamin to $[HgCl_n]^{2-n}$.[1606] ^{199}Hg and ^{207}Pb n.m.r. spectroscopy has been used to show that Hg^{2+} reacts with $[Me_3Pb]^+$ to give $[MeHg]^+$ and Pb^{2+}.[1607]

Boron. ^{11}B n.m.r. spectroscopy has been used to investigate the mechanism of the reaction of $BHCy_2$ with $Me_2NCH_2C_2H$.[1608] H_2 loss from some arachno-4-iridanonaboranes has been monitored by ^{11}B and ^{31}P n.m.r. spectroscopy, and $\Delta H^‡$ and $\Delta S^‡$ determined.[1609] ^1H n.m.r. spectroscopy has been used to monitor the exchange reactions of exo-nido-$(Ph_3P)_2Rh$-7-Me-8-Ph-7,8-$C_2B_9H_{10}$ and $C_2B_9H_{10}$. ^{11}B and ^{31}P

n.m.r. spectra were also recorded.[1610] ^{11}B and ^{31}P n.m.r. spectroscopy has been used to investigate the mechanism of hydrogenation of 1-butyl acrylate by some rhodacarboranes, e.g. closo-3,3-($Ph_3P)_2$-3-H-3,1,2-$RhC_2B_9H_{11}$.[1611] ^1H n.m.r. spectroscopy has been used to monitor scrambling reactions of hexene catalysed by closo-3,3-($Ph_3P)_2$-3-H-3,1,2-$RhC_2B_9H_{11}$ and similar compounds. ^{11}B and ^{31}P n.m.r. spectra were also recorded.[1612] ^{11}B n.m.r. spectroscopy has been used to determine $\Delta G^{\ddagger}$ for ester exchange in $ArB(OCH_2)_2CMe_2$.[1613]

Aluminium. Ligand exchange in $[AlMe_4]^-$ and $[AlEt_4]^-$ has been studied using ^1H-{^{27}Al} n.m.r. spectroscopy. The process is catalysed by Na^+.[1614] The reactions of $AlEt_3$ and $Al(CH_2Pr^i)_3$ with bipy, proceeded by complexation of the Al with the N has been followed by ^{13}C n.m.r. spectroscopy.[1615] ^{27}Al n.m.r. spectroscopy has been used to study the hydrolysis of Al(III) by NaOAc.[1616] ^{27}Al n.m.r spectroscopy has been used to investigate gelation behaviour of aluminosilicate solutions containing Na^+, K^+, Cs^+, and $[Me_4N]^+$.[1617]

Carbon. The exchange of hydrogen and carbon isotopes in the $H_2O/H_2/[CO_3H]^-/[HCO_2]^-$ system catalysed by Pd on charcoal has been studied using ^1H, ^2H, and ^{13}C n.m.r. spectroscopy.[1618]

Silicon, Germanium, and Tin. ^2H n.m.r. spectroscopy has been used to investigate the reaction of $[Me_2Si]_6$ with EtOD in the presence of light to give $Me_2SiD(OEt)$ and $MeSi(CH_2D)H(OEt)$.[1619] The insertion of Me_2Ge into the C-Br bond of $PhCH_2Br$ gives CIDNP effects in the ^1H n.m.r. spectrum.[1620] ^1H n.m.r. spectroscopy has been used to monitor the reaction between MeI and SnS to give methyl stannanes.[1621] The reaction between $R^1{}_2R^2MC_5Ph_4$ (M = Si, Ge, Sn) and $Ph_3P{=}CMe_2$ has been monitored by ^{13}C and ^{31}P n.m.r. spectroscopy.[1622] The effect of pressure on the polymerization kinetics of the $Si(OMe)_4$ sol-gel process has been studied using ^{29}Si n.m.r. spectroscopy. High pressures considerably enhance the reaction rates without altering the condensation mechanism during the initial stages of the polymerization process.[1623] ^1H n.m.r. spectroscopy has been used to study the formation of silica gels from $Si(OEt)_4$.[1624]

Phosphorus. The reaction between $ArP{=}CPh_2$ and S_8 to give $ArP(S){=}CPh_2$ has been investigated by ^{31}P n.m.r. spectroscopy.[1625] The alkaline hydrolysis of $[(H_2N)_3P{=}NP(NH_2)_3]^+$ has been monitored by ^{31}P n.m.r. spectroscopy.[1626] The kinetic process of the decomposition of (42) to give (43) has been monitored using ^{31}P n.m.r. spectroscopy, and an activation energy of 102.4 $\pm$ 6.0 kJ mol^{-1} determined.[1627] The kinetics of the cationic ring-opening polymerization

(42)

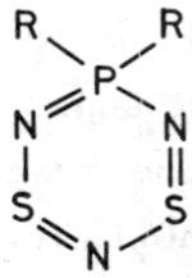

(43)

of 2-phenyl-1,2-oxaphospholane have been determined by ^{31}P n.m.r. spectroscopy.[1628] ^{31}P n.m.r. spectroscopy has been used to monitor the hydrolysis of $\overline{OCH_2CH_2OP}(O)(OR)$ to give $HOCH_2CH_2OPO_3H_2$.[1629] ^{31}P n.m.r. spectroscopy has been used to investigate the synthesis of oligodeoxyribonucleotides by the phosphotriester method.[1630] The reaction between P_4S_3 and R_2S_2 in light to give $SP(SR)_3$ and $P(SR)_3$ has been investigated by ^1H and ^{31}P n.m.r. spectroscopy.[1631]

4 Paramagnetic Complexes

In this section, compounds of d-block transition elements will be considered first and then those of the lanthanide and actinide elements. Papers concerning the use of paramagnetic complexes as 'shift reagents' are usually omitted.

Several reviews have appeared: 'N.m.r. of iron-porphyrin complexes',[1632] 'Nuclear magnetic resonance of paramagnetic metalloporphyrins',[1633] 'Basic knowledge of analytical reagents. N.m.r. shift reagents',[1634] and 'Nuclear magnetic resonance analysis using chiral shift reagents'.[1635]

Inhomogeneous broadening for ^1H, ^2H, and ^7Li n.m.r. signals has been studied in solutions of Co(II) and Ni(II) in H_2O, D_2O, and DMSO. No inhomogeneous broadening was observed in aqueous solutions of Fe(III), Cr(III), and Cu(II).[1636] The n.m.r. shift arising from the electron angular momentum and electron spin dipolar-nuclear spin angular momentum interactions has been investigated for a $4d^1$ system in a strong crystal-field environment of tetragonal symmetry.[1637] A similar study has also been performed for a $3d^1$ system in a strong crystal field of trigonal symmetry.[1638] A general expression using the non-multipole expansion method has been derived for the n.m.r. shift arising for a $3\underline{d}^2$ system in a strong crystal field of octahedral symmetry.[1639] Nuclear spin relaxation in paramagnetic systems in the slow-motion regime for the electron spin has been examined.[1640, 1641] Chelate complexes of transition metal salts have been investigated to determine their ability to reduce $\underline{T}_1$ and the toxicity of the metal. The EDTA complexes of $FeCl_2$, $GdCl_3$, and $MnCl_2$ are not as effective as the respective salts in reducing $\underline{T}_1$ of water protons.[1642] Paramagnetic metalloporphyrins have been investigated as potential contrast agents in n.m.r. imaging.[1643]

The Transition Metals.—Vanadium. ^1H and ^{13}C n.m.r. spectroscopy has been used to study the chemical shifts of $(\eta^5\text{-}C_5H_5)V(C_5H_5)Fe(C_5H_5)V(\eta^5\text{-}C_5H_5)$ as a function of temperature.[1644]

Chromium. Electronic effects have masked the small energy differences between the rotamers of the ethyl bond in $(\eta^5\text{-}C_5H_5)Cr(\eta^5\text{-}C_5Me_4Et)$.[1645] Trans-$[Cr(pn)_2(nic)_2]^+$ has been characterized in solution by ^2H n.m.r. spectroscopy.[1646] ^1H and ^{13}C $\underline{T}_1$ data for a variety of aromatic compounds in the presence of $Cr(acac)_3$ have been reported and discussed in terms of the possible interaction modes. If hydrogen-bond donating solvents are excluded, the interaction mode is dipole-dipole.[1647] The use of $Cr(acac)_3$ as a relaxation reagent to obtain quantitative ^{13}C n.m.r. spectra has been optimized.[1648]

Manganese. ^1H n.m.r. spectra have been reported for $(\eta^5\text{-}C_5H_5)Mn(tmed)$[1649] and $[(\eta^5\text{-}MeC_5H_4)MnI(PEt_3)]_2$.[1650] ^{13}C n.m.r. spectra have been measured for high-spin Mn(III) porphyrins to evaluate effects of axial ligand binding and to correlate isotropic shift patterns with $\underline{d}$-orbital occupation.[1651, 1652] Mn(II)-enhanced relaxation rates of ^1H and ^{13}C nuclei in 8-$\underline{R}$-quinoline-2-carboxylic acid have been measured in aqueous solution at various temperatures. Relative metal-ligand distances were calculated.[1653]

Rhenium. The ^1H n.m.r. spectra of $[ReCl_4(CNAr)_2]^-$ show substantial contact shifts.[1654] ^1H and ^{31}P n.m.r. spectra have been recorded for fac-$Re(bipy)(PMe_2Ph)Cl_3$.[1655]

Iron. ^{11}B and ^{13}C n.m.r. spectra have been recorded for $[M(C_2B_9H_{11})_2]^{2-}$.[1656] The ^1H n.m.r. hyperfine-shift patterns of quinine and alkoxide Fe(III) porphyrin are characteristic of a high-spin $S = 5/2$ state.[1657] For some Fe(III)(tetraphenylporphyrin) complexes, variable-temperature ^2H n.m.r. results are consistent with reversal in the ordering of 'pure' $S = 5/2$ and $S = 3/2$ states for the oxidized *versus* the parent complexes.[1658] Variable-temperature ^{13}C n.m.r. spectra have been measured in solution for μ-oxo dimeric Fe(III) tetraphenyl- and octaethyl-porphyrin species. Hyperfine chemical shifts were fitted to an antiferromagnetically coupled model.[1659] ^1H and ^{13}C n.m.r. spectra have been reported for Fe(tetraphenylporphyrin)(NNC_9H_{18}).[1660] The ^1H n.m.r. spectra of [Fe(protoporphyrin)]$^+$ and [Fe(tetraphenylporphyrin)]$^+$ have been used to discuss the ground-state electronic structure of the iron.[1661] The solution properties of a phenolate-bridged (tetraarylporphyrinato)iron(III) have been investigated by ^1H n.m.r. spectroscopy.[1662] The ^1H n.m.r. spectrum of Fe{5,10,15,20-(4-MeO-$C_6H_4)_4$porphyrinato}(SH) shows the pyrrole resonance at δ -18.7.[1663] ^1H and ^{13}C n.m.r. spectroscopy has been used to study the effect of the *meso* substituent on the $S = 3/2, 5/2$ spin state admixture of perchlorato(tetraarylporphinato)Fe(III) complexes.[1664] The vacancy of the $d_{x^2-y^2}$ in [Fe{[2,4,6-$(MeO)_3C_6H_2$]-porphine}]$^+$ is supported by the marked low-frequency shifts of the pyrrole β-^1H and β-^{13}C resonances.[1665] The ^1H n.m.r. spectra of the complexes of μ-oxo-bis(dicarbethoxytetramethylhaemin) and μ-oxo-bis(tricarbethoxytetramethylhaemin) have demonstrated the presence of several conformational forms.[1666] ^1H n.m.r. spectra were also recorded for some ferric isoporphyrins.[1667] The ^1H chemical shifts from the haem methyl groups in some proteins have given some information on the protein-haem interactions.[1668] Haem orientational heterogeneity in deuterohaemin-reconstituted horse and human haemoglobin has been studied by ^1H n.m.r. spectroscopy.[1669]

^1H n.m.r. investigations of $Fe_2O(OAc)(HBpz_3)_2$ and its deuterated analogue have been reported in which the paramagnetically shifted acetate-group methyl-proton resonances are identified at δ -10.5.[1670] ^1H n.m.r. spectra have also been reported for $[Fe_2S_2(OAr)_4]^{2-}$,[1671] $[MoFe_3S_4(SPh)_3(al_2cat)(NCMe)]^{2-}$,[1672] $[Fe_4S_4(SR)_4]^{3-}$,[1673] $[Fe_4S_4Cl_2(EPh)_2]^{2-}$ (E = O, S),[1674] $[Fe_6S_9(SR)_2]^{4-}$,[1675] and ferredoxin.[1676]

Ruthenium. The ^1H and ^2H n.m.r. spectra of Ru(II) and Co(III) octaethylporphyrin π-cation radicals have been reported. The signals are not enormously broadened but considerably shifted.[1677]

A detailed analysis of the paramagnetic shifts in the ^{1}H n.m.r. spectra has been presented, and a new model for referencing dipolar shifts, in binuclear ruthenium porphyrin dimers, was presented. The calculated contact shifts exhibit π symmetry and indicate that the dominant mode of spin transfer into the porphyrin ring is derived from porphyrinato $3\underline{e}(\pi) \longrightarrow$ Ru charge transfer.[1678]

Osmium. The ^{1}H n.m.r. spectrum of Os(salen)(py)$_2$ covers 80 p.p.m.[1679]

Cobalt. The ^{1}H n.m.r. spectra of $[(\eta^5\text{-}C_5Me_5)Co(\eta^6\text{-arene})]^+$ show a pronounced high-frequency shift of the methyl signal and a large low-frequency shift of the ring protons, in line with established spin-delocalization mechanisms.[1680] The ^{59}Co n.m.r. signal of the spin-crossover Co(III) in $[(\eta^5\text{-}C_5H_5)Co\{\mu\text{-}P(OR)_2O\}_3]_2Co$ has been observed. The ^{59}Co chemical shift exhibits an exceptionally large temperature dependence of ca. 113 p.p.m. K^{-1} at room temperature. Analysis of this temperature dependence, together with the thermodynamic parameters of the spin equilibrium taken from the temperature dependence of the optical spectrum, gave 17 650 $\pm$ 200 p.p.m. for the chemical shift of the diamagnetic low-spin species.[1681] The isotopic shifts in the ^{1}H n.m.r. spectra of N,N'-o-phenylene-bis(salicylideneiminato)cobalt(II) have been observed. The unpaired electron spin is delocalized onto the ligand through the highest-filled MO.[1682] ^{1}H n.m.r. spectra of $[Co(5\text{-nitrosalicylato})(NH_3)_4]^{2+}$,[1683] (N-salicylideneanthranilate)M(II) (M = Co, Ni),[1684] and Co(II) complexes of alkylpyrazoles have been recorded.[1685] The paramagnetic Co complex of $P(2,6\text{-}C_5H_3N)(CH{=}NO)_3BF$, L, has been used as a shift reagent to obtain structural information on $(\eta^3\text{-}C_3H_5)PdClL$ from ^{1}H and ^{2}H n.m.r. spectroscopy.[1686] Paramagnetically shifted n.m.r. resonances in Co(II) carbonic anhydrase have been assigned.[1687] The ^{1}H n.m.r. spectrum of $[CoCl_3(PPh_3)]^-$ has been reported.[1688]

Nickel. ^{1}H n.m.r. isotropic contact shifts have been studied for bis adducts of Ni(dpm)$_2$. A correlation was found between the lowest d-d transition energy and the ^{1}H chemical shift of Bu^t.[1689] Paramagnetic relaxation enhancement data have been reported for ^{1}H and ^{15}N in aniline in the presence of Ni(dpm)$_2$. For ^{15}N, T_1 and T_2 were reported. The experimental data were fitted to a new theoretical model, not invoking electron-spin relaxation times, which is valid in the slow motion regime for the electron spins.[1690]

Copper. The possibility has been examined of the combined use of data for rates of spin-lattice paramagnetic relaxation of protons and ^{13}C of ligands and quadrupolar spin-lattice relaxation of ^{2}H of the same but deuteriated ligands for studying the geometrical and electronic structures of paramagnetic metal complexes in solution. $Cu(acac)_2py$ and $Cu(hfac)_2(py)_2$ were studied.[1691]

Compounds of the Lanthanides and Actinides.—Lanthanides. By using a nonmultipolar expansion technique, the n.m.r. shift has been calculated as a function of the distance from an n.m.r. nucleus of the paramagnetic ion, and the temperature for an f^1 system where the f^1 ion is in a crystal-field environment of octahedral symmetry, and for the f^{13} system where the ion is in a crystal-field environment of icoshedral symmetry with a small distortion along a 3-fold axis.[1692] Lanthanide-induced shifts and extrapolation have been used to determine T_1 for overlapping ^{1}H n.m.r. signals and strongly coupled hydrogen atoms.[1693] N.m.r. data have also been reported for $[MMe_6]^{3-}$ (M = lanthanide; ^{13}C),[1694] $Eu(C_5H_5)(PhCOCHCOPh)_2$,[1695] $Yb_3(C_5Me_5)_4(C_2Ph)_4$,[1696] and $[Yb(C_5Me_5)_2(\mu\text{-}CO)Mn_n(CO)_{5-n}]_m$.[1697]

The structure of 1- and 2-ligand complexes of lanthanides with SALEN in solution has been studied as a result of the analysis of pseudocontact n.m.r. shifts of the ligand.[1698] A conformational equilibrium of the ligands in Eu(2,2',2"-terpyridyl)$_3$ has been observed.[1699] ^{1}H n.m.r. spectroscopy has been used to demonstrate that $Ln(NTA)_2$ (Ln = La, Pr, Nd, Sm, Eu, Tb, Ho, Er, Tm, Yb) are isostructural.[1700] N.m.r. data have also been reported for Ce{(p-tol)$_4$porphyrin},[1701] Yb complexes of polyazapolycarboxylic macrocycles,[1702] and $Dy\{(MeO)P(O)_2CH_2C(O)NEt_2\}_2NCS$ (^{13}C, ^{31}P).[1703]

The shifts induced by equimolar mixtures of CF_3CO_2Ag and $Ln(fod)_3$ have been analyzed.[1704] Larger shifts in the n.m.r. spectra of olefinic, aromatic, and halogenated substrates have been observed in the presence of binuclear lanthanide(III)-Ag(I) shift reagents, when pentane is used as the solvent.[1705] Secondary deuterium-isotope effects are observed in the presence of lanthanide(III)-Ag(I) n.m.r. shift reagents using C_6H_6, C_6H_5D, p-xylene, and p-$C_6H_4(CD_3)_2$.[1706]

The effect of a small amount of water on the ^{1}H and ^{13}C chemical shifts of $Pr(ClO_4)_3$ and 18-crown-6 has been studied. The chemical shifts showed a linear relationship with the amount of water.[1707]

$Yb(fod)_3$-induced ^1H and ^{13}C n.m.r. shifts exhibit regular differences if normalized by setting the shift at the nucleus closest to the binding site to relative shifts = 100%. Very flat minima were observed in the geometrical analysis.[1708] ^{13}C n.m.r. spectra of chondroitin 4- and 6-sulphates, chondroitin, ß-D-glucose-6-sulphate, and ß-D-glucuronate have been measured. The stuctures are similar to those reported for the Ca analogues.[1709] The shift and relaxation effects on the n.m.r. spectrum of AMP due to binding of Ln(III) cations have been studied.[1710]

[Eu{(R)-propylenediaminetetraacetate}]$^-$ has been shown to be a promising chiral shift reagent for ^1H n.m.r. spectroscopy in aqueous solution.[1711] The ability of $M(CF_3CF_2COCHCOC_6H_4F\text{-}4)_3$ (M = Pr, Sm) to act as a shift reagent has been examined.[1712] Induced shifts in ^{14}N chemical shifts for various nitrogenous compounds in the presence of a range of shift reagents have been measured.[1713] [M(triethylenetetra-aminehexa-acetate)]$^{3-}$ (M = Dy, Tm) has been examined as a shift reagent for $[^{14}NH_4]^+$, $^{23}Na^+$, $^{25}Mg^{2+}$, $^{39}K^+$, and $^{87}Rb^+$,[1714] and applied to $^{23}Na^+$ in human erythrocytes.[1715] Cellular Na^+ transport has been followed in vivo by ^{23}Na n.m.r. spectroscopy by using anionic Dy-based shift reagents to resolve internal and external $^{23}Na^+$ resonances.[1716] ^{39}K n.m.r. shifts in K_3(tripolyphosphate) have been investigated in the presence of Dy^{3+}, Tb^{3+}, and Yb^{3+}. Shifts were induced.[1717]

T_1 of pyridine, piperidine, and picoline has been measured as a function of added $Pr(dpm)_3$.[1718] The structure of the Gd(III)-pyridoxine complex has been studied by ^1H n.m.r. spectroscopy. The Gd(III) appears to extend out of the plane of the pyridine ring, and to form a 6-membered chelate.[1719] The effects have been determined of added Gd^{3+} on T_1 of ^{13}C of epi-inositol.[1720]

Actinides. The paramagnetic shift resulting from f-orbital covalency for lanthanide fluorides has been successfully treated theoretically in earlier studies. This treatment has been extended to the paramagnetic shift for ^1H and ^{13}C in uranocenes. The very large anisotropy in the magnetic moment makes treatments in the literature incorrect.[1721] The ^1H paramagnetic shift of dibenzouranocene has been separated into the contact and pseudo-contact components.[1722] UCl_4 has been postulated as a shift reagent in aqueous solution for amino acids.[1723] N.m.r. data have also been reported for $[U(BH_4)_6]^{2-}$ (^{11}B),[1724] $(\eta^5\text{-}C_5H_5)_3U(CyNC{=}CHPPh_2Me)$,[1725]

$(\eta^5-C_5H_5)_3U(Bu^nC{=}NR)$,[1726] $(\eta^5-C_5H_5)UNCMeCHPPh_2Me$,[1727] $(\eta^5-C_5H_5)_3$-UX,[1728] $(\eta^5-C_5H_5)_2U(trop)_2$,[1729] $[(\eta^5-C_5Me_5)_2U(OMe)]_2PH$ (^{31}P),[1730] $(\eta^5-C_5H_5)_2UX_2$,[1731] $UCl_{\underline{n}}(H_{\underline{m}}Bpz_{4-\underline{m}})_{4-\underline{n}}$ (^{11}B),[1732] and $[U_2(OBu^t)_9]^-$.[1733]

5 Solid-state N.M.R. Spectroscopy

This section consists of three main parts: 'Motions in Solids', 'Structure of Solids', and 'Molecules Sorbed Onto Solids'.

A number of reviews have appeared: 'High-resolution n.m.r. of solids',[1734] 'Solid-state n.m.r. studies of heterogeneous catalysts',[1735] 'Solid state high resolution n.m.r: ^{13}C and other nuclei',[1736] 'Macromolecular conformation studies by variable temperature ^{13}C and ^{29}Si m.a.s.n.m.r.',[1737] 'Solid state ^{29}Si and ^{27}Al n.m.r. studies of decationized and dealuminated zeolites',[1738] 'Structural characterization of zeolites by high resolution m.a.s. solid state ^{29}Si n.m.r. spectroscopy',[1739] 'Nuclear magnetic resonance studies of zeolites' (^{27}Al, ^{29}Si),[1740] 'Characterization of chemically modified silica gels by n.m.r.',[1741] 'Nuclear spin relaxation phenomena in amorphous semiconductors',[1742] 'E.s.r. and n.m.r. [for characterization of amorphous semiconductors]',[1743] 'Dislocations in solids investigated by means of n.m.r.',[1744] 'First and second order effects in n.m.r. spectral averaging',[1745] 'Gas hydrates as clathrate ices',[1746] 'N.m.r. studies of guest molecules in intercalation compounds',[1747] 'N.m.r. of organic compounds adsorbed on porous solids',[1748] and 'New n.m.r. methods for the study of hydroxyapatite surfaces'.[1749]

Motion in Solids.—Methyl tunnelling and reorientation in MMe_4 (M = Si, Ge, Sn, Pb) and $[MeLi]_4$ have been investigated.[1750] The diffusion of ^7Li in Li_3N has been measured by n.m.r. spectroscopy.[1751, 1752] Li^+ transport in $Li0.4MoO_3$ has been studied by ^7Li n.m.r. spectroscopy.[1753] Li movement and ^3H behaviour in Li_2O, $LiSiO_3$, and $LiAlO_2$ have been examined.[1754] Structure, ionic motion, and conductivity in some solid solutions of $LiCl$-MCl_2 (M = Mg, V, Mn) have been studied using ^7Li n.m.r. spectroscopy.[1755] From ^1H and ^7Li n.m.r. $\underline{T}_1$, $\underline{T}_2$, and $\underline{T}_1$ in Li_5NI_2, the diffusive motion of Li^+ has been studied.[1756] Li^+ diffusion and water-molecule reorientation in $LiI.H_2O$ have been investigated.[1757] ^{23}Na motion in $Na_2Ca(CO_3)_2$,[1758] ß-aluminas,[1759] and sodium ß-gallate[1760] has been studied. ^{13}C and ^{23}Na n.m.r. spectroscopy has been used to show molecular reorientations in NaCN and KCN.[1761] The temperature dependence of ^{87}Rb $\underline{T}_1$ in $Rb_{1-\underline{x}}(NH_4)_{\underline{x}}H_2PO_4$ indicates progressive slowing down of the proton intrabond jump motion.[1762]

The correlation times of the NH_3 reorientation in $[Mg(NH_3)_6]^+$ $[NO_3]^-_2$ have been derived.[1763] The librational motion of $[NH_4]^+$ in $[NH_4][MgF_3]$ has been studied.[1764] Ionic motion in $NaMgF_3$, $KMgF_3$, and $KCaF_3$ has been investigated by ^{19}F n.m.r. spectroscopy.[1765] N.m.r. spectroscopy has provided evidence for two ammonia rotational hindering potentials in $M(NH_3)_6$ (M = Sr, Ba, Yb).[1766] Fluorine mobility in $La_{1-\underline{x}}Ba_{\underline{x}}F_{3-\underline{x}}$ has been studied by ^{19}F n.m.r. spectroscopy.[1767-1769]

A ^1H n.m.r. study of proton self-diffusion in $ScH_{1.98}$ and $LuH_{2.03}$ has been reported.[1770] Hydrogen site occupancy and hydrogen diffusion in $LaNi_4BH_{1.5}$ have been studied.[1771] A ^{19}F n.m.r. study of anionic mobility in $LaO_{1-\underline{x}}F_{1+2\underline{x}}$ has been reported.[1772] Protonic diffusion in $H(UO_2PO_4).4H_2O$ has been investigated using a pulsed field gradient n.m.r. method.[1773]

A systematic ^2H n.m.r. study of $CoTiF_6.D_2O$ and $NiTiF_6.6D_2O$ single crystals has been undertaken to determine the effects of the diffusion of cations.[1774] A ^1H n.m.r. study of H diffusion in $ZrH_{\underline{x}}$ has been reported.[1775] ^{19}F n.m.r. spectroscopy has been used to study fluoride diffusion in Pb_3ZrF_{10}.[1776]

^7Li motion in $Li_{3+\underline{x}}V_{1-\underline{x}}Si_{\underline{x}}O_4$ has been studied.[1777] Motional narrowing in $HNbO_3$ has been demonstrated.[1778] The mobility of Tl^+ in $TlNb_2O_5F$ has been studied by ^{205}Tl n.m.r. spectroscopy.[1779] The quasi-reorientational dynamics of Li^+ in lithium-doped $KTaO_3$ have been studied using ^7Li, ^{39}K, and ^{181}Ta n.m.r. spectroscopy.[1780] The dynamics of random-site interacting dipoles in $(Na,K)TaO_3$ have been studied using ^{23}Na $\underline{T}_1$ values.[1781] Proton diffusion in some hydrogen rhenium bronzes has been shown.[1782] In the solid state, reorientational motion of $[(\eta^5-C_6H_7)Fe(CO)_3]^+$ results in temperature-dependent ^{13}C n.m.r. spectra.[1783] The exchange of bridging and terminal CO ligands in $Co_2(CO)_8$ has been studied using variable-temperature ^{13}C m.a.s.n.m.r. spectroscopy. $\underline{E}_a$ is 11.7 $\pm$ 0.6 kcal mol^{-1}.[1784] NH_3 rotation rates in $[Co(NH_3)_6]Cl_3$ have been determined.[1785]

^{19}F n.m.r. studies of thermochromic transitions in Cu(II) and Ni(II) complexes of dien have been reported.[1786] ^1H $\underline{T}_1$ values of $Cd(py)_2Ni(CN)_4$ have been measured to study motion.[1787] The reorientation of $[C(NH_3)_3]^+$ about its $\underline{C}_3$ axis has been studied with $[AuCl_4]^-$, $[PtCl_4]^{2-}$, and $[PdCl_4]^{2-}$ as anions.[1788] ^1H n.m.r. spectroscopy has been used to investigate dynamic disorder in Cu{Me_2C-$(CH_2NH_2)_2$}$(NO_3)_2$ [1789] and Cu$(pz)_2Br_2.H_2O$.[1790] The diffusion of Cu(I) in CsI has been studied by ^{127}I n.m.r. spectroscopy.[1791] The self-diffusion of Br^- in AgBr has been studied using ^{81}Br n.m.r. spectroscopy.[1792] The diffusion of Ag in MAg_4I_5 (M = K, Rb) has been measured by ^{109}Ag n.m.r. spectroscopy.[1793] NH_3 group rotation and ion tumbling have been studied in $[Cd(NH_3)_6][ClO_4]_2$.[1794] Internal motions in $[Me_4N]_2[HgCl_4]$ have been studied.[1795]

The molecular dynamics of H_3NBH_3 have been investigated.[1796] ^7Li, ^{17}O, and ^{19}F n.m.r. spectroscopy has been used to study ionic motion in the system Li_2O-Li_2Cl_2-B_2O_3.[1797] ^7Li, ^{11}B, and ^{19}F $\underline{T}_1$ values have been used to study molecular reorientations in $LiBF_4$.[1798] Co-operative Li^+ motion and relaxation of ^7Li by paramagnetic impurities in $LiAlSiO_4$ have been studied by ^7Li n.m.r. spectroscopy.[1799] The activation energies for electrical conductivity and $^7Li^+$ motion in Li_2SiO_3 and $LiAlO_2$ are in good agreement.[1800] ^1H and ^{27}Al n.m.r. spectroscopy has been used to follow the dehydroxylation of muscovite.[1801] Cement hydration has been studied using ^{27}Al m.a.s.n.m.r. spectroscopy.[1802] A ^{19}F, ^{203}Tl, and ^{205}Tl n.m.r. study of Tl^+ mobility in $TlNb_2O_5F$ has been reported.[1803]

Combined translational-rotational jumps in solid CO have been investigated using ^{13}C n.m.r. spectroscopy.[1804] Rotational tunnelling of the methyl group in $MeSiCl_3$ has been shown.[1805] The crystallization of zeolite Y from aqueous solution has been monitored by ^{27}Al and ^{29}Si n.m.r. spectroscopy.[1806] Solid-state ^{29}Si n.m.r. spectroscopy has been used to follow the kinetics of the hydration of tricalcium silicate.[1807] The Li mobility of $Li_{2\underline{x}}S_{\underline{x}}.GeS_2$ has been studied by ^7Li n.m.r. spectroscopy.[1808] The temperature dependence on ^1H $\underline{T}_1$ values for $[Me_2NH_2]_2[MCl_6]$ (M = Sn, Te) has been investigated.[1809]

The rotations of N_2 in the solid have been investigated by ^{14}N and ^{15}N n.m.r. spectroscopy.[1810] NF_3 reorientation in a chlathrate hydrate has been studied by ^{19}F n.m.r. spectroscopy.[1811] The phase transition in NH_4AlF_4 has been investigated by ^1H and ^{19}F n.m.r. spectroscopy.[1812] ^{14}N n.m.r. spectroscopy has indicated that the $[NO_3]^-$ group in NH_4NO_3 is undergoing rotations.[1813] Ammonium-ion reorientation in NH_4Cl has been studied using ^{35}Cl n.m.r. spectroscopy.[1814] Motion of the $[NH_4]^+$ and $[PF_6]^-$ groups in NH_4PF_6 has been investigated by ^{19}F n.m.r. spectroscopy as a function of pressure.[1815] The mobility of a solid benzene-cyclophosphazene inclusion compound has been studied using ^{2}H n.m.r. spectroscopy.[1816] ^{31}P chemical-shift anisotropies and molecular motions in phosphazene elastomers have been studied by ^{31}P solid-state n.m.r. spectroscopy.[1817] ^{13}C and ^{19}F n.m.r. studies of AsF_5 motion in graphite have been reported.[1818] According to ^{19}F n.m.r. data, the diffusional mobility of F^- in Bi_2OF_4 is greater than in other oxyfluorides.[1819]

^{17}O n.m.r. spectroscopy has been used to study the motion of oxygens in some oxysalts.[1820-1822] ^{1}H and ^{77}Se n.m.r. spectroscopy has been used to study motion in NH_4HSeO_4, $RbHSeO_4$,[1823] and $CsHSeO_4$.[1824]

Structure of Solids.—^{13}C shielding anisotropies have been reported for a series of common diamagnetic acetates.[1825] Well resolved solid-state ^{17}O m.a.s.n.m.r. spectra have been obtained for a variety of oxides and oxyanions. There is a wide range of ^{17}O quadrupolar-coupling constants, 0 to >5 MHz, chemical-shift anisotropies, 0 to >300 p.p.m., and linewidths, 0.3 to >40 p.p.m., in addition to a 1200 p.p.m. chemical-shift range.[1826]

^{13}C n.m.r spectra of solid $[Me^{6}Li]_4$ and $[CH_2{}^{6}Li]_{\underline{n}}$ have been discussed.[1827, 1828] Solid-state ^{1}H and ^{7}Li n.m.r. spectra have shown the presence of Li_2S in $[PhCLiCHCLiCHCLiPh]^-Li^+.3Li_2S$.[1829] Ternary graphite-lithium-thf compounds have been studied by ^{7}Li and ^{13}C n.m.r. spectroscopy.[1830] ^{7}Li n.m.r. $\underline{T}_1$ and $\underline{T}_2$ have been determined for polyvinyl acetate doped with $LiClO_4$.[1831] β-Ray-detected ^{8}Li n.m.r. signals and $\underline{T}_1$ have been observed for Li_3N.[1832] NaCN has been examined by n.m.r. spectroscopy.[1833] ^{23}Na n.m.r. investigations of $Na(CN)_{\underline{x}}Cl_{1-\underline{x}}$ have been reported.[1834] ^{23}Na chemical shifts of some inorganic and organic compounds in the solid state have been determined.[1835] Spurious signals caused by the piezoelectric ringing of $NaNO_2$ have been observed by ^{23}Na n.m.r. spectroscopy.[1836] ^{23}Na and ^{133}Cs n.m.r. spectra have been measured for $[M(crown\ ether)]^+Na^-$ (M = Na, Cs).[1837] ^{13}C and ^{87}Rb n.m.r. spectra have been observed for RbC_8.[1838]

Pulse locking has been applied to ^{1}H and ^{19}F in $(NH_4)_2BeF_4$.[1839] Polycondensation in calcium polyphosphate has been studied by ^{31}P n.m.r. spectroscopy.[1840] Spin-lattice relaxation at low temperatures in $Ca(OH)_2$ has been reported.[1841] Heteronuclear zero-field ^{13}C n.m.r. spectroscopy has been applied to ^{13}C-enriched $Ca(O_2CH)_2$.[1842] The ^{19}F n.m.r. spectrum of CaF_2 has been recorded.[1843] Dehydration of strontium hydrogen vanadate has been monitored by ^1H n.m.r. spectroscopy.[1844] A composite pulse sequence has been developed and applied to $Ba(ClO_3)_2.H_2O$.[1845]

Using solid-state ^{29}Si n.m.r. spectra of two modifications of $Y_2Si_2O_7$ with different Si-O-Si bond angles, it has been verified that increasing bond angles cause low-frequency isotropic ^{29}Si chemical shifts.[1846] The ionic conductivity of $Li_{0.05}Nb_{0.95}Ti_{0.05}O_3$ has been studied.[1847] Spin flip in $Dy_xY_{3-x}Fe_5O_{12}$ has been studied by ^{57}Fe n.m.r. spectroscopy.[1848] The interaction between H_2 and Rh/La_2O_3 and Pd/La_2O_3 catalysts has been investigated by ^1H n.m.r. spectroscopy.[1849] N.m.r. quadrupole effects of ^{139}La and ^{27}Al in $LaFe_xAl_{1-x}O_3$ have been reported.[1850] ^{169}Tm n.m.r. signals have been observed in $TmAsO_4$ [1851] and $Tm(EtSO_4)_3.9H_2O$.[1852]

^7Li and ^{23}Na n.m.r. spectra of M_xTiO_2 (M = Li, Na, K) have been reported.[1853] An extensive ^{23}Na n.m.r. study of TiS_2 has been performed.[1854] The tetragonal distortion in ZrH_x has been studied by ^1H n.m.r. spectroscopy.[1855] The OH group in $Zr_2(OH)_2(SO_4)_3.4H_2O$ has been characterized by ^1H n.m.r. spectroscopy.[1856] ^{77}Se n.m.r. spectroscopy has provided evidence for Li ordering in $Li_{0.29}ZrSe_2$.[1857]

^1H and ^{51}V n.m.r. spectra have been used to investigate the phase transition in $HfV_2H_{3.9}$.[1858] The electronic structures of VC_x, VN_x, VO_x, VC_xO_y, VN_xO_y, and VN_xC_y have been studied using ^{51}V n.m.r. spectroscopy.[1859, 1860] ^{51}V n.m.r. spectroscopy has been used to study the microscopic magnetic properties of V_6O_{13}.[1861] ^{27}Al, ^{29}Si, and ^{51}V n.m.r. studies of the interaction of the active component of vanadium catalysts for SO_2 oxidation with supports have been reported.[1862] The influence of SiO_2 on the properties of evaporated V-Mo oxide catalysts has been investigated by ^{51}V n.m.r. spectroscopy.[1863] A ^{51}V n.m.r. study of VO_2 and $V_{1-x}W_xO_2$ has been reported.[1864] The compounds $Al_2O_3.V_2O_5.7H_2O$, $Ga_2O_3.V_2O_5.6H_2O$, and $In_2O_3.V_2O_5.4H_2O$ have been studied by ^1H n.m.r. spectroscopy.[1865] The nature of the metal ions in $Tl_xV_6S_8$ has been investigated by ^{51}V and ^{205}Tl n.m.r. spectroscopy.[1866, 1867] ^{51}V relaxation in VX_2 (X = Cl, Br, I) has been investigated.[1868] The electronic structure of NbH_2 has been investigated by ^1H and ^{93}Nb n.m.r. spectroscopy.[1869] ^{93}Nb n.m.r. spectra of NbC have been used to study carbon vacancies.[1870] ^7Li, ^{39}K, and ^{181}Ta n.m.r. experiments have been carried out on $KTaO_3$ doped with Li, Nb, and Na.[1871] ^{23}Na spin-lattice relaxation has been reported for $(K,Na)TaO_3$.[1872]

The structure of $(Ph_3SnO)_2CrO_2$ has been studied by ^1H, ^{13}C, and ^{119}Sn.[1873] ^{35}Cl and ^{37}Cl n.m.r. spectra have been observed in

Rb_2CrCl_4.[1874] ^{13}C c.p.m.a.s.n.m.r. spectroscopy has been used to characterise species formed from reacting '$Mo(CO)_3$' with γ-alumina.[1875] The structure of $K_xH_{3-x}PMo_{12}O_{40}$ has been investigated by n.m.r. spectroscopy.[1876] Solid-state ^{31}P and ^{29}Si n.m.r. spectra have been used to characterise a number of heteropolyanions. Two inequivalent phosphorus sites are seen for $[P_2Mo_{18}O_{62}]^{6-}$, while at least 5 ^{31}P signals are observed for $[PMo_{11}WO_{40}]^{3-}$.[1877] ^{31}P m.a.s.n.m.r. spectroscopy has been used to characterize compounds of the type $(EtO)_3W_{12}PO_{37}$.[1878] The dependence of the Knight shift of ^{119}Sn in Mo_6S_8Sn with pressure and temperature has been reported.[1879]

The paramagnetic hydrates $MSO_4.H_2O$ (M = Mn, Fe, Ni) and $CoCl_2.6H_2O$ have been investigated by n.m.r. spectroscopy.[1880] ^1H n.m.r spectroscopy has been used to clarify the spin structure in $CsMn_{1-x}Co_xCl_3.2H_2O$.[1881] ^{54}Mn n.m.r. spectra of oriented nuclei in antiferromagnetic $MnCl_2.4H_2O$ have been studied.[1882] The quadrupolar interaction of ^{99}Tc in $KTcO_4$ has been reported.[1883]

The ^{13}C T_1 of $Fe(phen)_2(N^{13}CS)_2$ from 150 K to the spin-crossover temperature has been measured and shows non-exponential decay.[1884] ^{57}Fe n.m.r. spectra in haematite have been studied in the presence of mechanical stress.[1885] SiO_2-modified $LRu(CO)_4$ and $L_3Ru_3(CO)_9$ have been characterized by ^{13}C, ^{29}Si, and ^{31}P n.m.r. spectroscopy.[1886]

The effect of rare-earth-ion size on spin-state equilibrium has been studied by ^{59}Co n.m.r. spectroscopy in $LnCoO_3$.[1887] A ^2H n.m.r. study of $CoSiF_6.6D_2O$ shows a phase transition at 170 ± 2 K.[1888] ^1H and ^{19}F T_1 values have been measured in $CoSiF_6.6H_2O$.[1889] A ^{13}C n.m.r. investigation of the reaction between $[Rh(NH_3)_5Cl]^+$ and zeolite has suggested the formation of a $Rh(I)(CO)_2$ moiety.[1890]

Solid-state ^{13}C n.m.r. spectra of Ni(II) and Zn(II) derivatives of octaethylporphyrin have been reported.[1891] Local magnetizations in a manganese-doped K_2NiF_4 have been determined by ^{19}F n.m.r. spectroscopy.[1892]

A ^1H n.m.r. investigation of spin transfer in $K_2CuCl_4.2H_2O$ has been reported.[1893] ^{31}P n.m.r. spectra have been used to study a series of CuX-PPh_3 complexes.[1894]

^{11}B n.m.r. spectroscopy has been used to characterise several zinc borate crystalline compounds and glasses.[1895] ^{13}C spin diffusion has been used to determine the structure of ^{13}C-labelled $Zn(OAc)_2.2H_2O$.[1896] ^{87}Rb electric-field gradient tensors and symmetry have been investigated in Rb_2ZnBr_4 and Rb_2ZnCl_4.[1897, 1898]

^{113}Cd n.m.r. spectra of $Cd(bipy)_2X_2$ have been reported.[1899] CdO has been investigated using ^{113}Cd n.m.r. spectroscopy.[1900] ^{113}Cd shielding tensors have been determined in some oxo cadmium compounds.[1901, 1902] ^{113}Cd n.m.r. spectra have been reported for $[Cd_{19}(SCH_2CH_2OH)_{16}][SO_4]_2.4H_2O$ [1903] and $[C_{10}H_{16}NO][CdCl_3(OH_2)]$.[1904] The temperature dependence of $\underline{T}_1$ has been studied in $[Me_4N][CdCl_3]$.[1905]

The symmetry and local bonding of C in $B_{12}C_3$ have been studied using ^{13}C n.m.r. spectroscopy. $^1\underline{J}(^{13}C,^{11}B)$ was determined.[1906] The intensities in n.q.r.-n.m.r. spectra of $H_3{}^{10}BO_3$ have been examined.[1907] ^7Li n.m.r. studies on $Li_2B_2O_4$ have been reported.[1908] ^{11}B and ^{23}Na stimulated echoes have been measured in various glasses.[1909] ^{11}B m.a.s.n.m.r. spectra have been used to determine $BO_{\underline{n}}$ structural units in zeolites.[1910] N.m.r. spectroscopy has been used to investigate the effects of N on alkali borosilicate glasses.[1911] The fraction of 4-co-ordinate boron to total boron in $PbO-B_2O_3-Al_2O_3$ glasses has been measured by ^{11}B n.m.r. spectroscopy.[1912] Similar ^{11}B n.m.r. studies were applied to some borate glasses containing tellurium.[1913, 1914]

The ^{27}Al n.m.r. spectra of a $NiMoP/Al_2O_3$ hydrodesulphurization catalyst contain peaks for surface Al^{3+}, which can be distinguished from those of bulk γ-Al_2O_3.[1915] The $Na_2O-Al_2O_3-H_2O$ system has been studied by ^{27}Al n.m.r. spectroscopy.[1916] Spin-lattice relaxation of Fe^{3+} in $LiAl_5O_8$ and $LiGa_5O_8$ has been studied.[1917] ^7Li n.m.r. spectroscopy has been used to detect a highly conductive phase in $LiBr.H_2O-Al_2O_3$.[1918] ^7Li relaxation for LiI and LiI with 30 mol % Al_2O_3 is purely dipolar.[1919] N.m.r. imaging in solids has been applied to (Na,K) ß-alumina.[1920] The origin of the crystal field and its variation with temperature in several α-alums have been studied by ^{27}Al n.m.r. spectroscopy.[1921] Solid-state ^{27}Al n.m.r. spectra of some basic aluminium sulphate crystals have been reported.[1922] The pressure dependence of ^{27}Al n.m.r. spectra in $KAl(SO_4)_2.12H_2O$ has been examined.[1923]

An n.m.r. study of glasses has been published.[1924] Amorphous $Al_2O_3-SiO_2$ has been investigated by ^{27}Al n.m.r. spectroscopy.[1925] ^{27}Al and ^{29}Si m.a.s.n.m.r. spectra have been used to characterize a synthetic beidellite-smectite.[1926] The formation of non-framework $[Al(OH)_4]^-$ upon thermal treatment of Ca^{2+}-exchanged zeolite A has been followed by ^{27}Al and ^{29}Si m.a.s.n.m.r. spectroscopy.[1927] ^{27}Al m.a.s.n.m.r. spectroscopy has been used to characterize various natural and synthetic zeolites.[1928] ^{27}Al and ^{29}Si m.a.s.n.m.r.

spectroscopy has been used to assess the degree of dealumination of Na mordenite after heat treatment and acid leaching.[1929] The ^{27}Al and ^{29}Si n.m.r. spectra have been recorded for montmorillonite.[1930] Solid-state ^{27}Al and ^{23}Na n.m.r. spectroscopy has been used to follow the incorporation of Al^{3+} and Na^{+} into the mordenite lattice and channels.[1931] ^{27}Al m.a.s.n.m.r. spectroscopy has been used to distinguish between Al in the tetrahedral and octahedral sheets of phyllosilicates.[1932] A related study used both ^{27}Al and ^{29}Si m.a.s.n.m.r. spectroscopy.[1933] The crystallization of hydroxysodalite has been studied by ^{23}Na, ^{27}Al, and ^{29}Si m.a.s.n.m.r. spectroscopy.[1934] A statistical approach to the structure of faujasite has been used to obtain distributions of local structures that permit comparisons with the results of ^{29}Si n.m.r. experiments.[1935] ^{27}Al and ^{29}Si solid-state n.m.r. spectroscopy has been used to investigate the dehydroxylation of pyrophyllite. The ^{29}Si n.m.r. spectrum is consistent with the maintenance of a 2:1 layer structure. A change in the Al co-ordination from 6 to 5 on dehydroxylation results in the loss of the ^{27}Al resonance.[1936] ^{27}Al and ^{29}Si m.a.s.n.m.r. spectroscopy has been used to investigate Al in Sialon X-phase.[1937] The interpretation of the ^{29}Si n.m.r. spectrum of synthetic mazzite has been made using information about the sensitivity of ^{29}Si chemical shifts to the framework topology and Ga substitution.[1938] The ^{29}Si m.a.s.n.m.r. spectra of dealuminated zeolites have been studied, where the chemical-shift dispersion due to structurally distinct SiO_4 tetrahedra is comparable in magnitude to first coordination sphere effects. Some previous ^{29}Si peak assignments are incorrect.[1939] No silicon aggregation could be detected in the ^{29}Si m.a.s.n.m.r. spectrum of zeolite A.[1940] The ^{205}Tl n.m.r. spectrum of Tl-exchanged zeolites A, X, and Y shows 3 lines due to different cation positions.[1941] ^{13}C chemical shifts distinguish between $[Me_4N]^+$ trapped in the α and β cages of the zeolite A structure and are insensitive to the Al:Si ratio.[1942] The direct insertion of aluminium into high silica zeolite frameworks has been investigated by ^{27}Al n.m.r. spectroscopy.[1943] The dealumination of high silica zeolite A has been investigated by ^{29}Si m.a.s.n.m.r. spectroscopy.[1944] The hypothesis of Breck and Skeels concerning the reinsertion of Al^{3+} into the framework of dealuminated Y zeolites has been examined by ^{29}Si n.m.r. spectroscopy.[1945] The contributions to the ^{29}Si m.a.s.n.m.r. line widths of zeolites have been examined.[1946] ^{29}Si m.a.s.n.m.r. spectroscopy has proved to be a valuable tool for the quantitative identification of local silicon environments in zeolitic frameworks.[1947] Two possible interpretations of ^{29}Si n.m.r. spectra of kaolin-group minerals have been presented.[1948] ^{1}H n.m.r. spectroscopy has been used to detect protons in natural glasses.[1949] The structure has been determined for $Si_{6-\underline{x}}Al_{\underline{x}+\underline{y}}O_{\underline{x}}N_{\underline{y}+8-\underline{x}}$ by m.a.s.n.m.r. spectroscopy.[1950] The structure of some ruby crystals has been investigated by ^{27}Al n.m.r. spectroscopy.[1951] The Al coordination in disordered calcium hydrogen silicates containing Al_2O_3 has been studied by ^{27}Al n.m.r. spectroscopy.[1952] Electron-coupled nuclear spin-spin interactions have been observed in the ^{29}Si m.a.s.n.m.r. spectrum of low albite.[1953]

^{11}B m.a.s.n.m.r. spectroscopy has been used to investigate boron in tetrahedral sites of the ZSM-5 framework.[1954] ^{27}Al m.a.s.n.m.r. spectroscopy has been used to show the introduction of 4- and 6-co-ordinate Al occurs during treatment of highly siliceous ZSM-5/silicalite with $AlCl_3$ vapour at elevated temperatures.[1955] Several distinct Al tetrahedral sites in TPA-ZSM-5 have been observed by ^{27}Al n.m.r. spectroscopy.[1956] The state of Al and Si in zeolites Y, mordenite, and ZSM has been studied by ^{27}Al and ^{29}Si n.m.r. spectroscopy.[1957] The conversion of zeolite ZSM-5 by various dealumination procedures into silicalite has been studied by ^{27}Al and ^{29}Si m.a.s.n.m.r. spectroscopy.[1958] The stability and selectivity of ZSM-type zeolites in methanol transformation have been studied using ^{27}Al and ^{29}Si n.m.r. spectroscopy.[1959] The ^{29}Si m.a.s.n.m.r. spectrum and relaxation time have been measured for ZSM-39, the first synthetic zeolite for which it has been possible to uniquely correlate ^{29}Si n.m.r. resonances with crystallographic sites.[1960] ^{29}Si n.m.r. spectra of ZSM-5 indicate that the nature of the observed spectrum depends on the presence of SiOR, R = H, Na, or TPA, groups inside the zeolite lattice.[1961] ^{29}Si m.a.s.n.m.r. analysis of zeolite ZSM-39 has resolved the number of tetrahedral sites and the deviation from ideal symmetry.[1962] ^{27}Al and ^{29}Si m.a.s.n.m.r. spectroscopy has been used to characterize some clays.[1963-1965] Aluminium coordination in soils and clay fractions has been investigated using ^{27}Al m.a.s.n.m.r. spectroscopy.[1966] ^{27}Al and ^{31}P m.a.s.n.m.r. spectra of four aluminophosphate molecular sieves have been studied.[1967]

The ^{115}In n.m.r. spectrum of a single crystal of $(NH_4)_2InBr_5.H_2O$ has been measured to investigate electric-field gradients.[1968] ^{205}Tl n.m.r. spectra of $As_2S_3-Tl_2S$ glasses have been studied.[1969]

Diamonds of gem quality give narrow ^{13}C n.m.r. signals even in a spectrometer designed for use with liquids.[1970] ^{13}C n.m.r. spectroscopy has been used to determine C-C bond lengths in graphite intercalation compounds.[1971] ^{1}H and ^{19}F n.m.r. spectroscopy has been used to determine the ordering of $MeNO_2$ cointercalated with $[PF_6]^-$, $[AsF_6]^-$, or $[SbF_6]^-$ in graphite.[1972] Graphite intercalation compounds containing $[AsF_6]^-$ have been characterized by ^{19}F n.m.r. spectroscopy.[1973, 1974] ^{13}C n.m.r. spectra have been recorded for $SbCl_5$ intercalated graphite.[1975] The structure of SbF_5 intercalated graphite has been studied by ^{19}F n.m.r. spectroscopy.[1976, 1977] ^{121}Sb n.m.r. spectroscopy has been used to investigate a graphite/-$SbCl_5$ compound.[1978] ^{19}F n.m.r. spectroscopy has been used to study the action of $Br^+[FSO_3]^-$ on graphite to give $C_{12}BrSO_3F$.[1979]

Solid-state ^{29}Si n.m.r. spectroscopy distinguishes between various classes of network junctions in di-Me silicone rubber.[1980] ^{29}Si n.m.r. spectroscopy of polysilane has been used to characterize the chemical environment of Si in the backbone.[1981] The Si-O-Si bond angle distribution in vitreous silica has been determined by ^{29}Si m.a.s.n.m.r. spectroscopy.[1982] ^{29}Si m.a.s.n.m.r. spectroscopy has been used to study Si environments in lithium silicates.[1983] By using known n.m.r. spectroscopic and structural data of solid silicates, a linear correlation has been found between Si-O bond lengths and $\delta(^{29}Si)$.[1984] A semi-empirical quantum chemical rationalization of the correlation between Si-O-Si angles and ^{29}Si chemical shifts has been published.[1985] ^{29}Si n.m.r. shifts in silicates with various degrees of anion condensation and different kinds of cations have been studied.[1986] Nuclear spin lattice relaxation in $Li_2Si_2O_5$ glass has been studied using the β-decay radiation of polarized ^{8}Li nuclei.[1987] ^{29}Si n.m.r. studies of binary alkali glasses have been reported.[1988] The structure of lead silicate glasses has been studied by ^{29}Si n.m.r. spectroscopy.[1989]

The ^{13}C n.m.r. spectra of $\{Ge(hemiporphyrazine)\}_{\underline{n}}$ polymers have been assigned as cofacially linked polymer.[1990]

$Pb(OAc)_2.3H_2O$ undergoes solid-state transitions at room temperature resulting in anomalous and variable peaks in the ^{13}C c.p.m.a.s.n.m.r. spectra of powdered samples.[1991] The ^{19}F n.m.r. spectrum of PbF_2 has been investigated as a function of orientation and the ^{19}F chemical-shift anisotropy determined.[1992]

The state of proton-containing groups in the hydrates of some inorganic acids has been studied by ^{1}H n.m.r. spectroscopy.[1993] The ^{87}Rb electric-field gradient tensors and $\underline{T}_1$ of RbD_2PO_4 have been determined.[1994] ^{39}K and ^{75}As n.m.r. spectra of KH_2EO_4 (E = P, As) have been measured.[1995] The ^{31}P chemical-shielding anisotropies of six condensed phosphates have been measured.[1996] The ^{31}P chemical-shift tensors of ammonium phosphotellurate have been measured.[1997] The ^{19}F T_1, linewidths, and chemical shifts of $MAsF_6$ (M = Li, Na, K, Rb, Cs, and Ag) have been determined.[1998]

A number of studies of bistetramethyltetraselenafulvanium ion using ^{13}C[1999] and ^{77}Se[2000, 2001] n.m.r. spectroscopy have been published. Evidence for multipole echoes has been found in the ^{127}I n.m.r. spectrum of KI.[2002] $KIO_3.HIO_3.KCl$ has been studied by ^{127}I n.m.r. spectroscopy.[2003]

Molecules Sorbed Onto Solids.—This section is divided into two subsections: 'Water Sorbed Onto Solids' and 'Atoms and Other Molecules Sorbed Onto Solids'.

Water Sorbed Onto Solids. The presence of adsorbed water on $BaCl_2.H_2O$ has been shown by ^{1}H n.m.r. spectroscopy.[2004] The surface concentration of OH groups on $NiO\text{-}SiO_2$ has been determined using ^{2}H n.m.r. spectroscopy.[2005] ^{1}H n.m.r. spectroscopy has been used to investigate hydrogen chemisorption and water formation on Pt black catalysts.[2006] There have been several ^{1}H n.m.r. studies of water in zeolites,[2007-2011] and on silica.[2012, 2013]

Atoms and Other Molecules Sorbed Onto Solids. Species involved in the hydrogenation of CO to CH_4 over a Ru/SiO_2 catalyst have been identified by ^{13}C n.m.r. spectroscopy.[2014] The adsorption of benzene onto Rh/SiO_2[2015] and Pt/Al_2O_3[2016] has been investigated by ^{1}H n.m.r. spectroscopy. The structure of acetylene on Pt particles has been studied by ^{1}H and ^{13}C n.m.r. spectroscopy.[2017, 2018] The ^{129}Xe n.m.r. spectrum of Xe adsorbed on Pt supported on NaY zeolite consists of a single component.[2019]

Using ^{1}H and ^{13}C m.a.s.n.m.r. spectroscopy, propylene on Al_2O_3 has been studied.[2020] ^{1}H and ^{13}C n.m.r. studies of sodium decyl sulphate sorbed on $Al(OD)_3$ have been reported.[2021]

^{13}C n.m.r. spectroscopy has been used to characterize clay-organic systems.[2022] The effects of lanthanide, Ca^{2+}, and Mn^{2+} exchange and adsorbed substances on the hydrocarbon conversion activities of NaY zeolite catalysts have been studied by n.m.r. spectroscopy.[2023] N.m.r. measurements of adsorbed molecules have yielded information about the occupation of cationic sites in Y zeolites by paramagnetic lanthanide ions.[2024] The kinetics of adsorption of cyclohexane and CH_4 on zeolite NaX and Linde 13X have been studied by n.m.r. spectroscopy.[2025] Adsorption of CH_4 on H-ZSM zeolite causes a high-frequency shift of the ^{13}C resonance and a decrease in $^{1}\underline{J}(^{13}C,^{1}H)$.[2026] The activation energies for self-diffusion of oct-1-ene and octane on X-zeolites have been studied using the n.m.r. pulsed field gradient technique.[2027] The ^{29}Si m.a.s.n.m.r. spectrum of silicalite has been recorded to examine the effect of sorbates.[2028] The self-diffusion of paraffins and olefins in zeolite NaX under the influence of residual water molecules has been investigated.[2029] ^{13}C m.a.s.n.m.r. spectroscopy has been applied to the study of butene adsorbed on zeolites.[2030] An n.m.r. study of the reaction of methanol with isobutylene in the presence of ion-exchange catalysts has been reported.[2031] ^{13}C $\underline{T}_1$ of ^{13}C-enriched benzene has been determined in order to elucidate the adsorption properties of pillar interlayered montmorillonite.[2032] The molecular motion of C_6D_6 has been studied in NaX and Cs(Na)X zeolites by ^{2}H n.m.r. spectroscopy.[2033] ^{29}Si m.a.s.n.m.r. spectra of ZSM-5 have shown changes in the zeolite structure in the presence of sorbate molecules, e.g. p-xylene.[2034] Acid-base interactions of adsorbed MeCN in decationated zeolite Y have been studied by ^{1}H, ^{13}C, and ^{15}N n.m.r. spectroscopy.[2035] A similar study using ^{1}H and ^{15}N n.m.r. of pyridine has also been reported.[2036] The ^{31}P m.a.s.n.m.r. spectrum of Me_3P adsorbed on a dehydrated HY zeolite has been determined.[2037] The solidification of ^{3}He in a porous glass has been investigated.[2038] The void spaces in mordenites have been characterized by means of ^{129}Xe n.m.r. spectra of sorbed Xe.[2039] A similar study of ^{129}Xe n.m.r. of Xe on Y and A zeolites has also been reported.[2040]

The thermal and oxidative transformation of $C_{24}K(C_6H_6)_{2.8}$ has been studied by ^{1}H n.m.r. spectroscopy.[2041] $\underline{T}_1$ and $\underline{T}_2$ have been measured for CH_4 adsorbed on Graphoil.[2042] The ^{129}Xe n.m.r. spectra of Xe adsorbed on graphite has been observed in 3 different phases.[2043] ^{13}C n.m.r. spectroscopy has been used to study n-alkylchlorosilanes immobilized on reversed-phase liquid chromatographic surfaces.[2044] Nuclear polarization of ^{129}Xe has been studied on glass.[2045]

6 Group IIIB Compounds

Several relevant reviews have appeared: 'Boron-11',[2046] 'Aluminium-27',[2047] and 'Thallium N.M.R.'[2048]

Boron Hydrides and Carboranes.—A general technique has been developed for the direct determination of B-B atom connectivities in all types of boron cage compounds, based on *J*-correlated, 2D ^{11}B-^{11}B n.m.r. spectroscopy.[2049] ^{1}H, ^{11}B, and ^{13}C n.m.r. spectroscopy has been used to study the stereochemistry of some *N*-borane adducts of oxazotidines.[2050] The ^{13}C n.m.r. signals of $H_2B(B\text{-}pz)_2BR^1R^2$ have been assigned using 2D $^{1}H/^{13}C$ correlation.[2051] N.m.r. data have also been reported for $LiBH_3R$, $(MeO)_2AlH$ (^{11}B, ^{27}Al),[2052] 1-bora-4-stanna-2,5-cyclohexadienes (^{13}C, ^{119}Sn),[2053] $[BH_2(CN)_2]^-$ (^{11}B),[2054] $R^1R^2NHBH_2C(O)NHEt$ (^{11}B, ^{13}C),[2055] $Me_3NBH_2CO_2Et$ (^{11}B),[2056] $HC(CH_2\text{-}CH_2)_3NBH_2CO_3H$ (^{11}B),[2057] $ROBH_2$ (^{11}B),[2058] $Pr^i{}_2NBHNHN{=}CPr^i{}_2$ (^{11}B, ^{13}C),[2059] $[HB(Me_2pz)_3BH]^+[NbCl_6]^-$ (^{13}C),[2060] and $(C_8H_{14}B)_2(\mu\text{-}H)\text{-}(\mu\text{-}O_2CBu^t)$ (^{11}B, ^{13}C).[2061]

$^{1}J(^{11}B,^{11}B)$ has been observed in $[B_3H_6X^1X^2]^-$.[2062] ^{10}B and ^{11}B n.m.r. spectroscopy has been used to observe $^{1}J(^{11}B,^{10,11}B)$ in 2-centre, 2-electron B-B bonds linking coupled cage boranes and carboranes. $^{1}J(^{11}B,^{11}B)$ range from 79 to 151 Hz.[2063] 2D ^{11}B-^{11}B n.m.r. spectra have been used to characterize 2,2'-$(B_5H_8)_2$.[2064] Several types of previously unobserved long- and short-range couplings in *closo*-2,4-$C_2B_5H_7$ and some of its derivatives have been obtained from resolution-enhanced ^{1}H and ^{11}B n.m.r. spectra. $J(^{1}H,^{1}H)$ was derived from ^{1}H-{^{11}B,^{10}B} spectra.[2065] ^{11}B COSY n.m.r. spectra have been measured for $B_9H_{13}(SMe_2)$ and $B_{10}H_{14}$.[2066] The ^{11}B and ^{1}H n.m.r. spectra of $B_{10}H_{14}$ exhibit substantial solvent and temperature effects.[2067] Heteronuclear 2D n.m.r. spectroscopy has been applied to ^{11}B in decaborane.[2068] The ^{11}B quadrupole coupling constants for *o*-, *m*-, and *p*-$B_{10}C_2H_{12}$ have been determined and also calculated theoretically.[2069] N.m.r. data have also been reported for $[B_3H_6(PMe_3)_2]^+[B_3H_8]^-$ (^{11}B),[2070] $Ph_3PCHRB_3H_7$ (^{11}B, ^{31}P),[2071] $Me_2N(CH_2)_2B_3H_6$ (^{11}B),[2072] 1,2'-$B_4H_9B_5H_8$ (^{11}B),[2073] $Ir(\eta^4\text{-}B_4H_9)(CO)(PMe_2Ph)_2$ (^{11}B),[2074] 1-$(OC)_3FeB_4H_8$ (^{11}B, ^{13}C),[2075] $B_4H_8\text{-}PMe_3$ (^{11}B, ^{31}P),[2076] $(C_8H_8)Ti(Et_2C_2B_4H_4\text{-}\underline{n}\underline{n})$ (^{11}B),[2077] $(\eta^6\text{-}C_6Me_6)\text{-}Fe(2,3\text{-}Me_2C_2B_4H_4)$ (^{11}B),[2078] $Os(CO)_3\{(Me_3Si)_2C_2B_4H_4\}$ (^{11}B, ^{13}C, ^{29}Si),[2079] 1-Sn-2-(Me_3Si)-3-R-2,3-$C_2B_4H_4$ (^{11}B, ^{13}C, ^{29}Si, ^{119}Sn),[2080] RB_5H_6 (^{11}B, ^{19}F),[2081] 2,3,4-$Et_3B_5H_6$ (^{11}B),[2082] 1-X-2,2'-$(B_5H_7)(B_5H_8)$ (^{2}H, ^{11}B),[2083] $(HIrB_8H_{12})(CO)(PMe_3)_2$ (^{11}B),[2084] 4-(Et_2NCS_2)-*arachno*-4-AuB_8H_{12} (^{11}B),[2085] $B_9H_9SCNH_2Bu^t$ (^{11}B),[2086] $(\eta^6\text{-}C_6Me_3H_3)FeB_9H_{13}$ (^{11}B),[2087] 10-$(\eta^5\text{-}C_5H_5)Ni\text{-}\eta^4$-1-$CB_8H_9$ (^{11}B),[2088]

<u>arachno</u>-6,8-$C_2B_7H_{13}$ (^{11}B),[2089] $[(SCN)B_{10}H_9]^{2-}$ (^{11}B),[2090] $[X_{\underline{n}}B_{10}H_{10-\underline{n}}]^{2-}$ (^{11}B),[2091] <u>cis</u>-Pt($B_{10}H_{10}$)(PPh_3)$_2$ (^{11}B, ^{31}P),[2092] (Me_2PhP)$_2PtB_{10}H_{11}Cl$ (^{11}B, ^{31}P),[2093] μ-6,9-Pt(PPh_3)$_2$-6,9-$C_2B_8H_{10}$ (^{11}B),[2094] [<u>anti</u>-7,7':8,8'-bis(μ-dithio)bis(7,8-dicarbaundecaborate(10)]$^-$ (^{11}B),[2095] [N(PPh_3)$_2$][W(CR)(CO)$_2$(η-1,2-$C_2B_9H_9Me_2$)] (^{11}B, ^{13}C, ^{31}P),[2096] (OC)$_3RuC_2B_9H_{11}$ (^{11}B),[2097] 8-Br-8'-L-3-Co(1,2-$C_2B_9H_{10}$)$_2$ (^{11}B),[2098] <u>closo</u>-1-Me-2,2-(PEt_3)$_2$-2-H-8-Ph-2,1,8-$RhC_2B_9H_9$ (^{31}P),[2099] <u>closo</u>-1-R^1-2-R^2-3,3-(PPh_3)$_2$-3-H-3,1,2-$RhC_2B_9H_9$ (^{11}B, ^{31}P),[2100] <u>closo</u>-2,2-(PPh_3)$_2$-^{2}H-8-Ph-2,1,8-$IrC_2B_9H_{10}$ (^{11}B, ^{31}P),[2101] <u>exo</u>-<u>nido</u>-[(PPh_3)$_2$Rh][7,8-μ-Me_2-10-{(CH_2)$_2$C(O)O(CH_2)$_3$Me]-7,8-$C_2B_9H_9$ (^{31}P),[2102] $[X_{\underline{n}}B_{12}H_{12-\underline{n}}]^{2-}$ (^{11}B),[2103] $[1\text{-}CB_{11}H_{12}]^-$ (^{11}B),[2104] 1-$MeNH_2$-1-$CB_{11}H_{11}$ (^{11}B),[2105] $C_2H_{10}B_{10}(EH)_2$ (E = S, Se; ^{11}B),[2106] bis(acetylarylene)carboranes (^{13}C),[2107] ($C_2B_{10}H_{11}$)$_2S_2$ (^{11}B),[2108] $C_2B_{10}H_{11}C_5H_4Fe(CO)_2Br$ (^{11}B),[2109] and (Me_3Si)$_2C_4B_8H_{10}$ (^{11}B, ^{13}C, ^{29}Si).[2110]

Other Compounds of Boron.—A chiral Eu shift reagent has been used to show that (44) exists as enantiomers.[2111] N.m.r. data have

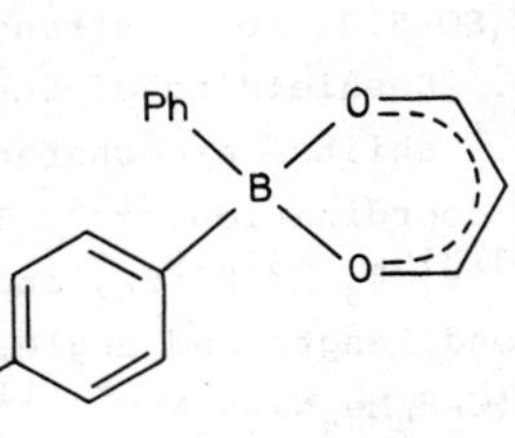

(44)

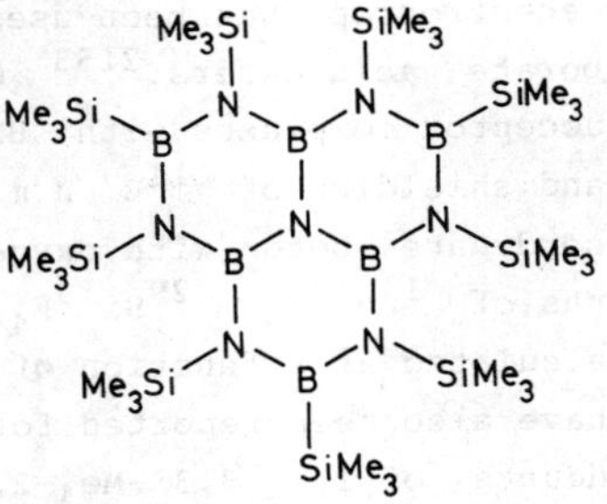

(45)

also been reported for Me_2BR (^{13}C),[2112] B($C_{10}H_{16}$R) (^{11}B, ^{13}C),[2113] BR($C_7H_8Me_3$)$_2$ (^{11}B),[2114] monoisopinocamphenylborane (^{11}B),[2115] $R^1C(PPh_3)BR^2{}_2$ (^{11}B, ^{13}C),[2116] RC(BBu^t)$_2CSiMe_3$ (^{11}B, ^{13}C),[2117] spiroboretenes (^{11}B, ^{13}C),[2118] CHR=CRB(C_8H_{13}Me) (^{13}C),[2119] $Me_3Sn\overline{C=C(BBu^t)}BClBu^t$ (^{11}B),[2120] 3-R_2B-pyridyl (^{13}C),[2121] $\overline{MeNCH_2CH_2NMe(BY)}Br_{\underline{n}}X_{3-\underline{n}}$ (^{11}B, ^{13}C, ^{14}N),[2122] $Me_2\overline{NCH_2C_6H_4B}(C_8H_{14})$ (^{13}C),[2123] ArCNB(C_9H_{18}) (^{11}B),[2124] (Me_3Si)$_2$C(BBu^t)$_2CEPh_3$ (E = P, As; ^{11}B, ^{13}C, ^{31}P),[2125] $Me_2BNR^1R^2$ (^{11}B),[2126] $Me_2B(NMe_2)AlX_3$ (^{11}B, ^{13}C, ^{14}N, ^{27}Al),[2127] $Cl_2P(O)NR^1BR^2{}_2$ (^{11}B, ^{13}C, ^{14}N, ^{15}N, ^{17}O, ^{31}P),[2128] $Et\overline{BCEt=CMeSiMe_2N}Me$ (^{11}B, ^{13}C, ^{14}N, ^{17}O, ^{29}Si),[2129] $Me_2NB(CBu^t)_2BNMe_2$ (^{11}B, ^{13}C),[2130] ($Ph_2\overline{BOC_6H_4CH=N}$)$_2$($CH_2$)$_{\underline{n}}$ (^{13}C),[2131] $R_2B(OMe)$, $AlH_2(OMe)$ (^{11}B, ^{27}Al),[2132] MeOB(C_9H_{15}Me) (^{13}C),[2133]

$R^1{}_2BOP(O)PhR_2$ (^{11}B, ^{13}C, ^{31}P),[2134] $Ph_2BONEt_2CH_2CHRO$ (^{11}B),[2135] $Bu^tC_6H_2N\{C(CF_3)_2O\}_2B$(2,2'-biphenyl) ($^{11}B$, ^{13}C, ^{19}F),[2136] $MeNCH_2CH_2N(BR)Me(MX_3)$ (M = Al, Ga; ^{11}B, ^{13}C, ^{14}N, ^{27}Al),[2137] $H_2C\{B(NMe_2)B(NMe_2)\}_2CH_2$ (^{11}B),[2138] $[CH_2(CH_2CMe_2)_2NBRNCR_3BX]^+[AlX_4]^-$ (^{11}B, ^{13}C, ^{27}Al),[2139] RB=NR (^{11}B),[2140] $Bu^tB{=}NBu^t$ (^{11}B, ^{13}C, ^{14}N),[2141] $S_2CR^1{=}CR^2BN(SiMe_3)_2$ (^{11}B, ^{13}C, ^{29}Si),[2142] $(mesityl)_2BC_6F_5$ (^{19}F),[2143] $R^1B(OR^2)_2$ (^{13}C),[2144] 1,3,2-dioxaborinanes (^{11}B, ^{13}C),[2145] azomethine boron complexes of semicarbazones and thiosemicarbazones (^{11}B, ^{13}C),[2146] $\underline{Z}$-$R^1CH{=}CHB(OR^2)_2$ (^{11}B, ^{13}C),[2147] $PhB(OCHR)_2P(O)Ph$ (^{31}P),[2148] 1,3,2-dioxaborinium cations (^{11}B),[2149] and $(Bu^tO)_3SiBNMeCH_2CH_2NMe$ (^{11}B, ^{13}C, ^{29}Si).[2150]

From the relative ^1H chemical shifts of $EtCNBX_3$, it has been suggested that the electron-donor abilities follow the order BBr_3 > BCl_3 > BF_3.[2151] The ^{19}F chemical shifts in $[BF_{4-\underline{n}}(OH)_{\underline{n}}]^-$ and $[BF_{4-\underline{n}}(OOH)_{\underline{n}}]^-$ have been calculated by a CNDO/2 method.[2152] ^{11}B n.m.r. spectroscopy has been used to determine the pH stability of some borate acid esters.[2153] $CF_3SO_3SiR_3$ forms strongly polarized donor-acceptor complexes with BX_3. Deshielding of the ^{29}Si n.m.r. sites and shielding of ^{11}B n.m.r. shifts are characteristic for tetracoordinate boron with oxygen coordination.[2154] The shielding constants of $^{11}B^{19}F_3$, $^{29}Si^{19}F_4$, $^{31}P^{19}F_3$, $^{31}P^{19}F_5$, and $^{31}P^1H_3$ have been calculated as a function of bond length and angle.[2155] N.m.r. data have also been reported for $(C_5H_6Me_4N)_2B_2NBu^t$ (^{11}B, ^{13}C),[2156] MX_3 adducts of 1,1',3,3'-Me_4-2,2'-bis(1,3,2-diazaborolidine) (M = B, Al, Ga; ^{11}B, ^{13}C, ^{27}Al),[2157] $MeNCH_2CH_2NMeBNMeCH_2CH_2NMe$-$BNMeCH_2CH_2NMe$ (^{11}B, ^{13}C, ^{14}N),[2158] $[MeNCH_2CH_2NMeBNC_{13}H_9]^+$ (^{11}B),[2159] $(Bu^tMe_2Si)NN(SiMe_2Bu^t)BN(SiMe_3)_2$ (^{13}C, ^{14}N, ^{29}Si),[2160] (45) (^{11}B, ^{13}C),[2161] $[MeNCH_2CH_2NMeBPPh_3]^+[AlCl_4]^-$ (^{11}B, ^{13}C, ^{27}Al, ^{31}P),[2162] $[(amine)_2BF_2]^+$ (^{11}B, ^{19}F),[2163] $Me_2NCH_2CH_2NMeBClBCl_2$ (^{11}B),[2164] BF_3-amine adducts (^{11}B, ^{19}F),[2165] BF_2NHAr (^{11}B, ^{13}C, ^{19}F),[2166] $(C_{12}H_{11}N_2)BX_2$ (^{11}B),[2167] $OCR^1R^2CHMeNR^3MeB(O_2C_6H_4)$ (^{11}B, ^{13}C),[2168] tetracoordinate compounds of boron with sulphur containing ligands (^{11}B),[2169] boric acid complexes of some hydroxylated diterpenoids (^{13}C),[2170] $(AcO)_2B(OCRCHCOCO_2Me)$ (^{11}B),[2171] $[B(PO_2F_2)_4]^-$ (^{11}B, ^{19}F, ^{31}P),[2172] $B(OR)_3$ (^{11}B),[2173]

$B_2O(OAc)_{4-\underline{n}}\{\overline{OCRC=CONPhNC}Me\}_{\underline{n}}$ (^{11}B),[2174] and $[B(SCH_2S)_4]^{5-}$ (^{11}B).[2175]

Complexes of Other Group IIIB Elements.—^{27}Al n.m.r. spectroscopy has been used to characterize species present in the system $LiAlH_4$-HOCHPhCHPhOH-EtOH.[2176] $^3\underline{J}(^{203,205}Tl,^1H)$ values have been used to determine the conformation of $(4\text{-}R\text{-}C_6H_{10})_2TlCl$ (R = H, Me, Bu^t).[2177] $(\underline{p}\text{-}Bu^tC_6H_4)_{\underline{n}}Tl(O_2CCF_3)_{3-\underline{n}}$ ($\underline{n}$ = 1 or 2) has a substantial $^5\underline{J}(^{203,205}Tl,^1H)$ coupling constant.[2178] N.m.r. data have also been reported for $(AlH_3)_6(Et_2O)_2$ (^{27}Al),[2179] $[AlH_{4-\underline{n}}(OR)_{\underline{n}}]^-$ (^{27}Al),[2180] (thf)(tmed)Li$(PMe_2CH_2)AlMe_3$ (^{31}P),[2181] $[Ga(C_6Me_6)]^+$ (^{13}C),[2182] $Tl(CF_3)_{\underline{n}}(C_2F_5)_{3-\underline{n}}$ (^{19}F),[2183] and $[Et_2MPEt_2]_3$ (M = Ga, In; ^{31}P).[2184]

$\underline{T}_1$ and $\underline{T}_2$ for ^{14}N, ^{27}Al, ^{69}Ga, and ^{71}Ga have been measured in $[M(NCMe)_6]^{3+}$ and $[MCl_{\underline{n}}Br_{4-\underline{n}}]^-$ (M = Al, Ga). ^{27}Al, ^{69}Ga, and ^{71}Ga relaxation involves both the quadrupolar and scalar mechanisms of the second kind, owing to the relaxation of ^{14}N, ^{35}Cl, ^{37}Cl, ^{79}Br, and ^{81}Br nuclei. In $[M(NCMe)_6]^{3+}$, $^1\underline{J}(^{27}Al,^{14}N)$ = 22 ± 2 Hz, $^1\underline{J}(^{69}Ga,^{14}N)$ = 59 ± 5 Hz, and $^1\underline{J}(^{71}Ga,^{14}N)$ = 74 ± 6 Hz.[2185] For K and Ba aluminates, ^{27}Al n.m.r. spectroscopy has shown the main solution species to be $[Al(OH)_4]^-$, while the species present with Na as a cation is $[Al(OH)_6]^{3-}$.[2186] The formation of AlX_3.solvent has been indicated from chemical-shift data at room temperature, but, on cooling, $[AlX_4]^-$ is observed.[2187] ^{27}Al n.m.r. spectroscopy has been used to investigate the complexing of AlX_3 with Et_2O and thf.[2188] N.m.r. data have also been reported for cyclic Tl(I) alkanediyl dithiophosphates (^{31}P),[2189] a Tl(I)-enniatin B complex (^{13}C, ^{205}Tl),[2190] and $[PhPC_5H_2Ph_3]^+[AlCl_4]^-$ (^{13}C, ^{27}Al, ^{31}P).[2191]

7 Group IVB Elements

Several reviews have appeared: 'Recent advances in ^{29}Si n.m.r. spectroscopy',[2192] 'N.m.r. investigations and analysis of carbon-functional organosilicon compounds',[2193] 'Applications of ^{29}Si n.m.r. spectroscopy',[2194] and 'Spectroscopic study of arylhydrosilanes'.[2195]

The ^1H, ^{13}C, and ^{15}N chemical shifts $^1\underline{J}(^{13}C,^1H)$, $^1\underline{J}(^{15}N,^{13}C)$, and $^1\underline{J}(^{15}N,^1H)$ have been determined in HCN in various liquid-crystal solvents. The ^{13}C shift anisotropy is 334 ± 20 p.p.m.[2196] A close correlation exists between ^{29}Si, ^{73}Ge, and ^{119}Sn chemical shifts. The slopes of the correlation lines correspond closely to

$\langle \underline{r}^{-3} \rangle_{\underline{np}}$ ratios for the element pairs.[2197] Polarization transfer has been used to observe ^{73}Ge n.m.r. signals.[2198] N.m.r. data have also been reported for CH_2=$CHCH_2SiH_2Me$ (^{13}C),[2199] $Me_3SiO(SiH_2O)_n$-$SiMe_3$ (^{29}Si),[2200] $H_2SiC_8H_6O_3$ (^{13}C),[2201] Me_2SiHR (^{13}C),[2202] $[Nu_2SiMe_2H]^+$ (^{29}Si),[2203] $HMeSiC_7H_6(CF_3)_2$ (^{13}C),[2204] $\overline{HMeSiCH_2CMe{=}CMeCH_2}$ (^{13}C),[2205] 1-Me_2SiH-adamantane (^{13}C),[2206] $(C_4H_3E)SiHR^1R^2$ (E = O, S; ^{29}Si),[2207] $H(SiCl_2)_{\underline{n}}H$ ($\underline{n}$ = 3 to 7; ^{29}Si),[2208] and $H_3MC_2CF_3$ (M = Si, Ge; ^{19}F).[2209]

δ(^{29}Si) chemical shifts in $Me_2\overline{SiCR^1{=}C}R^2$ are in the range -87 to -106 p.p.m.[2210] The ^{13}C n.m.r. chemical shifts of phenyl substituted Si, Ge, Sn, and Pb anions show that the remote phenyl positions are mainly affected by π-polarizing effects, and that conjugative effects are unimportant.[2211] The ^{13}C n.m.r. signals of the aromatic carbon atoms in $Ph_3MCH_2CH{=}CHMe$ (M = C, Si, Ge, Sn, Pb) show that polarization is present, which is consistent with a mesomeric electron-withdrawing effect for M = Si, Ge, Sn, and a mesomeric electron-donating effect for M = Pb.[2212] $^1\underline{J}(^{13}C,^{13}C)$ has been determined for R_3SiC_2X and found to depend on the electronegativity of X.[2213] ^{73}Ge $\underline{T}_1$ of $GeMe_4$ is due to quadrupolar relaxation and is solvent dependent.[2214] ^{73}Ge chemical shifts and $\underline{T}_1$ values have been determined for R_4Ge (R = Me, Et, Ph, 2-furyl, 2-thienyl, Cl).[2215] In the ^{13}C n.m.r. spectra of $XMe_2GeGeMe_3$, a correlation has been found between coupling and inductive parameters.[2216] The reaction of $R_{\underline{n}}SnCl_{4-\underline{n}}$ with $SbCl_5$ and $BiCl_3$ has been investigated by ^{11}B and ^{119}Sn n.m.r. spectroscopy.[2217] A linear correlation between $^2\underline{J}(^{119}Sn,^{13}C)$ and $^2\underline{J}(^{119}Sn,^{119}Sn)$ has been found in a wide variety of Me_3Sn derivatives.[2218] ^{2}H, ^{13}C, and ^{119}Sn n.m.r. spectroscopy has been used to show that trifluoroacetolysis of the cyclohexyl-Sn bond goes with retention of configuration at carbon.[2219] A correlation has been found between the calculated charge density and δ(^{119}Sn) in $[Sn_{9-\underline{n}}Pb_{\underline{n}}]^{4-}$, $[Sn_{9-\underline{n}}Ge_{\underline{n}}]^{4-}$, $[Sn_{8-\underline{n}}Pb_{\underline{n}}Tl]^{5-}$, $[Sn_2Bi_2]^{2-}$, and $[SnTe_4]^{4-}$.[2220] N.m.r. data have also been reported for Me_3MCN (M = C, Ge, Sn; ^{13}C),[2221] (allyl)$SiMe_3$ (^{13}C, ^{29}Si),[2222] $Me_{\underline{n}}Cl_{3-\underline{n}}Si$-$CH_2SiCl_{\underline{m}}Me_{3-\underline{m}}$ (^{29}Si),[2223] (cyclopropyl)$SiMe_3$ (^{29}Si),[2224] Me_3Si(norcarane) (^{29}Si),[2225] Me_3Si(bicyclo[$\underline{n}$.1.0]alkanes (^{29}Si),[2226] $\overline{(CH_2)_{\underline{n}}C}HSiMe_3$ (^{13}C, ^{29}Si),[2227] $[Li(tmed)]_2[\underline{p}$-$C_6H_4$-$(CHSiMe_3)_2]$ (^{13}C),[2228] $C_6H_4(CHSiMe_3Li)_2$ (^{13}C),[2229] Me_3Si-$\overline{CHCH{=}CHCH{=}CHSi}Bu^tCl$ (^{13}C, ^{29}Si),[2230] $(Me_3Si)_2CHSiFR^1NR^2SiMe_3$ (^{13}C,

^{19}F, ^{29}Si),[2231] $(Me_3Si)_2CHPHGeCl_3$ (^{31}P),[2232] $(Me_3Si)_2\overline{CN=NCR^1R^2P}Cl$ (^{13}C, ^{31}P),[2233] $(Me_3Si)_2C(SiFMePh)(SiMe_2F)$ (^{13}C, ^{19}F, ^{29}Si),[2234] $(Me_3Si)_2C(SiMe_2OMe)(SiMe_2SCN)$ (^{13}C),[2235] $(Me_3Si)_3CP(O)(O_2)_2P(O)C$-$(SiMe_3)_3$ (^{31}P),[2236] $(Me_3Si)_3CBPhO(CH_2)_4C(SiMe_3)_3$ (^{11}B, ^{13}C),[2237] $(Me_3Si)_3CP{=}PC(SiMe_3)_3$ (^{13}C, ^{31}P),[2238] $(Me_3Si)_3SiCO_2Me$ (^{29}Si),[2239] $(Me_3Si)_2(Me_3Ge)CSiMe_2F$ (^{19}F),[2240] $(Me_3Si)ClC(SiMe_2)_2CCl(SiMe_3)$ (^{13}C, ^{29}Si),[2241] poly(trimethylsilylacetylene) (^{13}C, ^{29}Si),[2242] $\{(Me_3Si)_2C_5H_3\}_2Sn$ (^{13}C),[2243] $Me_3Si\overline{C{=}CROSi}Me_2$ (^{13}C),[2244] $(Me_3Si)Ph$-$C{=}C(SiMe_3)_2C{=}CPhSiMe_3$ (^{13}C),[2245] $BuCF{=}CFSiMe_3$ (^{19}F),[2246] $\{(Me_3Si)_2$-$C{=}PPh_2\}_2PR$ (^{13}C, ^{31}P),[2247] $Bu^t{}_3C_6H_2P(O){=}C(SiMe_3)_2$ (^{31}P),[2248] (mesityl)$P(S){=}C(SiMe_3)_2$ (^{13}C, ^{31}P),[2249] $Bu^t{}_3C_6H_2P(O){=}CPh(SiMe_3)$ (^{13}C, ^{31}P),[2250] $Me_3SiC_2(CH_2)_{\underline{n}}C_2SiMe_3$ (^{13}C),[2251] $Me_3SiC_2P{=}C(SiMe_3)_2$ (^{13}C, ^{31}P),[2252] $N_3P_3F_{6-\underline{n}}(C_2SiMe_3)_{\underline{n}}$ (^{19}F, ^{31}P),[2253] Me_3SiR (^{13}C),[2254-2265] (^{13}C, ^{29}Si),[2266, 2267] $Me_2Si_2Cl_{\underline{n}}H_{4-\underline{n}}$ (^{13}C, ^{29}Si),[2268] $(Me_3Si)MeSi(CMe{=}CMe)_2SiMe(SiMe_3)$ (^{13}C),[2269] $(Me_3Si)_3Si$-CH_2CH_2OH (^{13}C, ^{29}Si),[2270] $P_4Si_4(SiMe_3)_8$ (^{29}Si, ^{31}P),[2271] 9,9,10,10-Me_4-9,10-H_2-10-M-9-silaanthracene (M = Si, Ge, Sn, Pb; ^{13}C),[2272] $Me_2SiR^1R^2$ (^{13}C),[2273] Me_2SiBu^tR (^{13}C),[2274] $Bu^tCH_2CH(SiMe_2CH{=}CH_2)$-$(SiMe_2F)$ (^{13}C),[2275] $Me_3SiCCl_2SiMe_2F$ (^{13}C, ^{19}F, ^{29}Si),[2276] $PhMe_2SiR$ (^{13}C),[2277] $Me_2\overline{MCH_2CR^1{=}CR^2CH_2M}Me_2$ (M = Si, Ge; ^{13}C, ^{29}Si),[2278] $(EtMeSi)_{\underline{n}}$ (^{13}C, ^{29}Si),[2279] 1-allyl-1,2,5-Me_3-1-silacyclopentane (^{13}C),[2280] $(PhOC_6H_4)_2SiMe(CH_2CH_2CH_2Cl)$ (^{13}C),[2281] $C_{12}H_8SiMeCH_2Pr^i$ (^{13}C),[2282] $Me(n\text{-}C_{10}H_{21})_2SiCH{=}CHC_6H_{13}\text{-}\underline{n}$ (^{13}C, ^{29}Si),[2283] copolymers containing methyl(ß-phenethyl)silylene or cyclohexyl(methyl)silylene units (^{13}C),[2284] $Me\overline{CHCH_2CH_2CH_2Si}MeSnR_3$ (^{119}Sn),[2285] Et_3SiR (^{13}C),[2286] $Et_3SiCH_2CH_2SiEt_3$ (^{13}C),[2287] $[(CH_2)_4Si]_{\underline{n}}$ (^{13}C, ^{29}Si),[2288] $R_fCH{=}CHSiEt_3$ (^{19}F),[2289] $R_3Si(CH_2)_{\underline{n}}NHCH_2CH_2NH(CH_2)_{\underline{n}}SiR_3$ (^{13}C),[2290] $Ph_3MCR{=}C{=}CMe_2$ (M = Si, Ge, Sn; ^{13}C),[2291] $Ph_3PbMPPh_3$ (M = Si, Ge, Sn, Pb; ^{13}C, ^{29}Si, ^{207}Pb),[2292] $[R_2Si]_{\underline{n}}$ (^{13}C),[2293] (mesityl)Bu^t-$Si{=}SiBu^t$(mesityl) (^{29}Si),[2294] (mesityl)$_2Si{=}Si$(mesityl)$_2$ (^{29}Si),[2295] $MeCCl{=}CHCH_2GeMe_3$ (^{13}C),[2296] $Me_2Ge(CH_2)_2CMe_2$ (^{13}C),[2297] $Ph_{\underline{n}}GeCl_{4-\underline{n}}$ (^{13}C),[2298] $[M(C_5Me_5)]^+$ (M = Ge, Sn; ^{13}C, ^{119}Sn),[2299] $(Ph_3SnCH_2)_3SnH$ (^{119}Sn),[2300] $Me_2C{=}CHCH_2SnMe_3$ (^{13}C),[2301] $Me_3SnCH_2MMe_3$ (M = C, Si, Ge, Sn; ^{13}C, ^{29}Si, ^{119}Sn),[2302] $Me_3SnCR^1R^2C(SnMe_3){=}CR^3R^4$ (^{13}C, ^{119}Sn),[2303] (naphthyl)$SnMe_3$ (^{13}C),[2304] $Et_2NCMe{=}CHC_2SnMe_3$ (^{13}C, ^{117}Sn, ^{119}Sn),[2305] Me_3SnR (^{13}C),[2306] $(Me^3Sn)_4Sn$ (^{13}C, ^{119}Sn),[2307] $Me_2Sn(CH_2CH_2CH_2)_2NMe$ (^{13}C, ^{119}Sn),[2308] $Me_2\overline{SnCH{=}CR^1SnMe_2C}R^2R^3$ {^{119}Sn, including $^1\underline{J}(^{119}Sn,^{119}Sn)$},[2309] (phen)$_2SnMe_2$ (^{119}Sn),[2310]

$R^1COCR^2R^3SnR^4{}_3$ (^{119}Sn),[2311] $[(PhCH_2)_2Sn]_6$ (^{119}Sn),[2312] Ph_3Sn(cyclohexyl) derivatives (^{13}C, ^{119}Sn),[2313] $[Ph_2Sn(CH_2)_{\underline{n}}]_4$ (^{13}C),[2314] $Ph_2Sn(CHRC_6H_4)_2$ (^{13}C),[2315] $PhSn\{(CH_2)_8\}SnPh$ (^{13}C),[2316] and $(C_5Me_5)_2Sn$ (^{13}C).[2317]

The ^{29}Si chemical shifts of $PhSO_2NHSiMe_3$ have been discussed in terms of S-N π-bonding.[2318] The ^{29}Si n.m.r. chemical shifts of organosiloxanes have been interpreted by excluding the contribution of Si 3*d*-orbitals.[2319] ^{19}F and ^{29}Si n.m.r. spectra have been used to provide qualitative and quantitative analysis of silylated hydroxyl functions in asphaltenes.[2320] ^{29}Si and ^{31}P n.m.r. spectra have been used to determine the chirality of silylated $[^{17}O,^{18}O]$ phosphatidylethanolamine.[2321] ^{13}C and ^{119}Sn n.m.r. spectra of a set of $Bu^n{}_3Sn$ compounds and their complexes in coordinating and non-coordinating solvents have been studied. $\delta(^{119}Sn)$, $\delta(^{13}C)$, and $^1\underline{J}(^{119}Sn,^{13}C)$ depend significantly on the coordination number of tin and the geometry of the coordination sphere.[2322] ^{13}C, ^{31}P, ^{119}Sn n.m.r. spectra have been used to indicate an intramolecular Sn-P interaction for $Me_2SnCl(CH_2)_{\underline{n}}PR^1R^2$.[2323] N.m.r. data have also been reported for $Me_2\overline{Si(OMe)NSiMe_2N(SiMe_3)Si}Me(OMe)$ (^{13}C, ^{29}Si),[2324] silylsulphonamides (^{29}Si),[2325] $(MeO_2C)(Me_3Si)^{15}N^{15}N$-$(SiMe_3)CO_2Me$ (^{15}N),[2326] $(Me_3Si)_2NCH_2OMe$ (^{13}C),[2327] $(Me_3SiN{=})_2S(N)$-$(O)C(CF_3)_2$ (^{19}F),[2328] $(Me_3Si)_2NP(CH_2CH{=}CH_2)_2$ (^{13}C, ^{31}P),[2329] Me_3Si-$N{=}PPh_2PPh_2$ (^{13}C, ^{29}Si, ^{31}P),[2330] $(Me_3Si)_2PC(S)NMe_2$ (^{13}C, ^{29}Si, ^{31}P),[2331] $P_4(SiMe_3)_3Bu^t$ (^{31}P),[2332] $Me_3Si(PBu^t)_3SiMe_3$ (^{31}P),[2333] $CH_2(CH_2NBu^t)_2PSiEt_3$ (^{13}C, ^{29}Si, ^{31}P),[2334] Me_3SiOR (^{13}C),[2335] (^{29}Si),[2336] Me_3SiOAr (^{13}C),[2337] $(C_5H_6)\{CH{=}C(OSiMe_3)PPh\}_2$ (^{13}C, ^{31}P),[2338] $RP{=}C(OSiMe_3)_2$ (^{13}C, ^{29}Si, ^{31}P),[2339] $N_3P_3F_{6-\underline{n}}(COSiMe_3)_{\underline{n}}$ (^{19}F, ^{31}P),[2340] $[S_2COSiMe_3]^-$ (^{13}C),[2341] Me_3SiOCS_2R (^{13}C, ^{29}Si),[2342] Me_3SiO(substituted cyclopropyl) (^{13}C),[2343, 2344] Bu^tMe_2Si-cyclodextrin (^{13}C),[2345] $(CH_2{=}CHSiMe_2)_2E$ (E = O, S, Se, Te; ^{13}C),[2346] $PhC(OSiMe_2Bu^t)CHNMe_2$ (^{13}C),[2347] $R_3SiOSO_2CF_3$ (^{29}Si),[2348] $R^1R^2\overline{SiSiR^1R^2OO}$ (^{29}Si),[2349] $RP{=}C(SSiMe_3)_2$ (^{13}C, ^{29}Si, ^{31}P),[2350] (^{13}C, ^{31}P),[2351] $Bu^t{}_3C_6H_2P{=}C(SSiMe_3)_2$ (^{13}C, ^{29}Si, ^{31}P),[2352] Me_3Si-TePh (^{29}Si, ^{125}Te),[2353] Me_3SiF (^{19}F, ^{29}Si),[2354] $(Me_3SiCH_2)_{\underline{n}}SiCl_{4-\underline{n}}$ (^{13}C),[2355] $(ClMe_2Si)_2C{=}PMe_2Cl$ (^{13}C, ^{29}Si, ^{31}P),[2356] $CH_2(SiCl_2CH_2)_2$-$SiCl(CH_2SiCl_3)$ (^{13}C, ^{29}Si),[2357] $Li(O_2COMMe_3)$ (M = Ge, Sn; ^{13}C),[2358] $(Ph_2XSn)_2CH_2$.HMPA (^{13}C, ^{31}P, ^{119}Sn),[2359] $(Bu_3SnO)_{\underline{n}}P(O)Ph_{3-\underline{n}}$ (^{31}P, ^{119}Sn),[2360] $\{(CF_3C_6H_4)_3Sn\}_2O$ (^{13}C),[2361] $Cy_3Sn\{PhC(O)CHC(O)Ph\}$ (^{13}C),[2362] $Bu^n{}_3Sn$(thio-semi carbazates) (^{13}C, ^{119}Sn),[2363] poly-

(Bu^n_3Sn methacrylate) (^{13}C),[2364] $[Bu_3Sn(OH_2)_2]^+$ (^{119}Sn),[2365] $(furyl)_nSnX_{4-n}$ (^{119}Sn),[2366] $P_7(SnMe_3)_3$ (^{31}P),[2367] $(Me_2SnCl)CHCH_2$-CH_2SOEt (^{13}C, ^{119}Sn),[2368] $(Ph_2XSn)_2CH_2$ (^{13}C, ^{119}Sn),[2369] Cl_nSn-$(CHMeCH_2COMe)_{4-n}$ (^{13}C),[2370] $(CH_2{=}CH)_3SnX$ (^{13}C),[2371] $CH_2\{PhSn$-$(SCH_2CH_2)_2NMe\}_2$ (^{13}C, ^{119}Sn),[2372] bis(R_3Sn) carbonates (^{119}Sn),[2373] and $Me_3SnOSPMe_2$ (^{31}P).[2374]

^{29}Si n.m.r. shifts of $Me_{2-n}Ph_nSiX_2$ indicate that the shifts on replacement of Me by Ph are largely determined by the electronegativity of X.[2375] ^{119}Sn n.m.r. spectra have indicated that the tin in $Me_2Sn(S_2COEt)_2$ is 4 coordinate in solution, but the X-ray structure shows it to be 6 coordinate in the solid state.[2376] The ^{17}O resonance in 4-$C_6H_4CO_2CH_2MF_3$ (M = C, Si) is more sensitive to R when M = Si than when M = C.[2377] N.m.r. data have also been reported for $R^1R^2SiFNHR^3$ (^{19}F),[2378] $PhO_2P(NHNHSiR_2)_2O$ (^{31}P),[2379] $RFSiCH_2CMe{=}NNBu^t$ (^{19}F, ^{29}Si, ^{31}P),[2380] $Bu^t_2Si(NH_2)OLi$ (^{13}C, ^{29}Si),[2381] poly(dimethylsiloxane) (^{13}C, ^{29}Si),[2382] $R^1R^2Si(OC_6H_4O)_2$-SiR^1R^2 (^{13}C, ^{29}Si),[2383] organotin derivatives of 2-thiouracil (^{13}C),[2384, 2385] $R^1_2Si(OH)OSiR^2R^3_2$ (^{19}F, ^{29}Si),[2386] $(Pr^i_2SiF)_2O$ (^{19}F, ^{29}Si),[2387] $FBu^t_2SiOSiBu^t_2OSiF_3$ (^{13}C, ^{19}F, ^{29}Si),[2388] Me_2Si-$(S)_2SiMe_2$ (^{29}Si),[2389] $(Pr^iO)_{4-n}Si(SEt)_n$ (^{29}Si),[2390] $Si_2Me_2Cl_{4-n}Br_n$ (^{29}Si),[2391] $ClMe_2P{=}C(SiCl_2CH_2)_2SiCl_2$ (^{13}C, ^{29}Si, ^{31}P),[2392] Me_2ClGe-(S_2CNMe_2) (^{13}C),[2393] $S(CH_2CMe_2CGeCl_2)_2$ (^{13}C),[2394] $(Ph_2GeO)_4$ (^{13}C),[2395] $BuPh(C_{10}H_{17}O)SnCl$ (^{119}Sn),[2396] $(Bu^t_2SnO)_3$ (^{119}Sn),[2397] diorganotin bis(O,O-alkylenedithiophosphates) (^{13}C, ^{31}P, ^{119}Sn),[2398] $Me(MeO)_2SiN(SiMe_3)_2$ (^{13}C, ^{29}Si),[2399] $[MeSiFO]_n$ (^{19}F, ^{29}Si),[2400] p-$R^1C_6H_4CO_2CH_2SiR^2_3$ (^{13}C),[2401] $RSiCl_n(OR)_{3-n}$ (^{13}C),[2402] $Cl_3CH_2CH_2CONH_2$, $Cl_3SnCH_2CH_2CO_2H$ (^{35}Cl),[2403] $Me_2NCH_2MCl_3$ (M = Si, Sn; ^{29}Si, ^{119}Sn),[2404] $MePPMe_2{=}C(SiMe_3)SiCl_2PMe$ (^{31}P),[2405] organotin phosphates (^{119}Sn),[2406] $MeN(CH_2CH_2S)_2SnCl(CHCl_2)$ (^{119}Sn),[2407] and $Cl_3SnCH_2CH_2CO_2H$ (^{13}C).[2408]

$^1J(^{29}Si,^{15}N)$ values for silylamines, *e.g.* $Bu^tNHSi(OEt)_3$, have been determined and interpreted in terms of the Fermi contact interaction.[2409] The ^{31}P n.m.r. spectrum of $(Bu^tP)_2Ge(Bu^tP)_4$ is $[AB]_2$ and X_2.[2410] $^1J(^{117,119}Sn,^{14}N)$ in $[Sn(NCS)_6]^{2-}$ is 150 Hz. The ^{14}N and ^{119}Sn n.m.r. spectra were recorded.[2411] The influence of the d-orbitals of the third-row elements and also the inductive and steric effects of the organooxy groups on the ^{29}Si chemical shift of compounds with 4 electronegative substituents attached to Si have been evaluated.[2412] Analogies have been shown to exist between the ^{29}Si chemical shifts in silicates and ^{31}P chemical shifts in phosphates.[2413] The properties of Na silicate in solution have been studied using ^{29}Si n.m.r. spectroscopy. A relationship between the concentration in solution and the degree of polymerization of the silicates has been described.[2414] A ^{29}Si n.m.r. study has been made to determine the composition of various silicate anions in solution.[2415] The structure-forming effects of cations in sodium tetramethylammonium silicate solutions have been studied by ^{29}Si n.m.r. spectroscopy.[2416] 2D n.m.r. techniques have been used to study the structures of species present in aqueous silicate solutions.[2417] ^{29}Si n.m.r. spectroscopy has been used to investigate the nature of the silicates in some silicate solutions.[2418-2421] A correlation has been found between ^{29}Si chemical shifts in silicates with orbital energy differences.[2422] The static and vibrational contributions to the quadrupolar relaxation rate of ^{73}Ge have been determined in $GeCl_{\underline{n}}Br_{4-\underline{n}}$. Relaxation in $GeBr_4$ is purely quadrupolar, but in $GeCl_4$ there are both quadrupolar and spin-rotation contributions.[2423] N.m.r. data have also been reported for $Sn(RNCH_2CH_2SCH_2CH_2NR)_2$ (^{13}C, ^{119}Sn),[2424] $N(CH_2CH_2O)_3SiOR$ (^{13}C, ^{29}Si),[2425] $[Si_7O_{21}]^{7-}$ (^{29}Si),[2426] $[Sn(SO_3F)_5]^-$ (^{19}F, ^{119}Sn),[2427] $SnCl_2$-diol systems (^{13}C, ^{119}Sn),[2428] $E(CH_2CH_2S)_2Sn(SR)_2$ (^{13}C, ^{119}Sn),[2429] $[SnTe_4]^{4-}$ (^{119}Sn, ^{125}Te),[2430] $[Pb(EPh)_3]^-$ (E = S, Se; ^{207}Pb),[2431] and $[SnX_{\underline{n}}F_{6-\underline{n}}]^{2-}$ (^{119}Sn).[2432]

8 Compounds of Group VB Elements

A book has appeared, 'Phosphorus-31 n.m.r.: principles and applications',[2433] which contains particularly relevant sections on 'Phosphorus-31 chemical shifts: principles and empirical observations',[2434] 'Two dimensional phosphorus-31 n.m.r.',[2435] 'Selective compilation of phosphorus-31 n.m.r. data',[2436] and 'Conformational analysis of phosphoryl compounds with phosphorus-carbon bonds'.[2437] A review entitled 'Application of optically active phosphinothioic acid as chiral shift reagent to the determination of enantiomeric purity' has appeared.[2438]

The effect of ^{2}H substitution on δ(^{14}N) n.m.r. and $^1\underline{J}(^{14}N,^1H)$ in $[NH_4]^+$ has been studied. Each additional replacement of ^{1}H by ^{2}H in $[NH_4]^+$ results in an approximate linear increase in the ^{14}N shielding.[2439] A theoretical study of the conformational dependence of the ^{31}P magnetic shielding tensor in P_2H_4 has been performed.[2440] N.m.r. data have also been reported for $H_2PCH_2PH_2$ (^{13}C,

^{31}P),[2441] $N_3P_3H_6$ (^{31}P),[2442] $PhPCBu^t{=}CBu^tPH$ (^{31}P),[2443] $P_5Bu^t{}_4H$ (^{31}P),[2444] $P_6Bu^t{}_5H$ (^{31}P),[2445] Li_2P_7H (^{31}P),[2446] $[Ph_2PH(OR)]^+$ (^{31}P),[2447] $(FC_6H_4)_2P(O)H$ (^{31}P),[2448] $(Et_2N)_2PH$ (^{31}P),[2449] $OCH_2CH_2OPH\{OCR^1R^2C(O)O\}$ (^{31}P),[2450] and (MeO)P(O)H(OR) (^{13}C).[2451]

^{31}P n.m.r. spectroscopy has been used to investigate the acceptor properties of $PR_{\underline{n}}Cl_{5-\underline{n}}$ with Cl^-, py, bipy, phen.[2452] Solvent effects of the ^{13}C n.m.r. spectra of polystyrene gels containing quaternary phosphonium ions have been examined.[2453] N.m.r. data have also been reported for $Me_3P{=}CHPBu^tCH_2CH_2CH{=}CH_2$ (^{13}C, ^{31}P),[2454] $(c\text{-}Pr)_2Pr^iP{=}CMe_2$ (^{13}C, ^{31}P),[2455] $[EtMeP(CH_2CH_2)_2\text{-}C(OH)_2]^+$ (^{13}C),[2456] [Ph-phospha-adamantane]$^+$ (^{13}C, ^{31}P),[2457] $[Et_2PhP(CH_2)_3PPhEt_2]^{2+}$ (^{31}P),[2458] $[PhRP(CH{=}CBu^t)_2O]^+$ (^{31}P),[2459] $[R_3PAr]^+$ (^{13}C),[2460] $[CH_2PPh_2C(PMe_2Ph)CH_2PPh_2CPPh_2Me]^{2+}$ (^{13}C, ^{31}P),[2461] $[Ph_3PCH_2R]^+$ (^{31}P),[2462] $[Ph_3PR]^+$ (^{13}C, ^{31}P),[2463, 2464] (^{13}C),[2465] $Me_{\underline{n}}Ph_{3-\underline{n}}P{=}C{=}PPh_3$ (^{13}C, ^{31}P),[2466] $[Ph_3PCR{=}CHNMe_2]^+$ (^{13}C, ^{31}P),[2467] $Ph_3P{=}C(SbPh_2)_2$ (^{31}P),[2468] $Ph_3P{=}CHPPr^i{}_2$ (^{13}C, ^{31}P),[2469] $[Ph_3P{=}CH(CH_2)_2O]^-$ (^{31}P),[2470] $Ph_2P(C_6H_4Li)CH_2$ (^{31}P),[2471] (mesityl)$_3$-$P{=}CH_2$ (^{31}P),[2472] $Bu^t(CH_2CH_2CH_2PMe_2)_2$ (^{13}C, ^{31}P),[2473] 7-phosphanorbornenes (^{13}C, ^{31}P),[2474] (^{31}P),[2475] $R^1R^2R^3P$ (^{31}P),[2476] $Pr^i{}_2PCH_2\text{-}PPr^i(CH_2)_{\underline{n}}PPr^iCH_2PPr^i{}_2$ (^{31}P),[2477] Cy_2PR (^{31}P),[2478] $PhP(CH_2\text{-}CH_2)_2PPh$ (^{13}C, ^{31}P),[2479] $Bu^t{}_2PCH(SiMe_3)_2$ (^{13}C, ^{31}P),[2480] $PhPCBu^t{=}CHCH{=}CHCH{=}CBu^t$ (^{13}C, ^{31}P),[2481] $PhPCPh{=}CHCH{=}CPhC(C_{12}H_8)$ (^{31}P),[2482] $Ph_2PCH_2CH_2OMe$ (^{13}C, ^{31}P),[2483] $Ph_3P{=}C(PPh_2)_2$ (^{13}C, ^{31}P),[2484] $CH_2\{CH_2N(CH_2PPh_2)CH_2CH_2N(CH_2PPh_2)CH_2\}_2CH_2$ (^{31}P),[2485] $RPhPC_6H_4SH$ (^{13}C, ^{31}P),[2486] $H_2N(CH_2)_3P(CH_2CH_2PPh_2)_2$ (^{13}C, ^{31}P),[2487] $Me(OCH_2CH_2)_{\underline{n}}OCH_2$-diop ($^{31}P$),[2488] Ph_2PR (^{13}C),[2489] $PhMePC_6H_4CHMeNMe_2$ (^{31}P),[2490] $Ph_2PCH{=}CHPPh_2$ (^{13}C, ^{31}P),[2491] 1,1-$(Ph_2P)_2$-cyclopropane (^{31}P),[2492] $[C_6H_4(CO)_2P]^-$ (^{13}C, ^{31}P),[2493] $P{=}CR^1NR^2N{=}CR^3$ (^{31}P),[2494] 2-pyridyl phosphabenzene (^{13}C, ^{31}P),[2495] 3-Bu-5-Ph-phosphabenzene (^{13}C, ^{31}P),[2496] $P{=}CPhNHN{=}CR$ (^{31}P),[2497] $P{=}CR^1NR^2N{=}C\text{-}Bu^t$ (^{13}C, ^{31}P),[2498] $OC_6H_4E{=}CBu^t$ (E = P, As; ^{13}C, ^{31}P),[2499] (mesityl)P=C(mesityl)$_2$ (^{31}P),[2500] (mesityl)P=CR$_2$ (^{31}P),[2501] $C_6H_2Bu^t{}_3P{=}CH\text{-}PXC_6H_2Bu^t{}_3$ (^{31}P),[2502] $C_6H_2Bu^t{}_3P{=}CHX$ (^{13}C, ^{31}P),[2503] $C_6H_2Bu^t{}_3P{=}CR\text{-}(NMe_2)$ (^{31}P),[2504] $C_6H_2Bu^t{}_3P{=}C{=}PC_6H_2Bu^t{}_3$ (^{13}C, ^{31}P),[2505, 2506] $C_6H_2Bu^t{}_3P{=}C{=}NPh$ (^{13}C, ^{31}P),[2507] $ArP{=}C(C_{12}H_8)$ (^{13}C, ^{31}P),[2508] $ArP{=}CPh_2$ (^{13}C, ^{31}P),[2509] $R^1P{=}C{=}NR^2$ (^{31}P),[2510] $CF_3E{=}CF_2$ (E = P, As; ^{19}F),[2511] $[PhPCN]^-$ (^{13}C, ^{31}P),[2512] and ArP=PAr (^{31}P).[2513]

Low-temperature ^{7}Li and ^{31}P n.m.r. spectra of lithiated organophosphorus compounds, e.g. Ph_2PLi, have resolved $^{1}J(^{31}P,^{7}Li)$ = 30 to 50 Hz.[2514] For $N_3P_3(NMe_2)_4(N{=}PPh_3)_2$ and related compounds, the ^{31}P n.m.r. parameters have been discussed in terms of the electron donating ability of the ligands.[2515] Plots of $^{3}J(^{31}P^{III}OC^{13}C)$, $^{3}J(^{31}P^{IV}OC^{13}C)$, and $^{3}J(^{31}PCC^{13}C)$ versus dihedral angle have been reported and take into account the steric orientation of the unshared electron pairs in the P=O and P=S groups.[2516] The origin of the oxygen in the oxidation of PPh_3 by $K_2S_2O_8$ has been studied using the ^{18}O isotope effect of the ^{31}P chemical shift.[2517] N.m.r. data have also been reported for $Ph_4BiO_2CCF_3$ (^{13}C),[2518] $RN{=}PPh_3$ (^{31}P),[2519 2520] (^{13}C),[2521] $Ph_3P{=}NCH_2CH_2CH_2CO_2Me$ (^{13}C, ^{31}P),[2522] $R^1{}_3P(OR)(NR^2{}_2)$ (^{31}P),[2523] $[(Ph_3P)_2P]^+$ (^{31}P),[2524] $MeP(O)CH_2OCH_2CH_2OCH_2$ (^{31}P),[2525] phosphoindole-1-oxide (^{13}C, ^{31}P),[2526] 2,2-dihydro-4,4,5,5-Me_4-2,2,2-trisubstituted-1,3,2-dioxaphospholanes (^{31}P),[2527] $R^1R^2P(O)C({=}NOH)CR^3{=}NNR^4CXNH_2$ (^{13}C, ^{31}P),[2528] $R^1R^2R^3PO$ (^{13}C),[2529] R^1R^2MePO (^{13}C, ^{31}P),[2530] R^1R^2PhPO (^{31}P),[2531] Ph_2RPO (^{13}C),[2532] $R^1Ph_2P(OR^2)_2$ (^{31}P),[2533] $[\{Ph_2P(O)\}_2$-$C(OH)]_2CR_2$ (^{13}C, ^{31}P),[2534] a bridgehead bicyclophosphine sulphide (^{13}C, ^{31}P),[2535] $Ph(S)P(CH_2CH_2)_2CH_2$ (^{13}C, ^{31}P),[2536] PhR^1R^2PS (^{13}C, ^{31}P),[2537] $[MeEPPh_3]^+$ (E = S, Se; ^{13}C, ^{31}P),[2538] $Ph_{3-n}PSeCl_n$ (^{31}P, ^{77}Se),[2539] $Me_2NCO_2PF_{4-n}(CF_3)_n$ (^{31}P),[2540] $(Cl_2P)_2C{=}PMe_2Cl$ (^{31}P),[2541] and $R^1ClPCR^2{=}CR^3$ (^{13}C, ^{31}P).[2542]

^{13}C and ^{15}N chemical shifts, $J(^{31}P,^{13}C)$, and $^{1}J(^{31}P,^{15}N)$, have been reported for a series of 10 4-substituted-1-(dimethylphosphono)piperazines.[2543] The ^{1}H and ^{31}P n.m.r. spectra of $[R^1R^2P$-$(X)]_2Y$ (X, Y = O, S) have been fully analyzed.[2544] The stereochemistry of the hydrolysis at an acyclic phosphoryl centre has been investigated using ^{18}O-induced ^{31}P chemical shifts.[2545] N.m.r. data have also been reported for $Me_3MPCHMeCH_2CH_2CH_2$ (M = Si, Ge; ^{13}C, ^{29}Si, ^{31}P),[2546] $PhN(OSiMe_3)PMe_2{=}NSiMe_3$ (^{13}C, ^{31}P),[2547] $C_8H_8PNEt_2$ (^{31}P),[2548] $(Pr^i{}_2N)_2PC_6H_8$ (^{31}P),[2549] $Bu^tPNRN{=}CMeCH_2$ (^{31}P),[2550] $Ph_2PNR^1R^2$ (^{13}C),[2551] $(R_f)_2P(NS)_2N$ (^{13}C, ^{19}F, ^{31}P),[2552] $(Ph_2PN)_2(NSPh)$ (^{31}P),[2553] $[(Ph_2PN)_4(SN)_2]^{2+}$ (^{13}C, ^{31}P),[2554] $P_3N_3Cl_4RAr$ (^{13}C, ^{19}F, ^{31}P),[2555] $P_3N_3R^1R^2R^3R^4R^5R^6$ (^{31}P),[2556] $(Ph_2PN)_2\{NS(O)NPPh_2NPPh_2NH_2\}$ (^{31}P),[2557] $(Et_2N)(O)PCH_2CMe{=}CHCH_2$ (^{13}C, ^{31}P),[2558] $MeNC_6H_4CH_2P(O)R$ (^{13}C, ^{31}P),[2559] (46) (^{13}C,

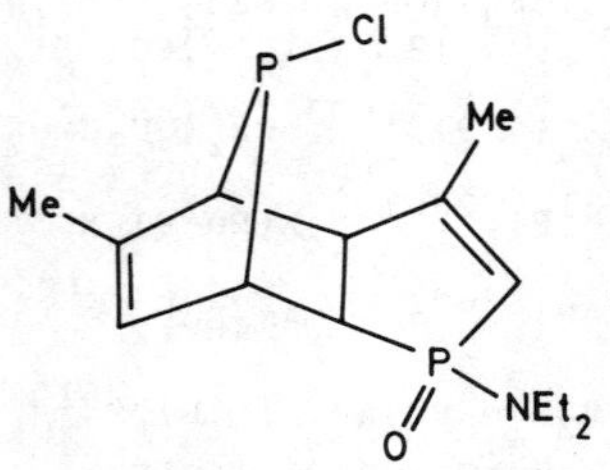

(46)

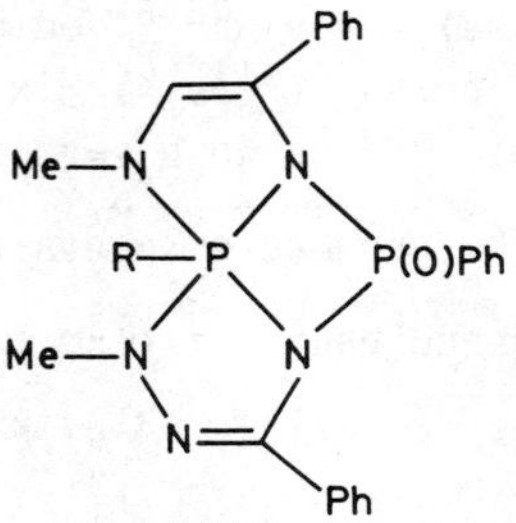

(47)

^{31}P),[2560] $RC_6H_4P(O)(C_3F_7)_2$ (^{19}F, ^{31}P),[2561] $RPhP(O)NHOSO_2Me$ (^{31}P),[2562] $R^1R^2P(Se)NMeP(E)(OR^3)_2$ (E = O, S, Se; ^{31}P, ^{77}Se),[2563] $R^1PR^1PCR^2$=$CHCR^3R^4CR^2CH$=CR^3R^4 (^{13}C, ^{31}P),[2564] $[OPR_2PPPR_2O]^{2-}$ (^{31}P),[2565] R_2C=$CPBu^tPBu^t$ (^{13}C, ^{31}P),[2566] $PhPCH_2CH_2PPh$ (^{31}P),[2567] Me_4Sb_2 (^{13}C),[2568] $Sb_2(C_6F_4)_3$ (^{13}C),[2569] MeC=CHCH=CMeBi-BiCMe=CHCH=CMe (^{13}C),[2570] $OC(CF_3)_2CHMePEt\{OCH(CF_3)_2\}2$ (^{19}F),[2571] $Ph(CH_2)_4P(CH_2CH_2CN)O_2H$ (^{13}C),[2572] $Ph(CH_2)_4PR^1(O)(OR^2)$ (^{31}P),[2573] $R^1{}_2P(O)OP(NR^2{}_2)_2$ (^{13}C, ^{31}P),[2574] $R^1R^2POSC_6H_{10}Cl$ (^{13}C, ^{31}P),[2575] $(HO)R^1C(CH$=$CR^2)_2P(O)OMe$ (^{13}C, ^{31}P),[2576] $OCH_2CH_2CH_2P(O)Ph$ (^{13}C, ^{31}P),[2577] $Ph_2POCHR^1CHR^2NMePPh_2$ (^{31}P),[2578] $[As(CF_3)_2O(OH)]_2$ (^{19}F),[2579] $(C_6F_5)_2BiO_2CR$ (^{19}F),[2580] $R^1R^2PSR^3$ (^{31}P),[2581] diphenylphosphino-N,N-dimethylthioformamide (^{13}C),[2582] and S=P=CHCMe=CMeCH=CPh (^{31}P).[2583]

^1H and ^{31}P n.m.r. spectra of $PhPCl_2$ dissolved in a nematic phase have been analyzed and the structure of the skeleton determined.[2584] The acceptor properties of $PBr_3(O_2C_6H_4)$, $PBr(O_2C_6H_4)_2$, $PhPBr_4$, and $[PhPBr_3]^+$ towards halides and pyridines have been investigated by ^{31}P n.m.r. spectroscopy.[2585] N.m.r. spectral data for a solution containing equivalent amounts of octanesulphonic acid, $SbCl_3$, and H_2O have been interpreted in terms of formation of $[H_3O][RSbCl_5]$.[2586] N.m.r. data have also been reported for $[RP(CN)_2X]^-$ (^{31}P),[2587] $(REN$=$)_2S$ (E = P, As, Sb, Bi; ^{13}C, ^{31}P),[2588] $[PhCH_2P(NMe_2)_3]^+$ (^{13}C),[2589] Bu^tC=$PNRN$=N (^{13}C, ^{31}P),[2590] (47) (^{13}C, ^{31}P),[2591] R^1NNCR^2NPMgX (^{31}P),[2592] $[Cl_2P(NPCl_2N)_2PR]_2$ (^{31}P),[2593] $OCH_2CH_2OPMeNHCMe_2C(O)O$ (^{31}P),[2594] $MeP(NMe_2)(OR)$ (^{31}P),[2595] ArP(E)-(NBu^t) (E = O, S, Se; ^{31}P),[2596] $C_6H_{10}(NMe)_2P(O)Et$ (^{31}P),[2597]

$PhP(O)(NHMe)_2$ (^{31}P),[2598] $OC(CF_3)_2C(CF_3)_2OPBu^tFNEt_2$ (^{13}C, ^{19}F),[2599] $N(SOPh{=}N)_2PClBu^t$ (^{31}P),[2600] $N_3P_3Cl_5Ar$ (^{13}C, ^{31}P),[2601] $R^1PF_2(NR^2{}_2)_2$ (^{19}F, ^{31}P),[2602] $Bu^t{}_3C_6H_2P{=}PC_6H_2Bu^t{}_3$ (^{31}P),[2603] $CH_2(CH_2CMe_2)_2NP{=}PBu^t$ (^{31}P),[2604] $PhPNMeC(S)NMeP(S)Ph$ (^{31}P),[2605] $X(Bu^tP)_3X$ (^{31}P),[2606] $Bu^tPPBu^tPPPBu^tPBu^t$ (^{13}C, ^{31}P),[2607] $As_4Bu^t{}_2$ (^{13}C),[2608] $MeNCHMeCHPhOPR$ (^{13}C, ^{31}P),[2609] $R_2NCH_2PO_3H_2$ (^{31}P),[2610] $[RPO_3]^{2-}$ (^{31}P),[2611] $[ArPO_3H]^-$ (^{31}P),[2612] $Bu^tCH(CH_2O)_2P(O)Bu^t$ (^{31}P),[2613] $RP(O)(OEt)_2$ (^{13}C),[2614] $RP(O)(OEt)_2$ (^{13}C, ^{31}P),[2615] $(EtO)_2OPC(CF_3)_2{=}C{=}C(CF_3)_2$ (^{19}F),[2616] $RP(O)(OEt)_2$ (^{13}C),[2617] $[OCHR^1CHR^2OPOCR^3MeCH{=}CH]^+$ (^{31}P),[2618] $\{(Pr^iO)_2P(O)\}_2CHCl$ (^{31}P),[2619] $(EtO)_2P(O)CHCH_2O$ (^{13}C, ^{31}P),[2620] Et_2O_3PR (^{13}C),[2621] $C_8H_{16}PO_2H$ (^{31}P),[2622] $OCH_2CHMeCH(OH)P(O)(OMe)$ (^{13}C),[2623] $(R^1O)_2P(O)CH{=}CBrCO_2R^2$ (^{13}C),[2624] $(R^1O)_2P(O)CH_2CR^2(OEt)(OLi)$ (^{31}P),[2625] $(RO)_2P(O)CHClCHO$ (^{31}P),[2626] $(Pr^iCH_2O)_2P(O)R$ (^{19}F, ^{31}P),[2627] polymers containing $RP(O)(OPh)_2$ (^{31}P),[2628] $(R^1O)_3P{=}CR^2R^3$ (^{31}P),[2629] $S(CH_2S)_2CHP(O)(OMe)_2$ (^{13}C),[2630] $Bu^t{}_3C_6H_2P(OMe)S_2H$ (^{31}P),[2631] $MePO_3H_2$ (^{13}C, ^{31}P),[2632] $Bu^t{}_3C_6H_2PECl_2$ (E = O, S; ^{13}C, ^{31}P),[2633] $PhP(S_2POCMe_2CMe_2O)_2$ (^{31}P),[2634] $(PhP)_3S_3$ (^{31}P),[2635] $(Me_3C_6H_2PS)_4$ (^{31}P),[2636] and $PhAs(S_2COR)$ (^{13}C).[2637]

The effect of anharmomic vibration and centrifugal distortion on ^{15}N and ^{13}C nuclear shielding in N_2O and CO_2 has been reported.[2638] Heteronuclear chemical-shift correlation has been applied to ^{13}C, ^{19}F, and ^{31}P in some phosphorus compounds.[2639] ^1H and ^{31}P n.m.r. spectroscopy has been used to determine the conformation of the ring in $(PhC_3H_5ONPh)(Me_2N)PO$ [2640] and $Me_2C_3H_4ONHP(O)NMe_2$.[2641] Molecular motion in phosphoserine and phosphothreonine has been investigated using ^{13}C $\underline{T}_1$ values.[2642] ^{31}P n.m.r. spectroscopy has been used to investigate the reaction of $C_6H_4O_2PNMe_2$ with halogen.[2643] For $(EtO)_{\underline{m}}Cl_{3-\underline{m}}P{=}NP(O)(OEt)_{\underline{n}}Cl_{2-\underline{n}}$, the ^{15}N and ^{31}P n.m.r. spectra have been recorded. $^1\underline{J}(^{31}P,^{15}N)$ and $^2\underline{J}(^{31}P,^{31}P)$ were discussed and interpreted qualitatively by semiempirical quantum chemical calculations.[2644] $(Me_2N)_3PE$ (E = Se, Te, BH_3) has been investigated using ^{11}B, ^{15}N, ^{31}P, ^{77}Se, and ^{125}Te n.m.r. spectroscopy. The ^{15}N relaxation is predominantly ^{15}N-^1H dipolar, while the ^{125}Te relaxation is controlled by nuclear screening anisotropy.[2645] Connectivity in $[P_{26}]^{4-}$ has been investigated using COSY 90 ^{31}P n.m.r.

spectroscopy.[2646] N.m.r. data have also been reported for $(R_2N)_2P(NMe_2)=NH$ (^{31}P),[2647] $(Me_2N)_3P=NR$ (^{31}P),[2648] $\{OC(NMe)_2\}_2PPh_2$ (^{31}P),[2649] $[(Me_2N)_2P=NNMe_2]^+$ (^{31}P),[2650] $P\{N=CSC(CF_3)_2SC(CF_3)_2\}_3$ (^{19}F, ^{31}P),[2651] $[(Me_2N)_2P=NNMe_2]^+$ (^{31}P),[2652] aminomethylenephosphines (^{13}C, ^{31}P),[2653] P=NC(CN)=C(CN)NMe (^{13}C, ^{31}P),[2654] $P=NC_6H_4NR$ (^{31}P),[2655] $(Pr^i{}_2N)_4P_2N_2$ (^{13}C, ^{31}P),[2656] $N_3P_3\{HN(CH_2)_3NH\}_3$ (^{31}P),[2657] $P_3N_3Cl_3Me(NHCH_2CH_2CH_2O)$ (^{31}P),[2658] N_3P_3(aziridinyl)$_4$-$(NH_2)_2$ (^{31}P),[2659] $N_3P_3Cl_5\{NH(CH_2)_4NH\}N_3P_3Cl_5$ (^{31}P),[2660] $P_4(NBu^t)_4$ (^{31}P),[2661] P(aziridinyl)$_2(NHC_5N_4MeNMe_3)O$ (^{13}C, ^{31}P),[2662] $[NPClO(NSOCl)_2]^-$ (^{31}P),[2663] $N_3P_3(NMeCH_2CH_2O)_2Cl_2$ (^{31}P),[2664] $N_3P_3Cl_{6-\underline{n}}$-(O-$\underline{p}$-tol)$_{\underline{n}}$ (^{13}C, ^{31}P),[2665] $N_3P_3Cl_2\{O(CH_2)_3O\}_2$ (^{31}P),[2666] $MeNCH_2CH_2NMeP(OR_f)_3$ (^{19}F),[2667] $N=CPhN=NMePO_2C_6H_2Bu^t{}_2$ (^{31}P),[2668] $(ClCH_2CH_2)_2NP(O)(NH_2)OCH_2CH_2CHO$ (^2H, ^{13}C, ^{31}P),[2669] $(ClCH_2CH_2)_2N$-P(O)OCH$_2$CH$_2$CHOHNH (^{31}P),[2670] $[C_6H_4(NH)_2PS_2]^-$ (^{31}P),[2671] $OCMe_2CMe_2OP(NMe_2)OCPh=CHCH=CPhO$ (^{31}P),[2672] OCMe=CMeP(O)(pyrrole) (^{31}P),[2673] 1,3,2-dioxaphospholanes (^{31}P),[2674] $[(RNHPO_2)_2O]^{2-}$ (^{31}P),[2675] ROP(OMe)(NMe$_2$) (^{31}P),[2676] $(R^1O)(R^2O)PNR^3R^4$ (^{31}P),[2677] $R_2NP\{OCH(CF_3)_2\}_2$ (^{19}F, ^{31}P),[2678] $Me_2NP(OCH_2)_2CMe_2$ (^{31}P),[2679] $MeN\{OC(CF_3)_2C(CF_3)_2OP\}_2NMe$ (^{13}C, ^{31}P),[2680] $O=CBu^tCH=NCH=CBu^tOP$ (^{13}C, ^{31}P),[2681] $(EtO)_3P=NP(O)(OEt)_2$ (^{31}P),[2682] a polymer with alternating N=PCl(OPh) and $N=P(OPh)(OCH_2CF_2CHF_2)$ units (^{31}P),[2683] $Cl_{3-\underline{n}}(RO)_{\underline{n}}P=NPOCl_{\underline{n}}(OR)_{2-\underline{n}}$ (^{31}P),[2684] $MeOP(O)(NH_2)(S^{13}CH_3)$ (^{13}C),[2685] $MeOP(O)SCH_2CH_2NCH_2CH_2Cl$ (^{31}P),[2686] $(PhO)_2P(O)NHP(S)$-$(OPh)_2$ (^{31}P),[2687] MeNCHMeCHPhOP(S)Cl (^{31}P),[2688] $pyPO_2F$ (^{19}F, ^{31}P),[2689] $[Cl_2P(O)N(O)PCl_2]^-$ (^{14}N, ^{31}P),[2690] $Na[N\{P(O)Cl_2\}_2]$ (^{23}Na, ^{31}P),[2691] $HN(POCl_2)_2$ (^{31}P),[2692] $\{(Et_2N)_2P(S)S\}_3P$ (^{13}C, ^{31}P),[2693] arsenazo (^{13}C),[2694] fluorinated diaminotetrachlorotriphosphazenes (^{19}F),[2695] $N_3P_3(NHCH_2CH_2CH)Cl_4$ (^{19}F, ^{31}P),[2696] $[R_2NPN=PClNR_2]^{2+}$ (^{13}C, ^{31}P),[2697] $\{(CF_3S)_4C_4N\}_2PCl$ (^{19}F, ^{31}P),[2698] $[PCl_{6-\underline{n}}(NCO)_{\underline{n}}]^-$ (^{31}P),[2699] $[P_7]^{3-}$ (^{31}P),[2700] and $[P_{21}]^{3-}$ (^{31}P).[2701]

The stereochemistry of $(MeO)_3P(O_2C_5H_8X)$ has been investigated by ^{13}C n.m.r. spectroscopy.[2702] The ^1H and ^{13}C n.m.r. spectra of oriented $(MeO)_3PO$ have shown solvent and temperature dependence of the conformer ratio.[2703] The technique of X-relayed ^1H-^1H correlated spectroscopy has been applied to diethylphosphonoacetate.[2704] The selective observation of phosphate ester protons by ^1H-$\{^{31}P\}$ spin-echo difference spectroscopy has been described.[2705] The interaction of nucleophiles with $(MeO)_2POH$ has been claimed to be the cause of the variability in the ^{31}P n.m.r. spectra.[2706] The principal values of the ^{31}P chemical-shielding tensors of substituted phosphonic acid diester compounds have been derived from their ^{31}P n.m.r. spectra, and the orientations of the principal axes were calculated.[2707] The attached proton test has been applied to the ^{31}P n.m.r. spectrum of $(R^1O)(R^2O)PS_2H$.[2708] A general procedure for assigning the ^{31}P n.m.r. spectra of nucleic acids has been published.[2709, 2710] Heteronuclear 2D n.O.e. has been used and applied to ^1H-^{31}P in ATP.[2711] Mononucleotide ^{31}P n.m.r. relaxation measurements have been reported.[2712] ^1H, ^{13}C, and ^{31}P n.m.r. spectra have been used to determine the conformation in 5'-P^V phosphorylated tetrahydrofurfuryl systems.[2713] The optimum pH for the analysis of inorganic phosphate mixtures by ^{31}P n.m.r. spectroscopy has been determined by a simplex method.[2714] $^2\underline{J}(^{31}P,^{31}P)$ has been determined for $[O_3POPO_2{}^{17}O]^{4-}$. The ^{17}O n.m.r. spectrum was also recorded.[2715] From a ^{19}F n.m.r. study, $^1\underline{J}(^{209}Bi,^{19}F) = 3810 \pm 10$ Hz in $[BiF_6]^-$.[2716] N.m.r. data have also been reported for $P(OEt)_5$ (^{31}P),[2717] $(C_6H_4O_2)_2POR$ (^{31}P),[2718] $\{OC(O)CHMeO\}P(OCH_2CH_2O)(OR)$ (^{31}P),[2719] $[\{OC(O)CMe_2O\}_2PS]^-$ (^{13}C, ^{31}P),[2720] $(MeO)_2P(O)SR$ (^{13}C, ^{31}P),[2721] $[(R^1O)(R^2O)POSe]^-$ (^{31}P),[2722] $(RO)_2P(O)X$ (^{31}P),[2723] $XP\{OC(CF_3)_2C(CF_3)_2O\}_2$ (^{19}F, ^{31}P),[2724] $(EtO)_2P(X)YP(OEt)_2$ (X, Y = O, S, Se; ^{31}P),[2725] $(OCR^1R^2CR^3R^4OPS_2)_3E$ (E = As, Sb, Bi; ^{31}P),[2726] S(Se)-phosphorylisothio(seleno)ureas (^{31}P),[2727] $[ROPO(S)OPO_3]^{3-}$ (^{31}P),[2728] $P_4O_{10-\underline{n}}S_{\underline{n}}$ (^{31}P),[2729] CH_3CSOPF_2 (^{13}C, ^{19}F, ^{31}P),[2730] $[As(OTeF_5)_3X]^-$ (^{19}F),[2731] $P(SCH_2)_2CHS$ (^{31}P),[2732] $P_4E_3I_2$ (E = S, Se; ^{31}P),[2733] $P_4S_{\underline{n}}Se_{3-\underline{n}}$ (^{31}P),[2734] $[AsF_5\{NHC(S)Me\}]^-$ (^{19}F),[2735] $[AsF_6]^-$ (^{19}F),[2736, 2737] and $[SbX_{\underline{n}}F_{6-\underline{n}}]^-$ (^{19}F, ^{75}As, ^{121}Sb).[2738]

9 Compounds of Group VIB, Iodine, and Xenon

Three relevant reviews have appeared: 'Oxygen-17 n.m.r.',[2739] 'Selenium-77 and tellurium-125 n.m.r.',[2740] and 'Sulphur, selenium, and tellurium fluorides'.[2741]

The fractional frequency shift in the ^1H n.m.r. signal from H_2O in a spherically shaped sample has been measured.[2742] ^{17}O chemical shifts have been reported for CO, CO_2, OCS, N_2O, and OF_2. An absolute ^{17}O shielding scale was proposed. In OF_2, $^1\underline{J}(^{19}F,^{17}O)$ =

300 ± 30 Hz.[2743] ^{33}S chemical shifts have been measured for SF_6, $HFSO_3$, X_2SO_2, SO_2, Cl_2SO, MeSH, H_2S, and OCS. The ^{33}S chemical shift range is ca. 1000 p.p.m.[2744]

^{13}C and ^{77}Se chemical shifts have been measured for some selenium heterocycles with a selenocarbonyl group. The relationship between the ^{77}Se chemical shift of the C=Se groups and the wavelengths of their $n \rightarrow \pi^*$ absorption was discussed.[2745] Cis-trans isomer assignments in substituted tetraselenafulvenes have been made on the basis of $^3J(^{77}Se,^{77}Se)$.[2746] For o-substituted seleno- and telluro-phenetoles, correlations exist between ^{77}Se and ^{125}Te chemical shifts and the halogen electronegativities or the ^{13}C chemical shifts.[2747] The ^1H n.m.r. spectrum of benzo[b]selenophene, including ^{77}Se satellites, has been investigated in liquid crystalline solvents.[2748] The ^{77}Se T_1 of several selenoketones and Ph_2Se_2 has been studied as a function of temperature at three different magnetic fields. Chemical-shift anisotropy relaxation is dominant at higher fields.[2749] N.m.r. data have also been reported for $[Me_2SF]^+$ (^{19}F),[2750] $MeSe(CH_2)_nSeMe$ (^{13}C, ^{77}Se),[2751] $[RC(SeMe)_2]^+$ (^{13}C),[2752] $(R^1Se)_nCH_{3-n}R^2$ (^{13}C),[2753] $R^1SeCO_2R^2$ (^{13}C, ^{77}Se),[2754] $R^1SeC(S)(OR^2)$ (^{13}C, ^{77}Se),[2755] $R^1SeC(S)(SR^2)$ (^{13}C, ^{77}Se),[2756] RSeC(S)Cl (^{13}C, ^{77}Se),[2757] $(CF^3Se)^3CH$ (^{13}C, ^{19}F, ^{77}Se),[2758] some selenium – substituted tetraphenylporphyrins (^{13}C),[2759] $C_2H_2Se_2C$=CHCH=$CSe_2C_2H_2$ (^{13}C),[2760] SeCPh=CPhOCRPh (^{13}C),[2761] $PhSeC(CH_2Br)$=CHCl (^{13}C),[2762] PhSeR (^{13}C),[2763] $(2\text{-pyridyl})_2Se$ (^{13}C),[2764] $(HOC_6H_4)_2Te(OH)_2$ (^{13}C),[2765] $Me_2C(C_2)_nCMe_2C$=Se (^{13}C),[2766] $C_{10}H_7S_2C$=Se (^{13}C),[2767] F_3CCSF_3 (^{19}F),[2768] $CF_3CFClSF_3$ (^{19}F),[2769] MeCH=SF_4 (^{19}F),[2770] $SF_5CFCF_2OSO_2$ (^{19}F),[2771] $CF_3CF(NO)SF_5$ (^{19}F),[2772] OCPh=CHC(=Te)CH=CPhO (^{13}C),[2773] and $C_6H_{10}Cl(TeCl_3)$ (^{13}C).[2774]

Solvent effects in the ^{19}F n.m.r. spectra in SF_5N=SF_2 have been examined.[2775] A ^{33}S n.m.r. study of $[S_2O_3]^{2-}$ has shown that the observed signal is due to the internal sulphur. T_1 and the activation energy for tumbling were also determined.[2776] The ^{19}F n.m.r. spectrum of gaseous SF_4 has been examined and the pressure dependence determined.[2777] The experimental dependence of the second virial coefficient of the screening constant of ^{19}F in gaseous SF_6 on temperature has been measured.[2778] N.m.r. data have also been reported for $[(Me_2N)_2SF_2]^{2+}$ (^{13}C, ^{19}F),[2779] F_4S=NF (^{19}F),[2780] SF_5-NR^1R^2 (^{13}C, ^{19}F),[2781] F_5ENCO (E = Se, Te; ^{19}F),[2782] $SF_4Cl(OSO_2CF_3)$

(^{19}F),[2783] $F_2E^1(OE^2F_5)_2$ (E^1, E^2 = Se, Te; ^{19}F, ^{77}Se, ^{125}Te),[2784] F_3SSSF (^{19}F),[2785] $MeOTeClF_4$ (^{19}F),[2786] and $[Te_2Se_6]^{2+}$ (^{77}Se, ^{125}Te).[2787]

^{131}Xe $\underline{T}_1$ values have been measured in a variety of solvents and the mechanism of relaxation discussed.[2788] ^{129}Xe chemical shifts are sensitive to the environment and have been used to probe the hydrophobic environment.[2789] ^{129}Xe n.m.r. spectroscopy has been used to investigate the intramolecular interaction between Xe and crown ethers.[2790] N.m.r. data have also been reported for $IF_{\underline{n}}(OR)_{5-\underline{n}}$ (^{19}F),[2791] and XeO_2F_2 (^{17}O, ^{129}Xe).[2792]

10 Appendix

This appendix contains a list of papers in which the use of nuclei other than ^1H has been described. The nuclei are ordered by increasing mass.

^2H 32, 54, 145, 147, 212, 447, 528, 1226, 1327-1329, 1345-1347, 1359, 1373, 1374, 1379, 1393, 1394, 1401, 1423, 1540, 1541, 1552, 1556, 1559, 1571, 1580-1582, 1593, 1603, 1618, 1619, 1636, 1646, 1658, 1677, 1687, 1691, 1774, 1816, 1888, 2005, 2033, 2083, 2219, and 2669.

^3H 1754.

^3He 2038.

^6Li 32, 33, 1065, and 1493.

^7Li 20, 32, 34, 35, 37, 40, 919, 1065, 1066, 1153, 1336, 1337, 1340, 1382, 1384, 1392, 1636, 1751-1753, 1755, 1756, 1777, 1780, 1797-1800, 1808, 1829-1831, 1853, 1871, 1908, 1918, 1919, and 2514.

^8Li 1832 and 1987.

^{10}B 1097, 2063, and 2065.

^{11}B 37, 54, 90, 233, 243, 438, 442, 783, 961, 963, 964, 1063, 1079, 1156, 1168, 1216, 1254, 1295, 1296, 1298, 1299, 1498, 1508, 1522, 1608-1613, 1656, 1724, 1732, 1798, 1895, 1906, 1909, 1910, 1912-1914, 1954, 2046, 2049, 2050, 2052, 2054-2059, 2061-2098, 2100, 2101, 2103-2106, 2108-2110, 2113-2118, 2120, 2122, 2124-2130, 2132, 2134-2142, 2145-2147, 2149, 2150, 2153-2159, 2161-2169, 2171-2176, 2217, 2237, and 2645.

^{13}C 24, 27, 28, 31, 33, 38, 39, 41, 43, 44, 51-53, 57, 63-72, 74,

76, 78, 83, 84, 86-88, 90-102, 105-107, 109-111, 113-115, 117, 119, 122, 126-130, 132, 137, 141, 144, 146, 148, 149, 154, 156, 158, 161-164, 166, 172-176, 178-185, 187-190, 192-211, 213, 216, 218, 219, 221-228, 230-233, 235-238, 241, 245, 246, 248, 249, 251, 253-260, 263, 264, 267, 268, 270, 272, 275, 278, 281, 287, 288, 292-294, 296, 297, 300-304, 307, 308, 310, 311, 313, 315, 316, 320, 327, 329, 351, 352, 360, 367-371, 375-382, 384-401, 403-407, 412-414, 416, 418-422, 425, 427, 429, 436, 437, 439-441, 443, 446, 451, 453-456, 459-463, 466-468, 471, 474-482, 485, 486, 487-492, 495-501, 503-508, 511, 513-516, 518, 520-522, 524-530, 532, 534-557, 559, 562, 563, 565, 567-573, 575, 576, 578-591, 593, 597, 598, 600-603, 605, 608, 609, 614-617, 620-622, 626, 629, 630, 633, 637, 638, 642, 643, 655-657, 659, 661, 662, 664, 668, 671, 673, 678, 679, 681, 683-688, 690, 693, 694, 697, 698, 700-703, 705-707, 709, 713-715, 718, 719, 723, 724, 727-732, 739-742, 746-754, 757, 761, 766, 767, 769, 771, 772, 779, 781, 783, 789, 791, 792, 794, 796, 802, 804-815, 818-820, 822, 824-827, 829-834, 836, 837, 839, 840, 845, 861-863, 877, 879-883, 886-892, 894-896, 898, 905, 907, 908, 911, 913, 919, 923, 924, 930, 934, 936, 937, 941-943, 946, 951, 952, 954-960, 962, 964, 965, 970, 972, 975, 981, 983-987, 991-995, 999, 1004, 1011, 1016-1018, 1030, 1032, 1039, 1050, 1059, 1062, 1064, 1070-1072, 1077, 1078, 1080-1083, 1085, 1086, 1093, 1096, 1097, 1099, 1100, 1103, 1107, 1108, 1110-1113, 1116, 1117-1123, 1128, 1130-1132, 1152, 1156, 1157, 1162, 1168-1173, 1176, 1177, 1179-1187, 1189-1194, 1198, 1199, 1202, 1206, 1211-1213, 1215-1217, 1220, 1222, 1223, 1225-1228, 1230-1232, 1234, 1236-1238, 1240-1242, 1244-1246, 1248-1252, 1255, 1256, 1260, 1264, 1265, 1268, 1269, 1275, 1279, 1280, 1282, 1283, 1288-1291, 1295, 1298-1300, 1305, 1306, 1308, 1314, 1319, 1320, 1325, 1326, 1331, 1340, 1341, 1354, 1365, 1366, 1377, 1378, 1400, 1415, 1416, 1419, 1422, 1424, 1430, 1434, 1435, 1442, 1443, 1454, 1456, 1461, 1466, 1476, 1480, 1482, 1485, 1490, 1493-1495, 1500, 1502, 1505, 1507, 1509, 1518, 1522, 1524, 1527, 1528, 1531, 1533, 1534, 1537, 1541-1543, 1545, 1546, 1549, 1550, 1552, 1555, 1557, 1558, 1562, 1563, 1573, 1575, 1576, 1582, 1588, 1600, 1605, 1606, 1615, 1618, 1622, 1644, 1647, 1648, 1651, 1653, 1656, 1659, 1660, 1664, 1665, 1691, 1694, 1703, 1707-1709, 1720, 1721, 1736, 1737, 1761, 1783,

1784, 1804, 1818, 1825, 1827, 1830, 1838, 1842, 1873, 1875, 1884, 1886, 1890, 1891, 1896, 1906, 1942, 1970, 1971, 1975, 1990, 1991, 1999, 2014, 2017, 2018, 2020-2022, 2026, 2030, 2032, 2035, 2044, 2050, 2051, 2053, 2055, 2059-2061, 2075, 2079, 2080, 2096, 2107, 2110, 2112, 2113, 2116-2119, 2121-2123, 2125, 2127-2134, 2136, 2137, 2139, 2141, 2142, 2144-2147, 2150, 2156-2158, 2160-2162, 2166, 2168, 2170, 2182, 2190, 2191, 2196, 2199, 2201, 2202, 2204-2206, 2211-2213, 2216, 2218, 2219, 2221, 2222, 2227-2231, 2233-2235, 2237, 2238, 2241-2245, 2247, 2249-2252, 2254-2270, 2272-2284, 2286-2288, 2290-2293, 2296-2299, 2301-2308, 2313-2317, 2322-2324, 2327, 2329-2331, 2234, 2335, 2337-2339, 2341-2347, 2350-2352, 2355-2359, 2361-2364, 2368-2372, 2381-2385, 2388, 2392-2395, 2398, 2399, 2401, 2402, 2408, 2424, 2425, 2428, 2429, 2441, 2451, 2453-2457, 2460, 2461, 2463-2467, 2469, 2473, 2474, 2479-2481, 2483, 2484, 2486, 2487, 2489, 2491, 2493, 2495, 2496, 2498, 2499, 2503, 2505-2509, 2512, 2515, 2518, 2521, 2522, 2526, 2528-2530, 2532, 2534-2538, 2542, 2543, 2546, 2547, 2551, 2552, 2554, 2555, 2558-2560, 2564, 2566, 2568-2570, 2572, 2574-2577, 2582, 2588-2591, 2599, 2601, 2607-2609, 2614, 2615, 2617, 2620, 2621, 2623, 2624, 2630, 2632, 2633, 2637-2639, 2642, 2653, 2654, 2656, 2662, 2665, 2669, 2680, 2681, 2685, 2693, 2694, 2697, 2702, 2703, 2713, 2720, 2721, 2730, 2745, 2747, 2751-2767, 2773, 2774, 2779, and 2781.

^{14}N 19, 218, 792, 1099, 1340, 1344, 1349, 1350, 1447, 1453, 1713, 1715, 1810, 1813, 2122, 2127-2129, 2137, 2141, 2158, 2160, 2185, 2411, 2439, and 2690.

^{15}N 19, 25, 73, 153, 184, 321-323, 435, 467, 606, 633, 636, 792, 794, 795, 1000, 1015, 1071-1073, 1115, 1124, 1125, 1131, 1320, 1400, 1435, 1537, 1690, 1810, 2035, 2036, 2128, 2196, 2326, 2409, 2543, 2638, 2644, and 2645.

^{17}O 24, 60, 89, 141, 267, 268, 339, 348, 350, 356, 401, 402, 531, 532, 847, 1203, 1222, 1231, 1327-1330, 1340, 1354, 1364, 1368, 1370, 1373, 1389, 1400, 1428, 1439, 1547, 1590, 1797, 1820-1822, 1826, 2128, 2129, 2377, 2715, 2739, 2743, and 2792.

^{19}F 48, 63, 75, 120, 143, 159, 191, 194, 207, 229, 232, 234, 249, 271, 275, 285, 307, 314, 331, 344, 347, 354, 358, 408, 434, 456, 458, 467, 470, 484, 499, 514, 555, 566, 574, 578, 669, 684, 695, 712, 716, 717, 723, 746, 784, 785, 788, 821,

821, 860, 864, 906, 912, 914, 924, 926, 938, 1058, 1069, 1071, 1072, 1098, 1114, 1153, 1216, 1279, 1288, 1305, 1313, 1316, 1317, 1322, 1323, 1332, 1397, 1421, 1433, 1458, 1479, 1489, 1498, 1499, 1520, 1548, 1561, 1765, 1767-1769, 1772, 1776, 1786, 1797, 1798, 1803, 1811, 1812, 1815, 1818, 1819, 1839, 1843, 1889, 1892, 1972-1974, 1976, 1977, 1979, 1992, 1998, 2081, 2136, 2143, 2152, 2155, 2163, 2165, 2166, 2172, 2183, 2209, 2231, 2234, 2240, 2246, 2253, 2276, 2289, 2320, 2328, 2340, 2354, 2378, 2380, 2386-2388, 2400, 2427, 2511, 2552, 2555, 2561, 2571, 2579, 2580, 2599, 2602, 2616, 2627, 2639, 2651, 2667, 2678, 2689, 2695, 2696, 2698, 2717, 2724, 2730, 2731, 2735-2738, 2744, 2750, 2758, 2768-2772, 2775, 2777, 2778-2786, and 2791.

^{23}Na 17, 20, 36, 1152, 1337, 1341-1343, 1348-1350, 1381, 1384, 1388-1391, 1394-1396, 1714-1716, 1758, 1761, 1781, 1834-1837, 1853, 1854, 1872, 1909, 1931, 1934, and 2691.

^{25}Mg 10, 42, 1389, 1401, and 1715.

^{27}Al 19, 194, 955, 1475, 1476, 1478, 1481, 1482, 1524, 1536, 1614, 1616, 1617, 1738, 1740, 1801, 1802, 1806, 1850, 1862, 1915, 1916, 1921-1923, 1925-1934, 1936, 1937, 1943, 1951, 1952, 1955-1959, 1963, 1966, 1967, 2047, 2052, 2127, 2132, 2137, 2139, 2157, 2162, 2176, 2177, 2180, 2185, 2186, 2188, and 2191.

^{29}Si 291, 340, 361, 511, 659, 1104, 1133, 1274, 1486-1488, 1623, 1737-1740, 1806, 1807, 1846, 1847, 1862, 1877, 1886, 1926, 1927, 1929, 1930, 1933-1940, 1944, 1946-1948, 1953, 1957-1965, 1980-1986, 1988, 1989, 2028, 2034, 2079, 2080, 2110, 2129, 2142, 2150, 2154, 2155, 2160, 2192, 2194, 2197, 2200, 2203, 2207, 2208, 2210, 2222-2227, 2230, 2231, 2234, 2239, 2241, 2242, 2266-2271, 2276, 2278, 2279, 2283, 2288, 2292, 2294, 2295, 2302, 2318-2321, 2324, 2325, 2330, 2331, 2334, 2336, 2339, 2342, 2348-2350, 2352-2354, 2356, 2357, 2374, 2380-2383, 2386-2392, 2399, 2400, 2404, 2409, 2412-2422, 2425, 2426, and 2546.

^{31}P 34, 38, 47, 56, 61, 66, 67, 69, 73, 77-79, 85, 97, 99, 101, 103, 104, 108, 112, 118, 125, 132, 138, 144, 145, 149-153, 155-160, 163-172, 177, 178, 186, 194, 195, 205, 208, 220-222, 230, 236, 238-248, 250-252, 254-256, 258, 261, 262, 265, 267-274, 276, 277, 279-286, 289-291, 295-299, 301, 304-307, 309-314, 316-319, 325-327, 330-337, 341, 345, 349, 354, 362-366, 374-377, 379, 382-385, 388-390, 396, 406, 407,

409-413, 415-418, 420, 422-424, 426, 428, 430-433, 436, 439, 444, 445, 448-454, 456, 457, 460-466, 469, 472, 483, 493, 494, 499, 501, 502, 508, 510, 514, 515, 517, 519, 523, 525, 544, 545, 556, 558, 561, 571, 574, 577, 578, 585, 589, 592-597, 601, 602, 604-607, 609-614, 617-619, 623-632, 634, 639-641, 644-651, 655, 656, 658, 660, 661, 663-667, 669, 670, 672-674, 676-678, 680, 682, 688, 689, 691, 692, 695, 696, 699, 704, 708, 711-714, 716, 717, 719-722, 733-738, 741-745, 753, 755, 756, 758-778, 780, 782-788, 790, 791, 797-801, 828, 838, 841-843, 846, 849-861, 863, 864, 867-875, 878-880, 882, 884, 885, 887, 891-893, 897-907, 909, 910, 914-922, 924, 925, 927-929, 931-939, 942, 944, 945, 947-950, 952, 953, 960, 966-969, 971-975, 977-980, 988, 989, 996-998, 1000, 1014-1016, 1020-1051, 1067, 1068, 1074, 1075, 1077-1079, 1084-1092, 1109, 1129, 1135, 1167, 1170, 1179-1181, 1183, 1191, 1192, 1196, 1207, 1208, 1212, 1220, 1223, 1224, 1227, 1228, 1230, 1235, 1238, 1245, 1247, 1249, 1255-1259, 1261, 1263, 1265, 1267, 1269, 1270, 1280, 1283, 1286, 1288, 1297, 1306, 1310, 1321, 1322, 1385-1387, 1400, 1403-1407, 1412, 1426, 1427, 1442, 1445, 1451, 1467, 1482, 1490, 1491, 1500, 1508, 1509, 1511, 1515-1517, 1529-1533, 1535, 1539, 1545, 1551, 1552, 1562, 1564, 1566, 1572-1575, 1579, 1582, 1584-1586, 1589, 1592, 1596, 1598, 1599, 1601, 1604, 1609-1612, 1622, 1625-1631, 1655, 1703, 1730, 1817, 1840, 1877, 1878, 1886, 1894, 1967, 1996, 1997, 2037, 2071, 2076, 2092, 2093, 2096, 2099-2102, 2125, 2128, 2134, 2148, 2155, 2162, 2172, 2181, 2184, 2189, 2191, 2232, 2233, 2236, 2238, 2247-2250, 2252, 2253, 2271, 2321, 2323, 2329-2334, 2338-2340, 2350-2352, 2356, 2359, 2360, 2367, 2374, 2379, 2380, 2392, 2398, 2405, 2410, 2413, 2440-2450, 2452, 2454, 2455, 2457-2459, 2461-2464, 2466-2488, 2490-2510, 2512-2517 2519, 2520, 2522-2528, 2530, 2531, 2533-2550, 2552-2567 2573-2578, 2581, 2583-2585, 2587, 2588, 2590-2598, 2600-2607 2609-2613, 2615, 2618-2620, 2622, 2625-2629, 2631-2636, 2639 2640, 2643-2666, 2668-2684, 2686-2693, 2697-2701, 2705-2715 2717-2730, and 2732-2734.

^{33}S 1330, 2744, and 2776.

^{35}Cl 848, 1136, 1337, 1340, 1373, 1814, 1874, 2185, and 2403.

^{37}Cl 1337, 1874, and 2185.

^{39}K 1388, 1390, 1714, 1717, 1780, 1871, and 1995.

^{43}Ca 10, 49, 50, 1408, and 1409.

^{45}Sc 54.
^{47}Ti 80 and 81.
^{49}Ti 80 and 81.
^{51}V 19, 121, 124, 126, 131, 133, 134, 136, 139, 140, 342, 1420, 1426, 1428, 1858-1864, and 1866-1868.
^{53}Cr 19.
^{54}Mn 1882.
^{55}Mn 359, 387, 396, 405, 416, 1216, and 1429.
^{57}Fe 435, 560, 1848, and 1885.
^{59}Co 22, 724, 792, 793, 795, 803, 816, 832, 835, 843, 844, 1436, 1567, 1581, and 1887.
^{63}Cu 1076.
^{67}Zn 1115 and 1465.
^{69}Ga 19 and 2185.
^{71}Ga 19 and 2185.
^{73}Ge 2197, 2198, 2214, 2215, and 2423.
^{75}As 1995 and 2738.
^{77}Se 33, 115, 116, 248, 669, 783, 1331, 1602, 1823, 1824, 1857, 2000, 2001, 2740, 2539, 2563, 2645, 2745-2749, 2751, 2754-2758, 2784, and 2787.
^{79}Br 2184.
^{81}Br 1792 and 2184.
^{85}Rb 1152.
^{87}Rb 36, 1152, 1349, 1350, 1714, 1762, 1838, 1897, and 1994.
^{89}Y 56.
^{91}Zr 82 and 98.
^{93}Nb 121, 1167, 1869, and 1870.
^{95}Mo 218, 266-268, 270, 324, 328, 338, 339, 343, 346, 353, 1172, 1202, and 1505.
^{99}Tc 19, 423, 424, 428, and 1883.
^{99}Ru 22, 512, and 1439.
^{103}Rh 22, 655, 657, 659, 676-678, 823, 846-848, 1268, and 1269.
^{105}Pd 1055.
^{113}Cd 10, 50, 1125-1127, 1133-1136, 1294, 1466, 1469, and 1899-1904.
^{115}In 1968.
^{117}Sn 512, 866, and 2305.
^{119}Sn 28, 512, 866, 902, 937, 939, 1103, 1175, 1182, 1298, 1456, 1489, 1514, 1515, 1525, 1526, 1598, 1873, 1879, 2053, 2080, 2197, 2217-2220, 2285, 2299, 2300, 2302, 2303, 2305, 2307-2313, 2322, 2323, 2359, 2360, 2363, 2365, 2366, 2368,

2369, 2372, 2373, 2376, 2396-2398, 2404, 2407, 2411, 2424, 2427-2430, and 2432.

^{121}Sb 1978 and 2738.

^{125}Te 1332, 2353, 2430, 2740, 2645, 2747, 2784, and 2787.

^{127}I 1352 and 1791.

^{129}Xe 2019, 2039, 2040, 2043, 2045, 2789, 2790, and 2792.

^{131}Xe 2788.

^{133}Cs 848, 1352, 1384, and 1837.

^{139}La 1850.

^{169}Tm 1851 and 1852.

^{181}Ta 1780 and 1871.

^{183}W 141, 142, 146, 272, 342, 343, and 355-357.

^{187}Os 360, 512, and 1229.

^{195}Pt 22, 164, 195, 208, 865-867, 872, 873, 878, 884, 903, 908, 910, 919, 921, 930, 937, 939, 951, 973, 976, 978, 979, 984, 1000, 1001, 1008, 1013, 1014, 1016, 1019-1022, 1039, 1041, 1045, 1050, 1052-1054, 1056, 1057, 1268, 1280, 1456, 1514-1517, 1596, and 1598.

^{199}Hg 28, 509, 677, 910, 1095-1097, 1099-1104, 1106, 1108, 1109 1521, and 1607.

^{203}Tl 1803 and 2048.

^{205}Tl 1484, 1485, 1779, 1803, 1866, 1867, 1941, 1969, 2048, 2177 2178, and 2190.

^{207}Pb 28, 1527, 1607, 2292, and 2431.

^{209}Bi 2716.

References

1 Nuclear Magnetic Resonance', ed. R.J. Abraham (Specialist Periodic Reports), The Royal Society of Chemistry, London, 1984, Vol. 13; 1985, Vo 14.

2 L.A. Fedorov, 'N.M.R. Spectroscopy of Organometallic Compounds: Testing t Use of N.M.R. in the Chemistry of Organic Compounds of Heavy Nontransiti Metals', 1984, 248pp. (Chem. Abstr., 1984, **101**, 171 490).

3 'N.M.R. of Newly Accesible Nuclei, Vol. 1: Chemical and Biochemic Applications', Ed. P. Laszlo, Academic Press, New York, 1983, 298 pp.

4 'N.M.R. of Newly Accessible Nuclei, Vol. 2: Chemically and Biochemical Important Elements', Ed. P. Laszlo, Academic Press, New York, 1983, 436pp.

5 P. Granger, 'N.M.R. of less common nuclei', N.M.R. Newly Accesib Nucl., 1984, **2**, 385.

6 R.F.M. White, Annu. Rep. Prog. Chem., Sect. B, 1983, **79**, 3.

7 H. Günther and P. Schmitt, Naturwissenschaften, 1984, **71**, 342 (Che Abstr., 1984, **101**, 142 448).

8 H. Iwamura and H. Tukada, Kagaku no Ryoiki, Zokan, 1983, 33 (Chem. Abstr 1984, **100**, 113 496).

9 C. T. Burt and J.A. Koutcher, J. Nucl. Med., 1984, **25**, 237 (Chem. Abstr 1984, **100**, 153 172).

10 S. Forsén, A. Andersson, T. Drakenberg, O. Teleman, E. Thulin, and H. Vogel, Dev. Biochem., 1983, **25**, 121 (Chem. Abstr., 1984, **100**, 135 144).

11 D. Rehder, Magn. Reson. Rev., 1984, **9**, 125 (Chem. Abstr., 1984, **101**, 64 555).
12 K. Murata, Bunseki, 1984, 357 (Chem. Abstr., 1984, **101**, 47 696).
13 K.M. Smith, Rev. Port. Quim., 1983, **25**, 138 (Chem. Abstr., 1984, **101**, 35 255).
14 K. I. Zamaraev, V.M. Mastikhin, and V.M. Nekipelov, 7 Sov.-yap. Seminar po Katalizu, Irkutsk, 1-7 Iyulya 1983. Sb. Dokl., Novosibirsk, 1983, 149. From Ref. Zh., Khim., abstr. no. 2B4051 (Chem. Abstr., 1984, **100**, 162 499).
15 V.N. Mirnyi, V.D. Prisyazhnyi, and T.A. Mirnaya, Termodin. Elektrokhim. Svoistva Ionnykh Rasplavov, 1984, 54 (Chem. Abstr., 1984, **101**, 221 017).
16 W. Jahn, K. Yunlu, W. Oroschin, H.D. Amberger, and R.D. Fischer, Inorg. Chim. Acta, 1984, **95**, 85.
17 A.A. Maudsley and S.K. Hilal, Br. Med. Bull., 1984, **40**, 165 (Chem. Abstr., 1984, **101**, 166 556).
18 R.K. Gupta, P. Gupta, and R.D. Moore, Annu. Rev. Biophys. Bioeng., 1984, **13**, 221 (Chem. Abstr., 1984, **101**, 85 728).
19 V.P. Tarasov, V.I. Privalov, Yu.A. Buslaev, and U. Eichhoff, Z. Naturforsch., Teil B, 1984, **39**, 1230 (Chem. Abstr., 1984, **101**, 239 164).
20 V.I. Chizhik, C. Cabal, A. Guzman, A. Miranda, and R. Novas, Rev. Cubana Fis., 1983, **3**, 33 (Chem. Abstr., 1984, **101**, 182 452).
21 C.J. Jameson and H.-J. Osten, J. Chem. Phys., 1984, **81**, 4300.
22 N. Juranic, J. Chem. Soc., Dalton Trans., 1984, 1537.
23 J.W. Akitt and W.S. McDonald, J. Magn. Reson., 1984, **58**, 401.
24 C.S. Hsu, H.T. Yin, and C. Chen, Hua Hsueh, 1983, **41**, 30 (Chem. Abstr., 1984, **101**, 97 898).
25 S. Donovan-Mtunzi, R.L. Richards, and J. Mason, J. Chem. Soc., Dalton Trans., 1984, 469.
26 V.A. Kuz'mitskii, Yu.B. Vysotskii, and K. Dzilinski, Zh. Strukt. Khim., 1983, **24**, 33 (Chem. Abstr., 1984, **100**, 95 404).
27 K.N. Ganesh, Proc. - Indian Acad. Sci., Chem. Sci., 1984, **93**, 647 (Chem. Abstr., 1984, **101**, 87 595).
28 L.A. Fedorov, Zh. Strukt. Khim., 1984, **25**, 43 (Chem. Abstr., 1984, **101**, 211 337).
29 C. Detellier, NMR Newly Accessible Nucl., 1983, **2**, 105. Ed. P. Laszlo, Academic, New York (Chem. Abstr., 1984, **100**, 166 564).
30 K.J. Neurohr, T. Drakenberg, and S. Forsén, NMR Newly Accessible Nucl., 1983, **2**, 229. Ed. P. Laszlo, Academic, New York (Chem. Abstr., 1984, **100**, 166 567).
31 H. Ahlbrecht, K. Zimmermann, G. Boche, and G. Decher, J. Organomet. Chem., 1984, **262**, 1.
32 L.M. Jackman and L.M. Scarmoutzos, J. Am. Chem. Soc., 1984, **106**, 4627.
33 D. Seebach, J. Gabriel, and R. Haessig, Helv. Chim. Acta, 1984, **67**, 1083.
34 P.B. Hitchcock, M.F. Lappert, P.P. Power, and S.J. Smith, J. Chem. Soc., Chem. Commun., 1984, 1669.
35 J.W. Robinson and T.A. Ekman, Anal. Lett., 1984, **17**, 565 (Chem. Abstr., 1984, **101**, 83 894).
36 D.M. Holton, P.P. Edwards, D.C. Johnson, C.J. Page, W. McFarlane, and B. Wood, J. Chem. Soc., Chem. Commun., 1984, 740.
37 C. Eaborn, M.N.A. El-Kheli, P.B. Hitchcock, and J.D. Smith, J. Chem. Soc., Chem. Commun., 1984, 1673.
38 L.M. Engelhardt, G.E. Jacobsen, C.L. Raston, and A.H. White, J. Chem. Soc., Chem. Commun., 1984, 220.
39 C. Najera, M. Yus, R. Haessig, and D. Seebach, Helv. Chim. Acta, 1984, **67**, 1100.
40 T. Fjeldberg, P.B. Hitchcock, M.F. Lappert, and A.J. Thorne, J. Chem. Soc., Chem. Commun., 1984, 822.
41 C.G. Screttas and M. Micha-Screttas, Organometallics, 1984, **3**, 904.
42 R. Benn, H. Lehmskuhl, K. Mehler, and A. Rufinska, Angew. Chem., 1984, **96**, 521.
43 B. Bogdanović, S.-t. Liao, R. Mynott, K. Schlichte, and U. Westeppe, Chem. Ber., 1984, **117**, 1378.
44 H. Lehmkuhl, K. Mehler, R. Benn, A. Rufińska, G. Schroth, and C. Krüger,

Chem. Ber., 1984, **117**, 389.
45 K.M. Smith, D.A. Goff, and R.J. Abraham, Org. Magn. Reson., 1984, **22**, 779.
46 R.P.H. Kooyman and T.J. Schaafsma, J. Am. Chem. Soc., 1984, **106**, 551.
47 P. Rösch, R.S. Goody, and M. Isakov, Phosphorus Sulphur, 1984, **21**, 9.
48 G.A. Smith, R.T. Hesketh, J.C. Metcalfe, J. Feeney, and P.G. Morris, Proc. Natl. Acad. Sci. USA., 1983, **80**, 7178 (Chem. Abstr., 1984, **100**, 64 354).
49 E. Bouhoutsos-Brown, C.H. Pletcher, G.L. Nelsestuen, and R.G. Bryant, J. Inorg. Biochem., 1984, **21**, 337.
50 T. Drakenberg and H.J. Vogel, Dev. Biochem., 1983, **25**, 73 (Chem. Abstr., 1984, **100**, 170 181).
51 S. Devarajan, M. Vijayan, and K.R.K. Easwaran, Int. J. Pept. Protein Res., 1984, **23**, 324 (Chem. Abstr., 1984, **101**, 171 714).
52 L. Rösch, C. Krüger, and A.P. Chiang, Z. Naturforsch., Teil B, 1984, **39**, 855.
53 D. Gordon and M.G.H. Wallbridge, Inorg. Chim. Acta, 1984, **88**, 15.
54 M. Mancini, P. Bougeard, R.C. Burns, M. Mlekuz, B.G. Sayer, J.I.A. Thompson, and M.J. McGlinchey, Inorg. Chem., 1984, **23**, 1072.
55 G.Z. Suleimanov, P.V. Petrovskii, Yu.S. Bogachev, I.L. Zhuravleva, E.I. Fedin, and I.P. Beletskaya, J. Organomet. Chem., 1984, **262**, C35.
56 S.O. Grim and S.A. Sangokoya, J. Chem. Soc., Chem. Commun., 1984, 1599.
57 A. Marzotto and H. Kozlowski, Congr. Naz. Chim. Inorg. [Atti], 15th, 1982, 245 (Chem. Abstr., 1984, **100**, 170 187).
58 A. Marzotto and H. Kozlowski, Congr. Naz. Chim. Inorg., [Atti], 16th, 1983, 200 (Chem. Abstr., 1984, **100**, 16 552).
59 A. Marzotto and H. Kozlowski, Inorg. Chim. Acta, 1984, **94**, 96.
60 W.S. Jung, Y. Ikeda, H. Tomiyasu, and H. Fukutomi, Bull. Chem. Soc. Jpn., 1984, **57**, 2317.
61 A.A. Pinkerton, D.L. Pisaniello, and F. Zonnevijlle, Inorg. Chim. Acta, 1984, **82**, 153.
62 H. Le Bail, C. Chachaty, P. Rigny, and R. Bougon, J. Phys. Lett., 1983, **44**, 1017 (Chem. Abstr., 1984, **100**, 44 255).
63 D.T. Haworth and M. Das, Synth. React. Inorg. Met.-Org. Chem., 1984, **14**, 283.
64 D.E. Fenton, U. Casellato, P.A. Vigato, and M. Vidali, Inorg. Chim. Acta, 1984, **95**, 187.
65 H. Schumann, I. Albrecht, J. Pickardt, and E. Hahn, J. Organomet. Chem., 1984, **276**, C5.
66 H. Schumann, I. Albrecht, F.-W. Reier, and E. Hahn, Angew. Chem., Int. Ed. Engl., 1984, **23**, 522.
67 H. Schumann and F.-W. Reier, Inorg. Chim. Acta, 1984, **95**, 43.
68 A.L. Wayda, J.L. Atwood, and W.E. Hunter, Organometalliics, 1984, **3**, 939.
69 H. Schumann and F.W. Reier, J. Organomet. Chem., 1984, **269**, 21.
70 A.L. Wayda, J.L. Dye, and R.D. Rogers, Organometallics, 1984, **3**, 1605.
71 T.D. Tilley, R.A. Andersen, and A. Zalkin, Inorg. Chem., 1984, **23**, 2271.
72 K.K. Abid and D.E. Fenton, Inorg. Chim. Acta, 1984, **95**, 119.
73 K. Bokolo, A. Courtois, J.J. Delpuech, E. Elkaim, J. Protas, D. Rinaldi, L. Rodehueser, and P. Rubini, J. Am. Chem. Soc., 1984, **106**, 6333.
74 S.C. Nayak, V. Chakravortty, and K.C. Dash, Indian J. Chem., Sect A, 1984, **23**, 602 (Chem. Abstr., 1984, **101**, 239 272).
75 J.H. Holloway, D. Laycock, and R. Bougon, J. Fluorine Chem., 1984, **26**, 281 (Chem. Abstr., 1984, **101**, 221 334).
76 C.M. Fendrick and T.J. Marks, J. Am. Chem. Soc., 1984, **106**, 2214.
77 P.G. Edwards, R.A. Andersen, and A. Zalkin, Organometallics, 1984, **3**, 293.
78 G.S. Girolami, G. Wilkinson, M. Thornton-Pett, and M.B. Hursthouse, J. Chem. Soc., Dalton Trans., 1984, 2789.
79 R. Choukroun and D. Gervais, J. Organomet. Chem., 1984, **266**, C37.
80 P.G. Gassman, W.H. Campbell, and D.W. Macomber, Organometallics, 1984, **3**, 385.
81 A. Dormond, M. Fauconet, J.C. Leblanc, and C. Moise, Polyhedron, 1984, **3**, 897.
82 R. Benn and A. Rufińska, J. Organomet. Chem., 1984, **273**, C51.
83 G. Erker, K. Berg, C. Krüger, G. Müller, K. Angermund, R. Benn, and G.

Schroth, Angew. Chem., Int. Ed. Engl., 1984, **23**, 455.
84 R. Beckhaus and K.-H. Thiele, J. Organomet. Chem., 1984, **268**, C7.
85 T. Fukumoto, K. Murata, and S. Ikeda, Anal. Chem., 1984, **56**, 929.
86 W.J. Sartain and J.P. Selegue, Organometallics, 1984, **3**, 1922.
87 G. Erker, U. Dorf, R. Benn, R.-D. Reinhardt, and J.L. Petersen, J. Am. Chem. Soc., 1984, **106**, 7649.
88 K. Mach, H. Antropiusová, V. Hanus, F. Tureček, and P. Sedmera, J. Organomet. Chem., 1984, **269**, 39.
89 G.L. Hillhouse and J.E. Bercaw, J. Am. Chem. Soc., 1984, **106**, 5472.
90 D. Männig and H. Nöth, J. Organomet. Chem., 1984, **275**, 169.
91 P.T. Barger, B.D. Santarsiero, J. Armantrout, and J.E. Bercaw, J. Am. Chem. Soc., 1984, **106**, 5178.
92 K.C. Ott, E.J.M. deBoer, and R.H. Grubbs, Organometallics, 1984, **3**, 223.
93 B.J.J. van de Heisteeg, G. Schat, O.S. Akkerman, and F. Bickelhaupt, Tetrahedron Lett., 1984, **25**, 5191.
94 W.R. Tikkanen, J.Z. Liu, J.W. Egan, jun., and J.L. Petersen, Organometallics, 1984, **3**, 825.
95 P.B. Mackenzie, K.C. Ott, and R.H. Grubbs, Pure Appl. Chem., 1984, **56**, 59.
96 M.F. Lappert, C.L. Raston, B.W. Skelton, and A.H. White, J. Chem. Soc., Dalton Trans., 1984, 893.
97 S.C.H. Ho, D.A. Straus, J. Armantrout, W.P. Schaefer, and R.H. Grubbs, J. Am. Chem. Soc., 1984, **106**, 2210.
98 P. Bougeard, J.J. McCullouugh, B.G. Sayer, and M.J. McGlinchey, Inorg. Chim. Acta, 1984, **89**, 133.
99 H.H. Karsch, G. Müller, and C. Krüger, J. Organomet. Chem., 1984, **273**, 195.
100 W.R. Tikkanen and J.L. Petersen, Organometallics, 1984, **3**, 1651.
101 F. Senocq, C. Randrianalimanana, A. Thorez, P. Kalck, R. Choukroun, and D. Gervais, J. Chem. Soc., Chem. Commun., 1984, 1376.
102 Y. Kai, N. Kanehisa, K. Miki, N. Kasai, M. Akita, H. Yasuda, and A. Nakamura, Bull. Chem. Soc. Jpn., 1983, **56**, 3735.
103 R. Choukroun, F. Dahan, and D. Gervais, J. Organomet. Chem., 1984, **266**, C33.
104 S.J. Young, H. Hope, and N.E. Schore, Organometallics, 1984, **3**, 1585.
105 R.M. Waymouth, B.D. Santarsiero, and R.H. Grubbs, J. Am. Chem. Soc., 1984, **106**, 4050.
106 C.P. Casey, R.F. Jordan, and A.L. Rheingold, Organometallics, 1984, **3**, 504.
107 M.E.E. Meijer-Veldman, J.L. de Boer, H.J. de L. Meijer, A.M.M. Schreurs, J. Kroon, and A.L. Spek, J. Organomet. Chem., 1984, **269**, 255.
108 G.S. Girolami, G. Wilkinson, M. Thornton-Pett, and M.B. Hursthouse, J. Chem. Soc., Dalton Trans., 1984, 2347.
109 M.D. Curtis, S. Thanedar, and W.M. Butler, Organometallics, 1984, **3**, 1855.
110 D.M. Giolando and T.B. Rauchfuss, J. Am. Chem. Soc., 1984, **106**, 6455.
111 S. Saxena, Y.P. Singh, and A.K. Rai, J. Organomet. Chem., 1984, **270**, 301.
112 J.W. Bruno, J.C. Huffman, M.A. Green, and K.G. Caulton, J. Am. Chem. Soc., 1984, **106**, 8310.
113 P.T. Barger and J.E. Bercaw, Organometallics, 1984, **3**, 278.
114 A. Antiñolo, M.F. Lappert, and D.J.W. Winterborn, J. Organomet. Chem., 1984, **272**, C37.
115 P. Granger, B. Gautheron, G. Tainturier, and S. Pouly, Org. Magn. Reson., 1984, **22**, 701.
116 B. Gautheron, G. Tainturier, S. Pouly,, F. Théobald, H. Vivier, and A. Laarif, Organometallics, 1984, **3**, 1495.
117 R.A. Jones, J.G. Hefner, and T.C. Wright, Polyhedron, 1984, **3**, 1121.
118 R.A. Jones, S.T. Schwab, and B.R. Wittlesey, Polyhedron, 1984, **3**, 505.
119 L.A. Fedorov, M.S. Zhukov, N.V. Korsakova, Yu.M. Dedkov, and A.N. Ermakov, Zh. Neorg. Khim., 1984, **29**, 2157 (Chem. Abstr., 1984, **101**, 182 608).
120 J.M. Canich, G.L. Gard, and J.M. Shreeve, Inorg. Chem., 1984, **23**, 441.
121 G.F.P. Warnock and J.E. Ellis, J. Am. Chem. Soc., 1984, **106**, 5016.
122 M.E. Eichner, H.G. Alt, and M.D. Rausch, J. Organomet. Chem., 1984, **264**, 309.
123 Yu.V. Skripkin, O.G. Volkov, A.A. Pasynskii, A.S. Antsyshkina, L.M. Dikareva, V.N. Ostrikova, M.A. Porai-Koshits, S.L. Davydova, and S.G.

Sakharov, J. Organomet. Chem., 1984, **263**, 345.
124 R. Talay and D. Rehder, J. Organomet. Chem., 1984, **262**, 25.
125 D. Rehder and H.C. Bechthold, Z. Naturforsch., Teil B, 1984, **39**, 323 (Chem. Abstr., 1984, **101**, 23 641).
126 G.F.P. Warnock, S.B. Philson, and J.E. Ellis, J. Chem. Soc., Chem. Commun., 1984, 893.
127 P.H.M. Budzelaar, A.A.H. van der Zeijden, J. Boersma, G.J.M. van der Kerk, A.L. Spek, and A.J.M. Duisenberg, Organometallics, 1984, **3**, 159.
128 J. Amaudrut, A. Kadmiri, J. Sala-Pala, and J.E. Guerchais, J. Organomet. Chem., 1984, **266**, 53.
129 L.R. Chamberlain, A.P. Rothwell, and I.P. Rothwell, J. Am. Chem. Soc., 1984, **106**, 1847.
130 P.A. Belmonte, F.G.N. Cloke, K.H. Theopold, and R.R. Schrock, Inorg. Chem., 1984, **23**, 2365.
131 D. Wenke and D. Rehder, J. Organomet. Chem., 1984, **273**, C43.
132 C.M. Bolinger, T.B. Rauchfuss, and S.R. Wilson, J. Am. Chem. Soc., 1984, **106**, 7800.
133 F. Naeumann, D. Rehder, and V. Pank, Inorg. Chim. Acta, 1984, **84**, 117.
134 E.A. Maatta, Inorg. Chem., 1984, **23**, 2560.
135 M.A. Fedotov, Fiz.-khim. Osnovy Prakt. Ispol'z. Izo-i Geteropolisoedin., Dnepropetrovsk, 1983, 38. From Ref. Zh., Khim., 1984, abstr. no. 11B1285 (Chem. Abstr., 1984, **101**, 102 748).
136 M. Hoch, D. Rehder, and C. Weidemann, Inorg. Chim. Acta, 1984, **92**, L5.
137 T.C. Jones, A.J. Nielson, and C.E.F. Rickard, J. Chem. Soc., Chem. Commun., 1984, 205.
138 J.L. Morancais and L.G. Hubert-Pfalzgraf, Transition Met. Chem., 1984, **9**, 130.
139 V.M. Mastikhin, O.B. Lapina, V.N. Krasil'nikov, and A.A. Ivakin, React. Kinet. Catal. Lett., 1984, **24**, 119 (Chem. Abstr., 1984, **100**, 162 513).
140 C. Madic, G.M. Begun, R.L. Hahn, J.P. Launay, and W.E. Thiessen, Inorg. Chem., 1984, **23**, 469.
141 C.J. Besecker, V.W. Day, W.G. Klemperer, and M.R. Thompson, J. Am. Chem. Soc., 1984, **106**, 4125.
142 R.G. Finke and M.W. Droege, J. Am. Chem. Soc., 1984, **106**, 7274.
143 M.E. Ignatov, E.G. Il'in, A.A. Shvets, and O.A. Osipov, Zh. Obshch. Khim., 1983, **53**, 1532 (Chem. Abstr., 1984, **100**, 6735).
144 D. Lyons, G. Wilkinson, M. Thornton-Pett, and M.B. Hursthouse, J. Chem. Soc., Dalton Trans., 1984, 695.
145 G.J. Kubas, R.R. Ryan, B.I. Swanson, P.J. Vergamini, and H.J. Wasserman, J. Am. Chem. Soc., 1984, **106**, 451.
146 W. Buchner and W.A. Schenk, Inorg. Chem., 1984, **23**, 132.
147 S.C. Kao, P.L. Gaus, K. Youngdahl, and M.Y. Darensbourg, Organometallics, 1984, **3**, 1601.
148 D.H. Gibson, K.R. Phillips, and F.U. Ahmed, J. Organomet. Chem., 1984, **276**, 363.
149 J. Powell, M.R. Gregg, and J.F. Sawyer, J. Chem. Soc., Chem. Commun., 1984, 1149.
150 R.H. Crabtree and G.G. Hlatky, Inorg. Chem., 1984, **23**, 2388.
151 F.G.N. Cloke, P.J. Fyne, V.C. Gibson, M.L.H. Green, M.J. Ledoux, R.N. Perutz, A. Dix, A. Gourdon, and K. Prout, J. Organomet. Chem., 1984, **277**, 61.
152 J. Febray, F. Casabianca, and J.G. Riess, J. Am. Chem. Soc., 1984, **106**, 7985.
153 R.A. Henderson, J. Chem. Soc., Dalton Trans., 1984, 2259.
154 M.J. Mockford, A.M. Soares, and W.P. Griffith, Transition Met. Chem., 1984, **9**, 40.
155 E.P. Kyba, J.D. Mather, K.L. Hassett, J.S. McKennis, and R.E. Davis, J. Am. Chem. Soc., 1984, **106**, 5371.
156 C.P. Casey and R.M. Bullock, Organometallics, 1984, **3**, 1100.
157 M.L.H. Green and G. Parkin, J. Chem. Soc., Chem. Commun., 1984, 1467.
158 M.L.H. Green, G. Parkin, K.J. Moynihan, and K. Prout, J. Chem. Soc., Chem. Commun., 1984, 1540.

159 M.L.H. Green, G. Parkin, C. Mingqin, and K. Prout, J. Chem. Soc., Chem. Commun., 1984, 1400.

160 R.H. Crabtree, G.G. Hlatky, C.P. Parnell, B.E. Segmüller, and R.J. Uriarte, Inorg. Chem., 1984, **23**, 354.

161 L. Arndt, T. Delord, and M.Y. Darensbourg, J. Am. Chem. Soc., 1984, **106**, 456.

162 M. Fajardo, M.P. Gómez-Sal, B.F.G. Johnson, J. Lewis, and P.R. Raithby, J. Organomet. Chem., 1984, **272**, C57.

163 W.J. Breen, P.M. Shulman, G.L. Geoffroy, A.L. Rheingold, and W.C. Fultz, Organometallics, 1984, **3**, 782.

164 J.C. Jeffery, I. Moore, and F.G.A. Stone, J. Chem. Soc., Dalton Trans., 1984, 1571.

165 T.S. Coolbaugh, B.D. Santarsiero, and R.H. Grubbs, J. Am. Chem. Soc., 1984, **106**, 6310.

166 G.A. Carriedo, J.A.K. Howard, F.G.A. Stone, and M.J. Went, J. Chem. Soc., Dalton Trans., 1984, 2545.

167 L. Weber, D. Wewers, W. Meyer, and R. Boese, Chem. Ber., 1984, **117**, 732.

168 L. Weber and D. Wewers, Chem. Ber., 1984, **117**, 1103.

169 L. Weber and D. Wewers, Chem. Ber., 1984, **117**, 3331.

170 A. Blagg, A.T. Hulton, P.G. Pringle, and B.L. Shaw, J. Chem. Soc., Dalton Trans., 1984, 1815.

171 H. Blau, K.-H. Griessmann, and W. Malisch, J. Organomet. Chem., 1984, **264**, C1.

172 M. Herberhold, W. Ehrenreich, K. Guldner, W. Jellen, U. Thewalt, and H.P. Klein, Z. Naturforsch., Teil B, 1983, **38**, 1383.

173 H. Fischer, S. Zeuner, and K. Ackermann, J. Chem. Soc., Chem. Commun., 1984, 684.

174 K.H. Dötz, W. Sturm, M. Popall, and J. Riede, J. Organomet. Chem., 1984, **277**, 267.

175 N.A. Ustynyuk, V.N. Vinogradova, V.G. Andrianov, and Yu.T. Struchkov, J. Organomet. Chem., 1984, **268**, 73.

176 R. Märkl and H. Fischer, J. Organomet. Chem., 1984, **267**, 277.

177 D. Rehder, R. Talay, and V. Pank, Inorg. Chim. Acta, 1984, **85**, L45.

178 G. Facchin, P. Uguagliati, and R.A. Michelin, Organometallics, 1984, **3**, 1818.

179 D.H. Gibson, K. Owens, and T.-S. Ong, J. Am. Chem. Soc., 1984, **106**, 1125.

180 S. Strutz and R.R. Shrock, Organometallics, 1984, **3**, 1600.

181 T. Desmond, F.J. Lalor, G. Ferguson, and M. Parvez, J. Chem. Soc., Chem. Commun., 1984, 75.

182 H. Adams, N.A. Bailey, and M.J. Winter, J. Chem. Soc., Dalton Trans., 1984, 273.

183 H. Adams, N.A. Bailey, J.T. Gauntlett, and M.J. Winter, J. Chem. Soc., Chem. Commun., 1984, 1360.

184 L. Blum, I.D. Williams, and R.R. Schrock, J. Am. Chem. Soc., 1984, **106**, 8316.

185 J.R. Shapley, J.T. Park, M.R. Churchill, J.W. Ziller, and L.R. Beanan, J. Am. Chem. Soc., 1984, **106**, 1144.

186 O. Rossell, M. Seco, and P. Braunstein, J. Organomet. Chem., 1984, **273**, 233.

187 M.F. Lappert, C.L. Raston, G.L. Rowbottom, B.W. Skelton, and A.H. White, J. Chem. Soc., Dalton Trans., 1984, 883.

188 M.R. Churchill, J.W. Ziller, S.F. Pedersen, and R.R. Schrock, J. Chem. Soc., Chem. Commun., 1984, 485.

189 I. Feinstein-Jaffe, D. Gibson, S.J. Lippard, R.R. Schrock, and A. Spool, J. Am. Chem. Soc., 1984, **106**, 6305.

190 M.H. Chisholm, D.M. Hoffman, and J.C. Huffman, J. Am. Chem. Soc., 1984, **106**, 6815.

191 J.L. Davidson, M. Shiralian, L. Manojlović-Muir, and K.W. Muir, J. Chem. Soc., Dalton Trans., 1984, 2167.

192 H.G. Alt, H.I. Hayen, H.-P. Klein, and U. Thewalt, Angew. Chem., Int. Ed. Engl., 1984, **23**, 809.

193 N.H.T. Huy, E.O. Fischer, J. Riede, U. Thewalt, and K.H. Dötz, J. Organomet. Chem., 1984, **273**, C29.

194 S.J. Holmes, R.R. Schrock, M.R. Churchill, and H.J. Wasserman, Organometallics, 1984, **3**, 476.
195 J.C. Jeffery, J.C.V. Laurie, I. Moore, H. Razay, and F.G.A. Stone, J. Chem. Soc., Dalton Trans., 1984, 1563.
196 F.R. Kreissl, W.J. Sieber, and M. Wolfgruber, J. Organomet. Chem., 1984, **270**, C45.
197 L.K. Fong and N.J. Cooper, J. Am. Chem. Soc., 1984, **106**, 2595.
198 M.P. Guy and D.W. Bennett, Inorg. Chim. Acta, 1984, **86**, L45.
199 J.T. Guy, jun. and D.W. Bennett, Transition Met. Chem., 1984, **9**, 43.
200 H. Fischer, S. Zeuner, K. Ackermann, and U. Schubert, J. Organomet. Chem., 1984, **263**, 201.
201 A. Mayr, K.C. Schaefer, and E.Y. Huang, J. Am. Chem. Soc., 1984, **106**, 1517.
202 M.R. Churchill, Y.-j. Li, K.H. Theopold, and R.R. Schrock, Inorg. Chem., 1984, **23**, 4472.
203 R.R. Schrock, S.F. Pedersen, M.R. Churchill, and J.W. Ziller, Organometallics, 1984, **3**, 1574.
204 M.R. Churchill, Y.-j. Li, L. Blum, and R.R. Schrock, Organometallics, 1984, **3**, 109.
205 H.G. Alt and U. Thewalt, J. Organomet. Chem., 1984, **268**, 235.
206 M. Green, J.A.K. Howard, S.J. Porter, F.G.A. Stone, and D.C. Tyler, J. Chem. Soc., Dalton Trans., 1984, 2553.
207 J.C. Jeffery, K.A. Mead, H. Razay, F.G.A. Stone, M.J. Went, and P. Woodward, J. Chem. Soc., Dalton Trans., 1984, 1383.
208 J.C. Jeffery, I. Moore, H. Razay, and F.G.A. Stone, J. Chem. Soc., Dalton Trans., 1984, 1581.
209 G.A. Carriedo, J.A.K. Howard, K. Marsden, F.G.A. Stone, and P. Woodward, J. Chem. Soc., Dalton Trans., 1984, 1589.
210 G.A. Carriedo, J.C. Jeffery, and F.G.A. Stone, J. Chem. Soc., Dalton Trans., 1984, 1597.
211 M. Green, J.A.K. Howard, A.P. James, A.N. de M. Jelfs, C.M. Nunn, and F.G.A. Stone, J. Chem. Soc., Chem. Commun., 1984, 1623.
212 J.W. Faller, K.H. Chao, and H.H. Murray, Organometallics, 1984, **3**, 1231.
213 M.R. Churchill and J.C. Fettinger, J. Am. Chem. Soc., 1984, **106**, 3356.
214 J. Brocard, A. Laconi, and D. Courturier, Org. Magn. Reson., 1984, **22**, 369.
215 C. Eischenbroich, B. Spangenberg, and H. Mellinghoff, Chem. Ber., 1984, **117**, 3165.
216 A. Ceccon, A. Gambaro, and A. Venzo, J. Organomet. Chem., 1984, **275**, 209.
217 K.R. Stewart, S.G. Levine, and A.T. McPhail, J. Organomet. Chem., 1984, **263**, 45.
218 M. Minelli, J.L. Hubbard, D.L. Lichtenberger, and J.H. Enemark, Inorg. Chem., 1984, **23**, 2721.
219 F.-W. Grevels and V. Skibbe, J. Chem. Soc., Chem. Commun., 1984, 681.
220 U. Kunze, H. Jawad, W. Conzelmann, and J. Strähle, Z. Anorg. Allg. Chem., 1984, **516**, 93.
221 M. Canestrari, M.L.H. Green, and A. Izquierdo, J. Chem. Soc., Dalton Trans., 1984, 2795.
222 E. Carmona, L. Sanchez, J.M. Marin, M.L. Poveda, J.L. Atwood, R.D. Priester, and R.D. Rogers, J. Am. Chem. Soc., 1984, **106**, 3214.
223 F.R. Kreissl, W.J. Sieber, and H.G. Alt, Chem. Ber., 1984, **117**, 2527.
224 H.G. Alt, H.I. Hayen, H.-P. Klein, and U. Thewalt, Angew. Chem., Int. Ed. Engl., 1984, **23**, 809.
225 H.P. Kim, S. Kim, R.A. Jacobson, and R.J. Angelici, Organometallics, 1984, **3**, 1124.
226 F.R. Kreissl, W.J. Sieber, M. Wolfgruber, and J. Riede, Angew. Chem., Int. Ed. Engl., 1984, **23**, 640.
227 J.R. Shapley, C.H. McAteer, M.R. Churchill, and L.V. Biondl, Organometallics, 1984, **3**, 1595.
228 B.J. Brisdon, M. Cartwright, and A.G. Hodson, J. Organomet. Chem., 1984, **277**, 85.
229 J.L. Davidson and L. Carlton, J. Chem. Soc., Chem. Commun., 1984, 964.
230 F.R. Kreissl, M. Wolfgruber, W.J. Sieber, and K. Ackermann, Organometallics, 1984, **3**, 777.

231 R.P. Hughes, J.W. Reisch, and A.L. Rheingold, Organometallics, 1984, **3**, 1761.

232 N.A. Ustynyuk, V.N. Vinogradova, V.N. Korneva, D.N. Kravtsov, V.G. Andrianov, and Yu. T. Struchkov, J. Organomet. Chem., 1984, **277**, 285.

233 G.A. Carriedo, G.P. Elliott, J.A.K. Howard, D.B. Lewis, and F.G.A. Stone, J. Chem. Soc., Chem. Commun., 1984, 1585.

234 L. Carlton, J.L. Davidson, J.C. Miller, and K.W. Muir, J. Chem. Soc., Chem. Commun., 1984, 11.

235 U. Behrens and F. Edelmann, J. Organomet. Chem., 1984, **263**, 179.

236 H.R. Allcock, G.H. Riding, and R.R. Whittle, J. Am. Chem. Soc., 1984, **106**, 5561.

237 F. Edelmann and U. Behrens, Chem. Ber., 1984, **117**, 3463.

238 R. Bender, P. Braunstein, J.-M. Jud, and Y. Dusausoy, Inorg. Chem., 1984, **23**, 4489.

239 A.S. Goldman and D.R. Tyler, J. Am. Chem. Soc., 1984, **106**, 4066.

240 E. Gross, K. Jörg, K. Fiederling, A. Göttlein, W. Malisch, and R. Boese, Angew. Chem., Int. Ed. Engl., 1984, **23**, 738.

241 O.J. Scherer, H. Sitzmann, and G. Wolmershäuser, J. Organomet. Chem., 1984, **268**, C9.

242 A.H. Cowley, N.C. Norman, and S. Quashie, J. Am. Chem. Soc., 1984, **106**, 5007.

243 R. Maisch, M. Barth, and W. Malisch, J. Organomet. Chem., 1984, **260**, C35.

244 H.P.M.M. Ambrosius, A.W. van Hemert, W.P. Bosman, J.H. Noordik, and G.J.A. Ariaans, Inorg. Chem., 1984, **23**, 2678.

245 L.D. Hutchins, E.N. Duesler, and R.T. Paine, Organometallics, 1984, **3**, 399.

246 O.J. Scherer, H. Sitzmann, and G. Wolmershäuser, Angew. Chem., Int. Ed. Engl., 1984, **23**, 968.

247 R. Janta, M. Luksza, W. Malisch, D. Kempf, and G. Künzel, J. Organomet. Chem., 1984, **266**, C22.

248 D.S. Gill, M. Green, K. Marsden, I. Moore, A.G. Orpen, F.G.A. Stone, I.D. Williams, and P. Woodward, J. Chem. Soc., Dalton Trans., 1984, 1343.

249 D. Lentz, I. Brüdgam, and H. Hartl, Angew. Chem., Int. Ed. Engl., 1984, **23**, 525.

250 R. Bartsch, P.B. Hitchcock, M.F. Meidine, and J.F. Nixon, J. Organomet. Chem., 1984, **266**, C41.

251 L.J. Farrugia, A.D. Miles, and F.G.A. Stone, J. Chem. Soc., Dalton Trans., 1984, 2415.

252 P. Braunstein, C.d.M.de Bellefon, M. Lanfranchi, and A. Tiripicchio, Organometallics, 1984, **3**, 1772.

253 M. Herberhold, W. Jellen, and H.H. Murray, J. Organomet. Chem., 1984, **270**, 65.

254 G.A. Carriedo, J.A.K. Howard, and F.G.A. Stone, J. Chem. Soc., Dalton Trans., 1984, 1555.

255 F.R. Kreissl, M. Wolfgruber, and W.J. Sieber, J. Organomet. Chem., 1984, **270**, C4.

256 M. Wolfgruber, W. Sieber, and F.R. Kreissl, Chem. Ber., 1984, **117**, 427.

257 O.A. Rajan, M. McKenna, J. Noordik, R.C. Haltiwanger, and M.R. DuBois, Organometallics, 1984, **3**, 831.

258 W.R. Jackson, I.D. Rae, and M.G. Wong, Aust. J. Chem., 1984, **37**, 1563.

259 C. Elschenbroich, R. Moeckel, and E. Bilger, Z. Naturforsch., Teil B, 1984, 39, 375 (Chem. Abstr., 1984, **101**, 23 654).

260 K. Dimroth, R. Thamm, and H. Kaletsch, Z. Naturforsch., Teil B, 1984, **39**, 207 (Chem. Abstr., 1984, **101**, 55 264).

261 R. Luck and R.H. Morris, Inorg. Chem., 1984, **23**, 1489.

262 R.L. Luck, R.H. Morris, and J.F. Sawyer, Organometallics, 1984, **3**, 247.

263 C.G. Kreiter and E. Michels, Chem. Ber., 1984, **117**, 344.

264 E. Vincent, L. Verdonck, D.F. van de Vondel, and G.P. van der Kelen, J. Mol. Struct., 1984, **112**, 119.

265 G. Huttner, J. Borm, and L. Zsolnai, J. Organomet. Chem., 1984, **263**, C33.

266 E.C. Alyea and A. Somogyvari, Inorg. Chim. Acta, 1984, **83**, L49.

267 G.M. Gray, R.J. Gray, and D.C. Berndt, J. Magn. Reson., 1984, **57**, 347.

268 G.M. Gray, Inorg. Chim. Acta, 1984, **81**, 157.

269 J.T. Mague and S.E. Dessens, J. Organomet. Chem., 1984, **262**, 347.
270 A.M. Bond, S.W. Carr, and R. Colton, Organometallics, 1984, **3**, 541.
271 J. Grobe and D. Le Van, Z. Anorg. Allg. Chem., 1984, **518**, 36.
272 I.J. Colquhoun, W. McFarlane, and R.L. Keiter, J. Chem. Soc., Dalton Trans., 1984, 455.
273 S.R. Bunkhall, H.D. Holden, B.F.G. Johnson, J. Lewis, G.N. Pain, P.R. Raithby, and M.J. Taylor, J. Chem. Soc., Chem. Commun., 1984, 25.
274 R.A. Michelin, G. Facchin, and P. Uguagliati, Inorg. Chem., 1984, **23**, 961.
275 D. Lentz, Chem. Ber., 1984, **117**, 415.
276 R.B. King and W.-K. Fu, J. Organomet. Chem., 1984, **272**, C33.
277 R.B. King and N.D. Sadanani, J. Chem. Soc., Chem. Commun., 1984, 955.
278 L. Verdonck, E. Vincent, and G.P. van der Kelen, J. Mol. Phys., 1984, **117**, 257.
279 H.P. Fritz and T. Schöttle, J. Organomet. Chem., 1984, **265**, 175.
280 J. Ellermann and A.A.M. Demuth, Angew. Chem., Int. Ed. Engl., 1984, **23**, 805.
281 S. Holand, C. Charrier, F. Mathey, J. Fischer, and A. Mitschler, J. Am. Chem. Soc., 1984, **106**, 826.
282 D.J. Chandler, R.A. Jones, A.L. Stuart, and T.C. Wright, Organometallics, 1984, **3**, 1830.
283 S. Al-jibori, M. Hall, A.T. Hutton, and B.L. Shaw, J. Chem. Soc., Dalton Trans., 1984, 863.
284 S. Al-jibori and B.L. Shaw, J. Organomet. Chem., 1984, **272**, 213.
285 J. Grobe, M. Köhne-Wächter, and D. Le Van, Z. Anorg. Allg. Chem., 1984, **519**, 67.
286 R.N. Bagchi, A.M. Bond, G. Brain, R. Colton, T.L.E. Henderson, and J.E. Kevekordes, Organometallics, 1984, **3**, 4.
287 G.J. Kruger, L. Linford, and H.G. Raubenheimer, J. Chem. Soc., Dalton Trans., 1984, 2337.
288 H. Werner and W. Paul, Angew. Chem., Int. Ed. Engl., 1984, **23**, 58.
289 A. Hengefeld and R. Nast, J. Organomet. Chem., 1984, **267**, 151.
290 R.B. King and K.S. RaghuVeer, Inorg. Chem., 1984, **23**, 2482.
291 W.A. Schenk and F.E. Baumann, J. Organomet. Chem., 1984, **260**, C6.
292 D. Sellmann, W. Weber, G. Liehr, and H.P. Beck, J. Organomet. Chem., 1984, **269**, 155.
293 T.J. McNeese, M.B. Cohen, and B.M. Foxman, Organometallics, 1984, **3**, 552.
294 J.W. Kolls, E.M. Holt, J.A. Hriljac, and D.F. Shriver, Organometallics, 1984, **3**, 496.
295 E.H. Wong, F.C. Bradley, L. Prasad, and E.J. Gabe, J. Organomet. Chem., 1984, **263**, 167.
296 F. Mercier and F. Mathey, J. Organomet. Chem., 1984, **263**, 55.
297 A. Brèque, C.C. Santini, F. Mathey, J. Fischer, and A. Mitschler, Inorg. Chem., 1984, **23**, 3463.
298 S.-G. Shyu and A. Wojcicki, Organometallics, 1984, **3**, 809.
299 A. Blagg, G.R. Cooper, P.G. Pringle, R. Robson, and B.L. Shaw, J. Chem. Soc., Chem. Commun., 1984, 933.
300 C.G. Young, J.A. Broomhead, and C.J. Boreham, J. Organomet. Chem., 1984, **260**, 91; **268**, C60.
301 C.Y. Chou and E.A. Maatta, Inorg. Chem., 1984, **23**, 2912.
302 J.A. Connor, E.J. James, C. Overton, and N. El Murr, J. Chem. Soc., Dalton Trans., 1984, 255.
303 R. Alvarez, E. Carmona, E. Gutierrez-Puebla, J.M. Marin, A. Monges, and M.L. Poveda, J. Chem. Soc., Chem. Commun., 1984, 1326.
304 R. Alvarez, E. Carmona, M.L. Poveda, and R. Sanchez-Delgado, J. Am. Chem. Soc., 1984, **106**, 2731.
305 E.Carmona, K. Doppert, J.M. Martin, M.L. Poveda, L. Sanchez, and R. Sanchez-Delgado, Inorg. Chem., 1984, **23**, 530.
306 H.P.M.M. Ambrosius, J. Willemse, J.A. Cras, W.P. Bosman, and J.H. Noordik, Inorg. Chem., 1984, **23**, 2672.
307 R.B. King, M. Shimura, and G.M. Brown, Inorg. Chem., 1984, **23**, 1398.
308 M.H. Chisholm, J.F. Corning, K. Folting, J.C. Huffmann, A.L. Ratermann, I.P. Rothwell, and W.E. Streib, Inorg. Chem., 1984, **23**, 1037.
309 F. Mercier and F. Mathey, J. Chem. Soc., Chem. Commun., 1984, 782.

310 A. Marinetti and F. Mathey, Organometallics, 1984, **3**, 1492.
311 A. Marinetti and F. Mathey, Organometallics, 1984, **3**, 456.
312 A. Marinetti and F. Mathey, Phosphorus Sulphur, 1984, **19**, 311.
313 B. Deschamps and F. Mathey, Phosphorus Sulphur, 1983, **17**, 317.
314 J. Chatt, P.B. Hitchcock, A. Pidcock, C.P. Warrens, and K.R. Dixon, J. Chem. Soc., Dalton Trans., 1984, 2237.
315 H.G. Raubenheimer, G.J. Krüger, and C.F. Marais, J. Chem. Soc., Chem. Commun., 1984, 634.
316 A. Marinetti, F. Mathey, J. Fischer, and A. Mitschler, J. Chem. Soc., Chem. Commun., 1984, 45.
317 S. Rosenberg, R.R. Whittle, and G.L. Geoffroy, J. Am. Chem. Soc., 1984, **106**, 5934.
318 G.L. Geoffroy, S. Rosenberg, P.M. Shulman, and R.R. Whittle, J. Am. Chem. Soc., 1984, **106**, 1519.
319 E.D. Morrison, A.D. Harley, M.A. Marcelli, G.L. Geoffroy, A.L. Rheingold, and W.C. Fultz, Organometallics, 1984, **3**, 1407.
320 E.O. Fischer, A.C. Filippou, and H.G. Alt, J. Organomet. Chem., 1984, **276**, 377.
321 S. Donovan-Mtunzi, R.L. Richards, and J. Mason, J. Chem. Soc., Dalton Trans., 1984, 2429.
322 S. Donovan-Mtunzi, R.L. Richards, and J. Mason, J. Chem. Soc., Dalton Trans., 1984, 2729.
323 S. Donovan-Mtunzi, R.L. Richards, and J. Mason, J. Chem. Soc., Dalton Trans., 1984, 1329.
324 M. Minelli, J.L. Hubbard, and J.H. Enemark, Inorg. Chem., 1984, **23**, 970.
325 R.H. Morris, J.M. Ressner, J.F. Sawyer, and M. Shiralian, J. Am. Chem. Soc., 1984, **106**, 3683.
326 S.N. Anderson, D.L. Hughes, and R.L. Richards, J. Chem. Soc., Chem. Commun., 1984, 958.
327 R. Luck, R.H. Morris, and J.F. Sawyer, Organometallics, 1984, **3**, 1009.
328 M. Minelli, K. Yamanouchi, J.H. Enemark, P. Subramanian, B.B. Kaul, and J.T. Spence, Inorg. Chem., 1984, **23**, 2554.
329 M.H. Chisholm, K. Folting, J.C. Huffman, and A.L. Ratermann, Inorg. Chem., 1984, **23**, 2303.
330 M.E. Noble, K. Folting, J.C. Huffman, and R.A.D. Wentworth, Inorg. Chem., 1984, **23**, 631.
331 P. Legzdins and J.C. Oxley, Inorg. Chem., 1984, **23**, 1053.
332 E. Carmona, A. Galindo, L. Sanchez, A.J. Nielson, and G. Wilkinson, Polyhedron, 1984, **3**, 347.
333 R.L. Luck and R.H. Morris, J. Am. Chem. Soc., 1984, **106**, 7978.
334 H.P.M.M. Ambrosius, F.A. Colton, L.R. Falvello, H.T.J.M. Hintzen, T.J. Melton, W. Schwotzer, M. Tomas, and J.G.M. van der Linden, Inorg. Chem., 1984, **23**, 1611.
335 A.D. Hunter and P. Legzdins, Inorg. Chem., 1984, **23**, 4198.
336 L.G. Hubert-Pfalzgraf, Inorg. Chim. Acta, 1984, **88**, 89.
337 D.C. Povey and R.L. Richards, J. Chem. Soc., Dalton Trans., 1984, 2585.
338 M.A. Fedotov, Izv. Akad. Nauk SSSR, Ser. Khim., 1984, 1166 (Chem. Abstr., 1984, **101**, 102 741).
339 S.F. Gheller, M. Sidney, A.F. Masters, R.T.C. Brownlee, M.J. O'Connor, and A.G. Wedd, Aust. J. Chem., 1984, **37**, 1825.
340 M.A. Fedotov and L.P. Kazanskii, Koord. Khim., 1984, **10**, 423 (Chem. Abstr., 1984, **100**, 166 829).
341 R.I. Maksimovskaya, Fiz.-khim. Osnovy Prakt. Ispol'z. Izo-i Geteropolisoedin., Dnepropetrovsk, 1983, 47. From Ref. Zh., Khim., 1984, abstr. no. 11B1294 (Chem. Abstr., 1984, **101**, 102 749).
342 P.J. Domaille, J. Am. Chem. Soc., 1984, **106**, 7677.
343 S.F. Gheller, T.W. Hambley, J.R. Rodgers, R.T.C. Brownlee, M.J. O'Connor, M.R. Snow, and A.G. Wedd, Inorg. Chem., 1984, **23**, 2519.
344 E.G. Il'in, M.E. Ignatov, B.V. Golovanov, V.D. Butskii, and Yu.A. Buslaev, Dokl. Akad. Nauk SSSR, 1984, **275**, 379 (Chem. Abstr., 1984, **101**, 47 653).
345 J.M. Berg and R.H. Holm, J. Am. Chem. Soc., 1984, **106**, 3035.
346 B. Piggott, R.N. Sheppard, and D.J. Williams, Inorg. Chim. Acta, 1984, **86**,

L65.
347 M.E. Ignatov, E.G. Il'in, B.V. Golovanov, V.D. Butskii, and Yu.A. Buslaev, Dokl. Akad. Nauk SSSR, 1984, **277**, 375 (Chem. Abstr., 1984, **101**, 221 331).
348 M.H. Chisholm, K. Folting, J.C. Huffman, and C.C. Kirkpatrick, Inorg. Chem., 1984, **23**, 1021.
349 S.F. Davison, B.E. Mann, and P.M. Maitlis, J. Chem. Soc., Dalton Trans., 1984, 1223.
350 N.Ya. Volovaya, R.I. Maksimovskaya, M.A. Fedotov, and S.A. Morosanova, Zh. Neorg. Khim., 1984, **29**, 1468 (Chem. Abstr., 1984, **101**, 64 830).
351 Z. Zhang, F. Wang, S. Li, C. Zhao, and Y. Li, Gaodeng Xuexiao Huaxue Xuebao, 1984, **5**, 276 (Chem. Abstr., 1984, **100**, 220 545).
352 M.M. Caldeira and V.M.S. Gil, Can. J. Chem., 1984, **62**, 2094.
353 M. Minelli, J.H. Enemark, J.R. Nicholson, and C.D. Garner, Inorg. Chem., 1984, **23**, 4384.
354 D.J. Santure, J.C. Huffman, and A.P. Sattelberger, Inorg. Chem., 1984, **23**, 938.
355 J. Martin-Frere and Y. Jeannin, Inorg. Chem., 1984, **23**, 3394.
356 M.A. Fedotov, L.P. Kazanskii, and V.I. Spitsyn, Dokl. Akad. Nauk SSSR, 1983, **272**, 1179 (Chem. Abstr., 1984, **100**, 44 225).
357 R. Acerate, C.F. Hammer, and L.C.W. Baker, Inorg. Chem., 1984, **23**, 1478.
358 S.G. Sakharov, Yu.V. Kokunov, M.P. Gustyakova, and Yu.A. Buslaev, Dokl. Akad. Nauk SSSR, 1984, **276**, 148 (Chem. Abstr., 1984, **101**, 162 658).
359 K. Kanda, H. Nakatsuji, and T. Yonezawa, J. Am. Chem. Soc., 1984, **106**, 5888.
360 A.A. Koridze, O.A. Kizas, N.E. Kolobova, V.N. Vinogradova, N.A. Ustynyuk, P.V. Petrovskii, A.I. Yanovsky, and Yu.T. Struchkov, J. Chem. Soc., Chem. Commun., 1984, 1158.
361 U. Schubert, G. Kraft, and C. Kalbas, Transition Metal Chem., 1984, **9**, 161.
362 R.B. King, W.-K. Fu, and E.M. Holt, J. Chem. Soc., Chem. Commun., 1984, 1439.
363 W.A. Skupiński, J.C. Huffman, J.W. Bruno, and K.G. Caulton, J. Am. Chem. Soc., 1984, **106**, 8128.
364 T.J. Emge, T.F. Koetzle, J.W. Bruno, and K.G. Caulton, Inorg. Chem., 1984, **23**, 4012.
365 J.D. Allison, F.A. Cotton, G.L. Powell, and R.A. Walton, Inorg. Chem., 1984, **23**, 159.
366 J.D. Allison and R.A. Walton, J. Am. Chem. Soc., 1984, **106**, 163.
367 J.A. Bandy, F.G.N. Cloke, M.L.H. Green, D. O'Hare, and K. Prout, J. Chem. Soc., Chem. Commun., 1984, 240.
368 M.L.H. Green, D. O'Hare, J.A. Bandy, and K. Prout, J. Am. Chem. Soc., 1984, **106**, 884.
369 C.G. Kreiter, K.-H. Franzreb, and W.S. Sheldrick, J. Organomet. Chem., 1984, **270**, 71.
370 G. Ciani, G. D'Alfonso, M. Freni, P. Romiti, and A. Sironi, Congr. Naz. Chim. Inorg., [Atti], 15th, 1982, 86 (Chem. Abstr., 1984, **100**, 166 991).
371 G.D. Vaughn and J.A. Gladysz, Organometallics, 1984, **3**, 1596.
372 D.F.R. Gilson, P.J. Fitzpatrick, and I.S. Butler, Organometallics, 1984, **3**, 387.
373 N. Suryaprakash, A.C. Kunwar, and C.L. Khetrapal, J. Organomet. Chem., 1984, **275**, 33.
374 P.V. Petrovskii, B.A. Kvasov, A.G. Ginzburg, V.N. Setkina, and D.N. Kursanov, Izv. Akad. Nauk SSSR, Ser. Khim., 1984, 1774 (Chem. Abstr., 1984, **101**, 230 723).
375 E. Lindner and A. Brösamle, Chem. Ber., 1984, **117**, 2730.
376 E. Deschamps, F. Mathey, C. Knobler, and Y. Jeannin, Organometallics, 1984, **3**, 1144.
377 H. Blau, K.-H. Griessmann, and W. Malisch, J. Organomet. Chem., 1984, **263**, C5.
378 K.C. Brinkman and J.A. Gladysz, Organometallics, 1984, **3**, 147.
379 N.E. Kolobova, L.L. Ivanov, O.S. Zhvanko, O.M. Khitrova, A.S. Batsanov, and Yu.T. Struchkov, J. Organomet. Chem., 1984, **265**, 271.
380 M.M. Singh and R.J. Angelici, Inorg. Chem., 1984, **23**, 2699.
381 K. Henrick, J.A. Iggo, M.J. Mays, and P.R. Raithby, J. Chem. Soc., Chem.

Commun., 1984, 209.
382 R.P. Rosen, J.B. Hoke, R.R. Whittle, G.L. Geoffroy, J.P. Hutchinson, and J.A. Zubieta, Organometallics, 1984, **3**, 846.
383 J.A. Iggo, M.J. Mays, P.R. Raithby, and K. Henrick, J. Chem. Soc., Dalton Trans., 1984, 633.
384 J.A. Iggo and M.J. Mays, J. Chem. Soc., Dalton Trans., 1984, 643.
385 M.L.H. Green, D. O'Hare, and J.M. Wallis, J. Chem. Soc., Chem. Commun., 1984, 233.
386 J.A. Clark and M. Kilner, J. Chem. Soc., Dalton Trans., 1984, 389.
387 T.G. Richmond, A.M. Crespi, and D.F. Shriver, Organometallics, 1984, **3**, 314.
388 F.B. McCormick, Organometallics, 1984, **3**, 1924.
389 J.H. Merrifield, J.M. Fernández, W.E. Buhro, and J.A. Gladysz, Inorg. Chem., 1984, **23**, 4022.
390 J.C. Jeffery, A.G. Orpen, W.T. Robinson, F.G.A. Stone, and M.J. Went, J. Chem. Soc., Chem. Commun., 1984, 396.
391 J.M. Mayer and T.H. Tulip, J. Am. Chem. Soc., 1984, **106**, 3878.
392 M. de los Angeles Paz-Sandoval, P. Powell, M.G.B. Drew, and R.N. Perutz, Organometallics, 1984, **3**, 1026.
393 P.G. Lenhert, C.M. Lukehart, and K. Srinivasan, J. Am. Chem. Soc., 1984, **106**, 124.
394 M. Leyendecker and C.G. Kreiter, J. Organomet. Chem., 1984, **260**, C67.
395 M. Brookhart and A. Lukacs, J. Am. Chem. Soc., 1984, **106**, 4161.
396 C. Arsenault, P. Bougeard, B.G. Sayer, S. Yeroushalmi, and M.J. McGlinchey, J. Organomet. Chem., 1984, **265**, 283.
397 P.L. Gaus, N. Marchant, M.A. Marsinek, and M.O. Funk, Inorg. Chem., 1984, **23**, 3269.
398 D.A. Lesch, J.W. Richardson, jun., R.A. Jacobson, and R.J. Angelici, J. Am. Chem. Soc., 1984, **106**, 2901.
399 I. Bernal, J.D. Korp, W.A. Herrmann, and R. Serrano, Chem. Ber., 1984, **117**, 434.
400 W.A. Herrmann, B. Koumbouris, T. Zahn, and M.L. Ziegler, Angew. Chem., Int. Ed. Engl., 1984, **23**, 812.
401 W.A. Herrmann, R. Serrang, A. Schäfer, U. Küsthardt, M.L. Ziegler, and E. Guggolz, J. Organomet. Chem., 1984, **272**, 55.
402 W.A. Herrmann, R. Serrano, and H. Bock, Angew. Chem., Int. Ed. Engl., 1984, **23**, 382.
403 P. Pasman and J.J.M. Snel, J. Organomet. Chem., 1984, **276**, 387.
404 S.R. Stobart and M.J. Zaworotko, J. Chem. Soc., Chem. Commun., 1984, 1700.
405 R.M. Nielson and S. Wherland, Inorg. Chem., 1984, **23**, 3265.
406 M. Lattman, B.N. Anand, S.S.C. Chu, and R.D. Rosenstein, Organometallics, 1984, **3**, 670.
407 M. Lattman, B.N. Anand, S.S.C. Chu, and R.R. Rosenstein, Phosphorus Sulphur, 1983, **18**, 303.
408 S.H. Strauss, K.D. Abney, K.M. Long, and O.P. Anderson, Inorg. Chem., 1984, **23**, 1994.
409 E. Lindner, K. Auch, W. Hiller, and R. Fawzi, Angew. Chem., Int. Ed. Engl., 1984, **23**, 320.
410 R. Fawzi, W. Hiller, I.-P. Lorentz, J. Mohyla, and C. Zeiher, J. Organomet. Chem., 1984, **262**, C43.
411 P. Jaitner and P. Peringer, Transition Met. Chem., 1984, **9**, 325.
412 A. Behr, U. Kanne, and G. Thelen, J. Organomet. Chem., 1984, **269**, Cl.
413 G.D. Vaugh, K.A. Krein, and J.A. Gladysz, Angew. Chem., Int. Ed. Engl., 1984, **23**, 245.
414 M.A. Urbancic, S.R. Wilson, and J.R. Shapley, Inorg. Chem., 1984, **23**, 2954.
415 O.J. Scherer, J. Kerth, and W.S. Sheldrick, Angew. Chem., Int. Ed. Engl., 1984, **23**, 156.
416 U. Kunze, A. Bruns, and D. Rehder, J. Organomet. Chem., 1984, **268**, 213.
417 J. Gimeno, V. Riera, M.A. Ruiz, A.M.M. Lanfredi, and A. Tiripicchio, J. Organomet. Chem., 1984, **268**, C13.
418 E. Lindner, C.-P. Krieg, W. Hiller, and D. Hübner, Chem. Ber., 1984, **117**, 489.
419 J. Keijsper, J. Mul, G. v. Koten, K. Vrieze, H.C. Ubbels, and C.H. Stam,

Organometallics, 1984, **3**, 1732.
420 B.F. Hoskins, R.J. Steen, and T.W. Turney, J. Chem. Soc., Dalton Trans., 1984, 1831.
421 M.G.B. Drew, B.J. Brisdon, and A.M. Watts, Polyhedron, 1984, **3**, 1059.
422 O.J. Scherer, R. Anselmann, and W.S. Sheldrick, J. Organomet. Chem., 1984, **263**, C26.
423 D.W. Wester, D.H. White, F.W. Miller, and R.T. Dean, Inorg. Chem., 1984, **23**, 1501.
424 M.J. Abrams, A. Davison, R. Faggiani, A.G. Jones, and C.J.L. Lock, Inorg. Chem., 1984, **23**, 3284.
425 R. Bhattacharyya and P.S. Roy, Transition Met. Chem., 1984, **9**, 281.
426 J.W. Buchler, C. Dreher, and K.-L. Lay, Chem. Ber., 1984, **117**, 2261.
427 N. Hebendanz, F.H. Köhler, and G. Müller, Inorg. Chem., 1984, **23**, 3043.
428 J.L. Vanderheyden, A.R. Ketring, K. Libson, M.J. Heeg, L. Roecker, P. Motz, R. Whittle, R.C. Elder, and E. Deutsch, Inorg. Chem., 1984, **23**, 3184.
429 P. Edwards and G. Wilkinson, J. Chem. Soc., Dalton Trans., 1984, 2695.
430 P.J. Blower, J.R. Dilworth, J. Hutchinson, T. Nicholson, and J. Zubieta, Inorg. Chim. Acta, 1984, **90**, L27.
431 T.J. Barder, F.A. Cotton, D. Lewis, W. Schwotzer, S.M. Tetrick, and R.A. Walton, J. Am. Chem. Soc., 1984, **106**, 2882.
432 T.J. Barder, F.A. Cotton, G.L. Powell, S.M. Tetrick, and R.A. Walton, J. Am. Chem. Soc., 1984, **106**, 1323.
433 P.G. Edwards, A.C. Skapski, A.M.Z. Slawin, and G. Wilkinson, Polyhedron, 1984, **3**, 1083.
434 A.M. Bol'shakov, M.A. Glushkova, and Yu.A. Buslaev, Dokl. Akad. Nauk SSSR, 1983, **273**, 1134.
435 L. Baltzer, E.D. Becker, B.A. Averill, J.M. Hutchinson, and O.A. Gansow, J. Am. Chem. Soc., 1984, **106**, 2444.
436 M.L.H. Green and L.-L. Wong, J. Chem. Soc., Chem. Commun., 1984, 1442.
437 P.L. Bogdan, K.H. Whitmire, J.W. Kolis, D.F. Shriver, and E.M. Holt, J. Organomet. Chem., 1984, **272**, 169.
438 M.V. Baker and L.D. Field, J. Chem. Soc., Chem. Commun., 1984, 996.
439 C. Lapinte, D. Catheline, and D. Astruc, Organometallics, 1984, **3**, 817.
440 D.K. Liu, C.G. Brinkley, and M.S. Wrighton, Organometallics, 1984, **3**, 1449.
441 D.J. Brauer, S. Hietkamp, H. Sommer, and O. Stelzer, Angew. Chem., 1984, **96**, 696.
442 J.C. Vites, C.E. Housecroft, G.B. Jacobsen, and T.P. Fehlner, Organometallics, 1984, **3**, 1591.
443 J. Vites and T.P. Fehlner, Organometallics, 1984, **3**, 491.
444 V.D. Patel, N.J. Taylor, and A.J. Carty, J. Chem. Soc., Chem. Commun., 1984, 99.
445 E.B. Boyar and S.D. Robinson, Inorg. Chim. Acta, 1984, **90**, L31.
446 P. Zanello, S. Aime, and D. Osella, Organometallics, 1984, **3**, 1374.
447 J.-c. Liu, E. Boyar, A.J. Deeming, and S. Donovan-Mtunzi, J. Chem. Soc., Chem. Commun., 1984, 1182.
448 B. Chaudret, J. Organomet. Chem., 1984, **268**, C33.
449 B. Chaudret, G. Commenges, and R. Poilblanc, J. Chem. Soc., Dalton Trans., 1984, 1635.
450 J.W. Bruno, J.C. Huffman, and K.G. Caulton, Inorg. Chim. Acta, 1984, **89**, 167.
451 W.-K. Wong, K.W. Chiu, J.A. Statler, G. Wilkinson, M. Motevalli, and M.B. Hursthouse, Polyhedron, 1984, **3**, 1255.
452 J.C. Briggs, C.A. McAuliffe, and G. Dyer, J. Chem. Soc., Dalton Trans., 1984, 423.
453 M.I. Bruce, T.W. Hambley, M.R. Snow, and A.G. Swincer, J. Organomet. Chem., 1984, **273**, 361.
454 G. Smith and D.J. Cole-Hamilton, J. Chem. Soc., Dalton Trans., 1984, 1203.
455 D.R. Saunders and R.J. Mawby, J. Chem. Soc., Dalton Trans., 1984, 2133.
456 M.A. Bennett, T.-N. Huang, and J.L. Latten, J. Organomet. Chem., 1984, **272**, 189.
457 R.H. Morris and M. Shiralian, J. Organomet. Chem., 1984, **260**, C47.
458 M.A. Bennett and J.P. Ennett, Organometallics, 1984, **3**, 1365.

459 T.V. Ashworth, D.C. Liles, and E. Singleton, J. Chem. Soc., Chem. Commun., 1984, 1317.
460 C.E. Graimann and M.L.H. Green, J. Organomet. Chem., 1984, **275**, C12.
461 M.J. Mays, P.R. Raithby, P.L. Taylor, and K. Henrick, J. Chem. Soc., Dalton Trans., 1984, 959.
462 S. Guesmi, G. Süss-Fink, P.H. Dixneuf, N.J. Taylor, and A.J. Carty, J. Chem. Soc., Chem. Commun., 1984, 1606.
463 D. Nucclaron, N.J. Taylor, and A.J. Carty, Organometallics, 1984, **3**, 177.
464 K. Kwek, N.J. Taylor, and A.J. Carty, J. Am. Chem. Soc., 1984, **106**, 4636.
465 J.W. Bruno, J.C. Huffman, and K.G. Caulton, J. Am. Chem. Soc., 1984, **106**, 1663.
466 R. Werner and H. Werner, Chem. Ber., 1984, **117**, 161.
467 D.E. Samkoff, J.R. Shapley, M.R. Churchill, and H.J. Wasserman, Inorg. Chem., 1984, **23**, 397.
468 M. Green, A.G. Orpen, and C.J. Schaverien, J. Chem. Soc., Chem. Commun., 1984, 37.
469 J.A. Clucas, D.F. Foster, M.M. Harding, and A.K. Smith, J. Chem. Soc., Chem. Commun., 1984, 949.
470 Z. Dawoodi and M.J. Mays, J. Chem. Soc., Dalton Trans., 1984, 1931.
471 L.Y. Hsu, W.L. Hsu, D.Y. Jan, A.G. Marshall, and S.G. Shore, Organometallics, 1984, **3**, 591.
472 E. Sappa, M. Valle, G. Predieri, and A. Tiripicchio, Inorg. Chim. Acta, 1984, **88**, L23.
473 E. Pelletier and J.F. Harrod, Organometallics, 1984, **3**, 1070.
474 B. Gregory, C.R. Jablonski, and Y.P. Wang, J. Organomet. Chem., 1984, **269**, 75.
475 M.E. Wright, J.F. Hoover, G.O. Nelson, C.P. Scott, and R.S. Glass, J. Org. Chem., 1984, **49**, 3059.
476 K.C. Brinkman, A.J. Blakeney, W. Krone-Schmidt, and J.A. Gladysz, Organometallics, 1984, **3**, 1235.
477 E. Lindner, E. Schauss, W. Hiller, and R. Fawzi, Angew. Chem., Int. Ed. Engl., 1984, **23**, 711.
478 R.S. Bly and G.S. Silverman, Organometallics, 1984, **3**, 1765.
479 S. Javaheri and W.P. Giering, Organometallics, 1984, **3**, 1927.
480 W. Keim, M. Röper, H. Strutz, and C. Krüger, Angew. Chem., Int. Ed. Engl., 1984, **23**, 433.
481 T. Forschner, K. Menard, and A. Cutler, J. Chem. Soc., Chem. Commun., 1984, 121.
482 M.F. Semmelhack, R. Tamura, W. Schnatter, and J. Springer, J. Am. Chem. Soc., 1984, **106**, 5363.
483 E. Lindner, C.-P. Krieg, W. Hiller, and R. Fawzi, Angew. Chem., Int. Ed. Engl., 1984, **23**, 523.
484 S.J. Doig, R.P. Hughes, R.E. Davis, S.M. Gadol, and K.D. Holland, Organometallics, 1984, **3**, 1921.
485 P. Vierling and J.G. Riess, J. Am. Chem. Soc., 1984, **106**, 2432.
486 G.-H. Kuo, P. Helquist, and K.C. Kerber, Organometallics, 1984, **3**, 806.
487 D.L. Reger and C.A. Swift, Organometallics, 1984, **3**, 876.
488 D. Catheline, E. Románe, and D. Astruc, Inorg. Chem., 1984, **23**, 4508.
489 I.R. Butler, W.E. Lindsell, and M.J.K. Thomas, J. Organomet. Chem., 1984, **262**, 59.
490 R.S. Bly, G.S. Silverman, M.M. Hossain, and R.K. Bly, Organometallics, 1984, **3**, 642.
491 D.L. Reger and E. Mintz, Organometallics, 1984, **3**, 1759.
492 A.J. Lindsay, S. Kim, R.A. Jacobson, and R.J. Angelici, Organometallics, 1984, **3**, 1523.
493 P. Robert, B. Demerseman, and P.H. Dixneuf, Organometallics, 1984, **3**, 1771.
494 H.H. Karsch, Chem. Ber., 1984, **117**, 3123.
495 C.P. Casey and W.H. Miles, Organometallics, 1984, **3**, 808.
496 G.L. Baird, S.G. Davies, and T.R. Maberly, Organometallics, 1984, **3**, 1764.
497 H. Berke, U. Grössmann, G. Huttner, and L. Zsolnai, Chem. Ber., 1984, **117**, 3432.
498 H. Umland and U. Behrens, J. Organomet. Chem., 1983, **273**, C39.

499 G. Grötsch and W. Malisch, J. Organomet. Chem., 1984, **262**, C38.
500 J.-P. Battioni, D. Dupre, P. Guerin, and D. Mansuy, J. Organomet. Chem., 1984, **265**, 53.
501 E.O. Fischer, J. Schneider, and D. Neugebauer, Angew. Chem., Int. Ed. Engl., 1984, **23**, 820.
502 H. Le Bozec, P.H. Dixneuf, and R.D. Adams, Organometallics, 1984, **3**, 1919.
503 H. Stolzenberg, W.P. Fehlhammer, M. Monari, V. Zanotti, and L. Busetto, J. Organomet. Chem., 1984, **272**, 73.
504 E.L. Hoel, G.B. Ansell, and S. Leta, Organometallics, 1984, **3**, 1633.
505 M.E. Wright and G.O. Nelson, J. Organomet. Chem., 1984, **263**, 371.
506 J. Ros, R. Mathieu, X. Solans, and M. Font-Altaba, J. Organomet. Chem., 1984, **260**, C40.
507 E.K. Lhadi, H. Patin, and A. Darchen, Organometallics, 1984, **3**, 1128.
508 I.D. Salter and F.G.A. Stone, J. Organomet. Chem., 1984, **260**, C71.
509 M.M. Kubicki, R. Kergoat, J.Y. De Vanssay, J.Y. Le Gall, and J.E. Guerchais, Polyhedron, 1984, **3**, 365.
510 H.-U. Wekel and W. Malisch, J. Organomet. Chem., 1984, **264**, C10.
511 G. Thum and W. Malisch, J. Organomet. Chem., 1984, **264**, C5.
512 H. Moriyama, P.S. Pregosin, Y. Saito, and T. Yamakawa, J. Chem. Soc., Dalton Trans., 1984, 2329.
513 E.D. Morrison, G.R. Steinmetz, G.L. Geoffroy, W.C. Fultz, and A.L. Rheingold, J. Am. Chem. Soc., 1984, **106**, 4783.
514 J.A. Statler, G. Wilkinson, M. Thornton-Pett, and M.B. Hursthouse, J. Chem. Soc., Dalton Trans., 1984, 1731.
515 T.D. Tilley, R.H. Grubbs, and J.E. Bercaw, Organometallics, 1984, **3**, 274.
516 R.P. Tooze, M. Motevalli, M.B. Hursthouse, and G. Wilkinson, J. Chem. Soc., Chem. Commun., 1984, 799.
517 M.F. Garbauskas, J.S. Kasper, and L.N. Lewis, J. Organomet. Chem., 1984, **276**, 241.
518 D.R. Saunders, M. Stephenson, and R.J. Mawby, J. Chem. Soc., Chem. Commun., 1984, 1153.
519 K.R. Grundy and J. Jenkins, J. Organomet. Chem., 1984, **265**, 77.
520 D.H. Gibson and T.-S. Ong, Organometallics, 1984, **3**, 1911.
521 J. Evans and G.S. McNulty, J. Chem. Soc., Dalton Trans., 1984, 79.
522 D.L. Davies, S.A.R. Knox, K.A. Mead, M.J. Morris, and P. Woodward, J. Chem. Soc., Dalton Trans., 1984, 2293.
523 R. Regragui, P.H. Dixneuf, N.J. Taylor, and A.J. Carty, Organometallics, 1984, **3**, 814.
524 M.M. Fleming, A.K. Pomeroy, and P. Rushman, J. Organomet. Chem., 1984, **273**, C33.
525 T. Behling, G.S. Girolami, G. Wilkinson, R.G. Somerville, and M.B. Hursthouse, J. Chem. Soc., Dalton Trans., 1984, 877.
526 P.A. Belmonte and Z.-Y. Own, J. Am. Chem. Soc., 1984, **106**, 7493.
527 J.A. Bandy, M.L.H. Green, D. O'Hare, and K. Prout, J. Chem. Soc., Chem. Commun., 1984, 1402.
528 G.R. Steinmetz, E.D. Morrison, and G.L. Geoffroy, J. Am. Chem. Soc., 1984, **106**, 2559.
529 C.M. Jensen, C.B. Knobler, and H.D. Kaesz, J. Am. Chem. Soc., 1984, **106**, 5926.
530 C.M. Jensen, Y.J. Chen, and H.D. Kaesz, J. Am. Chem. Soc., 1984, **106**, 4046.
531 L. Milone, S. Aime, R. Gobetto, and D. Osella, Congr. Naz. Chim. Inorg., [Atti], 15th, 1982, 69 (Chem. Abstr., 1984, **100**, 113 740).
532 S. Aime, R. Gobetto, D. Osella, G.E. Hawkes, and E.W. Randall, J. Chem. Soc., Dalton Trans., 1984, 1863.
533 S. Arumugam, A.C. Kunwar, and C.L. Khetrapal, J. Organomet. Chem., 1984, **265**, 73.
534 H. Fleckner, F.-W. Grevels, and D. Hess, J. Am. Chem. Soc., 1984, **106**, 2027.
535 V. Busetti, G. Granozzi, S. Aime, R. Gobetto, and D. Osella, Organometallics, 1984, **3**, 1510.
536 M. Crocker, M. Green, A.G. Orpen, and D.M. Thomas, J. Chem. Soc., Chem. Commun., 1984, 1141.
537 J.R. Bleeke and M.K. Hays, Organometallics, 1984, **3**, 506.

538 M.L.N. Marchino, E. Sappa, A.M.M. Lanfredi, and A. Tiripicchio, J. Chem. Soc., Dalton Trans., 1984, 1541.
539 T.V. Ashworth, A.A. Chalmers, E. Meintjies, H. Oosthuizen, and E. Singleton, J. Organomet. Chem., 1984, **276**, C19.
540 N.D. Feasey, N.J. Forrow, G. Hogarth, S.A.R. Knox, K.A. Macpherson, M.J. Morris, and A.G. Orpen, J. Organomet. Chem., 1984, **267**, C41.
541 N.J. Forrow and S.A.R. Knox, J. Chem. Soc., Chem. Commun., 1984, 679.
542 P. Eilbracht, R. Jelitte, and L. Walz, Chem. Ber., 1984, **117**, 3473.
543 A.M. Madonik, D. Mandon, P. Michaud, C. Lapinte, and D. Astruc, J. Am. Chem. Soc., 1984, **106**, 3381.
544 J.A.S. Howell, J.C. Kola, D.T. Dixon, P.M. Burkinshaw, and M.J. Thomas, J. Organomet. Chem., 1984, **266**, 83.
545 C.C. Santini and F. Mathey, J. Organomet. Chem., 1984, **266**, 285.
546 M.F. Semmelhack and J.W. Herndon, J. Organomet. Chem., 1984, **265**, C15.
547 Z. Goldschmidt, S. Antebi, and I. Goldberg, J. Organomet. Chem., 1984, **260**, 105.
548 R.G. Sutherland, A. Piorko, and C.C. Lee, Organometallics, 1984, **3**, 331.
549 H. Butenschön and A. de Meijere, Angew. Chem., Int. Ed. Engl., 1984, **23**, 707.
550 A.M. Madonik and D. Astruc, J. Am. Chem. Soc., 1984, **106**, 2437.
551 D.A. Brown, N.J. Fitzpatrick, W.K. Glass, and P.K. Sayal, Organometallics, 1984, **3**, 1137.
552 N. Broadley, N.G. Connelly, R.M. Mills, M.W. Whiteley, and P. Woodward, J. Chem. Soc., Dalton Trans., 1984, 683.
553 A.J. Birch, W.D. Raverty, and G.R. Stephenson, Organometallics, 1984, **3**, 1075.
554 Z. Goldschmidt, S. Antebi, D. Cohen, and I. Goldberg, J. Organomet. Chem., 1984, **273**, 347.
555 J. Ros, X. Solans, M. Font-Altaba, and R. Mathieu, Organometallics, 1984, **3**, 1014.
556 N.G. Connelly, A.R. Lucy, R.M. Mills, M.W. Whiteley, and P. Woodward, J. Chem. Soc., Dalton Trans., 1984, 161.
557 D.N. Cox and R. Roulet, Helv. Chim. Acta, 1984, **67**, 1365.
558 R.O. Rosete, D.J. Cole-Hamilton, and G. Wilkinson, J. Chem. Soc., Dalton Trans., 1984, 2067.
559 J.W. Hull, jun., and W.L. Gladfelter, Organometallics, 1984, **3**, 605.
560 T. Nozawa, M. Hatano, M. Sato, Y. Toida, and E. Batholdi, Bull. Chem. Soc. Jpn., 1983, **56**, 3837.
561 H. Nakazawa, W.E. Buhro, G. Bertrand, and J.A. Gladysz, Inorg. Chem., 1984, **23**, 3431.
562 R.J. Ranson and R.M.G. Roberts, J. Organomet. Chem., 1984, **260**, 307.
563 R.M.G. Roberts, J. Silver, and B. Yamin, J. Organomet. Chem., 1984, **270**, 221.
564 T.E. Bitterwolf and M.J. Golightly, Inorg. Chim. Acta, 1984, **84**, 153.
565 P.D. Beer, J. Elliot, P.J. Hammond, C. Dudman, and C.D. Hall, J. Organomet. Chem., 1984, **263**, C37.
566 A.A. Koridze and N.M. Astakhova, Izv. Akad. Nauk SSSR, Ser. Khim., 1984, 711 (Chem. Abstr., 1984, **101**, 91 191).
567 M. Moll, H. Behrens, W. Poff, and P. Würstl, Z. Anorg. Allg. Chem., 1984, **516**, 127.
568 U.S. Gill, C.C. Lee, and R.G. Sutherland, Synth. React. Inorg. Met.-Org. Chem., 1984, **14**, 953.
569 L.A. Paquette, R.G. Daniels, and R. Gleiter, Organometallics, 1984, **3**, 560.
570 M.M. Singh and R.J. Angelici, Inorg. Chem., 1984, **23**, 2691.
571 E. Lindner and C.-P. Krieg, J. Organomet. Chem., 1984, **269**, 65.
572 N. Kuhn and H. Schumann, J. Organomet. Chem., 1984, **276**, 55.
573 B.P. Gracey, S.A.R. Knox, K.A. Macpherson, A.G. Orpen, and S.R. Stobart, J. Organomet. Chem., 1984, **272**, C45.
574 J.M. Shreeve and S.M. Williamson, Organometallics, 1984, **3**, 1104.
575 D. Catheline and D. Astruc, J. Organomet. Chem., 1984, **272**, 417.
576 J.R. Matachek, R.J. Angelici, K.A. Schugart, K.J. Haller, and R.F. Fenske, Organometallics, 1984, **3**, 1038.

577 H.R. Allcock, K.D. Lavin, G.H. Riding, and R.R. Whittle, Organometallics, 1984, **3**, 663.
578 H.R. Allcock, K.D. Lavin, G.H. Riding, P.R. Suszko, and R.R. Whittle, J. Am. Chem. Soc., 1984, **106**, 2337.
579 E.I. Klimova, A.N. Pushin, and V.A. Sazonova, J. Organomet. Chem., 1984, **270**, 319.
580 M. Herberhold and L. Haumaier, Angew. Chem., Int. Ed. Engl., 1984, **23**, 521.
581 K.-P. Stahl, G. Boche, and W. Massa, J. Organomet. Chem., 1984, **277**, 113.
582 S. Akabori, Y. Habata, and M. Sato, Bull. Chem. Soc. Jpn., 1984, **57**, 68.
583 M. Hisatome, J. Watanabe, and K. Yamakawa, J. Organomet. Chem., 1984, **266**, 159.
584 M. Hisatome, J. Watanabe, K. Yamakawa, K. Kozawa, and T. Uchida, J. Organomet. Chem., 1984, **262**, 365.
585 B. Deschamps, F. Mathey, J. Fischer, and J.H. Nelson, Inorg. Chem., 1984, **23**, 3455.
586 M. Sato, M. Kubo, S. Ebine, and S. Akabori, Bull. Chem. Soc. Jpn., 1984, **57**, 421.
587 J.F. Helling and U.S. Gill, J. Organomet. Chem., 1984, **264**, 353.
588 V. Guerchais and D. Astruc, J. Chem. Soc., Chem. Commun., 1984, 881.
589 D. Catheline and D. Astruc, Organometallics, 1984, **3**, 1094.
590 C.C. Lee, U.S. Gill, and R.G. Sutherland, J. Organomet. Chem., 1984, **267**, 157.
591 T.V. Ashworth, A.A. Chalmers, D.C. Liles, E. Meintjies, H.E. Oosthuizen, and E. Singleton, J. Organomet. Chem., 1984, **276**, C49.
592 A. Stasunik, D.R. Wilson, and W. Malisch, J. Organomet. Chem., 1984, **270**, C18.
593 S.G. Davies and S.J. Simpson, J. Chem. Soc., Dalton Trans., 1984, 993.
594 E. Cesarotti, A. Chiesa, G.F. Ciani, A. Sironi, R. Vefghi, and C. White, J. Chem. Soc., Dalton Trans., 1984, 653.
595 A. Stasunik and W. Malisch, J. Organomet. Chem., 1984, **270**, C56.
596 F. Morandini, G. Consiglio, G. Ciani, and A. Sironi, Inorg. Chim. Acta, 1984, **82**, L27.
597 M.I. Bruce, M.G. Humphreys, J.M. Patrick, and A.H. White, Aust. J. Chem., 1983, **36**, 2065.
598 M.I. Bruce, M.G. Humphreys, A.G. Swincer, and R.C. Wallis, Aust. J. Chem., 1984, **37**, 1747.
599 G.J. Baird and S.G. Davies, J. Organomet. Chem., 1984, **262**, 215.
600 S. Akabori, Y. Habata, H. Muneguni, and M. Sato, Tetrahedron Lett., 1984, **25**, 1991.
601 M. Crocker, M. Green, A.G. Orpen, H.-P. Neumann, and C.J. Schaverien, J. Chem. Soc., Chem. Commun., 1984, 1351.
602 H. Werner, H. Kletzin, and Ch. Burschka, J. Organomet. Chem., 1984, **276**, 231.
603 R.O. Gould, T.A. Stephenson, and D.A. Tocher, J. Organomet. Chem., 1984, **263**, 375.
604 R. Regragui, P.H. Dixneuf, N.J. Taylor, and A.J. Carty, Organometallics, 1984, **3**, 1020.
605 G.-Y. Lin and J. Takats, J. Organomet. Chem., 1984, **269**, C4.
606 G. Albertin, S. Antoniutti, and E. Bordignon, J. Chem. Soc., Chem. Commun., 1984, 1688.
607 A.L. Rheingold and M.A. Fountain, Organometallics, 1984, **3**, 1417.
608 R.H. Neilson, R.J. Thoma, I. Vickovic, and W.H. Watson, Organometallics, 1984, **3**, 1132.
609 A.H. Cowley, J.E. Kilduff, J.G. Lasch, N.C. Norman, M. Pakulski, F. Ando, and T.C. Wright, Organometallics, 1984, **3**, 1044.
610 V. Riera, J. Ruiz, and F. Mayor, J. Organomet. Chem., 1984, **276**, 47.
611 D. Seyferth, T.G. Wood, J.P. Fackler, jun., and A.M. Mazany, Organometallics, 1984, **3**, 1121.
612 T. Rukachalsirikul, S. Arabi, F. Hartstock, N.J. Taylor, and A.J. Carty, Organometallics, 1984, **3**, 1587.
613 A.A. Bahsoun, J.A. Osborn, P.H. Bird, D. Nucciarone, and A.V. Peters, J. Chem. Soc., Chem. Commun., 1984, 72.

614 D. Touchard, P.H. Dixneuf, R.D. Adams, and B.E. Segmüller, Organometallics, 1984, **3**, 640.
615 H.-W. Frühauf and J. Breuer, J. Organomet. Chem., 1984, **277**, C13.
616 M. Herberhold and W. Bühlmeyer, Angew. Chem., Int. Ed. Engl., 1984, **23**, 80.
617 Y.-F. Yu, J. Gallucci, and A. Wojcicki, J. Chem. Soc., Chem. Commun., 1984, 653.
618 H.R. Allcock, P.R. Suszko, L.J. Wagner, R.R. Whittle, and B. Boso, J. Am. Chem. Soc., 1984, 4966.
619 S.I. Al-Resayes, P.B. Hitchcock, M.F. Meidine, and J.F. Nixon, J. Chem. Soc., Chem. Commun., 1984, 1080.
620 D. Seyferth, G.B. Womack, M. Cowie, and B.W. Hames, Organometallics, 1984, **3**, 1891.
621 I. Bruedgam, H. Hartl, and D. Lentz, Z. Naturforsch., Teil B, 1984, **39**, 721 (Chem. Abstr., 1984, **101**, 130 845).
622 K.H. Whitmore, C.B. Lagrone, M.R. Churchill, J.C. Fettinger, and L.V. Biondi, Inorg. Chem., 1984, **23**, 4227.
623 J.P. Fackler, jun., A.M. Mazany, D. Seyferth, H.P. Withers, jun., T.G. Wood, and C.F. Campana, Inorg. Chim. Acta, 1984, **82**, 31.
624 A.W. Coleman, D.F. Jones, P.H. Dixneuf, C. Brisson, J.-J. Bonnet, and G. Lavigne, Inorg. Chem., 1984, **23**, 952.
625 K. Joseph, S. Gopinathan, and C. Gopinathan, Synth. React. Inorg. Met.-Org. Chem., 1984, **14**, 1005.
626 M. Pizzotti, C. Crotti, and F. Demartin, J. Chem. Soc., Dalton Trans., 1984, 735.
627 E. Roland and H. Vahrenkamp, Chem. Ber., 1984, **117**, 1039.
628 A.G. Cowie, B.F.G. Johnson, J. Lewis, and P.R. Raithby, J. Chem. Soc., Chem. Commun., 1984, 1710.
629 B.P. Sulliivan, J.V. Caspar, S.R. Johnson, and T.J. Meyer, Organometallics, 1984, **3**, 1241.
630 A.J. Deeming, S. Donovan-Mtunzi, and S.E. Kabir, J. Organomet. Chem., 1984, **276**, C65.
631 R.D. Adams and T.S.A. Hor, Inorg. Chem., 1984, **23**, 4723.
632 R.D. Adams, T.S.A. Hor, and I.T. Horváth, Inorg. Chem., 1984, **23**, 4733.
633 A.R. Butler, C. Glidewell, A.R. Hyde, and J. McGinnis, Polyhedron, 1984, **3**, 1165.
634 S. Komiya, M. Katoh, T. Ikariya, R.H. Grubbs, T. Yamamoto, and A. Yamamoto, J. Organomet. Chem., 1984, **260**, 115.
635 J.K. Barton and J.S. Nowick, J. Chem. Soc., Chem. Commun., 1984, 1650.
636 J. Mason, D.M.P. Mingos, D. Sherman, and R.W.M. Wardle, J. Chem. Soc., Chem. Commun., 1984, 1223.
637 M. Le Plouzennec, A. Bondon, and G. Simmoneaux, Inorg. Chem., 1984, **23**, 4398.
638 R.R. Ruminski, K.D. Van Tassel, and J.D. Petersen, Inorg. Chem., 1984, **23**, 4380.
639 Y.-F. Yu, C.-N. Chau, A. Wojcicki, M. Calligaris, G. Nardin, and G. Balducci, J. Am. Chem. Soc., 1984, **106**, 3704.
640 G. Albertin and E. Bordignon, Inorg. Chem., 1984, **23**, 3822.
641 G.J. Organ, M.K. Cooper, K. Henrick, and M. McPartlin, J. Chem. Soc., Dalton Trans., 1984, 2377.
642 M.J. Cook, A.P. Lewis, and G.S.G. McAuliffe, Org. Magn. Reson., 1984, **22**, 388.
643 W.P. Griffith, N.T. McManus, and A.C. Skapski, J. Chem. Soc., Chem. Commun., 1984, 434.
644 K. Yoshinaga, T. Kito, and K. Ohkubo, J. Chem. Soc., Perkin Trans. 2, 1984, 469.
645 J. Gotzig, A.L. Rheingold, and H. Werner, Angew. Chem., Int. Ed. Engl., 1984, **23**, 814.
646 J. Gotzig, A.L. Rheingold, and H. Werner, Angew. Chem., Int. Ed. Engl., 1984, **23**, 814.
647 J.A. Ladd, H. Hope, and A.L. Balch, Organometallics, 1984, **3**, 1838.
648 M.O. Albers, D.C. Liles, E. Singleton, and J.E. Yates, J. Organomet. Chem., 1984, **272**, C62.

649 C.W. Jung, P.E. Garrou, P.R. Hoffman, and K.G. Caulton, Inorg. Chem., 1984, **23**, 726.

650 S.M. Bucknor, M. Draganjac, T.B. Rauchfuss, C.J. Ruffing, W.C. Fultz, and A.L. Rheingold, J. Am. Chem. Soc., 1984, **106**, 5379.

651 A.R. Chakravarty, F.A. Cotton, and W. Schwotzer, Inorg. Chem., 1984, **23**, 99.

652 P. Laszlo, 'Cobalt-59', NMR Newly Accessible Nucl., 1983, **2**, 253. Ed. P. Laszlo, Academic Press, New York (Chem. Abstr., 1984, **100**, 166 568).

653 B.E. Mann, 'Rhodium-103', NMR Newly Accessible Nucl., 1983, **2**, 301. Ed. P. Laszlo, Academic Press, New York (Chem. Abstr., 1984, **100**, 166 570).

654 J.J. Villafranca, Phosphorus-31 NMR, 1984, 155. Ed. D.G. Gorenstein, Academic Press, Orlando, Fla.

655 R. Bonnaire, D. Davoust, and N. Platzer, Org. Magn. Reson., 1984, **22**, 80.

656 V.V. Mainz and R.A. Andersen, Organometallics, 1984, **3**, 675.

657 J.S. Ricci, T.F. Koetzle, R.J. Goodfellow, P. Espinet, and P.M. Maitlis, Inorg. Chem., 1984, **23**, 1828.

658 R.J. Barton, D.G. Holah, H. Shengzhi, A.N. Hughes, S.I. Khan, and B.E. Robertson, Inorg. Chem., 1984, **23**, 2391.

659 M.J. Fernandez, P.M. Bailey, P.O. Bentz, J.S. Ricci, T.F. Koetzle, and P.M. Maitlis, J. Am. Chem. Soc., 1984, **106**, 5458.

660 X. You, R.S. Drago, and J.G. Miller, Huaxue Xuebao, 1984, **42**, 14 (Chem. Abstr., 1984, **100**, 131 445).

661 R.A. Periana and R.G. Bergman, J. Am. Chem. Soc., 1984, **106**, 7272.

662 W.P. Griffith, M.J. Mockford, and A.C. Skapski, J. Chem. Soc., Chem. Commun., 1984, 407.

663 A.M. Mueting, P. Boyle, and L.H. Pignolet, Inorg. Chem., 1984, **23**, 44.

664 M.D. Fryzuk, T. Jones, and F.W.B. Einstein, Organometallics, 1984, **3**, 185.

665 C. Woodcock and R. Eisenberg, Inorg. Chem., 1984, **23**, 4207.

666 H. Lehner, D. Matt, A. Togni, R. Thouvenot, L.M. Venanzi, and A. Albinati, Inorg. Chem., 1984, **23**, 4254.

667 L.F. Rhodes, J.C. Huffman, and K.G. Caulton, J. Am. Chem. Soc., 1984, **106**, 6874.

668 M.-J. Fernandez and P.M. Maitlis, J. Chem. Soc., Dalton Trans., 1984, 2063.

669 E.A.V. Ebsworth, P.G. Page, and D.W.H. Rankin, J. Chem. Soc., Dalton Trans., 1984, 2569.

670 E. Arpac, F. Mirzaei, A. Yardimcioglu, and L. Dahlenburg, Z. Anorg. Allg. Chem., 1984, **519**, 148.

671 M.J. Wax, J.M. Stryker, J.M. Buchanan, C.A. Kovac, and R.G. Bergman, J. Am. Chem. Soc., 1984, **106**, 1121.

672 T.R.B. Mitchell, J. Organomet. Chem., 1984, **270**, 245.

673 W.J. Youngs, B.L. Simms, and J.A. Ibers, J. Organomet. Chem., 1984, **272**, 295.

674 M.A. Lilga and J.A. Ibers, Inorg. Chem., 1984, **23**, 3538.

675 B. Kräutler and C. Caderas, Helv. Chim. Acta, 1984, **67**, 189.

676 M.J. Doyle, M.F. Lappert, P.L. Pye, and P. Terreros, J. Chem. Soc., Dalton Trans., 1984, 2355.

677 A.R. Sanger, Can. J. Chem., 1984, **62**, 822.

678 K. Isobe, S. Okeya, N.J. Meanwell, A.J. Smith, H. Adams, and P.M. Maitlis, J. Chem. Soc., Dalton Trans., 1984, 1215.

679 G.T. Bratt and H.P.C. Hogenkamp, Biochemistry, 1984, **23**, 5653 (Chem. Abstr., 1984, **101**, 211 442).

680 K.W. Chiu, G. Wilkinson, M. Thornton-Pett, and M.B. Hursthouse, Polyhedron, 1984, **3**, 79.

681 R.B. Cracknell, A.G. Orpen, and J.L. Spencer, J. Chem. Soc., Chem. Commun., 1984, 326.

682 H.H. Karsch, Chem. Ber., 1984, **117**, 783.

683 D.J. Pasto and D.A. Timmers, Inorg. Chem., 1984, **23**, 4115.

684 K.L. Brown, J.M. Hakimi, D.M. Nuss, Y.D. Montejano, and D.W. Jacobsen, Inorg. Chem., 1984, **23**, 1463.

685 E.N. Jacobsen and R.G. Bergman, Organometallics, 1984, **3**, 329.

686 D. Wormsbächer, R. Drews, F. Edelmann, and U. Behrens, J. Organomet. Chem., 1984, **270**, 93.

687 A.R. Siedle, R.A. Newmark, and L.H. Pignolet, Organometallics, 1984, **3**, 855.

688 J.M. Mayer and J.C. Calabrese, Organometallics, 1984, **3**, 1292.
689 H. Werner, W. Paul, and R. Zolk, Angew. Chem., Int. Ed. Engl., 1984, **23**, 626.
690 L.A. Oro, M. Esteban, R.M. Claramunt, J. Elguero, C. Foces-Foces, and F.H. Cano, J. Organomet. Chem., 1984, **276**, 79.
691 L. Dahlenburg and C. Prengel, Organometallics, 1984, **3**, 934.
692 L. Dahlenburg, A. Yardimcioglu, and N. Höck, Inorg. Chim. Acta, 1984, **89**, 213.
693 S. Sprouse, K.A. King, P.J. Spellane, and R.J. Watts, J. Am. Chem. Soc., 1984, **106**, 6647.
694 M. Nonoyama, J. Organomet. Chem., 1984, **262**, 407.
695 F.L. Barceló, J.C. Besteiro, P. Lahuerta, C. Foces-Foces, F.H. Cano, and M. Martinez-Ripoll, J. Organomet. Chem., 1984, **270**, 343.
696 M.L. Deem, J. Organomet. Chem., 1984, **267**, 191.
697 J.W. Suggs and C.-H. Jun, J. Am. Chem. Soc., 1984, **106**, 3054.
698 P.L. Burk, D. Van Engen, and K.S. Campo, Organometallics, 1984, **3**, 493.
699 N.G. Connelly, M.J. Freeman, I. Manners, and A.G. Orpen, J. Chem. Soc., Dalton Trans., 1984, 2703.
700 H. Werner, J. Wolf, G. Müller, and C. Krüger, Angew. Chem., Int. Ed. Engl., 1984, **23**, 431.
701 W.A. Herrmann, C. Weber, M.L. Ziegler, and C. Pahl, Chem. Ber., 1984, **117**, 875.
702 K. Isobe, A.V. de Miguel, A. Nutton, and P.M. Maitlis, J. Chem. Soc., Dalton Trans., 1984, 929.
703 R.H. Crabtree and R.P. Dion, J. Chem. Soc., Chem. Commun., 1984, 1260.
704 E. Arpac and L. Dahlenburg, J. Organomet. Chem., 1984, **277**, 127.
705 J.C. Calabrese, D.C. Roe, D.L. Thorn, and T.H. Tulip, Organometallics, 1984, **3**, 1223.
706 D.L. Thorne and J.C. Calabrese, J. Organomet. Chem., 1984, **272**, 283.
707 P.S. Braterman, G.A. Heath, A.J. Mackenzie, B.C. Noble, R.D. Peacock, and L.J. Yellowlees, Inorg. Chem., 1984, **23**, 3425.
708 D.A. Wrobleski, C.S. Day, B.A. Goodman, and T.B. Rauchfuss, J. Am. Chem. Soc., 1984, **106**, 5464.
709 S. Aime, D. Osella, E. Giamello, and G. Granozzi, J. Organomet. Chem., 1984, **262**, Cl.
710 M.P. Doyle, M.R. Colsman, and M.S. Chinn, Inorg. Chem., 1984, **23**, 3684.
711 J.T. Mague, C.L. Klein, R.J. Majeste, and E.D. Stevens, Organometallics, 1984, **3**, 1860.
712 W.J. Laws and R.J. Puddephatt, J. Chem. Soc., Chem. Commun., 1984, 116.
713 K. Kellner, W. Hanke, A. Tzschach, Z. Nagy-magos, and L. Markó, J. Organomet. Chem., 1984, **268**, 175.
714 C. Vaccher, A. Mortreux, F. Petit, J.-P. Picavet, H. Sliwa, N.W. Murrall, and A.J. Welch, Inorg. Chem., 1984, **23**, 3613.
715 J.W. Fitch and D. Westmoreland, J. Organomet. Chem., 1984, **268**, 269.
716 M. Cowie, R.S. Dickson, and B.W. Hames, Organometallics, 1984, **3**, 1879.
717 B.R. Sutherland and M. Cowie, Organometallics, 1984, **3**, 1869.
718 S. Padmanabhan and K.M. Nicholas, J. Organomet. Chem., 1984, **268**, C23.
719 P. Barabotti, P. Diversi, G. Ingrosso, A. Lucherini, and F. Nuti, J. Chem. Soc., Dalton Trans., 1984, 2517.
720 M.D. Fryzuk, T. Jones, and F.W.B. Einstein, J. Chem. Soc., Chem. Commun., 1984, 1556.
721 B.D. Murray and P.P. Power, Organometallics, 1984, **3**, 1199.
722 C. Bianchini, C. Mealli, A. Meli, and M. Sabat, Inorg. Chem., 1984, **23**, 4125.
723 R.S. Dickson, R.J. Nesbit, H. Pateras, J.M. Patrick, and A.H. White, J. Organomet. Chem., 1984, **265**, C25.
724 H. Bönnemann, W. Brijoux, R. Brinkmann, W. Meurers, R. Mynott, W. v. Philipsborn, and T. Egolf, J. Organomet. Chem., 1984, **272**, 271.
725 D.C. Dittmer, E.J. Parker, and J.R. Bodwell, Org. Magn. Reson., 1984, **22**, 609.
726 H.M. Colquhoun, S.M. Doughty, J.F. Stoddard, and D.J. Williams, Angew. Chem., Int. Ed. Engl., 1984, **23**, 235.

727 J. Elguero, M. Esteban, M.F. Grenier-Loustalot, L.A. Oro, and M.T. Pinillos, J. Chim. Phys. Phys.-Chim. Biol., 1984, **81**, 251.
728 R.H. Crabtree and C.P. Parnell, Organometallics, 1984, **3**, 1727.
729 G.A. Ville, K.P.C. Vollhardt, and M.J. Winter, Organometallics, 1984, **3**, 1177.
730 H. Bonnemann, W. Brijoux, R. Brinkmann, and W. Meurers, Helv. Chim. Acta, 1984, **67**, 1616.
731 K.-J. Jens and E. Weiss, Chem. Ber., 1984, **117**, 2469.
732 R. Aneja, B.T. Golding, and C. Pierpoint, J. Chem. Soc., Dalton Trans., 1984, 219.
733 B.D. Murray, H. Hope, J. Hvoslef, and P.P. Power, Organometallics, 1984, **3**, 657.
734 I. Ojima, Pure Appl. Chem., 1984, **56**, 99.
735 C. Claver, J.C. Rodriguez, and A. Ruiz, Transition Met. Chem., 1984, **9**, 83.
736 B.A. Boyce, A. Carroy, J.-M. Lehn, and D. Parker, J. Chem. Soc., Chem. Commun., 1984, 1586.
737 C. Valentini, E. Cernia, M. Fiorini, and G.M. Giongo, J. Mol. Catalysis, 1984, **23**, 81.
738 R.T. Smith, R.K. Ungar, L.J. Sanderson, and M.C. Baird, Ann. N.Y. Acad. Sci., 1983, **415**, 253 (Chem. Abstr., 1984, **101**, 12 751).
739 I.T. Chizhevsky, N.E. Kolobova, N.V. Rastova, V.F. Sizoi, P.V. Petrovskii, V.G. Andrianov, and Yu.T. Struchkov, J. Organomet. Chem., 1984, **262**, 81.
740 M.D. Jones, R.D.W. Kemmitt, A.W.G. Platt, D.R. Russell, and L.J.S. Sherry, J. Chem. Soc., Chem. Commun., 1984, 673.
741 J.L. Atwood, K.A. Beveridge, G.W. Bushnell, K.R. Dixon, D.T. Eadie, S.R. Stobart, and M.J. Zaworotko, Inorg. Chem., 1984, **23**, 4050.
742 P.G. Rasmussen, O.H. Bailey, J.C. Bayón, and W.M. Butler, Inorg. Chem., 1984, **23**, 343.
743 G.W. Bushnell, S.R. Stobart, R. Vefghi, and M.J. Zaworotko, J. Chem. Soc., Chem. Commun., 1984, 282.
744 M.J. Burk, R.H. Crabtree, and E.M. Holt, Organometallics, 1984, **3**, 638.
745 R. Uson, L.A. Oro, D. Carmona, M.J. Fernandez, and P. Lahuerta, Rev. Acad. Cienc. Exactas, Fis., Quim. Nat. Zaragoza, 1982, **37**, 61 (Chem. Abstr., 1984, **100**, 113 879).
746 M.E. Gross, C.E. Johnson, M.J. Maroney, and W.C. Trogler, Inorg. Chem., 1984, **23**, 2968.
747 E.J. Miller, A.L. Rheingold, and T.B. Brill, J. Organomet. Chem., 1984, **273**, 377.
748 W.A. Herrmann, C. Bauer, and J. Weichmann, Chem. Ber., 1984, **117**, 1271.
749 P. Powell, J. Organomet. Chem., 1984, **266**, 307.
750 S. Clamp, N.G. Connelly, J.A.K. Howard, I. Manners, J.D. Payne, and W.E. Geiger, J. Chem. Soc., Dalton Trans., 1984, 1659.
751 J. Jeffery, E.J. Probitts, and R.J. Mawby, J. Chem. Soc., Dalton Trans., 1984, 2423.
752 W.A. Herrmann, G.W. Kriechbaum, C. Bauer, H. Pfisterer, E. Guggolz, and M.L. Ziegler, J. Organomet. Chem., 1984, **262**, 89.
753 H. Werner, B. Klingert, R. Zolk, and P. Thometzek, J. Organomet. Chem., 1984, **266**, 97.
754 S.L. Grundy, A.J. Smith, H. Adams, and P.M. Maitlis, J. Chem. Soc., Dalton Trans., 1984, 1747.
755 J. Meunier-Piret, M. Van Meerssche, M. Gielen, and I. Vanden Eynde, Bull. Soc. Chim. Belg., 1984, **93**, 307.
756 A. Ceriotti, L. Garlaschelli, G. Longoni, M.C. Malatesta, D. Strumolo, A. Fumagalli, and S. Martinengo, J. Mol. Catal., 1984, **24**, 309.
757 D.J. Darensbourg, D.J. Zalewski, and T. Delord, Organometallics, 1984, **3**, 1210.
758 P. Braunstein, C.d.M. de Bellefon, and M. Ries, J. Organomet. Chem., 1984, **262**, C14.
759 A. Maisonnat and R. Poilblanc, C.R. Acad. Sci., Ser. 2, 1984, **298**, 69.
760 A.T. Teleshev, G.M. Grishina, A.A. Borisenko, N.N. Nevskii, and E.E. Nifant'ev, Zh. Obshch. Khim., 1984, **54**, 1710 (Chem. Abstr., 1984, **101**, 203 228).

761 A.R. Sanger, Can. J. Chem., 1984, **62**, 2168.
762 K. Joseph, S.A. Pardhy, S.K. Pandit, S. Gopinathan, and C. Gopinathan, Inorg. Chim. Acta, 1984, **84**, 149.
763 J.C.T.R.B. St. Laurent, J. Sinclair, and J.F. Nixon, J. Organomet. Chem., 1984, **262**, 379.
764 J. Jeanneaux, J.G. Riess, and J. Wachter, Inorg. Chem., 1984, **23**, 3036.
765 B. Giovannitti, M. Ghedini, and G. Dolcetti, Transition Met. Chem., 1984, **9**, 54.
766 D.W. Abbott and C. Woods, Inorg. Chem., 1984, **23**, 3626.
767 W. Willis and K.M. Nicholas, Inorg. Chim. Acta, 1984, **90**, L51.
768 A. Jegorov, B. Kratochvíl, V. Langer, and J. Podlahová, Inorg. Chem., 1984, **23**, 4288.
769 J.A.E. Gibson and M. Cowie, Organometallics, 1984, **3**, 984.
770 J.A. Ladd, M.M. Olmstead, and A.L. Balch, Inorg. Chem., 1984, **23**, 2318.
771 S.P. Deraniyagala and K.R. Grundy, Inorg. Chim. Acta, 1984, **84**, 205.
772 J.A.E. Gibson and M. Cowie, Organometallics, 1984, **3**, 722.
773 A.L. Balch, M.M. Olmstead, and R.R. Guimerans, Inorg. Chim. Acta, 1984, **84**, L21.
774 B.C. Whitmore and R. Eisenberg, Inorg. Chem., 1984, **23**, 1697.
775 J. Ellermann, G. Szucsányi, K. Geibel, and E. Wilhelm, J. Organomet. Chem., 1984, **263**, 297.
776 J. Wachter, F. Jeanneaux, G. Le Borgne, and J.G. Riess, Organometallics, 1984, **3**, 1034.
777 H. Schumann, G. Cielusek, S. Jurgis, E. Hahn, J. Pickardt, J. Blum, Y. Sasson, and A. Zoran, Chem. Ber., 1984, **117**, 2825.
778 R.R. Guimerans, F.E. Wood, and A.L. Balch, Inorg. Chem., 1984, **23**, 1308.
779 D. Galli, L. Garlaschelli, G. Ciani, A. Fumagalli, S. Martinengo, and A. Sironi, J. Chem. Soc., Dalton Trans., 1984, 55.
780 A.L. Balch, R.R. Guimerans, and M.M. Olmstead, J. Organomet. Chem., 1984, **268**, C38.
781 R.J. Crowte, J. Evans, and M. Webster, J. Chem. Soc., Chem. Commun., 1984, 1344.
782 D.W. Stephan, Inorg. Chem., 1984, **23**, 2207.
783 E.A.V. Ebsworth, R.O. Gould, N.T. McManus, N.J. Pilkington, and D.W.H. Rankin, J. Chem. Soc., Dalton Trans., 1984, 2561.
784 A.L. Bandini, G. Banditelli, F. Bonati, G. Minghetti, F. Demartin, and M. Manassero, J. Organomet. Chem., 1984, **269**, 91.
785 E.A.V. Ebsworth, N.J. Pilkington, D.W.H. Rankin, and J.H. Holloway, Angew. Chem., Int. Ed. Engl., 1984, **23**, 30.
786 B.R. Sutherland and M. Cowie, Inorg. Chem., 1984, **23**, 2324.
787 B.J. Fisher and R. Eisenberg, Inorg. Chem., 1984, **23**, 3216.
788 R. Ros, F. Canziani, A. Scrivanti, and L. Garlaschelli, Congr. Naz. Chim. Inorg., [Atti], 15th, 1982, 78 (Chem. Abstr., 1984, **100**, 166 990).
789 J.A. Clucas, M.M. Harding, B.S. Nicholls, and A.K. Smith, J. Chem. Soc., Chem. Commun., 1984, 319.
790 P.G. Rasmussen. O.H. Bailey, and J.C. Bayon, J. Am. Chem. Soc., 1984, **23**, 338.
791 R. Ros, F. Canziani, and R. Roulet, J. Organomet. Chem., 1984, **267**, C9.
792 V.P. Chacko and R.G. Bryant, J. Magn. Reson., 1984, **57**, 79.
793 S.H. Peterson, R.G. Bryant, and J.G. Russell, Anal. Chim. Acta, 1983, **154**, 211 (Chem. Abstr., 1984, **100**, 61 017).
794 Y. Yoshikawa, Y. Masuda, H. Yamatera, S. Utsuno, and Y. Nakashima, J. Magn. Reson., 1984, **58**, 473.
795 J. Bultitude, L.F. Larkworthy, J. Mason, D.C. Povey, and B. Sandell, Inorg. Chem., 1984, **23**, 3629.
796 J. Ribas, J.M. Julia, X. Solans, M. Font-Altaba, A. Isalgue, and X. Tejada, Transition Met. Chem., 1984, **9**, 57.
797 T.P. Haromy, P.F. Gilletti, R.D. Cornelius, and M. Sundaralingam, J. Am. Chem. Soc., 1984, **106**, 2812.
798 G. Brewer, S. Drucker, J. Girard, C. Grisham, and E. Sinn, Inorg. Chim. Acta, 1984, **89**, 105.
799 J. Reibenspies and R.D. Cornelius, Inorg. Chem., 1984, **23**, 1563.

800 I. Lin, A. Hsueh, and D. Dunaway-Mariano, Inorg. Chem., 1984, **23**, 1692.
801 I. Lin and D. Dunaway-Mariano, J. Am. Chem. Soc., 1984, **106**, 6074.
802 Y. Yamamoto and E. Toyota, Bull. Chem. Soc. Jpn., 1984, **57**, 2650.
803 S.C.F. Au-Yeung and D.R. Eaton, Can. J. Chem., 1984, **62**, 2127.
804 M. Strasak and F. Bachraty, J. Coord. Chem., 1984, **13**, 105.
805 R.J.Geue, M.G. McCarthy, and A.M. Sargeson, J. Am. Chem. Soc., 1984, **106**, 8282.
806 M. Yamaguchi, M. Saburi, and S. Yoshikawa, J. Am. Chem. Soc., 1984, **106**, 8293.
807 R. Saito and Y. Kidani, Bull. Chem. Soc. Jpn., 1984, **57**, 3430.
808 H. Ogino and K. Ohata, Inorg. Chem., 1984, **23**, 3312.
809 K. Yamanari and Y. Shimura, Chem. Lett., 1984, 761.
810 N.R. Brodsky, N.M. Nguyen, N.S. Rowan, C.B. Storm, R.J. Butcher, and E. Sinn, Inorg. Chem., 1984, **23**, 891.
811 A. Hammershoi, A.M. Sargeson, and W.L. Steffen, J. Am. Chem. Soc., 1984, **106**, 2819.
812 K. Yamanari, N. Takeshita, and Y. Shimura, Bull. Chem. Soc. Jpn., 1984, **57**, 1227.
813 Y. Yamamoto and E. Toyota, Bull. Chem. Soc. Jpn., 1984, **57**, 47.
814 K. Yamanari, N. Takeshita, and Y. Shimura, Bull. Chem. Soc. Jpn., 1984, **57**, 2852.
815 A. Watson, A.R. Gainsford, and D.A. House, Inorg. Chim. Acta, 1984, **86**, L11.
816 V.P. Tarasov, T.Sh. Kapanadze, I.B. Baranovskii, G.V. Tsintsadze, S.G. Drobyshev, and Yu.A. Buslaev, Koord. Khim., 1984, **10**, 368 (Chem. Abstr., 1984, **100**, 220 534).
817 R.J. Abraham, J. Plant, G.R. Bedford, and B. Wright, Org. Magn. Reson., 1984, **22**, 57.
818 L. Ernst, J. Chem. Soc., Perkin Trans. 1, 1984, 2267.
819 A.R. Battersby, C. Edlington, and C.J.R. Fookes, J. Chem. Soc., Chem. Commun., 1984, 527.
820 C.A. Chang, E.N. Duesler, and R.E. Tapscott, Inorg. Chim. Acta, 1984, **85**, 73.
821 K.L. Brown, E.P. Kirven, and L.Y. Lu, Inorg. Chim. Acta, 1984, **84**, 199.
822 M.F. Gargallo, R.E. Tapscott, and E.N. Duesler, Inorg. Chem., 1984, **23**, 918.
823 C.D. Garner, M. Berry, and B.E. Mann, Inorg. Chem., 1984, **23**, 1500.
824 L. Walder, G. Rytz, U. Vögeli, R. Scheffold, and P. Engel, Helv. Chim. Acta, 1984, **67**, 1801.
825 I.I. Creaser, J.D. Lydon, A.M. Sargeson, E. Horn, and M.R. Snow, J. Am. Chem. Soc., 1984, **106**, 5729.
826 R.J. Geue, T.W. Hambley, J.M. Harrowfield, A.M. Sargeson, and M.R. Snow, J. Am. Chem. Soc., 1984, **106**, 5478.
827 M. Kanesaka, K.-i. Okamoto, M. Nomoto, and J. Hidaka, Bull. Chem. Soc. Jpn., 1984, **57**, 3111.
828 C. Bianchini, D. Masi, C. Mealli, A. Meli, M. Sabat, and G. Scapacci, J. Organomet. Chem., 1984, **273**, 91.
829 J. Giusti, S. Chimichi, M. Ciampolini, M. Sabat, and D. Masi, Inorg. Chim. Acta, 1984, **88**, 51.
830 M. Suzuki, O. Arisato, K.-i. Okamoto, H. Einaga, and J. Hidaka, Bull. Chem. Soc. Jpn., 1984, **57**, 2751.
831 G.R. Brubaker and D.W. Johnson, Inorg. Chim. Acta, 1984, **83**, 155.
832 D.R. Eaton and J.D. Weegar, Inorg. Chim. Acta, 1984, **93**, 73.
833 M. Saburi, Y. Yokowo, and S. Yoshikawa, Bull. Chem. Soc. Jpn., 1984, **57**, 1235.
834 Y. Yamamoto and H. Kudo, Bull. Chem. Soc. Jpn., 1984, **57**, 287.
835 S.C.F. Au-Yeung and D.R. Eaton, Inorg. Chem., 1984, **23**, 1517.
836 N.E. Dixon, G.A. Lawrance, P.A. Lay, and A.M. Sargeson, Inorg. Chem., 1984, **23**, 2940.
837 X. Li, R. Irie, S. Shinoda, and Y. Saito, Nippon Kagaku Kaishi, 1984, 271 (Chem. Abstr., 1984, **101**, 22 767).
838 H. Brunner and A.F.M.M. Rahman, Chem. Ber., 1984, **117**, 710.
839 A. Benedetti, C. Preti, and G. Tosi, J. Mol. Struct., 1984, **116**, 397.
840 R. Usón, L.A. Oro, D. Carmona, M.A. Esteruelas, C. Foces-Foces, F.H. Cano,

S. Garcia-Blanco, and A. Vazquez de Miguel, J. Organomet. Chem., 1984, **273**, 111.
841 M. Ghedini and F. Neve, J. Chem. Soc., Dalton Trans., 1984, 1417.
842 F. Neve and M. Ghedini, Congr. Naz. Chim. Inorg., [Atti], 16th, 1983, 136 (Chem. Abstr., 1984, **100**, 78 833).
843 S.M. Socol and J.G. Verkade, Phosphorus Sulphur, 1983, **18**, 291.
844 N. Juranic, Inorg. Chim. Acta, 1984, **87**, L37.
845 T.J. Cardwell and T.H. Lorman, Inorg. Chim. Acta, 1984, **85**, 1.
846 G. Schmid, U. Giebel, W. Huster, and A. Schwenk, Inorg. Chim. Acta, 1984, **85**, 97.
847 M.A. Fedotov and A.V. Belyaev, Koord. Khim., 1984, **10**, 1236 (Chem. Abstr., 1984, **101**, 203 004).
848 A.V. Belyaev, M.A. Fedotov, V.I. Korsunskii, A.B. Venediktov, and S.P. Khranenko, Koord. Khim., 1984, **10**, 911 (Chem. Abstr., 1984, **101**, 142 802).
849 M. Di Vaira, M. Peruzzini, and P. Stoppioni, J. Chem. Soc., Dalton Trans., 1984, 259.
850 M. Di Vaira, S. Monet, M. Peruzzini, and P. Stoppioni, J. Organomet. Chem., 1984, **266**, C8.
851 Y. Koike, T. Takayama, and M. Watabe, Bull. Chem. Soc. Jpn., 1984, **57**, 3595.
852 T.L. Marxen, B.J. Johnson, P.V. Nilsson, and L.H. Pignolet, Inorg. Chem., 1984, **23**, 4663.
853 S. Slater and J.H. Wagenknecht, J. Am. Chem. Soc., 1984, **106**, 5367.
854 E. W. Burkhardt, W.C. Mercer, and G.L. Geoffroy, Inorg. Chem., 1984, **23**, 1779.
855 M. Bressan, F. Morandini, and P. Rigo, Congr. Naz. Chim. Inorg., [Atti], 16th, 1983, 76 (Chem. Abstr., 1984, **100**, 44 381).
856 J.M. Brown and L.R. Canning, J. Organomet. Chem., 1984, **267**, 179.
857 C. Bianchini, C. Mealli, A. Meli, and M. Sabat, J. Chem. Soc., Chem. Commun., 1984, 1647.
858 W.A. Schenk, J. Leissner, and C. Burschka, Angew. Chem., Int. Ed. Engl., 1984, **23**, 807.
859 W.A. Schenk, J. Leissner, and C. Burschka, Angew. Chem., Int. Ed. Engl., 1984, **23**, 806.
860 J. Telser and R.S. Drago, Inorg. Chem., 1984, **23**, 2599.
861 I. Lin, W.B. Knight, S.-J. Ting, and D. Dunaway-Mariano, Inorg. Chem., 1984, **23**, 988.
862 J. Telser and R.S. Drago, Inorg. Chem., 1984, **23**, 1798.
863 J.E. Hoots, D.A. Lesch, and T.B. Rauchfuss, Inorg. Chem., 1984, **23**, 3130.
864 P.B. Hitchcock, S. Morton, and J.F. Nixon, J. Chem. Soc., Chem. Commun., 1984, 603.
865 H.T. Stokes, ACS Symp. Ser., 1984, **248**, 385 (Chem. Abstr., 1984, **100**, 180 674).
866 P.S. Pregosin and H. Rüegger, Inorg. Chim. Acta, 1984, **86**, 55.
867 D.P. Arnold and M.A. Bennett, Inorg. Chem., 1984, **23**, 2117.
868 I. Tkatchenko, Phosphorus Sulphur, 1983, **18**, 311.
869 H. Riml and L.M. Venanzi, J. Organomet. Chem., 1984, **260**, C52.
870 D.P. Arnold and M.A. Bennett, Inorg. Chem., 1984, **23**, 2110.
871 D. Carmona, S. Chaloupka, J. Jans, R. Thouwenot, and L.M. Venanzi, J. Organomet. Chem., 1984, **275**, 303.
872 W.R. Meyer and L.M. Venanzi, Angew Chem., Int. Ed. Engl., 1984, **23**, 529.
873 S.W. Carr, R. Colton, and D. Dakternieks, Inorg. Chem., 1984, **23**, 720.
874 A.B. Goel and S. Goel, Inorg. Chim. Acta, 1984, **90**, L33.
875 R.J. Goodfellow, E.M. Hanson, J.A.K. Howard, J.L. Spencer, and D.G. Turner, J. Chem. Soc., Chem. Commun., 1984, 1604.
876 M.A. Andrews, T.C.-T. Chang, C.-W.F. Cheng, L.V. Kapustay, K.P. Kelly, and W.J. Zweifeil, Organometallics, 1984, **3**, 1479.
877 P.S. Pregosin and R. Rüedi, J. Organomet. Chem., 1984, **273**, 401.
878 H.A. Brune, M. Falck, R. Hemmer, and H.G. Alt, Chem. Ber., 1984, **117**, 2803.
879 C.A. Tolman, W.C. Seidel, J.D. Druliner, and P.J. Domaille, Organometallics, 1984, 3, 33.
880 K. Sano, T. Yamamoto, and A. Yamamoto, Bull. Chem. Soc. Jpn., 1984, **57**, 2741.

881 H. Hoberg, D. Schaefer, G. Burkhart, C. Krüger, and M.J. Romäo, J. Organomet. Chem., 1984, **266**, 203.
882 J. Carvajal, G. Muller, J. Sales, X. Solans, and C. Miravitlies, Organometallics, 1984, **3**, 996.
883 K.P. Pörschke and G. Wilke, Chem. Ber., 1984, **117**, 56.
884 R. Glaser, D.J. Kountz, R.D. Waid, J.C. Gallucci, and D.W. Meek, J. Am. Chem. Soc., 1984, **106**, 6324.
885 R. Tooze, K.W. Chiu, and G. Wilkinson, Polyhedron, 1984, **3**, 1025.
886 P. Diversi, D. Fasce, and R. Santini, J. Organomet. Chem., 1984, **269**, 285.
887 N. Ishio, Y. Nakamura, and S. Okeya, Bull. Chem. Soc. Jpn., 1984, **57**, 1012.
888 B. Åkermark and K. Zetterberg, J. Am. Chem. Soc., 1984, **106**, 5560.
889 G.R. Newkome, V.K. Gupta, H.C.R. Taylor, and F.R. Fronczek, Organometallics, 1984, **3**, 1549.
890 M.A. Andrews, T.C.-T. Chang, C.-W.F. Cheng, T.J. Emge, K.P. Kelly, and T.F. Koetzle, J. Am. Chem. Soc., 1984, **106**, 5913.
891 F. Ozawa, T. Sugimoto, Y. Yuasa, M. Santra, T. Yamamoto, and A. Yamamoto, Organometallics, 1984, **3**, 683.
892 F. Ozawa, T. Sugimoto, T. Yamamoto, and A. Yamamoto, Organometallics, 1984, **3**, 692.
893 J.-T. Chen and A. Sen, J. Am. Chem. Soc., 1984, **106**, 1506.
894 N. Barr, S.F. Dyke, and R.L.W. Frost, Org. Magn. Reson., 1984, **22**, 277.
895 P.W. Clark and S.F. Dyke, J. Organomet. Chem., 1984, **276**, 421.
896 T.E. Crossley, P. Davies, M. Louey, R. Robson, and T.N. Huckerby, Inorg. Chim. Acta, 1984, **85**, 199.
897 R. Bertani, C.B. Castellani, and B. Crociani, J. Organomet. Chem., 1984, **269**, C15.
898 J.W.L. Martin, J.A.L. Palmer, and S.B. Wild, Inorg. Chem., 1984, **23**, 2664.
899 B. Crociani, F. di Bianca, A. Giovenco, and A. Scrivanti, J. Organomet. Chem., 1984, **269**, 295.
900 R. Usón, J. Forniés, M.A. Usón, and M.A. Orta, Inorg. Chim. Acta, 1984, **89**, 175.
901 R. Usón, J. Forniés, R. Navarro, M.A. Usón, M.P. Garcia, and A.J. Welch, J. Chem. Soc., Dalton Trans., 1984, 345.
902 M.L. Luetkens, F. Teixidor, and R.W. Rudolph, Inorg. Chim. Acta, 1984, **83**, L13.
903 M. Crespo, O. Rossell, J. Sales, and M. Seco, J. Organomet. Chem., 1984, **273**, 415.
904 A.B. Goel and S. Goel, Inorg. Chim. Acta, 1984, **82**, 41.
905 R.J. Puddephatt, M.C. Rendle, and C.F.H. Tipper, J. Organomet. Chem., 1984, **269**, 305.
906 G. Strukul, R.A. Michelin, J.D. Orbell, and L. Randaccio, Inorg. Chem., 1983, **22**, 3706.
907 H.A. Brune, J. Unsin, H.G. Alt, G. Schmidtberg, and K.H. Spohn, Chem. Ber., 1984, **117**, 1606.
908 A.N. Kitaigorodskii, V.M. Nekipelov, A.T. Nikitaev, and G.B. Shul'pin, J. Organomet. Chem., 1984, **275**, 295.
909 S. Takahashi, Y. Takai, H. Morimoto, and K. Sonogashira, J. Chem. Soc., Chem. Commun., 1984, 3.
910 J. Calvet, O. Rossell, and M. Seco, Transition Met. Chem., 1984, **9**, 208.
911 H.C. Clark, G. Ferguson, V.K. Jain, and M. Parvez, J. Organomet. Chem., 1984, **270**, 365.
912 V.F. Sutcliffe and G.B. Young, Polyhedron, 1984, **3**, 87.
913 P.K. Monaghan and R.J. Puddephatt, Organometallics, 1984, **3**, 444.
914 H.E. Bryndza, J.C. Calabrese, and S.S. Wreford, Organometallics, 1984, **3**, 1603.
915 A.B. Goel, S. Goel, and D. Vanderveer, Polyhedron, 1984, **3**, 373.
916 A.T. Hutton, B. Shebanzadeh, and B.L. Shaw, J. Chem. Soc., Chem. Commun., 1984, 549.
917 K.A. Azam, R.H. Hill, and R.J. Puddephatt, Can. J. Chem., 1984, **62**, 2029.
918 K.A. Azam, M.P. Brown, R.H. Hill, R.J. Puddephatt, and A. Yavari, Organometallics, 1984, **3**, 697.
919 A. Sebald, B. Wrackmeyer, C.R. Theocharis, and W. Jones, J. Chem. Soc.,

Dalton Trans., 1984, 747.
920 T.C. Flood and J.A. Statler, Organometallics, 1984, **3**, 1795.
921 P.G. Pringle and B.L. Shaw, J. Chem. Soc., Dalton Trans., 1984, 849.
922 A.B. Goel, S. Goel, and D. Vanderveer, Inorg. Chim. Acta, 1984, **82**, L9.
923 S. Komiya and T. Shindo, J. Chem. Soc., Chem. Commun., 1984, 1672.
924 Z.-Y. Yang and G.B. Young, J. Chem. Soc., Dalton Trans., 1984, 2019.
925 K. Kundu, J. Bangladesh Acad. Sci., 1983, **7**, 111 (Chem. Abstr., 1984, **100**, 139 327).
926 T.G. Appleton, J.R. Hall, D.W. Neale, and M.A. Williams, J. Organomet. Chem., 1984, **276**, C73.
927 A. Mantovani, M. Pelloso, G. Bandoli, and B. Crociani, J. Chem. Soc., Dalton Trans., 1984, 2223.
928 S. Cenini, F. Porta, M. Pizzotti, and G. La Monica, J. Chem. Soc., Dalton Trans., 1984, 355.
929 R.M. Ceder and J. Sales, J. Organomet. Chem., 1984, **276**, C31.
930 L. Chassot, E. Müller, and A. von Zelewsky, Inorg. Chem., 1984, **23**, 4249.
931 R.J. Cross and J. Gemmill, J. Chem. Res. (S), 1984, 30.
932 R. Usón, J. Forniés, M. Tomás, and R. Fandos, J. Organomet. Chem., 1984, **263**, 253.
933 G.R. Cooper, A.T. Hutton, C.R. Langrick, D.M. McEwan, P.G. Pringle, and B.L. Shaw, J. Chem. Soc., Dalton Trans., 1984, 855.
934 R.J. Cross and J. Gemmill, J. Chem. Soc., Dalton Trans., 1984, 199.
935 R.J. Cross and J. Gemmill, J. Chem. Soc., Dalton Trans., 1984, 205.
936 D. Afzal, P.G. Lenhert, and C.M. Lukehart, J. Am. Chem. Soc., 1984, **106**, 3050.
937 A. Albinati, P.S. Pregosin, and H. Rüegger, Angew. Chem., Int. Ed. Engl., 1984, **23**, 78.
938 H.C. Clark, G. Ferguson, A.B. Goel, and B.L. Ruhl, Organometallics, 1984, **3**, 15.
939 A. Albinati, P.S. Pregosin, and H. Rüegger, Inorg. Chem., 1984, **23**, 3223.
940 V.G. Albano, F. Demartin, A. De Renzi, N. Masciocchi, G. Morelli, and A. Panunzi, Congr. Naz. Chim. Inorg., [Atti], 16th, 1983, 47.
941 S.S. Hall and B. Åkermark, Organometallics, 1984, **3**, 1745.
942 K.R. Poerschke, R. Mynott, C. Krüger, and M.J. Romao, Z. Naturforsch., Teil B, 1984, **39**, 1076 (Chem. Abstr., 1984, **101**, 211 430).
943 R. del Rosario and L.S. Stuhl, J. Am. Chem. Soc., 1984, **106**, 1160.
944 E. Carmona, E. Gutiérrez-Puebla, A. Monge, J.M. Marín, M. Paneque, and M.L. Poveda, Organometallics, 1984, **3**, 1438.
945 A.H. Cowley, R.A. Jones, J.G. Lasch, N.C. Norman, C.A. Stewart, A.L. Stuart, J.L. Atwood, W.E. Hunter, and H.M. Zhang, J. Am. Chem. Soc., 1984, **106**, 7015.
946 A. Lavery and S.M. Nelson, J. Chem. Soc., Dalton Trans., 1984, 615.
947 D.H. Farrar, R.R. Gukathasan, and S.A. Morris, Inorg. Chem., 1984, **23**, 3258.
948 D.H. Farrar, R.R. Gukathasan, and K. Won, J. Organomet. Chem., 1984, **275**, 263.
949 R. Usón, J. Forniés, and M.A. Usón, Inorg. Chim. Acta, 1984, **81**, L27.
950 R. Usón, J. Forniés, and M.A. Usón, Synth. React. Inorg. Met.-Org. Chem., 1984, **14**, 355.
951 I.M. Al-Najjar, S.S. Al-Showiman, and H.M. Al-Hazimi, Inorg. Chim. Acta, 1984, **89**, 57.
952 P.G. Gassman and I.G. Cesa, Organometallics, 1984, **3**, 119.
953 R. Taube and J.P. Gehrke, Z. Chem., 1983, **23**, 437 (Chem. Abstr., 1984, **100**, 139 680).
954 H. Lehmkuhl, A. Rufińska, C. Naydowski, and R. Mynott, Chem. Ber., 1984, **117**, 376.
955 H. Hoberg, D. Schaefer, and B.W. Oster, J. Organomet. Chem., 1984, **266**, 313.
956 B. Åkermark, S. Hansson, B. Krakenberger, A. Vitagliano, and K. Zetterberg, Organometallics, 1984, **3**, 679.
957 M.K. Andari, O.G. Levanda, and A.P. Belov, Koord. Khim., 1984, **10**, 1100 (Chem. Abstr., 1984, **101**, 211 435).
958 R.J.P. Corriu, N. Escudié, and C. Guerin, J. Organomet. Chem., 1984, **271**, C7.

959 J. Powell and N.I. Dowling, J. Organomet. Chem., 1984, **264**, 387.
960 N. Kuhn, U. Schwenk, M. Winter, and R. Mynott, J. Organomet. Chem., 1984, **260**, C43.
961 G. Schmid, D. Kampmann, U. Höhner, D. Bläser, and R. Boese, Chem. Ber., 1984, **117**, 1052.
962 D.W. Slocum, M. Moronski, R. Gooding, and S. Duraj, J. Organomet. Chem., 1984, **260**, C26.
963 T. Kuhlmann, H. Pritzkow, U. Zenneck, and W. Siebert, Angew. Chem., Int. Ed. Engl., 1984, **23**, 965.
964 H. Wadepohl and W. Siebert, Z. Naturforsch., Teil B, 1984, **39**, 50 (Chem. Abstr., 1984, **100**, 175 043).
965 F. Canziani, C. Allevi, L. Garlaschelli, M.C. Malatesta, A. Albinati, and F. Ganazzoli, J. Chem. Soc., Dalton Trans., 1984, 2637.
966 F. Cecconi, C.A. Ghilardi, P. Innocenti, C. Mealli, S. Midollini, and A. Orlandini, Inorg. Chem., 1984, **23**, 922.
967 V.K. Polovnyak, V.Sh. Slobodina, T.E. Busygina, and A.E. Ivantsov, J. Mol. Liq., 1984, **28**, 123 (Chem. Abstr., 1984, **101**, 32 326).
968 H. Hope, M.M. Olmstead, P.P. Power, and M. Viggiano, Inorg. Chem., 1984, **23**, 326.
969 G. Elbaze, F. Dahan, M. Dartiguenave, and Y. Dartiguenave, Inorg. Chim. Acta, 1984, **85**, L3.
970 P.F. d.S.Filho and U. Schuchardt, J. Organomet. Chem., 1984, **263**, 385.
971 B.R. Lloyd and R.J. Puddephatt, Inorg. Chim. Acta, 1984, **90**, L77.
972 M.L. Kullberg and C.P. Kublak, Organometallics, 1984, **3**, 632.
973 C.R. Langrick, P.G. Pringle, and B.L. Shaw, J. Chem. Soc., Dalton Trans., 1984, 1233.
974 R. Bender, P. Braunstein, A. Tiripicchio, and M. Tiripicchio-Camellini, J. Chem. Soc., Chem. Commun., 1984, 42.
975 J.M. Brown, S.J. Cook, and S.J. Kimber, J. Organomet. Chem., 1984, **269**, C58.
976 D.B. Dell'Amico, F. Calderazzo, C.A. Veracini, and N. Zandonà, Inorg. Chem., 1984, **23**, 3030.
977 S.A. Bryan, M.K. Dickson, and D.M. Roundhill, J. Am. Chem. Soc., 1984, **106**, 1882.
978 V.F. Yudanov, N.K. Eremenko, E.G. Mednikov, and S.P. Gubin, Zh. Strukt. Khim., 1984, **25**, 49 (Chem. Abstr., 1984, **101**, 32 333).
979 A. Moor, P.S. Pregosin, L.M. Venanzi, and A.J. Welch, Inorg. Chim. Acta, 1984, **85**, 103.
980 C.E. Briant, R.W.M. Wardle, and D.M.P. Mingos, J. Organomet. Chem., 1984, **267**, C49.
981 N.W. Alcock, P. Moore, and C. Pierpoint, J. Chem. Soc., Dalton Trans., 1984, 2371.
982 R. Ocampo, H.J. Callot, P. Albrecht, and J.P. Kintzinger, Tetrahedron Lett., 1984, **25**, 2589.
983 A. Ulman, Org. Magn. Reson., 1984, **22**, 114.
984 J.-M. Bret, P. Castan, and J.-P. Laurent, J. Chem. Soc., Dalton Trans., 1984, 1975.
985 U. Kohler, H.-D. Hausen, and J. Weidlein, J. Organomet. Chem., 1984, **272**, 337.
986 N.W. Alcock, R.G. Kingston, P. Moore, and C. Pierpoint, J. Chem. Soc., Dalton Trans., 1984, 1937.
987 B. Korybut-Daszkiewicz, M. Kojima, J.H. Cameron, N. Herron, M.Y. Chavan, A.J. Jircitano, B.K. Coltrain, G.L. Neer, N.W. Alcock, and D.H. Busch, Inorg. Chem., 1984, **23**, 903.
988 G. Elbaze, F. Dahan, M. Dartiguenave, and Y. Dartiguenave, Inorg. Chim. Acta, 1984, **87**, 91.
989 D.G. Holah, A.N. Hughes, and N.I. Khan, Can. J. Chem., 1984, **62**, 1016.
990 S.H. Kim and R.B. Martin, Inorg. Chim. Acta, 1984, **91**, 11.
991 L.R. Milgrom, Polyhedron, 1984, **3**, 879.
992 K. Nakayama, T. Komorita, and Y. Shimura, Bull. Chem. Soc. Jpn., 1984, **57**, 1240.
993 U. Taubald, U. Nagel, and W. Beck, Chem. Ber., 1984, **117**, 1003.
994 G. Pneumatikakis, Polyhedron, 1984, **3**, 9.

995 G. Thiel and W. Beck, Z. Naturforsch., Teil B, 1983, **38**, 1081 (Chem. Abstr., 1984, **100**, 44 337).
996 P. Braunstein, J. Fischer, D. Matt, and M. Pfeffer, J. Am. Chem. Soc., 1984, **106**, 410.
997 T.-W. Lai and A. Sen, Organometallics, 1984, **3**, 866.
998 A. Sen and T.-W. Lai, Inorg. Chem., 1984, **23**, 3257.
999 H.C. Hufmans, H.O. Desseyn, and R.A. Dommisse, Transition Met. Chem., 1984, **9**, 213.
1000 T.G. Appleton, R.D. Berry, C.A. Davis, J.R. Hall, and H.A. Kimlin, Inorg. Chem., 1984, **23**, 3514.
1001 A. Woollins and B. Rosenberg, J. Inorg. Biochem., 1984, **20**, 23.
1002 A.T.M. Marcelis, C. Erkelens, and J. Reedijk, Inorg. Chim. Acta, 1984, **91**, 129.
1003 J.H.J. den Hartog, H. van den Elst, and J. Reedijk, J. Inorg. Biochem., 1984, **21**, 83.
1004 W. Tang, S. Zhang, C. Yuan, and A. Tai, Gaodeng Xuexiao Huaxue Xuebao, 1984, **5**, 1.
1005 D.N. Bochkov, N.A. Smorygo, L.L. Shcherbakova, and B.A. Ivin, Zh. Obshch. Khim., 1984, **54**, 900 (Chem. Abstr., 1984, **101**, 64 873).
1006 J.P. Girault, J.C. Chottard, J.M. Neumann, Son Tran Dinh, Tam Huynh Dinh, and J. Igolen, Nouv. J. Chim., 1984, **8**, 7 (Chem. Abstr., 1984, **101**, 16 914).
1007 J.H.J. Den Hartog, C. Altona, J.H. Van Boom, G.A. Van der Marel, C.A.G. Haasnoot, and J. Reedijk, J. Am. Chem. Soc., 1984, **106**, 1528.
1008 R. Kuroda, I.M. Ismail, and P.J. Sadler, J. Inorg. Biochem., 1984, **22**, 103.
1009 J.H.J. Den Hartog, M.L. Salm, and J. Reedijk, Inorg. Chem., 1984, **23**, 2001.
1010 D. Hermann, G.V. Fazakerley, and W. Guschlbauer, Biopolymers, 1984, **23**, 973.
1011 P. Arrizabalaga, P. Castan, and J.-P. Laurent, J. Am. Chem. Soc., 1984, **106**, 1300.
1012 P. Arrizabalaga, P. Castan, and J.-P. Laurent, J. Am. Chem. Soc., 1984, **106**, 4814.
1013 R.J. Brandon and J.C. Dabrowiak, J. Med. Chem., 1984, **27**, 861 (Chem. Abstr., 1984, **101**, 16 188).
1014 J.G. Kraaijkamp, G. v. Koten, K. Vrieze, D.M. Grove, G. Abbel, C.H. Stam, and A. Schmidpeter, Inorg. Chim. Acta, 1984, **85**, L33.
1015 L. Olgemoeller and W. Beck, Chem. Ber., 1984, **117**, 1241.
1016 A. Albinati, C.G. Anklin, and P.S. Pregosin, Inorg. Chim. Acta, 1984, **90**, L37.
1017 J. Auffret, P. Courtot, R. Pichon, R. Rumin, and J.Y. Salaun, J. Organomet. Chem., 1984, **262**, C19.
1018 A. Shamim, A. Adeyemo, M.T. Shamim, J. Wheeler, A. Turner, and A.B. Hussein, Inorg. Chim. Acta, 1984, **91**, 179.
1019 I. Georgii, B.E. Mann, B.F. Taylor, and A. Musco, Inorg. Chim. Acta, 1984, **86**, L81.
1020 O.J. Scherer, H. Jungmann, C. Krüger, and G. Wolmershäuser, Chem. Ber., 1984, **117**, 2382.
1021 G.W. Bushnell, K.R. Dixon, R. Ono, and A. Pidcock, Can. J. Chem., 1984, **62**, 696.
1022 H.A. Brune, M. Falck, R. Hemmer, G. Schmidtberg, and H.G. Alt, Chem. Ber., 1984, **117**, 2791.
1023 D. Fenske and K. Merzweiler, Angew. Chem., Int. Ed. Engl., 1984, **23**, 635.
1024 M. Di Vaira, S. Moneti, M. Peruzzini, and P. Stoppioni, Congr. Naz. Chim. Inorg., [Atti], 16th, 1983, 177.
1025 R. Bartsch, S. Hietkamp, H. Peters, and O. Stelzer, Inorg. Chem., 1984, **23**, 3304.
1026 C. Bianchini, C.A. Ghilardi, A. Meli, and A. Orlandini, J. Organomet. Chem., 1984, **270**, 251.
1027 M. Ciampolini, N. Nardi, P.L. Orioli, S. Mangani, and F. Zanobini, J. Chem. Soc., Dalton Trans., 1984, 2265.
1028 M. Ciampolini, N. Nardi, P. Dapporto, P. Innocenti, and F. Zanobini, J. Chem. Soc., Dalton Trans., 1984, 575.
1029 T. Yamamoto and Y. Sekine, Inorg. Chim. Acta, 1984, **83**, 47.
1030 H.M. Büch, P. Binger, and C. Krüger, Organometallics, 1984, **3**, 1504.

1031 G. Salem and S.B. Wild, Inorg. Chem., 1984, **23**, 2655.
1032 F. Mercier, F. Mathey, J. Fischer, and J.H. Nelson, J. Am. Chem. Soc., 1984, **106**, 425.
1033 J. Podlahová, L. Kavan, J. Silha, and J. Podlaha, Polyhedron, 1984, **3**, 963.
1034 E. Pfeiffer, M.L. Pasquier, and W. Marty, Helv. Chim. Acta, 1984, **67**, 654.
1035 F. Cecconi, C.A. Ghilardi, and S. Midollini, Inorg. Chem., 1983, **22**, 3802.
1036 H. Urata, T. Tsukui, H. Suzuki, Y. Morooka, and T. Ikawa, Chem. Lett., 1984, 191.
1037 J.R. Fisher and D.J. Cole-Hamilton, J. Chem. Soc., Dalton Trans., 1984, 809.
1038 F. Morandini and M. Bressan, Congr. Naz. Chim. Inorg. [Atti], 15th, 1982, 26 (Chem. Abstr., 1984, **100**, 202 358).
1039 R. Schierl, U. Nagel, and W. Beck, Z. Naturforsch., Teil B, 1984, **39**, 649 (Chem. Abstr., 1984, **101**, 32 334).
1040 J. Kozelka, H.-P. Lüthi, E. Dubler, and R.W. Kunz, Inorg. Chim. Acta, 1984, **86**, 155.
1041 D.W. Bruce, R.F. Jones, and D.J. Cole-Hamilton, J. Chem. Soc., Dalton Trans., 1984, 2249.
1042 M. Ciampolini, P. Dapporto, P. Innocenti, N. Nardi, and F. Zanobini, Congr. Naz. Chim. Inorg., [Atti], 16th, 1983, 123.
1043 P. Boron-Rettore, D.M. Grove, and L.M. Venanzi, Helv. Chim. Acta, 1984, **67**, 65.
1044 H.C. Clark, P.N. Kapoor, and I.J. McMahon, J. Organomet. Chem., 1984, **265**, 107.
1045 R.D. Waid and D.W. Meek, Inorg. Chem., 1984, **23**, 778.
1046 J.M. Solar, R.D. Rogers, and W.R. Mason, Inorg. Chem., 1984, **23**, 373.
1047 S. Muralidharan and J.H. Espenson, J. Am. Chem. Soc., 1984, **106**, 8104.
1048 M. Shimura and J.H. Espenson, Inorg. Chem., 1984, **23**, 4069.
1049 A.B. Goel and S. Goel, Angew. Chem., Int. Ed. Engl., 1984, **23**, 375.
1050 C.E. Briant, C.J. Gardner, T.S.A. Hor, N.D. Howells, and D.M.P. Mingos, J. Chem. Soc., Dalton Trans., 1984, 2645.
1051 P. Braunstein, H. Lehner, D. Matt, A. Tiripicchio, and M. Tiripicchio-Camellini, Angew. Chem., Int. Ed. Engl., 1984, **23**, 304.
1052 C. Carr, P.L. Goggin, and R.J. Goodfellow, Inorg. Chim. Acta, 1984, **81**, L25.
1053 J.J. Dechter and J. Kowalewski, J. Magn. Reson., 1984, **59**, 146.
1054 I.M. Oglezneva and I.K. Igumenov, Koord. Khim., 1984, **10**, 313 (Chem. Abstr., 1984, **101**, 14 208).
1055 M.A. Fedotov and V.A. Likholobov, Izv. Akad. Nauk SSSR, Ser. Khim., 1984, 1917 (Chem. Abstr., 1984, **101**, 162 447).
1056 T.G. Appleton, J.R. Hall, S.F. Ralph, and C.S.M. Thompson, Inorg. Chem., 1984, **23**, 3521.
1057 R. Colton and R.A. Stephenson, Polyhedron, 1984, **3**, 231.
1058 W.N. Hunter, K.W. Muir, and D.W.A. Sharp, Acta Crystallogr., 1984, **C40**, 37 (Chem. Abstr., 1984, **100**, 112 596).
1059 C. Bellitto, G. Dessy, V. Fares, and A. Flamini, Congr. Naz. Chim. Inorg., [Atti], 16th, 1983, 138 (Chem. Abstr., 1984, **100**, 44 387).
1060 P.M. Henrichs, 'Silver-109', NMR Newly Accessible Nucl., 1983, **2**, 319. Ed. P. Laszlo, Academic, New York (Chem. Abstr., 1984, **100**, 166 571).
1061 H. Nakatsuji, K. Kand, K. Endo, and T. Yonezawa, J. Am. Chem. Soc., 1984, **106**, 4653.
1062 G. Hallnemo, T. Olsson, and C. Ullenius, J. Organomet. Chem., 1984, **265**, C22.
1063 S.A. Snow, M. Shimoi, C.D. Ostler, B.K. Thompson, G. Kodama, and R.W. Parry, Inorg. Chem., 1984, **23**, 511.
1064 Y. Yamamoto, Bull. Chem. Soc. Jpn., 1984, **57**, 43.
1065 G. van Koten, J.T.B.H. Jastrzebski, C.H. Stam, and N.C. Niemann, J. Am. Chem. Soc., 1984, **106**, 1880.
1066 C. Eaborn, P.B. Hitchcock, J.D. Smith, and A.C. Sullivan, J. Chem. Soc., Chem. Commun., 1984, 870.
1067 R. Usón, A. Laguna, M. Laguna, B.R. Manzano, P.G. Jones, and G.M. Sheldrick, J. Chem. Soc., Dalton Trans., 1984, 839.
1068 F. Bonati, A. Burini, B.R. Pietroni, and M. Felici, J. Organomet. Chem.,

1984, **273**, 275.
1069 R. Usón, A. Laguna, M. Laguna, I. Colera, and E. de Jesús, J. Organomet. Chem., 1984, **263**, 121.
1070 J.G. Hefner, P.M. Zizelman, L.D. Durfee, and G.S. Lewandos, J. Organomet. Chem., 1984, **260**, 369.
1071 G.C. Van Stein, G. Van Koten, K. Vrieze, C. Brevard, and A.L. Spek, J. Am. Chem. Soc., 1984, **106**, 4486.
1072 G.C. Van Stein, G. Van Koten, K. Vrieze, and C. Brevard, Inorg. Chem., 1984, **23**, 4269.
1073 G.C. Van Stein, G. Van Koten, B. De Bok, L.C. Taylor, K. Vrieze, and C. Brevard, Inorg. Chim. Acta, 1984, **89**, 29.
1074 A.H. Cowley, N.C. Norman, and M. Pakulski, J. Chem. Soc., Chem. Commun., 1984, 1054.
1075 S.M. Socol and J.G. Verkade, Inorg. Chem., 1984, **23**, 3487.
1076 S. Kitagawa and M. Munakata, Inorg. Chem., 1984, **23**, 4388.
1077 A. Bell and D.A. Edwards, J. Chem. Soc., Dalton Trans., 1984, 1317.
1078 N.K. Homsy, M. Noltemeyer, H.W. Roesky, H.-G. Schmidt, and G.M. Sheldrick, Inorg. Chim. Acta, 1984, **90**, L59.
1079 P.G. Egan, D.G. Holah, and K.W. Morse, Inorg. Chem., 1984, **23**, 2203.
1080 J. Perron and A.L. Beauchamp, Can. J. Chem., 1984, **62**, 1287.
1081 J. Perron and A.L. Beauchamp, Inorg. Chem., 1984, **23**, 2853.
1082 Y. Kojima, T. Yamashita, Y. Ishino, T. Hirashima, and T. Miwa, Bull. Chem. Soc. Jpn., 1983, **56**, 3841.
1083 G. Banditelli, F. Bonati, S. Calogero, and G. Valle, J. Organomet. Chem., 1984, **275**, 153.
1084 D.A. Wrobleski, T.B. Rauchfuss, A.L. Rheingold, and K.A. Lewis, Inorg. Chem., 1984, **23**, 3124.
1085 S.M. Socol, R.A. Jacobson, and J.G. Verkade, Inorg. Chem., 1984, **23**, 88.
1086 S.O. Grim, Phosphorus Sulphur, 1983, **18**, 283.
1087 M. Camalli, F. Caruso, S. Chaloupka, P.N. Kapoor, P.S. Pregosin, and L.M. Venanzi, Helv. Chim. Acta, 1984, **67**, 1603.
1088 S.J. Berners-Price, M.A. Mazid, and P.J. Sadler, J. Chem. Soc., Dalton Trans., 1984, 969.
1089 A.K.H. Al-Sa'ady, C.A. McAuliffe, K. Moss, R.V. Parish, and R. Fields, J. Chem. Soc., Dalton Trans., 1984, 491.
1090 J.W.A. Van der Velden, P.T. Beurskens, J.J. Bour, W.P. Bosman, J.H. Noordik, M. Kolenbrander, and J.A.K.M. Buskes, Inorg. Chem., 1984, **23**, 146.
.091 J.W.A. Van der Velden, J.J. Bour, W.P. Bosman, J.H. Noordik, and P.T. Beurskens, Recl.: J.R. Neth. Chem. Soc., 1984, **103**, 13.
.092 C.E. Briant, K.P. Hall, and D.M.P. Mingos, J. Chem. Soc., Chem. Commun., 1984, 290.
093 M.S. Hussain and A.A. Isab, Transition Met. Chem., 1984, **9**, 398.
094 I.M. Armitage and Y. Boulanger, 'Cadmium-113 NMR', NMR Newly Accessible Nucl., 1983, **2**, 337. Ed. P. Laszlo, Academic, New York
095 R.J. Gillespie, P. Granger, K.R. Morgan, and G.J. Schrobilgen, Inorg. Chem., 1984, **23**, 887.
096 A.A. Solomin, V.I. Mstislavskii, T.A. Obukhova, Yu.K. Grishin, and Yu.A. Ustynyuk, Vestn. Mosk. Univ., Ser. 2: Khim., 1984, **25**, 394 (Chem. Abstr., 1984, **101**, 192 107).
097 C. Cauletti, C. Furlani, M.N. Piancastelli, A. Sebald, and B. Wrackmeyer, Inorg. Chem., 1984, **23**, 1113.
098 A.S. Peregudov, V.F. Ivanov, D.N. Kravtsov, and E.I. Fedin, Izv. Akad. Nauk SSSR, Ser. Khim., 1984, 849 (Chem. Abstr., 1984, **101**, 72 862).
099 F. De Sarlo, A. Brandi, L. Fabrizi, A. Guarna, and N. Niccolai, Org. Magn. Reson., 1984, **22**, 372.
100 E. Buncel, B.K. Hunter, R. Kumar, and A.R. Norris, J. Inorg. Biochem., 1984, **20**, 1771.
101 A.R. Norris and R. Kumar, Inorg. Chim. Acta, 1984, **93**, L63.
102 A.A. Solimin, V.I. Mstislavskii, A.G. Shakhatuni, Yu.K. Grishin, and Yu.A. Ustynyuk, Vestn. Mosk. Univ., Ser. 2: Khim., 1984, **25**, 78 (Chem. Abstr., 1984, **100**, 210 041).
103 T.N. Mitchell and B. Fabisch, J. Organomet. Chem., 1984, **269**, 249.

1104 M.F. Larin, D.V. Gendin, V.A. Pestunovich, L.I. Rybin, O.A. Vyazankina, and N.S. Vyazankin, Izv. Akad. Nauk SSSR, Ser. Khim., 1983, 2139 (Chem. Abstr., 1984, **100**, 22 745).
1105 A. Osman, R.G. Steevensz, D.G. Tuck, H.A. Meinema, and J.G. Noltes, Can. J. Chem., 1984, **62**, 1698.
1106 G.B. Deacon, G.N. Stretton, and M.J. O'Connor, J. Organomet. Chem., 1984, **277**, C1.
1107 A.R. Norris, R. Kumar, and E. Buncel, J. Inorg. Biochem., 1984, **22**, 11.
1108 A.R. Norris and R. Kumar, Inorg. Chim. Acta, 1984, **93**, 33.
1109 M. Lusser and P. Peringer, Organometallics, 1984, **3**, 1916.
1110 R.S. Varma, S.A. Kunda, and G.W. Kabalka, J. Organomet. Chem., 1984, **276**, 311.
1111 R.S. Varma, S.A. Kunda, and G.W. Kabalka, J. Organomet. Chem., 1984, **272**, 331.
1112 R.C. Larock and K. Narayanan, J. Org. Chem., 1984, **49**, 3411.
1113 F. de Sarlo, A. Guarna, A. Goti, and A. Brandi, J. Organomet. Chem., 1984, **269**, 115.
1114 N.A.A. Al-Jabar and A.G. Massey, J. Organomet. Chem., 1984, **275**, 9.
1115 B.P. Bammel and R.F. Evilia, Inorg. Chem., 1984, **23**, 1574.
1116 L. Casella, M. Gullotti, and A. Rockenbauer, J. Chem. Soc., Dalton Trans., 1984, 1033.
1117 R.J. Abraham, P.S. Clezy, and Le Van Thuc, Aust. J. Chem., 1984, **37**, 335.
1118 H.B. Stegmann, E. Nagel, R. Merkle, and K. Scheffler, Fresenius' Z. Anal. Chem., 1983, **316**, 678 (Chem. Abstr., 1984, **100**, 79 206).
1119 R. Shunmugam and D.N. Sathyanarayana, Indian J. Chem., Sect. A, 1983, **22**, 784 (Chem. Abstr., 1984, **100**, 44 223).
1120 R. Shunmugam and D.N. Sathyanarayana, Indian J. Chem., Sect. A, 1984, **23**, 742 (Chem. Abstr., 1984, **101**, 162 646).
1121 T. Tatarowski, M. Kubiak, T. Glowiak, J.P. Morawiec, H. Kozlowski, and M. Gosalvez, Inorg. Chim. Acta, 1984, **93**, L3.
1122 L.V. Sudha and D.N. Sathyanarayana, J. Coord. Chem., 1984, **13**, 207.
1123 S. Muralidharan, K.S. Nagaraja, and M.R. Udupa, Polyhedron, 1984, **3**, 619.
1124 B.P. Bammel and R.F. Evilia, Inorg. Chim. Acta, 1984, **81**, L5.
1125 P. Gettins and J.E. Coleman, J. Biol. Chem., 1984, **259**, 4991 (Chem. Abstr., 1984, **100**, 205 682).
1126 M.F. Summers and L.G. Marzilli, Inorg. Chem., 1984, **23**, 521.
1127 R.A. Bulman and J.K. Nicholson, J. Am. Chem. Soc., 1984, **106**, 1118.
1128 A. Adeyamo, A. Shamin, and A. Turner, Inorg. Chim. Acta, 1984, **91**, L23.
1129 G. Crisponi, R. Caminiti, S. Biagini, M. Casu, and A. Lai, Polyhedron, 1984, **3**, 1105.
1130 N.W. Alcock, E.H. Curzon, P. Moore, and C. Pierpoint, J. Chem. Soc., Dalton Trans., 1984, 605.
1131 N.W. Alcock, E.H. Curzon, and P. Moore, J. Chem. Soc., Dalton Trans., 1984, 2813.
1132 E.S. Raper, A.R.W. Jackson, and D.J. Gardiner, Inorg. Chim. Acta, 1984, **84**, L1.
1133 K.J. Fisher and E.C. Alyea, Polyhedron, 1984, **3**, 509.
1134 P. Gettins and J.E. Coleman, J. Biol. Chem., 1984, **259**, 4987 (Chem. Abstr., 1984, **100**, 205 681).
1135 K.M. Welsh and B.S. Cooperman, Biochemistry, 1984, **23**, 4947.
1136 P. Gettins and J.E. Coleman, J. Biol. Chem., 1984, **259**, 11036 (Chem. Abstr., 1984, **101**, 186 810).
1137 C.T. Hunt, Y. Boulanger, S.W. Fesik, and I.M. Armitage, Bull. Magn. Reson. 1984, **6**, 55 (Chem. Abstr., 1984, **101**, 2519).
1138 D. Neuhaus, G. Wagner, M. Vasak, J.H.R. Jeremias, and K. Wüthrich, Eur. J Biochem., 1984, **143**, 659 (Chem. Abstr., 1984, **101**, 206 284).
1139 A. Andersson, T. Drakenberg, E. Thulin, H. Vogel, and S. Forsén, Dev Biochem., 1983, **25**, 165 (Chem. Abstr., 1984, **100**, 152 662).
1140 E. Thulin, A. Andersson, T. Drakenberg, S. Forsen, and H.J. Vogel Biochemistry, 1984, **23**, 1862 (Chem. Abstr., 1984, **100**, 134 510).
1141 H.R. Engeseth, D.R. McMillin, and J.D. Otvos, J. Biol. Chem., 1984, **259** 4822 (Chem. Abstr., 1984, **100**, 187 620).

1142 J.L. Marshall, J.E. Booth, and J.W. Williams, J. Biol. Chem., 1984, **259**, 3033 (Chem. Abstr., 1984, **100**, 152 998).
1143 D. Dakternieks, Inorg. Chim. Acta, 1984, **89**, 209.
1144 T. Allman and R.G. Goel, Can. J. Chem., 1984, **62**, 615.
1145 M. Camalli, F. Caruso, L. Zambonelli, R.W. Kunz, and P.S. Pregosin, Congr. Naz. Chim. Inorg., [Atti], 16th, 1983, 174 (Chem. Abstr., 1984, **100**, 85 846).
1146 J. Eichbichler and P. Peringer, Chem. Ber., 1984, **117**, 1215.
1147 P.A.W. Dean and J.J. Vittal, J. Am. Chem. Soc., 1984, **106**, 6436.
1148 T. Allman, R.C. Goel, N.K. Jha, and A.L. Beauchamp, Inorg. Chem., 1984, **23**, 914.
1149 N.A. Bell, M. Goldstein, L.A. March, and I.W. Nowell, Inorg. Chim. Acta, 1984, **83**, 75.
1150 D. Dakternieks and C.L. Rolls, Inorg. Chim. Acta, 1984, **87**, 5.
1151 T. Allman and R.G. Goel, Can. J. Chem., 1984, **62**, 621.
1152 P.P. Edwards, J. Phys. Chem., 1984, **88**, 3772.
1153 H. Ohno, Proc. - Electrochem. Soc´, 1984, **84**, 31 (Chem. Abstr., 1984, **100**, 161 887).
1154 E.W. Abel, S.K. Bhargava, and K.G. Orrell, Prog. Inorg. Chem., 1984, **32**, 1
1155 H. Matsui, A. Yoshino, T. Yoshida, and K. Takahashi, Bull. Chem. Soc. Jpn., 1984, **57**, 1052.
1156 E. Kalbarczyk and S. Pasynkiewicz, J. Organomet. Chem., 1984, **273**, C23.
1157 L. Echegoyen, A. Kaifer, H.D. Durst, and G.W. Gokel, J. Org. Chem., 1984, **49**, 688.
1158 G. Laurenczy, E. Bruecher, and L. Radics, Magy. Kem. Foly., 1983, **89**, 519 (Chem. Abstr., 1984, **100**, 57 378).
1159 M.R. Spirlet, J. Rebizant, J.F. Desreux, and M.F. Loncin, Inorg. Chem., 1984, **23**, 359.
1160 J.M. Boncella and R.A. Andersen, J. Chem. Soc., Chem. Commun., 1984, 809.
1161 Y. Ikeda, H. Tomiyasu, and H. Fukutomi, Inorg. Chem., 1984, **23**, 1356.
1162 G.S. Bristow, M.F. Lappert, T.R. Martin, J.L. Atwood, and W.F. Hunter, J. Chem. Soc., Dalton Trans., 1984, 399.
1163 A. Shaver and J.M. McCall, Organometallics, 1984, **3**, 1823.
1164 D.M. Giolando and T.B. Rauchfuss, Organometallics, 1984, **3**, 487.
1165 R.C. Fay, J.R. Weir, and A.H. Bruder, Inorg. Chem., 1984, **23**, 1079.
1166 K. Berg and G. Erker, J. Organomet. Chem., 1984, **263**, 37.
1167 K. Bachmann and D. Rehder, J. Organomet. Chem., 1984, **276**, 177.
1168 D.L. Reger, C.A. Swift, and L. Lebioda, Inorg. Chem., 1984, **23**, 349.
1169 K.A. Mahmoud, A.J. Rest, H.G. Alt, M.E. Eiichner, and B.M. Jansen, J. Chem. Soc., Dalton Trans., 1984, 175.
1170 M. Green, A.G. Orpen, I.D. Salter, and F.G.A. Stone, J. Chem. Soc., Dalton Trans., 1984, 2497.
1171 J.T. Park, J.R. Shapley, M.R. Churchill, and C. Bueno, Inorg. Chem., 1984, **23**, 4476.
1172 M. Green, N.C. Norman, A.G. Orpen, and C.J. Schaverien, J. Chem. Soc., Dalton Trans., 1984, 2455.
1173 W.A. Herrmann, J.L. Hubbard, I. Bernal, J.D. Korp, B.L. Haymore, and G.L. Hillhouse, Inorg. Chem., 1984, **23**, 2978.
1174 M.H. Chisholm, J.C. Huffman, and R.J. Tatz, J. Am. Chem. Soc., 1984, **106**, 5385.
1175 M.H. Chisholm, H.T. Chiu, K. Folting, and J.C. Huffman, Inorg. Chem., 1984, **23**, 4097.
1176 M.H. Chisholm, J.A. Heppert, and J.C. Huffman, J. Am. Chem. Soc., 1984, **106**, 1151.
1177 C.P. Casey, N.W. Vollendorf, and K.J. Haller, J. Am. Chem. Soc., 1984, **106**, 3754.
1178 A.M. Kook, P.N. Nickias, J.P. Selegue, and S.L. Smith, Organometallics, 1984, **3**, 499.
1179 S.R. Allen, M. Green, G. Moran, A.G. Orpen, and G.E. Taylor, J. Chem. Soc., Dalton Trans., 1984, 441.
1180 S.R. Allen, T.H. Glauert, M. Green, K.A. Mead, N.C. Norman, A.G. Orpen, C.J. Schaverien, and P. Woodward, J. Chem. Soc., Dalton Trans., 1984, 2747.

1181 E. Lindner, E.U. Küster, W. Hiller, and R. Fawzi, Chem. Ber., 1984, **117**, 127.
1182 J.M. Maher, J.R. Fox, B.M. Foxman, and N.J. Cooper, J. Am. Chem. Soc., 1984, **106**, 2347.
1183 P.R. Sharp, Organometallics, 1984, **3**, 1217.
1184 M.H. Chisholm, D.M. Hoffman, and J.C. Huffman, J. Am. Chem. Soc., 1984, **106**, 6806.
1185 M.H. Chisholm, K. Folting, D.M. Hoffman, and J.C. Huffman, J. Am. Chem. Soc., 1984, **106**, 6794.
1186 S.R. Allen, S.G. Barnes, M. Green, G. Moran, L. Trollope, N.W. Murrall, A.J. Welch, and D.M. Sharaiha, J. Chem. Soc., Dalton Trans., 1984, 1157.
1187 R.D. Barr, M. Green, T.B. Marder, and F.G.A. Stone, J. Chem. Soc., Dalton Trans., 1984, 1261.
1188 M.L.H. Green and P.A. Newman, J. Chem. Soc., Chem. Commun., 1984, 816.
1189 R.D. Barr, M. Green, J.A.K. Howard, T.B. Marder, A.G. Orpen, and F.G.A. Stone, J. Chem. Soc., Dalton Trans., 1984, 2757.
1190 J. Wachter, J.G. Riess, and A. Mitschler, Organometallics, 1984, **3**, 714.
1191 L. Carlton, W.E. Lindsell, K.J. McCullough, and P.N. Preston, J. Chem. Soc., Dalton Trans., 1984, 1693.
1192 J.F. Blount, G. Hunter, and K. Mislow, J. Chem. Soc., Chem. Commun., 1984, 170.
1193 G. Hunter and K. Mislow, J. Chem. Soc., Chem. Commun., 1984, 172.
1194 M.J. McGlinchey, P. Bougeard, B.G. Sayer, R. Hofer, and C.J.L. Lock, J. Chem. Soc., Chem. Commun., 1984, 789.
1195 E.M. Campi, B.M.K. Gatehouse, W.R. Jackson, I.D. Rae, and M.G. Wong, J. Chem. Soc., Chem. Commun., 1984, 175.
1196 R.A. Jones, J.G. Lasch, N.C. Norman, A.L. Stuart, T.C. Wright, and B.R. Whittlesey, Organometallics, 1984, **3**, 114.
1197 E.W. Abel, M.A.Z. Chowdhury, K.G. Orrell, and V. Sik, J. Organomet. Chem., 1984, **262**, 293.
1198 D.J. Darensbourg and D.J. Zalewski, Organometallics, 1984, **3**, 1598.
1199 C.P. Horwitz, E.M. Holt, and D.F. Shriver, Inorg. Chem., 1984, **23**, 2491.
1200 J.P. Collman and L.K. Woo, Proc. Natl. Acad. Sci. U.S.A., 1984, **81**, 2592 (Chem. Abstr., 1984, **101**, 32 327).
1201 T.W. Coffindaffer, I.P. Rothwell, and J.C. Huffman, Inorg. Chem., 1984, **23**, 1433.
1202 I. Buchanan, C.D. Garner, and W. Clegg, J. Chem. Soc., Dalton Trans., 1984, 1333.
1203 J.L. Corbin, K.F. Miller, N. Pariyadath, J. Heinecke, A.E. Bruce, S. Wherland, and E.I. Stiefel, Inorg. Chem., 1984, **23**, 3404.
1204 H.M. Colquhoun and D.J. Williams, J. Chem. Soc., Dalton Trans., 1984, 1675.
1205 M.H. Chisholm, K. Folting, J.C. Huffman, and A.L. Ratermann, Inorg. Chem., 1984, **23**, 613.
1206 P.J. Manning and L.K. Peterson, Inorg. Chim. Acta, 1984, **88**, L5.
1207 N.G. Connelly, J.A.K. Howard, J.L. Spencer, and P.K. Woodley, J. Chem. Soc., Dalton Trans., 1984, 2003.
1208 N.J. Hazel, J.A.K. Howard, and J.L. Spencer, J. Chem. Soc., Chem. Commun., 1984, 1663.
1209 D. Baudry, M. Ephritikhine, H. Felkin, and J. Zakrzewski, J. Organomet. Chem., 1984, **272**, 391.
1210 T. Beringhelli, E. Cesarotti, G. Ciiani, G. D'Alfonso, and A. Sironi, J. Organomet. Chem., 1984, **268**, C18.
1211 T. Beringhelli, G. Ciani, G. D'Alfonso, P. Romiti, A. Sironi, and M. Freni, Inorg. Chem., 1984, **23**, 2849.
1212 P.O. Nubel and T.L. Brown, J. Am. Chem. Soc., 1984, **106**, 644.
1213 P.O. Nubel and T.L. Brown, Organometallics, 1984, **3**, 29.
1214 K.H. Franzreb and C.G. Kreiter, Z. Naturforsch., Teil B, 1984, **39**, 81 (Chem. Abstr., 1984, **100**, 156 753).
1215 W.A. Kiel, W.E. Buhro, and J.A. Gladysz, Organometallics, 1984, **3**, 879.
1216 T.G. Richmond and D.F. Shriver, Organometallics, 1984, **3**, 305.
1217 K.-W. Lee, W.T. Pennington, A.W. Cordes, and T.L. Brown, Organometallics 1984, **3**, 404.

1218 E.W. Abel, M.Z.A. Chowdhury, K.G. Orrell, and V. Sik, Polyhedron, 1984, **3**, 331.
1219 E.W. Abel, S.K. Bhargava, K. Kite, K.G. Orrell, V. Sik, and B.L. Williams, J. Chem. Soc., Dalton Trans., 1984, 365.
1220 S.G. Davies, J. Hibbard, S.J. Simpson, S.E. Thomas, and O. Watts, J. Chem. Soc., Dalton Trans., 1984, 701.
1221 J. Vites, C. Eigenbrot, and T.P. Fehlner, J. Am. Chem. Soc., 1984, **106**, 4633.
1222 C.P. Horwitz and D.F. Shriver, Organometallics, 1984, **3**, 756.
1223 T.V. Ashworth, A.A. Chalmers, E. Meintjies, H.E. Oosthuizen, and E. Singleton, Organometallics, 1984, **3**, 1485.
1224 K. Osakada, T. Yamamoto, and A. Yamamoto, Inorg. Chim. Acta, 1984, **90**, L5.
1225 C.E. Kampe, N.M. Boag, C.B. Knobler, and H.D. Kaesz, Inorg. Chem., 1984, **23**, 1390.
1226 E. Rosenberg, E.V. Anslyn, C. Barner-Thorsen, S. Aime, D. Osella, R. Gobetto, and L. Milone, Organometallics, 1984, **3**, 1790.
1227 C.E. Kampe and H.D. Kaesz, Inorg. Chem., 1984, **23**, 4646.
1228 J.A.K. Howard, I.D. Slater, and F.G.A. Stone, Polyhedron, 1984, **3**, 567.
1229 A.A. Koridze, O.A. Kizas, N.E. Kolobova, P.V. Petrovskii, and E.I. Fedin, J. Organomet. Chem., 1984, **265**, C33.
1230 B.F.G. Johnson, J. Lewis, P.R. Raithby, S.N. Azman, B. Syed-Mustaffa, M.J. Taylor, K.H. Whitmore, and W. Clegg, J. Chem. Soc., Dalton Trans., 1984, 2111.
1231 G.E. Hawkes, E.W. Randall, S. Aime, D. Osella, and J.E. Elliott, J. Chem. Soc., Dalton Trans., 1984, 279.
1232 A.A. Koridze, O.A. Kizas, N.E. Kolobova, and P.V. Petrovskii, Izv. Akad. Nauk SSSR, Ser. Khim., 1984, 472 (Chem. Abstr., 1984, **101**, 55 303).
1233 Y.-C. Lin, A. Mayr, C.B. Knobler, and H.D. Kaesz, J. Organomet. Chem., 1984, **272**, 207.
1234 D.L. Reger, K.A. Belmore, E. Mintz, and P.J. McElligott, Organometallics, 1984, **3**, 134.
1235 G. Consiglio, F. Bangerter, C. Darpin, F. Morandini, and V. Lucchini, Organometallics, 1984, **3**, 1446.
1236 G.J. Kruger, L. Linford, H.G. Raubenheimer, and A.A. Chalmers, J. Organomet. Chem., 1984, **262**, 69.
1237 Z. Goldschmidt and Y. Bakal, J. Organomet. Chem., 1984, **269**, 191.
1238 J.A.S. Howell, D.T. Dixon, and J.C. Kola, J. Organomet. Chem., 1984, **266**, 69.
1239 M. Moll, H. Behrens, W. Popp, G. Liehr, and W.P. Fehlhammer, Z. Naturforsch., Teil B, 1983, **38**, 1446 (Chem. Abstr., 1984, **100**, 139 289).
1240 S. Aime, G. Nicola, D. Osella, A.M.M. Lanfredi, and A. Tiripicchio, Inorg. Chim. Acta, 1984, **85**, 161.
1241 L. Harland, G.R. Stephenson, and M.J. Whittaker, J. Organomet. Chem., 1984, **263**, C30.
1242 B.E. Mann, C.M. Spencer, B.F. Taylor, and P. Yavari, J. Chem. Soc., Dalton Trans., 1984, 2027.
1243 K.G. Orrell, V. Šik, C.H. Brubaker, jun., and B. McCulloch, J. Organomet. Chem., 1984, **276**, 267.
1244 B. McCulloch and C.H. Brubaker, jun., Organometallics, 1984, **3**, 1707.
1245 H. Werner, R. Werner, and C. Burschka, Chem. Ber., 1984, **117**, 152.
1246 K. Itoh, H. Nagashima, T. Ohshima, N. Oshima, and H. Nishiyama, J. Organomet. Chem., 1984, **272**, 179.
1247 H. Werner and R. Werner, Chem. Ber., 1984, **117**, 142.
1248 J. Keijsper, L. Polm, G. van Koten, K. Vrieze, G. Abbel, and C.H. Stam, Inorg. Chem., 1984, **23**, 2142.
1249 A. Mayr, Y.C. Lin, N.M. Boag, C.E. Kampe, C.B. Knobler, and H.D. Kaesz, Inorg. Chem., 1984, **23**, 4640.
1250 A. Fumagalli and G. Ciani, J. Organomet. Chem., 1984, **272**, 91.
1251 B.F.G. Johnson, J. Lewis, J.M. Mace, P.R. Raithby, R.E. Stevens, and W.L. Gladfelter, Inorg. Chem., 1984, **23**, 1600.
1252 Y. Masuda and H. Yamatera, J. Phys. Chem., 1984, **88**, 3425.
1253 H. t. Dieck, W. Kollvitz, and I. Kleinwächter, Inorg. Chem., 1984, **23**,

2685.
1254 D.J. Wink and N.J. Cooper, J. Chem. Soc., Dalton Trans., 1984, 1257.
1255 J.R. Bleeke and W.-J. Peng, Organometallics, 1984, **3**, 1422.
1256 L.C. Ananias de Carvalho, M. Dartiguenave, Y. Dartiguenave, and A.L. Beauchamp, J. Am. Chem. Soc., 1984, **106**, 6848.
1257 B.R. Sutherland and M. Cowie, Inorg. Chem., 1984, **23**, 1290.
1258 O.W. Howarth, C.H. McAteer, P. Moore, and G.E. Morris, J. Chem. Soc., Dalton Trans., 1984, 1171.
1259 G.B. Robertson and P.A. Tucker, Aust. J. Chem., 1984, **37**, 257.
1260 S. Okeya, N.J. Meanwell, B.F. Taylor, K. Isobe, A.V. de Miguel, and P.M. Maitlis, J. Chem. Soc., Dalton Trans., 1984, 1453.
1261 W.D. Jones and F.J. Feher, Inorg. Chem., 1984, **23**, 2376.
1262 B.M. Louie, S.J. Rettig, A. Storr, and J. Trotter, Can. J. Chem., 1984, **62**, 1057.
1263 P.A. Chaloner, J. Organomet. Chem., 1984, **266**, 191.
1264 M.J. Freeman, A.D. Miles, M. Murray, A.G. Orpen, and F.G.A. Stone, Polyhedron, 1984, **3**, 1093.
1265 C.H. Bushweller, C.D. Rithner, and D.J. Butcher, Inorg. Chem., 1984, **23**, 1967.
1266 Y. Yamamoto and H. Yamazaki, Bull. Chem. Soc. Jpn., 1984, **57**, 297.
1267 R.J. Crowte and J. Evans, J. Chem. Soc., Chem. Commun., 1984, 1332.
1268 A. Fumagalli, S. Martinengo, P. Chini, D. Galli, B.T. Heaton, and R. Della Pergola, Inorg. Chem., 1984, **23**, 2947.
1269 A.R. Siedle, R.A. Newmark, L.H. Pignolet, and R.D. Howells, J. Am. Chem. Soc., 1984, **106**, 1510.
1270 J.A.S. Duncan, T.A. Stephenson, M.D. Walkinshaw, D. Hedden, and D.M. Roundhill, J. Chem. Soc., Dalton Trans., 1984, 801.
1271 B.M. Mattson, A.E. Madera, and M.C. Palazzotto, J. Coord. Chem., 1984, **13**, 321.
1272 D.R. Eaton, M.J. McGlinchey, K.A. Moffat, and R.J. Buist, J. Am. Chem. Soc., 1984, **106**, 8110.
1273 B.M. Trzcinska, J.P. Fackler, jun., and A.B. Anderson, Organometallics, 1984, **3**, 319.
1274 M.D. Fryzuk and P.A. MacNeil, J. Am. Chem. Soc., 1984, **106**, 6993.
1275 H. Lehmkuhl, C. Naydowski, F. Danowski, M. Bellenbaum, R. Benn, A. Rufińska, G. Schroth, R. Mynott, and S. Pasynkiewicz, Chem. Ber., 1984, **117**, 3231.
1276 E.W. Abel, P.K. Mittal, K.G. Orrell, V. Šik, and T.S. Cameron, J. Chem. Soc., Chem. Commun., 1984, 1312.
1277 E.W. Abel, G.D. King, K.G. Orrell, V. Šik, T.S. Cameron, and K. Jochem, J. Chem. Soc., Dalton Trans., 1984, 2047.
1278 L. Manojlovic-Muir, K.W. Muir, A.A. Frew, S.S.M. Ling, M.A. Thomson, and R.J. Puddephatt, Organometallics, 1984, **3**, 1637.
1279 J.D. Scott and R.J. Puddephatt, J. Chem. Soc., Chem. Commun., 1984, 193.
1280 D.A. Clark, R.D.W. Kemmitt, M.A. Mazid, P. McKenna, D.R. Russell, M.D. Schilling, and L.J.S. Sherry, J. Chem. Soc., Dalton Trans., 1984, 1993.
1281 M. Pfeffer, J. Fischer, and A. Mitschler, Organometallics, 1984, **3**, 1531.
1282 H. Tanaka, K. Isobe, S. Kawaguchi, and S. Okeya, Bull. Chem. Soc. Jpn., 1984, **57**, 1850.
1283 H. Kurosawa, N. Asada, A. Urabe, and M. Emoto, J. Organomet. Chem., 1984, **272**, 321.
1284 R Lazzaroni, G. Uccello-Barretta, S. Bertozzi, and P. Salvadori, J. Organomet. Chem., 1984, **275**, 145.
1285 H. Meyer and A. Zschunke, J. Organomet. Chem., 1984, **269**, 209.
1286 R.A. Head and M.I. Tabb, J. Mol. Catal., 1984, **26**, 149.
1287 M. Onishi, K. Hiraki, A. Ueno, Y. Yamaguchi, and Y. Ohama, Inorg. Chim. Acta, 1984, **82**, 121.
1288 S. Okeya, T. Miyamoto, S. Ooi, Y. Nakamura, and S. Kawaguchi, Bull. Chem. Soc. Jpn., 1984, **57**, 395.
1289 H. Werner, H. Otto, Tri Ngo Khac, and C. Burschka, J. Organomet. Chem., 1984, **262**, 123.
1290 A.K.H. Al-Sa'ady, K. Moss, C.A. McAuliffe, and R.V. Parish, J. Chem. Soc.,

Dalton Trans., 1984, 1609.
1291 P.H.M. Budzelaar, J. Boersma, G.J.M. van der Kerk, and A.L. Spek, Organometallics, 1984, **3**, 1187.
1292 P. Mirti, Transition Met. Chem., 1984, **9**, 357.
1293 L.E. Nivorozhkin, Yu.E. Chernysh, G.S. Borodkin, R.Ya. Olekhnovich, Yu.A. Zhdanov, and V.I. Minkin, Dokl. Akad. Nauk SSSR, 1983, **273**, 144 (Chem. Abstr., 1984, **100**, 74 697).
1294 M. Vašák, J. Mol. Catal., 1984, **23**, 293.
1295 T.L. Venable, R.B. Maybard, and R.N. Grimes, J. Am. Chem. Soc., 1984, **106**, 6187.
1296 G.B. Jacobsen, J.H. Morris, and D. Reed, J. Chem. Soc., Dalton Trans., 1984, 415.
1297 J.A. Long, T.B. Marder, P.E. Behnken, and M.F. Hawthorne, J. Am. Chem. Soc., 1984, **106**, 2979.
1298 B. Wrackmeyer, Organometallics, 1984, **3**, 1.
1299 R. Koester, G. Seidel, and B. Wrackmeyer, Angew. Chem., Int. Ed. Engl., 1984, **23**, 512.
1300 B.L. Li, M.A. Goodman, and R.H. Neilson, Inorg. Chem., 1984, **23**, 1368.
1301 W. Kliegel, S.J. Rettig, and J. Trotter, Can. J. Chem., 1984, **62**, 515.
1302 P.A. Odorisio, S.D. Pastor, J.D. Spivack, D. Bini, and R.K. Rodebaugh, Phosphorus Sulphur, 1984, **19**, 285.
1303 D.L. Grossmann and D.T. Haworth, Inorg. Chim. Acta, 1984, **84**, L17.
1304 C. Eaborn, P.B. Hitchcock, P.D. Lickiss, A. Pidcock, and K.D. Safa, J. Chem. Soc., Dalton Trans., 1984, 2015.
1305 G. Glebe, M. Nix, and K. Hensen, Chem. Ber., 1984, **117**, 797.
1306 H.R. O'Neal and R.H. Neilson, Inorg. Chem., 1984, **23**, 1372.
1307 V.V. Negrebetskii, O.V. Balitskaya, V.I. Kal'chenko, and M.Yu. Kornilov, Zh. Obshch. Khim., 1984, **54**, 76 (Chem. Abstr., 1984, **100**, 174 936).
1308 S.D. Pastor, J.D. Spivack, R.K. Rodebaugh, and D. Bini, J. Org. Chem., 1984, **49**, 1297.
1309 M. Oki and M. Ohira, Bull. Chem. Soc. Jpn., 1984, **57**, 3117.
1310 K. Issleib, E. Leissring, and M. Riemer, Z. Anorg. Allg. Chem., 1984, **519**, 75.
1311 J. Grobe, G. Henkel, B. Krebs, and N. Voulgarakis, Z. Naturforsch., Teil B, 1984, **39**, 341.
1312 V.M. Litvinov, B.D. Lavrukhin, V.S. Papkov, and A.A. Zhdanov, Dokl. Akad. Nauk SSSR, 1983, **271**, 900 (Chem. Abstr., 1984, **100**, 52 290).
1313 C.-h. Yu, S.-C. Chang, and C.-s. Liu, J. Org. Chem., 1983, **48**, 5228.
1314 J.L. Atwood, A.D. McMaster, R.D. Rogers, and S.R. Stobart, Organometallics, 1984, **3**, 1500.
1315 M. Bee, A.J. Dianoux, and F. Volino, Mol. Phys., 1984, **51**, 221.
1316 J.C. Martin and W.H. Stevenson, tert., Phosphorus Sulphur, 1983, **18**, 81.
1317 R.J.P. Corriu, A. Kpoton, M. Poirier, G. Royo, and J.Y. Corey, J. Organomet. Chem., 1984, **277**, C25.
1318 P. Jutzi and H. Saleske, Chem. Ber., 1984, **117**, 222.
1319 W.B. Jennings and M. Rutherford, Tetrahedron Lett., 1984, **25**, 3131.
1320 W.B. Jennings, M.S. Tolley, J.H. Hargis, S.D. Worley, and L. Chang, J. Chem. Soc., Perkin Trans. 2, 1984, 1207.
1321 V.V. Negrebetskii, V.I. Kal'chenko, R.B. Rudyi, and L.N. Markovskii, Zh. Obshch. Khim., 1984, **54**, 71 (Chem. Abstr., 1984, **100**, 174 935).
1322 A. Connelly and R.K. Harris, J. Chem. Soc., Dalton Trans., 1984, 1547.
1323 D.B. Denney, D.Z. Denney, and L. Liu, Huaxue Xuebao, 1984, **42**, 293 (Chem. Abstr., 1984, **101**, 55 215).
1324 L. Shen, J. Qiu, L. Li, and Y. Xiao, Fenzi Kexue Yu Huaxue Yanjiu, 1983, **3**, 47 (Chem. Abstr., 1984, **100**, 121 243).
1325 D.W. Aksnes and K. Ramstad, Acta Chem. Scand., Ser. A, 1984, **38**, 781.
1326 F.Y. Fujiwara and C. Airoldi, J. Phys. Chem., 1984, **88**, 1640.
1327 H.D. Luedemann and E.W. Lang, J. Phys., Colloq., 1984, (C7), 41 (Chem. Abstr., 1984, **101**, 239 172).
1328 E.W. Lang, H.D. Luedemann, and L. Piculell, J. Chem. Phys., 1984, **81**, 3820.
1329 C. Chachaty and J.P. Quaegebeur, Mol. Phys., 1984, **52**, 1081.
1330 R.E. Wasylishen, J.B. Macdonald, and J.O. Friedrich, Can. J. Chem., 1984,

62, 1181.
1331 J.A. Anderson, J.D. Odom, and A.J. Zozulin, Organometallics, 1984, **3**, 1458.
1332 R.S. Michalak, S.R. Wilson, and J.C. Martin, J. Am. Chem. Soc., 1984, **106**, 7529.
1333 V.I. Chizhik, Termodinam. Solvatats. Protsessov, Ivanovo, 1983, 6. From Ref. Zh., Khim., 1984, abstr. no. 2B3277 (Chem. Abstr., 1984, **100**, 162 665).
1334 A.M. Sazonov, A.S. Kurakin, G.A. Shmuilovich, and A.B. Porai-Koshits, Izv. Vyssh. Uchebn. Zaved., Khim. Khim. Tekhnol., 1984, **27**, 318 (Chem. Abstr., 1984, **100**, 181 021).
1335 R.K. Mazitov, Radiospektroskopiya (Perm), 1983, **15**, 44 (Chem. Abstr., 1984, **101**, 82 846).
1336 Y. Nakamura, M. Niibe, and M. Shimoji, J. Phys. Chem., 1984, **88**, 3755.
1337 S. Engstroem, B. Joensson, and R.W. Impey, J. Chem. Phys., 1984, **80**, 5481.
1338 A.Yu. Efimov, L.S. Lilich, M.K. Khripun, and A.G. Shmatko, Teor. Eksp. Khim., 1984, **20**, 590 (Chem. Abstr., 1984, **101**, 239 177).
1339 A. Rauhvargers, T.I. Gorbovitskaya, I. Kreitus, and J. Tiliks, Latv. PSR Zinat. Akad. Vestis, Kim. Ser., 1984, 207 (Chem. Abstr., 1984, **100**, 216 646).
1340 D.W. James and R.E. Mayes, J. Phys. Chem., 1984, **88**, 637.
1341 A.S.N. Murthy and A.P. Bhardwaj, J. Chem. Soc., Perkin Trans. 2, 1984, 727.
1342 A. Delville, L. Herwats, and P. Laszlo, Nouv. J. Chim., 1984, **8**, 557 (Chem. Abstr., 1984, **101**, 236 003).
1343 M. Niibe, Y. Nakamura, and M. Shimoji, J. Phys. Chem., 1984, **88**, 3555.
1344 M. Niibe and Y. Nakamura, J. Phys. Chem., 1984, **88**, 5608.
1345 O.A. El Seoud, M.I. El Seoud, P.A.R. Pires, and J.P.S. Farah, J. Phys. Chem., 1984, **88**, 2669.
1346 M.I. El Seoud and O.A. El Seoud, Ber. Bunsen-Ges. Phys. Chem., 1984, **88**, 742.
1347 E.W. Lang, W. Fink, and H.D. Luedemann, J. Phys. Colloq., 1984, (C7), 173 (Chem. Abstr., 1984, **101**, 221 194).
1348 U. Braendle, E. Kammerer, D. Koehnlein, and O. Lutz, Z. Naturforsch., Teil A, 1984, **39**, 615 (Chem. Abstr., 1984, **101**, 126 194).
1349 B.M. Braun and M. Holz, J. Solution Chem., 1983, **12**, 685.
1350 B.M. Braun and M. Holz, Z. Phys. Chem. (Wiesbaden), 1983, **135**, 77.
1351 Yu.G. Vlasov and P.P. Antonov, Izv. Vyssh. Uchebn. Zaved., Khim. Khim. Tekhnol., 1984, **27**, 755 (Chem. Abstr., 1984, **101**, 117 633).
1352 W.W. Warren, jun., S. Sotier, and G.F. Brennert, Phys. Rev. B: Condens. Matter, 1984, **30**, 65 (Chem. Abstr., 1984, **101**, 82 854).
1353 R. McIntyre, J. Electroanal. Chem. Interfacial Electrochem., 1984, **167**, 157.
1354 D.L. Pisaniello, L. Helm, P. Meier, and A.E. Merbach, Inorg. Chim. Acta, 1984, **94**, 91.
1355 I. Ursu, D.E. Demco, M. Bogdan, P. Fitori, and A. Darabont, Report, 1983, IFIN-CS-7-1983, 17 pp. Avail. INIS. From INIS Atomindex, 1984, 15, abstr. no. 15:043282 (Chem. Abstr., 1984, **101**, 203 023).
1356 Y. Ikeda, H. Tomiyusu, and H. Fukutomi, Bull. Chem. Soc. Jpn., 1984, **57**, 2925.
1357 M. Harada, Y. Ikeda, H. Tomiyasu, and H. Fukutomi, Chem. Lett., 1984, 1195.
1358 A. Saito, Y. Ikeda, H. Tomiyasu, and H. Fukuto-i, Bull. Res. Lab. Nucl. React. (Tokyo Inst. Technol.), 1984, **9**, 43 (Chem. Abstr., 1984, **101**, 44 114).
1359 T. Kushnir and G. Navon, J. Magn. Reson., 1984, **56**, 373.
1360 S.F. Lincoln, A.M. Hounslow, D.L. Pisaniello, B.G. Doddridge, J.H. Coates, A.E. Merbach, and D. Zbinden, Inorg. Chem., 1984, **23**, 1090.
1361 Y. Masuda and H. Yamatera, Bull. Chem. Soc. Jpn., 1984, **57**, 58.
1362 S. Chopra, R.E.D. McClung, and R.B. Jordan, J. Magn. Reson., 1984, **59**, 361.
1363 K.E. Newman, Inorg. Chim. Acta, 1984, **89**, L3.
1364 P. Moore, J. Sachinidis, and G.R. Willey, J. Chem. Soc., Dalton Trans., 1984, 1323.
1365 S.F. Lincoln, J.H. Coates, D.A. Hadi, and D.L. Pisaniello, Inorg. Chim. Acta, 1984, **81**, L9.
1366 S.F. Lincoln, T.W. Hambley, D.L. Pisaniello, and J.H. Coates, Aust. J. Chem., 1984, **37**, 713.

1367 S.F. Lincoln, J.H. Coates, B.G. Doddridge, and D.L. Pisaniello, Aust. J. Chem., 1984, **37**, 947.
1368 L. Helm, L.I. Elding, and A.E. Merbach, Helv. Chim. Acta, 1984, **67**, 1453.
1369 I. Nagypal and F. Debreczeni, Inorg. Chim. Acta, 1984, **81**, 69.
1370 I. Bertini, G. Lanini, and C. Luchinat, Inorg. Chim. Acta, 1984, **93**, 51.
1371 L. Guinand, K.L. Hobt, E. Mittermaier, E. Roessler, A. Schwenk, and H. Schneider, Z. Naturforsch., Teil A, 1984, **39**, 83 (Chem. Abstr., 1984, **100**, 92 223).
1372 A. Kamars and E. Svarcs, Latv. PSR Zinat. Akad. Vestis, Kim. Ser., 1984, 294 (Chem. Abstr., 1984, **101**, 98 468).
1373 H.G. Hertz and R. Maurer, Z. Phys. Chem. (Wiesbaden), 1983, **135**, 107.
1374 M. Saunders, S. Saunders, and C.A. Johnson, J. Am. Chem. Soc., 1984, **106**, 3098.
1375 G.V. Lagodzinskaya, I.Yu. Kozyreva, N.G. Yunda, and G.B. Manelis, Izv. Akad. Nauk SSSR, Ser. Khim., 1984, 2212 (Chem. Abstr., 1984, **101**, 239 178).
1376 T.K. Chandrashekar and V. Krishnan, Bull. Soc. Chim. Fr., 1984, 42.
1377 R. Caminiti, P. Cucca, M. Monduzzi, G. Saba, and G. Crisponi, J. Chem. Phys., 1984, **81**, 543.
1378 V.A. Maroni and M.L. Maciejewski, Proc. - Electrochem. Soc., 1984, **84**, 359 (Chem. Abstr., 1984, **100**, 182 496).
1379 G. Doucet Ladeveze, L. Rodehuser, P.R. Rubini, and J.J. Delpuech, Nouv. J. Chim., 1984, **8**, 93 (Chem. Abstr., 1984, **101**, 12 811).
1380 R.J.M. Nolte and D.J. Cram, J. Am. Chem. Soc., 1984, **106**, 1416.
1381 R.W. Hay and D.M.S. Clark, Inorg. Chim. Acta, 1984, **83**, L23.
1382 C.S. Chen, S.J. Wang, and S.C. Wu, Inorg. Chem., 1984, **23**, 3901.
1383 W. Offermann and E. Weber, Chem. Ber., 1984, **117**, 234.
1384 R.R. Rhinebarger, J.W. Rovang, and A.I. Popov, Inorg. Chem., 1984, **23**, 2557.
1385 K. Matsushita, O. Kamo, Y. Terada, T. Yoshida, and H. Okabayashi, Chem. Scr., 1984, **23**, 228 (Chem. Abstr., 1984, **101**, 102 782).
1386 P. Haake and R.V. Prigodich, Inorg. Chem., 1984, **23**, 457.
1387 J.A. Walmsley, R.G. Barr, E. Bouhoutsos-Brown, and T.J. Pinnavaia, J. Phys. Chem., 1984, **88**, 2599.
1388 W.H. Braunlin and L. Nordenskioeld, Eur. J. Biochem., 1984, **142**, 133 (Chem. Abstr., 1984, **101**, 85 884).
1389 L. Piculell, B. Lindman, and R. Einarsson, Biopolymers, 1984, **23**, 1683.
1390 W.H. Braunlin, H.J. Vogel, and S. Forsén, Eur. J. Biochem., 1984, **142**, 139 (Chem. Abstr., 1984, **101**, 86 009).
1391 J.W. Pettegrew, D.E. Woessner, N.J. Minshew, and T. Glonek, J. Magn. Reson., 1984, **57**, 185.
1392 B. Lindman, M.C. Puyal, N. Kamenka, R. Rymden, and P. Stilbs, J. Phys. Chem., 1984, **88**, 5048.
1393 N.A.P. Vaz and J.W. Doane, Phys. Rev. A, 1983, **28**, 3663 (Chem. Abstr., 1984, **100**, 43 449).
1394 K. Rendall, G.J.T. Tiddy, and M.A. Trevethan, J. Colloid Interface Sci., 1984, **98**, 565 (Chem. Abstr., 1984, **100**, 145 467).
1395 D. Perahia, D. Goldfarb, and Z. Luz, Mol. Cryst. Liq. Cryst., 1984, **108**, 107 (Chem. Abstr., 1984, **101**, 64 777).
1396 P.K. Kilpatrick and W.G. Miller, J. Phys. Chem., 1984, **88**, 1649.
1397 H. Ohno, Nippon Genshiryoku, [Rep.] JAERI-M 1984, JAERI-M-84-060, 35 pp. (Chem. Abstr., 1984, **101**, 235 898).
1398 A.B. Goel and E.C. Ashby, Inorg. Chim. Acta, 1984, **87**, 61.
1399 H. Kondo, K. Miura, S. Uno, and J. Sunamoto, J. Inorg. Biochem., 1984, **21**, 93.
1400 S. Khazaeli and R.E. Viola, J. Inorg. Biohem., 1984, **22**, 33.
1401 A. Khan, K. Fontell, and B. Lindman, Surfactants Solution, [Proc. Int. Symp.] 4th, 1982 (Pub. 1984), **1**, 193 (Chem. Abstr., 1984, **101**, 28 783).
1402 T. Huang, S. Jiang, L. Chen, S. Lian, K. Ruan, and Y. Li, Jiegou Huaxue, 1983, **2**, 175 (Chem. Abstr., 1984, **101**, 165 701).
1403 V.L. Pecoraro, J.D. Hermes, and W.W. Cleland, Biochemistry, 1984, **23**, 5262 (Chem. Abstr., 1984, **101**, 186 300).
1404 R.K. Gupta, P. Gupta, W.D. Yushok, and Z.B. Rose, Biochem. Biophys. Res. Commun., 1983, **117**, 210 (Chem. Abstr., 1984, **100**, 47 236).

1405 R.K. Gupta, P. Gupta, W.D. Yushok, and Z.B. Rose, Physiol. Chem. Phys., 1983, **15**, 265 (Chem. Abstr., 1984, **100**, 153 313).
1406 C.J. Turner and P.B. Garlick, J. Magn. Reson., 1984, **57**, 221.
1407 K.V. Vasavada, B.D. Ray, and B.D. Nageswara Rao, J. Inorg. Biochem., 1984, **21**, 323.
1408 T. Drakenberg, T. Andersson, S. Forsén, and T. Wieloch, Biochemistry, 1984, **23**, 2387 (Chem. Abstr., 1984, **100**, 205 529).
1409 H.J. Vogel, T. Andersson, W.H. Braunlin, T. Drakenberg, and S. Forsén, Biochem. Biophys. Res. Commun., 1984, **122**, 1350.
1410 H.B. Amin, S.S. Al-Showiman, and I.M. Al-Najjar, J. Chem. Soc. Pak., 1983, **5**, 201 (Chem. Abstr., 1984, **100**, 78 693).
1411 L. Shen, Y. Gao, Y. Xiao, and J. Ni, Zhongguo Kexueyuan Changchun Yingyong Huaxue Yanjiuso Jikan, 1983, **20**, 1 (Chem. Abstr., 1984, **101**, 98 650).
1412 C.D. Eads, P. Mulqueen, W.D. Horrocks, jun., and J.J. Villafranca, J. Biol. Chem., 1984, **259**, 9379 (Chem. Abstr., 1984, **101**, 125 135).
1413 V.V. Ustyak and Yu.I. Sal'nikov, Zh. Neorg. Khim., 1984, **29**, 806 (Chem. Abstr., 1984, **100**, 145 882).
1414 V.V. Ustyak and Yu.I. Sal'nikov, Zh. Neorg. Khim., 1984, **29**, 679 (Chem. Abstr., 1984, **100**, 145 879).
1415 K. Dill, M.E. Daman, R.L. Batstone-Cunningham, J.M. Lacombe, and A.A. Pavia, Carbohdr. Res., 1983, **123**, 123 (Chem. Abstr., 1984, **100**, 86 102).
1416 K. Dill, M.E. Daman, R.L. Batstone-Cunningham, M. Denarie, and A.A. Pavia, Carbohydr. Res., 1983, **123**, 137 (Chem. Abstr., 1984, **100**, 121 573).
1417 N.A. Kostromina, V.V. Dudko, and N.I. Salo, Ukr. Khim. Zh. (Russ. Ed.), 1984, **50**, 158 (Chem. Abstr., 1984, **100**, 162 743).
1418 Y. Ikeda, H. Tomiyasu, and H. Fukutomi, Inorg. Chem., 1984, **23**, 3197.
1419 M.T. Nunes, V.M.S. Gil, and A.V. Xavier, Inorg. Chim. Acta, 1984, **95**, 13.
1420 L. Pettersson, B. Hedman, I. Andersson, and N. Ingri, Chem. Scr., 1983, **22**, 254.
1421 A.I. Nikolaev, N.I. Kasikova, E.G. Il'in, and M.E. Ignatov, Zh. Neorg. Khim., 1984, **29**, 2062 (Chem. Abstr., 1984, **101**, 117 706).
1422 A.M. Bond, S.W. Carr, and R. Colton, Inorg. Chem., 1984, **23**, 2343.
1423 W.D. Wheeler and J.I. Legg, Inorg. Chem., 1984, **23**, 3798.
1424 A.M.V.S.V. Cavaleiro, V.M.S. Gil, J.D. Pedrosa de Jesus, R.D. Gillard, and P.A. Williams, Transition Met. Chem., 1984, **9**, 62.
1425 A.M.V. Cavaleiro, J.D. Pedrosa de Jesus, R.D. Gillard, and P.A. Williiams, Transition Met. Chem., 1984, **9**, 81.
1426 L.G. Detusheva, E.N. Yurchenko, and S.V. Tkachev, Koord. Khim., 1983, **9**, 1220 (Chem. Abstr., 1984, **100**, 40 571).
1427 E.N. Yurchenko and L.G. Detusheva, Koord. Khim., 1984, **10**, 61 (Chem. Abstr., 1984, **100**, 113 987).
1428 M.K. Kotvanova, M.A. Fedotov, L.P. Kazanskii, and E.A. Torchenkova, Koord. Khim., 1984, **10**, 1062 (Chem. Abstr., 1984, **101**, 162 529).
1429 R. Nielson and S. Wherland, Inorg. Chem., 1984, **23**, 1338.
1430 J.J. Led, J. Phys. Chem., 1984, **88**, 5531.
1431 E.E. Zaev and L.M. Khalilov, Kolloidn. Zh., 1984, **46**, 427 (Chem. Abstr., 1984, **101**, 60 582).
1432 C.F. Barrientos-Penna, C.F. Campana, F.W.B. Einstein, T. Jones, D. Sutton, and A.S. Tracey, Inorg. Chem., 1984, **23**, 363.
1433 E. Horn and M.R. Snow, Aust. J. Chem., 1984, **37**, 1375.
1434 D.L. Davies, B.P. Gracey, V. Guerchais, S.A.R. Knox, and A.G. Orpen, J. Chem. Soc., Chem. Commun., 1984, 841.
1435 D.E. Fjare and W.L. Gladfelter, J. Am. Chem. Soc., 1984, **106**, 4799.
1436 E. Ragg and G.R. Moore, J. Inorg. Biochem., 1984, **21**, 253.
1437 J.D. Satterlee and J.A. Shelnutt, J. Phys. Chem., 1984, **88**, 5487.
1438 K.S. Hagen and R.H. Holm, Inorg. Chem., 1984, **23**, 418.
1439 P. Bernhard, L. Helm, I. Rapaport, A. Ludi, and A.E. Merbach, J. Chem. Soc., Chem. Commun., 1984, 302.
1440 I.T. Fanchiang and J.J. Pignatello, Inorg. Chim. Acta, 1984, **91**, 147.
1441 M.F. Summers, L.G. Marzilli, N. Bresciani-Pahor, and L. Randaccio, J. Am. Chem. Soc., 1984, **106**, 4478.
1442 K.L. Brown and J.M. Hakimi, Inorg. Chem., 1984, **23**, 1756.

1443 J.G. Russell and R.G. Bryant, J. Phys. Chem., 1984, **88**, 4298.
1444 E.V. Makhonina, N.V. Sharova, M.B. Rozenkevich, and Yu.A. Sakharovskii, Koord. Khim., 1983, **9**, 1683 (Chem. Abstr., 1984, **100**, 57 384).
1445 K.L. Brown, J.M. Hakima, and D.W. Jacobsen, J. Am. Chem. Soc., 1984, **106**, 7894.
1446 H. Slebocka-Tilk, J.L. Cocho, Z. Frackman, and R.S. Brown, J. Am. Chem. Soc., 1984, **106**, 2421.
1447 S. Ito and Y. Yano, Bull. Chem. Soc. Jpn., 1984, **57**, 2824.
1448 J.R. Vriesenga and G.H. Fairchild, Inorg. Chim. Acta, 1984, **82**, L21.
1449 F.L. Dickert and W. Gumbrecht, Z. Naturforsch., Teil B, 1984, **39**, 69 (Chem. Abstr., 1984, **100**, 162 578).
1450 Yu.I. Sal'nikov, F.V. Devyatov, N.E. Zhuravleva, and D.V. Golodnitskaya, Zh. Neorg. Khim., 1984, **29**, 2273 (Chem. Abstr., 1984, **101**, 178 608).
1451 E.S. Stoyanov and V.A. Mikhailov, Koord. Khim., 1984, **10**, 518 (Chem. Abstr., 1984, **100**, 198 541).
1452 R.R. Amirov and Z.A. Saprykova, Zh. Neorg. Khim., 1984, **29**, 2138 (Chem. Abstr., 1984, **101**, 117 744).
1453 J. Sachinidis and M.W. Grant, Aust. J. Chem., 1983, **36**, 2019.
1454 W.A. Donaldson, J. Organomet. Chem., 1984, **269**, C25.
1455 K. Miki, K. Yamamoto, N. Kasai, H. Kurosawa, M. Emoto, and A. Urabe, J. Chem. Soc., Chem. Commun., 1984, 1520.
1456 L.Yu. Alt, A.S. Belyi, V.K. Duplyakin, V.I. Perevalova, and N.P. Shitova, React. Kinet. Catal. Lett., 1983, **23**, 13 (Chem. Abstr., 1984, **100**, 92 074).
1457 N.B. Shitova, L.Ya. Alt, V.I. Perevalova, A.S. Belyi, and V.K. Duplyakin, React. Kinet. Catal. Lett., 1983, **23**, 17 (Chem. Abstr., 1984, **100**, 92 203).
1458 K.C. Lee and F. Aubke, Inorg. Chem., 1984, **23**, 2124.
1459 E. Dubler, U.K. Haering, K.H. Scheller, P. Baltzer, and H. Sigel, Inorg. Chem., 1984, **23**, 3785.
1460 E.J. Pulliam and D.R. McMillin, Inorg. Chem., 1984, **23**, 1172.
1461 K. Kobe and J.C. Valdes, Vestn. Slov. Kem. Drus., 1984, **31**, 71 (Chem. Abstr., 1984, **101**, 83 606).
1462 E.L. Gogolashvili, A.V. Zakharov, and Z.A. Saprykova, React. Kinet. Catal. Lett., 1983, **23**, 249 (Chem. Abstr., 1984, **101**, 12 783).
1463 M.A. Fedotov, O.P. Krivoruchko, A.V. Golovin, and R.A. Buyanov, Termodin. i Struktura Gidoksokompleksov v Rastvorakh. Materialy III Vses. Soveshch., L., 1983, 165. From Ref. Zh., Khim., 1984, abstr. no. 7V35 (Chem. Abstr., 1984, **101**, 32 387).
1464 K.B. Yatsimirskii, V.A. Bidzilya, G.G. Talanova, Z.Z. Rozhkova, T.N. Kudrya, A.A. Chaikovskaya, and I.E. Boldeskul, Teor. Eksp. Khim., 1983, **19**, 500 (Chem. Abstr., 1984, **100**, 16 657).
1465 D. Koehnlein, O. Lutz, and P. Ruppert, J. Magn. Reson., 1984, **57**, 486.
1466 G.C. Baumann, M.F. Summers, J.P. Hutchinson, J. Zubieta, and L.G. Marzilli, Inorg. Chem., 1984, **23**, 3104.
1467 B. Szpoganicz and A.E. Martell, Inorg. Chem., 1984, **23**, 4442.
1468 H. Sigel, K.H. Scheller, and R.M. Milburn, Inorg. Chem., 1984, **23**, 1933.
1469 R. Sommer and D. Beyersmann, J. Inorg. Biochem., 1984, **20**, 131.
1470 P.D. Ellis, P. Strang, and J.D. Potter, J. Biol. Chem., 1984, **259**, 10 348 (Chem. Abstr., 1984, **101**, 186 346).
1471 D.L. Rabenstein and R.S. Reid, Inorg. Chem., 1984, **23**, 1246.
1472 A.A. Isab, Inorg. Chim. Acta, 1984, **91**, L35.
1473 A. Kamars and E. Svarcs, PSR Zinat. Akad. Vestis, Kim. Ser., 1984, 299 (Chem. Abstr., 1984, **101**, 60 953).
1474 E.G. Ippolitov, B.N. Chernyshov, G.P. Shchetinina, V.V. Kon'shin, and N.A. Didenko, Deposited Doc., 1982, VINITI 140-83, 27 pp. Avail. VINITI (Chem. Abstr., 1984, **100**, 113 984).
1475 G.A.R. Van Dijk and H.C.A.M. Smoorenburg, J. Electrochem. Soc., 1984, **131**, 345 (Chem. Abstr., 1984, **100**, 114 104).
1476 H.P. Erb and T. Bluhm, J. Phys. Chem., 1984, **88**, 4158.
1477 A.A. Fannin, jun., L.A. King, J.A. Levisky, and J.S. Wilkes, J. Phys. Chem., 1984, **88**, 2609.
1478 T. Matsumoto and K. Ichikawa, J. Am. Chem. Soc., 1984, **106**, 4316.
1479 Yu.A. Buslaev and S.P. Petrosyants, Polyhedron, 1984, **3**, 265.

1480 M. Yokoi, Y. Usami, and E. Kubota, Chem. Lett., 1984, 1477.
1481 V.E. Larchenko, K.I. Popov, A.I. Grigor'ev, L.V. Nikitina, and N.M. Dyatlova, Koord. Khim., 1984, **10**, 492 (Chem. Abstr., 1984, **101**, 32 313).
1482 J.P. Laussac and G. Commenges, Nouv. J. Chim., 1983, **7**, 579 (Chem. Abstr., 1984, **100**, 98 487).
1483 G. Pan, Gaodeng Xuexiao Huaxue Xuebao, 1984, **5**, 395 (Chem. Abstr., 1984, **101**, 117 800).
1484 Y. Nakamura, K. Baba, M. Shimoji, and S. Shimokawa, Ber. Bunsen-Ges. Phys. Chem., 1984, **88**, 398.
1485 I. Bertini, E. Borghi, C. Luchinat, and A. Scozzafava, J. Mol. Struct., 1984, **113**, 191.
1486 R.K. Harris, J. Jones, T.G. Knight, and R.H. Newman, J. Mol. Liq., 1984, **29**, 63 (Chem. Abstr., 1984, **101**, 217 217).
1487 C.J. Creswell, R.K. Harris, and P.T. Jageland, J. Chem. Soc., Chem. Commun., 1984, 1261.
1488 G. Engelhardt and D. Hoebbel, J. Chem. Soc., Chem. Commun., 1984, 514.
1489 T. Birchall and G. Denes, Can. J. Chem., 1984, **62**, 591.
1490 T.G. Appleton, J.R. Hall, A.D. Harris, H.A. Kimlin and I.J. McMahon, Aust. J. Chem., 1984, **37**, 1833.
1491 C. Wang and A.F. Janzen, Can. J. Chem., 1984, **62**, 1563.
1492 T.K. Chandrashekar and V. Krishnan, Can. J. Chem., 1984, **62**, 475.
1493 G. Fraenkel, M. Henrichs, M. Hewitt, and B.M. Su, J. Am. Chem. Soc., 1984, **106**, 255.
1494 P.E.M. Allen, S. Hagias, C. Mair, and E.H. Williams, Ber. Bunsenges. Phys. Chem., 1984, **88**, 623.
1495 K.G. Moloy and T.J. Marks, J. Am. Chem. Soc., 1984, **106**, 7051.
1496 N. Fujiwara, H. Tomiyasu, and H. Fukutomi, Bull. Chem. Soc. Jpn., 1984, **57**, 1576.
1497 R.D. Fischer and K. Yunlu, Z. Naturforsch., Teil B, 1983, **38**, 1369.
1498 P.B. Hitchcock, M.F. Lappert, and R.G. Taylor, J. Chem. Soc., Chem. Commun., 1984, 1082.
1499 A.J. Downs and C.J. Gardner, J. Chem. Soc., Dalton Trans., 1984, 2127.
1500 G.A. Zank and T.B. Rauchfuss, Organometallics, 1984, **3**, 1191.
1501 G. Erker, K. Berg, R. Benn, and G. Schroth, Angew. Chem., Int. Ed. Engl., 1984, **23**, 625.
1502 N.A. Ustynyuk, L.N. Novikova, Yu.F. Oprunenko, S.G. Malyugina, and Yu.A. Ustynyuk, J. Organomet. Chem., 1984, **277**, 75.
1503 A.G. Stepanov, V.M. Nekipelov, and K.I. Zamaraev, Dokl. Akad. Nauk SSSR, 1984, **274**, 1144 (Chem. Abstr., 1984, **101**, 47 603).
1504 A.G. Stepanov and V.M. Nekipelov, React. Kinet. Catal. Lett., 1984, **24**, 361 (Chem. Abstr., 1984, **100**, 216 597).
1505 I. Buchanan, M. Minelli, M.T. Ashby, T.J. King, J.H. Enemark, and C.D. Garner, Inorg. Chem., 1984, **23**, 495.
1506 H. Fischer, S. Zeuner, and J. Riede, Angew. Chem., Int. Ed. Engl., 1984, **23**, 726.
1507 F.W.B. Einstein, T. Jones, R.K. Pomeroy, and P. Rushman, J. Am. Chem. Soc., 1984, **106**, 2707.
1508 L. Mordenti, J.-L. Roustan, and J.G. Riess, Inorg. Chem., 1984, **23**, 4503.
1509 K.L. Chalk and R.K. Pomeroy, Inorg. Chem., 1984, **23**, 444.
1510 L.G. Marzilli, M.F. Summers, J.H. Ramsden, jun., N. Bresciani-Pahor, and L. Randaccio, J. Chem. Soc., Dalton Trans., 1984, 511.
1511 J.M. Brown and A.R. Lucy, J. Chem. Soc., Chem. Commun., 1984, 914.
1512 R. Knorr and F. Ruf, Angew. Chem., Int. Ed. Engl., 1984, **23**, 368.
1513 N.W. Alcock, J.M. Brown, and T.D. MacLean, J. Chem. Soc., Chem. Commun., 1984, 1689.
1514 L.Ya. Alt, V.P. Mel'nikova, A.N. Startsev, and V.K. Duplyakin, React. Kinet. Catal. Lett., 1983, **23**, 21 (Chem. Abstr., 1984, **100**, 92 204).
1515 M.S. Holt, J.J. MacDougall, F. Mathey, and J.H. Nelson, Inorg. Chem., 1984, **23**, 449.
1516 T.A. Van der Knaap, F. Bickelhaupt, J.G. Kraaykamp, G. Van Koten, J.P.C. Bernards, H.T. Edzes, W.S. Veeman, E. De Boer, and E.J. Baerends, Organometallics, 1984, **3**, 1804.

1517 T.A. van der Knaap and F. Bickelhaupt, Phosphorus Sulphur, 1983, **18**, 47.

1518 E.G. Hoffmann, H. Nehl, H. Lehmkuhl, K. Seevogel, and W. Stempfle, Chem. Ber., 1984, **117**, 1364.

1519 C.B. Storm, A.H. Turner, and M.B. Swann, Inorg. Chem., 1984, **23**, 2743.

1520 D.N. Kravtsov, A.S. Peregudov, V.F. Ivanov, and E.I. Fedin, Izv. Akad. Nauk SSSR, Ser. Khim., 1984, 1626 (Chem. Abstr., 1984, **101**, 230 694).

1521 A.M. Bond, R. Colton, M.L. Dillon, J.E. Moir, and D.R. Page, Inorg. Chem., 1984, **23**, 2883.

1522 G.E. Herberich, H. Ohst, and H. Mayer, Angew. Chem., Int. Ed. Engl., 1984, **23**, 969.

1523 R.L. Shuler, R.A. De Marco, and A.D. Berry, Inorg. Chim. Acta, 1984, **85**, 185.

1524 H.P. Erb, H. Felbecker, and T. Bluhm, Ber. Bunsenges. Phys. Chem., 1984, **88**, 122.

1525 A.G. Stepanov and V.M. Nekipelov, React. Kinet. Catal. Lett., 1984, **24**, 351 (Chem. Abstr., 1984, 100, 220 605).

1526 C.T.G. Knight and A.E. Merbach, J. Am. Chem. Soc., 1984, **106**, 804.

1527 M.M. Kubicki, R. Kergoat, J.-E. Guerchais, and P. L'Haridon, J. Chem. Soc., Dalton Trans., 1984, 1791.

1528 T.V. Zykova, V.M. Ismailov, V.V. Moskva, and R.A. Salakhutdinov, Zh. Obshch. Khim., 1984, **54**, 1288 (Chem. Abstr., 1984, **101**, 171 368).

1529 A.-M. Caminade, M. Verrier, C. Ades, N. Paillous, and M. Koenig, J. Chem. Soc., Chem. Commun., 1984, 875.

1530 A. Belforte, F. Calderazzo, A. Morville, G. Pelizzi, and D. Vitali, Inorg. Chem., 1984, **23**, 1504.

1531 J. Pretula, K. Kaluzynski, and S. Penczek, J. Polym. Sci., Polym. Chem. Ed., 1984, **22**, 1251 (Chem. Abstr., 1984, **101**, 55 615).

1532 C. Blonski, M.B. Gasc, A.F. Hegarty, A. Klaebe, and J.J. Perie, J. Am. Chem. Soc., 1984, **106**, 7523).

1533 J.J.C. Van Lier, R.J.M. Hermans, and H.M. Buck, Phosphorus Sulphur, 1984, **19**, 173.

1534 P. Sobota, M. Nowak, and W. Kramarz, J. Organomet. Chem., 1984, **275**, 161.

1535 M. Watanabe, Y. Morii, and S. Sato, Bull. Chem. Soc. Jpn., 1984, **57**, 2914.

1536 E.A. Fushman, A.N. Shupik, V.E. L'vovskii, and L.F. Borisova, Acta Polym., 1983, **34**, 674 (Chem. Abstr., 1984, **100**, 139 672).

1537 P. Sobota and Z. Janas, J. Organomet. Chem., 1984, **276**, 171.

1538 K.I. Arsenin, K.F. Karlysheva, L.A. Malinko, and I.A. Shcheka, Termodin. i Struktura Gidroksokompleksov v Rastvorakh. Materialy III Vses. Soveshch., L., 1983, 109. From Ref. Zh., Khim., 1984, abstr. no. 8B3204 (Chem. Abstr., 1984, **101**, 61 118).

1539 Q.-Z. Shi, T.G. Richmond, W.C. Trogler, and F. Basolo, J. Am. Chem. Soc., 1984, **106**, 71.

1540 P.L. Gaus, S.C. Kao, M.Y. Darensbourg, and L.W. Arndt, J. Am. Chem. Soc., 1984, **106**, 4752.

1541 J.T. Gauntlett, B.F. Taylor, and M.J. Winter, J. Chem. Soc., Chem. Commun., 1984, 420.

1542 M.R. Churchill, J.W. Ziller, J.H. Freudenberger, and R.R. Schrock, Organometallics, 1984, **3**, 1554.

1543 J.H. Freudenberger, R.R. Shrock, W.R. Churchill, A.L. Rheingold, and J.W. Ziller, Organometallics, 1984, **3**, 1563.

1544 T.G. Traylor and K. Stewart, Organometallics, 1984, **3**, 324.

1545 D.T. Dixon, J.C. Kola, and J.A.S. Howell, J. Chem. Soc., Dalton Trans., 1984, 1307.

1546 D.J. Darensbourg and R.L. Gray, Inorg. Chem., 1984, **23**, 2993.

1547 J.L. Corbin, K.F. Miller, N. Pariyadath, S. Wherland, A.E. Bruce, and E.I. Stiefel, Inorg. Chim. Acta, 1984, **90**, 41.

1548 E.G. Il'in, B.V. Golovanov, M.E. Ignatov, V.D. Butskii, and Yu.A. Buslaev, Dokl. Akad. Nauk SSSR, 1984, 276, 612 (Chem. Abstr., 1984, **101**, 158 509).

1549 L.-N. Ji, M.E. Rerek, and F. Basolo, Organometallics, 1984, **3**, 740.

1550 D. Brenner, A. Davison, J. Lister-James, and A.G. Jones, Inorg. Chem., 1984, **23**, 3793.

1551 P.C. Heah and J.A. Gladysz, J. Am. Chem. Soc., 1984, **106**, 7636.

1552 E.H.K. Zeiher, D.G. DeWit, and K.G. Caulton, J. Am. Chem. Soc., 1984, **106**, 7006.
1553 J.W. Connolly, Organometallics, 1984, **3**, 1333.
1554 J.S. Holmgren and J.R. Shapley, Organometallics, 1984, **3**, 1322.
1555 G. Cardaci, G. Reichenbach, and G. Bellachioma, Inorg. Chem., 1984, **23**, 2936.
1556 C.P. Casey, M.W. Meszaros, S.R. Marder, and P.J. Fagan, J. Am. Chem. Soc., 1984, **106**, 3680.
1557 S. Aime, R. Gobetto, D. Osella, S. Pellegrino, and O. Gambino, Congr. Naz. Chim. Inorg., [Atti], 16th, 1983, 375 (Chem. Abstr., 1984, **100**, 44 392).
1558 D.J. Darensbourg, R.L. Gray, and M. Pala, Organometallics, 1984, **3**, 1928.
1559 D.L. Hickman and H.M. Goff, J. Am. Chem. Soc., 1984, **106**, 5013.
1560 A.L. Balch, Y.-W. Chan, R.-J. Cheng, G.N. La Mar, L. Latos-Grazynski, and M.W. Renner, J. Am. Chem. Soc., 1984, **106**, 7779.
1561 N.E. Schore, C.S. Ilenda, M.A. White, H.E. Bryndza, M.G. Matturro, and R.G. Bergman, J. Am. Chem. Soc., 1984, **106**, 7451.
1562 Y.T. Fanchiang, G.T. Bratt, and H.P.C. Hogenkamp, Proc. Natl. Acad. Sci. U.S.A., 1984, **81**, 2698 (Chem. Abstr., 1984, **101**, 152 240).
1563 Y.-T. Fanchiang, Inorg. Chem., 1984, **23**, 3983.
1564 S.J. Landon and T.B. Brill, Inorg. Chem., 1984, **23**, 1266.
1565 S.J. Landon and T.B. Brill, Inorg. Chem., 1984, **23**, 4177.
1566 C.D. Wood and P.E. Garrou, Organometallics, 1984, **3**, 170.
1567 J.G. Russell, R.G. Bryant, and M.M. Kreesoy, Inorg. Chem., 1984, **23**, 4565.
1568 R.K. Harris and R.J. Morrow, J. Chem. Soc., Faraday Trans. 1, 1984, **80**, 3071.
1569 S. Balt, H.J. Gamelkoorn, C.R.P. Mac-Coll, and W.E. Renkema, Transition Met. Chem., 1984, **9**, 224.
1570 S. Balt and H.J. Gamelkoorn, Inorg. Chim. Acta, 1984, **86**, L61.
1571 J. Lichtig, M.E. Sosa, and M.L. Tobe, J. Chem. Soc., Dalton Trans., 1984, 581.
1572 I.I. Creaser, G.P. Haight, R. Peachey, W.T. Robinson, and A.M. Sargeson, J. Chem. Soc., Chem. Commun., 1984, 1568.
1573 D.R. Jones, L.F. Lindoy, and A.M. Sargeson, J. Am. Chem. Soc., 1984, **106**, 7807.
1574 W.D. Jones and F.J. Feher, J. Am. Chem. Soc., 1984, **106**, 1650.
1575 E.A.V. Ebsworth, N.T. McManus, and D.W.H. Rankin, J. Chem. Soc., Dalton Trans., 1984, 2573.
1576 J.A. Kampmeier and S. Mahalingam, Organometallics, 1984, **3**, 489.
1577 B.C. Whitmore and R. Eisenberg, Report, 1983, UROCHESTER/DC/83/TR-5; Order No. AD-A135508, 11 pp. Avail. NTIS. From Gov. Rep. Announce. Index (U.S.), 1984, **84**, 104 (Chem. Abstr., 1984, **100**, 180 818).
1578 B.C. Whitmore and R. Eisenberg, J. Am. Chem. Soc., 1984, **106**, 3225.
1579 A. Miyashita, H. Takaya, T. Souchi, and R. Noyori, Tetrahedron, 1984, **40**, 1245.
1580 M.J. Auburn and S.R. Stobart, J. Chem. Soc., Chem. Commun., 1984, 281.
1581 M.J. Auburn, R.D. Holmes-Smith, and S.R. Stobart, J. Am. Chem. Soc., 1984, **106**, 1314.
1582 R.H. Crabtree, M.F. Mellea, and J.M. Quirk, J. Am. Chem. Soc., 1984, **106**, 2913.
1583 J.M. Brown, L.R. Canning, and A.R. Lucy, J. Chem. Soc., Chem. Commun., 1984, 915.
1584 P. Hendry and A.M. Sargeson, J. Chem. Soc., Chem. Commun., 1984, 164.
1585 E.G. Samsel and J.R. Norton, J. Am. Chem. Soc., 1984, **106**, 5505.
1586 F.E. Wood, J. Hvoslef, H. Hope, and A.L. Balch, Inorg. Chem., 1984, **23**, 4309.
1587 L.S. Hegedus, B. Akermark, K. Zetterberg, and L.F. Olsson, J. Am. Chem. Soc., 1984, **106**, 7122.
1588 A. Benefiel, D.M. Roundhill, W.C. Fultz, and A.L. Rheingold, Inorg. Chem., 1984, **23**, 3316.
1589 G. Guglielmo, V. Ricevuto, P. Bergamini, S. Sostero, and O. Traverso, J. Organomet. Chem., 1984, **276**, 281.
1590 E. Curzon, B.T. Golding, C. Pierpoint, and B.W. Waters, J. Organomet. Chem.,

1984, **262**, 263.
1591 N. Gregor, K. Zaw, and P.M. Henry, Organometallics, 1984, **3**, 1251.
1592 R.H. Reamey and G.M. Whitesides, J. Am. Chem. Soc., 1984, **106**, 81.
1593 T.C. Flood and S.P. Bitler, J. Am. Chem. Soc., 1984, **106**, 6076.
1594 S. Lanza, D. Minniti, R. Romeo, P. Moore, J. Sachinidis, and M.L. Tobe, J. Chem. Soc., Chem. Commun., 1984, 542.
1595 S. Lanza, D. Minniti, P. Moore, J. Sachinidis, R. Romeo, and M.L. Tobe, Inorg. Chem., 1984, **23**, 4428.
1596 I.M. Al-Najjar, Inorg. Chim. Acta, 1984, **83**, L53.
1597 G. Guillot-Edelheit and J.-C. Chottard, J. Chem. Soc., Dalton Trans., 1984, 169.
1598 S.W. Carr, R. Colton, D. Dakternieks, B.F. Hoskins, and R.J. Steen, Inorg. Chem., 1983, **22**, 3700.
1599 R.N. Bose, R.E. Viola, and R.D. Cornelius, J. Am. Chem. Soc., 1984, **106**, 3336.
1600 S. Shinoda, T. Nishikimi, S.-i. Uchino, Y. Koie, and Y. Saito, J. Chem. Soc., Dalton Trans., 1984, 2689.
1601 R.N. Bose, R.D. Cornelius, and R.E. Viola, Inorg. Chem., 1984, **23**, 1181.
1602 M. Clarembeau, J.L. Bertrand, and A. Krief, Isr. J. Chem., 1984, **24**, 125.
1603 J.P. Fackler, jun., H.H. Murray, and J.D. Basil, Organometallics, 1984, **3**, 821.
1604 E.M. Kinsch and D.W. Stephan, Inorg. Chim. Acta, 1984, **91**, 263.
1605 B. Fabisch and T.N. Mitchell, J. Organomet. Chem., 1984, **269**, 219.
1606 Y.-T. Fanchiang, Inorg. Chem., 1984, **23**, 3428.
1607 A.M. Bond, J.R. Bradbury, P.J. Hanna, G.N. Howell, H.A. Hudson, S. Strother, and M.J. O'Connor, Anal. Chem., 1984, **56**, 2392.
1608 J.L. Torregrosa, M. Baboulene, V. Speziale, and A. Lattes, J. Organomet. Chem., 1984, **277**, 159.
1609 J. Bould, N.N. Greenwood, and J.D. Kennedy, J. Chem. Soc., Dalton Trans., 1984, 2477.
1610 J.A. Long, T.B. Marder, and M.F. Hawthorne, J. Am. Chem. Soc., 1984, **106**, 3004.
1611 P.E. Behnken, D.C. Busby, M.S. Delaney, R.E. King, tert., C.W. Kreimendahl, T.B. Marder, J.J. Wilczynski, and M.F. Hawthorne, J. Am. Chem. Soc., 1984, **106**, 7444.
1612 P.E. Behnken, J.A. Belmont, D.C. Busby, M.S. Delaney, R.E. King, tert., C.W. Kreimendahl, T.B. Marder, J.J. Wilczynski, and M.F. Hawthorne, J. Am. Chem. Soc., 1984, **106**, 3011.
1613 G. Wulff, M. Lauer, and H. Böhnke, Angew. Chem., Int. Ed. Engl., 1984, **23**, 741.
1614 N. Ahmad, F.K. Cartledge, M.C. Day, and J.H. Medley, Organometallics, 1984, **3**, 389.
1615 A.Yu. Fisenko and B.A. Ershov, Deposited Doc., 1982, SPSTL 1069 Khp-D82, 16 pp. Avail. SPSTL (Chem. Abstr., 1984, **101**, 55 149).
1616 J.W. Akitt and N.B. Milic, J. Chem. Soc., Dalton Trans., 1984, 981.
1617 L.S.D. Glasser and G. Harvey, J. Chem. Soc., Chem. Commun., 1984, 1250.
1618 S. Chao, C.J. Stalder, D.P. Summers, and M.S. Wrighton, J. Am. Chem. Soc., 1984, **106**, 2723.
1619 I.S. Alnaimi, W.P. Weber, A.S. Nazran, and D. Griller, J. Organomet. Chem., 1984, **272**, C10.
1620 J. Koecher and M. Lehnig, Organometallics, 1984, **3**, 937.
1621 W.F. Manders, G.J. Olson, F.E. Brinckman, and J.M. Bellama, J. Chem. Soc., Chem. Commun., 1984, 538.
1622 I.V. Borisova, Yu.N. Luzikov, N.N. Zemlyansky, Yu.A. Ustynyuk, and I.P. Beletskaya, J. Organomet. Chem., 1984, **268**, 11.
1623 I. Artaki, S. Sinha, and J. Jonas, Mater. Lett., 1984, **2**, 448 (Chem. Abstr., 1984, **101**, 137 954).
1624 C.J. Brinker, K.D. Keefer, D.W. Schaefer, R.A. Assink, B.D. Kay, and C.S. Ashley, J. Non-Cryst. Solids, 1984, **63**, 45 (Chem. Abstr., 1984, **100**, 160 800).
1625 T.A. van der Knaap and F. Bickelhaupt, J. Organomet. Chem., 1984, **277**, 351.
1626 W. Toepelmann, B. Thomas, H. Schadow, and L. Meznik, Z. Chem., 1984, **24**, 134

(Chem. Abstr., 1984, **101**, 151 940).
1627 N. Burford, T. Chivers, R.T. Oakley, and T. Oswald, Can. J. Chem., 1984, **62**, 712.
1628 S. Kobayashi, M. Suzuki, and T. Saegusa, Macromolecules, 1984, **17**, 107 (Chem. Abstr., 1984, **100**, 34 877).
1629 K. Taira, T. Fanni, and D.G. Gorenstein, J. Org. Chem., 1984, **49**, 4531.
1630 S. Chandrasegaran, A. Murakami, and L.-s. Kan, J. Org. Chem., 1984, **49**, 4951.
1631 F.H. Musa and B.W. Tattershall, J. Chem. Soc., Dalton Trans., 1984, 1517.
1632 M. Nakamura, Kagaku no Ryoiki, 1983, **37**, 976 (Chem. Abstr., 1984, **100**, 184 441).
1633 L.B. Dugad and S. Mitra, Proc. - Indian Acad. Sci., Chem. Sci., 1984, **93**, 295 (Chem. Abstr., 1984, **101**, 32 029).
1634 A. Yamasaki, Bunseki, 1983, 916 (Chem. Abstr., 1984, **100**, 95 561).
1635 R.R. Fraser, Asymmetric Synth., 1983, **1**, 173. Ed. J.D. Morrison, Academic, New York (Chem. Abstr., 1984, **100**, 208 656).
1636 C. Cabal, Rev. Cubana Fis., 1983, **3**, 97 (Chem. Abstr., 1984, **101**, 182 440).
1637 S. Ahn, E. Park, and S.W. Oh, Bull. Korean Chem. Soc., 1984, **5**, 55 (Chem. Abstr., 1984, **100**, 184 619).
1638 S. Ahn, E. Park, and K.H. Lee, Bull. Korean Chem. Soc., 1983, **4**, 240 (Chem. Abstr., 1984, **100**, 44 266).
1639 S. Ahn, S.W. Oh, and K.H. Lee, Bull. Korean Chem. Soc., 1984, **5**, 93 (Chem. Abstr., 1984, **101**, 47 523).
1640 N. Benetis, J. Kowalewski, L. Nordenskioeld, H. Wennerstroem, and P.O. Westlund, J. Magn. Reson., 1984, **58**, 261.
1641 P.O. Westlund, H. Wennerstroem, L. Nordenskioeld, J. Kowalewski, and N. Benetis, J. Magn. Reson., 1984, **59**, 91.
1642 M.A. Brown and G.A. Johnson, Med. Phys., 1984, **11**, 67 (Chem. Abstr., 1984, **100**, 170 968).
1643 C.W. Chen, J.S. Cohen, C.E. Myers, and M. Sohn, FEBS Lett., 1984, **168**, 70 (Chem. Abstr., 1984, **101**, 20 038).
1644 F.H. Köhler, W.A. Geike, P. Hofmann, U. Schubert, and P. Stauffert, Chem. Ber., 1984, **117**, 904.
1645 F.H. Köhler, R. De Cao, and G. Manlik, Inorg. Chim. Acta, 1984, **91**, L1.
1646 C.A. Green, R.J. Bianchini, and J.I. Legg, Inorg. Chem., 1984, **23**, 2713.
1647 H. Grahn, U. Edlund, and G.C. Levy, J. Magn. Reson., 1984, **56**, 61.
1648 D.J. Cookson and B.E. Smith, J. Magn. Reson., 1984, **57**, 355.
1649 J. Heck, W. Massa, and P. Weinig, Angew. Chem., Int. Ed. Engl., 1984, **23**, 722.
1650 F.H. Köhler, N. Hebendanz, U. Thewalt, B. Kanellakopulos, and R. Klenze, Angew. Chem., Int. Ed. Engl., 1984, **23**, 721.
1651 H.M. Goff and A.P. Hansen, Inorg. Chem., 1984, **23**, 321.
1652 A.P. Hansen and H.M. Goff, Inorg. Chem., 1984, **23**, 4519.
1653 C. Moberg and M. Weber, Polyhedron, 1984, **3**, 491.
1654 C.J. Cameron, S.M. Tetrick, and R.A. Walton, Organometallics, 1984, **3**, 240 (Chem. Abstr., 1984, **100**, 60 824).
1655 J.V. Caspar, B.P. Sullivan, and T.J. Meyer, Inorg. Chem., 1984, **23**, 2104.
1656 J.Plešek, K. Baše, F. Mareš, F. Hanousek, B. Štír, and S. Heřmánek, Collect. Czech. Chem. Commun., 1984, **49**, 2776.
1657 D.V. Behere and H.M. Goff, J. Am. Chem. Soc., 1984, **106**, 4945.
1658 A.D. Boersma and H.M. Goff, Inorg. Chem., 1984, **23**, 1671.
1659 A.D. Boersma, M.A. Phillippi, and H.M. Goff, J. Magn. Reson., 1984, **57**, 197.
1660 J.-P. Mahy, P. Battioni, D. Mansuy, J. Fisher, R. Weiss, J. Mispelter, I. Morgenstern-Badarau, and P. Gans, J. Am. Chem. Soc., 1984, **106**, 1699.
1661 D.V. Behere, R. Birdy, and S. Mitra, Inorg. Chem., 1984, **23**, 1978.
1662 H.M. Goff, E.T. Shimomura, Y.J. Lee, and W.R. Scheidt, Inorg. Chem., 1984, **23**, 315.
1663 D.R. English, D.N. Hendrickson, K.S. Suslick, C.W. Eigenbrot, jun., and W.R. Scheidt, J. Am. Chem. Soc., 1984, **106**, 7258.
1664 G.E. Toney, A. Gold, J. Savrin, L.W. TerHaar, R. Sangaiah, and W.E. Hatfield, Inorg. Chem., 1984, **23**, 4350.
1665 G.E. Toney, L.W. TerHaar, J.E. Savrin, A. Gold, W.E. Hatfield, and R.

Sangaiah, Inorg. Chem., 1984, **23**, 2561.
1666 T.A. Babushkina, R.E. Garibov, L.B. Lazukova, and V.V. Khrapov, Koord. Khim., 1984, **10**, 77 (Chem. Abstr., 1984, **100**, 131 210).
1667 A. Gold, W. Ivey, G.E. Toney, and R. Sangaiah, Inorg. Chem., 1984, **23**, 2932.
1668 A. Mayer and H. Eicher, J. Mol. Catal., 1984, **23**, 151 (Chem. Abstr., 1984, **100**, 170 210).
1669 T. Jue and G.N. La Mar, Biochem. Biophys. Res. Commun., 1984, **119**, 640 (Chem. Abstr., 1984, **100**, 170 230).
1670 W.H. Armstrong, A. Spool, G.C. Papaefthymiou, R.B. Frankel, and S.J. Lippard, J. Am. Chem. Soc., 1984, **106**, 3653.
1671 W.E. Cleland, jun., and B.A. Averill, Inorg. Chem., 1984, **23**, 4192.
1672 R.E. Palermo, R. Singh, J.K. Bashkin, and R.H. Holm, J. Am. Chem. Soc., 1984, **106**, 2600.
1673 K.S. Hagen, A.D. Watson, and R.H. Holm, Inorg. Chem., 1984, **23**, 2984.
1674 M.G. Kanatzidis, N.C. Baenziger, D. Coucouvanis, A. Simopoulos, and A. Kostikas, J. Am. Chem. Soc., 1984, **106**, 4500.
1675 H. Strasdeit, B. Krebs, and G. Henkel, Inorg. Chem., 1984, **23**, 1816.
1676 I. Bertini, G. Lanini, and C. Luchinat, Inorg. Chem., 1984, **23**, 2731.
1677 I. Morishima, Y. Takamuri, and Y. Shiro, J. Am. Chem. Soc., 1984, **106**, 766.
1678 J.P. Collman, C.E. Barnes, P.N. Swepston, and J.A. Ibers, J. Am. Chem. Soc., 1984, **106**, 3500.
1679 F.C. Anson, J.A. Christie, T.J. Collins, R.J. Coots, T.T. Furatani, S.L. Gipson, J.T. Keech, T.E. Krafft, B.D. Santarsiero, and G.H. Spies, J. Am. Chem. Soc., 1984, **106**, 4460.
1680 U. Koelle, B. Fuss, M.V. Rajasekharan, B.L. Ramakrishna, J.H. Ammeter, and M.C. Boehm, J. Am. Chem. Soc., 1984, **106**, 4152.
1681 G. Navron and W. Klaeui, Inorg. Chem., 1984, **23**, 2722.
1682 K. Migita, C.T. Migita, and M. Iwaizumi, J. Chem. Soc., Dalton Trans., 1984, 1195.
1683 Y. Yamamoto, S. Shimokawa, and E. Yamada, Bull. Chem. Soc. Jpn., 1984, **57**, 1153.
1684 G.E. Jackson and L.G. Scott, S. Afr. J. Chem., 1983, **36**, 120 (Chem. Abstr., 1984, **100**, 78 658).
1685 A.V. Kessenikh, A.S. Lomov, and O.V. Ivanov, Radiospektroskopiya (Perm), 1983, **15**, 54 (Chem. Abstr., 1984, **101**, 202 982).
1686 J.W. Faller, C. Blankenship, and S. Sena, J. Am. Chem. Soc., 1984, **106**, 793.
1687 P.J. Morley, S.B. Carroll, P. Harmon, and R.G. Khalifah, J. Am. Chem. Soc., 1984, **106**, 1830.
1688 D. Fenske, R. Basoglu, J. Hachgenei, and F. Rogel, Angew. Chem., Int. Ed. Engl., 1984, **23**, 160.
1689 Y.Y. Lim, Inorg. Chim. Acta, 1984, **86**, L1.
1690 N. Benetis, J. Kowalewski, L. Nordenskioeld, and U. Edlund, J. Magn. Reson., 1984, **58**, 282.
1691 V.M. Nekipelov and K.I. Zamaraev, Khim. Fiz., 1983, 917 (Chem. Abstr., 1984, **101**, 82 868).
1692 R.M. Golding, R.O. Pascual, and I.C. Hoare, J. Magn. Reson., 1984, **58**, 219.
1693 M.D. Johnston, jun., G.H. Caines, and A.S. Zektzer, J. Magn. Reson., 1984, **60**, 415.
1694 H. Schumann, J. Müller, N. Bruncks, H. Lauke, J. Pickardt, H. Schwarz, and K. Eckhart, Organometallics, 1984, **3**, 69.
1695 S. Zhu, K. Sha, H. Tan, and Z. Gao, Gaodeng Xuexiao Huaxue Xuebao, 1984, **5**, 657 (Chem. Abstr., 1984, **101**, 211 353).
1696 J.M. Boncella, T.D. Tilley, and R.A. Andersen, J. Chem. Soc. Chem. Commun., 1984, 710.
1697 J.M. Boncella and R.A. Andersen, Inorg. Chem., 1984, **23**, 432.
1698 T.A. Babushkina, V.F. Zolin, and L.G. Koreneva, Zh. Strukt. Khim., 1984, **25**, 28 (Chem. Abstr., 1984, **101**, 182 556).
1699 R.D. Chapman, R.T. Loda, J.P. Riehl, and R.W. Schwartz, Inorg. Chem., 1984, **23**, 1652.
1700 N.I. Grebenshchikov, L.V. Ruzaikina, L.A. Fedorov, and A.N. Ermakov, Zh. Neorg. Khim., 1984, **29**, 917 (Chem. Abstr., 1984, **100**, 184 614).
1701 J.W. Buchler, H.G. Kapellmann, M. Knoff, K.L. Lay, and S. Pfeifer, Z.

Naturforsch., Teil B, 1983, **38**, 1339 (Chem. Abstr., 1984, **100**, 78 856).
1702 H.G. Brittain and J.F. Desreux, Inorg. Chem., 1984, **23**, 4459.
1703 S.M. Bowen, E.N. Duesler, and R.T. Paine, Inorg. Chim. Acta, 1984, **84**, 221.
1704 A. Dambska and A. Janowski, J. Magn. Reson., 1984, **59**, 13.
1705 T.J. Wenzel, J. Org. Chem., 1984, **49**, 1834.
1706 T.J. Wenzel, Spectrosc. Lett., 1984, **17**, 77 (Chem. Abstr., 1984, **101**, 32 180).
1707 L. Shen, Y. Gao, Y. Xiao, and J. Ni, Zhongguo Kexueyuan Changchun Yingyong Huaxue Yanjiuso Jikan, 1982, **19**, 43 (Chem. Abstr., 1984, **101**, 98 647).
1708 H.J. Schneider, U. Buchheit, and P.K. Agrawal, Tetrahedron, 1984, **40**, 1017.
1709 S. Balt, M.W.G. De Bolster, and G. Visser-Luirink, Carbohydr. Res., 1983, **121**, 1 (Chem. Abstr., 1984, **100**, 7024).
1710 C.F.G.C. Geraldes and J.R. Ascenso, J. Chem. Soc., Dalton Trans., 1984, 267.
1711 K. Kabuto and Y. Sasaki, J. Chem. Soc., Chem. Commun., 1984, 316.
1712 K.C. Joshi, V.N. Pathak, and V. Grover, J. Indian Chem. Soc., 1983, **60**, 802 (Chem. Abstr., 1984, **100**, 44 237).
1713 H. Januszewski, L. Stefaniak, M. Witanowski, and G.A. Webb, Bull. Pol. Acad. Sci., Chem., 1984, **32**, 53 (Chem. Abstr., 1984, **101**, 121 777).
1714 S.C. Chu, M.M. Pike, E.T. Fossel, T.W. Smith, J.A. Balschi, and C.S. Springer, jun., J. Magn. Reson., 1984, **56**, 33.
1715 M.M. Pike, E.T. Fossel, T.W. Smith, and C.S. Springer, jun., Am. J. Physiol., 1984, **246**, C528 (Chem. Abstr., 1984, **101**, 86 661).
1716 L.O. Sillerud and J.W. Heyser, Plant Physiol., 1984, **75**, 269 (Chem. Abstr., 1984, **101**, 51 093).
1717 M.K. Hayer and F.G. Riddell, Inorg. Chim. Acta, 1984, **92**, L37.
1718 V.N. Torocheshnikov and N.M. Sergeev, Zh. Strukt. Khim., 1983, **24**, 121 (Chem. Abstr., 1984, **100**, 131 202).
1719 V.D. Buikliskii and V.T. Panyushkin, Zh. Strukt. Khim., 1983, **24**, 49 (Chem. Abstr., 1984, **100**, 64 068).
1720 M.E. Daman and K. Dill, J. Magn. Reson., 1984, **60**, 118.
1721 B.R. McGarvey, Can. J. Chem., 1984, **62**, 1349.
1722 A. Streitwieser, jun., Inorg. Chim. Acta, 1984, **94**, 171.
1723 Z. Yang, Huaxue Shiji, 1984, **6**, 222 (Chem. Abstr., 1984, **101**, 142 656).
1724 G.V. Fazakerley, G. Folcher, and H. Marquet-Ellis, Polyhedron, 1984, **3**, 457.
1725 R.E. Cramer, K. Panchanatheswaran, and J.W. Gilje, Angew. Chem., Int. Ed. Engl., 1984, **23**, 912.
1726 A. Dormond, A.A. Elbouadili, and C. Moise, J. Chem. Soc., Chem. Commun., 1984, 749.
1727 R.E. Cramer, K. Panchanatheswaran, and J.W. Gilje, J. Am. Chem. Soc., 1984, **106**, 1853.
1728 G. Paolucci, G. Rossetto, P. Zanella, K. Yünlü, and R.D. Fischer, J. Organomet. Chem., 1984, **272**, 363.
1729 P. Zanella, G. Rossetto, A. Berton, and G. Paolucci, Inorg. Chim. Acta, 1984, **95**, 263.
1730 M.R. Duttera, V.W. Day, and T.J. Marks, J. Am. Chem. Soc., 1984, **106**, 2907.
1731 G. Paolucci, P. Zanella, G. Rossetto, and U. Croatto, Inorg. Chim. Acta, 1984, **94**, 55.
1732 N. Marques, A.P. de Matos, and K.W. Bagnall, Inorg. Chim. Acta, 1984, **95**, 75.
1733 F.A. Cotton, D.O. Marler, and W. Schwotzer, Inorg. Chem., 1984, **23**, 4211.
1734 G.E. Maciel, Science, 1984, **226**, 282 (Chem. Abstr., 1984, **101**, 202 781).
1735 G.E. Maciel, Heterog. Catal., Proc. Symp. Ind.-Univ. Coop. Chem. Program Dep. Chem., Texas A&M Univ., 1984, 349 (Chem. Abstr., 1984, **101**, 221 022).
1736 H. Shindo and T. Fujito, Kagaku no Ryoiki, Zokan, 1983, 29 (Chem. Abstr., 1984, **100**, 184 444).
1737 H.J. Cantow, D. Emeis, W. Gronski, A. Hasenhindl, D. Lausberg, M. Moeller, and Y. Shahab, Macromol. Chem. Phys., Suppl., 1984, **7**, 63 (Chem. Abstr., 1984, **101**, 192 721).
1738 G. Engelhardt, U. Lohse, M. Magi, and E. Lippmaa, Stud. Surf. Sci., Catal., 1984, **18**, 23 (Chem. Abstr., 1984, **101**, 117 449).
1739 Z. Gabelica, J.B. Nagy, P. Bodart, G. Debras, E.G. Derouane, and P.A. Jacobs, NATO ASI Ser., Ser. E, 1984, **80**, 193 (Chem. Abstr., 1984, **100**, 180

481).
1740 J. Klinowski, Prog. Nucl. Magn. Reson., Spectrosc., 1984, **16**, 237 (Chem. Abstr., 1984, **101**, 235 972).
1741 S. Shinoda and Y. Saito, Hyomen, 1983, **21**, 722 (Chem. Abstr., 1984, **100**, 140 034).
1742 P.C. Taylor, J. Non-Cryst. Solids, 1983, **59**, 109 (Chem. Abstr., 1984, **100**, 95 298).
1743 T. Shimizu, Jpn. Annu. Rev. Electron. Comput. Telecommun., 1983, **6**, 62 (Chem. Abstr., 1984, **100**, 149 011).
1744 J.T.M. De Hosson, O. Kanert, and A.W. Sleeswyk, Dislocat. Solids, 1983, **6**, 441 (Chem. Abstr., 1984, **100**, 59702).
1745 J.L. Bjorkstam, J. Mol. Struct., 1983, **111**, 135 (Chem. Abstr., 1984, **100**, 78 500).
1746 D.W. Davidson, Nat. Gas Hydrates, 1983, 1 (Chem. Abstr., 1984, **100**, 177 357).
1747 W. Mueller-Warmuth, Colloids Surf., 1984, **11**, 1 (Chem. Abstr., 1984, **101**, 221 021).
1748 H. Pfeifer, W. Meiler, and D. Deininger, Annu. Rep. NMR Spectrosc., 1983, **15**, 291.
1749 J.P. Yesinowski, R.A. Wolfgang, and M.J. Mobley, Adsorpt. Surf. Chem. Hydroxyapatite, [Proc. Symp.] 1982 (Pub. 1984), 151 (Chem. Abstr., 1984, **100**, 198 152).
1750 W. Mueller-Warmuth, K.H. Dupree, and M. Prager, Z. Naturforsch., Teil A, 1984, **39**, 66.
1751 E. Bechtold-Schweickert, M. Mali, J. Roos, and D. Brinkmann, Phys. Rev. B: Condens. Matter, 1984, **30**, 2891 (Chem. Abstr., 1984, **101**, 158146).
1752 K. Nishida, T. Asai, and S. Kawai, Solid State Commun., 1983, **48**, 701 (Chem. Abstr., 1984, **100**, 43 520).
1753 A. Hardwick, P.G. Dickens, and R.C.T. Slade, Solid State Commun., 1984, **50**, 705 (Chem. Abstr., 1984, **101**, 41 027).
1754 H. Ohno, S. Konishi, K. Noda, H. Takeshita, H. Katsuta, H. Yoshida, H. Watanabe, and T. Matsuo, Report, 1982, JAERI-M-82-146, 37 pp. Avail. INIS. From INIS Atomindex, 1983, 14, abstr. no. 793 962 (Chem. Abstr., 1984, **100**, 95 016).
1755 C. Cros, L. Hanebali, L. Latie, G. Villeneuve, and G. Wang, Solid State Ionics, 1983, **9**, 139 (Chem. Abstr., 1984, **100**, 77 963).
1756 T. Asai, S. Kawai, R. Nagai, and S. Mochizuki, J. Phys. Chem. Solids, 1984, **45**, 173 (Chem. Abstr., 1984, **100**, 166 037).
1757 J.L. Bjorkstam, M. Villa, M. Rahman, and P.M. Skarstad, Solid State Ionics, 1983, **9**, 111 (Chem. Abstr., 1984, **100**, 78 722).
1758 C. Jaeger and K. Heide, Phys. Status Solidi A, 1984, **81**, K7 (Chem. Abstr., 1984, **100**, 131 233).
1759 K.R. Carduner, G.C. Farrington, D. White, and M. Villa, Solid State Ionics, 1983, **9**, 339 (Chem. Abstr., 1984, **100**, 78 723).
1760 L. Bottyan, R. Dupree, and G.V. Chandrashekhar, Solid State Ionics, 1983, **9**, 347 (Chem. Abstr., 1984, **100**, 78 724).
1761 H.T. Stokes, D.C. Ailion, and T.A. Case, Phys. Rev. B: Condens. Matter, 1984, **30**, 4925 (Chem. Abstr., 1984, **101**, 220 209).
1762 J. Slak, R. Kind, R. Blinc, E. Courtens, and S. Zumer, Phys. Rev. B: Condens. Matter, 1984, **30**, 85.
1763 J.A. Janik, J.M. Janik, A. Migdal-Mikuli, E. Mikuli, M. Rachwalska, T. Stanek, K. Otnes, B.O. Fimland, and I. Svare, Physica B+C (Amsterdam), 1983, **122**, 315 (Chem. Abstr., 1984, **100**, 28 269).
1764 E. Palacios, J. Bartolome, R. Navarro, J. Garcia, D. Gonsalez, and H.B. Brom, Ferroelectrics, 1984, **55**, 955 (Chem. Abstr., 1984, **100**, 201 249).
1765 A.V. Chadwick, J.H. Strange, G.A. Ranieri, and M. Terenzi, Solid State Ionics, 1983, **9**, 555 (Chem. Abstr., 1984, **100**, 77 977).
1766 K.B. Rawlings, R.F. Marzke, and W.S. Glaunsinger, J. Phys. Chem., 1984, **88**, 3880.
1767 A.F. Aalders, A. Polman, A.F.M. Arts, and H.W. Wijn, Solid State Ionics, 1983, **9**, 539 (Chem. Abstr., 1984, **100**, 77 976).
1768 A. Roos, A.F. Aalders, J. Schoonman, A.F.M. Arts, and H.W. Wijn, Solid State

Ionics, 1983, **9**, 571 (Chem. Abstr., 1984, **100**, 77 978).
1769 A. Roos, F.C.M. Van de Pol, R. Keim, and J. Schoonman, Solid State Ionics, 1984, **13**, 191 (Chem. Abstr., 1984, **101**, 82 223).
1770 O.J. Zogal, C. Jaeger, H. Duehler, and B. Schnabel, Phys. Status Solidi A, 1984, **82**, K153.
1771 F.E. Spada, H. Oesterreicher, R.C. Bowman, jun., and M.P. Guse, Phys. Rev. B: Condens. Matter, 1984, **30**, 4909 (Chem. Abstr., 1984, **101**, 235 919).
1772 A.I. Livshits, V.M. Buznik, P.P. Fedorov, and B.P. Sobolev, Zh. Strukt. Khim., 1983, **24**, 44 (Chem. Abstr., 1984, **100**, 15 850).
1773 Y.T. Tsai, W.P. Halperin, and D.H. Whitmore, J. Solid State Chem., 1983, **50**, 263 (Chem. Abstr., 1984, **100**, 57 157).
1774 M. Bose, A. Ghoshray, and K. Roy, Proc. Nucl. Phys. Solid State Phys. Symp. 1980 (Pub. 1983), **23**, 834 (Chem. Abstr., 1984, **101**, 47 509).
1775 R.C. Bowman, jun. and B.D. Craft, J. Phys. C, 1984, **17**, L477 (Chem. Abstr., 1984, **101**, 198 532).
1776 J. Senegas, J.P. Laval, and B. Frit, J. Solid State Chem., 1984, **53**, 344 (Chem. Abstr., 1984, **101**, 102 752).
1777 L. Chen, K. Wang, G. Che, G. Wang, and Z. Li, Solid State Ionics, 1983, **9**, 149 (Chem. Abstr., 1984, **100**, 77 964).
1778 J.L. Fourquet, M.F. Renou, R. De Pape, H. Theveneau, P.R. Man, O. Lucas, and J. Pannetier, Solid State Ionics, 1983, **9**, 1011 (Chem. Abstr., 1984, **100**, 77 720).
1779 P.P. Man, H. Theveneau, P. Papon, and J.L. Fourquet, Solid State Ionics, 1983, **9**, 73 (Chem. Abstr., 1984, **100**, 77 960).
1780 J.J. Van der Klink and F. Borsa, Phys. Rev. B: Condens. Matter, 1984, **30**, 52 (Chem. Abstr., 1984, 101, 82 853).
1781 M. Corti, U.T. Hoechli, and A. Rigamonti, Ferroelectrics, 1984, **54**, 397 (Chem. Abstr., 1984, **100**, 184 233).
1782 M.T. Weller and P.G. Dickens, Solid State Ionics, 1983, **9**, 1081 (Chem. Abstr., 1984, **100**, 113 842).
1783 M.J. Buckingham, B.W. Fitzsimmons, and I. Sayer, J. Chem. Soc., Chem. Commun., 1984, 339.
1784 B.E. Hanson, M.J. Sullivan, and R.J. Davis, J. Am. Chem. Soc., 1984, **106**, 251.
1785 J.A.J. Lourens and E.C. Reynhardt, J. Magn. Reson., 1984, **57**, 123.
1786 R.J. Pylkki, R.D. Willett, and H.W. Dodgen, Inorg. Chem., 1984, **23**, 594.
1787 E. Oztekin, Hacettepe Bull. Nat. Sci. Eng., 1984, **13**, 41 (Chem. Abstr., 1984, **101**, 202 988).
1788 S. Gima, Y. Furukawa, R. Ikeda, and D. Nakamura, J. Mol. Struct., 1983, **111**, 189.
1789 L. Menabue, G.C. Pellacani, L.P. Battaglia, A.B. Corradi, F. Sandrolini, A. Motori, R.J. Pylkki, and R.D. Willett, J. Chem. Soc., Dalton Trans., 1984, 2187.
1790 K.V.G. Reddy, K.V.R. Chary, B.A. Sastry, G. Ponticelli, and R. Pinna, Indian J. Pure Appl. Phys., 1983, **21**, 635 (Chem. Abstr., 1984, **100**, 78 674).
1791 K. Budde and H. Richtering, Phys. Status Solidi B, 1983, **120**, 595 (Chem. Abstr., 1984, **100**, 74 392).
1792 O.A.P. Tavares, B. D'Aguanno, and L.G. Conti, Gazz. Chim. Ital., 1983, **113**, 529 (Chem. Abstr., 1984, **100**, 198 120).
1793 H. Looser, M. Mali, J. Roos, and D. Brinkmann, Solid State Ionics, 1983, **9**, 1237 (Chem. Abstr., 1984, **100**, 74 389).
1794 L. Piekara-Sady, S. Idziak, M. Krupski, E. Dynowska, M. Mackowiak, and J. Stankowski, Physica B+C (Amsterdam), 1984, **123**, 211 (Chem. Abstr., 1984, **100**, 148 825).
1795 B.S. Sarma and J. Ramakrishna, Indian J. Pure Appl. Phys., 1984, **22**, 380 (Chem. Abstr., 1984, **101**, 142 628).
1796 E.C. Reynhardt and C.F. Hoon, J. Phys. C, 1983, **16**, 6137 (Chem. Abstr., 1984, **100**, 131 198).
1797 P.J. Bray, D.E. Hintenlang, M.L. Lui, and R.V. Mulkern, Glastech. Ber., 1983, **56**, 892 (Chem. Abstr., 1984, **100**, 112 960).
1798 E.C. Reynhardt and J.A.J. Lourens, J. Chem. Phys., 1984, **80**, 6240.
1799 E. Schweickert, M. Mali, J. Roos, D. Brinkmann, P.M. Richards, and R.M.

Biefeld, Solid State Ionics, 1983, **9**, 1317 (Chem. Abstr., 1984, **100**, 95 403).
1800 S. Konishi and H. Ohno, J. Am. Ceram. Soc., 1984, **67**, 418.
1801 A.S. Litovchenko and V.V. Mazykin, Phys. Status Solidi A, 1984, **81**, K47 (Chem. Abstr., 1984, **100**, 130 535).
1802 D. Mueller, A. Rettel, W. Gessner, and G. Scheler, J. Magn. Reson., 1984, **57**, 152.
1803 P.P. Man, H. Theveneau, P. Papon, and J.L. Fourquet, J. Chem. Phys., 1984, **80**, 1272.
1804 S.B. Liu and M.S. Conradi, Phys. Rev. B: Condens. Matter, 1984, **30**, 24 (Chem. Abstr., 1984, **101**, 64 771).
1805 S. Takeda and H. Chihara, J. Magn. Reson., 1984, **56**, 48.
1806 S. Ueda, N. Kageyama, and M. Koizumi, Proc. Int. Symp. Hydrotherm. React., 1st 1982 (Pub. 1983), 695.
1807 N.J. Clayden, C.M. Dobson, C.J. Hayes, and S.A. Rodger, J. Chem. Soc., Chem. Commun., 1984, 1396.
1808 J. Senegas and J. Olivier-Fourcade, J. Phys. Chem. Solids, 1983, **44**, 1033 (Chem. Abstr., 1984, **100**, 112 785).
1809 H. Ishida, R. Ikeda, and D. Nakamura, Ber. Bunsenges. Phys. Chem., 1984, **88**, 546.
1810 M.A. Doverspike and M.S. Conradi, Phys. Rev. B: Condens. Matter, 1984, **30**, 4905 (Chem. Abstr., 1984, **101**, 221 198).
1811 D.W. Davidson, S.K. Garg, C.I. Ratcliffe, J.S. Tse, and S.R. Gough, Can. J. Chem., 1984, **62**, 1229.
1812 M. Mackowiak and R.J.C. Brown, Mater. Res. Bull., 1984, **19**, 249 (Chem. Abstr., 1984, **100**, 183 598).
1813 R.E. Wasylishen, Spectrochim. Acta, Part A, 1984, **40**, 115 (Chem. Abstr., 1984, **100**, 131 209).
1814 K. Morimoto, K. Shimomura, and M. Yoshida, J. Phys. Soc. Jpn., 1983, **52**, 3927 (Chem. Abstr., 1984, **100**, 16 564).
1815 R. Kaliaperumal, K.V. Ramanathan, and R. Srinivasan, Proc. Nucl. Phys. Solid State Phys. Symp. 1980 (Pub. 1983), **23**, 784 (Chem. Abstr., 1984, **101**, 32 191).
1816 E. Meirovitch, I. Belsky, and S. Vega, J. Phys. Chem., 1984, **88**, 1522.
1817 R.G. Pratt and J.L. Ackerman, Polym. Mater. Sci. Eng., 1983, **49**, 293 (Chem. Abstr., 1984, **100**, 193 319).
1818 L. Mattix, G.R. Miller, J. Milliken, M.J. Moran, H.A. Resing, and D.C. Weber, Polym. Prepr. (Am. Chem. Soc., Div. Polym. Chem.), 1984, **25**, 354 (Chem. Abstr., 1984, **100**, 220 410).
1819 A.I. Livshits, V.M. Buznik, M.P. Borzenkova, and F.V. Kalinchenko, Izv. Akad. Nauk SSSR, Neorg. Mater., 1984, **20**, 495 (Chem. Abstr., 1984, **101**, 98 103).
1820 K. Morimoto, Solid State Commun., 1984, **49**, 1169 (Chem. Abstr., 1984, **100**, 184 613).
1821 K. Morimoto and K. Shimomura, J. Phys. Soc. Jpn., 1984, **53**, 59 (Chem. Abstr., 1984, **100**, 113 723).
1822 K. Morimoto and K. Shimomura, J. Phys. Soc. Jpn., 1984, **53**, 2752 (Chem. Abstr., 1984, **101**, 142 646).
1823 Yu.N. Moskvich, A.A. Sukhovskii, and O.V. Rozanov, Fiz. Tverd. Tela (Leningrad), 1984, **26**, 38 (Chem. Abstr., 1984, **100**, 95 424).
1824 R. Blinc, J. Dolinsek, G. Lahajnar, I. Zupancic, L.A. Shuvalov, and A.I. Baranov, Phys. Status Solidi B, 1984, **123**, K83 (Chem. Abstr., 1984, **100**, 219 764).
1825 S. Ganapathy, V.P. Chacko, and R.G. Bryant, J. Magn. Reson., 1984, **57**, 239.
1826 S. Schramm and E. Oldfield, J. Am. Chem. Soc., 1984, **106**, 2502.
1827 J.A. Gurak, J.W. Chinn, jun., R.J. Lagow, H. Steinfink, and C.S. Yannoni, Inorg. Chem., 1984, **23**, 3717.
1828 R.J. Lagow and J.A. Gurak, Chem. Future, Proc. IUPAC Congr., 29th 1983 (Pub. 1984), 107 (Chem. Abstr., 1984, **100**, 209 928).
1829 B. Bogdanovic, A. Cordi, H. Stepowska, P. Locatelli, B. Wermeckes, U. Wilczok, K. Haertel, and H. Richtering, Chem. Ber., 1984, **117**, 42.

1830 F. Beguin, H. Estrade-Szwarckopf, J. Conard, P. Lauginie, P. Marceau, D. Guerard, and L. Facchini, Synth. Met., 1983, **7**, 77 (Chem. Abstr., 1984, **100**, 131 295).

1831 M.C. Wintersgill, J.J. Fontanella, J.P. Calame, S.G. Greenbaum, and C.G. Andeen, J. Electrochem. Soc., 1984, **131**, 2208 (Chem. Abstr., 1984, **101**, 161 876).

1832 B. Bader, P. Heitjans, H. Ackermann, P. Freilaender, F. Fujara, G. Kiese, H.J. Stoeckmann, and C. Van der Marel, Hyperfine Interact., 1983, **16**, 593 (Chem. Abstr., 1984, **100**, 44 260).

1833 P.W. Gash, J. Appl. Phys., 1983, **54**, 6900 (Chem. Abstr., 1984, **100**, 43 350).

1834 J. Albers, S. Elschner, A. Kloepperpieper, and J. Petersson, Ferroelectrics, 1984, **55**, 769 (Chem. Abstr., 1984, **100**, 184 622).

1835 R. Tabeta and H. Saito, Chem. Lett., 1984, 293 (Chem. Abstr., 1984, **100**, 113 739).

1836 D.G. Hughes and L. Pandey, J. Magn. Reson., 1984, **56**, 428.

1837 A. Ellaboudy, M.L. Tinkham, B. Van Eck, J.L. Dye, and P.B. Smith, J. Phys. Chem., 1984, **88**, 3852.

1838 Y. Chabre, P. Segransan, C. Berthier, F.J. Di Salvo, and J.E. Fischer, Synth. Met., 1983, **8**, 7.

1839 L.N. Erofeev, V.V. Laiko, V.P. Tarasov, and B.N. Provotorov, Khim. Fiz., 1982, 387 (Chem. Abstr., 1984, **101**, 82 863).

1840 V.A. Urikh and O.Yu. Fishbein, Zh. Neorg. Khim., 1984, **29**, 56 (Chem. Abstr., 1984, **100**, 78 948).

1841 L.J. De Haas, C.M.B. Van der Zon, W.T. Wenckebach, and N.J. Poulis, Physica B+C (Amsterdam), 1983, **123**, 35 (Chem. Abstr., 1984, **100**, 44 238).

1842 D.B. Zax, A. Bielecki, K.W. Zilm, and A. Pines, Chem. Phys. Lett., 1984, **106**, 550.

1843 A.E. Mefed, Zh. Eksp. Teor. Fiz., 1984, **86**, 302 (Chem. Abstr., 1984, **100**, 131 213).

1844 M. Drabik, L. Zurkova, and I. Goljer, J. Therm. Anal., 1983, **26**, 73 (Chem. Abstr., 1984, **100**, 60 910).

1845 R. Tycko, E. Schneider, and A. Pines, J. Chem. Phys., 1984, **81**, 680.

1846 A.R. Grimmer, F. Von Lampe, M. Mägi, and E. Lippmaa, Monatsh. Chem., 1984, **115**, 561.

1847 L. Latie, G. Villeneuve, D. Conte, and G. Le Flem, J. Solid State Chem., 1984, **51**, 293 (Chem. Abstr., 1984, **100**, 166 052).

1848 G.A. Babushkin, V.A. Borodin, V.D. Doroshev, R.Z. Levitin, V.N. Orlov, A.I. Popov, and T.N. Tarasenko, Zh. Eksp. Teor. Fiz., 1984, **87**, 989 (Chem. Abstr., 1984, **101**, 202 931).

1849 V.L. Kuznetsov, I.L. Mudrakovskii, A.V. Romanenko, A.V. Pashis, V.M. Mastikhin, and Yu.I. Ermakov, React. Kinet. Catal. Lett., 1984, **25**, 137 (Chem. Abstr., 1984, **101**, 60 889).

1850 B.Ya. Brach, S.P. Gabuda, A.M. Panich, and R.M. Rakhmankulov, Zh. Strukt. Khim., 1983, **24**, 126 (Chem. Abstr., 1984, **100**, 131 203).

1851 B. Bleaney, J.F. Gregg, M.J.M. Leask, and M.R. Wells, Proc. R. Soc. London, [Ser.] A, 1984, **394**, 69 (Chem. Abstr., 1984, **101**, 102 758).

1852 A.V. Egorov, A.A. Kudryashov, M.S. Tagirov, and M.A. Teplov, Fiz. Tverd. Tela (Leningrad), 1984, **26**, 2223 (Chem. Abstr., 1984, **101**, 121 772).

1853 L. Brohan and R. Marchand, Solid State Ionics, 1983, **9**, 419 (Chem. Abstr., 1984, **100**, 95 441).

1854 P. Molinie, L. Trichet, J. Rouxel, C. Berthier, Y. Chabre, and P. Segransan, J. Phys. Chem. Solids, 1984, **45**, 105 (Chem. Abstr., 1984, **100**, 145 794).

1855 J.S. Cantrell, R.C. Bowman, jun., and D.B. Sullenger, J. Phys. Chem., 1984, **88**, 918.

1856 E.P. Bochkarev, N.M. Kolpachkova, Yu.B. Muravlev, L.G. Nekhamkin, N.V. Ovsyannikova, and E.L. Sokolova, Dokl. Akad. Nauk SSSR, 1984, **274**, 619 (Chem. Abstr., 1984, **101**, 63 958).

1857 Y. Chabre, C. Berthier, P. Segransan, L. Trichet, and P. Deniard, Solid State Ionics, 1983, **9**, 467 (Chem. Abstr., 1984, **100**, 77 655).

1858 M.Yu. Belyaev, A.V. Skripov, V.N. Kozhanov, A.P. Stepanov, and E.V. Galoshina, Solid State Commun., 1983, **48**, 1049 (Chem. Abstr., 1984, **100**,

60 678).
1859 A.V. Dmitriev, R.N. Pletnev, V.A. Gubanov, Yu.G. Zainulin, and G.P. Shveikin, Dokl. Akad. Nauk SSSR, 1984, **275**, 906 (Chem. Abstr., 1984, **101**, 43 773).
1860 A.V. Dmitriev, R.N. Pletnev, Yu.G. Zainulin, and V.A. Gubanov, Izv. Akad. Nauk SSSR, Neorg. Mater., 1984, **20**, 807 (Chem. Abstr., 1984, **101**, 102 781.
1861 M. Itoh, H. Yasuoka, Y. Ueda, and K. Kosuge, J. Phys. Soc. Jpn., 1984, **53**, 1847 (Chem. Abstr., 1984, **101**, 16 120).
1862 V.M. Mastikhin, O.B. Lapina, and L.G. Simonova, React. Kinet. Catal. Lett., 1984, **24**, 127 (Chem. Abstr., 1984, **100**, 162 514).
1863 T.P. Gorshkova, R.I. Maksimovskaya, D.V. Tarasova, N.N. Chumachenko, and T.A. Nikoro, React. Kinet. Catal. Lett., 1984, **24**, 107 (Chem. Abstr., 1984, **100**, 162 511).
1864 K. Takanashi, H. Yasuoka, Y. Ueda, and K. Kosuage, J. Phys. Soc. Jpn., 1983, **52**, 3953 (Chem. Abstr., 1984, **100**, 16 566).
1865 V.G. Pitsyuga, A.P. Nakhodnova, and L.F. Chernysh, Zh. Neorg. Khim., 1984, **29**, 84 (Chem. Abstr., 1984, **100**, 149 971).
1866 H. Eckert, W. Mueller-Warmuth, W. Schramm, and R. Schoelhorn, Solid State Ionics, 1984, **13**, 1 (Chem. Abstr., 1984, **101**, 15 657).
1867 W. Schramm, R. Schoelhorn, H. Eckert, and W. Mueller-Warmuth, Mater. Res. Bull., 1983, **18**, 1283 (Chem. Abstr., 1984, **100**, 128 640).
1868 M. Itoh, T. Tsuda, and H. Yasuoka, J. Phys. Soc. Jpn., 1984, **53**, 1923 (Chem. Abstr., 1984, **101**, 47 525).
1869 B. Nowak and M. Minier, J. Phys. F, 1984, **14**, 1291.
1870 A.A. Rempel, M.Yu. Belyaev, L.B. Dudrovskaya, and G.P. Shveikin, Zh. Neorg. Khim., 1984, **29**, 2451 (Chem. Abstr., 1984, **101**, 202 487).
1871 F. Borsa and J.J. Van der Klink, Ferroelectrics, 1984, **53**, 117 (Chem. Abstr., 1984, **100**, 184 609).
1872 U.T. Hoechli and A. Rigamonti, J. Phys. C, 1983, **16**, 6321 (Chem. Abstr., 1984, **100**, 131 199).
1873 K. Handlir, J. Holecek, M. Nadvornik, A. Lycka, and M. Hucl., Collect. Czech. Chem. Commun., 1984, **49**, 1497.
1874 H. Kubo, K. Hirakawa, M. Tanimoto, and K. Katsumata, Kogaku Shuho - Kyushu Daigaku, 1983, **56**, 805 (Chem. Abstr., 1984, **100**, 166 835).
1875 B.E. Hanson, G.W. Wagner, R.J. Davis, and E. Motell, Inorg. Chem., 1984, **23**, 1635.
1876 P.L. Gai, N.J. Clayden, J. Black, and J.B. Goodenough, Conf. Ser. - Inst. Phys., 1984, **68**, 253 (Chem. Abstr., 1984, **100**, 127 387).
1877 J.B. Black, N.J. Clayden, L. Griffiths, and J.D. Scott, J. Chem. Soc., Dalton Trans., 1984, 2765.
1878 W.H. Knoth and R.D. Farlee, Inorg. Chem., 1984, **23**, 4765.
1879 N.E. Alekseevskii and E.G. Nikolaev, Pis'ma Zh. Eksp. Teor. Fiz., 1984, **39**, 23 (Chem. Abstr., 1984, **100**, 113 708).
1880 J. Kassovicova, J. Murin, and M. Rakos, Czech. J. Phys., 1984, **B 34**, 1015 (Chem. Abstr., 1984, **101**, 221 175).
1881 H. Kubo, N. Uryu, H. Nishihara, and H. Yasuoka, J. Phys. Soc. Jpn., 1984, **53**, 2099 (Chem. Abstr., 1984, **101**, 47 526).
1882 A. Kotlicki, B.A. McLeod, M. Shott, and B.G. Turrell, Phys. Rev. B: Condens. Matter, 1984, **29**, 26 (Chem. Abstr., 1984, **100**, 60 698).
1883 V.P. Tarasov, V.I. Privalov, S.A. Petrushin, G.A. Kirakosyan, S.V. Kryuchkov, K.E. German, and Yu.A. Buslaev, Dokl. Akad. Nauk SSSR, 1983, **272**, 919 (Chem. Abstr., 1984, **100**, 16 558).
1884 B. Maiti and B.R. McGarvey, J. Magn. Reson., 1984, **58**, 37.
1885 R.S. Nasibullin, M.S. Setchenkov, and I.L. Maksimov, Izv. Vyssh. Uchebn. Zaved., Fiz., 1983, **26**, 114 (Chem. Abstr., 1984, **100**, 60 673).
1886 D.K. Liu, M.S. Wrighton, D.R. McKay, and G.E. Maciel, Inorg. Chem., 1984, **23**, 212.
1887 M. Bose, A. Ghoshray, A. Basu, and C.N.R. Rao, Proc. Nucl. Phys. Solid State Phys. Symp. 1980 (Pub. 1983), **23**, 831 (Chem. Abstr., 1984, **101**, 32 194).
1888 M. Bose, K. Roy, and A. Ghoshray, J. Phys. C, 1984, **17**, L757 (Chem. Abstr., 1984, **101**, 220 190).

1889 H. Rager, Z. Naturforsch., Teil A, 1984, **39**, 111 (Chem. Abstr., 1984, **100**, 95 411).
1890 F. Lefebvre and Y. Ben Taarit, Nouv. J. Chim., 1984, **8**, 387 (Chem. Abstr., 1984, **101**, 137 823).
1891 M. Okazaki and C.A. McDowell, J. Am. Chem. Soc., 1984, **106**, 3185.
1892 H. Van der Vlist, A.F.M. Arts, and H.W. De Wijn, Phys. Rev. B: Condens. Matter, 1984, **30**, 5270 (Chem. Abstr., 1984, **101**, 221 113).
1893 T.J. Han and S.H. Choh, Can. J. Phys., 1983, **61**, 1627 (Chem. Abstr., 1984, **100**, 95 408).
1894 P.F. Barron, J.C. Dyason, L.M. Engelhardt, P.C. Healy, and A.H. White, Inorg. Chem., 1984, **23**, 3766.
1895 I.A. Harris, jun. and P.J. Bray, Phys. Chem. Glasses, 1984, **25**, 69 (Chem. Abstr., 1984, **101**, 77 540).
1896 P.M. Henrichs and M. Linder, J. Magn. Reson., 1984, **58**, 458.
1897 B.W. Van Beest, A. Janner, and R. Blinc, J. Phys. C, 1983, **16**, 5409 (Chem. Abstr., 1984, **100**, 43 364).
1898 J. Emery, S. Hubert, and J.C. Fayet, J. Phys. Lett., 1984, **45**, 693.
1899 P.F. Rodesiler, R.W. Turner, N.G. Charles, E.A.H. Griffith, and E.L. Amma, Inorg. Chem., 1984, **23**, 999.
1900 M.P. Shatlock and G.E. Maciel, J. Chem. Phys., 1984, **81**, 895.
1901 R.S. Honkonen and P.D. Ellis, J. Am. Chem. Soc., 1984, **106**, 5488.
1902 S. Ganapathy, V.P. Chacko, and R.G. Bryant, J. Chem. Phys., 1984, **81**, 661.
1903 S. Lacelle, W.C. Stevens, D.M. Kurtz, jun., J.W. Richardson, jun., and R.A. Jacobson, Inorg. Chem., 1984, **23**, 930.
1904 N.G. Charles, P.F. Rodesiler, E.A.H. Griffith, and E.L. Amma, Acta Crystallogr., 1984, **C40**, 1676 (Chem. Abstr., 1984, **101**, 201 981).
1905 B.S. Sarma and J. Ramakrishna, Curr. Sci., 1984, **53**, 459 (Chem. Abstr., 1984, **101**, 64 730).
1906 T.M. Duncan, J. Am. Chem. Soc., 1984, **106**, 2270.
1907 G.S. Beloglazov and V.S. Grechishkin, Zh. Fiz. Khim., 1984, **58**, 2254 (Chem. Abstr., 1984, **101**, 221 165).
1908 G. Wang, Z. Li, L. Chen, and L. Wang, Wuli Xuebao, 1983, **32**, 1104 (Chem. Abstr., 1984, **100**, 60 674).
1909 M. Engelsberg and N. Mendes Borges, J. Phys. C, 1984, **17**, 3633 (Chem. Abstr., 1984, **101**, 202 986).
1910 Z. Gabelica, J.B. Nagy, P. Bodart, and G. Debras, Chem. Lett., 1984, 1059.
1911 M.I. Nieto, H. Arribart, P. Sapoval, and J.L. Oteo, Glastech. Ber., 1983, **56**, 468 (Chem. Abstr., 1984, **100**, 108 016).
1912 J.W. Khang, Nonmunjip - Cheju Taehak, 1983, **16**, 187 (Chem. Abstr., 1984, **100**, 72 757).
1913 I.A. Harris, jun. and P.J. Bray, Phys. Chem. Glasses, 1984, **25**, 44 (Chem. Abstr., 1984, **100**, 202 209).
1914 H. Buerger, W. Vogel, V. Kozhukharov, and M. Marinov, J. Mater. Sci., 1984, **19**, 403 (Chem. Abstr., 1984, **100**, 108 029).
1915 G. Haller, B. McMillan, and J. Brinen, Prepr. - Am. Chem. Soc., Div. Pet. Chem., 1984, **29**, 939 (Chem. Abstr., 1984, **101**, 178 276).
1916 M. Weinberger, D. Mueller, and W. Gessner, Z. Chem., 1983, **23**, 420.
1917 S.C. Bhargava and A.M. Nene, Proc. Nucl. Phys. Solid State Phys. Symp. 1980 (Pub. 1983), **23**, 806.
1918 T. Asai and S. Kawai, Chem. Lett., 1984, 1125 (Chem. Abstr., 1984, **101**, 121 108).
1919 R. Dupree, J.R. Howells, A. Hooper, and F.W. Poulsen, Solid State Ionics, 1983, **9**, 131 (Chem. Abstr., 1984, **100**, 88 702).
1920 B.H. Suits and D. White, Solid State Commun., 1984, **50**, 291 (Chem. Abstr., 1984, **101**, 16 114).
1921 S. Sinha and R. Srinivasan, Pramana, 1984, **22**, 345 (Chem. Abstr., 1984, **100**, 220 074).
1922 A.C. Kunwar, A.R. Thompson, H.S. Gutowsky, and E. Oldfield, J. Magn. Reson., 1984, **60**, 467.
1923 S.V. Bhat, S. Sinha, and R. Srinivasan, Curr. Sci., 1984, **53**, 27 (Chem. Abstr., 1984, **101**, 142 630).
1924 P.J. Bray, I.A. Harris, jun., F. Bucholtz, and A.E. Geissberger,

Stekloobraz. Sostoyanie. Materialy 7 Vses. Soveshch., Leningrad, 13-15 Okt., 1981, L., 1983, 55. From Ref. Zh., Fiz. (A-D), 1983, abstr. no. 11E117 (Chem. Abstr., 1984, **100**, 40 027).
1925 D. Müller, P. Starke, M. Jank, K.-P. Wendlandt, H. Bremer, and G. Scheler, Z. Anorg. Allg. Chem., 1984, **517**, 167.
1926 P.A. Diddams, J.M. Thomas, W. Jones, J.A. Ballantine, and J.H. Purnell, J. Chem. Soc., Chem. Commun., 1984, 1340.
1927 D.R. Corbin, R.D. Farlee, and G.D. Stucky, Inorg. Chem., 1984, **23**, 2920.
1928 J.B. Nagy, Z. Gabelica, G. Debras, E.G. Derouane, J.P. Gilson, and P.A. Jacobs, Zeolites, 1984, **4**, 133 (Chem. Abstr., 1984, **100**, 198 202).
1929 J.A. Ripmeester, A. Majid, and R.E. Hawkins, J. Inclusion Phenom., 1983, **1**, 193 (Chem. Abstr., 1984, **100**, 145 560).
1930 T.M. Duncan, P. Winslow, and A.T. Bell, Prepr. - Am. Chem. Soc., Div. Pet. Chem., 1984, 13 (Chem. Abstr., 1984, **101**, 195 390).
1931 P. Bodart, Z. Gabelica, J.B. Nagy, and G. Debras, NATO ASI Ser., Ser. E, 1984, **80**, 211 (Chem. Abstr., 1984, **100**, 183 462).
1932 J. Sanz and J.M. Serratosa, Clay. Miner., 1984, **19**, 113 (Chem. Abstr., 1984, **101**, 26 267).
1933 J. Sanz and J.M. Serratosa, J. Am. Chem. Soc., 1984, **106**, 4790.
1934 S. Hayashi, K. Suzuki, S. Shin, K. Hayamizu, and O. Yamamoto, Chem. Phys. Lett., 1984, **110**, 54.
1935 A.W. Peters, ACS Symp. Ser., 1984, **248**, 201 (Chem. Abstr., 1984, **100**, 180 650).
1936 R.L. Frost and P.F. Barron, J. Phys. Chem., 1984, **88**, 6206.
1937 J. Klinowski, J.M. Thomas, D.P. Thompson, P. Korgul, K.H. Jack, C.A. Fyfe, and G.C. Gobbi, Polyhedron, 1984, **3**, 1267.
1938 R.H. Jarman, A.J. Jacobson, and M.T. Melchior, J. Phys. Chem., 1984, **88**, 5748.
1939 C.A. Fyfe, G.C. Gobbi, G.J. Kennedy, C.T. De Schutter, W.J. Murphy, R.S. Ozubko, and D.A. Slack, Chem. Lett., 1984, 163.
1940 K. Seff and M.D. Mellum, J. Phys. Chem., 1984, **88**, 3560.
1941 M. Alla, E. Lippmaa, D. Freude, and D. Prager, Z. Phys. Chem. (Leipzig), 1984, **265**, 825.
1942 R.H. Jarman and M.T. Melchior, J. Chem. Soc., Chem. Commun., 1984, 414.
1943 C.D. Chang, C.T.-W. Chu, J.N. Miale, R.F. Bridger, and R.B. Calvert, J. Am. Chem. Soc., 1984, **106**, 8143.
1944 C.A. Fyfe, G.J. Kennedy, G.T. Kokotailo, and C.T. De Schutter, J. Chem. Soc., Chem. Commun., 1984, 1093.
1945 G. Engelhardt and U. Lohse, J. Catal., 1984, **88**, 513 (Chem. Abstr., 1984, **101**, 98 381).
1946 C.A. Fyfe, G.C. Gobbi, W.J. Murphy, R.S. Ozubko, and D.A. Slack, J. Am. Chem. Soc., 1984, **106**, 4435.
1947 S. Ramdas and J. Klinowski, Nature (London), 1984, **308**, 521.
1948 J.G. Thompson, Clays Clay Miner., 1984, **32**, 233 (Chem. Abstr., 1984, **101**, 57 868).
1949 P.J. Bray and E.J. Holupka, J. Non-Cryst. Solids, 1984, **67**, 119 (Chem. Abstr., 1984, 101, 214 255).
1950 N.D. Butler, R. Dupree, and M.H. Lewis, J. Mater. Sci. Lett., 1984, **3**, 469 (Chem. Abstr., 1984, **100**, 219 462).
1951 Yu.A. Grinchenko, Yu.N. Savvin, and B.L. Timan, Sb. Nauchn. Tr. - Vses. Nauchno-Issled. Inst. Monokrist., Stsintill. Mater. Osobo Chist. Khim. Veshchestv, 1982, **8**, 36 (Chem. Abstr., 1984, **100**, 201 178).
1952 H. Stade, D. Mueller, and G. Scheler, Z. Anorg. Allg. Chem., 1984, **510**, 16.
1953 D.E. Woessner and J.C. Trewella, J. Magn. Reson., 1984, **59**, 352.
1954 Z. Gabelica, G. Debras, and J.B. Nagy, Stud. Surf. Sci. Catal., 1984, **19**, 113 (Chem. Abstr., 1984, **101**, 238 424).
1955 M.W. Anderson, J. Klinowski, and L. Xinsheng, J. Chem. Soc., Chem. Commun., 1984, 1596.
1956 K.F.M.G.J. Scholle, W.S. Veeman, P. Frenken, and G.P.M. Van der Velden, J. Phys. Chem., 1984, **88**, 3395.
1957 V.G. Stepanov, A.A. Shubin, K.G. Ione, V.M. Mastikhin, and K.I. Zamaraev, Kinet. Katal., 1984, **25**, 1225 (Chem. Abstr., 1984, **101**, 236 172).

1958 C.A. Fyfe, G.C. Gobbi, and G.J. Kennedy, J. Phys. Chem., 1984, **88**, 3248.

1959 K.G. Ione, G.V. Echevskii, and G.N. Nosyreva, J. Catal., 1984, **85**, 287 (Chem. Abstr., 1984, **100**, 36 651).

1960 J.B. Higgins, D.E. Woessner, J.C. Trewella, and J.L. Schlenker, Zeolites, 1984, **4**, 112.

1961 G. Boxhoorn, A.G.T.G. Kortbeek, G.R. Hays, and N.C.M. Alma, Zeolites, 1984, **4**, 15 (Chem. Abstr., 1984, **100**, 95 429).

1962 G.T. Kokotailo, C.A. Fyfe, G.C. Gobbi, G.J. Kennedy, and C.T. De Schutter, J. Chem. Soc., Chem. Commun., 1984, 1208.

1963 N.C.M. Alma, G.R. Hays, A.V. Samoson, and E.T. Lippmaa, Anal. Chem., 1984, **56**, 729.

1964 J.G. Thompson, Clay Miner., 1984, **19**, 229 (Chem. Abstr., 1984, **101**, 155 016).

1965 M. Lipsicas, R.H. Raythatha, T.J. Pinnavaia, I.D. Johnson, R.F. Giese, jun., P.M. Costanzo, and J.L. Robert, Nature (London), 1984, **309**, 604 (Chem. Abstr., 1984, **101**, 41 228).

1966 M.A. Wilson, P.F. Barron, and A.S. Campbell, J. Soil Sci., 1984, **35**, 201 (Chem. Abstr., 1984, **101**, 109 555).

1967 C.S. Blackwell and R.L. Patton, J. Phys. Chem., 1984, **88**, 6135.

1968 K. Yamada, N. Weiden, and A. Weiss, J. Mol. Struct., 1983, **111**, 217.

1969 V.A. Ananichev, A.N. Kudryavtsev, L.A. Bailakov, and L.N. Blinov, Fiz. Khim. Stekla, 1984, **10**, 93 (Chem. Abstr., 1984, **100**, 143 738).

1970 P.M. Henrichs, M.L. Cofield, R.H. Young, and J.M. Hewitt, J. Magn. Reson., 1984, **58**, 85.

1971 G.R. Miller, C.F. Poranski, jun., and H.A. Resing, J. Chem. Phys., 1984, **80**, 1708.

1972 G.R. Miller, H.A. Resing, M.J. Moran, L. Banks, F.L. Vogel, A. Pron, and D. Billaud, Synth. Met., 1983, **8**, 77 (Chem. Abstr., 1984, **100**, 113 696).

1973 M.J. Moran, G.R. Miller, R.A. DeMarco, and H.A. Resing, J. Phys. Chem., 1984, **88**, 1580.

1974 S. Karunanithy and F. Aubke, J. Fluorine Chem., 1984, **25**, 339 (Chem. Abstr., 1984, **101**, 102 830).

1975 K. Kume, Y. Maniwa, H. Suematsu, Y. Iye, and S. Tanuma, Synth. Met., 1983, **8**, 69 (Chem. Abstr., 1984, **100**, 95 406).

1976 G. Roth, K. Lueders, and H.J. Güntherodt, Synth. Met., 1983, **8**, 91 (Chem. Abstr., 1984, **100**, 95 407).

1977 S. Karunanity and F. Aubke, J. Fluorine Chem., 1983, **23**, 541 (Chem. Abstr., 1984, **100**, 113 820).

1978 L.B. Ebert, D.R. Mills, and J.C. Scanlon, Mater. Res. Bull., 1983, **18**, 1505 (Chem. Abstr., 1984, **100**, 60 663).

1979 S. Karunanithy and F. Aubke, Can. J. Chem., 1984, **61**, 2638.

1980 K. Beshah, J.E. Mark, and J.L. Ackerman, Polym. Mater. Sci. Eng., 1983, **49**, 288 (Chem. Abstr., 1984, **100**, 193 318).

1981 W.J. Welsh, K. Beshah, J.L. Ackerman, J.E. Mark, L.D. David, and R. West, Polym. Prepr. (Am. Chem. Soc., Div. Polym. Chem.), 1983, **24**, 131 (Chem. Abstr., 1984, **100**, 192 616).

1982 E. Dupree and R.F. Pettifer, Nature (London), 1984, **308**, 523 (Chem. Abstr., 1984, **100**, 179 100).

1983 B.H.W.S. De Jong, C.M. Schramm, and V.E. Parziale, J. Am. Chem. Soc., 1984, **106**, 4396.

1984 A.R. Grimmer and R. Radeglia, Chem. Phys. Lett., 1984, **106**, 262.

1985 G. Engelhardt and R. Radeglia, Chem. Phys. Lett., 1984, **108**, 271.

1986 M. Magi, E. Lippmaa, A. Samoson, G. Engelhardt, and A.R. Grimmer, J. Phys. Chem., 1984, **88**, 1518.

1987 P. Heitjans, B. Bader, H.J. Stoeckmann, K. Doerr, G. Kiese, H. Ackermann, P. Freilaender, W. Mueller-Warmuth, and K. Meise-Gresch, Hyperfine Interact., 1983, **16**, 597 (Chem. Abstr., 1984, **100**, 44 261).

1988 A.R. Grimmer, M. Magi, M. Haehnert, H. Stade, A. Samoson, W. Wieker, and E. Lippmaa, Phys. Chem. Glasses, 1984, **25**, 105 (Chem. Abstr., 1984, **101**, 135 652).

1989 T. Fujiu and M. Ogino, J. Non-Cryst. Solids, 1984, **64**, 287 (Chem. Abstr., 1984, **101**, 11 376).

1990 C.W. Dirk and T.J. Marks, Inorg. Chem., 1984, **23**, 4325.
1991 R.G. Bryant, V.P. Chacko, and M.C. Etter, Inorg. Chem., 1984, **23**, 3580.
1992 R.E.J. Sears, Q.Z. Guo, and H.J. Mackey, J. Chem. Phys., 1984, **80**, 5448.
1993 A.B. Yaroslavtsev, V.F. Chuvaev, Z.N. Prozorovskaya, and I.I. Baskin, Zh. Neorg. Khim., 1983, **28**, 2746 (Chem. Abstr., 1984, **100**, 44 304).
1994 B. Topic, R. Blic, and L.A. Shuvalov, Phys. Status Solidi A, 1984, **85**, 409 (Chem. Abstr., 1984, **101**, 221 186).
1995 M. Mali, J. Roos, E. Courtens, and K.A. Müller, Ferroelectrics, 1984, **53**, 215 (Chem. Abstr., 1984, **100**, 184 610).
1996 T.M. Duncan and D.C. Douglass, Chem. Phys., 1984, **87**, 339.
1997 J. Gaillard, J. Gloux, P. Gloux, B. Lamotte, and G. Ruis, Ferroelectrics, 1984, **54**, 421 (Chem. Abstr., 1984, **100**, 219 939).
1998 Q.Z. Guo and R.E.J. Sears, J. Magn. Reson., 1984, **57**, 143.
1999 L.J. Azevedo, J.E. Schriber, and E.M. Engler, Phys. Rev. B: Condens. Matter, 1984, **29**, 464 (Chem. Abstr., 1984, **100**, 78 676).
2000 T. Takahashi, D. Jerome, and K. Bechgaard, J. Phys. (Les Ulis, Fr.), 1984, **45**, 945 (Chem. Abstr., 1984, **100**, 219 899).
2001 C. Bourbonnais, F. Creuzet, D. Jerome, K. Bechgaard, and A. Moradpour, J. Phys., Lett., 1984, **45**, 755 (Chem. Abstr., 1984, **101**, 102 779).
2002 C.W.B. Lee and B.C. Sanctuary, J. Magn. Reson., 1984, **59**, 189.
2003 A.M. Petrosyan, Kristallografiya, 1984, **29**, 1024 (Chem. Abstr., 1984, **101**, 221 387).
2004 J. Andersson, M. Azoulay, and J. De Pablo, Thermochim. Acta, 1983, **70**, 291 (Chem. Abstr., 1984, **100**, 127 438).
2005 G. Wendt, B. Staudte, K.H. Steinberg, R.V. Dmitriev, and C.M. Minacev, Z. Chem., 1984, **24**, 33 (Chem. Abstr., 1984, **100**, 162 470).
2006 N. Reinecke and R. Haul, Ber. Bunsenges. Phys. Chem., 1984, **88**, 1232.
2007 P.H. Kasai and P.M. Jones, J. Mol. Catal., 1984, **27**, 81.
2008 W. Oehme, D. Michel, H. Pfeifer, and S.P. Zhdanov, Zeolites, 1984, **4**, 120 (Chem. Abstr., 1984, **100**, 162 418).
2009 K.F.M.G.J. Scholle, A.P.M. Kentgens, W.S. Veeman, P. Frenken, and G.P.M. Van der Velden, J. Phys. Chem., 1984, **88**, 5 (Chem. Abstr., 1984, **100**, 40 118).
2010 O.V. Veselova, E.I. Mal'ko, B.P. Nikolaev, L.N. Petrov, and A.M. Shlyakov, Kolloidn. Zh., 1983, **45**, 1053 (Chem. Abstr., 1984, **100**, 40 235).
2011 D. Freude, M. Hunger, H. Pfeifer, G. Scheler, J. Hoffmann, and W. Schmitz, Chem. Phys. Lett., 1984, **105**, 427.
2012 F. Hanus and P. Gillis, J. Magn. Reson., 1984, **59**, 437.
2013 P. Stilbs and B. Lindman, J. Colloid Interface Sci., 1984, **99**, 290 (Chem. Abstr., 1984, **100**, 162 249).
2014 T.M. Duncan, P. Winslow, and A.T. Bell, Prepr. - Am. Chem. Soc., Div. Pet. Chem., 1984, **29**, 1135 (Chem. Abstr., 1984, **101**, 194 809).
2015 S.J. Decanio, J.H. Onuferko, H.C. Foley, B.C. Gates, and C. Dybowski, Surf. Sci., 1984, **136**, L67 (Chem. Abstr., 1984, **100**, 191 238).
2016 C.F. Tirendi, G.A. Mills, and C. Dybowski, J. Phys. Chem., 1984, **88**, 5765.
2017 P.K. Wang, Report, 1984, DOE/ER/01198-1406; Order No. DE84008661, 93 pp. From Energy Res. Abstr., 1984, 9, abstr. no. 22862 (Chem. Abstr., 1984, **101**, 98 238).
2018 P.K. Wang, C.P. Slichter, and J.H. Sinfelt, Phys. Rev. Lett., 1984, **53**, 82 (Chem. Abstr., 1984, **101**, 60 650).
2019 T. Ito, L.C. De Menorval, and J.P. Fraissard, J. Chim. Phys. Phys.-Chim. Biol., 1983, **80**, 573.
2020 V.M. Mastikhin, I.L. Mudrakovskii, A.G. Pel'menchikov, and G.M. Zhidomirov, Surf. Sci., 1984, **144**, 550 (Chem. Abstr., 1984, **101**, 216 881).
2021 W. Abraham, T.M. Harris, and D.J. Wilson, Sep. Sci. Technol., 1984, 389 (Chem. Abstr., 1984, **101**, 117 400).
2022 D.T.B. Tennakoon, R. Schloegl, T. Rayment, J. Klinowski, W. Jones, and J.M. Thomas, Clay Miner., 1983, **18**, 357 (Chem. Abstr., 1984, **100**, 91 973).
2023 K.H. Steinberg, F. Roessner, H. Winkler, B. Staudte, Kh. Dimitrov, Z. Popova, R. Dmitriev, and Kh. Minachev, Geterog. Katal., 1983, 5th, Pt. 1, 491 (Chem. Abstr., 1984, **100**, 180 697).
2024 H. Winkler, K.H. Steinberg, and G. Kapphahn, J. Colloid Interface Sci., 1984, **98**, 144 (Chem. Abstr., 1984, **100**, 106 067).

2025 A. Zikanova, M. Kocirik, and A.G. Bezus, Adsorbts. Mikroporakh. [Tr. Konf. Teor. Vopr. Adsorbts.]. 5th 1979 (Pub. 1983), 119 (Chem. Abstr., 1984, **100**, 162 274).
2026 I.L. Mudrakovskii, S.S. Shepelev, K.G. Ione, K.I. Zamaraev, E. Paukstis, and G.K. Boreskov, Kinet. Katal., 1984, **25**, 766 (Chem. Abstr., 1984, **101**, 110 244).
2027 H. Herden, W.D. Einicke, U. Messow, E. Volkmann, R. Schoellner, and J. Kaerger, J. Colloid Interface Sci., 1984, **102**, 227 (Chem. Abstr., 1984, **101**, 198 756).
2028 G.W. West, Aust. J. Chem., 1984, **37**, 455.
2029 A. Germanus, J. Kaerger, and H. Pfeifer, Zeolites, 1984, **4**, 188 (Chem. Abstr., 1984, **100**, 162 345).
2030 E.G. Derouane and J.B. Nagy, ACS Symp. Ser., 1984, **248**, 101 (Chem. Abstr., 1984, **101**, 54 477).
2031 V.V. Volynets, O.P. Yablonskii, L.F. Titova, and D.N. Chaplits, Issled. i Razrab. Tekhnol. Pr-va Monomerov i Sintet. Kauchukov, M., 1983, 52. From Ref. Zh., Khim., 1984, abstr. no. 2N20 (Chem. Abstr., 1984, **100**, 208 781).
2032 M. Matsumoto, S. Shinoda, H. Takahashi, and Y. Saito, Bull. Chem. Soc. Jpn., 1984, **57**, 1795.
2033 D.L. Hasha, V.W. Miner, J.M. Garces, and S.C. Rocke, Prepr. - Am. Chem. Soc., Div. Pet. Chem., 1984, **29**, 953 (Chem. Abstr., 1984, **101**, 158 331).
2034 C.A. Fyfe, G.J. Kennedy, C.T. De Schutter, and G.T. Kokotailo, J. Chem. Soc., Chem. Commun., 1984, 541.
2035 V. Bosacek, D. Freude, W. Gruender, W. Meiler, and H. Pfeifer, Z. Phys. Chem. (Leipzig), 1984, **265**, 241.
2036 D. Freude, H. Pfeifer, A. Schmidt, and B. Staudte, Z. Phys. Chem. (Leipzig), 1984, **265**, 250.
2037 W.P. Rothwell, W.X. Shen, and J.H. Lunsford, J. Am. Chem. Soc., 1984, **106**, 2452.
2038 M. Shimoda, T. Mizusaki, T. Suzuki, A. Hirai, and K. Eguchi, Phys. Lett. A, 1984, **102**, 426 (Chem. Abstr., 1984, **101**, 28 470).
2039 J.A. Ripmeester, J. Magn. Reson., 1984, **56**, 247.
2040 M.A. Springuel-Huet, T. Ito, and J.P. Fraissard, Stud. Surf. Sci. Catal., 1984, **18**, 13 (Chem. Abstr., 1984, **101**, 137 619).
2041 A.L. Blyumenfel'd, Yu.V. Isaev, Yu.N. Novikov, and E.I. Fedin, Zh. Fiz. Khim., 1984, **58**, 2094 (Chem. Abstr., 1984, **101**, 203 140).
2042 J.H. Quateman and M. Bretz, Phys. Rev. B: Condens. Matter, 1984, **29**, 1159 (Chem. Abstr., 1984, **100**, 95 422).
2043 T. Shibanuma, H. Asada, S. Ishi, and T. Matsui, Jpn. J. Appl. Phys. Part 1, 1983, **22**, 1656 (Chem. Abstr., 1984, **100**, 16 574).
2044 R.K. Gilpin and M.E. Gangoda, Anal. Chem., 1984, **56**, 1470.
2045 A.T. Nicol, Phys. Rev. B: Condens. Matter, 1984, **29**, 2397 (Chem. Abstr., 1984, **100**, 131 234).
2046 R.G. Kidd, 'Boron-11' in NMR Newly Accessible Nucl., 1983, **2**, 49. Ed. P. Laszlo, Academic, New York (Chem. Abstr., 1984, **100**, 166 562).
2047 J.J. Delpuech, 'Aluminium-27' in NMR Newly Accessible Nucl., 1983, **2**, 153. Ed. P. Laszlo, Academic, New York (Chem. Abstr., 1984, **100**, 166 565).
2048 J.F. Hinton and K.R. Metz, 'Thallium NMR' in NMR Newly Accessible Nucl., 1983, **2**, 367. Ed. P. Laszlo, Academic, New York (Chem. Abstr., 1984, **100**, 166 573).
2049 T.L. Venable, W.C. Hutton, and R.N. Grimes, J. Am. Chem. Soc., 1984, **106**, 29.
2050 R. Contreras, F. Santiesteban, M.A. Paz-Sandoval, and B. Wrackmeyer, Tetrahedron, 1984, **40**, 3829.
2051 K. Niedenzu and P.M. Niedenzu, Inorg. Chem., 1984, **23**, 3713.
2052 B. Singaram, T.E. Cole, and H.C. Brown, Organometallics, 1984, **3**, 774.
2053 S. Kerschl and B. Wrackmeyer, Z. Naturforsch., Teil B, 1984, **39**, 1037 (Chem. Abstr., 1984, **101**, 211325).
2054 B.F. Spielvogel, F.U. Ahmed, M.K. Das, and A.T. McPhail, Inorg. Chem., 1984, **23**, 3263.
2055 B.F. Spielvogel, F.U. Ahmed, K.W. Morse, and A.T. McPhail, Inorg. Chem., 1984, **23**, 1776.

2056 B.F. Spielvogel, F.U. Ahmed, G.L. Silvey, P. Wisian-Neilson, and A.T. McPhail, Inorg. Chem., 1984, **23**, 4322.
2057 B. Kemp, S. Kalbag, and R.A. Geanangel, Inorg. Chem., 1984, **23**, 3063.
2058 L. Mendoza, M. Beatriz Pena, L. Cuellar, and R. Contreras, Rev. Latinoam. Quim., 1984, **15**, 57 (Chem. Abstr., 1984, **101**, 211 227).
2059 H. Nöth and S. Weber, Chem. Ber., 1984, **117**, 2504.
2060 D.C. Bradley, M.B. Hursthouse, J. Newton, and N.P.C. Walker, J. Chem. Soc., Chem. Commun., 1984, 188.
2061 P. Idelmann, G. Müller, W.R. Scheidt, W. Schüssler, and R. Köster, Angew. Chem., Int. Ed. Engl., 1984, **23**, 153.
2062 M. Arunchaiya, J.H. Morris, S.J. Andrews, D.A. Welch, and A.J. Welch, J. Chem. Soc., Dalton Trans., 1984, 2525.
2063 J.A. Anderson, R.J. Astheimer, J.D. Odom, and L.G. Sneddon, J. Am. Chem. Soc., 1984, **106**, 2275.
2064 C.T. Brewer and R.N. Grimes, J. Am. Chem. Soc., 1984, **106**, 2722.
2065 W. Nam, M. Soltis, C. Gordon, S. Lee, and T. Onak, J. Magn. Reson., 1984, **59**, 399.
2066 D. Reed, J. Chem. Res. (S), 1984, 198.
2067 D.F. Gaines, C.K. Nelson, J.C. Kunz, J.H. Morris, and D. Reed, Inorg. Chem., 1984, **23**, 3252.
2068 D.P. Burum, J. Magn. Reson., 1984, **59**, 430.
2069 A.J. Leffler, J. Chem. Phys., 1984, **81**, 2574.
2070 M. Kameda and G. Kodama, Inorg. Chem., 1984, **23**, 3710.
2071 P.C. Choi and J.H. Morris, J. Chem. Soc., Dalton Trans., 1984, 2119.
2072 N.E. Miller, J. Organomet. Chem., 1984, **269**, 123.
2073 E.W. Corcoran, jun. and L.G. Sneddon, J. Am. Chem. Soc., 1984, **106**, 7793.
2074 S.K. Boocock, M.A. Toft, K.E. Inkrott, L.Y. Hsu, J.C. Huffman, K. Folting, and S.G. Shore, Inorg. Chem., 1984, **23**, 3084.
2075 T.L. Venable, E. Sinn, and R.N. Grimes, J. Chem. Soc., Dalton Trans., 1984, 2275.
2076 M. Kameda, M. Shimoi, and G. Kodama, Inorg. Chem., 1984, **23**, 3705.
2077 R.G. Swisher, E. Sinn, and R.N. Grimes, Organometallics, 1984, **3**, 599.
2078 R.P. Micciche, J.J. Briguglio, and L.G. Sneddon, Organometallics, 1984, **3**, 1496.
2079 N.S. Hosmane and N.N. Sirmokadam, Organometallics, 1984, **3**, 1119.
2080 N.S. Hosmane, N.N. Sirmokadam, and R.H. Herber, Organometallics, 1984, **3**, 1665.
2081 R.J. Astheimer and L.G. Sneddon, Inorg. Chem., 1984, **23**, 3207.
2082 D.F. Gaines and G.A. Steehler, J. Chem. Soc., Chem. Commun., 1984, 1127.
2083 J.A. Heppert, M.A. Kulzick, and D.F. Gaines, Inorg. Chem., 1984, **23**, 14.
2084 J. Bould, J.E. Cook, N.N. Greenwood, and J.D. Kennedy, J. Chem. Soc., Dalton Trans., 1984, 1903.
2085 M.A. Beckett, J.E. Crook, N.N. Greenwood, and J.D. Kennedy, J. Chem. Soc., Dalton Trans., 1984, 1427.
2086 A. Arafat, G.D. Friesen, and L.J. Todd, Inorg. Chem., 1983, **22**, 3721.
2087 R.P. Micciche, J.J. Briguglio, and L.G. Sneddon, Inorg. Chem., 1984, **23**, 3992.
2088 B. Štíbr, Z. Janoušek, K. Baše, J. Plešek, K.A. Solntsev, L.A. Butman, I.I. Kuznetsov, and N.T. Kuznetsov, Collect. Czech. Chem. Commun., 1984, **49**, 1660.
2089 K. Baše, S. Heřmánek, and F. Hanousek, J. Chem. Soc., Chem. Commun., 1984, 299.
2090 H.G. Srebny and W. Preetz, Z. Anorg. Allg. Chem., 1984, **513**, 7.
2091 W. Preetz, H.G. Srebny, and H.C. Marsmann, Z. Naturforsch., Teil B, 1984, **39**, 6.
2092 Yu.L. Gaft, Yu.A. Ustynyuk, A.A. Borisenko, and N.T. Kuznetsov, Zh. Neorg. Khim., 1983, **28**, 2234 (Chem. Abstr., 1984, **100**, 28 830).
2093 J.E. Crook, N.N. Greenwood, J.D. Kennedy, and W.S. McDonald, J. Chem. Soc., Dalton Trans., 1984, 2487.
2094 B. Štíbr, Z. Janoušek, K. Base, S. Heřmánek, J. Plešek, and I.A. Zakharova, Collect. Czech. Chem. Commun., 1984, **49**, 1891.
2095 C. Vinas, W.M. Butler, F. Teixidor, and R.W. Rudolph, Organometallics, 1984,

3, 503.
2096 M. Green, J.A.K. Howard, A.P. James, C.M. Nunn, and F.G.A. Stone, J. Chem. Soc., Chem. Commun., 1984, 1113.
2097 P.E. Behnken and M.F. Hawthorne, Inorg. Chem., 1984, **23**, 3420.
2098 J. Plešek, B. Štíbr, and S. Heřmánek, Collect. Czech. Chem. Commun., 1984, **49**, 1492.
2099 C.B. Knobler, T.B. Marder, E.A. Mizusawa, RG. Teller, J.A. Long, P.E. Behnken, and M.F. Hawthorne, J. Am. Chem. Soc., 1984, **106**, 2990.
2100 R.T. Baker, M.S. Delaney, R.E. King, tert., C.B. Knobler, J.A. Long, T.B. Marder, T.E. Paxson, R.G. Teller, and M.F. Hawthorne, J. Am. Chem. Soc., 1984, **106**, 2965.
2101 J.A. Doi, E.A. Mizusawa, C.B. Knobler, and M.F. Hawthorne, Inorg. Chem., 1984, **23**, 1482.
2102 J.D. Hewes, C.W. Kreimendahl, T.B. Marder, and M.F. Hawthorne, J. Am. Chem. Soc., 1984, **106**, 5757.
2103 H.G. Srebny, W. Preetz, and H.C. Marsmann, Z. Naturforsch., Teil B, 1984, **39**, 189.
2104 J. Plešek, T. Jelinek, E. Drdáková, S. Heřmánek, and B. Štíbr, Collect. Czech. Chem. Commun., 1984, **49**, 1559.
2105 J. Plešek, T. Jelíek, and B. Štíbr, Polyhedron, 1984, **3**, 1351.
2106 L.I. Zakharkin and I.V. Pisareva, J. Organomet. Chem., 1984, **267**, 73.
2107 M.M. Teplyakov, I.A. Khotina, Ts.L. Gelashvili, and V.V. Korshak, Dokl. Akad. Nauk SSSR, 1983, **271**, 874 (Chem. Abstr., 1984, **100**, 52 097).
2108 L.I. Zakharkin and I.V. Pisareva, Phosphorus Sulphur, 1984, **20**, 357.
2109 L.I. Zakharkin, A.I. Kovredov, V.A. Olshevskaya, and V.A. Antonovich, J. Organomet. Chem., 1984, **267**, 81.
2110 N.S. Hosmane, M. Dehghan, and S. Davies, J. Am. Chem. Soc., 1984, **106**, 6435.
2111 C. Uncuta, I. Bally, C. Draghici, F. Chiraleu, and A.T. Balaban, Rev. Roum. Chim., 1984, **29**, 121.
2112 M.E. Gurskü, S.V. Baranin, A.I. Lutzenko, and B.M. Mikhailov, J. Organomet. Chem., 1984, **270**, 17.
2113 M.E. Gurskü, D.G. Pershin, and B.M. Mikhailov, J. Organomet. Chem., 1984, **260**, 17.
2114 H.C. Brown, P.K. Jadhav, and M.C. Desai, Tetrahedron, 1984, **40**, 1325.
2115 H.C. Brown and A.K. Mandal, J. Org. Chem., 1984, **49**, 2558.
2116 H.-J. Bestmann and T. Arenz, Angew. Chem., Int. Ed. Engl., 1984, **23**, 381.
2117 R. Wehrmann, C. Pues, H. Klusik, and A. Berndt, Angew. Chem., Int. Ed. Engl., 1984, **23**, 372.
2118 R. Wehrmann, H. Klusik, and A. Berndt, Angew. Chem., Int. Ed. Engl., 1984, **23**, 369.
2119 L.S. Vasilyev, O.D. Smirnova, M.I. Struchkova, and B.M. Mikhailov, J. Organomet. Chem., 1984, **275**, 19.
2120 C. Pues and A. Berndt, Angew. Chem., Int. Ed. Engl., 1984, **23**, 313.
2121 M. Terashima, H. Kakimi, M. Ishikura, and K. Kamata, Chem. Pharm. Bull., 1983, **31**, 4573.
2122 K. Anton, H. Nöth, and H. Pommerening, Chem. Ber., 1984, **117**, 2479.
2123 E. Kalbarczyk and S. Pasynkiewicz, J. Organomet. Chem., 1984, **262**, 11.
2124 B.M. Mikhailov and T.K. Baryshnikova, J. Organomet. Chem., 1984, **260**, 25.
2125 R. Wehrmann, H. Klusik, and A. Berndt, Angew. Chem., Int. Ed. Engl., 1984, **23**, 826.
2126 W. Maringgele, G.M. Sheldrick, A. Meller, and M. Noltemeyer, Chem. Ber., 1984, **117**, 2112.
2127 K. Anton, P. Konrad, and H. Nöth, Chem. Ber., 1984, **117**, 863.
2128 H. Nöth and W. Storch, Chem. Ber., 1984, **117**, 2140.
2129 R. Köster and G. Seidel, Angew. Chem., Int. Ed. Engl., 1984, **23**, 155.
2130 M. Hildenbrand, H. Pritzkow, U. Zenneck, and W. Siebert, Angew. Chem., Int. Ed. Engl., 1984, **23**, 371.
2131 E. Hohaus, Fresenius' Z. Anal. Chem., 1983, **315**, 696 (Chem. Abstr., 1984, **100**, 68 348).
2132 B. Singaram, T.E. Cole, and H.C. Brown, Organometallics, 1984, **3**, 1520.
2133 M.E. Gurskii, S.V. Baranin, and B.M. Mikhailov, J. Organomet. Chem., 1984, **270**, 9.

2134 L. Synoradzki, R. Mynott, J. Anbei, C. Krüger, Y.-h. Tsay, and R. Köster, Chem. Ber., 1984, **117**, 2863.
2135 W. Kliegel and E. Ahlenstiel, J. Organomet. Chem., 1984, **277**, 173.
2136 D.Y. Lee and J.C. Martin, J. Am. Chem. Soc., 1984, **106**, 5745.
2137 K. Anton, C. Euringer, and H. Nöth, Chem. Ber., 1984, **117**, 1222.
2138 H. Fisch, H. Pritzkow, and W. Siebert, Angew. Chem., Int. Ed. Engl., 1984, **23**, 608.
2139 H. Nöth and S. Weber, Angew. Chem., Int. Ed. Engl., 1984, **23**, 994.
2140 H.-U. Meier, P. Paetzold, and E. Schröder, Chem. Ber., 1984, **117**, 1954.
2141 P. Paetzold, C. v. Plotho, G. Schmid, R. Boese, B. Schrader, D. Bougeard, U. Pfeiffer, R. Gleiter, and W. Schäfer, Chem. Ber., 1984, **117**, 1089.
2142 C. Habben and A. Meller, Chem. Ber., 1984, **117**, 2531.
2143 J.G. Morse and W.K. Glanville, Inorg. Chem., 1984, **23**, 11.
2144 D.S. Matteson and K.M. Sadhu, Organometallics, 1984, **3**, 614.
2145 V.V. Kuznetsov, A.I. Gren, I.V. Petrovskii, and V.A. Antonovich, Deposited Doc., 1982, VINITI 89, 22 pp. Avail. VINITI (Chem. Abstr., 1984, **100**, 209 932).
2146 L. Bhal, R.V. Singh, and J.P. Tandon, Acta Chim. Hung., 1984, **115**, 251 (Chem. Abstr., 1984, **101**, 91 023).
2147 H.C. Brown and T. Imai, Organometallics, 1984, **3**, 1392.
2148 B.A. Arbuzov, O.A. Erastov, G.N. Nikonov, I.P. Romanova, R.P. Arshinova, and O.V. Ovodova, Izv. Akad. Nauk SSSR, Ser. Khim., 1983, 2535.
2149 C.K. Narula and H. Nöth, Z. Naturforsch., Teil B, 1983, **38**, 1161.
2150 J. Pfeiffer, W. Maringgele, and A. Meller, Z. Anorg. Allg. Chem., 1984, **511**, 185.
2151 S.K. Tandon, Indian J. Chem., Sect. A, 1984, **23**, 65 (Chem. Abstr., 1984, **101**, 14 225).
2152 B.N. Chernyshov, G.P. Shchetinina, V.L. Pershin, and E.G. Ippolitov, Koord. Khim., 1984, **10**, 171 (Chem. Abstr., 1984, **100**, 197 978).
2153 M. van Duin, J.A. Peters, A.P.G. Kieboom, and H. van Bekkum, Tetrahedron, 1984, **40**, 2901.
2154 G.A. Olah, K. Laali, and O. Farooq, Organometallics, 1984, **3**, 1337.
2155 B.T. Hamdi, D.J. Reynolds, and G.A. Webb, Org. Mag. Res., 1984, **22**, 90.
2156 F. Dirschl, H. Nöth, and W. Wagner, J. Chem. Soc., Chem. Commun., 1984, 1533.
2157 K. Anton, H. Nöth, and H. Pommerening, Chem. Ber., 1984, **117**, 2495.
2158 K. Anton, H. Fussstetter, and H. Nöth, Chem. Ber., 1984, **117**, 2542.
2159 C.K. Narula and H. Nöth, J. Chem. Soc., Chem. Commun., 1984, 1023.
2160 U. Klingebiel, Angew. Chem., Int. Ed. Engl., 1984, **23**, 815.
2161 T. Gasparis-Ebeling and H. Nöth, Angew. Chem., Int. Ed. Engl., 1984, **23**, 303.
2162 C.K. Narula and H. Nöth, Inorg. Chem., 1984, **23**, 4147.
2163 M.J. Farquharson and J.S. Hartman, J. Chem. Soc., Chem. Commun., 1984, 256.
2164 W. Haubold, J. Hrebicek, and G. Sawitzki, Z. Naturforsch., Teil B, 1984, **39**, 1027 (Chem. Abstr., 1984, **101**, 203 250).
2165 J.A. Happe, R.J. Morgan, and C.M. Walkup, Natl. SAMPE Symp. Exhib., [Proc.], 1984, 29th.(Technol. Vectors), 921 (Chem. Abstr., 1984, **101**, 91 984).
2166 P.B. Hitchcock, H.A. Jasim, M.F. Lappert, and H.D. Williams, J. Chem. Soc., Chem. Commun., 1984, 662.
2167 G. Klebe and D. Tranqui, Inorg. Chim. Acta, 1984, **81**, 1.
2168 F. Santiesteban, M.A. Campos, H. Morales, R. Contreras, and B. Wrackmeyer, Polyhedron, 1984, **3**, 589.
2169 K.K. Chaturvedi, R.V. Singh, and J.P. Tandon, Synth. React. Inorg. Met-Org. Chem., 1984, **14**, 1085.
2170 F. Fernández-Gadea, M.L. Jimeno, and B. Rodriguez, Org. Magn. Reson., 1984, **22**, 515.
2171 Y. Singh and A.K. Rai, Proc. - Indian Acad. Sci., Chem. Sci., 1984, **93**, 177 (Chem. Abstr., 1984, **100**, 202 429).
2172 M.F.A. Dove, R.C. Hibbert, and N. Logan, J. Chem. Soc., Dalton Trans., 1984, 2719.
2173 H.C. Brown, J.S. Cha, and B. Nazer, Inorg. Chem., 1984, **23**, 2929.
2174 Y.P. Singh and A.K. Rai, Polyhedron, 1984, **3**, 929.

2175 W. Diamantikos, H. Heinzelmann, E. Rath, and H. Binder, Z. Anorg. Allg. Chem., 1984, **517**, 111.
2176 J. Yamashita, S. Tomiyama, H. Hashimoto, K. Kitahara, and H. Sato, Chem. Lett., 1984, 749.
2177 B.V. Cheesman and R.F.M. White, Org. Magn. Reson., 1984, **22**, 764.
2178 B.V. Cheesman and R.F.M. White, Can. J. Chem., 1984, **62**, 521.
2179 B.M. Bulychev and S.E. Tokareva, Zh. Obshch. Khim., 1984, **54**, 1092 (Chem. Abstr., 1984, **101**, 102 978).
2180 V.V. Gavrilenko, M.I. Vinnikova, V.A. Antonovich, and K.I. Zakharkin, Izv. Akad. Nauk SSSR, Ser. Khim., 1984, 865 (Chem. Abstr., 1984, **101**, 110 067).
2181 H.H. Karsch, A. Appelt, and G. Müller, J. Chem. Soc., Chem. Commun., 1984, 1415.
2182 H. Schmidbaur, U. Thewalt, and T. Zafiropoulos, Angew. Chem., Int. Ed. Engl., 1984, **23**, 76.
2183 T.R. Bierschenk, T.J. Juhlke, W.I. Bailey, jun., and R.J. Lagow, J. Organomet. Chem., 1984, **277**, 1.
2184 F. Maury and G. Constant, Polyhedron, 1984, **3**, 581.
2185 V.P. Tarasov, V.I. Privalov, G.A. Kirakosyan, and Yu.A. Buslaev, Z. Naturforsch., Teil B, 1983, **38**, 1173 (Chem. Abstr., 1984, **100**, 121 157).
2186 J.W. Akitt and W. Gessner, J. Chem. Soc., Dalton Trans., 1984, 147.
2187 V.P. Tarasov, G.A. Kirakosyan, S.B. Randarevich, and Yu.A. Buslaev, Koord. Khim., 1984, **10**, 487 (Chem. Abstr., 1984, **100**, 198 540).
2188 V.P. Tarasov, G.A. Kirakosyan, S.B. Randarevich, and Yu.A. Buslaev, Koord. Khim., 1984, **10**, 950.
2189 R. Ahmad, G. Srivastava, and R.C. Mehrotra, Inorg. Chim. Acta, 1984, **89**, 41.
2190 J.F. Hinton and G.L. Turner, J. Magn. Reson., 1984, **59**, 262.
2191 T.N. Dave, H. Kaletsch, and K. Dimroth, Angew. Chem., Int. Ed. Engl., 1984, **23**, 989.
2192 E.A. Williams, Annu. Rep. NMR Spectrosc., 1983, **15**, 235 (Chem. Abstr., 1984, **101**, 72 776).
2193 J. Schraml, Carbon-Funct. Organosilicon Compd., 1984, 121 (Chem. Abstr., 1984, **101**, 23 540).
2194 B. Coleman, 'Applications of silicon-29 NMR spectroscopy', in NMR Newly Accessible Nucl., 1983, **2**, 197 (Chem. Abstr., 1984, **100**, 166 566).
2195 E. Lukevics and V.F. Matorykina, Latv. PSR Zinat. Akad. Vestis, Kim. Ser., 1983, 643 (Chem. Abstr., 1984, **100**, 156 653).
2196 G. Dombi, P. Diehl, J. Lounila, and R. Wasser, Org. Magn. Reson., 1984, **22**, 573.
2197 P.J. Watkinson and K.M. Mackay, J. Organomet. Chem., 1984, **275**, 39.
2198 K.M. Mackay, P.J. Watkinson, and A.L. Wilkins, J. Chem. Soc., Dalton Trans., 1984, 133.
2199 G. Maier, G. Mihm, and H.P. Reisenauer, Chem. Ber., 1984, **117**, 2351.
2200 D. Seyferth and C.C. Prud'homme, Inorg. Chem., 1984, **23**, 4412.
2201 G. Maier, G. Mihm, G.O.W. Baumgärtner, and H.P. Reisenauer, Chem. Ber., 1984, **117**, 2337.
2202 T.J. Barton and A. Revis, J. Am. Chem. Soc., 1984, **106**, 3802.
2203 A.R. Bassindale and T. Stout, J. Chem. Soc., Chem. Commun., 1984, 1387.
2204 R.T. Collin and Y.-W. Kwak, Organometallics, 1984, **3**, 918.
2205 G. Maier, G. Mihm, H.P. Reisenauer, and D. Littmann, Chem. Ber., 1984, **117**, 2369.
2206 Y.-M. Pai, E. Wanek, and W.P. Weber, J. Organomet. Chem., 1984, **270**, 271.
2207 E. Lukevics and M. Dzintara, J. Organomet. Chem., 1984, **271**, 307.
2208 E. Hengge and G. Miklau, Z. Anorg. Allg. Chem., 1984, **508**, 33.
2209 D.W.W. Anderson, S. Cradock, E.A.V. Ebsworth, A.R. Green, D.W.H. Rankin, and A.G. Robiette, J. Organomet. Chem., 1984, **271**, 235.
2210 D. Seyferth, D.C. Annarelli, and S.C. Vick, J. Organomet. Chem., 1984, **272**, 123.
2211 E. Buncel, T.K. Venkatachalam, U. Edlund, and B. Eliasson, J. Chem. Soc., Chem. Commun., 1984, 1479.
2212 E. Matarasso-Tchiroukhine, Can. J. Chem., 1984, **62**, 133.
2213 K. Kamienska-Trela, Z. Biedrzycka, R. Machinek, B. Knieriem, and W. Luettke, Org. Magn. Reson., 1984, **22**, 317.

2214 I. Sekacis, E. Liepins, I. Zicmane, and E. Lukevics, Zh. Obshch. Khim., 1983, **53**, 2064 (Chem. Abstr., 1984, **100**, 51 743).
2215 Y. Takeuchi, T. Harazono, and N. Kakimoto, Inorg. Chem., 1984, **23**, 3835.
2216 C.H. Yoder, J.E. Crouse, and D.A. Wilson, J. Organomet. Chem., 1984, **276**, 9.
2217 K.B. Dillon and G.F. Hewitson, Polyhedron, 1984, **3**, 957.
2218 B. Wrackmeyer, J. Magn. Reson., 1984, **59**, 141.
2219 H.A. Olszowy and W. Kitching, Organometallics, 1984, **3**, 1676.
2220 L.L. Lohr, jun., Int. J. Quantum Chem., 1984, **25**, 211 (Chem. Abstr., 1984, **100**, 57 099).
2221 L. D'Ornellas, J. Cadenas, B. Mendez, and A.C. Rojas, Acta Cient. Venez., 1982, **33**, 382 (Chem. Abstr., 1984, **101**, 7317).
2222 V. Alberts, M.J. Cuthbertson, D.W. Hawker, and P.R. Wells, Org. Magn. Reson., 1984, **22**, 556.
2223 G. Délèris, M. Birot, J. Dunoguès, B. Barbe, M. Pétraud, and M. Lefort, J. Organomet. Chem., 1984, **266**, 1.
2224 M. Grignon-Dubois, M. Laguerre. B. Barbe, and M. Pétraud, Organometallics, 1984, **3**, 1060.
2225 M. Grignon-Dubois and M. Laguerre, Actual. Chim., 1983, 21 (Chem. Abstr., 1984, **100**, 191 938).
2226 M. Grignon-Dubois, M. Laguerre, B. Barbe, and M. Pétraud, Organometallics, 1984, **3**, 359.
2227 M. Grignon-Dubois, A. Marchand, J. Dunoguès, B. Barbe, and M. Pétraud, J. Organomet. Chem., 1984, **272**, 19.
2228 W.-P. Leung, C.L. Raston, B.W. Skelton, and A.H. White, J. Chem. Soc., Dalton Trans., 1984, 1801.
2229 R.v.R. Schleyer, A.J. Kos, D. Wilhelm, T. Clark, G. Boche, G. Decher, H. Etzrodt, H. Dietrich, and W. Mahdi, J. Chem. Soc., Chem. Commun., 1984, 1495.
2230 P. Jutzi, C. Otto, and T. Wippermann, Chem. Ber., 1984, **117**, 1885.
2231 U. Klingebiel and L. Skoda, Z. Anorg. Allg. Chem., 1984, **510**, 175.
2232 J. Escudie, C. Couret, H. Ranaivonjatovo, and J. Satgé, J. Chem. Soc., Chem. Commun., 1984, 1621.
2233 Y.Y.C.Y.L. Ko and R. Carrié, J. Chem. Soc., Chem. Commun., 1984, 1640.
2234 S. Pohlmann and U. Klingebiel, Z. Anorg. Allg. Chem., 1984, **510**, 169.
2235 C. Eaborn and M.N. Romanelli, J. Chem. Soc., Chem. Commun., 1984, 1616.
2236 A.-M. Caminade, C. Couret, J. Escudie, and M. Koenig, J. Chem. Soc., Chem. Commun., 1984, 1622.
2237 C. Eaborn, M.N.A. El-Kheli, P.B. Hitchcock, and J.D. Smith, J. Organomet. Chem., 1984, **272**, 1.
2238 A.H. Cowley, J.E. Kilduff, J.G. Lasch, S.K. Mehrotra, N.C. Norman, M. Pakulski, B.R. Whittlesey, J.L. Atwood, and W.E. Hunter, Inorg. Chem., 1984, **23**, 2582.
2239 A.G. Brook and L. Yau, J. Organomet. Chem., 1984, **271**, 9.
2240 C. Eaborn and A.K. Saxena, J. Chem. Soc., Chem. Commun., 1984, 1482.
2241 G. Fritz, J. Thomas, K. Peters, E.-M. Peters, and H.G. v. Schnering, Z. Anorg. Allg. Chem., 1984, **514**, 61.
2242 Y. Okano, T. Masuda, and T. Higashimura, J. Polym. Sci., Polym. Chem. Ed., 1984, **22**, 1603 (Chem. Abstr., 1984, **101**, 73 157).
2243 A.H. Cowley, P. Jutzi, F.X. Kohl, J.G. Lasch, N.C. Norman, and E. Schlüter, Angew. Chem., Int. Ed. Engl., 1984, **23**, 616.
2244 A. Sekiguchi and W. Ando, J. Am. Chem. Soc., 1984, **106**, 1487.
2245 M. Ishikawa, S. Matsuzawa, K. Hirotsu, S. Kamitori, and T. Higuchi, Organometallics, 1984, **3**, 1930.
2246 S. Martin, R. Sauvêtre, and J.-F. Normant, J. Organomet. Chem., 1984, **264**, 155.
2247 R. Appel, G. Haubrich, and F. Knoch, Chem. Ber., 1984, **117**, 2063.
2248 R. Appel and C. Casser, Tetrahedron Lett., 1984, **25**, 4109.
2249 M. Caira, R.H. Neilson, W.H. Watson, P. Wisian-Neilson, and Z.-M. Xie, J. Chem. Soc., Chem. Commun., 1984, 698.
2250 R. Appel, F. Knoch, and H. Kunze, Angew. Chem., Int. Ed. Engl., 1984, **23**, 157.
2251 P.C. Auderset, A.S. Dreiding, and E.R.F. Gesing, Synth. Commun., 1983, **13**,

881 (Chem. Abstr., 1984, **100**, 68 360).
2252 R. Appel, C. Casser, and F. Knoch, Chem. Ber., 1984, **117**, 2693.
2253 C.W. Allen, J.L. Desorcie, and K. Ramachandran, J. Chem. Soc., Dalton Trans., 1984, 2843.
2254 H. Duddeck and M.R. Islam, Chem. Ber., 1984, **117**, 554.
2255 L.D. Boardman, V. Bagheri, H. Sawada, and E.-i. Negishi, J. Am. Chem. Soc., 1984, **106**, 6105.
2256 B. Halton, C.J. Randall, and P.J. Stang, J. Am. Chem. Soc., 1984, **106**, 6108.
2257 T. Hayama, S. Tomoda, Y. Takeuchi, and Y. Nomura, J. Org. Chem., 1984, **49**, 3235.
2258 J.A. Walker, S.P. Bitler, and F. Wudl, J. Org. Chem., 1984, **49**, 4733.
2259 K. Tamao, T. Nakajima, and M. Kumada, Organometallics, 1984, **3**, 1655.
2260 H. Lehmkuhl, K. Hauschild, and M. Bellenbaum, Chem. Ber., 1984, **117**, 383.
2261 I. Reichelt and H.-U. Reissig, Liebigs Ann Chem., 1984, 828.
2262 J.A. Miller and E.-i. Negishi, Isr. J. Chem., 1984, **24**, 76.
2263 O. Tsuge, S. Kanemasa, and K. Matsuda, J. Org. Chem., 1984, **49**, 2689.
2264 L.M. Engelhardt, R.I. Papasergio, C.L. Raston, and A.H. White, J. Chem. Soc., Dalton Trans., 1984, 311.
2265 L.M. Engelhardt, W.-P. Leung, C.L. Raston, P. Twiss, and A.H. White, J. Chem. Soc., Dalton Trans., 1984, 321.
2266 J.P. Picard, A. Aziz-Elyusufi, R. Calas, J. Dunoguès, and N. Duffaut, Organometallics, 1984, **3**, 1660.
2267 M. Ahra, M. Grignon-Dubois, and J. Dunoguès, J. Organomet. Chem., 1984, **271**, 15.
2268 H. Schmoelzer and E. Hengge, J. Organomet. Chem., 1984, **260**, 31.
2269 J.D. Rich, F. Shafiee, K.J. Haller, S.G. Harsy, and R. West, J. Organomet. Chem., 1984, **264**, 61.
2270 A.G. Brook and J.J. Chrusciel, Organometallics, 1984, **3**, 1317.
2271 K. Hassler, J. Organomet. Chem., 1984, **266**, Cl.
2272 W.Z. McCarthy, J.Y. Corey, and E.R. Corey, Organometallics, 1984, **3**, 255.
2273 D. Seyferth, D.P. Duncan, M.L. Shannon, and E.W. Goldman, Organometallics, 1984, **3**, 574.
2274 M.E. Scheller and B. Frei, Helv. Chim. Acta, 1984, **67**, 1734.
2275 P.R. Jones, A.H.-B. Cheng, and T.E. Albanesi, Organometallics, 1984, **3**, 78.
2276 G. Fritz and J. Thomas, J. Organomet. Chem., 1984, **271**, 107.
2277 W. Bernhard and I. Fleming, J. Organomet. Chem., 1984, **271**, 281.
2278 A. Marchand, P. Gerval, F. Duboudin, M.-H. Gaufryau, M. Joanny, and P. Mazerolles, J. Organomet. Chem., 1984, **267**, 93.
2279 A. Katti, C.W. Carlson, and R. West, J. Organomet. Chem., 1984, **271**, 353.
2280 F. Franke, M.J. Cuthbertson, and P.R. Wells, J. Org. Chem., 1984, **49**, 1258.
2281 J.Y. Corey, E.R. Corey, V.H.T. Chang, M.A. Hauser, M.A. Leiber, T.E. Reinsel, and M.E. Riva, Organometallics, 1984, **3**, 1051.
2282 M. Ishikawa, T. Tabohashi, M. Kumada, and J. Iyoda, J. Organomet. Chem., 1984, **264**, 79.
2283 A. Onopchenko, E.T. Sabourin, and D.L. Beach, J. Org. Chem., 1984, **49**, 3389.
2284 X.H. Zhang and R. West, J. Polym. Sci., Polym. Chem. Ed., 1984, **22**, 225 (Chem. Abstr., 1984, **100**, 140 569).
2285 J.-P. Quintard, G. Dumartin, C. Guerin, J. Dubac, and A. Laporterie, J. Organomet. Chem., 1984, **266**, 123.
2286 K. Ohga, U.C. Yoon, and P.S. Mariano, J. Org. Chem., 1984, **49**, 213.
2287 M.G. Voronkov, N.I. Ushakova, I.I. Tsykhanskaya, and V.B. Pukhnarevich, J. Organomet. Chem., 1984, **264**, 39.
2288 C.W. Carlson, K.J. Haller, X.-H. Zhang, and R. West, J. Am. Chem. Soc., 1984, **106**, 5521.
2289 I. Ojima, T. Fuchikami, and M. Yatabe, J. Organomet. Chem., 1984, **260**, 335.
2290 C. Hu, J.-G. He, D.H. O'Brien, and K.J. Irgolic, J. Organomet. Chem., 1984, **268**, 31.
2291 K. Ruitenberg, H. Westmijze, H. Kleijn, and P. Vermeer, J. Organomet. Chem., 1984, **277**, 227.
2292 N. Kleiner and M. Drager, J. Organomet. Chem., 1984, **270**, 151.
2293 H. Watanabe, T. Muraoka, M.-a. Kageyama, K. Yoshizumi, and Y. Nagai, Organometallics, 1984, **3**, 141.

2294 M.J. Michalczyk, R. West, and J. Michl, J. Am. Chem. Soc., 1984, **106**, 821.
2295 M.J. Fink, M.J. Michalczyk, K.J. Haller, R. West, and J. Michl, Organometallics, 1984, **3**, 793.
2296 B. Mauze, P. Ongoka, and L. Miginiac, J. Organomet. Chem., 1984, **264**, 1.
2297 J.W.F.L. Seetz, B.J.J. van de Heisteeg, G. Schat, O.S. Akkerman, and F. Bickelhaupt, J. Organomet. Chem., 1984, **277**, 319.
2298 D. Simon, K. Häberle, and M. Dräger, J. Organomet. Chem., 1984, **267**, 133.
2299 P. Jutzi, F.-X. Kohl, E. Schlüter, M.B. Hursthouse, and N.P.C. Walker, J. Organomet. Chem., 1984, **271**, 393.
2300 Y. Ducharme, S. Latour, and J.D. Wuest, Organometallics, 1984, **3**, 208.
2301 F.K. Sheffy, J.P. Godschalx, and J.K. Stille, J. Am. Chem. Soc., 1984, **106**, 4833.
2302 D.W. Hawker and P.R. Wells, Org. Magn. Reson., 1984, **22**, 280.
2303 H. Killing and T.N. Mitchell, Organometallics, 1984, **3**, 1318.
2304 C.D. Schaeffer, jun., J.L. Lefferts, and J.J. Zuckerman, Org. Magn. Reson., 1984, **22**, 125.
2305 B. Wrackmeyer and C. Bihlmayer, J. Organomet. Chem., 1984, **266**, 33.
2306 H. Duddeck and M.R. Islam, Chem. Ber., 1984, **117**, 565.
2307 D.W. Hawker and P.R. Wells, J. Organomet. Chem., 1984, **266**, 37.
2308 K. Jurkschat and A. Tzschach, J. Organomet. Chem., 1984, **272**, C13.
2309 H. Killing and T.N. Mitchell, Organometallics, 1984, **3**, 1917.
2310 W.P. Neumann and J. Fu, J. Organomet. Chem., 1984, **273**, 295.
2311 K. Kobayashi, M. Kawanisi, T. Hitomi, and S. Kozima, Chem. Lett., 1984, 497.
2312 H. Puff, C. Bach, H. Reuter, and W. Schuh, J. Organomet. Chem., 1984, **277**, 17.
2313 H.A. Olszowy and W. Kitching, Organometallics, 1984, **3**, 1670.
2314 Y. Azuma and M. Newcomb, Organometallics, 1984, **3**, 9.
2315 L.M. Engelhardt, W.-P. Leung, C.L. Raston, P. Twiss, and A.H. White, J. Chem. Soc., Dalton Trans., 1984, 331.
2316 M. Newcomb, M.T. Blanda, Y. Azuma, and T.J. Delord, J. Chem. Soc., Chem. Commun., 1984, 1159.
2317 W.J. Evans, L.A. Hughes, and T.P. Hanusa, J. Am. Chem. Soc., 1984, **106**, 4270.
2318 J. Iley, A.R. Bassindale, and P. Patel, J. Chem. Soc., Perkin Trans. 2, 1984, 77.
2319 A.I. Ermakov, E.A. Kirichenko, N.A. Makrushin, and N.I. Pimkin, Zh. Fiz. Khim., 1984, **58**, 1049 (Chem. Abstr., 1984, **101**, 43 819).
2320 J.-M. Dereppe and B. Parbhoo, Anal. Chem., 1984, **56**, 2740.
2321 K. Bruzik and M.-D. Tsai, J. Am. Chem. Soc., 1984, **106**, 747.
2322 M. Nádvornik, J. Holeček, K. Handlř, and A. Lyčka, J. Organomet. Chem., 1984, **275**, 43.
2323 H. Weichmann, J. Organomet. Chem., 1984, **262**, 279.
2324 S.A. Kazoura and W.P. Weber, J. Organomet. Chem., 1984, **268**, 19.
2325 J. Iley, A.R. Bassindale, and P. Patel, J. Chem. Soc., Perkin Trans. 2, 1984, 77.
2326 B.N. Khasapov, A.V. Kalinin, A.Ya. Shteinshneider, A.A. Blyumenfel'd, S.L. Ioffe, and V.A. Tartakovskii, Izv. Akad. Nauk SSSR, Ser. Khim., 1984, 1296 (Chem. Abstr., 1984, **101**, 192 018).
2327 T. Morimoto, T. Takahashi, and M. Sekiya, J. Chem. Soc., Chem. Commun., 1984, 794.
2328 F.M. Tesky and R. Mews, Z. Anorg. Allg. Chem., 1984, **518**, 197.
2329 R.R. Ford, M.A. Goodman, R.H. Neilson, A.K. Roy, U.G. Wettermark, and P. Wisian-Neilson, Inorg. Chem., 1984, **23**, 2063.
2330 H. Schmidbaur, S. Lauteschläger, and F.H. Köhler, J. Organomet. Chem., 1984, **271**, 173.
2331 G. Becker, W. Becker, and G. Uhl, Z. Anorg. Allg. Chem., 1984, **518**, 21.
2332 G. Fritz and K. Stoll, Z. Anorg. Allg. Chem., 1984, **514**, 69.
2333 M. Baudler, J. Hellmann, and G. Reuschenbach, Z. Anorg. Allg. Chem., 1984, **509**, 38.
2334 E.E. Nifantiev, N.S. Vyazankin, S.F. Sorokina, L.A. Vorobieva, O.A. Vyazankina, D.A. Bravo-Zhivotovsky, and A.R. Bekker, J. Organomet. Chem., 1984, **277**, 211.

2335 R. Hässig, D. Seebach, and H. Siegel, Chem. Ber., 1984, **117**, 1877.
2336 J. Schraml, J. Magn. Reson., 1984, **59**, 515.
2337 D.T. Plummer, B.A. Karcher, R.A. Jacobson, and R.J. Angelici, J. Organomet. Chem., 1984, **260**, 347.
2338 R. Appel and S. Korte, Z. Anorg. Allg. Chem., 1984, **510**, 123.
2339 R. Appel, B. Laubach, and M. Siray, Tetrahedron Lett., 1984, **25**, 4447.
2340 C.W. Allen, J.L. Desorcie, and K. Ramachandran, Report, 1984, TR-14, Order No. AD-A141429, 18 pp. Avail. NTIS. From Gov. Rep. Announce. Index (U.S.), 1984, **84**, 47 (Chem. Abstr., 1984, **101**, 211 307).
2341 G. Gattow and G. Schubert, Z. Anorg. Allg. Chem., 1984, **515**, 173.
2342 G. Gattow and G. Schubert, Z. Anorg. Allg. Chem., 1984, **515**, 182.
2343 I. Reichelt and H.U. Reissig, Liebigs Ann. Chem., 1984, 531.
2344 E. Kunkel, I. Reichelt, and H.U. Reissig, Liebigs Ann. Chem., 1984, 512.
2345 T.J. Michalski, A. Kendler, and M.L. Bender, J. Inclusion Phenom., 1983, **1**, 125 (Chem. Abstr., 1984, **100**, 210 299).
2346 J.E. Drake, A. Vahradian, and B.M. Glavincevski, Can. J. Chem., 1984, **61**, 2712.
2347 M. Wada, Y. Nishihara, and K. Akiba, Tetrahedron Lett., 1984, 25, 5405.
2348 A.R. Bassindale and T. Stout, J. Organomet. Chem., 1984, **271**, C1.
2349 M.J. Michalczyk, R. West, and J. Michl, J. Chem. Soc., Chem. Commun., 1984, 1525.
2350 G. Becker, O. Mundt, and G. Uhl, Z. Anorg. Allg. Chem., 1984, **517**, 89.
2351 G. Becker, W. Massa, R.E. Schmidt, and G. Uhl, Z. Anorg. Allg. Chem., 1984, **517**, 75.
2352 R. Appel, P. Fölling, L. Krieger, M. Siray, and F. Knoch, Angew. Chem., Int. Ed. Engl., 1984, **23**, 970.
2353 C.H.W. Jones and R.D. Sharma, J. Organomet. Chem., 1984, **268**, 113.
2354 A.R. Bassindale and T. Stout, Tetrahedron Lett., 1984, **25**, 1631.
2355 K.R. Pope and P.R. Jones, Organometallics, 1984, **3**, 354.
2356 G. Fritz and W. Schick, Z. Anorg. Allg. Chem., 1984, **511**, 108.
2357 G. Fritz and A. Wörsching, Z. Anorg. Allg. Chem., 1984, **512**, 131.
2358 H. Yildirimyan and G. Gattow, Z. Anorg. Allg. Chem., 1984, **519**, 213.
2359 M. Gielen, K. Jurkschat, J. Meunier-Piret, and M. Van Meerssche, Bull. Soc. Chim. Belg., 1984, **93**, 379 (Chem. Abstr., 1984, **101**, 171 391).
2360 S.J. Blunden, R. Hill, and D.G. Gillies, J. Organomet. Chem., 1984, **270**, 39.
2361 Th. Butters, K. Scheffler, H.B. Stegmann, U. Weber, and W. Winter, J. Organomet. Chem., 1984, **272**, 351.
2362 S.K. Brahma, W.H. Nelson, and W.F. Howard, jun., J. Organomet. Chem., 1984, **273**, 47.
2363 A. Saxena and J.P. Tandon, Polyhedron, 1984, **3**, 681.
2364 B.B. Dambatta and J.R. Ebdon, Br. Polym. J., 1984, **16**, 69 (Chem. Abstr., 1984, **101**, 192 757).
2365 S.J. Blunden and R. Hill, Inorg. Chim. Acta, 1984, **87**, 83.
2366 D.W. Allen, D.J. Derbyshire, J.S. Brooks, S.J. Blunden, and P.J. Smith, J. Chem. Soc., Dalton Trans., 1984, 1889.
2367 H.-P. Abicht, W. Hönle, and H.G. v. Schnering, Z. Anorg. Allg. Chem., 1984, **519**, 7.
2368 K. Swami, J.P. Hutchinson, H.G. Kuivila, and J.A. Zubieta, Organometallics, 1984, **3**, 1687.
2369 M. Gielen and K. Jurkschat, J. Organomet. Chem., 1984, **273**, 303.
2370 J.W. Burley, P. Hope, and A.G. Mack, J. Organomet. Chem., 1984, **277**, 37.
2371 E. Vincent, L. Verdonck, L. Naessens, and G.P. van der Kelen, J. Organomet. Chem., 1984, **277**, 235.
2372 R. Willem, M. Gielen, J. Meunier-Piret, M. van Meerssche, K. Jurkschat, and A. Tzschach, J. Organomet. Chem., 1984, **277**, 335.
2373 S.J. Blunden and J.N.R. Ruddick, J. Organomet. Chem., 1984, **267**, C5.
2374 A.F. Shihada, I.A.A. Jassim, and F. Weller, J. Organomet. Chem., 1984, **268**, 125.
2375 R.H. Cragg and R.D. Lane, J. Organomet. Chem., 1984, **277**, 199.
2376 D. Dakternieks, B.F. Hoskins, E.R.T. Tiekink, and G. Winter, Inorg. Chim. Acta, 1984, **85**, 215.
2377 E. Liepins, I. Ziimane, E. Ignatovics, L.I. Gubanova, and M.G. Voronkov, Zh.

Obshch. Khim., 1983, **53**, 1789 (Chem. Abstr., 1984, **100**, 68 362).
2378 W. Clegg, U. Klingebiel, G.M. Sheldrick, and D. Stalke, J. Organomet. Chem., 1984, **265**, 17.
2379 U. Engelhardt and T. Bünger, Z. Anorg. Allg. Chem., 1984, **517**, 177.
2380 O. Graalmann, M. Hesse, U. Klingebiel, W. Clegg, M. Haase, and G.M. Sheldrick, Z. Anorg. Allg. Chem., 1984, **514**, 49.
2381 O. Graalmann, U. Klingebiel, W. Clegg, M. Haase, and G.M. Sheldrick, Angew. Chem., Int. Ed. Engl., 1984, **23**, 891.
2382 R.A. Newmark and B.C. Copley, Macromolecules, 1984, **17**, 1973 (Chem. Abstr., 1984, **101**, 172 064).
2383 R.H. Cragg and R.D. Lane, J. Organomet. Chem., 1984, **270**, 25.
2384 G.C. Stocco, L. Pellerito, A. Girasolo, and A.G. Osborne, Congr. Naz. Chim. Inorg., [Atti], 15th, 1982, 329 (Chem. Abstr., 1984, **100**, 192 005).
2385 G.C. Stocco, L. Pellerito, M.A. Girasolo, and A.G. Osborne, Inorg. Chim. Acta, 1984, **83**, 79.
2386 O. Graalmann, U. Klingebiel, W. Clegg, M. Haase, and G.M. Sheldrick, Chem. Ber., 1984, **117**, 2988.
2387 O. Graalmann and U. Klingebiel, J. Organomet. Chem., 1984, **275**, C1.
2388 O. Graalmann, U. Klingebiel, W. Clegg, M. Haase, and G.M. Sheldrick, Z. Anorg. Allg. Chem., 1984, **519**, 87.
2389 H.G. Horn and M. Hemeke, Chem.-Ztg., 1983, **107**, 367 (Chem. Abstr., 1984, **100**, 191 936).
2390 W. Wojnowski and J. Pikies, Z. Anorg. Allg. Chem., 1984, **508**, 201.
2391 H. Schmölzer and E. Hengge, Monatshefte, 1984, **115**, 1125.
2392 G. Fritz, W. Schick, W. Hönle, and H.G. v. Schnering, Z. Anorg. Allg. Chem., 1984, **511**, 95.
2393 R.K. Chadha, J.E. Drake, and A.B. Sarkar, Inorg. Chem., 1984, **23**, 4769.
2394 A.A. Espenbetov, Yu.T. Struchkov, S.P. Kolesnikov, and O.M. Nefedov, J. Organomet. Chem., 1984, **275**, 33.
2395 L. Ross and M. Draeger, Z. Naturforsch., Teil B, 1984, **39**, 868 (Chem. Abstr., 1984, **101**, 211 321).
2396 H. Schumann, B. Pachaly, and B.C. Schütze, J. Organomet. Chem., 1984, **265**, 145.
2397 H. Puff, W. Schuh, R. Sievers, W. Wald, and R. Zimmer, J. Organomet. Chem., 1984, **260**, 271.
2398 R.J. Rao, G. Srivastava, R.C. Mehrotra, B.S. Saraswat, and J. Mason, Polyhedron, 1984, **3**, 485.
2399 S.A. Kazoura and W.P. Weber, J. Organomet. Chem., 1984, **271**, 47.
2400 I. Ruppert, K. Schlich, and P. Hardt, J. Organomet. Chem., 1984, **267**, C29.
2401 A.I. Albanov, M.G. Voronkov, L.I. Gubanova, M.F. Larin, E. Liepins, and V.A. Pestunovich, Izv. Akad. Nauk SSSR, Ser. Khim., 1983, 2402 (Chem. Abstr., 1984, **100**, 85 780).
2402 V.M. Kopylov, I.S. Shragin, T.I. Rybkina, A.I. Chernyshev, M.I. Shkol'nik, and E.N. Ukhina, Zh. Obshch. Khim., 1984, **54**, 359 (Chem. Abstr., 1984, **101**, 7259).
2403 V.P. Feshin, P.A. Nikitin, M.G. Voronkov, T.K. Gar, N.A. Viktorov, S.N. Gurkova, A.I. Gusev, and V.I. Shiryaev, Zh. Obshch. Khim., 1984, **54**, 646 (Chem. Abstr., 1984, **101**, 91 103).
2404 A. Tzschach, W. Uhlig, and K. Kellner, J. Organomet. Chem., 1984, **266**, 17.
2405 G. Fritz and W. Schick, Z. Anorg. Allg. Chem., 1984, **518**, 14.
2406 J. Otera, T. Yano, E. Kunimoto, and T. Nakata, Organometallics, 1984, **3**, 426.
2407 A. Tzschach, M. Scheer, and K. Jurkschat, Z. Anorg. Allg. Chem., 1984, **512**, 177.
2408 E.S. Paterson, J.L. Wardell, and J.W. Burley, J. Organomet. Chem., 1984, **273**, 313.
2409 E. Kupce, E. Liepins, O. Pudova, and E. Lukevics, J. Chem. Soc., Chem. Commun., 1984, 581.
2410 M. Baudler, L. de Riese-Meyer, and U. Schings, Z. Anorg. Allg. Chem., 1984, **519**, 24.
2411 S.J. Blunden, P.A. Cusack, and D.G. Gillies, J. Magn. Reson., 1984, **60**, 114.
2412 J. Pikies and W. Wojnowski, Z. Anorg. Allg. Chem., 1984, **511**, 219.

2413 A.R. Grimmer, Z. Chem., 1984, **24**, 194 (Chem. Abstr., 1984, **101**, 82 858).
2414 J. Wang, R. Chen, Y. Mao, C. Li, J. Wei, H. Liu, and A. Tai, Huaxue Xuebao, 1984, **42**, 308 (Chem. Abstr., 1984, **100**, 220 422).
2415 F. Schlenkrich, O. Ziemens, O. Rademacher, and H. Scheler, Z. Chem., 1984, **24**, 130 (Chem. Abstr., 1984, **101**, 79 410).
2416 G. Engelhardt and O. Rademacher, J. Mol. Liq., 1984, **27**, 125.
2417 R.K. Harris, M.J. O'Connor, E.H. Curzon, and O.W. Howarth, J. Magn. Reson., 1984, **57**, 115.
2418 O. Rademacher, O. Ziemens, and H. Scheler, Z. Anorg. Allg. Chem., 1984, **519**, 165.
2419 F. Schlenkrich, E. Beil, O. Rademacher, and H. Scheler, Z. Anorg. Allg. Chem., 1984, **511**, 41.
2420 D. Hoebbel, A. Vargha, G. Engelhardt, and K. Ujszaszy, Z. Anorg. Allg. Chem., 1984, **509**, 85.
2421 O. Rademacher, F. Schlenkrich, O. Ziemens, and H. Scheler, Z. Chem., 1984, **24**, 131 (Chem. Abstr., 1984, **101**, 81 789).
2422 J.A. Tossell, Phys. Chem. Miner., 1984, **10**, 137 (Chem. Abstr., 1984, **100**, 149 869).
2423 V.P. Tarasov, V.I. Privalov, S.G. Drobyshev, and Yu.A. Buslaev, Dokl. Akad. Nauk SSSR, 1983, **272**, 1176 (Chem. Abstr., 1984, **100**, 28 775).
2424 A. Tzschach, M. Scheer, and K. Jurkschat, Z. Anorg. Allg. Chem., 1984, **515**, 147.
2425 J. Schraml, A.M. Krapivin, A.P. Luzin, V.M. Kilesso, and V.A. Pestunovich, Collect. Czech. Chem. Commun., 1984, **49**, 2897.
2426 A. Winkler, D. Hoebbel, A.R. Grimmer, W. Wieker, K. Ujszaszy, and M. Magi, Rev. Chim. Miner., 1983, **20**, 801 (Chem. Abstr., 1984, **100**, 149 984).
2427 S.P. Marella, L.C. Lee, and F. Aubke, Inorg. Chem., 1984, **23**, 653.
2428 S.J. Blunden, P.A. Cusack, P.J. Smith, and D.G. Gillies, Inorg. Chim. Acta, 1984, **84**, 35.
2429 A. Tzschach, M. Scheer, and K. Jurkschat, Z. Anorg. Allg. Chem., 1984, **508**, 73.
2430 J.C. Huffman, J.P. Haushalter, A.M. Umarji, G.K. Shenoy, and R.C. Haushalter, Inorg. Chem., 1984, **23**, 2312.
2431 P.A.W. Dean, J.J. Vittal, and N.C. Payne, Inorg. Chem., 1984, **23**, 4232.
2432 K.B. Dillon and A. Marshall, J. Chem. Soc., Dalton Trans., 1984, 1245.
2433 'Phosphorus-31 NMR: Principles and Applications', ed. D.G. Gorenstein, Academic Press: Orlando, Fla.), 1984, 604 pp. (Chem. Abstr., 1984, **101**, 68 907).
2434 D.G. Gorenstein, Phosphorus-31 NMR, 1984, 7.
2435 W.C. Hutton, Phosphorus-31 NMR, 1984, 479.
2436 D.G. Gorenstein and D.O. Shah, Phosphorus-31 NMR, 1984, 549.
2437 O.A. Raevskii and T.I. Ignat'eva, Russ. Chem. Rev., 1983, **52**, 1156.
2438 M. Moriyama, Yuki Gosei Kagaku Kyokaishi, 1984, **42**, 355 (Chem. Abstr., 1984, **100**, 220 220).
2439 R.E. Wasylishen and J.O. Friedrich, J. Chem. Phys., 1984, **80**, 585.
2440 V. Galasso, J. Magn. Reson., 1984, **58**, 311.
2441 S. Hietkamp, H. Sommer, and O. Stelzer, Chem. Ber., 1984, **117**, 3400.
2442 C. Mihart, S. Raduly, and Z. Simon, Rev. Roum. Chim., 1983, **28**, 857 (Chem. Abstr., 1984, **101**, 23 595).
2443 C. Charrier, F. Mathey, F. Robert, and Y. Jeannin, J. Chem. Soc., Chem. Commun., 1984, 1707.
2444 M. Baudler and H. Tschäbunin, Z. Anorg. Allg. Chem., 1984, **511**, 77.
2445 M. Baudler and H. Tschäbunin, Z. Anorg. Allg. Chem., 1984, **516**, 7.
2446 M. Baudler, R. Heumüller, and K. Langerbeins, Z. Anorg. Allg. Chem., 1984, **514**, 7.
2447 H.R. Hudson, A. Kow, and J.C. Roberts, Phosphorus Sulphur, 1984, **19**, 375.
2448 P.N. Kapoor, D.D. Pathak, G. Gaur, and M. Kutty, J. Organomet. Chem., 1984, **276**, 167.
2449 R.B. King, N.D. Sadanani, and P.M. Sundaram, Phosphorus Sulphur, 1983, **18**, 125.
2450 B. Garrigues, D. Boyer, and A. Munoz, Can. J. Chem., 1984, **62**, 2170.
2451 K. Troev and M. Simeonov, Phosphorus Sulphur, 1984, **19**, 367.

2452 R.M.K. Deng and K.B. Dillon, J. Chem. Soc., Dalton Trans., 1984, 1911.
2453 W.T. Ford, Org. Coat. Appl. Polym. Sci. Proc., 1983, **48**, 464 (Chem. Abstr., 1984, **100**, 121 979).
2454 R. Benn, R. Mynott, W.J. Richter, and G. Schroth, Tetrahedron, 1984, **40**, 3273.
2455 A. Schier and H. Schmidbaur, Chem. Ber., 1984, **117**, 2314.
2456 B.R. Bailey, tert., K.D. Berlin, and E.M. Holt, Phosphorus Sulphur, 1984, **20**, 131.
2457 H.J. Meeuwissen, T.A. van der Knaap, and F. Bickelhaup, Phosphorus Sulphur, 1983, **18**, 109.
2458 F.R. Benn, J.C. Briggs, and C.A. McAuliffe, J. Chem. Soc., Dalton Trans., 1984, 293.
2459 J. Skolimowski, R. Skowronski, and M. Simalty, Phosphorus Sulphur, 1984, **19**, 159.
2460 D.W. Allen, I.W. Nowell, L.A. March, and B.F. Taylor, J. Chem. Soc., Perkin Trans. 1, 1984, 2523.
2461 H. Schmidbaur, R. Herr, and J. Riede, Angew.Chem., Int. Ed. Engl., 1984, **23**, 247.
2462 A.B. Reitz, M.S. Mutter, and B.E. Maryanoff, J. Am. Chem. Soc., 1984, **106**, 1873.
2463 H.J. Cristau, B. Chabaud, and H. Christol, J. Org. Chem., 1984, **49**, 2023.
2464 R.M. Moriarty, I. Prakash, O. Prakash, and W.A. Freeman, J. Am. Chem. Soc., 1984, **106**, 6082.
2465 W.v.E. Doering and J.C. Schmidhauser, J. Am. Chem. Soc., 1984, **106**, 5025.
2466 H. Schmidbaur, R. Herr, and C.E. Zybill, Chem. Ber., 1984, **117**, 3374.
2467 H.J. Bestmann, G. Schmid, H. Oechsner, and P. Ermann, Chem. Ber., 1984, **117**, 1561.
2468 H. Schmidbaur, B. Milewski-Mahrla, G. Müller, and C. Krüger, Organometallics, 1984, **3**, 38.
2469 H. Schmidbaur, A. Schier, S. Lauteschläger, J. Riede, and G. Müller, Organometallics, 1984, **3**, 1906.
2470 A.B. Reitz and B.E. Maryanoff, J. Chem. Soc., Chem. Commun., 1984, 1548.
2471 B. Schaub, T. Jenny, and M. Schlosser, Tetrahedron Lett., 1984, **25**, 4097.
2472 D.G. Gilheany, D.A. Kennedy, J.F. Malone, and B.J. Walker, J. Chem. Soc., Chem. Commun., 1984, 1217.
2473 M. Antberg, C. Prengel, and L. Dahlenburg, Inorg. Chem., 1984, **23**, 4170.
2474 L.D. Quin, K.C. Caster, J.C. Kisalus, and K.A. Mesch, J. Am. Chem. Soc., 1984, **106**, 7021.
2475 L.D. Quin, K.C. Caster, and J.C. Kisalus, Phosphorus Sulphur, 1983, **18**, 105.
2476 T. Butters and W. Winter, Chem. Ber., 1984, **117**, 990.
2477 H. Sommer, S. Hietkamp, and O. Stelzer, Chem. Ber., 1984, **117**, 3414.
2478 K. Tani, T. Ise, Y. Tatsuno, and T. Saito, J. Chem. Soc., Chem. Commun., 1984, 1641.
2479 D.M. Schubert and A.D. Norman, Inorg. Chem., 1984, **23**, 4130.
2480 B.-L. Li and R.H. Neilson, Inorg. Chem., 1984, **23**, 3665.
2481 G. Märkl and W. Burger, Angew. Chem., Int. Ed. Engl., 1984, **23**, 894.
2482 D.G. Gilheany and B.J. Walker, Phosphorus Sulphur, 1983, **18**, 183.
2483 G.K. Anderson and R. Kumar, Inorg. Chem., 1984, **23**, 4064.
2484 H. Schmidbaur, R. Herr, and J. Riede, Chem. Ber., 1984, **117**, 2322.
2485 H. Hope, M. Viggiano, B. Moezzi, and P.P. Power, Inorg. Chem., 1984, **23**, 2550.
2486 E.P. Kyba and C.N. Clubb, Inorg. Chem., 1984, **23**, 4766.
2487 D.L. DuBois, Inorg. Chem., 1984, **23**, 2047.
2488 Y. Amrani and D. Sinou, J. Mol. Catal., 1984, **24**, 231.
2489 R.A. Rossi, A.N. Santiago, and S.M. Palacios, J. Org. Chem., 1984, **49**, 3387.
2490 L. Horner and G. Simons, Phosphorus Sulphur, 1984, **19**, 65.
2491 S. Hietkamp and O. Stelzer, Inorg. Chem., 1984, **23**, 258.
2492 H. Schmidbaur and T. Pollock, Helv. Chim. Acta, 1984, **67**, 2175.
2493 C.L. Liotta, M.L. McLaughlin, D.G. Van Derveer, and B.A. O'Brien, Tetrahedron Lett., 1984, **25**, 1665.
2494 A. Schmidpeter and A. Willhalm, Angew. Chem., Int. Ed. Engl., 1984, **23**, 903.
2495 J.-M. Alcaraz, E. Deschamps, and F. Mathey, Phosphorus Sulphur, 1984, **19**,

45.
2496 G. Märkl, K. Hock, and L. Merz, Chem. Ber., 1984, **117**, 763.
2497 G. Märkl and I. Trötsch, Angew. Chem., Int. Ed. Engl., 1984, **23**, 901.
2498 W. Rösch and M. Regitz, Angew. Chem., Int. Ed. Engl., 1984, **23**, 900.
2499 J. Heinicke and A. Tzschach, Phosphorus Sulphur, 1984, **20**, 347.
2500 T.A. van der Knaap and F. Bickelhaupt, Phosphorus Sulphur, 1984, **21**, 227.
2501 R.H. Neilson, Phosphorus Sulphur, 1983, **18**, 43.
2502 H.H. Karsch, F.H. Köhler, and H.-U. Reisacher, Tetrahedron Lett., 1984, **25**, 3687.
2503 R. Appel, C. Casser, M. Immenkeppel, and F. Knoch, Angew. Chem., Int. Ed. Engl., 1984, **23**, 895.
2504 J. Navech, J.P. Majoral, A. Meriem, and R. Kraemer, Phosphorus Sulphur, 1983, **18**, 27.
2505 R. Appel, P. Fölling, B. Josten, M. Siray, V. Winkhaus, and F. Knoch, Angew. Chem., Int. Ed. Engl., 1984, **23**, 619.
2506 M. Yoshifuji, K. Toyota, and N. Inamoto, J. Chem. Soc., Chem. Commun., 1984, 689.
2507 M. Yoshifuji, K. Toyota, K. Shibayama, and N. Inamoto, Tetrahedron Lett., 1984, **25**, 1809.
2508 Th.A. van der Knaap and F. Bickelhaupt, Chem. Ber., 1984, **117**, 915.
2509 Th.A. van der Knaap, Th.C. Klebach, F. Visser, F. Bickelhaupt, P. Ros, E.J. Baerends, C.H. Stam, and M. Konijn, Tetrahedron, 1984, **40**, 765.
2510 O.I. Kolodiazhnyi, Phosphorus Sulphur, 1983, **18**, 39.
2511 J. Grobe and D. Le Van, Angew. Chem., Int. Ed. Engl., 1984, **23**, 710.
2512 A. Schmidpeter, K.-H. Zirzow, G. Burget, G. Huttner, and I. Jibril, Chem. Ber., 1984, **117**, 1695.
2513 M. Yoshifuji, K. Shibayama, I. Shima, and N. Inamoto, Phosphorus Sulphur, 1983, **18**, 11.
2514 I.J. Colquhoun, H.C.E. McFarlane, and W. McFarlane, Phosphorus Sulphur, 1983, **18**, 61.
2515 C. Lemsink, B. de Ruiter, and J.C. van de Grampel, J. Chem. Soc., Dalton Trans., 1984, 1521.
2516 Yu.Yu. Samitov and F.Kh. Karataeva, Zh. Obshch. Khim., 1984, **54**, 805 (Chem. Abstr., 1984, **101**, 55 230).
2517 J.M. Risley and R.L. Van Etten, Int. J. Chem. Kinet., 1984, **16**, 1167 (Chem. Abstr., 1984, **101**, 211 292).
2518 D.H.R. Barton, B. Charpiot, E.T.H. Dau, W.B. Motherwell, C. Pascard, and C. Pichon, Helv. Chim. Acta, 1984, **67**, 586.
2519 P. Laszlo and E. Polla, Tetrahedron Lett., 1984, **25**, 4651.
2520 A.M. Pinchuk, V.A. Zasorina, A.S. Shtepanek, and Z.Z. Rozhkova, Zh. Obshch. Khim., 1983, **53**, 2012 (Chem. Abstr., 1984, **100**, 22 720).
2521 H. Wamhoff and G. Haffmanns, Chem. Ber., 1984, **117**, 585.
2522 J. Garcia, F. Urpí, and J. Vilarrasa, Tetrahedron Lett., 1984, **25**, 4841.
2523 K. Tanaka, T.-r. Igarashi, S. Maeno, and K. Mitsuhashi, Bull. Chem. Soc. Jpn., 1984, **57**, 2689.
2524 A. Schmidpeter, S. Lochschmidt, G. Burget, and W.S. Sheldrick, Phosphorus Sulphur, 1983, **18**, 23.
2525 G. Borisov, V. Vasileva, E.N. Tsvetkov, T.E. Kron, and M.I. Kabachnik, Phosphorus Sulphur, 1984, **21**, 59.
2526 L.D. Quin, K.C. Caster, and S.B. Soloway, J. Org. Chem., 1984, **49**, 2627.
2527 A.L. Baumstark and P.C. Vasquez, J. Org. Chem., 1984, **49**, 793.
2528 V.A. Pavlov, T.A. Zyablikova, A.V. Il'yasov, B.G. Liorber, I.E. Ismaev, G.N. Bukhmina, and T.A. Tarzivolova, Zh. Obshch. Khim., 1984, **54**, 784 (Chem. Abstr., 1984, **101**, 91 072).
2529 P. Eisenbarth and M. Regitz, Chem. Ber., 1984, **117**, 445.
2530 C. Laurenco, L. Villien, and G. Kaufmann, Tetrahedron, 1984, **40**, 2731.
2531 A.J. Macpherson and D.J.H. Smith, J. Chem. Res. (S), 1984, 32.
2532 M. Regitz and S.G. Khbeis, Chem. Ber., 1984, **117**, 2233.
2533 P.D. Beer, R.C. Edwards, C.D. Hall, J.R. Jennings, and R.J. Cozens, Phosphorus Sulphur, 1983, **17**, 283.
2534 E. Lindner and H. Kern, Chem. Ber., 1984, **117**, 355.
2535 J.-M. Alcaraz and F. Mathey, J. Chem. Soc., Chem. Commun., 1984, 508.

2536 N. Satyamurthy, K.D. Berlin, M.B. Hossain, and D. van der Helm, Phosphorus Sulphur, 1984, **19**, 113.
2537 E. Deschamps and F. Mathey, J. Chem. Soc., Chem. Commun., 1984, 1214.
2538 R.K. Haynes and C. Indorato, Aust. J. Chem., 1984, **37**, 1183.
2539 K.M. Enikeev, E.V. Bayandina, I.E. Ismaev, N.A. Buina, A.V. Il'yasov, and I.A. Nuretdinov, Zh. Obshch. Khim., 1983, **53**, 2143 (Chem. Abstr., 1984, **100**, 22721).
2540 R.G. Cavell and L. Vande Griend, Phosphorus Sulphur, 1983, **18**, 89.
2541 G. Fritz and W. Schick, Z. Anorg. Allg. Chem., 1984, **511**, 132.
2542 R. Breslow and L.A. Deuring, Tetrahedron Lett., 1984, **25**, 1345.
2543 G.W. Buchanan and S.H. Preusser, Org. Magn. Reson., 1984, **22**, 127.
2544 G. Hägele, G. Tossing, W. Kückelhaus, J. Seega, and R.K. Harris, J. Chem. Soc., Dalton Trans., 1984, 2803.
2545 S. Trippett and C.L. White, J. Chem. Soc., Chem. Commun., 1984, 251.
2546 C. Couret, J. Escudie, and S.A. Thaoubane, Phosphorus Sulphur, 1984, **20**, 81.
2547 J.D. Buynak, K.P. Jadhav, and D.L. Lively, Phosphorus Sulphur, 1984, **20**, 145.
2548 G. Märkl and B. Alig, J. Organomet. Chem., 1984, **273**, 1.
2549 A.H. Cowley, C.A. Stewart, B.R. Whittlesey, and T.C. Wright, Tetrahedron Lett., 1984, **25**, 815.
2550 W. Rösch and M. Regitz, Phosphorus Sulphur, 1984, **21**, 97.
2551 J.M.A. Al-Rawi, M.A. Sheat, and N. Ayed, Org. Mag. Reson., 1984, **22**, 336.
2552 H.W. Roesky, J. Lucas, M. Noltemeyer, and G.M. Sheldrick, Chem. Ber., 1984, **117**, 1583.
2553 T. Chivers and M.N.S. Rao, Inorg. Chem., 1984, **23**, 3605.
2554 N. Burford, T. Chivers, M.N.S. Rao, and J.F. Richardson, Inorg. Chem., 1984, **23**, 1946.
2555 P.J. Harris, K.B. Williams, and B.L. Fisher, J. Org. Chem., 1984, **49**, 406.
2556 S.A. Ali, E. Herrmann, and B. Thomas, Z. Chem., 1984, **24**, 133 (Chem. Abstr., 1984, **101**, 111 046).
2557 T. Chivers, M.N.S. Rao, and J.F. Richardson, Inorg. Chem., 1984, **23**, 1531.
2558 L.D. Quin and J. Szewczyk, Phosphorus Sulphur, 1984, **21**, 161.
2559 D.J. Collins, P.F. Drygala, and J.M. Swan, Aust. J. Chem., 1984, **37**, 1009.
2560 L.D. Quin and J. Szewczyk, J. Chem. Soc., Chem. Commun., 1984, 1551.
2561 L.M. Yagupol'skii, N.V. Pavlenko, N.V. Ignat'ev, G.I. Matyushecheva, and V.Ya. Semenii, Zh. Obshch. Khim., 1984, **54**, 334 (Chem. Abstr., 1984, **101**, 7298).
2562 M.J.P. Harger and A. Smith, J. Chem. Soc., Chem. Commun., 1984, 1140.
2563 L.K. Nikonorova, K.M. Enikeev, N.P. Grechkin, I.E. Ismaev, A.V. Il'yasov, and I.A. Nuretdinov, Zh. Obshch. Khim., 1983, **53**, 2452.
2564 R. Appel, F. Knoch, and H. Kunze, Chem. Ber., 1984, **117**, 3151.
2565 A. Schmidpeter, G. Burget, H.G. von Schnering, and D. Weber, Angew. Chem., Int. Ed. Engl., 1984, **23**, 816.
2566 M. Baudler, F. Saykowski, M. Hintze, K.-F. Tebbe, T. Heinlein, A. Vissers, and M. Fehér, Chem. Ber., 1984, **117**, 1542.
2567 T. Kauffmann, E. Antfang, and J. Olbrich, Tetrahedron Lett., 1984, **25**, 1963.
2568 A.J. Ashe, tert., E.G. Ludwig, jun., J. Oleksyszyn, and J.C. Huffman, Organometallics, 1984, **3**, 337.
2569 N.A.A. Al-Jabar and A.G. Massey, J. Organomet. Chem., 1984, **276**, 331.
2570 A.J. Ashe, tert., and F.J. Drone, Organometallics, 1984, **3**, 495.
2571 U. v. Allwörden, I. Tseggai, and G.V. Röschenthaler, Phosphorus Sulphur, 1984, **21**, 177.
2572 J.K. Thottathil, D.E. Ryono, C.A. Przybyla, J.L. Moniot, and R. Neubeck, Tetrahedron Lett., 1984, **25**, 4741.
2573 J.K. Thottathil, C.A. Przybyla, and J.L. Moniot, Tetrahedron Lett., 1984, **25**, 4737.
2574 J. Szewczyk, J.R. Lloyd, and L.D. Quin, Phosphorus Sulphur, 1984, **21**, 155.
2575 G. Hägele, M. Engelhardt, W. Peters, A. Skowronska, J. Gwara, and D. Wendisch, Phosphorus Sulphur, 1984, **21**, 53.
2576 K. Dimroth, H. Kaletsch, and T.N. Dave, Angew. Chem., Int. Ed. Engl., 1984, **23**, 990.
2577 G. Singh, Phosphorus Sulphur, 1983, **18**, 217.

2578 C. Pracejus and H. Pracejus, J. Mol. Catal., 1984, **24**, 227.
2579 R. Bohra, H.W. Roesky, M. Noltemeyer, and G.M. Sheldrick, J. Chem. Soc., Dalton Trans., 1984, 2011.
2580 G.B. Deacon, W.R. Jackson, and J.M. Pfeiffer, Aust. J. Chem., 1984, **37**, 527.
2581 J. Omebańczuk and M. Mikołajczyk, Tetrahedron Lett., 1984, **25**, 2493.
2582 A. Bruns, W. Hiller, and U. Kunze, Z. Naturforsch., Teil B, 1984, **39**, 14.
2583 J.-M. Alcaraz and F. Mathey, Tetrahedron Lett., 1984, **25**, 4659.
2584 K. Saito, A. Yoshino, T. Yoshida, and K. Takahashi, Bull. Chem. Soc. Jpn., 1983, **56**, 3585.
2585 R.M.K. Deng and K.B. Dillon, J. Chem. Soc., Dalton Trans., 1984, 1917.
2586 N.S. Golubev, Khim. Fiz., 1984, **3**, 772 (Chem. Abstr., 1984, **101**, 64 973).
2587 R.M.K. Deng, K.B. Dillon, and A.W.G. Platt, Phosphorus Sulphur, 1983, **18**, 93.
2588 M. Herberhold, W. Ehrenreich, and K. Guldner, Chem. Ber., 1984, **117**, 1999.
2589 F. Costanzi, M. Pianca, M. Tatò, and G. Moggi, Org. Magn. Reson., 1984, **22**, 705.
2590 Y.Y.C.Y.L. Ko, R. Carrié, A. Muench, and G. Becker, J. Chem. Soc., Chem. Commun., 1984, 1634.
2591 R.O. Day, A. Schmidpeter, and R.R. Holmes, Inorg. Chem., 1983, **22**, 3696.
2592 M. Haddad, Th. N'Gando M'Pondo, C. Malavaud, L. Lopez, and J. Barrans, Phosphorus Sulphur, 1984, **20**, 333.
2593 H. Winter and J.C. van de Grampel, Recl. Trav. Chim. Pays-Bas, 1984, **103**, 241.
2594 B. Garrigues and A. Munoz, Can. J. Chem., 1984, **62**, 2179.
2595 A. Jäger and J. Engels, Tetrahedron Lett., 1984, **25**, 1437.
2596 L.N. Markovski, V.D. Romanenko, A.V. Ruban, and A.B. Drapailo, J. Chem. Soc., Chem. Commun., 1984, 1692.
2597 S. Hanessian, D. Delorme, S. Beaudoin, and Y. Leblanc, J. Am. Chem. Soc., 1984, **106**, 5754.
2598 S. Kleemann and E. Fluck, Z. Anorg. Allg. Chem., 1984, **516**, 79.
2599 R. Francke, V. Roeschenthaler, R. Di Giacomo, and D. Dakternieks, Phosphorus Sulphur, 1984, **20**, 107.
2600 H. Winter and J.C. van de Grampel, J. Chem. Soc., Chem. Commun., 1984, 489.
2601 H.R. Allcock, R.A. Nissan, P.J. Harris, and R.R. Whittle, Organometallics, 1984, **3**, 432.
2602 J. Svara and E. Fluck, Synth. React. Inorg. Met.-Org. Chem., 1984, **14**, 869.
2603 A.H. Cowley and M.K. Pakulski, J. Am. Chem. Soc., 1984, **106**, 1491.
2604 L.N. Markovski, V.D. Romanenko, and A.V. Kirsanov, Phosphorus Sulphur, 1983, **18**, 31.
2605 H. Riffel, S. Kleemann, H. Hess, and E. Fluck, Z. Anorg. Allg. Chem., 1984, **508**, 61.
2606 M. Baudler and J. Hellmann, Z. Anorg. Allg. Chem., 1984, **509**, 53.
2607 M. Baudler and B. Makowka, Angew. Chem., Int. Ed. Engl., 1984, **23**, 986.
2608 M. Baudler and S. Wietfeldt-Haltenhoff, Angew. Chem., Int. Ed. Engl., 1984, **23**, 379.
2609 W.J. Richter, Chem. Ber., 1984, **117**, 2328.
2610 P.J. Diel and L. Maier, Phosphorus Sulphur, 1984, **20**, 311.
2611 K.C. Calvo and F.H. Westheimer, J. Am. Chem. Soc., 1984, **106**, 4205.
2612 J.J. Bulot, E.E. Aboujaoude, N. Collignon, and P. Savignac, Phosphorus Sulphur, 1984, **21**, 197.
2613 R.O. Day, W.G. Bentrude, K.C. Lee, W.N. Setzer, J.A. Deiters, and R.R. Holmes, J. Am. Chem. Soc., 1984, **106**, 103.
2614 F. Nicotra, L. Panza, and G. Russo, J. Chem. Soc., Chem. Commun., 1984, 5.
2615 H. Ahlbrecht, W. Farnung, and H. Simon, Chem. Ber., 1984, **117**, 2622.
2616 H. Collet, P. Calas, and A. Commeyras, J. Chem. Soc., Chem. Commun., 1984, 1152.
2617 H. Ahlbrecht and W. Farnung, Chem. Ber., 1984, **117**, 1.
2618 C.M. Angelov, C.Zh. Christov, and D.M. Mondeshka, Phosphorus Sulphur, 1984, **21**, 249.
2619 D.W. Hutchinson and G. Semple, Phosphorus Sulphur, 1984, **21**, 1.
2620 G. Sturtz and A. Pondaven-Raphalen, Phosphorus Sulphur, 1984, **20**, 35.
2621 J.M. McIntosh and L.Z. Pillon, Can. J. Chem., 1984, **62**, 2089.

2622 J. Emsley, A.P. Dunning, R.J. Parker, J.K. Williams, S. Brown, S. Earnshaw, and D.S. Moore, Polyhedron, 1984, **3**, 325.
2623 A.E. Wróblewski and W.T. Konieczko, Monatshefte, 1984, **115**, 785.
2624 E. Ohler and E. Zbiral, Monatshefte, 1984, **115**, 493.
2625 E.E. Aboujaoude, N. Collignon, and P. Savignac, J. Organomet. Chem., 1984, **264**, 9.
2626 V.M. Ismailov, V.V. Moskva, and T.V. Zykova, Zh. Obshch. Khim., 1983, **53**, 2793 (Chem. Abstr., 1984, **100**, 139 221).
2627 V.I. Shibaev, A.V. Garabadzhiu, and A.A. Rodin, Zh. Obshch. Khim., 1983, **53**, 1743 (Chem. Abstr., 1984, **100**, 6 679).
2628 S. Sosnowski, A. Duda, S.Slomkowski, and S. Penczek, Makromol. Chem., Rapid Commun., 1984, **5**, 551 (Chem. Abstr., 1984, **101**, 192 555).
2629 J.C. Caesar, D.V. Griffiths, J.C. Tebby, and S.E. Willetts, J. Chem. Soc., Perkin Trans. 1, 1984, 1627.
2630 M. Mikołajczyk, P. Bałczewski, K. Wroblewski, J. Karolak-Wojciechowska, A. Miller, M.W. Wieczorek, M.Y. Antipin, and Y.T. Struchkov, Tetrahedron, 1984, **40**, 4885.
2631 J. Navech, M. Revel, and R. Kraemer, Phosphorus Sulphur, 1984, **21**, 105.
2632 O.S. Sadikov, A.A. Abduvakhobov, Sh.K. Kosimov, and A.G. Dobren'kov, Uzb. Khim. Zh., 1983, 23 (Chem. Abstr., 1984, **100**, 60 662).
2633 M. Yoshifuji, I. Shima, K. Shibayama, and N. Inamoto, Tetrahedron Lett., 1984, **25**, 411.
2634 R.K. Gupta, A.K. Rai, R.C. Mehrotra, and V.K. Jain, Inorg. Chim. Acta, 1984, **88**, 201.
2635 C. Lensch, W. Clegg, and G.M. Sheldrick, J. Chem. Soc., Dalton Trans., 1984, 723.
2636 C. Lensch and G.M. Sheldrick, J. Chem. Soc., Dalton Trans., 1984, 2855.
2637 R.K. Gubta, A.K. Rai, R.C. Mehrotra, and V.K. Jain, Polyhedron, 1984, **3**, 721.
2638 C.J. Jameson and H.J. Osten, J. Chem. Phys., 1984, **81**, 2556.
2639 V. Rutar, Chem. Phys. Lett., 1984, **106**, 258.
2640 W.G. Bentrude, R.O. Day, J.M. Holmes, G.S. Quin, W.N. Setzer, A.E. Sopchik, and R.R. Holmes, J. Am. Chem. Soc., 1984, **106**, 106.
2641 R.R. Holmes, R.O. Day, W.N. Setzer, A.E. Sopchik, and W.G. Bentrude, J. Am. Chem. Soc., 1984, **106**, 2353.
2642 L. Pogliani, N. Niccolai, and C. Rossi, Spectrosc. Lett., 1984, **17**, 159 (Chem. Abstr., 1984, **101**, 111 374).
2643 J. Gloede, Z. Anorg. Allg. Chem., 1984, **513**, 193.
2644 A. John, B. Thomas, A. Pfuetzner, and G. Grossmann, Z. Anorg. Allg. Chem., 1984, **511**, 7.
2645 B. Wrackmeyer, Z. Naturforsch., Teil B, 1984, **39**, 533 (Chem. Abstr., 1984, **100**, 202 207).
2646 M. Baudler, R. Heumüller, D. Düster, J. Germeshausen, and J. Hahn, Z. Anorg. Allg. Chem., 1984, **518**, 7.
2647 G. Sicard, A. Baceiredo, G. Bertrand, and J.-P. Majoral, Angew. Chem., Int. Ed. Engl., 1984, **23**, 459.
2648 Yu.P. Egorov and A.A. Kudryavtsev, Zh. Obshch. Khim., 1983, **53**, 2222 (Chem. Abstr., 1984, **100**, 121 195).
2649 G. Bettermann, H. Buhl, R. Schmutzler, D. Schomburg, and U. Wermuth, Phosphorus Sulphur, 1983, **18**, 77.
2650 J.P. Majoral, G. Bertrand, A. Baceiredo, M. Mulliez, and R. Schmutzler, Phosphorus Sulphur, 1983, **18**, 221.
2651 H.W. Roesky and K.S. Dhathathreyan, J. Chem. Soc., Chem. Commun., 1984, 1053.
2652 M. Mulliez, J.-P. Majoral, and G. Bertrand, J. Chem. Soc., Chem. Commun., 1984, 284.
2653 V.D. Romanenko, L.K. Polyachenko, and L.N. Markovskii, Phosphorus Sulphur, 1984, **19**, 189.
2654 L. Lopez, J.-P. Majoral, A. Meriem, T. N'G. M'Pondo, J. Navech, and J. Barrans, J. Chem. Soc., Chem. Commun., 1984, 183.
2655 C. Malavaud, T. N'G. M'Pondo, L. Lopez, J. Barrans, and J.-P. Legros, Can. J. Chem., 1984, **62**, 43.

2656 A. Baceiredo, G. Bertrand, J.-P. Majoral, G. Sicard, J. Jaud, and J. Galy, J. Am. Chem. Soc., 1984, **106**, 6088.

2657 N. el Murr, R. Lahana, J.-F. Labarre, and J.-P. Declercq, J. Mol. Struct., 1984, **117**, 73.

2658 P.J. Harris and K.B. Williams, Inorg. Chem., 1984, **23**, 1495.

2659 A. Mahmoun, P. Castera, and J.-F. Labarre, Inorg. Chim. Acta, 1984, **86**, L41.

2660 F. Sournies, J.-F. Labarre, P.J. Harris, and K.B. Williams, Inorg. Chim. Acta, 1984, **90**, L61.

2661 D. DuBois, E.N. Duesler, and R.T. Paine, J. Chem. Soc., Chem. Commun., 1984, 488.

2662 K.F. Turchin, Yu.N. Sheinker, and T.S. Safonova, Khim.-Farm. Zh., 1984, **18**, 366 (Chem. Abstr., 1984, **101**, 43 452).

2663 B. de Ruiter, H. Winter, T. Wilting, and J.C. van de Grampel, J. Chem. Soc., Dalton Trans., 1984, 1027.

2664 V. Chandrasekhar, S.S. Krishnamurthy, H. Manohar, A.R.V. Murthy, R.A. Shaw, and M. Woods, J. Chem. Soc., Dalton Trans., 1984, 621.

2665 S. Karthikeyan and S.S. Krishnamurthy, Z. Anorg. Allg. Chem., 1984, **513**, 231.

2666 S.R. Contractor, M.B. Hursthouse, H.G. Parkes, L.S. Shaw, R.A. Shaw, and H. Yilmaz, J. Chem. Soc., Chem. Commun., 1984, 675.

2667 D.B. Denney, D.Z. Denney, and L. Liu, Gaodeng Xuexiao Huaxue Xuebao, 1984, **5**, 509 (Chem. Abstr., 1984, **101**, 211 284).

2668 O. Diallo, M.T. Boisdon, C. Malavaud, L. Lopez, M. Haddad, and J. Barrans, Tetrahedron Lett., 1984, **25**, 5521.

2669 G. Zon, S.M. Ludeman, J.A. Brandt, V.L. Boyd, G. Ozkan, W. Egan, and K.L. Shao, J. Med. Chem., 1984, **27**, 466 (Chem. Abstr., 1984, **100**, 120 238).

2670 R.F. Borch, T.R. Hoye, and T.A. Swanson, J. Med. Chem., 1984, **27**, 490.

2671 M. Meisel and C. Donath, Phosphorus Sulphur, 1983, **18**, 159.

2672 Y. Leroux, D. El Manouni, B.B. Jaafar, and R. Burgada, Phosphorus Sulphur, 1983, **18**, 205.

2673 F. Ramirez, J.S. Ricci, jun., H. Okazaki, J.F. Marecek, and M. Lewy, Phosphorus Sulphur, 1984, **20**, 279.

2674 J. Nielsen and O. Dahl, J. Chem. Soc., Perkin Trans. 2, 1984, 553.

2675 H.-A. Lehmann and W. Schaffrath, Z. Anorg. Allg. Chem., 1984, **508**, 145.

2676 S.L. Beaucage, Tetrahedron Lett., 1984, **25**, 375.

2677 M.W. Schwarz and W. Pfleiderer, Tetrahedron Lett., 1984, **25**, 5513.

2678 R. Francke, R. Di Giacomo, D. Dakternieks, and G.-V. Röschenthaler, Z. Anorg. Allg. Chem., 1984, **519**, 141.

2679 D.E. Schiff, J.W. Richardson, jun., R.A. Jacobson, A.H. Cowley, J. Lasch, and J.G. Verkade, Inorg. Chem., 1984, **23**, 3373.

2680 A. Baceiredo, G. Bertrand, J.-P. Majoral, U. Wermuth, and R. Schmutzler, J. Am. Chem. Soc., 1984, **106**, 7065.

2681 S.A. Culley and A.J. Arduengo, tert., J. Am. Chem. Soc., 1984, **106**, 1164.

2682 J. Steinbach, E. Herrmann, and L. Riesel, Z. Anorg. Allg. Chem., 1984, **511**, 51.

2683 E.G. Bulycheva, P.V. Petrovskii, M.A. Andreeva, V.V. Korshak, and S.V. Vinogradova, Vysokomol. Soedin., Ser. A, 1984, **26**, 135 (Chem. Abstr., 1984, **100**, 139 732).

2684 L. Riesel, A. Pfützner, and E. Herrmann, Z. Anorg. Allg. Chem., 1984, **511**, 33.

2685 C.M. Thompson, S. Castellino, and T.R. Fukoto, J. Org. Chem., 1984, **49**, 1696.

2686 M.-P. Teulade and P. Savignac, Phosphorus Sulphur, 1984, **21**, 23.

2687 O. Navratil, M. Fofana, and J. Smola, Z. Chem., 1984, **24**, 30 (Chem. Abstr., 1984, **100**, 216 572).

2688 C.R. Johnson, R.C. Elliott, and T.D. Penning, J. Am. Chem. Soc., 1984, **106**, 5019.

2689 M. Meisel, C. Donath, M. Cernik, G.U. Wolf, K. Dostal, and H. Grunze, Z. Anorg. Allg. Chem., 1984, **512**, 72.

2690 J. Heubel, T. Abouchakra, E. Puskaric, R. de Jaeger, and S.P. Narula, Z. Anorg. Allg. Chem., 1984, **511**, 212.

2691 R. De Jaeger, J. Belky, T. Abou Chakra, and J. Heubel, Rev. Chim. Miner.,

1983, **20**, 463 (Chem. Abstr., 1984, **100**, 149 975).
2692 J. Heubel, R. De Jaeger, and T. Abou Chakra, Rev. Chim. Miner., 1984, **21**, 67 (Chem. Abstr., 1984, **100**, 216 398).
2693 V.A. Al'fonsov, I.E. Ismaev, D.A. Pudovik, K.M. Enikeev, M.A. Pudovik, V.K. Khairullin, E.S. Batyeva, A.V. Il'yasov, and A.N. Pudovik, Zh. Obshch. Khim., 1984, **54**, 1258 (Chem. Abstr., 1984, **101**, 192 063).
2694 L.A. Fedorov, Zh. Anal. Khim., 1984, **39**, 10 (Chem. Abstr., 1984, **100**, 109 882).
2695 T.T. Bamgboye, Spectrochim. Acta, Part A, 1984, **40**, 329.
2696 B.S. Suresh, V. Chandrasekhar, and D.K. Padma, J. Chem. Soc., Dalton Trans., 1984, 1787.
2697 M.-R. Marre, M. Sanchez, and R. Wolf, J. Chem. Soc., Chem. Commun., 1984, 566.
2698 A. Haas and F. Weiss, Z. Anorg. Allg. Chem., 1984, **513**, 234.
2699 K.B. Dillon and A.W.G. Platt, Phosphorus Sulphur, 1984, **19**, 299.
2700 W. Hönle, H.G. v. Schnering, A. Schmidpeter, and G. Burget, Angew. Chem., Int. Ed. Engl., 1984, **23**, 817.
2701 M. Baudler, D. Duester, K. Langerbeins, and J. Germeshausen, Angew. Chem., 1984, **96**, 309.
2702 P.J.J.M. Van Ool and H.M. Buck, Recl., J.R. Neth. Chem. Soc., 1984, **103**, 119.
2703 C.L. Khetrapal, G. Govil, and H.J.C. Yeh, J. Mol. Struct., 1984, **116**, 303.
2704 D. Neuhaus, G. Wider, G. Wagner, and K. Wüthrich, J. Magn. Reson., 1984, **57**, 164.
2705 J.S. Cohen, C.W. Chen, and A. Bax, J. Magn. Reson., 1984, **59**, 181.
2706 L.V. Nesterov and N.E. Krepysheva, Zh. Obshch. Khim., 1984, **54**, 50 (Chem. Abstr., 1984, **100**, 156 718).
2707 T. Weller, U. Franck, G. Klose, and R. Lochmann, Stud. Biophys., 1983, **96**, 275 (Chem. Abstr., 1984, **100**, 31 176).
2708 H. Jancke, R. Radeglia, J. Neels, and A. Porzel, Org. Magn. Reson., 1984, **22**, 376.
2709 A.P. Joseph and P.H. Bolton, J. Am. Chem. Soc., 1984, **106**, 437.
2710 M. Petersheim, S. Mehdi, and J.A. Gerlt, J. Am. Chem. Soc., 1984, **106**, 439.
2711 C. Yu, and G.C. Levy, J. Am. Chem. Soc., 1984, **106**, 6533.
2712 A.M. Avedikian, D. Besserre, and M. Delepierre, J. Chim. Phys.-Chim. Biol., 1983, **80**, 831.
2713 L.H. Koole, E.J. Lanters, and H.M. Buck, J. Am. Chem. Soc., 1984, **106**, 5451.
2714 I.B. Rubin, Anal. Lett., 1984, **17**, 1259.
2715 G. Lowe and B.V.L. Potter, J. Chem. Soc., Chem. Commun., 1984, 877.
2716 F.G. Il'in, V.F. Sukhoverkhov, A.V. Sharabarin, and Yu.A. Buslaev, Dokl. Akad. Nauk SSSR, 1983, **271**, 652 (Chem. Abstr., 1984, **100**, 44 220).
2717 K. Taira, W.L. Mock, and D.G. Gorenstein, J. Am. Chem. Soc., 1984, **106**, 7831.
2718 J. Gloede, Phosphorus Sulphur, 1984, **20**, 15.
2719 S. Kobayashi, Y. Narukawa, and T. Saegusa, Macromolecules, 1984, **17**, 134 (Chem. Abstr., 1984, **100**, 68 850).
2720 L. Lamande, A. Munoz, and D. Boyer, J. Chem. Soc., Chem. Commun., 1984, 225.
2721 B. Młotkowska and A. Markowska, Ann., 1984, 1.
2722 W.J. Stec and G. Zon, Tetrahedron Lett., 1984, **25**, 5279.
2723 L. Riesel, E. Herrmann, and J. Steinbach, Phosphorus Sulphur, 1983, **18**, 253.
2724 R. Bohlen and G.-V. Röschenthaler, Z. Anorg. Allg. Chem., 1984, **513**, 199.
2725 A. Skowrońska, E. Krawczyk, and J. Burski, Phosphorus Sulphur, 1983, **18**, 233.
2726 H.P.S. Chauhan, G. Srivastava, and R.C. Mehrotra, Polyhedron, 1984, **3**, 1337.
2727 M. Mikolajczyk, P. Kielbasinski, and W. Basinski, J. Org. Chem., 1984, **49**, 899.
2728 R. Iyengar, H.-T. Ho, R.D. Sammons, and P.A. Frey, J. Am. Chem. Soc., 1984, **106**, 6038.
2729 G.U. Wolf and M. Meisel, Z. Anorg. Allg. Chem., 1984, **509**, 101.
2730 E.A.V. Ebsworth, C.M. Huntley, and D.W.H. Rankin, J. Chem. Soc., Dalton Trans., 1984, 2301.
2731 S.H. Strauss and K.D. Abney, Inorg. Chem., 1984, **23**, 515.

2732 A.H. Cowley, M. Lattman, R.A. Montag, and J.G. Verkade, Inorg. Chem., 1984, **23**, 3378.
2733 R. Blachnik, G. Kurz, and U. Wickel, Z. Naturforsch., Teil B, 1984, **39**, 778.
2734 R. Blachnik, U. Wickel, and P. Schmitt, Z. Naturforsch., Teil B, 1984, **39**, 1135.
2735 E.G. Il'in, Yu.A. Buslaev, U. Kalov, and L. Kolditz, Dokl. Akad. Nauk SSSR, 1984, **276**, 371 (Chem. Abstr., 1984, **101**, 121 891).
2736 A.V. Kopityin, P. Gabor Klatsmanyi, and E. Pungor, Magy. Kem. Foly., 1983, **89**, 561 (Chem. Abstr., 1984, **100**, 78 987).
2737 D.D. Desmarteau, W.Y. Lam, B.A. O'Brien, and S.C. Chang, J. Fluorine Chem., 1984, **25**, 387 (Chem. Abstr., 1984, **101**, 221 369).
2738 M.F.A. Dove, J.C.P. Sanders, E.L. Jones, and M.J. Parkin, J. Chem. Soc., Chem. Commun., 1984, 1578.
2739 J.P. Kintzinger, 'Oxygen-17 NMR' in 'NMR Newly Accessible Nucl.', 1983, **2**, 79. Ed. P. Laszlo, Academic, New York (Chem. Abstr., 1984, **100**, 166 563).
2740 H.C.E. McFarlane and W. McFarlane, 'Selenium-77 and tellurium-125', in 'NMR Newly Accessible Nucl.', 1983, **2**, 275 (Chem. Abstr., 1984, **100**, 166 569).
2741 Yu.A. Buslaev and Yu.V. Kokunov, Koord. Khim., 1984, **10**, 435 (Chem. Abstr., 1984, **100**, 202 297).
2742 B.W. Petley and R.W. Donaldson, Metrologia, 1984, **20**, 81 (Chem. Abstr., 1984, **101**, 239 176).
2743 R.E. Wasylishen, S. Mooibroek, and J.B. Macdonald, J. Chem. Phys., 1984, **81**, 1057.
2744 R.E. Wasylishen, C. Connor, and J.O. Friedrich, Can. J. Chem., 1984, **62**, 981.
2745 H. Poleschner, R. Radeglia, and H. Meyer, Org. Magn. Reson., 1984, **22**, 480.
2746 I. Johannsen and H. Eggert, J. Am. Chem. Soc., 1984, **106**, 1240.
2747 M. Baiwir, G. Llabres, A. Luxen, L. Christiaens, and J.L. Piette, Org. Magn. Reson., 1984, **22**, 312.
2748 N. Suryaprakash, S. Raghothama, C.L. Khetrapal, A.C. Kunwar, M. Baiwir, and G. Llabres, J. Organomet. Chem., 1984, **277**, 359.
2749 T.C. Wong, T.T. Ang, F.S. Guziec, jun., and C.A. Moustakis, J. Magn. Reson., 1984, **57**, 463.
2750 A.M. Forster and A.J. Downs, J. Chem. Soc., Dalton Trans., 1984, 2827.
2751 D.J. Gulliver, E.G. Hope, W. Levason, S.G. Murray, D.M. Potter, and G.L. Marshall, J. Chem. Soc., Perkin Trans. 2, 1984, 429.
2752 L. Hevesi, S. Desauvage, B. Georges, G. Evrard, P. Blanpain, A. Michel, S. Harkema, and G.J. Van Hummel, J. Am. Chem. Soc., 1984, **106**, 3784.
2753 J. Gabriel and D. Seebach, Helv. Chim. Acta, 1984, **67**, 1070.
2754 B. Sturm and G. Gattow, Z. Anorg. Allg. Chem., 1984, **513**, 183.
2755 B. Sturm and G. Gattow, Z. Anorg. Allg. Chem., 1984, **514**, 120.
2756 B. Sturm and G. Gattow, Z. Anorg. Allg. Chem., 1984, **514**, 129.
2757 B. Sturm and G. Gattow, Z. Anorg. Allg. Chem., 1984, **509**, 61.
2758 A. Haas and K.W. Kempf, Tetrahedron, 1984, **40**, 4963.
2759 R.J. Abraham, P. Leonard, and R. Ulman, Org. Magn. Reson., 1984, **22**, 561.
2760 Z.-i. Yoshida, H. Awaji, and T. Sugimoto, Tetrahedron Lett., 1984, **25**, 4227.
2761 J. Gramza, R.B. Mitchell, and D.C. Dittmer, J. Org. Chem., 1984, **49**, 2057.
2762 S. Halazy and L. Hevesi, J. Org. Chem., 1983, **48**, 5242.
2763 A. Toshimitsu, S. Uemura, M. Okano, and N. Watanabe, J. Org. Chem., 1984, **48**, 5246.
2764 A. Toshimitsu, H. Owada, K. Terao, S. Uemura, and M. Okano, J. Org. Chem., 1984, **49**, 3796.
2765 B.L. Khandelwal, A.K. Singh, R. Mehta, and K. Kumar, Synth. React. Inorg. Met.-Org. Chem., 1984, **14**, 921.
2766 A. Krebs, W. Rüger, B. Ziegenhagen, M. Hebold, I. Hardtke, R. Müller, M. Schütz, M. Wietzke, and M. Wilke, Chem. Ber., 1984, **117**, 277.
2767 E. Fanghänel, A.M. Richter, and G. Schukat, J. Prakt. Chem., 1984, **326**, 479.
2768 B. Pötter and K. Seppelt, Angew. Chem., Int. Ed. Engl., 1984, **23**, 150.
2769 T. Krügerke and K. Seppelt, Z. Anorg. Allg. Chem., 1984, **517**, 59.
2770 B. Potter, G. Kleemann, and K. Seppelt, Chem. Ber., 1984, **117**, 3255.
2771 J.M. Canich, M.M. Ludvig, G.L. Gard, and J.M. Shreeve, Inorg. Chem., 1984,

23, 4403.
2772 S.A. Kinkead and J.M. Shreeve, Inorg. Chem., 1984, **23**, 4174.
2773 M.R. Detty and H.R. Luss, J. Org. Chem., 1984, **48**, 5147.
2774 S. Uemura and S.-i. Fukuzawa, J. Organomet. Chem., 1984, **268**, 223.
2775 J.S. Thrasher, J.L. Howell, D.E. Maurer, and A.F. Clifford, J. Fluorine Chem., 1984, **24**, 205 (Chem. Abstr., 1984, **100**, 131 207).
2776 J.F. Hinton and D. Buster, J. Magn. Reson., 1984, **58**, 324.
2777 C.A. Spring and N.S. True, J. Magn. Reson., 1984, **58**, 338.
2778 S.N. Terent'ev, P.G. Zykov, and V.M. Zbrod'ko, Deposited Doc., 1983, VINITI 4086, 10 pp. Avail. VINITI (Chem. Abstr., 1984, **101**, 162 435).
2779 H. Henle, R. Hoppenheit,and R. Mews, Angew. Chem., Int. Ed. Engl., 1984, **23**, 507.
2780 B.A. O'Brien and D.D. DesMarteau, Inorg. Chem., 1984, **23**, 2188.
2781 J.S. Thrasher, J.L. Howell, and A.F. Clifford, Chem. Ber., 1984, **117**, 1707.
2782 P. Huppmann, G. Klöter, J.S. Thrasher, K. Seppelt, and D.D. DesMarteau, Inorg. Chem., 1984, **23**, 2217.
2783 B.A. O'Brien and D.D. DesMarteau, Inorg. Chem., 1984, **23**, 644.
2784 R. Damerius, P. Huppmann, D. Lentz, and K. Seppelt, J. Chem. Soc., Dalton Trans., 1984, 2821.
2785 H. Willner, Z. Anorg. Allg. Chem., 1984, **514**, 171.
2786 L.J. Lawlor and J. Passmore, Can. J. Chem., 1984, **62**, 1477.
2787 M.J. Collins and R.J. Gillespie, Inorg. Chem., 1984, **23**, 1975.
2788 T.R. Stengle, N.V. Reo, and K.L. Williamson, J. Phys. Chem., 1984, **88**, 3225.
2789 T.R. Stengle, S.M. Hosseini, H.G. Basiri, and K.L. Williamson, J. Soln. Chem., 1984, **13**, 779.
2790 M. Claessens, O. Fabre, D. Zimmermann, and J. Reisse, Bull. Soc. Chim. Belg., 1984, **93**, 983.
2791 H.J. Frohn and W. Pahlmann, J. Fluorine Chem., 1984, **24**, 219 (Chem. Abstr., 1984, **100**, 167 074).
2792 G.A. Schumacher and G.J. Schrobilgen, Inorg. Chem., 1984, **23**, 2923.

2

Nuclear Quadrupole Resonance Spectroscopy

BY K. B. DILLON

1 Introduction

This chapter reports on the pure nuclear quadrupole resonance (n.q.r.) spectra of quadrupolar ($I > \frac{1}{2}$) nuclei in inorganic or organometallic solids. The study of structural phase transitions by n.m.r. and n.q.r. methods has been reviewed.[1] A review of some n.q.r. studies of structural and magnetic phase transitions, particularly in methylammonium and tetramethylammonium hexahalometallates(IV), methylammonium halides, and copper(II) complexes, has appeared,[2] as well as a more specialised review of ^{35}Cl n.q.r. work at low temperature on the one-dimensional XY system $PrCl_3$.[3]

The usual format is adopted in the more detailed sections, i.e. results for main-group elements (^{2}H, Groups I, III, V, VI and VII), followed by those for transition metals and lanthanides.

2 Main-group Elements

2.1 Deuterium.- ^{2}H nuclear quadrupole double resonance from partially deuterated powdered samples of KH_2PO_4 has been studied as a function of temperature and of percentage deuteration.[4] Results were obtained from samples containing 15, 50 and 90% deuterium. An isotope effect was observed on the quadrupolar coupling constant e^2Qq/h, but the asymmetry parameter η remained independent of percentage deuteration. Double transitions were also observed from the 50% and 90% deuterated materials, enabling the orientation of the ferroelectric electric field gradient (e.f.g.) tensor to be determined. ^{2}H and ^{14}N n.q.r. frequencies at 77K have been determined for deuterated phenylphosphonic acid, and the related ternary cocrystals of the pyridinium salt of 2-hydroxyphenyl phenylphosphonate.catechol, and pyridinium phenylphosphonate.2 catechol.[5] The crystal structures and hydrogen bonding in the compounds were discussed. Deuterium sites which were not located in the X-ray diffraction study could be assigned from the n.q.r. data.

2.2 Group I (Lithium-7, Sodium-23, and Potassium-40).- The observation has been reported of Fourier transform pure n.q.r. with high field detection by pulsed field cycling.[6] The advantages of the method for quadrupolar nuclei with low resonance frequencies were discussed, and its use was exemplified by showing the ^{7}Li spectrum at room temperature from polycrystalline $Li_2SO_4.H_2O$, with the two inequivalent Li sites clearly resolved. A phase coherent nuclear quadrupole pulse and double resonance spectrometer has been described, and used to record the spin echo double resonance (SEDOR) spectrum of ^{23}Na nuclei in $NaClO_3$ at 300K.[7] In a more recent paper by the same authors, a scheme for the detection of signals from rare quadrupolar nuclei based on off-resonant pulse excitation of abundant quadrupolar nuclei coupled to the rare ones has been proposed.[8] It was applied to ^{23}Na n.q.r. from a polycrystalline sample of $NaClO_3$ at room temperature, and the advantages of this double resonance technique were discussed. The expected relative intensities of lines in the n.q.r. spectra of nuclei with spins of 4, 5 or 6 have been calculated theoretically, and compared with experimental data for ^{40}K in $KClO_3$.[9] Good agreement was obtained.

<u>2.3 Group III (Boron-10, Aluminium-27, Gallium-69 and -71, and Indium-115)</u>.- The predictions of simple level crossing theory for the signal intensities in n.q.r. spectra obtained by double resonance by level crossing (d.r.l.c.) for a nucleus of I = 3 have been compared with experimental results for ^{10}B in borazine, $B_3N_3H_6$.[10] This method was found to give much better agreement than conventional matrix elements. Fourier transform n.q.r. with pulsed field cycling has been used to obtain ^{27}Al data from potassium alum, ammonium alum and a mixed sample (56 mol % potassium alum, 44 mol % ammonium alum).[11] In contrast with n.m.r. by magic angle spinning, where the chemical shifts were very similar, the different aluminium sites, including those in the mixture, could be readily identified, suggesting important applications of this technique to aluminium-containing solids such as zeolites. Fine structure has been observed in the ^{69}Ga and ^{71}Ga n.q.r. spectra of GaS and GaSe which persisted from 293-77K, although the spectra for the different gallium isotopes differed qualitatively in each of the compounds.[12] This structure was attributed to an indirect spin-spin interaction between nearest neighbour gallium atom pairs, and a model giving reasonable agreement with the experimental results was proposed. Other ^{69}Ga, ^{71}Ga results are described in the subsection on Group VII (^{35}Cl),[13] and ^{115}In data in the subsection on Group V (^{123}Sb, ^{209}Bi).[14]

<u>2.4 Group V (Nitrogen-14, Arsenic-75, Antimony-121 and -123, and Bismuth-209)</u>.- A two-frequency double n.m.r.-n.q.r. method has been

used to obtain ^{14}N n.q.r. frequencies for a series of mainly organic compounds, but including $C(NH_2)_3.HCl$ and $H_6TeO_6.2H_2O.$-$2(CH_2)_6N_4$.[15] The results were compared with data obtained by a direct n.q.r. method. ^{14}N n.q.r. by the technique of double resonance by level crossing (d.r.l.c.) has been recorded for several adducts of amines with BF_3, mainly at 77K but including data for $F_3B.NH_2Et$ at 298K.[16] The results were compared with corresponding data for the free amines and amine-borane adducts, and charge distribution in the compounds was discussed. ^{14}N n.q.r. at 120K has been obtained by the d.r.l.c. method for the compounds $4\text{-}H_2NC_6H_4SO_2NHR$ (R = CH_2SO_3Na or $NH_2.H_2O$).[17] A combination of the results with ^1H n.m.r. data enabled the structures of the compounds to be deduced. Temperature (T)-dependence studies on the ^{14}N n.q.r. signals from RbTCNQ and NaTCNQ (TCNQ = tetracyanoquinodimethanide) showed small but sharp discontinuities at the regular-alternant phase transition (214K and 346K respectively).[18] No ν^- lines were detected from the sodium compound. The n.q.r. spectra revealed that both phases can coexist, particularly in the Na compound where the range of coexistence was about 8°. Some estimates of relaxation rates were also obtained. Other ^{14}N results have been described in the subsection on ^2H.[5]

The ^{75}As n.q.r. spectrum at 77K of the yellow crystalline modification of arsenic consisted of eight lines between 14.8 and 16.3 MHz.[19] It is known to consist of tetrahedral As_4 molecules, and the results were in agreement with two crystallographically inequivalent common positions of As_4 molecules in the unit cell. ^{75}As n.q.r. lineshape measurements on the amorphous (glassy) mixed system $As_2S_xSe_{3-x}$ showed an asymmetric resonance ca. 8 MHz. in width.[20] The resonance frequency increased approximately linearly with x. The results indicated the presence of mixed $As_2(S,Se)_3$ pyramidal units, and were not consistent with anion subsite segregation, as has been proposed in some models. Measurements on crystalline As_2SSe_2 supported the structural model proposed for the glasses. The T-dependence of the n.q.r. parameters of ^{75}As nuclei in a single crystal of proustite, Ag_3AsS_3, has been determined from 75-20K; some specific heat measurements were also reported.[21] The results indicated that not one (as deduced in previous work) but two phase transitions occur in this region, one of second order at 60.5 ± 0.5K and one of first order at 49 ± 0.5K, the latter accompanied by thermal hysteresis. Behaviour in the temperature range between these values was explained by postulating the existence of an incommensurate phase, and a model was developed for the appearance of such a phase. A model has been proposed for the description of precursor order clusters at ferroelectric phase transitions.[22] ^{75}As n.q.r. data for the low temperature first order ferroelectric transition in Ag_3AsS_3, including pressure-dependence and spin-lattice relaxation results, were analysed on this basis. Temperature- and pressure-dependence studies have been carried out on the n.q.r. (^{75}As or ^{35}Cl) from the ferroelectric compounds KH_2AsO_4, KD_2AsO_4, $PbHAsO_4$, $PbDAsO_4$ and $NH_4H(ClCH_2COO)_2$.[23] The phase transition temperatures were determined, and their behaviour was discussed on the basis of the pseudospin-lattice coupling model. T-dependence measurements in the range 77-300K on the ^{63}Cu and ^{75}As signals from $[Et_3As.CuX]_4$ (X = Cl, Br or I) have shown the occurrence of structural phase transitions at 93 and 173K for X = Cl, 152K (first order) for X = Br, and 86K for X = I.[24] The signals from both ^{63}Cu and ^{75}As nuclei in the iodo-compound disappeared at 86$\pm$1K on cooling, and no resonances could be detected at 77K. The chloro- and bromo-compounds showed an increase in the number of resonances (from one to two) for both nuclei in the lowest temperature phase.

Acoustic n.q.r. spectra and spin-lattice relaxation times of ^{121}Sb and ^{123}Sb in a single crystal of antimony metal at 4.2K have been measured by means of a superconducting quantum interference device (SQUID) acoustomagnetic spectrometer.[25] Novel features of the spectra were described, including the first determination of magnetic field-dependent nuclear spin relaxation rates in a bulk metal. Structural phase transitions in K_2SbF_5 have been followed by various physical means, including the T-dependence of the ^{123}Sb n.q.r. frequencies for the $\frac{1}{2}\leftrightarrow\frac{3}{2}$ and $\frac{3}{2}\leftrightarrow\frac{5}{2}$ transitions in the range 77-400K.[26] There was one type of Sb site only at 77K, but a sequence of anomalies in the physical parameters was found in the regions 110.9-112.6, 123-124.8, 134.7-135, 140.3, 145.1, 267.2 and 280-315K, including the appearance between 112.6 and 267K of more than one type of Sb site. A first order transition, probably consisting of several phases, was suggested in the interval 113-140.3K, and a second order phase transition at 267K, both non-ferroelectric. Possible mechanisms for the phase transitions, including reorientation movements, rotation of the octahedron, and slight lattice disordering above room temperature, were discussed. Similar methods have been employed to investigate phase transitions in $KSbF_4$ between 77 and 380K.[27] The results showed four different Sb positions in the unit cell at 77K. On heating, the signals disappeared at 184K, and two new lines appeared at 194K. These persisted up to 380K, when the signals again disappeared, in agreement with the room temperature crystal structure which showed two crystallographically inequivalent Sb sites. It was deduced from this and other work that phase transitions occurred at 190K and 483K, the latter not detectable by n.q.r. Anomalies in physical properties in the vicinity of the phase transitions were discussed. The high temperature phase transition was a polymorphic irreversible transition into a highly conducting state ($\alpha \rightarrow \beta KSbF_4$). Other results for Sb are described in the subsection on Group VII (^{127}I).[28,29]

A theoretical description of the shape of the spin-echo envelope obtained on applying a pulsed external electric field to a powder specimen has been given.[30] The existence of the maximum shift direction for a quadrupole system was established theoretically, and shown experimentally from data for ^{121}Sb, ^{123}Sb nuclei in $SbCl_3$ and for ^{209}Bi nuclei in $BiBr_3$, $BiCl_3$ and $Bi(Ph)_3$. A new and improved pulse technique for studying electric effects in n.q.r. was described. An expression has been derived for the calculation of the thermal expansion coefficient and isothermal compressibility from n.q.r. data.[14] Gruneisen parameters, heat capacities and other thermophysical parameters could also be estimated. Its use was illustrated for ^{209}Bi (298K) or ^{123}Sb (288K) data, as appropriate, from $BiCl_3$, $SbCl_3$, $2SbCl_3.C_{10}H_8$, $2SbCl_3.C_6H_4Me_2$, $SbCl_3$.PhEt and the α and β phases of $SbBr_3$, for ^{127}I results from HgI_2 and HIO_3 (273K), and for ^{115}In data from InI_3 (308K). ^{209}Bi n.q.r. frequencies have been recorded for $Bi_2(SO_4)_3.3H_2O$, $KBiCl_2SO_4$, and $NH_4BiCl_2SO_4$ at 77 and 298K, and for $Rb_3Bi_2Cl_5(SO_4)_2.3H_2O$ at 77K.[31] No signals were obtained from the last compound at 298K. The first three compounds had one bismuth environment only, but $Rb_3Bi_2Cl_5(SO_4)_2.3H_2O$ had two inequivalent positions of the Bi atoms in the unit cell. On the basis of n.q.r. and i.r. results it was concluded that the sulphato-chloride complexes were probably polymeric.

<u>2.5 Group VI (Oxygen-17)</u>.- A double resonance technique has been used to obtain ^{17}O n.q.r. spectra from CsH_2PO_4 at various temperatures between 120K and room temperature, including measurements in both the paraelectric and ferroelectric phases (T_c 153K).[32] The results showed that the protons in the short one-dimensionally linked O-H(2)---O hydrogen bonds move between two equilibrium sites above T_c, and freeze into one of the two off-centre sites below T_c. The protons in the asymmetric O-H(1)---O bonds appeared to be static both above and below T_c, and to play a minor role only in the transition mechanism. The reorientational frequencies of the CO

dipoles in the α-phase of carbon monoxide have been derived from the temperature behaviour of the ^{17}O n.q.r. linewidth in the range 25-60K and from dielectric measurements.[33] The extreme slowing down of the reorientations and the orientational disorder in the "frozen" phase were shown by a progressive broadening of the resonance over a few degrees on cooling below 30K. A metastable phase was expected to occur below 16K, reminiscent of orientational dipolar glasses.

<u>2.6 Group VII (Chlorine-35 and -37, Bromine-79 and -81, and Iodine-127)</u>.- The synthesis of chlorine chlorosulphate, $ClOSO_2Cl$, has been reported, from reaction between Cl_2 and SO_3 in CH_2Cl_2 at 195K.[34] The compound gave two ^{35}Cl signals at 77K, at 36.17 (chlorosulphate) and 54.0 MHz. ^{35}Cl n.q.r. data at 77K for the complex $K_2[RuCl_5(NO)]$ have been interpreted as indicating that the Cl atoms in cis positions have a greater ionicity and double bond character with Ru than the trans Cl atom.[35] The preparation and some properties of $[SCl_3][AlCl_4]$ have been described, including ^{35}Cl n.q.r. frequencies of 42.014, 42.104 and 43.400 MHz. for the cation at 77K.[36] A ^{35}Cl n.q.r. frequency of 26.838 MHz. at 77K has been recorded for the chlorines bound to phosphorus in $PCl_2(O)(NCH_2CH_2Cl)_2$, in a study by ^{14}N and ^{35}Cl n.q.r. of some nitrogen mustards.[37] ^{35}Cl n.q.r. at 77 and 296K (where observable) has been used, together with vibrational spectroscopy, to ascertain the structures of $[PdCl_2.(ECl_2)_2]$ (E = S or Se), $[PtCl_4.(ECl_2)_2]$ and $[PtCl_4.(E'Cl_4)_2]$ (E' = Se or Te).[38] The complexes were deduced to have trans structures, with E coordinated directly to Pd or Pt, in the ECl_2 adducts, but the $E'Cl_4$ derivatives had ionic structures $[E'Cl_3]_2PtCl_6$. Coordination by $SeCl_2$, which is unstable under normal conditions, was achieved for the first time. The Zeeman ^{35}Cl n.q.r. spectrum of a single crystal of Me_2SnCl_2 has been re-investigated, in view of an improved crystal structure determination.[39] Two zero-splitting loci for the chlorine atoms were found. The asymmetry parameter η was evaluated as 0.32, and e^2Qq/h as -30.92 MHz. The principal e.f.g. Z-axes lay along the Sn-Cl bonds, while the Y-axes were parallel to the crystallographic b-axis and the X-axes lay in the cb-plane. Structure and π-bonding in the compound were discussed on the basis of Townes-Dailey theory.

The T-dependence of ^{35}Cl resonances from Cs_2HgCl_4 has been determined from 293.3-77K.[40] Two equally intense lines were found at room temperature, which showed a normal T-dependence on cooling to 245K. At this point the amplitude of the lines decreased rapidly, and then increased on further cooling, attributed to a phase transition of the order-disorder type. A first order phase transition was detected at 170K, when an additional signal was observed, showing three inequivalent Cl positions in the unit cell. At 145K a second order phase transition occurred, causing further splitting, and nine lines were seen below this temperature, some with anomalous (positive) T-coefficients. Crystals synthesised under different conditions showed that the number of resonances at 293K could vary from two to five, ascribed to the presence of more than one phase. Various physical methods, including the T-dependence of the ^{35}Cl n.q.r. frequency from 77-186K, have been used to study phase transitions in $[Cd(NH_3)_6](ClO_3)_2$.[41] No apparent anomalies were found in the n.q.r. plot, although anomalies were detected in the proton spin-lattice relaxation time and thermal expansion coefficient between 162 and 166K. This low temperature phase transition was suggested to be associated with a change of orientation of the ClO_3 groups, with a change in the e.f.g. direction at the chlorine sites but no change in its value. A second phase transition was detected at 248K, although this was not followed by n.q.r. Molecular motion in the various phases was discussed. The T-dependence of the frequency and signal-to-noise ratio of the high frequency ^{35}Cl resonance from K_2ZnCl_4 has been followed.[42] Measurements close to the incommensurate-commensurate phase transition allowed calculation of the T-dependence of the soliton density n_s, and showed the existence of a metastable chaotic state close to the transition temperature T_c, which extended on cooling far below T_c into the C phase. The results of similar studies on ^{35}Cl n.q.r. from Rb_2ZnCl_4 showed that thermal unpinning and floating of the incommensurate modulation wave produced a partial motional averaging of the splitting between the two edge singularities in the spectrum, close to the incommensurate-paraelectric transition T_1.[43] An expression was derived for the T-dependence of the mean square fluctuations, the results of which agreed well with those from ^{87}Rb n.m.r. data.

^{35}Cl n.q.r. frequencies at 77K have been measured for several trichloroorgano-germanes and -stannanes.[44] The results were compared with X-ray data, where known, and used to establish the structures of the compounds and the coordination number of Ge or Sn. A linear correlation was found between the ^{35}Cl n.q.r. frequencies and the Ge-Cl bond lengths in the germanes. ^{35}Cl n.q.r. frequencies at 77K for a series of organochlorogermanes $Cl_3GeCR^1R^2CH_m(Me)_{2-m}COZ$ have shown that the oxygen atom can coordinate to germanium to give a five-coordinate structure.[45] The frequency difference between axial and equatorial chlorines was particularly striking in

$Cl_3GeCH_2CH_2CONH_2$ and $Cl_3GeCH_2CH(Me)CONMe_2$, where crystal structure determination showed the largest variation in Ge-Cl bond lengths. Conditions for the formation of pentacoordinate compounds of Si, Ge and Sn were discussed. A series of correlational equations has been derived by a least squares method, relating the ^{35}Cl n.q.r. frequency to the Taft (σ*) and Thomson (E) parameters for various organo-derivatives of Si, Ge, Sn, P and As chlorides, as well as for carbon compounds.[46] Good correlation coefficients were obtained. ^{35}Cl n.q.r. spectra at 77K have been recorded for 34 compounds, mainly organic but including the phosphorus compounds $RPOCl_2$ (R = $MeCHClCH_2$, $Me_2C{=}C(Me)CCl{=}CH$, PhCCl=CH, or $Cl_3P{=}N$), $R'PSCl_2$ (R' = PhCCl=CH or PhCH=CH), $CCl_3PF_2Cl_2$, and $3,4\text{-}Me_2C_6H_3PCl_2$.[47] The T-dependence from 77K to the melting point of the ^{35}Cl n.q.r. frequency and spin-lattice relaxation time T_1 has been determined for $(Cl_3PNR)_2$, where R = Me, Ph, or $4\text{-}ClC_6H_4$, and the molecular dynamics were analysed on the basis of the results.[48] Intra-molecular rearrangements of the pseudorotation type were deduced to occur, with positional exchange of the axial and equatorial chlorine atoms. The activation energies of these rearrangements were evaluated as 25.0, 15.6 and 21.0 kcal mol^{-1} for R = Me, Ph and $4\text{-}ClC_6H_4$ respectively.

A 1:1 complex between $SnCl_4$ and $2\text{-}MeOC_6H_4COCl$ had a five-coordinate structure immediately after preparation and crystallisation, as shown by its ^{35}Cl n.q.r. spectrum at 77K (1 signal for C-Cl at 32.590 MHz., 1 signal for equatorial chlorines at 23.305 MHz., and 1 signal for the axial chlorine at 19.994 MHz., with relative intensities 1:3:1).[49] The ligand coordinated in an axial position via the carbonyl oxygen. When the solid compound was kept at low temperature, new signals appeared, with 5 additional lines in the spectrum after 5 h., and these were the only components of the spectrum after 3 days. The carbon-bonded chlorine was higher in frequency, but the other signals were closer together (between 19.662 and 21.420 MHz.) and of approximately equal intensity. The change was attributed to isomerisation to a six-coordinate cis complex, with coordination by the methoxy-group. The ^{35}Cl resonance from $MeNH_3^+HgCl_3^-$ showed an increase in frequency from 9.98 MHz. at 123K to 10.32 MHz. at 331K, followed by a discontinuous drop to 9.71 MHz. at 334K.[50] A further increase in the temperature caused the frequency to increase again. The results provided clear evidence for a first order phase transition around 333K, probably related to an order-disorder transition of the $MeNH_3$ groups. It was thought that motions of these groups might also account for the anomalous T-coefficient of the ^{35}Cl frequency. ^{35}Cl n.q.r. results including T-dependence data from 77-300K have been obtained for $N_3P_3Cl_4Ph_2$(I) and $N_3P_3Cl_4(NMe_2)_2$(II).[51] Both compounds gave four lines at room temperature, but one signal from compound II could not be clearly observed below 250K. All signals showed a normal T-dependence, and the results were analysed using the Bayer-Kushida-Brown approach. Torsional frequencies in the range 10-27 cm^{-1} were calculated; these were found to be only slightly temperature-dependent. In an extension of previous work, the T-dependence of the spin-lattice relaxation time T_1 of the ^{35}Cl nuclei has been determined from 77-300K for both crystalline phases of 4-chlorobenzenesulphonyl chloride.[52] It was concluded that below 160K the relaxation time depended mainly on molecular torsional oscillations, but above this temperature the triggering of wagging semi-external movements took place.

A ^{35}Cl n.q.r. spectrum has been obtained at 77K from a layer compound of graphite-$SnCl_4$-Cl_2 of formula $C_{54}SnCl_6$.[53] Weak lines were detected corresponding to an excess of $SnCl_4$ and Cl_2, showing the formation of crystals where the intercalants had penetrated into defects. A further signal was found at 15.635 MHz., characteristic of the $SnCl_6^{2-}$ anion in the interlayer space of the compound, with all six chlorines equivalent. Hence it was considered that complete electron transfer had taken place from the carbon network to the $SnCl_6^{2-}$ anion in the compound. The effect of hydrostatic pressure p ($0 < p < 5$ kbar) on the ^{35}Cl frequencies and spin-lattice relaxation rates for chlorine nuclei in A_2SnCl_6 (A = K, NH_4, Rb or Cs) has been investigated in the temperature range 20-450K.[54] In the cubic phase the pressure coefficients of the n.q.r. frequencies were negative. This behaviour was interpreted by considering variations in the static and dynamic parts of the e.f.g. At high temperatures the relaxation rate decreased on applying pressure, attributed to the increase in the enthalpy of activation of the $SnCl_6^{2-}$ motion. The activation volume was determined for K_2SnCl_6, enabling the height of the barrier to rotation to be evaluated. The effect of substitutional impurities on the upper structural phase transition temperature T_{c1} has been followed by recording the frequency and intensity of the ^{35}Cl resonances in the range 100-300K for the mixed crystals $(A_xK_{1-x})_2SnCl_6$ (A = Rb or NH_4) and $K_2(Re_xSn_{1-x})Cl_6$.[55] The mixed compounds could be investigated up to 10% impurity concentration in K_2SnCl_6 and up to 20% impurity concentration in K_2ReCl_6. The observed change of T_c as a function of concentration x could be fitted by the equation

$$T_{c(x)} = T_{c(o)} \frac{(1+ax)}{(1+bx)} \quad (1)$$

This relationship was explained theoretically on the assumption that Landau theory was still valid in the mixed crystals, and that the virtual crystal approximation could be applied to soft phonons. Further work on correlation times from ^{35}Cl relaxation measurements in the ferroelectric phase of HCl-DCl mixed crystals has confirmed the conclusions reported in last year's review (p.166, ref. 69).[56]

The dependence of the spin-lattice relaxation time T_1 for ^{35}Cl nuclei on uniaxial pressure applied along the [110] direction has been measured at five temperatures from 278-313K for single crystals of $NaClO_3$.[57] Comparison of the results with similar data for p-dichlorobenzene showed that $NaClO_3$ had a lower elastic constant, and its torsional frequencies were less affected by pressure. The construction of a spectrometer for n.q.r. measurements, including a temperature controller, has been described, and its performance tested by observing the ^{35}Cl resonance from $NaClO_3$ in the range 77-300K.[58] The temperature of the sample could be controlled within $\pm 0.2^{o}$ during the experiment. A simple search spectrometer for n.q.r. employing a field effect transistor has been constructed, and used to obtain the ^{35}Cl resonance from a crystal of $NaClO_3$ at room temperature.[59] Spin thermodynamics have been used to predict the relations between various aspects of pulsed n.q.r. for nuclei of spin $\frac{3}{2}$, and compared with experimental results for ^{35}Cl from single crystals of $NaClO_3$ at 313K.[60] Very satisfactory overall agreement was obtained, although the experiments suggested that the choice of quasi-invariants related to spin-spin interactions in $NaClO_3$ should be reconsidered. The development of an approximation to a "double-beam" super-regenerative n.q.r. spectrometer has been described, and its use illustrated by means of the room temperature ^{35}Cl spectrum of polycrystalline $NaClO_3$.[61]

The T-dependence has been determined of the ^{35}Cl n.q.r. from $[C(NH_2)_3][MCl_4]$ (M = Au, Pt or Pd) from 77K to 360, 167 and 216K respectively.[62] The gold compound gave four lines, two of which crossed at 239K; two signals faded out at 345K and the others at ca. 360K. The plot suggested a phase transition near to the fade-out temperature, and DTA indicated an unusual phase transition at 363K. The high temperature phase was easily supercooled. The Pt and Pd complexes gave two equally intense signals at 77K, which faded out at 169 and 220K respectively, but no heat anomalies were apparent between 77K and room temperature from DTA measurements. T-dependence studies on the ^{1}H n.m.r. spin-lattice relaxation times for the complexes were also reported, and molecular motion in the compounds was discussed on the basis of the results. A theory proposed previously by Negita, Nakamura and Chihara for the effect of rotation of ions or molecules on the n.q.r. frequencies of nuclei in adjacent ions has been applied to ^{35}Cl data from $[Ca.6H_2O][SnCl_6]$ and ^{187}Re results from NH_4ReO_4.[63] The theory provided an adequate explanation of the T-dependence in $[Ca.6H_2O][SnCl_6]$, identifying the flipping motion of the water molecules as responsible for the anomaly observed, but failed to explain details of pressure-dependence measurements on NH_4ReO_4. The contributions of the dipolar, electric and dynamic linewidth second moments to the ^{35}Cl, ^{69}Ga and ^{71}Ga n.q.r. linewidths in $NaClO_3$, $KClO_3$, $HgCl_2$, $SbCl_3$, $CrCl_3$, $GaCl_3$ and p-$C_6H_4Cl_2$ have been calculated, and compared with experimental data.[13] It was concluded that the electric term was usually the dominant contribution well below the melting point. The dipolar second moment was constant between 77 and 295K for the compounds studied. The dynamic contribution accounted for motion-related effects, for example at temperatures near the melting point, or where such motions occurred spontaneously because of phase transitions. The effects on spin-lattice relaxation times and line-shapes were considered. Comparison of experimental n.q.r. dipolar second moments for very pure single crystals of $NaClO_3$, $KClO_3$ and $SbCl_3$ with theoretical values has shown that the degeneracy in the $\pm\frac{3}{2}$ and $\pm\frac{1}{2}$ levels is not removed by internal dipolar fields, as suggested by Abragam and Kambe.[64] Better agreement was obtained by using the degenerate equations of Bize et al.

In an extension of previous work, a group theoretical classification of the patterns of halogen n.q.r. lines resulting from distortions of some tetragonal R_2MX_6 structures has been given.[65] Charts were presented to permit identification of the space groups and primary order parameters compatible with particular spectra, and the sequences of structural transitions in K_2ReCl_6 and $(NH_4)_2SnBr_6$ were discussed as illustrations of the use of these charts. A similar treatment has been applied to predict possible low symmetry space groups for a commensurate phase transition associated with a broken symmetry in hexagonal crystals of space group $P6_3/m$.[66] The T-dependence of ^{35}Cl n.q.r. from $PrCl_3$ in the vicinity of T_c = 0.4K was considered as an example of this approach.[66,67] Analysis of the n.q.r. data and other information identified the low temperature phase as a C^{I}_{3i} structure, in which chains of Pr ions were dimerised. The phase transition was thus considered to represent a novel type of Peierls dimerisation, induced by a Jahn-Teller coupling.[67] Other ^{35}Cl results have been described in the subsection on Group V (^{75}As).[23]

^{35}Cl n.q.r. frequencies have been measured at 77 and 273K for several 2,4,6-trichlorophenolates of copper(II), where X-ray crystallography had revealed secondary bonding interactions between the metal and the ortho chlorine.[68] Analogous 4-bromo-2,6-dichloro- and 2,6-dichlorophenolate complexes were also investigated to aid in frequency assignment. ^{81}Br frequencies at 308K were reported for some of the bromo-compounds. The secondary-bonding chlorines were found to have n.q.r. frequencies from 0.8-2.2 MHz. lower than corresponding ortho chlorines not involved in secondary bonding. Analysis of the data by an extended Townes-Dailey approach suggested the transfer of about 0.06 electrons from Cl to (or towards) Cu. The results were compared with literature data on pyridine complexes. The n.q.r. frequencies of the secondary-bonding chlorines also showed an anomalous T-dependence. In a similar study, ^{35}Cl n.q.r. spectra have been recorded at various temperatures from 77-273K for some cobalt(II) 2,4,6-trichloro-, 2,6-dichloro- and 4-bromo-2,6-dichlorophenolates.[69] Some ^{81}Br frequencies at 308K were reported for the bromo-derivatives. The difference in frequency between the coordinated and non-coordinated ortho chlorines was found to increase (to as high as 3 MHz., corresponding to the donation of about 0.1 electrons) as the metal-chlorine bond distance decreased. When this frequency difference exceeded ca. 1.3 MHz. at 77K it decreased at higher temperatures. ^{35}Cl or ^{79}Br, ^{81}Br n.q.r. frequencies (as appropriate) have been measured at 77K for some compounds in the series $ClCH_2(CH_3)_{3-n}Si(OCOCF_3)_n$ and $BrCH_2(CH_3)_{3-n}$-$Si(OCOCH_3)_n$.[70] The effects of substituents on the n.q.r. frequency were considered, and the results were correlated with data for other silanes and analogous carbon compounds. ^{35}Cl or ^{79}Br, ^{81}Br frequencies have been measured for $MoS_2Cl_3.SeCl_2$, $MoS_2Br_3.SeBr_2$ and $MoBr_2$ at 77 and 296K, and for $SCl_3^+MoCl_6^-$ and $MoBr_3$ at 77K only.[71] The structures of the compounds were discussed, and the selenium dihalides were thought to coordinate as ligands to molybdenum. The e.f.g. parameters at the halogen (^{35}Cl, ^{79}Br or ^{127}I) sites in some metal halates have been calculated by both CNDO/2 and INDO methods, and the results compared with experimental n.q.r. data.[72] It was concluded that the CNDO/2 method was better suited to this type of compound than the INDO approach.

Two ^{79}Br signals were observed from a commercial sample of $ThBr_4$, and these were shown by thermal treatment to arise from the α and β phases.[73] The corresponding ^{81}Br resonances were also detected. The T-dependence of the ^{81}Br signal from α-$ThBr_4$ between 63-300K was shown to be almost linear. Two ^{79}Br signals were detected from UBr_4 at 77K, and these showed opposite temperature coefficients between 77 and 300K. It was not known whether these arose from two crystallographic sites or from two different phases. Zeeman studies at 293K on the ^{81}Br n.q.r. from a single crystal of $(NH_4)_2InBr_5.H_2O$ have shown four resonances, and the quadrupolar coupling constants and asymmetry parameters were calculated for each of them.[74] The e.f.g. axes were found to be almost parallel to the In-Br bond directions. Some ^{115}In and ^{14}N n.m.r. results for the compound were also given, as well as ^{2}H n.m.r. data for $(ND_4)_2InBr_5.D_2O$, which was correlated with the structure of the compound and motions of the D_2O molecule. Molecular orbital calculations using CNDO/2 and INDO methods have been used to compute the e.f.g. at the bromine site in some metal bromates $M(BrO_3)_2.6H_2O$ (M = Zn, Mg or Ni).[75] The covalent f.g. was found to contribute > 80% of the total f.g. The effect of the water molecules was also considered. The data were compared with experimental results for e^2Qq/h and η obtained at room temperature from Zeeman studies on single crystal samples. Four physically inequivalent Br sites were found in each of the compounds. The calculations were improved by inclusion of the bromine d orbitals, indicating that the extended basis set should be used for the MO calculations. The T-dependence from 77-305K of ^{79}Br n.q.r. from $Cd(BrO_3)_2.2H_2O$, $M(BrO_3)_2.H_2O$ (M = Sr or Ba) and $M'(BrO_3)_2.6H_2O$ (M' = Zn, Mg or Ni) has been determined.[76] ^{81}Br frequencies were also measured at 77 and 305K. The Cd compound showed two bromine sites, but the other compounds gave single resonances only, and there was no evidence for phase transitions. The T-dependence of the resonance frequency was analysed on the basis of the torsional motion of the BrO_3^- groups.

As part of the characterisation of complexes containing mononuclear $[CuI_3]^{2-}$, $[CuBr_3]^{2-}$ and $[AgI_3]^{2-}$ ions, the n.q.r. frequencies (^{63}Cu, ^{65}Cu, ^{79}Br, ^{81}Br or ^{127}I as appropriate) have been measured at ca. 298K for $[PMePh_3]_2[CuI_3]$, $[PMePh_3]_2[AgI_3]$ and $[PMePh_3]_2[CuBr_2][Br]$.[77] The structure of the bromo-complex was deduced from vibrational spectroscopy, and its n.q.r. results were similar to those for other $[CuBr_2]^-$ salts. Structure and bonding in the complexes was discussed. The T-dependence of the halogen resonance from 4-cyanoanilinium bromide and from the corresponding iodide has been determined from 77-390K.[78] A phase transition was found for the iodide at 99.5±1.0K, characterised by a sudden change in the slope of the T-dependence plot. The asymmetry parameter η, which had a value of zero in the high temperature phase, showed a small but finite increase below the transition, which was deduced to be similar to those observed in anilinium, 4-fluoro- and 4-chloro-anilinium iodides. No phase transition was found for the bromide in this temperature range. Similar studies on $4\text{-MeC}_6H_4NH_3^+X^-$ (X = Br or I) and $4\text{-MeC}_6H_4ND_3^+I^-$ showed two reversible phase transitions when X = I, one between 247 and 258K and one between 174 and 179K.[79] Deuteration produced a slight shift of these transition temperatures and a considerable lowering of the ^{127}I frequencies and e^2Qq/h, but a slight increase in η. X-ray single crystal measurements on the iodide at 295K indicated an orthorhombic space group (P mmn or P $mn2_1$, Z = 2). A phase transition near 281K was detected for X = Br.

A good linear relationship has been found between the n.q.r. frequencies of each of the terminal iodine atoms I_a and I_b and the corresponding bond distance r_{ac} or r_{bc} in the I_3^- anions $I_a\text{-}I_c\text{-}I_b$.[80] A simple extended Hückel MO calculation suggested that this correlation could be used to check the validity of wavefunctions derived for materials containing I-I bonds. Zeeman studies on the $\frac{1}{2}\leftrightarrow\frac{3}{2}$ transition for ^{127}I nuclei in a single crystal of H_5IO_6 at 296K have shown two physically inequivalent iodine sites in the unit cell, with an average η value of 0.25.[81] The maximum field gradient was found to lie along the short bond I-O(2), as suggested in previous work. The e.f.g. parameters calculated by MO (CNDO/2 and INDO) methods as well as those from point charge calculations were compared with the experimental results. N.q.r. data have been given for four crystalline modifications of $KIO_3.HIO_3$.[82] Results for the α and β forms were obtained at 77 and 300K, and for $\gamma\text{-}KIO_3.HIO_3$, $\gamma\text{-}KIO_3.DIO_3$ and $\delta\text{-}KIO_3.HIO_3$ at 77K. The T-dependence of η and the components of the quadrupole coupling tensor from 77-300K was also determined for the α form. The results, together with i.r. spectra and thermograms, were discussed with relation to the crystal structures of the different modifications. The effect of crystal defects on the linewidths of the two resonances for nuclei of $I = \frac{5}{2}$ has been examined theoretically by considering the roles of both the principal component of the field gradient and the asymmetry component.[83] These affected the linewidths for the two transitions to different extents. Comparison with experimental data at 297K for ^{127}I nuclei in 1,4-diiodobenzene, iodic acid, 4-iodobenzoic acid, periodic acid, mercury(II) iodide and sodium periodate, and for ^{187}Re nuclei in potassium perrhenate, showed various ways in which defects interacted with quadrupolar nuclei. The structure and symmetry of intracrystalline fields in some iodates under hydrostatic pressures have been investigated by n.q.r.[84] Similar studies have been reported on the effects of hydrostatic pressure from 0-5 kbar on the n.q.r. parameters of $NaIO_3$, NH_4IO_3, α-HIO_3 and α-DIO_3 (both transitions) in the temperature range 4.2-300K.[85] The main effect of the pressure in HIO_3 appeared to be a decrease in the length of the intermolecular hydrogen bond and an increase in the length of the covalent O-H bond. The effect of applied pressure on the bonding in $LiIO_3$ and Ag_3SbS_3 has been followed by n.q.r.[28,29] The soft ionic bonds between the cations and pyramidal anions in $LiIO_3$ were easily distorted, but almost no stress was transferred to the anions. The pyramid in Ag_3SbS_3, where the covalent bond between the metal atom and the pyramidal group of atoms is similar to bonds within the pyramid, underwent greater distortion. Other results for ^{127}I have been described in the subsection on Group V.[14]

3 Transition Metals and Lanthanides

3.1 Manganese-55 and Cobalt-59.- A coordination sphere model based on electrostatic charges has been described for the e.f.g. in organometallic complexes with fac-a_3b_2c, fac-a_3b_3 or cis-a_3b_2 geometries.[86] The theory was tested by means of e^2Qq/h and η data for a number of cobalt and manganese complexes having these structures. The model was surprisingly successful where b and c were single-atom donor ligands, but yielded poor results for π-olefin and some other strong π-acceptor ligands. It was regarded as a useful addition to the Townes-Dailey approach.

3.2 Copper-63 and -65.- Low temperature (0-1.5K) studies on ^{63}Cu n.q.r. from normal and superconducting specimens of $CeCu_2Si_2$ have shown that in the superconducting sample (T_c ≈0.6K) the T-dependence of the spin-lattice relaxation rate is consistent with a conventional quasiparticle excitation spectrum below T_c.[87] A possible structural phase transition between T_c and 1.2K was deduced. The results from the normal sample were consistent with extensive disorder in the Cu site occupation. Spin-lattice relaxation times as a function of T have been measured for $CeCu_2Si_2$ in both the normal state (1-77K) and the superconducting state below T_c (0.67K) down to 65mK.[88] The relaxation rate decreased drastically just below T_c down to $0.5T_c$, but was almost temperature-independent below $0.3T_c$. Analysis based on the triplet pairing model with an anisotropic energy gap described well the behaviour from T_c to $0.5T_c$, but the results below

$0.3T_c$ remained unexplained. Other results for ^{63}Cu have been described in the subsections on Group V (^{75}As)[24] and Group VII (^{79}Br, ^{127}I).[77]

3.3 Niobium-93.- Spin-echo methods have been used to observe ^{93}Nb n.q.r. from $NbSe_3$ in the frequency range 2.5-5.3 MHz. at various temperatures between 100 and 300K, particularly in the region near to the charge-density-wave transition at 145K.[89] Only six of the twelve resonances expected from the crystal structure above 145K were observed, corresponding to two of the three types of Nb site. These were deduced to arise from two distinct quadrupole fields, with one axial or nearly axial ($\eta \approx 0$), and the other nearly "anti-axial" ($\eta = 1$). Below 145K the former became non-axial (η 0.55). Analysis of relaxation and lineshape data indicated that changes in the spectra could be due principally to a change in the dynamics of the quadrupole field at the nearly axial site, and that dynamics were relatively unimportant at the "anti-axial" site.

3.4 Europium-151 and -153.- Optically detected n.q.r. in the ground state has been observed for ^{151}Eu and ^{153}Eu ions in $KEu(WO_4)_2$ at 4.2K.[90] Ground state splittings of 20 and 31 MHz. for ^{151}Eu and of 51 and 84 MHz. for ^{153}Eu were found. The results were used in conjunction with hole-burning experiments to determine the hyperfine splitting in the ground (^7F_o) and excited (^5D_o) states of Eu^{3+} in the compound. Values of η were also calculated.

3.5 Tantalum-181.- ^{181}Ta n.q.r. from $1\underline{T}$-TaS_2 has been observed by a pulsed method in the commensurate phase at 4.2K, and in the frequency range 75-235 MHz.[91] Three distinct groups of Ta sites were found, one with axial and two with non-axial symmetry. The former was assigned to the α site, and the others to β and γ sites. The results were interpreted on the basis of the star-shaped cluster model, including the effect of three-dimensional ordering of the charge density wave.

3.6 Rhenium-187.- Quasi-elastic neutron scattering experiments at high scattering vectors have shown that the geometry of the re-orientation of the NH_4^+ ion in NH_4ReO_4 does not change between 120 and 300K.[92] Hence the anomalous T-dependence of the n.q.r. frequency could not be explained by a direct effect of reorientation on the e.f.g. at the Re nucleus, and a model was proposed in which the re-orientation of the NH_4^+ ion induced a small rearrangement of the oxygen atoms around the Re with change of temperature. This model gave qualitative agreement with the observed n.q.r. anomaly. Other results for ^{187}Re have been described in the subsection on Group VII (^{35}Cl[63] and ^{127}I[83]).

References

1 A. Rigamonti, Adv. in Phys., 1984, 33, 115.
2 D. Nakamura, J. Mol. Struct., 1983, 111, 341.
3 R.L. Armstrong and D.R. Taylor, Bull. Magn. Reson., 1984, 6, 123.
4 C.R. Brett and D.T. Edmonds, J. Mol. Struct., 1983, 111, 357.
5 R.O. Day, N. Hadipour and J.L. Ragle, J. Magn. Reson., 1984, 59, 373.
6 A. Bielecki, J.B. Murdoch, D.P. Weitekamp, D.B. Zax, K.W. Zilm, H. Zimmerman and A. Pines, J. Chem. Phys., 1984, 80, 2232.
7 R. Ramachandran and P.T. Narasimhan, Proc. Nucl. Phys. Solid State Phys. Symp. 1980 (publ. 1983), 23C, 946.
8 R. Ramachandran and P.T. Narasimhan, J. Mol. Struct., 1983, 111, 53.
9 G.S. Beloglazov and V.S. Grechishkin, Izv. Vyssh. Uchebn. Zaved., Fiz., (Engl. Transl.), 1983, 26, 582.
10 A. Lötz and J. Voitländer, J. Mol. Struct., 1983, 111, 61.
11 D.B. Zax, A. Bielecki, A. Pines and S.W. Sinton, Nature, 1984, 312, 351.
12 T.J. Bastow, I.D. Campbell and S.N. Stuart, J. Mol. Struct., 1983, 111, 71.
13 S.J. Melnick and M.A. Whitehead, J. Mol. Struct., 1983, 111, 397.
14 N.E. Ainbinder and A.S. Azheganov, Radiospektroskopiya (Perm), 1983, 15, 155.
15 S.V. Anferova and V.S. Grechishkin, Russian J. Phys. Chem., 1983, 57, 1534.
16 A. Lötz and J. Voitländer, J. Magn. Reson., 1984, 58, 235.
17 V.S. Grechishkin, R.V. Grechishkina and O.V. Starovoitov, Zh. Strukt. Khim. (Engl. Transl.), 1983, 24, 917.
18 R. Ambrosetti, R. Angelone, A. Colligiani and J. Murgich, J. Mol. Struct., 1983, 111, 255.
19 A.V. Parygin, G.V. Chikvaidze, G.K. Semin, B.V. Lokshin and Ya.A. Eiduss, Phys. Status Solidi A, 1984, 84, K101.
20 D.J. Treacy, S.G. Greenbaum, U. Strom and P.C. Taylor, J. Non-Cryst. Solids, 1983, 59-60, 847.
21 A.V. Bondar', V.S. Vikhnin, S.M. Ryabchenko and V.E. Yachmenev, Sov. Phys. Solid State, 1983, 25, 1497.
22 A. Gordon and J. Genossar, Physica B+C, 1984, 125, 53.
23 J. Stankowski and M. Zdanowska-Frączek, J. Mol. Struct., 1983, 111, 11.
24 J. Ribas, M. Durand and G. Jugie, J. Mol. Struct., 1983, 111, 281.
25 K.S. Pickens, D.I. Bolef, M.R. Holland and R.K. Sundfors, Phys. Rev. B, 1984, 30, 3644.
26 L.A. Zemnukhova, R.L. Davidovich, P.S. Gordienko, J. Grigas, A.N. Kovrianov, S.I. Kuznetsov, T.A. Kaidalova and V. Urbonavičius, Phys. Status Solidi A, 1983, 80, 553.
27 R.L. Davidovich, P.S. Gordienko, J. Grigas, T.A. Kaidalova, V. Urbonavičius and L.A. Zemnukhova, Phys. Status Solidi A, 1984, 84, 387.
28 D.F. Baisa and S.V. Mal'tsev, Sov. J. Chem. Phys., 1982 (2), 301.
29 D.F. Baisa and S.V. Mal'tsev, Sovrem. Metody YaMR EPR Khim. Tverd. Tela [Mater. Vses. Koord. Soveshch.] 3rd 1982, 105; Chem. Abstr., 1983, 99, 146436.
30 I.Yu. Amiantov, A.A. Boguslavsky and G.K. Semin, J. Mol. Struct., 1984, 117, 275.
31 R.L. Davidovich, M.A. Medkov, E.B. Bogomolova and L.A. Zemnukhuva, Izv. Akad. Nauk SSSR, Ser. Khim., (Engl. Transl.), 1984 (5), 895.
32 J. Seliger, V. Žagar and R. Blinc, J. Chem. Phys., 1984, 81, 3247.
33 J. Brookeman, A. Rigamonti and N. Sullivan, Ferroelectrics, 1984, 54, 393.
34 N.S. Zefirov, A.S. Koz'min and V.D. Sorokin, J. Org. Chem., 1984, 49, 4086.
35 B.V. Dresvyankin and A.B. Nikol'skii, Radiospektroskopiya (Perm), 1983, 15, 145.
36 A. Radde, L. Kolditz, S.I. Troyanov, Yu.T. Struchkov and M.Yu. Antipin, Z. Chem., 1984, 24, 101.
37 M.L. Buess, P.J. Bray and D.W. Sheppard, Org. Magn. Reson., 1984, 22, 67.

38 S.V. Volkov, Z.A. Fokina and N.I. Timoshchenko, Rev. Chim. Miner., 1983, 20, 776.
39 J.D. Graybeal, J. Mol. Struct., 1983, 111, 329.
40 V.V. Petrov, V.G. Pitsyuga, V.A. Gordeev, A.V. Bogdanova, M.A. Bagina and A.Yu. Khalakhan, Sov. Phys. Solid State, 1983, 25, 1995.
41 L. Piekara-Sady, S. Idziak, M. Krupski, E. Dynowska, M. Máckowiak and J. Stankowski, Physica B+C, 1984, 123, 211.
42 F. Milia, Phys. Lett. A, 1984, 102, 317.
43 F. Milia, R. Blinc and S. Žumer, Solid State Commun., 1984, 50, 1019.
44 V.P. Feshin, P.A. Nikitin, M.G. Voronkov, T.K. Gar, N.A. Viktorov, S.N. Gurkova, A.I. Gusev and V.I. Shiryaev, J. Gen. Chem. USSR, 1984, 54, 572.
45 V.P. Feshin, P.A. Nikitin, M.G. Voronkov, T.K. Gar, N.A. Viktorov, S.N. Gurkova and A.I. Gusev, Dokl. Akad. Nauk SSSR, (Engl. Transl.), 1984, 274, 112.
46 I.P. Biryukov, Dokl. Akad. Nauk SSSR, (Engl. Transl.), 1983, 271, 571.
47 A.D. Gordeev, I.A. Kyuntsel' and G.B. Soifer, Radiospektroskopiya (Perm), 1983, 15, 174.
48 G.E. Kibrik, É.S. Kozlov, I.A. Kyuntsel', V.A. Mokeeva and G.B. Soifer, J. Gen. Chem. USSR, 1984, 54, 702.
49 G.V. Dolgushin, I.M. Lazarev, V.P. Feshin and M.G. Voronkov, J. Gen. Chem. USSR, 1984, 54, 1276.
50 F. Milia, Solid State Commun., 1984, 51, 625.
51 K.R. Sridharan, J. Ramakrishna, K. Ramachandran and S.S. Krishna Murty, Polyhedron, 1984, 3, 867.
52 D.J. Pusiol and A.H. Brunetti, J. Phys. C, 1984, 17, 4487.
53 G.K. Semin, S.I. Kuznetsov, T.L. Khotsyanova, V.L. Solozhenko and Ya.A. Kalashnikov, Chem. Phys. Lett., 1984, 109, 500.
54 C. Dimitropoulos and J. Pelzl, J. Mol. Struct., 1983, 111, 163.
55 J. Pelzl, V. Waschk, Y.M. Seo and C. Dimitropoulos, J. Mol. Struct., 1983, 111, 363.
56 M. Crowley, J. Brookeman and A. Rigamonti, Ferroelectrics, 1984, 54, 405.
57 R.C. Zamar, A.H. Brunetti and D.J. Pusiol, J. Mol. Struct., 1983, 111, 171.
58 H. Niki, Y. Ikushima, Y. Kikuchi and R. Igei, Bull. Coll. Sci. Univ. Ryukyus, 1983, 36, 27.
59 H. Nabeshima, Physics Reports Kumamoto Univ., 1983, 6, 33.
60 C. Segebarth and J. Jeener, Phys. Rev. B, 1984, 29, 1176.
61 E.P.A. Sullivan, J. Mol. Struct., 1983, 111, 103.
62 S.-E. Gima, Y. Furukawa, R. Ikeda and D. Nakamura, J. Mol. Struct., 1983, 111, 189.
63 M. Mackowiak and R.J.C. Brown, J. Mol. Struct., 1983, 111, 233.
64 S.J. Melnick and M.A. Whitehead, J. Mol. Struct., 1984, 116, 217.
65 R.M. Morra and R.L. Armstrong, J. Magn. Reson., 1984, 60, 1.
66 R.M. Morra and R.L. Armstrong, Can. J. Phys., 1984, 62, 1152.
67 D.R. Taylor, R.M. Morra and R.L. Armstrong, J. Mol. Struct., 1983, 111, 239.
68 G. Wulfsberg, J. Yanisch, R. Meyer, J. Bowers and M. Essig, Inorg. Chem., 1984, 23, 715.
69 R. Meyer, J. Gagliardi Jr., and G. Wulfsberg, J. Mol. Struct., 1983, 111, 311.
70 V.P. Feshin, M.G. Voronkov, L.S. Romanenko, L.P. Ignat'eva and G.V. Dolgushin, J. Gen. Chem. USSR, 1984, 54, 1175.
71 Z.A. Fokina, S.I. Kuznetsov, N.I. Timoshchenko, V.L. Kolesnichenko, E.V. Bryukhova and G.K. Semin, Izv. Akad. Nauk SSSR, Ser. Khim., (Engl. Transl.), 1984 (4), 830.
72 R. Valli, S. Ramaprabhu and K.V.S. Rama Rao, J. Mol. Struct., 1983, 111, 317.
73 M. Gourdji, A. Péneau, M. Genet and L. Guibé, J. Mol. Struct., 1983, 111, 27.
74 K. Yamada, N. Weiden and A. Weiss, J. Mol. Struct., 1983, 111, 217.
75 S. Ramaprabhu and K.V.S. Rama Rao, Cryst. Lattice Defects Amorphous Mater., 1984, 10, 229.
76 S. Ramaprabhu, R. Valli and K.V.S. Rama Rao, Pramana, 1984, 23, 141.
77 G.A. Bowmaker, G.R. Clark, D.A. Rogers, A. Camus and N. Marsich, J. Chem. Soc., Dalton Trans., 1984, 37.

78 V.G. Krishnan and A. Weiss, J. Mol. Struct., 1983, 111, 379.
79 V. Seshu Bai and A. Weiss, Z. Naturforsch. A, 1984, 39A, 366.
80 Y. Yoshioka, N. Nakamura and H. Chihara, J. Mol. Struct., 1983, 111, 151.
81 S. Ramaprabhu, K.V.S. Rama Rao and A. Weiss, J. Mol. Struct., 1983, 111, 295.
82 A.M. Petrosyan, A.F. Volkov, V.A. Shishkin and Yu.N. Venevtsev, Sov. Phys. Crystallogr., 1983, 28, 542.
83 P.J. Bryant and S. Hacobian, J. Mol. Struct., 1983, 111, 201.
84 A.I. Barabash, Spektroskopiya Molekul i Kristallov, Kiev, 1983, 243; Ref. Zh. Khim., 1984, Abstr. No. 4B1305; Chem. Abstr., 1984, 101, 121788.
85 D.F. Baisa, A.I. Barabash and I.G. Vertegel, Sov. J. Chem. Phys., 1982 (3), 550.
86 T.B. Brill, J. Mol. Struct., 1983, 111, 269.
87 D.E. MacLaughlin, C. Tien, L.C. Gupta, J. Aarts, F.R. de Boer and Z. Fisk, Phys. Rev. B, 1984, 30, 1577.
88 Y. Kitaoka, K. Ueda, T. Kohara and K. Asayama, Solid State Commun., 1984, 51, 461.
89 B.H. Suits and C.P. Slichter, Phys. Rev. B, 1984, 29, 41.
90 A.J. Silversmith and N.B. Manson, J. Phys. C, 1984, 17, L97.
91 H. Nishihara and S. Tanaka, J. Phys. Soc. Jpn., 1984, 53, 1610.
92 R.M. Richardson and J. Howard, Chem. Phys., 1984, 86, 235.

3

Rotational Spectroscopy

BY S. CRADOCK

1 Introduction

There seems to have been a significant upsurge in activity in this area in 1984, with about 50% more reports found than for 1983. This has required rather more selectivity in the handling of the material, and I have tried to emphasise reports with some bearing on molecular structure at the expense of those dotting i's and crossing t's concerning well known and perhaps over-exposed molecules. One technical highlight is the direct observation using a diode laser spectrometer of dipole-allowed pure rotation lines of HD in the IR (see Table 1).

2 van der Waals and Hydrogen-Bonded Complexes

Further studies have been reported on rare gas/molecule complexes such as Ar.HF[1], where the equilibrium structure was established using new data for ^{36}Ar and ^{38}Ar isotopes, as well as existing data for ^{40}Ar.HF and DF, HCl complexes of Ne and Kr[2], Ar.HCN[3], which gives a very intense spectrum using molecular beam electric resonance spectroscopy (mbers) techniques, suggesting strong bonding, but clearly has a very non-rigid structure from the observed spectrum, and Ar.C_2N_2[4], whose structure is T-shaped with equivalent N atoms (from the observed nuclear quadrupole couplings). We may also note the reporting of more detailed studies by pulsed-nozzle/Fourier Transform microwave spectroscopy of complexes of Hg atoms with HCl + DCl[5], which have very similar structures to those found for the rare gas/hydrogen halide complexes.

1. B.L. Cousins, S.C. O'Brien and J.M. Lisy, J.Phys.Chem., 1984, 88, 5142.
2. E.J. Campbell, W.G. Read and J.A. Shea, Mol.Phys., 1984, 51, 1493.
3. K.R. Leopold, G.T. Fraser, F.J. Lin, D.D. Nelson Jr., and W. Klemperer, J.Chem.Phys., 1984, 81, 4922.
4. W.L. Ebenstein and J.S. Muenter, J.Chem.Phys., 1984, 80, 1417.
5. J.A. Shea and E.J. Campbell, J.Chem.Phys., 1984, 81, 5326.

A fascinating array of complexes between two molecules is now being studied, and both mbers and pulsed-nozzle/Fourier Transform microwave spectroscopy are being used. Mbers measurements of the pure rotational spectra of complexes of BF_3 with cyanogen, carbon dioxide and nitrous oxide[6] are interpreted in terms of a symmetric top structure for the C_2N_2 complex, but an asymmetric top structure for the CO_2 complex and (probably) the N_2O species. Dimers of CO_2 and N_2O have been detected in the infrared using a tunable colour centre laser[7], being identified by the rotational structure of a band near $\nu_1 + \nu_3$ of the monomer in each case.

The $CO_2.NH_3$ complex (studied by mbers[8]) has the nitrogen atom co-ordinated to the carbon of CO_2, but apparently slightly asymmetrically, with $<NCO = 85°$. The $CO_2.H_2O$ complex has also been reported[9]; it is planar, with an oxygen-carbon "bond".

Many complexes are clearly hydrogen-bonded, unlike those mentioned so far. They include OC..HCN, which is linear[10], and $^{15}N_2$..DF, for which ^2D hyperfine structure has been resolved using FT microwave spectroscopy.[11] HCN complexes with NH_3[12], PH_3[13] ($HC^{15}N$ isotope) and Me_3P[14] have all been studied; they have essentially linear Z..H-C hydrogen bonds (Z = N,P). The structure of CH_3OH..HCl has been deduced[15] from its MW spectrum; the C-O..(H)-Cl angle is close to the tetrahedral angle, as expected, but the three bonds at oxygen are close to coplanar, probably with a small inversion barrier.

Acetylene acts as a hydrogen-bonding ligand in complexes with NH_3[16] and H_2O[17], but methyl acetylene..HF[18] has the HF hydrogen-bonded to the triple bond. Complexes of H_2S with HF[19], HCl[20] and HCN[21] have the stronger acid H-bonded to S more or less perpendicular to the plane of the H_2S molecule.

6. K.R. Leopold, G.T. Fraser and W. Klemperer, J.Amer.Chem.Soc., 1984, 106, 897.
7. R.E. Miller and R.O. Watts, Chem.Phys. Letters, 1984, 105, 409.
8. G.T. Fraser, K.R. Leopold and W. Klemperer, J.Chem.Phys., 1984, 81, 2577.
9. K.I. Peterson and W. Klemperer, J.Chem.Phys., 1984, 80, 2439.
10. E.J. Goodwin and A.C. Legon, Chem.Phys., 1984, 87, 81.
11. A.C. Legon and L.C. Willoughby, Chem.Phys. Letters, 1984, 109, 502.
12. G.T. Fraser, K.R. Leopold, D.D. Nelson Jr., A. Turig and W. Klemperer, J.Chem.Phys., 1984, 80, 3073.
13. A.C. Legon and L.C. Willoughby, Chem.Phys. Letters, 1984, 111, 566.
14. H.L. Hirani, A.C. Legon, D.J. Millen and L.C. Willoughby, J.Mol.Struct., 1984, 125, 171.
15. P. Cope, A.C. Legon and D.J. Millen, Chem.Phys. Letters, 1984, 112, 59.
16. G.T. Fraser, K.R. Leopold and W. Klemperer, J.Chem.Phys., 1984, 80, 1423.
17. K.I. Peterson and W. Klemperer, J.Chem.Phys., 1984, 81, 3842.
18. J.A. Shea, R.E. Bumgarner and G. Henderson, J.Chem.Phys., 1984, 80, 4605.
19. L.C. Willoughby, A.J. Fillery-Travis and A.C. Legon, J.Chem.Phys., 1984, 81, 20.
20. E.J. Goodwin and A.C. Legon, J.C.S. Faraday II, 1984, 80, 51.
21. E.J. Goodwin and A.C. Legon, J.C.S. Faraday II, 1984, 80, 1669.

The HF dimer has been studied[22] by mbers, and also a more detailed study of the IR bands of $(HF)_2$ and $(DF)_2$ using a difference-frequency IR laser[23] has allowed an analysis of the tunnelling motion which makes the two components of the apparently unsymmetrical dimer effectively equivalent (Scheme 1). A similar dynamic structure is found[24] for $(HCl)_2$.

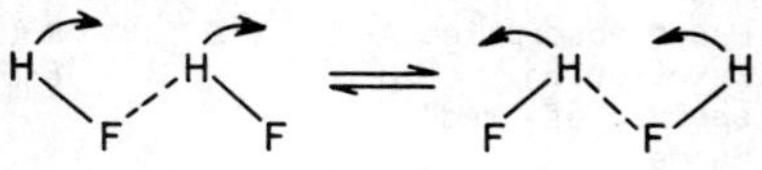

Scheme 1

References

22. B.J. Howard, T.R. Dykes and W. Klemperer, J.Chem.Phys., 1984, 81, 5417.
23. A.S. Pine, W.J. Lafferty and B.J. Howard, J.Chem.Phys., 1984, 81, 2939.
24. N. Ohashi and A.S. Pine, J.Chem.Phys., 1984, 81, 73.

3 Diatomic Species

Table 1

Species	Ref.	Technique	State(s)	Information derived
a) 2-electron species				
HD	a	Diode laser IR absorption	$X\ ^1\Sigma^+$ V=0	R(3),R(5),R(6) line frequencies
$^{6,7}Li^{1,2}H$	b	Sub-mmw (~260 GHz) (~520 GHz)	$X\ ^1\Sigma^+$ v=0 and 1	$1 \leftarrow 0$ and $2 \leftarrow 1$ pure rotation lines; unified molecular constants
KH	c	Diode Laser IR absorption	$X\ ^1\Sigma^+$ $v=0 \rightarrow 1$	Rot. anal. of vib. bands r_e = 224.1152(16)pm
CsH	d	Laser-induced fluorescence (l.i.f.)	$A^1\Sigma^+ \rightarrow X\ ^1\Sigma^+$	Dunham coefficients, dissociation energy
Na_2	e	Laser absorption in supersonic jet	$A\ ^1\Sigma_u^+ \leftarrow X\ ^1\Sigma_g^+$	Rot. anal.
K_2	f	L.i.f.	$B^1\Pi_u\ \ X\ ^1\Sigma_g^+$	Revised assignments, constants for both states
Cs_2	g	L.i.f.	$E\ ^1\Sigma_u^+,(1)^1\Pi_g$	r_e for both states r(E) = 534.0733pm r(1) = 569.7722pm

(a) P. Essenwanger and H.P. Gush, Can.J.Phys., 1984, 62, 1680.
(b) G.M. Plummer, E. Herbst and F.C. De Lucia, J.Chem.Phys., 1984, 81, 4893.
(c) N.N. Haese, Di-Jia Liu and R.S. Altman, J.Chem.Phys., 1984, 81, 3766.
(d) C. Crepin, J. Verges and C. Amiot, Chem.Phys.Letters, 1984, 112, 10.
(e) R.N. Ahmad-Bitar and A.S. Al-Ayash, J.Mol.Spectrosc., 1984, 106, 299.
(f) F. Engelke, H. Hage and U. Schühle, Chem.Phys.Letters, 1984, 106, 535.
(g) C. Amiot, C. Crepin and J. Verges, J.Mol.Spectrosc., 1984, 107, 28.

Species	Ref.	Technique	State(s)	Information derived
NaCs	h	L.i.f. and sub-Doppler polarisation spectra	$D^1\Pi \rightarrow X^1\Sigma^+$	Ground-state constants; $r_e(X)$=385pm, $r_e(D)$=422.6pm
b) 3-electron species				
CaH	i	Laser absorption /excitation spectra of "red" bands	$A^2\Pi$, $B^2\Sigma$ $\leftarrow X^2\Sigma$	Rot. anal.; equil. consts. for X
$Hg^{1,2}H$	j	L.i.f. in supersonic jet	$A_1\ ^2\Pi_{\frac{1}{2}}$	Equilib. constants, r_e=158.3(2)pm
c) 4-electron species				
GeH^+	k	Emission	$a^3\Pi - X^1\Sigma^+$	Rot.consts. for a and X
Be_2	l	L.i.f.	$A^1\Sigma_u^+ \leftrightarrow X^1\Sigma_g^+$	r_o for both states (r(x)=246, r(A)=216pm)
Sr_2	m	L.i.f.	$A^1\Sigma_u^+ \rightarrow X^1\Sigma_g^+$	r_e for both states (r(x)=445, r(A)=395pm)
d) 5-electron species				
CH	n	IR absorption (difference frequency tunable laser)	$X\ ^2\Pi$ v=0 $\rightarrow$ 1	Rot. anal. B_e; ω_e=2858.38(9)cm^{-1} r_e=111.8096(44)pm
SiH	o	Laser magnetic resonance (lmr) near 5 microns	$X\ ^2\Pi$ v = 0 $\rightarrow$1	ν_o, α_e
SiH	p	Far. I.R.l.m.r.	$X\ ^2\Pi$ v=0	Rotation, spin-rotation and Zeeman parameters
e) 6-electron species				
NH	q	L.i.f.	$A^3\Pi \leftarrow X^3\Sigma^-$	Hyperfine and n.q. coupling constants
PH	r	Diode laser IR absorption	$X\ ^3\Sigma^-$ v=0 $\rightarrow$1	Rot.anal. B_e, ω_e, D_e, α_e, β_e

(h) U. Diemer, H. Weickenmeier, M. Wahl and W. Demtröder, Chem.Phys.Letters, 1984, 104, 489.
(i) H. Martin, J.Mol.Spectrosc., 1984, 108, 66.
(j) S. Mayama, S. Hiraoka and K. Obi, J.Chem.Phys., 1984, 81, 4760.
(k) M. Tsuji, S. Yamagushi, S. Shimada and Y. Nishimura, J.Mol.Spectrosc., 1984, 103, 498.
(l) V.E. Bondybey and J.H. English, J.Chem.Phys., 1984, 80, 568.
(m) G. Gerber, R. Möller and H. Schneider, J.Chem.Phys., 1984, 81, 1538.
(n) K.G. Lubic and T. Amano, J.Chem.Phys., 1984, 81, 1655.
(o) J.M. Brown and D. Robinson, Mol.Phys., 1984, 51, 883.
(p) J.M. Brown, R.F. Curl and K.M. Evenson, J.Chem.Phys., 1984, 81, 2884.
(q) W. Ubachs, J.J. ter Meulen and A. Dymanus, Can.J.Phys., 1984, 62, 1374.
(r) J.R. Anacona, P.B. Davies and P.A. Hamilton, Chem.Phys.Letters., 1984, 104, 269.

Species	Ref.	Technique	State(s)	Information derived
$P^{1,2}H$	s	Far-IR l.m.r.	$X\ {}^3\Sigma^-$ v=0,1	Rot. consts. r_e=142.140(22)pm
PH	t	Emission	$b\ {}^1\Sigma^+ \rightarrow X^3\Sigma^-$	r_e(b)=141.78pm
AsH	u	Far-IR l.m.r.	$X\ {}^3\Sigma^-$	Ground-state rot. constants & hyperfine constants
$S^{1,2}H^+$	v	Emission spectra	$A\ {}^3\Pi \rightarrow X^3\Sigma^-$	Rot.anal. equil. consts r_e(X)=136.38pm r_e(A)=150.08pm
f) 7-electron species				
$O^{1,2}H$	w	IR absorption (difference frequency tunable laser)	$X\ {}^2\Pi$ v=0 → 1	Many constants, r_e= 96.9628(9)pm(OH) 96.9680(10)pm(OD)
HF^+	x	L.m.r.	$X\ {}^2\Pi$ v=0 → 1	Ground-state constants (B_o, D_o, etc.)
DCl^+	y	L.m.r.	$X\ {}^2\Pi$ v=0 → 1	B_1, ΔG_{vib}
g) 8-electron species				
DBr	z	IR absorption (heterodyne frequency measurements)	$X\ {}^1\Sigma^+$ v=0 → 1	Improved mol. constants
ArH^+, KrH^+	aa	Emission (FTIR)	$X\ {}^1\Sigma^+$ v=1 → 0	Rot. anal. B_e, ω_e, etc.
KrD^+	bb	Microwave absorption	$X\ {}^1\Sigma^+$ v=0	1 ← 0 rot. line; r_e=142.11911(7)pm
BeO	cc	Reanalysis of emission spectra	$A\ {}^1\Pi - X\ {}^1\Sigma^+$	Perturbation analysis, many mol. constants
MgO	dd	L.i.f. and microwave optical double resonance (modr) spectra	$B\ {}^1\Sigma^+ - X\ {}^1\Sigma^+$	Molecular constants for both states

(s) N. Ohashi, K. Kawaguchi and E. Hirota, J.Mol.Spectrosc., 1984, 103, 337.
(t) A.T. Droege and P.C. Engelking, J.Chem.Phys., 1984, 80, 5926.
(u) K. Kawaguchi and E. Hirota, J.Mol.Spectrosc., 1984, 106, 423.
(v) J. Rostas, M. Horani, J. Bion, D. Daumont and J. Malicet, Mol. Phys., 1984, 52, 1431.
(w) T. Amano, J.Mol.Spectrosc., 1984, 103, 436.
(x) D.C. Hovde, E. Schäfer, S.E. Strahan, C.A. Ferrari, D. Ray, K.G. Lubic and D.J. Saykally, Mol.Phys., 1984, 52, 245.
(y) A. Hinz, W. Bohle, D. Zeitz, J. Werner, W. Seeban and W. Urban, Mol. Phys., 1984, 53, 1017.
(z) J.S. Wells, D.A. Jennings and A.G. Maki, J.Mol.Spectrosc., 1984, 107, 48.
(aa) J.W.C. Johns, J.Mol.Spectrosc., 1984, 106, 124.
(bb) H.E. Warner, W.T. Conner and R.C. Woods, J.Chem.Phys., 1984, 81, 5413.
(cc) H. Lavendy, B. Pouilly and J.M. Robbe, J.Mol.Spectrosc., 1984, 103, 379.
(dd) Y. Azuma, T.R. Dyke, G.K. Gerke and T.C. Steimle, J.Mol.Spectrosc., 1984, 108, 137.

Species	Ref.	Technique	State(s)	Information derived
MgO	ee	Millimeter-wave (mmw) absorption	X $^1\Sigma^+$ v=0,1	Molecular constants
KCl	ff	Molecular beam electric resonance spectroscopy (mbers)	X $^1\Sigma^+$ v=0,1	Hyperfine structure; rot. and vib. dependence of n.q. couplings
h) 9-electron species				
CO^+	gg	L.i.f.	A $^2\Pi \rightarrow$ X $^2\Sigma^+$ v'=0 $\rightarrow$ v"=0	Rot.anal; B_o for both states
N_2^+	hh	L.i.f.	A $^2\Pi_u \rightarrow$ X $^2\Sigma_g^+$	Rot.anal; B_e
BO	ii	Visible emission	A $^2\Pi \rightarrow$ X $^2\Sigma^+$	Rot.anal.; equil. consts.
$^{13}C^{14}N$	jj	Mmw absorption (109 & 218GHz)	X $^2\Sigma^+$ v=0	1 $\leftarrow$ 0 and 2 $\leftarrow$ 1 lines, with fine and hyperfine structure constants
SiN	kk	Emission	B $^2\Sigma^+ \rightarrow$ A $^2\Pi_i$	Equilib. consts; T_e(A) = 994 cm^{-1}
CaF	ll	Molecular beam laser/radio-frequency double-resonance Stark spectroscopy	X $^2\Sigma^+$ v=0	Dipole moment = 3.07(7)D
CaI	mm	Molecular beam laser/radio-frequency double-resonance spectroscopy	B $^2\Sigma^+$-X $^2\Sigma^+$	Spin-rotation and hyperfine constants
SrCl	nn	Sub-Doppler polarisation spectroscopy	B $^2\Sigma^+$-X $^2\Sigma^+$	Rot.anal.; Dunham coefficients
SrCl	oo	Microwave/optical polarisation spectroscopy	X $^2\Sigma^+$ v=0	Magnetic-dipole and electric-quadrupole coupling constants
BaI	pp	Emission	C $^2\Pi \rightarrow$ X $^2\Sigma^+$ v'=0 $\rightarrow$ v"=0	Rot.anal.; molecular constants

(ee) T.C. Steimle, Y. Azuma and P.G. Carrick, Astrophys.J., 1984, 277, L21.
(ff) D. Nitz, J. Cederberg, A. Kotz, K. Hetzler, T. Aakve and T. Walhout., J.Mol.Spectrosc., 1984, 108, 6.
(gg) R.D. Brown, R.G. Dittman and D.C. McGilvery, J.Mol.Spectrosc., 1984, 104, 337.
(hh) T.A. Miller, T. Suzuki and E. Hirota, J.Chem.Phys., 1984, 80, 4671.
(ii) J.A. Coxon, S.C. Foster and S. Naxakis, J.Mol.Spectrosc., 1984, 105, 465.
(jj) M. Bogey, C. Demuynck and J.L. Destombes, Can.J.Phys., 1984, 62, 1248.
(kk) S.C. Foster, J.Mol.Spectrosc., 1984, 106, 369.
(ll) W.J. Childs, L.S. Goodman, U. Nielsen and V. Pfeufer, J.Chem.Phys., 1984, 80, 2283.
(mm) W.J. Childs, G.L. Goodman, L.S. Goodman and V. Pfeufer, J.Mol.Spectrosc., 1984, 107, 94.
(nn) W.E. Ernst and J.O. Schröder, J.Chem.Phys., 1984, 81, 136.
(oo) W.E. Ernst, J.O. Schröder and T. Toerring, Chem. Phys. Letters, 1984, 109, 175.
(pp) M.A. Johnson, C. Noda, J.S. McKillop and R.N. Zare, Can.J.Phys., 1984, 62, 1467.

Species	Ref.	Technique	State(s)	Information derived
i) 10-electron species				
VO^+	qq	Emission in near IR (FTIR)	$^1\Sigma^+ \rightarrow {}^1\Sigma^+$ $^1\Delta \rightarrow {}^1\Delta$	Rot.anal.; r_e for all 4 states
ThO	rr	Visible emission	Yellow bands	Rot. anal.
N_2	ss	UV absorption (1st positive bands)	$B\ {}^3\Pi_g \leftarrow A\ {}^3\Sigma_u^+$ $v'=12-15 \leftarrow v''=8-11$	Rot. anal.
^{14}CO	tt	IR absorption (2.4-3.0)(FTIR)	$X\ {}^1\Sigma^+$ $\Delta v=2$, $v'=0-13$	Rot.anal. using calculated distortion constants
$^{13}C^{18}O$	uu	Emission	$B\ {}^1\Sigma^+ \rightarrow A\ {}^1\Pi$	Rot.anal.; $r_e(A)=123.50$pm $r_e(B)=111.98$pm
CS	vv	IR emission	$X\ {}^1\Sigma^+$ $\Delta v=2$, $v''=0-6$	Vib. and rot. constants and Dunham coefficients
GeS	ww	Microwave data reanalysed	$X\ {}^1\Sigma^+$	Improved molecular constants
GaF	xx	UV absorption	$C\ {}^1\Pi \leftarrow X^1\Sigma$	Equil. constants
AlBr	yy	Emission and Absorption	$a\ {}^3\Pi \leftrightarrow X\ {}^1\Sigma^+$ $A\ {}^1\Pi$	Improved molecular constants
V_2	zz	2-photon ionisation laser spectroscopy	$A\ {}^3\Pi_u \leftarrow X\ {}^3\Sigma_g^-$	Rot.anal.; $r''_e=177$pm
j) 11-electron species				
O_2^+	aaa	Emission	$A\ {}^2\Pi_u \rightarrow X\ {}^2\Pi_g$	Rot.anal.; mol. constants
SO^+	bbb	Emission	$A\ {}^2\Pi \rightarrow X\ {}^2\Pi$	Rot.anal. equil. consts; $r_e(X)=142.50(3)$pm $r_e(A)=165.78(1)$pm
SO^+	ccc	Emission	$A\ {}^2\Pi \rightarrow X\ {}^2\Pi$	Rot.anal.; mol. constants

(qq) A.J. Merer, A.S-C. Cheung and A.W. Taylor, J.Mol.Spectrosc., 1984, 108, 343.
(rr) G. Edvinsson and A. Lagerqvist, Phys.Scr., 1984, 30, 309.
(ss) R.D. Verma, Can.J.Phys., 1984, 62, 414.
(tt) C. Amiot, J. Verges and C.R. Vidal, J.Mol.Spectrosc., 1984, 103, 364.
(uu) C.V.V. Prasad, G.L. Bhale and S. Paddi Reddy, J.Mol.Spectrosc., 1984, 104, 165.
(vv) R.J. Winkel, Jr., S.P. Davis, R. Peeyner and J.W. Brault, Can.J.Phys., 1984, 62, 1414.
(ww) A.C. Le Flock and J. Masson, J.Mol.Spectrosc., 1984, 103, 408.
(xx) W.B. Griffith, Jr., G.A. Bickel, H. D. McSwiney and C. Weldon Matthews, J.Mol.Spectrosc., 1984, 104, 343.
(yy) W.B. Griffith, Jr., C. Weldon Matthews, J.Mol.Spectrosc., 1984, 104, 347.
(zz) P.R.R. Langridge-Smith, M.D. Morse, G.P. Hansen, R.E. Smalley and A.J. Merer, - J.Chem.Phys., 1984, 80, 593.
(aaa) J.A. Coxon and M.P. Haley, J.Mol.Spectrosc., 1984, 108, 119.
(bbb) J.L. Hardwick, Yin Luo, D.H. Winicur and J.A. Coxon, Can.J.Phys., 1984, 62, 1792.
(ccc) J.A. Coxon and S.C. Foster, J.Mol.Spectrosc., 1984, 103, 281.

Species	Ref.	Technique	State(s)	Information derived
SbO	ddd	Emission	$A\ ^2\Pi \rightarrow X\ ^2\Pi$	Rot.anal.; equil. constants; $r_e(X)$=182.3pm $r_e(A)$=203.4 pm
NS	eee	Far-IR l.m.r.	$X\ ^2\Pi_{\frac{3}{2}}$	Hyperfine constants, Λ-doubling
SiCl	fff	Microwave	$X\ ^2\Pi_{\frac{1}{2},\frac{3}{2}}$ v=0	Ground-state constants
CeF	ggg	Absorption (568 nm band)	$X\ ^2\Phi_{\frac{5}{2}}$?	Ground-state consts. r_o=204.84(4)pm
k) 12-electron species				
O_2	hhh	L.m.r.	$X\ ^3\Sigma_g^-$	Ground-state consts.
$^{17}O_2$	iii	Mmw absorption	$X\ ^3\Sigma_g^-$	Ground-state consts.
S_2	jjj	L.i.f.	$B^3\Sigma_u^- \rightarrow X^3\Sigma_g^-$ v'=2 → v"=3	Analysis of perturbations by B" $^3\Pi_u$
S_2	kkk	L.i.f. from supersonic jet	$B\ ^3\Sigma_u^- \rightarrow X^3\Sigma_g^-$ and $B''^3\Pi_u \rightarrow X^3\Sigma_g^-$	Rot. consts. for B", v=3-18
CrO	lll	FTIR emission	$A\ ^5\Sigma \rightarrow X\ ^5\Pi$ v'=0 → v"=0	Constants for both states $r_o(X)$=162.1pm, $r_o(A)$=166.2pm
l) 13-electron species				
ClO	mmm	Mmw absorption (up to 723 GHz)	$X\ ^2\Pi_{\frac{1}{2},\frac{3}{2}}$	Rot., distortion and hyperfine consts.
BrO	nnn	Diode laser IR absorption	$X\ ^2\Pi$ v=0 → 1	Rot. & dist. consts. for both vib. states; r_e = 172.072(51)pm
ReO	ooo	Emission	Several bands, states not identified	Rot. analysis

(ddd) W.J. Balfour and R.S. Ram, J.Mol.Spectrosc., 1984, 105, 246.
(eee) J.R. Anacona and P.B. Davies, Chem.Phys.Letters., 1984, 108, 128.
(fff) M. Tanimoto, S. Saito, Y. Endo and E. Hirota, J.Mol.Spectrosc., 1984, 103, 330.
(ggg) R.M. Clements and R.F. Barrow, J.Mol.Spectrosc., 1984, 107, 119.
(hhh) M. Mizushima, L.R. Zink and K.M. Evenson, J.Mol.Spectrosc., 1984, 107, 395.
(iii) A. Cazzoli, C. degli Esposti and B.M. Landsberg, Nuovo.Cim.Soc.Ital.Fis.D., 1984, 3D, 341.
(jjj) M. Heaven, T.A. Miller and V.E. Bondybey, J.Chem.Phys., 1984, 80, 51.
(kkk) Y. Matsumi, T. Munakata and T. Kasuya, J.Chem.Phys., 1984, 81, 1108.
(lll) A.S.-C. Cheung, W. Zyrnicki and A.J. Merer., J.Mol.Spectrosc., 1984, 104, 315.
(mmm) E.A. Cohen, H.M. Pickett and M. Geller, J.Mol.Spectrosc., 1984, 106, 430.
(nnn) J.E. Butler, K. Kawaguchi and E. Hirota, J.Mol.Spectrosc., 1984, 104, 372.
(ooo) W.J. Balfour and R.S. Ram, Can.J.Phys., 1984, 62, 1524.

Species	Ref.	Technique	State(s)	Information derived
m) 14-electron species				
BrF	ppp	Emission 340-360 nm	D'(2) → A'(2)	Rot. analysis; r_e(A')=207pm, r_e(D')=266pm
IF	qqq	L.i.f.	$B^3\Pi(0^+) \rightarrow X^1\Sigma^+$ v'=0-4 → v"=3-9	Rot. analysis, Dunham coefficients
ICl	rrr	Optical/optical double resonance	$E(0^+) \leftarrow B$ $f(0^+) \leftarrow B$	Dunham coefficients
Br_2	sss	Optical/optical double resonance	$D(0_u^+)$ ion-pair state	Dunham coefficients $r_e(D) \sim 317.5$pm
Br_2	ttt	Optical/optical double resonance	$F(0_u^+)$ ion-pair state	r_e=327.6(8)pm
FeO	uuu	Mmw absorption (154 and 184GHz)	$X\ ^5\Delta_i$	Rotation constant
OsO	vvv	Emission in visible/near IR	Several bands, states not identified	Rot. analysis
n) 18-electron species				
AgI	www	Microwave (8 GHz)	$X\ ^1\Sigma^+$ v=0,1,2,3,4	3 ← 2 rot. lines, μ_0, vib. variation of n.q. coupling
o) 20-electron species				
Ni_2	xxx	2-photon ionisation laser spectroscopy	$X(^1\Gamma_g$ or $^3\Gamma_g)$	Rot. analysis; r_0=220.0(7)pm

(ppp) S.D. Henderson and J. Tellinghuisen, Chem.Phys.Letters, 1984, 112, 543.
(qqq) T. Trickl and J. Wanner, J.Mol.Spectrosc., 1984, 104, 174.
(rrr) J.C.D. Brand, A.R. Hoy and S.M. Jaywant, J.Mol.Spectrosc., 1984, 106, 388.
(sss) T. Ishiwata, A. Tokunaga, T. Shinzawa and I. Tanaka, J.Mol.Spectrosc., 1984, 108, 314.
(ttt) T. Shinzawa, A. Tokunaga, T. Ishiwata, I. Tanaka, K. Kasatani, M. Kawasaki and H. Sato, J.Chem.Phys., 1984, 80, 5909.
(uuu) Y. Endo, S. Saito and E. Hirota, Astrophys.J., 1984, 278, L131.
(vvv) W.J. Balfour and R.S. Ram, J.Mol.Spectrosc., 1984, 105, 360.
(www) K.P.R. Nair and J. Hoeft, Phys.Rev.A, 1984, 29, 1889.
(xxx) M.D. Morse, G.P. Hanson, P.R.R. Langridge-Smith, Lan-Sum Zheng, M.E. Geusic, D.L. Michalopoulos and R.E. Smalley, J.Chem.Phys., 1984, 80, 5400.

4 Triatomic Molecules and Ions

Several papers have reported analyses of IR spectra of isotopes of H_3^+: 46 lines of ν_2 of $^1H_3^+$ itself[25], some lines of ν_1 of H_2D^+ [26] and 35 b-type lines of ν_1 of HD_2^+ [27]. A pure rotation line ($1_{10} \leftarrow 1_{11}$) of H_2D^+ has been observed[28] at 372,421.380 MHz, and a start has been made[29] on the analysis of the many thousands of lines in the predissociation spectrum of very highly excited vibrational states of H_3^+, using a model based on a loose complex $H_2...H^+$.

The rotation and distortion constants of the $v_2 = 3$ level of the ground electronic state of NH_2 have been calculated from a simultaneous analysis of emission spectra and laser-induced fluorescence (l.i.f.) spectra[30], and l.i.f. spectra of NH_2 in a supersonic jet have been used[31] to derive constants for the $\tilde{A}\ ^2A_1$ excited state. Submillimeter-wave spectra (300-1100 GHz) of H_2O, HDO, and D_2O have been reported[32] and analysed to give constants for the $v_2 = 1$ state as well as the vibrational ground state. A tunable F_2^- : LiF centre laser has been used[33] to study the $\nu_2 + 2\nu_3$ band of HDO between 8558 and 8774 cm^{-1}. The ν_2 band of T_2O [34] has been reported, and the report of the IR spectrum of $T_2{}^{18}O$ mentioned last year has now appeared in a more accessible form[35]. FTIR spectra of H_2S between 2150 and 2950 cm^{-1}, with 0.01 cm^{-1} resolution, have been reported[36] and analysed, in terms of ν_1, ν_3, and $2\nu_2$ bands of several isotopes present at natural abundance. Mbers results on H_2S, HDS, and D_2S include dipole moments (to six decimal places!) and nuclear hyperfine coupling constants[37]. A simultaneous analysis of the ν_2 bands for all isotopes of H_2Se (observed by FTIR at 0.025 cm^{-1} resolution) using mass-adjusted parameters has been described[38]. Velocity-modulated laser absorption spectroscopy in the 3080-3520 cm^{-1} range has been used to observe transitions of ν_1 and ν_3 bands of H_2F^+ [39,40].

25. J.K.G. Watson, S.C. Foster, A.R.W. McKellar, P. Bernath, T. Amano, F.S. Pan, M.W. Crofton, R.S. Altman and T. Oka, Can.J.Phys., 1984, 62, 1875.
26. T. Amano and J.K.G. Watson, J.Chem.Phys., 1984, 81, 2869.
27. K.G. Luki and T.A. Amano, Can.J.Phys., 1984, 62, 1886.
28. H.E. Warner, W.T. Conner, R.H. Petrmichl and R.G. Woods, J.Chem.Phys., 1984, 81, 2514.
29. A. Carrington and R.A. Kennedy, J.Chem.Phys., 1984, 81, 91.
30. M.F. Merienne-Lafore and M. Vervloet, J.Mol.Spectrosc., 1984, 108, 160.
31. S. Mayama, S. Hiraoka and K. Obi, J.Chem.Phys., 1984, 80, 7.
32. J.K. Messer, F.C. De Lucia and P. Helminger, J.Mol.Spectrosc., 1984, 105, 139.
33. A.D. Bykov, Y.S. Makushkin, V.I. Serdyukov, L.N. Sinitsa, O.N. Ulenikov and G.A. Ushakova, J.Mol.Spectrosc., 1984, 105, 397.
34. H.A. Fry, L.H. Jones and J.E. Barefield, J.Mol.Spectrosc., 1984, 103, 41.
35. I. Kanesaka, M. Tsuchida, K. Kawai and T. Takeuchi, J.Mol.Spectrosc., 1984, 104, 405.
36. L. Lechuga-Fossat, J.-M.Flaud, C. Camy-Peyret and J.W.C. Johns, Can.J.Phys., 1984, 62, 18
37. R. Viswanathan and T.R. Dyke, J.Mol.Spectrosc., 1984, 103, 231.
38. W.C. Lane, T.H. Edwards, J.R. Gillis, F.S. Bonomo, and F.C. Murcray, J.Mol.Spectrosc., 1984, 107, 306.
39. E. Schäfer and R.J. Saykally, J.Chem.Phys., 1984, 80, 2973.
40. E. Schäfer and R.J. Saykally, J.Chem.Phys., 1984, 81, 4189.

The rotational structure of the 000-000 and 100-100 bands of the $\tilde{B}^2\Sigma^+ - \tilde{X}^2\Sigma^+$ transition of CaOH has been studied with a tunable dye laser, giving upper- and lower-state rotation constants[41]. The l-type doublet transitions of HCN, DCN, and $DC^{15}N$ with $v_2 = 1$ have been measured by FT microwave techniques in the 4-16 GHz region[42], and line intensities and widths have been measured[43] for ν_3 of HCN broadened by N_2, by FTIR spectroscopy. The ν_2 band of DCP has been observed by FTIR and analysed[44].

The ν_3 band of HCO^+ near 2200 cm^{-1} has been observed using diode lasers, and the line positions analysed to give B_0 and B_3[45,46]; the ν_2 band has also been studied near 830 cm^{-1} [47]. The ν_3 bands of HN_2^+, DN_2^+, and DCO^+ have also been studied using tunable IR diode lasers[48], and the ν_1 region of DN_2^+ analysed (including a hot-band) from difference-frequency lower spectra[49]. The millimeter wave (mmw) spectrum of HCS^+ (between 128 and 300 GHz) has been measured, giving B and D for the ground vibrational state[50]. The results confirm the identification from interstellar sources. Mmw and sub-mmw spectra of HCO have been reported[51]; the lines are of interest to astrophysicists for identification of this species, which may play a significant role in the chemistry in molecular clouds.

A similar study of HNO and DNO gives observed and predicted line positions up to 500 GHz, for states with $J \leq 10$ [52]. Microwave-optical double resonance spectra of HNO[53] and DNO[54] give information about the vibrational ground state of the $\tilde{A}\ ^1A''$ electronic state, which has also been studied by l.i.f. spectroscopy of DNO. In this study[55], as well as a rotational analysis, the identification of the onset of predissociation allowed the measurement of the dissociation energy of the ground state ($\tilde{X}\ ^1A'$) as 17,030(10) cm^{-1}. L.i.f. spectra of HCF have allowed some vibration and rotation constants of $\tilde{X}\ ^1A'$ to be defined[56]. The

41. P.F. Bernath and S. Kinsey-Nielsen, Chem.Phys.Letters, 1984, 105, 663.
42. E. Fliege, H. Dreizler, A.P. Cox and S.D. Hubbard, Z.Nat.A., 1984, 39A, 1104.
43. M.A.H. Smith, G.A. Harvey, G.L. Pellett, A. Goldman and D.J. Richardson, J.Mol.Spectrosc., 1984, 105, 105.
44. J. Lavigne, C. Pepin and A. Cabana, J.Mol.Spectrosc., 1984, 104, 49.
45. S.C. Foster, A.R.W. McKellar and T.J. Sears, J.Chem.Phys., 1984, 81, 578.
46. P.B. Davies, P.A. Hamilton and W.J. Rothwell, J.Chem.Phys., 1984, 81, 1598.
47. P.B. Davies and W.J. Rothwell, J.Chem.Phys., 1984, 81, 5239.
48. S.C. Foster and A.R.W. McKellar, J.Chem.Phys., 1984, 81, 3424.
49. D.J. Nesbitt, H. Petek, C.S. Gudeman, C.B. Moore and R.J. Saykally, J.Chem.Phys., 1984, 81, 5281.
50. M. Bogey, C. Demuynck, J.L. Destombes and B. Lemoine, J.Mol.Spectrosc., 1984, 107, 417.
51. G.A. Blake, K.V.L.N. Sastry and F.C. De Lucia, J.Chem.Phys., 1984, 80, 95.
52. K.V.L.N. Sastry, P. Helminger, G.M. Plummer, E. Herbst and F.C. De Lucia, Astrophys.J.Suppl.Ser., 1984, 55, 563.
53. J.C. Petersen, S. Saito, T. Amano and D.A. Ramsay, Can.J.Phys., 1984, 62, 1731.
54. J.C. Petersen, T. Amano and D.A. Ramsey, J.Chem.Phys., 1984, 81, 5449.
55. R.N. Dixon and C.A. Rosser, Chem.Phys.Letters, 1984, 108, 323.
56. K. Hakuta, J.Mol.Spectrosc., 1984, 106, 56.

ν_1 band of DO_2 has been studied using tunable IR lasers[57], allowing an r_e structure of HO_2 to be defined; parameters are r OH 97.07(20) pm, r OO 133.05(9) pm, <HOO 104.3(3)°. The microwave and mmw spectrum of HOCl has been reported and analysed[58].

The alkali-metal cyanides have been studied as gas-phase monomers by mbers, confirming that Li forms a linear isocyanide LiNC[59] whereas Na[60] and K[61] both have a non-rigid T-shaped cyanide. Many molecular constants have been reported. SiC_2 has been observed in supersonic jets, and its visible spectrum probed using laser photoionisation spectroscopy[62]. Analysis shows it to be triangular in the ground state. Dye-laser excited emission spectra of CCN ($\tilde{A} \rightarrow \tilde{X}$) have been analysed[63] to give constants for v_2 = 1 and 2 levels of $\tilde{X}\ ^2\Pi_r$.

FCP^+ has been detected [64] by its emission ($\tilde{A}^2\Sigma^+ \rightarrow \tilde{X}^2\Pi_i$) following electron impact on FCP. Rotation constants of NCO in some vibrationally excited states of $\tilde{X}\ ^2\Pi_i$ have been reported[65,66].

Emission spectra of rare isotopes of CO_2 in the ν_3 region have been reported[67-69], and lines in the 002-000/201-000/12°1-000 regions of the IR spectrum of N_2O measured extremely precisely by heterodyne techniques[70]. Mmw spectra of OCS in vibrationally excited states (ν_1, $n\nu_2$ for n = 2-4, $\nu_1 + n\nu_2$ for n = 1-3) have been reported[71]. The difference band $\nu_3 - \nu_1$ of CS_2 has been studied using a diode laser[72].

57. K.G. Lubic, T. Amano, H. Uehara, K. Kawaguchi and E. Hirota, J.Chem.Phys., 1984, 81, 4826.
58. H.E. Gillis Singbeil, W.D. Anderson, R. Wellington Davis, M.C.L. Gerry, E.A. Cohen, H.M. Pickett, F.J. Lovas and R.D. Suenram, J.Mol.Spectrosc., 1984, 103, 466.
59. J.J. Van Waals, W.L. Meerts and A. Dymanus, Chem.Phys., 1983, 82, 385.
60. J.J. Van Waals, W.L. Meerts and A. Dymanus, Chem.Phys., 1984, 86, 147.
61. J.J. Van Waals, W.L. Meerts and A. Dymanus, J.Mol.Spectrosc., 1984, 106, 280.
62. D.L. Michalopoulos, M.E. Gensic, P.R.R. Langridge-Smith and R.E. Smalley, J.Chem.Phys., 1984, 80, 3556.
63. K. Kawaguchi, T. Suzuki, S. Saito, E. Hirota and T. Kasuya, J.Mol.Spectrosc., 1984, 106, 320.
64. M.A. King, D. Klapstein, H.W. Kroto, R. Kuhn, J.P. Maier and J.F. Nixon, J.Chem.Phys., 1984, 80, 2332.
65. R.A. Copeland and D.R. Crosby, Can.J.Phys., 1984, 62, 1488.
66. K.N. Wong, W.R. Anderson, A.J. Kotlar and J.A. Vanderhoff, J.Chem.Phys., 1984, 81, 2970.
67. D. Bailly and C. Rossetti, J.Mol.Spectrosc, 1984, 105, 215.
68. D. Bailly and C. Rossetti, J.Mol.Spectrosc, 1984, 105, 229.
69. D. Bailly and C. Rossetti, J.Mol.Spectrosc, 1984, 105, 331.
70. C.R. Pollock, F.R. Petersen, D.A. Jennings, J.S. Wells and A.G. Maki, J.Mol.Spectrosc., 1984, 107, 62.
71. K. Tanaka, H. Ito and T. Tanaka, J.Mol.Spectrosc., 1984, 107, 324.
72. E. Baetun, G. Blanquet, J. Walrand and C.P. Courtoy, Can.J.Phys., 1984, 62, 1286.

The far-IR spectrum of NO_2 (from 8-80 cm^{-1}) measured at 0.003 cm^{-1} resolution by FT methods has been described[73], and microwave lines of $^{15}NO_2$ reported[74]. The ν_1 band of NO_2 has been studied by FTIR[75], and excitation spectra of NO_2 at 463 and 474 nm have been reported[76]. Far-IR FT spectra of O_3 [77] and SO_2[78] have been reported, and microwave spectra of ^{18}O isotopes of SO_2 used to define an r_e structure[79,80]. FTIR spectra of SO_2 in the ν_1 and ν_3 bands at 0.005 cm^{-1} resolution[81], and sub-Doppler laser spectra of SO_2 in a supersonic jet[82] have also been reported. Microwave spectra of FNO in $v_n = 1$ (n = 1,2,3) states have been used[83] to define some cubic potential constants in the force field, and FTIR methods have been used to measure the ν_2 and ν_3 bands of FNO[84]. A diode laser has been used to study the ν_2 band of FO_2[85].

5 Tetra-atomic Molecules

Among the many papers relating to the NH_3 molecule and its isotopes[86-93] we may note a study of the laser Stark spectrum giving sub-Doppler resolution, from which an expression for the J and K dependence of the "ground-state dipole moment" is derived[86], and a study of "forbidden" IR transitions of $^{15}NH_3$[91].

73. B. Carli, M. Carlotti, F. Mencaraglia, A. Mahmoudi, C. Camy-Peyret and J.-M. Flaud, Mol.Phys., 1984, 51, 1505.
74. N. Tanaka, Y. Hamada, M. Tsuboi, S. Saito, Y. Endo and E. Hirota, J.Mol.Spectrosc., 1984, 103, 87.
75. A. Perrin, J.-Y. Mandin, C. Camy-Peyret, J.-M. Flaud, J.-P. Chevillard and G. Guelachvili, J.Mol.Spectrosc., 1984, 103, 417.
76. N. Sugimoto and S. Takezawa, J.Mol.Spectrosc., 1984, 106, 307.
77. M. Calotti, G. Di Lonardo, L. Fusina, A. Trombetti, A. Bonetti, B. Carli and F. Mencaraglia, J.Mol.Spectrosc., 1984, 107, 84.
78. M. Carlotti, G. Di Lonardo, L. Fusina, B. Carli and F. Mencaraglia, J.Mol.Spectrosc., 1984, 106, 235.
79. R. Van Riet and G. Steenbeckeliers, Ann.Soc.Sci.Bruxelles, Ser.1, 1983, 97, 117.
80. R. Van Riet and G. Steenbeckaliers, Ann.Soc.Sci.Bruxelles, Ser. 1, 1983, 97, 243.
81. G. Guelachvili, O.N. Ulenikov and G.A. Ushakova, J.Mol.Spectrosc., 1984, 108, 1.
82. R. Kullmer and W. Demtröder, J.Chem.Phys., 1984, 81, 2919.
83. G. Cazzoli, C. Degli Esposti, P.G. Favero and P. Palmieri, Nuovo Cim.Soc.Ital.Fis.D., 1984, 3D, 627.
84. S.C. Foster and J.W.C. Johns, J.Mol.Spectrosc., 1984, 103, 176.
85. C. Yamada and E. Hirota, J.Chem.Phys., 1984, 80, 4694.
86. W.H. Weber, J.Mol.Spectrosc., 1984, 107, 405.
87. R.L. Poynter and J.S. Margolis, Mol.Phys., 1984, 51, 393.
88. G. Magerl, W. Schupita, J.M. Frye, W.A. Kreiner and T. Oka, J.Mol.Spectrosc., 1984, 107, 72.
89. G. Baldacchini, S. Marchetti and V. Montelatici, J.Mol.Spectrosc., 1984, 103, 257.
90. W. Hüttner and W. Major, Mol.Phys., 1984, 52, 631.
91. S. Urban, R. D'Cunha and K.N. Rao, J.Mol.Spectrosc., 1984, 106, 64.
92. A. Guarnieri, G.T. Tolkmit, M. Bester, S. Urban, K. Yamada and G. Winnewisser, J.Mol.Spectrosc., 1984, 106, 22.
93. S. Urban, D. Papousek, M. Bester, K. Yamada, M. Winnewisser and A. Guarnieri, J.Mol.Spectrosc., 1984, 106, 29.

IR/radio-frequency double-resonance spectra of PH_3 have been studied[94], and some unexpected lines in its laser Stark spectrum explained[95]. The ν_2 and ν_4 bands of AsH_3 have been reported and analysed[96], and a misassignment in the laser Stark spectrum has been corrected[97]. Lines of the ν_2[98] and ν_3[99] bands of H_3O^+ in the gas phase have been measured and assigned; the inversion barrier is found to be 904 cm^{-1}, by a combined analysis[99] of the ν_3 data and earlier measurements on ν_1.

N-protonated HCN ($HCNH^+$) has been detected by its IR spectrum[100], and line positions measured in the ν_1 and ν_2 bands using a difference frequency IR laser[101]. Similar measurements are reported for ν_1 of $DCNH^+$ [102]. Far-IR laser magnetic resonance (lmr) spectra are reported[103] for NH_2O.

HCCN is shown[104] to be a linear molecule in its ground electronic state ($\tilde{X}\ ^3\Sigma^-$) by analysis of its microwave spectrum. Microwave spectra of HBF_2 have been analysed[105], giving rotation constants, distortion constants including sextic terms, and boron nuclear quadrupole coupling constants.

The far-IR spectrum of HNCO has been measured from 8-80 cm^{-1} using FT methods at 0.004 cm^{-1} resolution[106], and the ν_1 band of $D^{14}N_3$ has been studied using both FTIR[107] and Raman[108] spectra. Mmw spectra of HNCSe, including several isotopes, have given[109] an r_s structure, with rNC 119.2 pm, r CSe 171.9 pm, <HNC 140°, and <NCSe (assumed) 175°. The rotational structures of the vibrational bands of DONO in the mid-IR have been investigated by FTIR methods[110].

Two groups report a partial rotational analysis of the $\tilde{A}\ ^1A'' \rightarrow \tilde{X}\ ^1A'$ absorption bands of NCNO in the near IR[111,112]. The

94. S. Tanimura and K. Tagachi, J.Mol.Spectrosc., 1984, 104, 414.
95. K. Takagi, Chem.Phys.Letters, 1984, 112, 302.
96. G. Di Lonardo, L. Fusina and J.W.C. Johns, J.Mol.Spectrosc., 1984, 104, 282.
97. F. Scappini, L. Fusina and G. Di Lonardo, J.Mol.Spectrosc., 1984, 107, 419.
98. N.N. Haese and T. Oka, J.Chem.Phys., 1984, 80, 572.
99. P.R. Bunker, T. Amano and V. Spirko, J.Mol.Spectrosc., 1984, 107, 208.
100. R.S. Altman, M.W. Crofton and T. Oka, J.Chem.Phys., 1984, 80, 3911.
101. R.S. Altman, M.W. Crofton and T. Oka, J.Chem.Phys., 1984, 81, 4255.
102. T. Amano, J.Chem.Phys., 1984, 81, 3350.
103. P.B. Davies, P. Dransfeld, F. Temps and H.G. Wagner, J.Chem.Phys., 1984, 81, 3763.
104. S. Saito, Y. Endo and E. Hirota, J.Chem.Phys., 1984, 80, 1427.
105. N.P.C. Westwood, W. Lewis-Bevan and M.C.L. Gerry, J.Mol.Spectrosc., 1984, 106, 227.
106. L. Fusina, M. Carlotti and B. Carli, Can.J.Phys., 1984, 62, 1452.
107. J. Bendtsen and G. Guelachvili, J.Mol.Spectrosc., 1984, 107, 385.
108. J. Bendtsen, J.Raman Spectrosc., 1984, 15, 113.
109. J. Vogt and M. Winnewisser, Ber.Bunsen-Ges.Phys.Chem., 1984, 88, 444.
110. L.O. Halonen, C.M. Deeley, I.M. Mills and V.-M. Horneman, Can.J.Phys., 1984, 62, 1300.
111. S. Bell and E.R. McNee, Chem.Phys.Letters, 1984, 111, 105.
112. M. Noble, I. Nadler, H. Reisler and C. Wittig, J.Chem.Phys., 1984, 81, 4333.

transition is accompanied by a dramatic change in the A rotation constant because of a large change in the bond angle CNO. The microwave spectrum of BrNCO has been analysed, giving A_0 and the bromine nuclear quadrupole coupling term χ_{ab}[113]. Microwave spectra of $COCl_2$ in vibrationally excited states[114] and diode laser spectra of the ν_1 and ν_5 IR bands[115] have given upper-state rotation constants and allowed correction of ground-state constants to equilibrium constants.

6 Penta-atomic Molecules

Velocity-modulated laser absorption spectra of ν_3 of NH_4^+ in the gas phase have been reported and analysed[116]. Ground-state constants of SiH_4 to sixth order have been derived by combination-difference methods from FTIR spectra at 0.002 cm^{-1} resolution of the ν_2 and ν_4 bands[117]. Two studies report data on the Q-branch of ν_2 of GeH_4 using a pulsed IR diode laser, first on isotopically pure $^{74}GeH_4$[118] and then on a natural isotopic mix[119]. The main Schüler band of ND_4[120] has been analysed, confirming the assignment as the 0-0 band of the 3p $^2F_2$ - 3s $^2A_1$ electronic transition[121].

A systematic assault on the IR spectra of silyl and germyl halides continues. Reports appearing in 1984 include studies of ν_6 of SiH_3Cl[122], of the ν_2 and ν_5 bands of SiH_3Cl[123], of $SiH_3{}^{79}Br$[124], of $^{74}GeH_3F$[125] (together with the ν_3 and ν_6 bands) and $^{74}GeH_3Br$[126], where the $\nu_3 + \nu_6$ combination complicates the observed bands, and of the ν_1 and ν_4 bands of $^{74}GeH_3{}^{35}Cl$[127] and $^{74}GeH_3Br$[128]. (The last report also covers the ν_6 and $2\nu_6$ bands

113. H.M. Jemson, W. Lewis-Bevan, N.P.C. Westwood, and M.C.L. Gerry, Chem.Phys.Letters, 1984, 108, 496.
114. S. Yamamoto, M. Nakata, M. Sugie, H. Takeo, C. Matsumura and K. Kuchitsu, J.Mol.Spectrosc., 1984, 105, 299.
115. S. Yamamoto, T. Nakanaga, H. Takeo, C. Matsumura, M. Nakata and K. Kuchitsu, J.Mol.Spectrosc., 1984, 106, 376.
116. E. Schäfer, R.J. Saykally and A.G. Robiette, J.Chem.Phys., 1984, 80, 3969.
117. G. Pierre, A. Valentin and L. Henry, Can.J.Phys., 1984, 62, 254.
118. A.E. Cheglokov, Y.A. Kuritsyn, E.P. Snegirev, O.N. Ulenikov and G.V. Vedeneeva, J.Mol.Spectrosc., 1984, 105, 385.
119. A.E. Cheglokov, Yu. A. Kuritsyn, E.P. Snegirev, O.N. Ulenikov and G.V. Vedeneeva, Mol.Phys., 1984, 53, 287.
120. F. Alberti, K.P. Huber and J.K.G. Watson, J.Mol.Spectrosc., 1984, 107, 133.
121. J.K.G. Watson, J.Mol.Spectrosc., 1984, 107, 124.
122. A. Ruoff, H. Essig, H. Bürger, S. Dobos and P. Schulz, Mol.Phys., 1984, 51, 499.
123. H. Bürger, A. Rahner, P. Schulz, J. Dupre, G. Graner, P.B. Davies and D.P. Stern, J.Mol.Spectrosc., 1984, 106, 337.
124. H. Bürger, H. Beckers and J. Kauppinen, J.Mol.Spectrosc., 1984, 108, 215.
125. S. Cradock, Mol.Phys., 1984, 51, 697.
126. F. Lattanzi, C. Di Lauro, H. Bürger, R. Eujen, P. Schulz and S. Cradock, Mol. Phys., 1984, 51, 81.
127. F. Lattanzi, C.Di Lauro, H. Bürger, R. Eujen and S. Cradock, J.Mol.Spectrosc., 1984, 107, 229.
128. H. Bürger, R. Eujen, A. Rahner, P. Schulz and S. Cradock, Z.Nat.A., 1984, 39A, 871.

of germyl bromide). All these reports concern FTIR spectra as a basis, but in several cases spectra of Doppler-limited resolution (FT, diode laser, or SISAM) are also employed. Of particular interest is the demonstration that A_0 and D_K for a symmetric top such as SiH_3Cl can be systematically derived from a simultaneous analysis of ν_6, $2\nu_6$, and $2\nu_6-\nu_6$ bands[129]; no accidental resonances or perturbation-allowed transitions are involved. Pure rotation spectra of SiH_3Cl in the mmw region[130] give ground-state and vibrationally excited-state rotation and distortion constants in agreement with the IR studies, though it has been shown[125] that some care is needed in the analysis of pure rotation spectra if the results are to be compatible with IR data for GeH_3F.

The OH stretching band of NH_2OH in the IR[131], microwave and FTIR spectra of $CH_2=S=O$[132], and microwave spectra of $CH_2=P-Cl$[133] have been reported. Sub-mmw spectra of HNO_3 (550-800 GHz)[134] and diode laser IR spectra of the ν_5 band[135] have been measured, and FT microwave spectra of $D^{15}NO_3$ analysed[136] to give ^2D nuclear quadrupole constants.

The manifold of bands associated with ν_5 of C_3O_2, which includes many hot-bands and sum and difference bands due to the very low-frequency band ν_7, has been re-analysed[137]. The mmw spectrum of chlorine nitrate ($ClONO_2$) from 80-228 GHz has been reported[138] and analysed to give rotation constants and line position predictions up to 300 GHz, to assist in the monitoring of this species in the atmosphere. IR/radiofrequency double resonance spectra of OsO_4 have allowed tetrahedral splitting constants for the ground state and the ν_3 = 1 state to be defined[139].

7 Molecules with Six or More Atoms

A careful analysis of rotation constants and vibrational corrections for various isotopic forms of B_2H_6 has allowed the definition of r_z and r_e structures that are consistent with electron-diffraction results, though rather more precise[140]. The

129. H. Bürger, G. Schippel, A. Ruoff, H. Essig and S. Cradock, J.Mol.Spectrosc., 1984, 106, 349.
130. D. Barrett and J.H. Carpenter, J.Mol.Spectrosc., 1984, 107, 153.
131. M.E. Coles, A.J. Merer and R.F. Curl, J.Mol.Spectrosc., 1984, 103, 300.
132. Y. Koga, H. Takeo, S. Kondo, C. Matsumura, Bull.Chem.Soc.Japan, 1984, 57, 2400.
133. H.W. Kroto, J.F. Nixon, O. Ohashi, K. Ohno and N.P.C. Simmons, J.Mol.Spectrosc., 1984, 103, 113.
134. J.K. Messer, F.C. De Lucia and P. Helminger, J.Mol.Spectrosc., 1984, 104, 417.
135. A.G. Maki and J.S. Wells, J.Mol.Spectrosc., 1984, 108, 17.
136. E. Fliege and H. Dreizler, Z.Nat.A., 1984, 39A, 630.
137. P. Jensen, J.Mol.Spectrosc., 1984, 104, 59.
138. R.D. Suenram and F.J. Lovas, J.Mol.Spectrosc., 1984, 105, 351.
139. F. Scappini, W.A. Kreiner, J.M. Frye and T. Oka, J.Mol.Spectrosc., 1984, 106, 436.
140. J.L. Duncan and J. Harper, Mol.Phys., 1984, 51, 371.

pure rotation spectrum of $SiH_3C\equiv CH$ in the $v_{10} = 3$ excited states has been analysed[141]; some anomalies in the $l = 1$ lines are attributed to perturbations by states with $v_n = 1$, $n = 5$, 8 or 9. Microwave spectra of CH_3SnH_2D, CH_3SnHD_2, and CH_3SnD_3 have been used[142] to derive a partial substitution structure (r_s SnH = 171.3(5) pm) and an effective barrier to internal rotation (V_3 = 595(8) cal/mole ≡ 2.49(4) kJ $mole^{-1}$).

A new model, in which the 2-dimensional anharmonic bend of a pseudo-linear skeleton is combined with all three rotational degrees of freedom in a variational calculation, is very successful in accounting for the microwave spectra of CH_3NCS[143] and SiH_3NCO[144], with only a small number of refining potential and structural parameters.

Another study of the 10 micron IR bands of N_2H_4 using FTIR (at 0.003 cm^{-1} resolution) and diode laser methods has been reported[145].

Diode laser spectra of ν_3 bands of PF_5[146] and WF_6[147] in supersonically cooled jets (where hot-bands are drastically reduced in intensity) have been recorded and two-photon Doppler-free spectra of SF_6 in the $2\nu_2$ region[148] reported. Microwave spectra of $(CH_2SH)_2$ have given a dipole moment and rotation constants[149], consistent with a skeletal torsion angle of -65° measured from the eclipsed conformation. There is no evidence of any intra-molecular hydrogen bonding. Internal rotation parameters have been deduced for microwave spectra of Me_2Se in torsionally and vibrationally excited states[150]. The mmw spectrum of IOF_5 (30-75 GHz) has been analysed[151], giving centrifugal distortion constants and data on the K-doubling in the ground state (for which $\Delta l = 0$, $\Delta k = \pm 4$).

141. A. Bauer and J. Carlier, J.Mol.Spectrosc., 1984, 106, 155.
142. J.R. Durig, C.M. Whang, G.M. Attia and Y.S. Li, J.Mol.Spectrosc., 1984, 108, 240.
143. M. Kreglewski, Chem.Phys.Letters, 1984, 112, 275.
144. M. Kreglewski and P. Jensen, J.Mol.Spectrosc., 1984, 103, 312.
145. M. Tsuboi, Y. Hamada, L. Henry, J. Chezelas and A. Valentin, J.Mol.Spectrosc., 1984, 108, 328.
146. M. Takami and H. Kuze, J.Chem.Phys., 1984, 80, 2314.
147. M. Takami and H. Kuze, J.Chem.Phys., 1984, 80, 5994.
148. C.W. Patterson, F. Herlemont, M. Azizi and J. Lemaire, J.Mol.Spectrosc., 1984, 108, 31.
149. R.M. Nandi, Chun-Fu Su and M.D. Harmony, J.Chem.Phys., 1984, 81, 1051.
150. W. Lalowski and H. Dreizler, Z.Nat.A., 1983, 38A, 1346.
151. P.N. Brier and M.J. Winrow, J.Mol.Spectrosc., 1984, 107, 21.

4

Characteristic Vibrations of Compounds of Main-group Elements

BY G. DAVIDSON

1 Group I

The v= 1←0 band of HD has been investigated at 77K over a range of pressures.[1]

I.r. and Raman spectra of THF and 1,3-dioxolane solutions of MNCS (M = Li, Na or K) give evidence for the formation of $LiNa(NCS)_2$, Li_2NCS^+, Na_2NCS^+ and $(MNCS)_2$.[2]

A force field has been calculated *ab initio* for $LiBO_2$.[3] I.r. spectra of matrices containing MBO_2, $LiCO_2$ or $MClO_2$ (M = Li-Cs) have been assigned by means of $^6Li/^7Li$, $^{10}B/^{11}B$, $^{12}C/^{13}C$, $^{16}O/^{18}O$ and $^{35}Cl/^{37}Cl$ substitutions. The MBO_2 molecules have C_s symmetry, with linear O-B-O and M-$\hat{O}$-B greater than 100°.[4] The Hartree-Fock-Roothaan method was used to calculate a force-field for $LiNO_3$ *ab initio*.[5] νLi-O is at 354cm^{-1} in LiCl.DMF.0.5H_2O, with νMg-O at 406cm^{-1} and νCa-O at 240cm^{-1} in related Group II dimethylformamide complexes.[6]

$^6Li/^7Li$ substitution was used to analyse the infrared and Raman spectra of Li_2SnO_3 and Li_2TiO_3.[7] A band due to νLi-O was seen at 460cm^{-1} in $Li_2O.2Al_2O_3.0{\cdot}9CO_2.8{\cdot}8H_2O$ (with νAl-O at 760cm^{-1}).[8] Both tetrahedral and octahedral Li are present in $Li_2Ti_3O_7$ (from i.r. data, with $^6Li/^7Li$ shifts).[9] Several modes involving cation motion were detected in the i.r. spectrum of $LiNaSO_4$.[10]

The band centres for the 1←0 vibrational transition in LiF are at 894.34104 ± 0.00020cm^{-1}(^{7}Li), 946.12074 ±0.00024cm^{-1}(^{6}Li).[11]

νK-O stretches gave 6 bands in the range 183-240cm^{-1} for KOX, where X = H or D, at 77K and at room temperature. δKOX were in the range 80-124cm^{-1}.[12] The vibrational spectra of KF.$(CH_2CO_2X)_2$ (X = H or D) contain a band due to νK-F near 330 cm^{-1}.[13]

Vapours from heated alkali-metal perchlorates were trapped in N_2 matrices and studied by i.r. spectroscopy. $^{16}O/^{18}O$ substitution shows that $RbClO_4$ contains bidentate perchlorate, Rb(μ-O)$_2$ClO$_2$, but $CsClO_4$ has three-fold symmetry.[14]

2 Group II

There have been ab initio calculations of vibrational spectra for $LiBeH_3$, $NaBeH_3$, $LiMgH_3$,[15] Be(OH)F,[16] $LiBeF_2^+$ and $LiBF_3^+$.[17] I.r. and Raman spectra of alkaline beryllate solutions contain bands due to $Be(OH_2)_4^{2+}$ (520cm^{-1}, 870-900cm^{-1}) and $Be(OH)_4^{2-}$ (700-750cm^{-1}).[18]

Some vibrational assignments have been proposed for the first organogermyl-magnesium compound, $Mg(GeMe_3)_2$.2DME (νMgO 386cm^{-1}, νMgO or $\nu MgGe_2$ 294, 269cm^{-1}, $\nu MgGe_2$ 280cm^{-1}).[19] Isotopic shift data ($^{24}Mg/^{26}Mg$, $^{40}Ca/^{44}Ca$ and $^{28}Si/^{30}Si$) were used to assign the vibrational spectrum of monticellite, $MgCaSiO_4$.[20] I.r. and Raman spectra were reported for Ln_2MgTiO_6, where Ln = La-Yb, or Y. $^{24}Mg/^{26}Mg$ substitution showed that the magnesium atoms were octahedrally coordinated.[21]

3 Group III

3.1 Boron.- Fermi resonance interactions ($\nu_1/2\nu_4$; $\nu_3/2\nu_4/\nu_2 + \nu_4$) were investigated in the Raman spectra of BH_4^- or BD_4^- isolated in alkali halide crystals.[22] νB-H and νB-O were assigned for K[$(RO)_3$-BH] in THF (where R = ^{i}Pr, ^{s}Bu, ^{t}Bu, Ph, etc.).[23] νCN of Na[$BH_2(CN)_2$] was at 2200cm^{-1}, compared to 2179cm^{-1} in BH_3CN^-. There was no evidence for B-NC bonding.[24]

114 spectroscopic data from 6 isotopic species of diborane were used to give an empirical harmonic force field, with 30 of the 33 independent force constants determined with numerical significance.[25] A force field has been calculated ab initio for methyldiborane.[26]

νB-H and cage modes in 1- and 2-$(SCN)B_{10}H_9^{2-}$ and 1,10-$(SCN)_2$-$B_{10}H_8^{2-}$ are very similar to those in the halo-analogues.[27] I.r. and Raman spectra gave detailed assignments for $B_{10}X_{10}^{2-}$, where X = H, D, Cl, Br or I.[28] A broad Raman band near 2080cm^{-1} in the Cs^+ salt of o-dicarba-nido-undecaborane and several weak bands (1770-2125 cm^{-1}) for the m-isomer all shift on deuteriation. They were assigned to vibrations of the 'extra hydrogen'.[29] A detailed vibrational assignment has been given for divinylfluoroborane, $(CH_2{=}CH)_2BF$. There is only one conformer (C_s) in the solid, but there are at least two conformers in the gas and liquid phases.[30]

Raman spectra of vitreous B_2O_3 have been recorded in the range 8-700K, and at pressures up to 8 kbar at room temperature. At room temperature, vitreous B_2O_3 comprises an equal proportion of boroxole rings (B_3O_6) and BO_3 triangles.[31] I.r. spectra of boron species in solutions containing B_2O_3, Na_2O and Al_2O_3 were reported. Bands were seen due to $B(OH)_4^-$, $Al(OH)_4^-$, Al-O-Al and Al-O-B species.[32] Raman spectra of B_2O_3-TeO_2 and K_2O-B_2O_3-TeO_2 systems reveal changes in the coordination of the B-O and Te-O units.[33]

Raman spectra of solid BF_3 at 80 and 20K contain bands due to lattice modes and the ν_3 mode - seen for the first time.[34] I.r. evidence has been found for the formation of weakly bound species BF_3...L, where L = CO or N_2, by the controlled condensation of appropriate mixtures on to a cold CsI window.[35] Some i.r. bands have been assigned for solid B_2I_4: νB-I 680-710(e), 484-503(b_2) cm^{-1}; δBI_2 240(e) cm^{-1}.[36] νB-X and δB-X modes have been assigned from the i.r. and Raman spectra of 2-X-$B_{10}H_9^{2-}$, 1-X-$B_{12}H_{11}^{2-}$, where X = Cl, Br or I.[37,38]

3.2 Aluminium.- There is some i.r. evidence for both monomer, $AlMe_3$, and dimer, Al_2Me_6, produced in an argon matrix using a Knudsen effusion technique.[39]

I.r. spectra of α-, β- and γ-modifications of $LiAlO_2$ confirm conclusions from ^{27}Al n.m.r. about the proportions of octahedral and tetrahedral aluminium coordination.[40] νAl-O modes were assigned from the i.r. and Raman spectra of $LnAlO_3$, where Ln = Sm or Eu.[41] ν_{as}(M-O-M) modes (M = Al or Si) have been assigned in i.r. spectra of Tschermak's pyroxene, gehlenite and glasses of composition $CaAl_2SiO_6$.[42] Polyalumoxanes and alumina aggregates have been identified (i.r. and Raman spectra) as intermediates formed during the hydrolysis of Al_2Me_6 in C_6H_6.[43] A Raman band at 900cm^{-1}, due to the Al-O stretch of AlO_6 octahedra, is seen for β-Al_2TiO_5.[44] νAl-O bands were assigned for $Cu[Al(OR)_4]_2$ (R = Me, Et, ^{i}Pr or ^{n}Bu).[45] $Al(OH)_6^{3-}$ has νAl-O at 520 and 750 cm^{-1}; $Ga(OH)_6^{3-}$ has νGa-O at 500 and 650cm^{-1}.[46]

Raman and i.r. spectra of AlSX and AlSeX, where X = Cl,Br or I, show that they are dimers, with planar, four-membered Al_2S_2 or Al_2Se_2 rings and short terminal Al-X bonds. These dimers are linked *via* long X...Al interactions to give a chain structure.[47]

Ab initio calculations have been made of i.r. wavenumbers and transition probabilities for AlF and AlF^+; for AlF the calculated wavenumber is close to the experimental value.[48] νM-F in $M(C_2O_4)_2F_2^{3-}$ are at 610cm^{-1} (M = Al), 480cm^{-1} (Ga).[49]

The i.r. spectra of Al_2Cl_6 and $AlCl_3$ (including data on $Al_2{}^{35}Cl_6$ and $Al_2{}^{37}Cl_6$) in the temperature range 473-843K, Raman spectra of Al_2Cl_6 vapour at ca. 500K, and $AlCl_3$ at 1075K (at pressures of 0.3 to 10 atm.), including polarisation measurements and $^{35}Cl/^{37}Cl$ isotopic data, have been reported. 9 Raman- and 7 i.r.-active fundamentals of the dimer were assigned, and all 4 fundamentals of the D_{3h} monomer.[50] Dimer and monomer bands were assigned from the high-temperature i.r. spectra of AlX_3 (X = Cl, Br or I) and $GaCl_3$ (300-860K). The i.r. emission spectrum of $Al_2Cl_7^-$, as Li^+, Na^+, K^+ or Cs^+ salts, can be assigned in terms of D_{3d} symmetry for the anion, with linear Al-O-Al.[51] I.r. emission spectra were also measured and assigned for MCl_4^-, where M = Al, Ga or In, and $M_2Cl_7^-$, where M = Al or Ga. In MCl_4^- there was some distortion from tetrahedral geometry, depending on the cation ($Li^+ > Na^+ > K^+ > Rb^+ > Cs^+$). For $M_2Cl_7^-$ it was suggested that the bridge is linear, except for the Li^+ species.[52]

R-B$(NMe_2)_2ECl_3$, where R = Me or Ph, E = Al or Ga, give νE-Cl in the expected regions, e.g. for R = Me, E = Al, $\nu_{as}AlCl_3$ is at 523 cm^{-1}, ν_sAlCl_3 at $420cm^{-1}$.[53]

I.r. spectra of aluminium bromide and iodide and gallium chloride vapours ($50-700cm^{-1}$) can be assigned in terms of D_{2h} dimers (7 of the 8 i.r.-active fundamentals seen) and D_{3h} monomers (all 3 i.r.-active fundamentals assigned).[54]

<u>3.3 Gallium</u>.- νGa-H is at $1620cm^{-1}$ in $KGa(CH_2SiMe_3)_2H_2$ ($1170cm^{-1}$ on deuteriation).[55] $(Et_2Ga-PEt_2)_3$ has νGa-C at 498 and $527cm^{-1}$.[56]

Raman spectra of M_2O, where M = Ga, In or Tl, isolated in argon matrices, gave assignments for the symmetric M_2O stretches. These were in good agreement with calculations for the gaseous molecules.[57] A complete assignment has been given for the Raman-active modes of $M(OH_2)_6^{3+}$, where M = Ga or In. The totally symmetric stretches were at higher wavenumbers than had previously been suggested ($537cm^{-1}$ for M = Ga, $505cm^{-1}$ for In).[58] I.r. and Raman spectra were reported for $CaGa_4O_7$, $CaGa_2O_4$, $Ca_3Ga_4O_9$, $Ca_3Ga_2O_6$,[59] and single crystals of $Nd_3Ga_5O_{12}$.[60]

Skeletal vibrational modes have been assigned for a large number of organothio-gallanes, e.g. MeGaCl(SR), $PhGa(SR)_2$, $Me_2Ga(SR)$, and $X_2Ga(SR)$, where X = Cl, Br or I, and R = various alkyl or aryl groups.[61-64] Raman and far-i.r. spectra of $TlGaS_2$, $TlGaSe_2$ and β-$TlInS_2$ can be interpreted in terms of the space group C_{2h}^{6}.[65] I.r. spectra of $CdGa_2S_4$ and $CdGa_2Se_4$ were also reported.[66]

The i.r. and Raman spectra of $GaBr_3$-MBr_2 systems, where M = Zn, Cd, Hg, Sn(II) or Pb(II), can be assigned in terms of $GaBr_4^-$ and $Ga_2Br_7^-$ vibrations. In $GaBr_3$-MBr_3 (M = Sb or Bi) there was also evidence for Ga-Br-M units.[67]

<u>3.4 Indium and Thallium</u>.- I.r. and Raman spectra of $TlIn_{1-x}Tl_xSe_2$ solid solutions contain vibrations of In atoms in a TlSe lattice at 190 and 187cm^{-1}.[68] Assignments of Raman spectra of $TlInSe_2$, $TlInTe_2$ and $TlGaTe_2$ were made by analogy with $TlInS_2$.[69]

The stretching wavenumbers of InX, where X = F, Cl, Br or I, are linearly dependent on bond length.[70]

$TlI_3(MeC_5H_4NO)_2$ gives νTl-O at 305 and 290cm^{-1}, and νTl-I at 165 and 150cm^{-1}.[71] Solid-phase i.r. and Raman spectra of pyH^+ (and related) salts of $TlCl_5^{2-}$ show that it is a genuine five-coordinate species.[72]

<u>4 Group IV</u>

<u>4.1 Carbon</u>.- The ν_2 band (CH stretch) of $HCNH^+$ is at 3187.86382 (39)cm^{-1}, in good agreement with <u>ab initio</u> calculations. ν_1 (NH stretch) is at 3482.844cm^{-1}.[73] I.r. and Raman spectra were assigned for $CHCl_2SiH_3$, $CHCl_2SiD_3$,[74] CD_3SiF_3 and CD_3SiCl_3.[75] I.r. spectra of CHD_2GeH_3 and CH_3GeHD_2 were used to study the correlation of 'isolated' νCH and νGeH with molecular parameters.[76] All of the fundamentals of $CHFCl_2$ were assigned from gas-phase i.r. spectra, except ν_6 (estimated from combination bands to be <u>ca</u>. 277cm^{-1}). $^{35}Cl/^{37}Cl$ isotopic shifts were measured in a number of cases.[77] Evidence was found for the formation of two different 1:1 complexes between CH_3CN and HCN (from far-i.r. spectra).[78]

Vibrational relaxation in MMe_4, where M = C, Si, Ge or Sn, was studied as a function of temperature and density, using line-shapes of the ν_3 M-C stretch.[79] Detailed assignments were given of the i.r. spectra of dimethylsilylene, Me_2Si, and 1-methylsilene, CH_3Si=CH_2, in an argon matrix at 28K; νC=Si in the latter is at 988cm^{-1}.[80]

The νC≡N region in $Me_3N.DCN$ can be assigned in terms of the formulation $(Me_3ND)^+CN^-$. The CN^- rotates, interacting with a translational mode.[81] Semi-empirical calculations have been made of the vibrational wavenumbers of H_2CN^+.[82] The Raman and i.r. spectra of polycrystalline C_2N_2, at 20K, can be assigned in terms of the known orthorhombic crystal structure.[83] A full assignment has been made of the i.r. and Raman bands of CF_3NC (e.g. νC-NC at $550cm^{-1}$).[84] A revised assignment of fundamental modes has been given for $ClN{=}CCl_2$ (using $Cl^{15}N{=}CCl_2$ and $ClN{=}^{13}CCl_2$ data).[85] νCN stretching wavenumbers in $S_2CN\text{-}NHCS_2^{3-}$ are at 1400 and $1310cm^{-1}$. Non-solvated salts of $S_2CNH\text{-}NHCS_2^{2-}$ were shown to have rather low symmetry.[86]

The i.r. vibration/rotation spectrum of DCP has been analysed in the region of the ν_2 fundamental.[87] Unstable $CH_3C{\equiv}P$ has been detected by i.r. spectroscopy in the gas phase. νC≡P is centred at $1558.724(25)cm^{-1}$ ($1554.962(25)cm^{-1}$ in $CD_3C{\equiv}P$).[88]

I.r. emission studies gave values of vibrational parameters for v = 0 to 13 for ^{14}CO.[89] <u>Ab initio</u> calculations of i.r. wavenumbers and intensities for the fundamentals of CO^+, CO_2^+ and CS_2^+ gave wavenumber values within 10% of the experimental values, where these are known.[90] The vibrational spectra (J = 0) of CO_2 and CS_2 have been compared with those calculated from a potential model for two-valued triatomic surfaces.[91] The ν_3 fundamental (CO stretch) of HCO^+ has its band origin at $2183.9496(2)cm^{-1}$, close to some recent <u>ab initio</u> calculations.[92] I.r. and Raman spectra of FC(O)SCl show that <u>trans</u>- and <u>cis</u>-isomers are both present (with the <u>trans</u>-form more abundant at room temperature).[93]

I.r. spectra of isotopic carbon disulphide, $^{12}C^{32}S^{34}S$, $^{12}C^{32}S^{33}S$, $^{12}C^{32}S_2$ and $^{13}C^{32}S_2$, gave improved molecular constants.[94] The observed i.r. intensity of ν_1 of CS_2 is largely due to coupling of ν_1 with ν_2 or ν_3 <u>via</u> a dimer model.[95] The Raman spectrum of CS_2 shows that ν_1 shifts very little with increased pressure, because of the strong S=C=S bonding.[96] The first C≡S triple bond has been characterised: νC≡S in $F_3CC{\equiv}SF_3$ is at $1740cm^{-1}$.[97]

The ν_3 band centre of $:CF_2$ was found to be at $1114.4435(3)cm^{-1}$ from i.r. diode laser spectroscopy.[98] The rotational fine structure of the ν_1 band of $CF_3{}^{37}Cl$ has been analysed, to give molecular constants for the band.[99] I.r. and Raman spectra of solid CF_3Br at 77K gave assignments for all 6 fundamentals.[100] Overtone and combination bands of CF_3Br and CF_3I can be assigned using continuously variable CO_2 laser absorptions.[101] A detailed study of the i.r. spectrum of solid $CFBr_3$ shows that the a_1 modes give doublets, the e modes triplets. This can be rationalised in terms of an orthorhombic crystal structure, based on the C_{2v} factor group.[102] High-resolution i.r. spectra of CCl_4 in the ν_3 and $\nu_1+\nu_4$ region, in a krypton matrix, show that there are two dominant sites in the matrix, one of T_d and one of C_{3v} symmetry.[103,104]

4.2 Silicon.- Methods of determining 'isolated' M-H stretching modes, where M = Si or Ge, have been reviewed.[105] The 850cm^{-1} i.r. band in Si/H alloy films suggests the presence of $(SiH_2)_n$ groups.[106] The thermal decomposition of polysilane, $-(SiH_2)_n-$, can be followed by the reduction in intensity of the SiH_2 stretching and bending modes.[107] A detailed analysis has been made of the rotation/vibration spectrum of $^{28}SiH_4$ (840-1040cm^{-1}).[108] A series of papers reported very high resolution i.r. studies on $H_3Si^{35}Cl$ and $H_3Si^{37}Cl$ in the ν_1/ν_3, ν_2/ν_5, ν_6 and $2\nu_6$ regions.[109-112]

A full vibrational assignment for Ph_4Si was based on i.r. and Raman spectra of solutions, melts and crystalline samples.[113] The Raman spectrum (Ar matrix, 10K) of the pyrolysis products of tetramethyldisila-1,2-cyclohexenes-4 shows the presence of tetramethyldisilene, $Me_2Si{=}SiMe_2$, with νSi=Si at 630cm^{-1}.[114]

Isotopic shifts and force constant calculations on Si_2O_2 show that it is cyclic, with D_{2h} symmetry.[115] I.r. and Raman spectra of $Me_3SiOSiMe_3$ and $Me_3SiNHSiMe_3$ are consistent with D_{3d} symmetry and a C_3 axis along Si-X-Si, where X = NH or O.[116] Vibrational spectra were reported for crystalline methyl, ethyl and vinyloctasiloxanes (100-1000cm^{-1}).[117] A band at 875cm^{-1} in the i.r. spectrum of evaporated silicon oxides, SiO_x, is due to a structural unit $Si-(O_ySi_{4-y})$, y = 2-4.[118] Shifts of νSi-O bands with pressure in vitreous silica can be related to Si-O-Si angle changes. The angles decrease with increased pressure, but to a smaller extent than in quartz.[119] Vibrational assignments for MO_4^{4-}, where M = ^{28}Si, ^{30}Si or Ge, are summarised in Table 1. A normal coordinate analysis was performed.[120]

Table 1 Vibrational assignments for MO_4^{4-} ions (/cm^{-1})

M =	Si	Ge
ν_1	780	710
ν_2	450	330
ν_3	935	735
ν_4	605	360

A comparative study established the following characteristic absorption regions in the i.r. spectra of calcium silicates: SiO_4 810-865cm^{-1}, $-SiO_3$ 850- 900cm^{-1}, $=SiO_2$ 910-1000cm^{-1}, ≡SiO >1000cm^{-1}, ≡Si-O-Si≡ (chains) 630-700cm^{-1}, (rings) 570-600cm^{-1}.[121] An i.r. band at 515cm^{-1} due to interstitial oxygen in silicon was assigned to ν_sSi-O-Si.[122] Raman spectra of the high-pressure polymorph of $MgSiO_3$ contain bands ca. 800cm^{-1} due to SiO_6 groups in this ilmenite structure.[123] ν(≡Si-O-Si) and ν(≡Si-O-Ga) modes were identified on very thin films of SiO_2 deposited on a ZnSe window after exposure to $(Me_3Si)_2NH$ or $GaMe_3$ respectively.[124] Raman spectra have been reported for the clinopyroxenes $CaScAlSiO_6$ (I)/ $CaTiAl_2O_6$ (II). νSi-O-Si was very weak in 70%I/ 30%II, suggesting non-random Si/Al distribution in tetrahedral sites.[125] Detailed vibrational analyses have been proposed for Cl_2MeSi-O-$SiCl_2Me$ [126] and $Cl_3SiO(SiCl_2O)_nSiCl_3$, where n = 1 or 2.[127] Some skeletal mode assignments have been made for M(hp)Cl_2, M(hp)$(OH)_2$ and $[M(hp)O]_2$, where hp = hemiporphyrazine, M = Si or Ge. Thus, νM-Cl are at 493, 446cm^{-1} (Si), 347, 317cm^{-1} (Ge); in the polymers νM-O are at 990cm^{-1} (Si), 903cm^{-1} (Ge).[128]

I.r. spectra have been listed and assigned for Tl_2XS_3, where X = Si or Ge.[129] The Raman spectra of Si_xS_{1-x} glasses, 0.25 > x, give bands due to isolated $Si(S_{\frac{1}{2}})_4$ tetrahedral units.[130] I.r. bands in $H(SiCl_2)_nH$, where n = 3-7, are very similar to those in perchloro-oligosilanes.[131]

<u>4.3 Germanium</u>.- Accurate values for the band origins of ν_2 (929.9081 ± 0.0010cm^{-1}) and ν_4 (820.7326 ± 0.0082cm^{-1}) have been reported for $^{74}GeH_4$.[132] A high-resolution i.r. spectrum of GeH_3F shows that ν_3/ν_6 and ν_2/ν_5 give Coriolis-coupled band pairs.[133] Gas-phase i.r. spectra of monoisotopic $H_3{}^{74}Ge{}^{35}Cl$ gave $\nu_1(a_1)$ = 2119.97703(19)cm^{-1} and ν_2(e) = 2128.48465(8)cm^{-1}.[134] I.r. spectra of monoisotopic $H_3{}^{74}Ge{}^{79}Br$ and $H_3{}^{74}Ge{}^{81}Br$ have been reported in the regions of ν_1, ν_4 and ν_6 [135] and ν_2, ν_5 and $\nu_3+\nu_6$.[136]

I.r. and Raman spectra of $(Ph_2GeO)_x$, where x = 3, 4 or n, contain ν_sGeO ca. 500cm^{-1}, ν_{as}GeO ca. 800-900cm^{-1}.[137] Skeletal mode assignments have been proposed for $GeF_2(SO_3F)_2$, $SnF_2(SO_3F)_2$, $Sn(SO_3F)_4$, $Ge(SO_3F)_6^{2-}$ and $Sn(SO_3F)_6^{2-}$.[138]

A detailed vibrational assignment has been given for GeF_5^-, as XeF_5^+, ClO_2^+, NO_2^+ and NBu_4^+ salts. Except for the last, polymeric species are present, but the NBu_4^+ salt contains D_{3h} monomers.[139]

$GeF_4.L$, where L = H_2O, MeOH, Me_2O or Me_2CO, have been prepared in matrices - the first such 1:1 complexes of GeF_4 with oxygen donors. νGe-F modes were as expected for a five-coordinate system.[140]

4.4 Tin.- The i.r. spectra (ca. 1900cm^{-1}) have been recorded for natural H_3SnCl and monoisotopic $H_3{}^{116}Sn^{35}Cl$. Band centres for the latter were at ν_1 1917.1932(7)cm^{-1}, ν_4 1924.6790(4)cm^{-1}. The isotopic shifts for ν_4 were -0.170(6), -0.326(6)cm^{-1} respectively on going to ^{118}Sn, ^{120}Sn.[141] Skeletal wavenumbers have been assigned in $(CH_2{=}CH)_3SnX$, where X = Cl, Br or I.[142] νSn-C were listed for $R_2Sn[SP(S)O_2G]_2$, where R = Me, Et or ^{n}Bu, G = $-CH_2CMe_2CH_2-$, $-CMe_2CMe_2-$ or $-CMe_2CH_2CHMe-$.[143] ν M-C and νM-S, where M = Sn or Pb, were assigned in various organotin and organolead 1,3-dithiolates, e.g. for $Me_2Sn(C_6H_3ClS_2)$ ν_{as}Sn-C 550cm^{-1}, ν_sSn-C 523cm^{-1} and νSn-S 361 and 330cm^{-1}.[144] The skeletal mode assignments (ν_s, ν_{as}Sn-C, νSn-O) for $R_3Sn(L-L)$, where R = Me, Ph or C_6H_{11}, L-L = 1,3-diphenyl-1,3-propanedionato, or R = Ph or C_6H_{11}, L-L = tropolonato, are all consistent with monomeric, five-coordinate tin species.[145] νSnC_3 is 503-548cm^{-1} in $Me_3SnOSPRR'$, where R,R' = Me or Ph.[146] νSn-C (500-600cm^{-1}) and νSn-NCO (349-380cm^{-1}) were assigned in N-di- and -triorganotin(IV) derivatives of saccharin, $R_nSn[NC(O)C_6H_4SO_2]_{4-n}$, where n = 2 or 3, R = Me, Et, ^{n}Pr, ^{n}Bu or C_6H_{11}.[147] The i.r. and Raman spectra of $(Ph_2Sn)_6$ reveal strong coupling between Sn_6 ring modes and the mass-sensitive vibrations of the phenyl groups.[148]

The skeletal (νSn-N, νSn-X) modes of SnX_4L_2, where X = Cl or Br and L = N-(phenyl)benzaldimine and derivatives, are all assignable in terms of trans geometry.[149] $(R_2SnO)_3$, where R = ^{t}Bu or ^{t}amyl, give i.r. and Raman spectra indicative of planar six-membered ring structures.[150] K_4SnTe_4 in ethylenediamine solution gives Raman bands at 180 and 130cm^{-1}, assigned to tetrahedral $SnTe_4^{4-}$.[151]

Skeletal mode assignments for $SnCl_2Br_2(MeCN)_2$ were consistent with C_1 symmetry. For $SnCl_2Br_2L_2$, where L = THF or OEt_2, C_{2v} symmetry appeared to be appropriate (trans L ligands, cis Cl_2 and Br_2).[152] The skeletal modes of $SnCl_4$.18-crown-6.$2H_2O$ and related $SnCl_4$ adducts suggest a cis-configuration of the non-halide ligand atoms. The Me_2SnCl_2 analogues give spectra consistent with an all-trans octahedral geometry at the tin.[153] $S_2N_2.SnCl_4$ gives νSnCl at 354cm^{-1} (i.r.), 355cm^{-1}(Raman). The coincidence suggests C_{2v}

symmetry, with distorted octahedral coordination at the tin.[154] ν_{as} and ν_sSn-X were assigned in $X_{4-n}Sn[Mn(CO)_4PPh_3]_n$, n = 2, X = Cl, Br or I, and νSn-X for n = 3. The values are given in Table 2.[155]

Table 2 Sn-X stretches in $X_{4-n}Sn[Mn(CO)_4PPh_3]_n$ (/cm^{-1})

n = 2	X =	Cl	Br	I
ν_{as}		304	213	176
ν_s		287	202	168
n = 3				
ν		278	195	160

4.5 Lead.- Detailed assignments of skeletal modes have been given for: $(Ph_3PbXY)^-$, where X, Y = Cl, Br or I; $(Ph_6Pb_2X_2Y)^-$, where X, Y = Cl or Br; $(Ph_2PbX_4)^{2-}$, where X = Cl or Br; $(Ph_2PbCl_3)^-$; Ph_3PbX and Ph_2PbX_2, where X = Cl, Br or I. I.r. and Raman spectra show that all the triphenyl-lead systems, except Ph_3PbI, have five-coordinate lead, while all the diphenyl-lead systems, except Ph_2PbI_2, have six-coordinate lead.[156] The effects of pressure on the Raman spectrum of crystalline $PbCl_2$ show that the anharmonic effects are greater for motions involving Cl than for those only involving Pb.[157]

The Raman spectra of PbI_2X^-, where X = Cl, Br or I, have been assigned - Table 3. All of the halides are bridging.[158]

Table 3 Vibrational assignments for PbI_2X^- (/cm^{-1})

X =	I	Br	Cl
νPbX		156	184
νPbI	126	124	116
νPbI	105	112	93
δPbICl			72
δPbIBr		67	
δPbI_2	55		
δPbI_2	36	36	39

5 Group V

5.1 Nitrogen.- Pulsed multichannel laser Raman Stokes spectroscopy gave information on the vibrational states V = 1-5 of N_2 $(X'\Sigma_g^+)$.[159] H_2N^+=N^- is formed in a glass at 80K or an Ar matrix at 10K by the irradiation of carbamoyl azide. $\nu N{=}N$ was seen at $1574cm^{-1}$, cf. $1529cm^{-1}$ for trans-HN=NH.[160] Ab initio calculations gave values for the vibrational wavenumbers and intensities of HN_2^+. ν_1 and ν_3 were predicted to be at 3234, $2254cm^{-1}$.[161] The rotational structure of the ν_1 band of $D^{14}N_3$ has been analysed - the band centre is at $2478.32(1)cm^{-1}$.[162] The ν_1 fundamental (νNH) has been observed in $DCNH^+$, with a band origin at 3473.10392(21) cm^{-1}.[163] The CARS spectrum of NH_3 and the free radical NH_2 have been recorded in the νNH region (3334, $3220cm^{-1}$ respectively).[164] A detailed analysis has been made of the rotational fine structure of the ν_2 band of NH_3.[165] The ν_2 band origins for $^{14}ND_3$ and $^{15}ND_3$ are at 745.5993(15) and $739.5349(15)cm^{-1}$.[166] I.r. spectra of strongly hydrogen-bonded complexes $RNH_2.HCl$ in Ar or N_2 matrices (R = H, alkyl) have been analysed.[167,168] The i.r. spectrum of the crystalline dihydrate of ammonia, $NH_3.2H_2O$, is consistent with the space group $P2_12_12_1$.[169] The Raman spectra of $CX_3OX.NX_3$, where X = H or D, at 77K can be assigned in terms of a polymeric chain, of C_s symmetry.[170]

The i.r. and Raman spectra of KNCS contain bands due to a' modes at 2024, 961, 950, 750 and 490 cm^{-1}, with an a" mode at $485cm^{-1}$.[171] I.r. spectra of gaseous and matrix-isolated HNCSe contain bands assigned as follows: ν_1 3535, ν_2 1957, ν_3 625, ν_4 508, ν_5 $419cm^{-1}$.[172] A normal-coordinate analysis has been performed for Me_3ECl^+, where E = N, P, As or Sb, from literature data.[173] The Raman and i.r. spectra of polycrystalline $Li^+NH_2SO_3^-$ are assigned on the basis of the space group C_{2h}^1, with four molecules per unit cell.[174]

Some skeletal modes have been assigned from the Raman spectra of silatranes, $RSi(OCH_2CH_2)_3N$.[175] Normal coordinate analyses on Me_2NPF_2, Me_2NPCl_2 and $(CD_3)_2NPCl_2$ showed that "PNC_2" modes near 700, 1000 and $1300cm^{-1}$ are in fact due to complex mixtures of internal coordinates. The results suggest a PN bond order of ca. 1.6.[176] $\nu N{=}P$ in $F_5S\text{-}N{=}PCl_3$ is at $1315cm^{-1}$, νNSe in $F_5S\text{-}NSeCl_2$ at $1045cm^{-1}$.[177]

High-resolution FT i.r. for $^{14}N^{16}O_2$ gave precise measurements of line positions and intensities for the ν_1 band in the gas phase.[178] Raman spectra of the solid NO dimer in the isotopic forms $^{16}O^{14}N^{14}N^{16}O$, $^{16}O^{15}N^{15}N^{16}O$ and $^{18}O^{15}N^{15}N^{18}O$ have been reported. Observed isotopic shifts gave an unambiguous assignment for ν_1 to ν_6.[179] An estimate was also given for ν_6 from ν_4-ν_6 in the i.r. spectrum.[180] Absorption intensities were measured (FT i.r.) for ν_1 and ν_3 of N_2O_3.[181] Raman spectra of solid N_2O_4 show that deposition at 80K gives O_2NNO_2, at 15K O_2NNO_2 and also NO_2 monomer, together with the isomer O_2NONO. Warming to 180K leads to auto-ionisation, to $NO^+NO_3^-$, which is then stable in the range 15-180K.[182] Measurements of the temperature and pressure dependence of the i.r. spectrum of pure HNO_3 have been made in the ranges 236-294K, 0.05 to 1.1 torr.[183] $Li^+NO_3^-$ ion-pairs give $\nu_3$of NO_3^- split by 262cm^{-1}. Solvation of the ion-pairs by amines can be followed by the decreased splitting $\Delta\nu_3$ as the solvation proceeds.[184] I.r. and Raman spectra of NO_3^- isolated in RbX (X = Cl, Br or I) crystals show that the ions occupy sites of D_3 symmetry at room temperature, but C_{3v} at low temperature.[185]

Theoretical calculations were made of the vibrational spectra of NSF and SNF using a 6-31G* basis set. For NSF, 6-31G*/MP2 optimisation gave reasonable agreement with experiment. The so far unknown SNF was predicted to have a very different spectrum.[186] A normal coordinate analysis was performed on NS_2^+ using literature data.[187] Detailed analysis of the ν_1 and ν_3 bands of NF_2 (X^2B_1) was reported.[188,189]

5.2 Phosphorus.- The resonance Raman spectrum of bis(2,4,6-tri-*tert*-butylphenyl)diphosphene gave a value for νP=P for the first time (610cm^{-1}).[190] The same mode was found, with a harmonic wavenumber of 595cm^{-1}, from the resonance Raman spectrum of $(Me_3Si)_3CP{=}PC(SiMe_3)_3$.[191] I.r. diode laser spectroscopy on the gaseous PH radical ($^3\Sigma^-$ state) gave a band origin for the stretch at 2276.2106(29)cm^{-1}.[192] $P_5(^tBu)_4H$ has νP-H at 2285cm^{-1}.[193] I.r. and Raman spectra of $Me_2P(H){=}O$ were assigned, assuming C_s symmetry.[194]

Assignments to C_2PX skeletal modes (C_s symmetry) were made from the Raman and i.r. spectra of Ph_2PX, where X = Cl, Br or I.[195] The vibrational spectrum of $CH_2(PCl_2)_2$ showed that two conformers were present at room temperature.[196] Vibrational assignments were reported for $P(SiR_3)_3$, where R = Ph, CH_3 or CD_3.[197] $P_6Si_4(SiMe_3)_8$ has a very complex vibrational spectrum, i.e. it has very low symmetry.[198]

The i.r. spectra of a number of new phosphazenes containing amine substituents, e.g. $P_3N_3Cl_4(NEt_2)_2$, $P_3N_3Cl_2F_2(NEt_2)_2$, etc., were reported and νPN assigned.[199]

Force constants have been calculated for PO_4^{3-} present in MPO_4 lattices, where M = Am, Cm, Bk, Cf or Es.[200] I.r. and Raman spectra of $Mg_xCa_{10-x}(PO_4)_6F_2$, where x = 1-4, show that Mg incorporation has negligible effect on the P-O bond strength.[201] The i.r. spectra of $P_2O_5F_2^-$ and $P_2O_3F_4$ have been assigned.[202] The Raman and i.r. spectra of triclinic $Ca_2P_2O_7.2H_2O$ can be interpreted by factor group analysis. 8 P-O stretches can all be assigned unambiguously.[203] Comparison with P_4O_6 and P_4O_{10} was used to assign the i.r. and Raman spectra of P_4O_7.[204] The presence of chain-type polyphosphate anions in $AgCeP_4O_{12}$ was suggested from its i.r. spectrum.[205] General assignments of the vibrational spectrum of $K_{13}(KP_2W_{20}O_{72}).xH_2O$ have been proposed. The PO_4 groups appeared to have C_{3v} symmetry.[206]

Gas-phase i.r. spectra of $CH_3OP(O)Cl_2$ and $CD_3OP(O)Cl_2$ show that both gauche and trans conformers were present, but with much more of the former.[207] I.r. spectra of gaseous $(CH_3)_2P(O)OCH_3$, $(CH_3)_2P(O)OCD_3$, $(CD_3)_2P(O)OCH_3$, $(CD_3)_2P(O)OCD_3$ and $(CH_3)_2P(O)^{13}CH_3$ have been assigned. The methyl dimethylphosphinate is in the gauche conformation.[208] Vibrational assignments have been proposed for $(Me_2PS)_2O$ and $(Me_2PS)_2S$ in terms of C_{2v} symmetry; ν_sP-O-P is at 675cm^{-1}, ν_sP-S-P at 431cm^{-1}.[209] There is i.r. evidence for the presence of both O- and S-bonded isomers in difluorophosphino monothioacetate: $CH_3C(=S)OPF_2$ and $CH_3C(=O)SPF_2$, with 80% of the O-bonded form at room temperature.[210]

I.r. and Raman spectra of matrix-isolated S=P-Cl contain the following bands: νPS 716.1cm^{-1}, νPCl 462.4cm^{-1}, and δSPCl 229cm^{-1}. ^{34}S-enriched samples give data showing that the phosphorus atom is central, with $\hat{SPCl}$ ca.110°.[211] C_{2v} symmetry proved adequate for assigning i.r. and Raman spectra of $PS_2X_2^-$, where X = Cl or Br. For $PS_2Cl_2^-$ the assignments are as follows: $\nu_{as}PS_2(b_1)$ 704cm^{-1}, $\nu_sPS_2(a_1)$ 569cm^{-1}, $\nu_{as}PCl_2(b_2)$ 421cm^{-1}, $\nu_sPCl_2(a_1)$ 384cm^{-1}, $\delta PS_2(a_1)$ 239cm^{-1}, $\delta PSCl(b_1, b_2)$ 234cm^{-1}, $\delta PCl_2(a_1)$ 166cm^{-1}.[212] Vibrational assignments were given for crystalline $P(CH_2CH_2CN)_3$ and $P(CH_2CH_2CN)_3E$, where E = O, S or Se. Both νP=S and νP=Se interact with skeletal bending modes.[213] Raman data for PS_4^{3-} in $PrPS_4$ were assigned by factor group analysis.[214] I.r. and Raman spectra for $Tl_2P_2S_6$ can be assigned on the basis of $P_2S_6^{2-}$ with D_{2h} symmetry.[215] Molten $Tl_2P_2S_6$, however, involves the equilibrium $P_2S_6^{2-} \rightleftharpoons 2PS_3^-$; the monomer is planar, with D_{3h} symmetry.[216] $P_2Se_6^{4-}$

in $Pb_2P_2Se_6$ appears to have C_{2h} symmetry.[217]

I.r. and Raman spectra of $PF_2(NH_2)$ have been assigned in gas, liquid and solid phases. $\nu_s PF_2$ is at 806cm^{-1}, $\nu_{as} PF_2$ 760cm^{-1}, νPN 875cm^{-1}.[218] $PF_2(^{14}NH_2)$ in a nitrogen matrix is present as two different species. These are differentiated by the NH_2 group conformations.[219] The ν_3 band of PF_5 has been studied at low rotational and vibrational temperatures in a pulsed supersonic free jet of PF_5. No splitting due to Berry pseudorotation was observed in either the fundamental or in the hot band $\nu_3 + \nu_7 - \nu_7$.[220] An <u>ab initio</u> calculation of the quadratic force field of PF_5 gave good agreement with the observed vibrational wavenumbers.[221]

<u>5.3 Arsenic</u>.- A high-resolution study of the ν_2 and ν_4 bands of AsH_3 gave the following values for the band centres: ν_2 906.7518314(702)cm^{-1}; ν_4 999.2245023(589)cm^{-1}.[222] νAs-C is near 470cm^{-1} and νAs-S in the range 350-400cm^{-1} in $PhAs(S_2COR)_2$, where R = Me, Et, ^{n}Pr, ^{i}Pr, allyl or ^{n}Bu.[223] AsOCl, from the reactions of Ag with $AsCl_3/O_2$ mixtures at 1300K, has νAs=O at 984.4cm^{-1}, νAs-Cl at 378.7cm^{-1}, confirmed by $^{16}O/^{18}O$ and $^{35}Cl/^{37}Cl$ shifts.[224] I.r. matrix spectra of $M^+AsO_3^-$, where M = K, Rb or Cs, and of the ^{18}O-enriched Cs^+ salt, can be assigned in terms of C_{2v} symmetry for AsO_3^-. Analogous data for $M^+ClO_3^-$, where M = Li, Na, K, Rb or Cs, show that ClO_3^- has C_{3v} symmetry.[225] Several fundamental modes for As_2O_5 and $AsSbO_5$ were assigned using factor group analysis in the i.r. and Raman spectra.[226] νAs-S bands due to the AsS_3 unit were assigned in the Raman spectra of glassy and liquid $As_xS_yBr_z$ samples.[227] The Raman spectra of glassy samples from the As-S-I system can be interpreted using a model comprising a disordered network of AsI_3 pyramids connected by -S- and (-S-S-)S bridges.[228] The vibrational spectrum of $AsCl_2F_3$ can be assigned under C_{2v} symmetry (equatorial chlorine atoms).[229] Analysis of the chemiluminescence spectra of AsI and SbI gave the following values for ω_e: 257 $\pm$ 2cm^{-1} (AsI), 206 $\pm$ 6cm^{-1} (SbI).[230]

<u>5.4 Antimony</u>.- Some general assignments were given from i.r. and Raman spectra of $Sb(NRR')_{3-m}Cl_m$, for M = 2 or 1, R,R' = Me or Et, and of ClSb(NRR')OR", where R, R', R" = Me or Et.[231] Dimeric $[Cl_4Sb(NCNSiMe_3)]_2$ gives i.r. and Raman spectra consistent with

a centrosymmetric structure of C_{2h} symmetry.[232] The vibrational spectra of $Sb(OEt)_nX_{3-n}$ (where X = Cl, n = 1-3; X = Br, n = 1,2; X = I, n = 2) were assigned by analogy with related species.[233] I.r. spectra and some assignments have been given for (1), where R, R^1, R^2, R^3 = H or alkyl, X = Cl, OMe or O^tBu.[234]

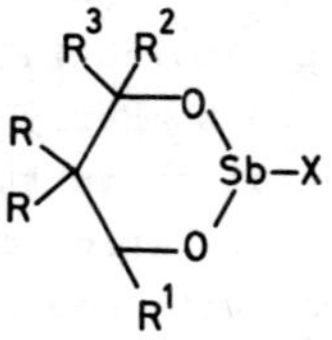

(1)

The compounds Me_2Sb-E-$SbMe_2$ have νSb_2Me bands at 230 and 215 cm^{-1} for E = Se, νSb_2Te at 180cm^{-1} for E = Te.[235] Polarised Raman spectra of a single crystal of $CaSbF_6$ suggest that it has a rhombohedral structure, with some distortions from O_h symmetry for SbF_6^-.[236] Tentative skeletal mode assignments were proposed for $SbX_3L_{1.5}$, where X = Cl or Br, $BiCl_3L_2$ and SnX_4L, where X = Cl or Br, for L = RNHC(S)C(S)NHR, R = Me, Et, ^iPr, ^nBu, C_6H_{11} or CH_2Ph.[237] The Raman spectrum of $SbCl_3.S_8$ is a superposition of the component spectra.[238] Raman spectra were reported for substituted pyridinium bromoantimonate(III) compounds containing $SbBr_4^-$, $SbBr_5^{2-}$ or $Sb_2Br_9^{3-}$.[239]

6 Group VI

6.1 Oxygen.- The temperature dependence of the i.r. spectrum of OH^- in NaCl crystals has been measured (1.45-4.2K). A band at 3653.2cm^{-1} was assigned to dimers of OH^-.[240] I.r. absorptions due to the radical $HOSO_2$ were detected in an argon matrix at 10K. The vibrational wavenumbers are: νOH 3539.9, $\nu_{as}SO_2$ 1309.2, ν_sSO_2 1097.3, νS-OH 759.5cm^{-1}. $^{16}O/^{18}O$ shifts were used in the assignment.[241] A theoretical analysis has been carried out on the overtone and combination vibrational spectrum of H_2O.[242] High-resolution far i.r. (10-60cm^{-1}) of $H_2{}^{16}O$, $H_2{}^{18}O$, $HD^{16}O$, $HD^{18}O$, $D_2{}^{16}O$ and $D_2{}^{18}O$ in Ar or Kr matrices can be interpreted in terms of the rotational/translational coupling model.[243] Low-wavenumber Raman bands of supercooled water have been measured down to -20°C. The O-O stretch (190cm^{-1}) changes markedly with temperature, but the O-O-O bend (60cm^{-1}) does not.[244] I.r. bands were identified due to two 1:1 complexes of H_2O and H_2S ($H_2O..HSH$, $H_2S..HOH$) in Ar matrices. There was also evidence for a third 1:1 species, with a bifurcated hydrogen bond between water and hydrogen sulphide, and up to 5 possible $H_2S(H_2O)_2$ species.[245] Several assignments to normal modes were made for $H_2O...HF$, including O...H stretching at 176 ± $15cm^{-1}$.[246] Various components of the first Raman overtone of νOH in liquid water (20-95°C) can be attributed to non-hydrogen-bonded OH, triply and quadruply hydrogen-bonded H_2O species.[247] Refined vibrational data were given for H_2O isolated in D_2O cubic ice.[248] The pressure dependence of the wavenumber and i.r. intensity of the OH stretch in a dilute solution of HDO in D_2O ice VII was investigated (23.6-189kbar).[249] The transformation of ice I_h to ice IX was observed from the Raman spectra of H_2O or D_2O ice at 88K and <u>ca</u>.3 kbar.[250] I.r. absorption intensities were measured for H_2O molecules in hydrated β-alumina.[251] νOH modes of H_2O in molten alkali metal bromides can be related to the radius of the solvent cation and to temperature.[252] A high-resolution study of the ν_2 band of T_2O vapour showed the band centre to be at 995.3258(29)cm^{-1}.[253] I.r. spectra gave the following fundamental wavenumbers for $T_2{}^{18}O$: ν_1 2220, ν_2 988, ν_3 2346cm^{-1}.[254]

Full vibrational assignments have been given for OH_3^+ and OD_3^+ (as AsF_6^- and SbF_6^- salts) - Table 4.[255] The ν_2 ($1^- \leftarrow 0^+$) inversion mode of H_3O^+ was observed for the first time using high-resolution i.r. spectroscopy. The band centre was at 954.417(14)cm^{-1}.[256]

Table 4 Vibrational assignments for OH_3^+ and OD_3^+ (/cm^{-1})

	OD_3^+	OH_3^+
$\nu_1(a_1)$	2300	3150
$\nu_2(a_1)$	715	900
$\nu_3(e)$	2450	3300
$\nu_4(e)$	1182	1620

A Raman study of hydrogen-bonding in H_2O_2 in the range 40 to -75°C is explicable in terms of a two-state model with a continuum of H_2O_2 molecules, and close-packed OH groups in two kinds of environment.[257] $M^+O_3^-$, where M = Ga, In or Tl, were detected in matrix systems. ν_3 of O_3^- gave several components, due to site-splitting; these were 830-870cm^{-1} for M = Ga or In,

787-819cm^{-1} for M = Tl. The difference was thought to be due to a covalent M-O_3 contribution in the Ga and In cases.[258] I.r. spectra of matrices containing Ozone and ICl contain bands due to the 1:1 complex O_3.ICl. This in turn decomposes to give bands of OICl and O_2ICl.[259]

6.2 Sulphur, Selenium, and Tellurium Rings and Chains.- I.r. and Raman spectra of SF_5(SSF) contain bands due to $\nu_{as}S_3$ (585cm^{-1}) and ν_sS_3 (472cm^{-1}).[260] The Raman spectrum of S_6.S_{10} is a simple superposition of those of S_6 and S_{10} themselves.[261] The Raman spectrum of liquid sulphur to 300°C can be explained in terms of the ratio between S_8 and S chains.[262] Se_2S_5 gives a Raman spectrum consistent only with the isomer (2). SeS_5 gives the spectrum expected for a six-membered ring.[263]

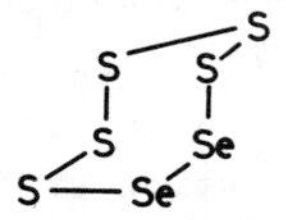

(2)

The resonance Raman spectrum of Te_3 in a nitrogen matrix gives an overtone series for the symmetric stretch. For $^{126}Te_3$ ω_1 = 206.3 ± 0.15cm^{-1}; for $^{130}Te_3$ ω_1 = 203.1 ± 0.2cm^{-1}. Force constants suggest a bond order greater than 1 in the molecule.[264] The Raman spectrum of liquid tellurium is different from that of liquid selenium, and shows no evidence for the presence of two-coordinate chains.[265]

6.3 Sulphur-Nitrogen Compounds.- Sulphur atoms react with monomeric and dimeric NO in Ar matrices (13-20K) to form SNO and SN_2O_2. The i.r. spectra of $S^{14}N^{16}O$, $S^{15}N^{16}O$ and $S^{14}N^{18}O$ suggest the following fundamental wavenumbers for $S^{14}N^{16}O$: νNO 1554.9, νNS 790.0, bending mode ca.567cm^{-1}. SN_2O_2 appears to exist as two isomeric forms, S(O)N.NO and ONSNO.[266] A novel sulphur(VI) dication has been prepared: $(Me_2N)_2SF_2^{2+}$, with $\nu_{as}SF_2$ 945, ν_s910 cm^{-1}.[267] The i.r. spectrum of $S(NAsR_2)_2$, where R = CH_3, C_6H_{11} or C_6H_5, contains ν_{as},ν_s of the NSN skeleton ca.1125-1176cm^{-1}, ca. 1000-1070cm^{-1} respectively.[268] SO_2.amine complexes have been studied by i.r. spectroscopy. $\nu_{as}SO_2$ (1350cm^{-1} in free SO_2) is at 1338cm^{-1} in SO_2NH_3, ca.1270cm^{-1} in $SO_2.NMe_3$, confirmed by ^{18}O

substitution.[269] I.r. bands were reported for $(H_2N.SO_2.NH_3)^+$ HSO_4^-,[270] and Raman spectra for $(CF_3SO_2)_2NH$.[271] Vibrational assignments were given for the cation $(S_3N_2)_2N^+$,(3), e.g. $\nu_{as}S_2N$ 865cm^{-1}, ν_sS_2N 738cm^{-1}.[272] The cyclic radical cation $S_4N_4^+$, as the $FeCl_4^-$ salt, has i.r. bands at 952, 941, 763, 750 and 715cm^{-1}, i.e. lower than in $S_4N_4^{2+}$. The large number of bands suggests that $S_4N_4^+$ has very low symmetry.[273] (4) gives νS=N at 1171cm^{-1}, (5) νS-Cl at 440cm^{-1}.[274]

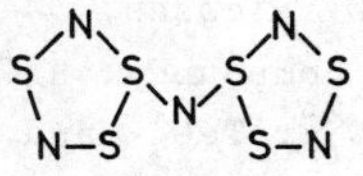

(3)

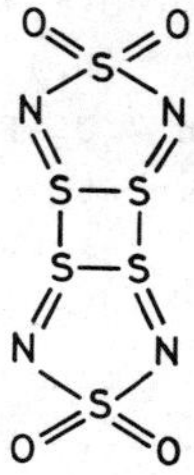

(4)

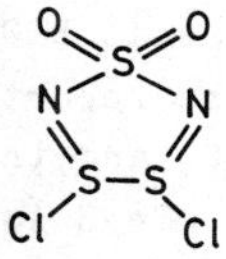

(5)

6.4 Other Sulphur and Selenium Compounds.- F.T.i.r. spectra of the transient molecule sulphine, CH_2SO, show that it has ν_5 at 1175.75, ν_9 at 762.07cm^{-1}.[275] The Raman spectrum of solid SO_2 provides good evidence to support an orthorhombic structure, with the acentric space group C_{2v}^{17}.[276] High-resolution F.T.i.r. spectroscopy of gaseous SO_3 shows that the ν_3 band origin is at 1391.5205(15)cm^{-1}.[277] νS-O bands are seen at 962 (a_1), 1116.9 (b_2) and 1098.6(e)cm^{-1} in the i.r. and Raman spectra of matrix-isolated Cs_2SO_4.[278]

Me_2SF^+ has been prepared and its vibrational spectrum reported.[279] The i.r.-forbidden ν_6 band of SF_6 is seen at 351 cm^{-1} in the gas phase, activated by Coriolis interactions.[280] The vibrational overtone spectrum of SF_6(gas) at 193K, in the range 2800-4200cm^{-1}, shows 10 quaternary combination bands, and one band of 5 vibrational quanta, $2\nu_2 + 3\nu_3$.[281] A full vibrational assignment was proposed for CF_3SBr.[282]

The band centre for ν_2 of H_2Se is at 1034.1652(19)cm^{-1}.[283] The i.r. spectra of R_2SeX_2, where R = $PhCOCH_2$, X = Cl or Br, and their ^{74}Se isotopic analogues show that νSeC is ca.554cm^{-1}, ν_sSeCl_2 282cm^{-1}, $\nu_{as}SeCl_2$ 241cm^{-1} ($\nu SeBr_2$ modes were not seen).

There was no evidence for Se-O bonding.[284] Force constants and vibrational mean square amplitudes were calculated for $SeOX_2$, where X = F, Cl or Br.[285] The proton in the anion of $NO_2^+HSe_2O_7^-$ leads to splitting of ν_sSe-O-Se (544, 560cm^{-1}).[286]

6.5 Tellurium.- I.r. spectra of Et_3Te^+ with various anions have been reported. νTe-C is in the range 470-550cm^{-1}.[287] νTe-OH (644cm^{-1}) and $\nu TeCl_4$ (283cm^{-1}) were seen in the Raman spectrum of $Te(OH)Cl_4^-$.[288] The vibrational modes of trans-$(MeO)TeClF_4$ can be assigned under C_{4v} symmetry: νTeO(a_1) 599cm^{-1}, νTeCl(a_1) 398cm^{-1}, νTeF_4(e) 692cm^{-1}, (a_1) 619cm^{-1}, (b_1) 583cm^{-1}.[289] A band due to νTeO_2 occurs at 810-820cm^{-1} in V_2O_5-P_2O_5-TeO_2 glasses.[290] I.r. spectra were reported for a variety of Te/O compounds: H_2TeO_4, H_6TeO_6, TeO_3 and the Na^+ and K^+ salts of TeO_4^{2-}, TeO_3^{2-} and $Te_2O_7^{4-}$.[291] νTe-S is at 265cm^{-1} in diphenyltellurium β-mercaptopropionate.[292] In the resonance Raman spectra of TeX_6^{2-}, where X = Cl or Br, the most enhanced band is ν_2 (e_g), in resonance with $^3T_{1u} \leftarrow {}^1A_{1g}$ and $^1T_{1u} \leftarrow {}^1A_{1g}$ transitions. This indicates that the ions are tetragonally distorted in these excited states.[293]

7 Group VII

The ν_1/ν_3 region of H_2F^+ (3080-3500cm^{-1}) has been investigated by velocity modulation i.r. laser spectroscopy. The band origins are: ν_1 3348.710(20)cm^{-1}, ν_3 3334.6718(32)cm^{-1}.[294] I.r. and Raman spectra of FHF^- and FDF^- isolated in single crystals of alkali halides have been obtained over a range of temperature and pressure. Low temperatures lead to a decrease in ν_3 and an increase in ν_2. The reverse changes occur on increasing the pressure.[295] I.r. spectra of H_2F_2 in Ar, N_2, CO and several mixed matrices have been reported. All 6 fundamentals are seen in the Ar matrix: ν_1 3826, ν_2 3702, ν_3 561, ν_4 400, ν_5 263 and ν_6 446cm^{-1}. The relative intensities of ν_1 and ν_2 depend strongly on the matrix environment. Higher oligomers (stretching modes 3430-3175cm^{-1}) were also detected.[296] Different results on $(HF)_2$ have been given, i.e. H-F stretches at 3896 and 3825cm^{-1}. The strongest absorption of an open trimer is at 3702cm^{-1}. Four different groups of bands (3427-3118cm^{-1}) were assigned to cyclic $(HF)_n$, where n = 3,4,5 or 6.[297] I.r. and Raman spectra of $BaF(HF_2)$

and its 10% and 50% deuteriated derivatives at 90 and 300K show that the $(F-H-F)^-$ ion is not centrosymmetric, despite the short F...F distance.[298] νF-H in hydrogen-bonded complexes RX-HF, where R = alkyl, X = Cl, Br or I, decreases slightly with increased base proton affinity and with increased size of R and/or X, i.e. increased hydrogen-bond strength in that order. CH_3X-HF complexes give two non-degenerate H-F librational modes, i.e. they are bent.[299]

Assignments for F-NCO in an argon matrix are consistent with C_s symmetry. ^{15}N substitution was used to confirm this assignment.[300] The ν_2 and ν_3 fundamentals of F-N=O are at 765.3551(4) cm^{-1} and 519.5980(4)cm^{-1} respectively.[301] The ν_2 mode of the FO_2 radical is at 579.31839(35)cm^{-1}; the rotational fine structure suggested O-F and O-O bond lengths of ca. 1.65 and 1.20Å.[302] Other i.r. bands due to FO_2 were also reported.[303]

The values of ω_e for the transient radical anions $Cl_2^{\dot{-}}$ and $Br_2^{\dot{-}}$ are at 286.4, 187.2cm^{-1} respectively (from resonance Raman spectra in aqueous solution).[304] Raman spectra of Cl_3^- show that it is stable in acetonitrile, nitromethane and N,N-dimethylformamide, but not in water, DMSO, dioxan, THF or HMPA.[305] I.r. spectra have been recorded for HCl, DCl, HBr and DBr in liquid Ar, Kr, Xe, O_2 and N_2.[306]

A full vibrational assignment for planar $ClONO_2$ in N_2 and Ar matrices gave no evidence for the presence of any isomers.[307] The Raman spectra of $KClO_3$ at room temperature have been recorded at pressures up to 60kbar. The phase change I ⟶ II is very sluggish at ca. 7kbar (the two forms co-exist over a range of ca. 5 kbar).[308] Isotopic substitution was used to give a detailed vibrational assignment of anion wavenumbers for matrix-isolated $MClO_3$ and $MClO_4$, where M = Li, Na, K or Cs, e.g. ^{18}O substitution on $Li^{35}ClO_4$, $K^{35}ClO_4$ and $Cs^{35}ClO_3$. The ClO_4^- ions are very slightly distorted tetrahedra, while $M^+ClO_3^-$ have C_{3v} symmetry.[309] It is possible to detect species $(ClO_4^-).(H_{aq}^+)$ in the Raman spectra of aqueous perchloric acid at high concentrations.[310] I.r. spectra of mixed crystals $KClO_4$-$KMnO_4$, $KClO_4$-KBF_4 (as well as K_2SO_4-K_2XO_4, where X = Se, Cr or Mn) show that all have symmetry D_{2h}^{16}, with tetrahedral anions.[311]

Surface-enhanced Raman spectra (SERS) of Br_2 and Br_3^- at a platinum electrode show a band at 161cm^{-1} due to Br_3^- adsorbed on the Pt, and another ca. $305cm^{-1}$ due to Br_2 molecularly bound to the Pt electrode.[312] Br_2 is adsorbed at preferred sites on graphite and KBr, but not on fused quartz, as shown by Raman spectra.[313] Fundamental modes of BrO_3^- (in $KBrO_3$) are at 790, 360 cm^{-1} (a_1), 835, $413cm^{-1}$ (e).[314]

Excitation of the Raman spectrum of ICl by an argon ion laser gives strong resonance enhancement.[315] The resonance Raman spectrum of I_3^-, generated by the 488nm Ar^+ laser line, contains 11 overtones of the symmetric stretch (fundamental at $112cm^{-1}$). Bands at 153, 170, 264 and $304cm^{-1}$ were assigned as ν_3, $2\nu_2$, $\nu_1 + \nu_3$ and $2\nu_3$ (Σ^+) respectively.[316]

8 Group VIII

I.r. emission spectra of $^{40}ArH^+$ and $^{84}KrH^+$ gave the following parameters: ω_e (Ar) $2710.92cm^{-1}$, (Kr) $2494.65cm^{-1}$; $\omega_e x_e$ (Ar) $61.64cm^{-1}$, (Kr) $48.53cm^{-1}$.[317] A calculation of the vibrational characteristics of XeH^+ gave $\omega_e = 2313 \pm 50cm^{-1}$, $\omega_e x_e = 41 \pm 5$ cm^{-1}.[318] The Raman spectra of solid $O_2Xe(OTeF_5)_2$, $OXe(OTeF_5)_4$ and $Xe(OTeF_5)_4$ were reported. Reasonably detailed assignments were proposed, e.g. for the XeO_2 species ν_{as} XeO_2 889/880 cm^{-1}, $\nu_s XeO_2$ $820\text{-}830cm^{-1}$, for the XeO species νXeO 887, $873cm^{-1}$. The strongly coupled Te-O-Xe stretches lay in the range 415-500 cm^{-1}.[319]

References

1 N.H.Rich and A.R.W.McKellar, Can. J. Phys., 1983, 61, 1648.
2 M.Chabanel and Z.Wang, J. Phys. Chem., 1984, 88, 1441.
3 S.P.Konovalov and V.G.Solomonik, Russ. J. Inorg. Chem., 1984, 29, 948.
4 R.Teghil, B.Janis, and L.Bencivenni, Inorg. Chim. Acta, 1984, 88, 115.
5 V.G.Solomonik and S.P.Konovalov, Zh. Strukt. Khim., 1983, 24, 10.
6 C.P.Rao, A.M.Rao, and C.N.R.Rao, Inorg. Chem., 1984, 23, 2080
7 N.V.Porotnikov, N.G.Chaban, and K.I.Petrov, Russ. J. Inorg. Chem., 1983, 28, 1402.
8 A.P.Nemudryi, V.P.Isupov, and N.P.Kotsupalo, Russ. J. Inorg. Chem., 1983, 28, 1574.
9 E.V.Proskuryakova, N.V.Porotnikov, and N.G.Chaban, Russ. J. Inorg. Chem., 1984, 29, 780.
10 E.A.Secco, J.Chem. Phys., 1983, 79, 5208.
11 A.G.Maki, J.Mol. Spectrosc., 1983, 102, 361.
12 I.Kanesaka, H.Kawahara, and K.Kawai, J. Raman Spectrosc., 1984, 15, 165.
13 D.J.Jones, J.Emsley, and J.Rozière, J. Chem. Soc., Dalton Trans., 1984, 1625.
14 I.R.Beattie and J.E.Parkinson, J. Chem. Soc., Dalton Trans., 1984, 1363.

15 L.P.Sukhanov and A.I.Boldyrev, Zh. Fiz. Khim., 1984, 58, 654.
16 T.P.Pogrebnya and V.G.Solomonik, Zh. Fiz. Khim., 1983, 57, 2754.
17 V.M.Ozerova and V.G.Solomonik, Russ. J. Inorg. Chem., 1983, 28, 1408.
18 A.V.Kulikova, E.A.Kopyleva and M.A.Kolenkova, Russ. J. Inorg. Chem., 1984, 29, 965.
19 L.Rösch, C.Krüger and A.-P.Chiang, Z. Naturforsch., Teil B, 1984, 39, 855.
20 M.Handke, M.Kosinski, and P.Tarte, J. Mol. Struct., 1984, 115, 401.
21 N.V.Porotnikov, M.German and L.M.Kovba, Russ. J. Inorg. Chem. 1984, 29, 825.
22 M.I.Memon, W.F.Sherman, and G.R.Wilkinson, J. Mol. Struct., 1984, 115, 213.
23 H.C.Brown, J.S.Cha, and B.Nazer, Inorg., Chem., 1984, 23, 2929.
24 B.F.Spielvogel, F.U.Ahmed, M.K.Das, and A.T.McPhail, Inorg. Chem., 1984, 23, 3263.
25 J.L.Duncan, J.Harper, E.Hamilton, and G.D.Nivellini, J.Mol. Spectrosc., 1983, 102, 416.
26 D.D.Keeports and D.F.Eggers, Inorg. Chem., 1984, 23, 2305.
27 H.G.Srebny and W.Preetz, Z. Anorg. Allg. Chem., 1984, 513, 7.
28 L.A.Leites, A.P.Kurbakova, M.M.Kaganskii, Yu.L.Gaft, I.A. Zakharova, and N.T.Kuznetsov, Izv. Akad. Nauk SSSR, Ser. Khim 1983, 2284.
29 L.A.Leites, S.S.Bukalov, L.E.Vinogradova, V.N.Kalinin, N.I. Kobel'kova, and L.I.Zakharkin, Izv. Akad. Nauk SSSR, Ser. Khim., 1984, 954.
30 J.D.Odom, J.A.Smooter-Smith, J.R.Durig, and E.J.Stampf, J. Mol. Struct., 1984, 112, 19.
31 S.Guha and G.E.Walrafen, J. Chem. Phys., 1984, 80, 3807.
32 B.K.Nauryzbaeva, E.A.Kopylova, L.P.Ni, and T.V.Solenko, Kompleksn. Ispol'z. Miner. Syr'ya, 1983, 42 (Chem. Abstr., 1984, 100, 26589).
33 K.Kneipp, H.Bürger, D.Fassler, and W.Vogel, J. Non-Cryst. Solids, 1984, 65, 223.
34 O.S.Binbrek and A.Anderson, Can. J. Spectrosc., 1983, 28, 173
35 J.Gebicki and J.Liang, J.Mol. Struct., 1984, 117, 283.
36 W.Haubold and P.Jacob, Z. Anorg. Allg. Chem., 1983, 507, 231.
37 W.Preetz, H.G.Srebny, and H.C.Marsmann, Z. Naturforsch., Teil B, 1984, 39, 6.
38 H.G.Srebny, W.Preetz, and H.C.Marsmann, Z. Naturforsch., Teil B, 1984, 39, 189.
39 S.Kvisle and E.Rytter, J. Mol. Struct., 1984, 117, 51.
40 W.Gessner and D.Müller, Z. Anorg. Allg. Chem., 1983, 505, 195
41 M.C.Saine and E.Husson, Spectrochim. Acta, Part A, 1984, 40, 733.
42 S.K.Sharma, B.Simons, and H.S.Yoder, Am. Mineral.,1983, 68, 1113.
43 S.Lasserre and J.Derouault, Nouv. J. Chim., 1983, 7, 659.
44 S.A.Azimov, D.D.Gulamova, N.N.Mel'nik, M.Kh.Sarkisova, S.Kh.Suleimanov, and L.M.Tsapen'ko, Izv. Akad. Nauk SSSR, Neorg. Mater., 1984, 20, 469.
45 R.C.Mehrotra and J.V.Singh, Z.Anorg. Allg. Chem., 1984, 512, 221.

46 B.N.Ivanov-Emin, G.Z.Kaziev, T.Yu.Gerasimova, and S. Ol'gin-Kin'ones, Russ. J. Inorg. Chem., 1984, 29, 523.
47 L.Zhengyan, H.Janssen, R.Mattes, H.Schnöckel, and B.Krebs, Z.Anorg. Allg. Chem., 1984, 513, 67.
48 R.Klein and P.Rosmus, Theor. Chim. Acta, 1984, 66, 21.
49 A.K.Sengupta and K.Sinha, J.Fluorine Chem., 1984, 24, 153.
50 T.Tomita, C.E.Sjøgren, P.Klaeboe, G.N.Papatheodorou, and E.Rytter, J. Raman Spectrosc., 1983, 14, 415.
51 P.Klaeboe, E.Rytter, and C.E.Sjøgren, J.Mol. Struct., 1984, 113, 213.
52 J.Hvistendahl, P.Klaeboe, E.Rytter, and H.A.Øye, Inorg. Chem., 1984, 23, 706.
53 K.Anton, P.Konrad, and H.Nöth, Chem. Ber., 1984, 117, 863.
54 C.E.Sjøgren, P.Klaeboe, and E.Rytter, Spectrochim. Acta, Part A, 1984, 40, 457.
55 O.T.Beachley and R.B.Hallock, Organometallics,1984, 3, 199.
56 F.Maury and G.Constant, Polyhedron, 1984, 3, 581.
57 I.V.Ovchinnikov, L.V.Serebrennikov, and A.A.Mal'tsev, Vestn. Mosk. Univ., Ser. 2, Khim., 1984, 25, 157.
58 S.P.Best, J.K.Beattie, and R.S.Armstrong, J.Chem. Soc., Dalton Trans., 1984, 2611.
59 N.V.Porotnikov, V.G.Savenko, and L.N.Margolin, Russ. J. Inorg Chem., 1983, 28, 1098.
60 A.E.Nosenko, A.I.Bilyi, and V.V.Artamanov, Zh. Prikl. Khim., 1984, 57, 413.
61 G.G.Hoffmann, J.Organomet. Chem., 1984, 273, 187.
62 G.G.Hoffmann, J.Organomet. Chem., 1984, 277, 189.
63 G.G.Hoffmann, Z. Naturforsch., Teil B, 1984, 39, 1216.
64 G.G.Hoffmann, Chem. Ber., 1983, 116, 3858.
65 K.R.Allakhverdiev, E.A.Vinogradov, R.Kh.Nani, E.Yu. Salaev, R.M.Sardarly, and N.Yu.Safarov, Fiz. Svoistva Slozhnykh Poluprovodn., 1982, 55 (Chem. Abstr.,1984, 100, 164535).
66 T.G.Kerimova, R.Kh.Nani, A.Sh.Khidirov, E.Yu.Salaev, and V.Ya.Shteinshraiber, Fiz. Svoistva Slozhnykh Poluprovodn., 1982, 132 (Chem. Abstr., 1984, 100, 164537).
67 A.Yu.Tsivadze, M.Rabbani, and A.G.Dudareva, Russ. J. Inorg. Chem., 1983, 28, 1414.
68 N.M.Gasanly, N.N.Mel'nik, A.S.Ragimov, and V.I.Tagirov, Fiz. Tverd. Tela., 1984, 26, 558.
69 N.M.Gasanly, B.M.Dzhavadov, and V.I.Tagirov, Fiz. Svoistva Slozhnykh Poluprovodn., 1982, 86 (Chem. Abstr., 1984, 100, 164536).
70 S.V.Volkov, V.F.Kozin, I.A.Sheka and N.I.Buryak, Ukr. Khim. Zh. (Russ. Ed.), 1983, 49, 1123.
71 M.R.Bermejo, A.Castineiras, M.Gayoso, W.Hiller, U.Englert, and J.Strähle, Z. Naturforsch., Teil B, 1984, 39, 1159.
72 M.B.Millikan and B.D.James, Inorg. Chim. Acta,1984, 81, 109.
73 R.S.Altman, M.W.Crofton, and T.Oka, J.Chem. Phys., 1984, 80, 3911.
74 S.V.Sin'ko, Yu.A.Pentin, G.M.Kuramshina, V.P.Kozyukov, G.S. Gol'din, and N.L.Sin'ko,Vestn. Mosk. Univ.,Ser. 2, Khim., 1984, 25, 148.
75 S.V.Sin'ko, G.M.Kuramshina, E.M.Protasov, G.S.Gol'din, Yu.A. Pentin, and S.A.Bashkirova, Vestn. Mosk. Univ., Ser. 2, Khim. 1983, 24, 449.
76 D.C.McKean, M.W.MacKenzie, and A.R.Morrisson, J. Mol. Struct. 1984, 116, 331.

77 A.Baldacci, A.Passerini, and S.Ghersetti, Spectrochim. Acta, Part A, 1984, 40, 165.
78 E.Knözinger and R.Wittenbeck, J.Chem. Phys., 1984, 80, 5979.
79 S.Perry and J.Jonas, J.Chem. Phys., 1983, 79, 6308.
80 C.A.Arrington, K.A.Klingensmith, and R.West, J. Am. Chem. Soc., 1984, 106, 525.
81 I.Kanesaka, Y.Kiyokawa, and K.Kawai, J. Chem. Phys., 1984, 80, 645.
82 T.J.Lee and H.F.Schaefer, J.Chem. Phys., 1984, 80, 2977.
83 B.Andrews, A.Anderson and B.Torrie, J. Raman Spectrosc., 1984, 15, 67.
84 D.Lentz, J. Fluorine Chem., 1984, 24, 523.
85 K.Kalcher, E.Nachbaur, and A.Popitsch, J. Mol. Spectrosc., 1983, 102, 174.
86 R.Mattes and B.Füsser, Z. Naturforsch., Teil B,1984, 39, 1.
87 J.Lavigne, C.Pepin, and A.Cabana, J. Mol. Spectrosc., 1984, 104, 49.
88 K.Ohno, Y.Yamamoto, M.Matsuura, and H.Murata, Chem. Lett., 1984, 413.
89 C.Amiot, J.Verges, and C.R.Vidal, J. Mol. Spectrosc., 1984, 103, 364.
90 S.Chin and W.B.Person, J.Phys. Chem., 1984, 88, 553.
91 S.Carter and J.N.Murrell, Croat. Chim. Acta,1984, 57, 355.
92 S.C.Foster, A.R.W.McKellar, and T.J.Sears, J.Chem. Phys., 1984, 81, 578.
93 C.O.Della Vedova, E.L.Varetti, and P.J.Aymonino, Can. J. Spectrosc., 1983, 28, 107.
94 R.D'Cunha, C.J.Seliskar, J.R.Manheim, and K.N.Rao, J. Phys. Chem., 1984, 88, 401.
95 I.Kanesaka, T.Miyajima, and K.Kawai, J. Mol. Struct., 1984, 118, 189.
96 H.Shimizu, Ferroelectrics, 1983, 52, 141.
97 B.Pötter and K.Seppelt, Angew. Chem., Int. Ed. Engl., 1984, 23, 150.
98 P.B.Davies, P.A.Hamilton, J.M.Elliott, and M.J.Rice, J. Mol. Spectrosc., 1983, 102, 193.
99 S.Giorgianni, R.Visinoni, A.Gambi, S.Ghersetti, F.Cappelani, and G.Restelli, J. Mol. Spectrosc., 1984, 104, 40.
100 Y.S.Park and H.F.Shurvell, J. Raman Spectrosc., 1984, 15, 194
101 E.Borsella, U. Del Bello, R.Fantoni, A.Ferretti, and A.Giardini-Guidoni, Contrib.-Symp. At. Surf. Phys., 1984, 345 (Chem. Abstr., 1984, 100, 129062).
102 H.F.Shurvell and J.R.Durig, Can. J. Spectrosc., 1984, 29, 1.
103 L.H.Jones and B.I.Swanson, J. Chem. Phys.,1984, 80, 3050.
104 L.H.Jones, B.I.Swanson, and S.A.Ekberg, J. Phys. Chem., 1984, 88, 5560.
105 D.C.McKean, J. Mol. Struct., 1984, 113, 251.
106 S.Furukawa and N.Matsumoto, Solid Stat Comm., 1983, 48, 539.
107 P.John, B.C.Cowie, and I.M.Odeh, Philos. Mag.,B, 1984, 49, 559.
108 G.Pierre, A.Valentin, and L.Henry, Can. J. Phys.,1984, 62, 254.
109 H.Bürger, A.Rahner, P.Schulz, J.Dupre, G.Graner, P.B.Davies, and D.P.Stern, J. Mol. Spectrosc., 1984, 106, 337.
110 A.Ruoff, H.Essig, H.Bürger, S.Dobos, and P.Schulz, Mol. Phys., 1984, 51, 499.

111 H.Bürger, G.Scippel, A.Ruoff. H.Essig, and S.Cradock, J.Mol. Spectrosc., 1984, 106, 349.
112 H.Bürger and P.Schulz, J.Mol. Spectrosc., 1983, 102, 160.
113 G.Varsányi, B.Zelei, S.Dobos, and M.Gál, Spectrochim. Acta, Part A,1984, 40, 529.
114 A.Marchand, P.Gerval, F.Duboudin, M.-H.Gaufryau, M.Joanny, and P.Mazerolles, J.Organomet. Chem., 1984, 267, 93.
115 V.V.Sliznev and V.G.Solomonik, Koord. Khim., 1984, 10, 1199.
116 K.Hamada and H.Morishita, Spectrosc. Lett.,1983, 16, 717.
117 V.G.Koslovskii, A.N.Kanev, and L.E.Garsh, Zh. Prikl. Spektrosk.,1984, 40, 399.
118 M.Nakamura, Y.Mochizuki, K.Usami, Y.Itoh, and T.Nozaki, Solid State Comm., 1984. 50, 1079.
119 F.A.Seifert, B.O.Mysen, and D.Virgo, Phys. Chem. Glasses, 1983, 24, 141.
120 M.Handke, J.Mol. Struct., 1984, 114, 187.
121 R.Bertram, S.Schönherr, and H.Görz, Z. Chem.,1984, 24, 225.
122 M.Stavola, Appl. Phys. Lett., 1984, 44, 514.
123 H.L.Ross and P.McMillan, Amer. Mineral., 1984, 69, 719.
124 B.A.Morrow, C.P.Tripp, and R.A.McFarlane, J.Chem. Soc., Chem. Comm., 1984, 1282.
125 H.Ohashi and M.Sekita, Ganseki Kobutsu Kosho Gakkaishi, 1983, 78, 239 (Chem. Abstr.,1984, 100, 27631).
126 A.A.Morozov, L.V.Khristenko, Yu.A.Pentin, S.V.Krasnoshchenko, E.A.Chernyshev, and T.L.Krashova, Zh. Prikl. Spektrosk. 1984, 40, 970.
127 A.A.Morozov, L.V.Khristenko, Yu.A.Pentin, S.V.Krasnoshchekov, G.S.Gol'din, L.S.Baturina, and T.N.Pushakova, Vestn. Mosk. Univ., Ser. 2, Khim., 1984, 25, 153.
128 C.W.Dirk and T.J.Marks, Inorg. Chem., 1984, 23, 4325.
129 Y.Nakamura, I.Nakai, and K.Nagashima, Nippon Kagaku Kaishi, 1984, 508.
130 M.Tenhover, M.A.Hazle, and R.K.Grasselli, Phys. Rev.,B, Condens. Mater., 1984, 29, 6732.
131 E.Hengge and G.Miklau, Z. Anorg. Allg. Chem.,1984, 508, 43.
132 A.E.Cheglokov, Yu.A.Kuritsyn, E.P.Snegirev, and O.N. Ulenikov, J. Mol. Spectrosc.,1984, 105, 385.
133 S.Cradock, Mol. Phys., 1984, 51, 697.
134 F.Lattanzi, C. DiLauro, H.Bürger, R.Eujen, and S.Cradock, J. Mol. Spectrosc., 1984, 107, 229.
135 H.Bürger, R.Eujen, A.Rahner, P.Schulz, and S.Cradock, Z.Naturforsch., Teil A, 1984, 39, 871.
136 F.Lattanzi, C.DiLauro, H.Bürger, R.Eujen, P.Schulz, and S.Cradock, Mol. Phys., 1984, 51, 81.
137 L.Ross and M.Dräger, Z.Naturforsch., Teil B, 1984, 39, 868.
138 S.P.Mallela, K.C.Lee, and F.Aubke, Inorg. Chem., 1984, 23, 653.
139 T.E.Mallouk, B.Desbat, and N.Bartlett, Inorg. Chem., 1984, 23, 3160.
140 A.M.Walther and B.S.Ault, Inorg. Chem., 1984, 23, 3892.
141 M.Betzel, H.Bürger, and P.Schulz, Z. Naturforsch., Teil A, 1984, 39, 155.
142 E.Vincent, L.Verdonck, L.Naessens, and G.P.Van der Kelen, J.Organomet. Chem., 1984, 277, 235.
143 R.J.Rao, G.Srivastava, R.C.Mehrotra, B.S.Saraswat, and J.Mason, Polyhedron, 1984, 3, 485.
144 K.Grätz, F.Huber, A.Silvestri, and R.Barbieri, J. Organomet. Chem., 1984, 273, 283.

145 S.K.Brahma, W.H.Nelson, and W.F.Howard, J. Organomet. Chem., 1984, 273, 47.
146 A.-F.Shihada, I.A.-A.Jassim, and F.Weller, J.Organomet. Chem., 1984, 268, 125.
147 F.A.K.Nasser and J.J.Zuckerman, J.Organomet. Chem., 1984, 266, 225.
148 M.Dräger, B.Mathiasch, L.Ross, and M.Ross, Z. Anorg. Allg. Chem., 1983, 506, 99.
149 J.A.García-Vázquez, M.López-Becerra, and J.R.Masaguer, Z. Anorg. Allg. Chem., 1984, 508, 187.
150 H.Puff, W.Schuh, R.Sievers, W.Wald, and R.Zimmer, J. Organomet. Chem., 1984, 260, 271.
151 J.C.Huffman, J.P.Haushalter, A.N.Umarji, G.K.Shenoy, and R.C.Haushalter, Inorg. Chem., 1984, 23, 2312.
152 D.Tudela, V.Fernandez, and J.Tornero, Z. Anorg. Allg. Chem., 1984, 509, 174.
153 U.Russo, A.Cassol, and A.Silvestri, J.Organomet. Chem., 1984, 260, 69.
154 H.W.Roesky, J.Anhaus, and W.S.Sheldrick, Inorg. Chem., 1984, 23, 75.
155 H.-J.Haupt, P.Balsaa, B.Scwab, U.Flörke, and H.Preut, Z. Anorg. Allg. Chem., 1984, 513, 22.
156 I.Wharf, R.Cuenca, E.Besso, and M.Onyszchuk, J. Organomet. Chem., 1984, 277, 245.
157 F.E.A.Melo, J. Mendes Filho, J.E.Moreira, V.Lemos, and F.Cerdeira, J. Raman Spectrosc., 1984, 15, 128.
158 J.G.Contreras, G.V.Seguel, B.Ungerer, W.F.Maier, and F.J. Hollander, J. Mol. Struct., 1983, 102, 295.
159 B.Mutel, M.Bridoux, M.Crunelle-Cras, O.Dessaux, F.Grasse, P. Goudmand, and G.Moreau, Chem. Phys. Lett., 1984, 104, 290.
160 A.P.Sylvester and P.B.Dervan, J. Am. Chem. Soc., 1984, 106, 4648.
161 P.Botschwina, Chem. Phys. Lett., 1984, 107, 535.
162 J.Bendtsen, J. Raman Spectrosc., 1984, 15, 113.
163 T.Amano, J.Chem. Phys., 1984, 81, 3350.
164 T.Dreier and J.Wulfrum, Appl. Phys.,B, 1984, 33, 213.
165 R.L.Poynter and J.S.Margolis, Mol. Phys., 1984, 51, 393.
166 S.Urban, D.Papoušek, M.Bester, K.Yamada, G.Winnewisser, and A.Guarnieri, J. Mol. Spectrosc., 1984, 106, 29.
167 A.J.Barnes, T.R.Beech, and Z.Mialke, J. Chem. Soc., Farad. Trans.,2, 1984, 80, 455.
168 A.J.Barnes, N.S.Kuzniarski, and Z.Mialke, J.Chem. Soc., Farad. Trans.,2, 1984, 80, 465.
169 J.E.Bertie and M.R.Shehata, J. Chem. Phys.,1984, 81, 27.
170 I.Kanesaka, H.Shozen, and H.Kawai, Spectrochim. Acta, Part A, 1984, 40, 383.
171 S,Mohan, S.Gunesekaran, and N.G.Ravikumar, Ind. J. Pure Appl. Phys., 1984, 22, 424.
172 J.Vogt and M.Winnewisser, Ber. Bunsenges. Phys. Chem., 1984, 88, 439.
173 W.Dorsch and A.Schmidt, Z. Anorg. Allg. Chem., 1984, 508, 149.
174 P.Muthusubramanian, P.S.Santos, and O.Sala, J. Mol. Struct., 1984, 112, 233.
175 P.Hencsei, M.Gál, and L.Bihátsi, J. Mol. Struct., 1984, 114, 391.

176 S.A.Katsyubova and R.R.Shagidullin, Zh. Obshch. Khim., 1984, 54, 694.
177 J.S.Thrasher and K.Seppelt, Z. Anorg. Allg. Chem.,1983, 507, 7
178 A.Perrin, J.Y.Mandin, C.Camy-Peyret, J.M.Flaud, J.P.Chevillard, and G.Guelachvili, J. Mol. Spectrosc., 1984, 103, 417.
179 E.M.Nour, L.H.Chen, M.M.Strube, and J.Laane, J. Phys. Chem., 1984, 88, 756.
180 V.Menoux, R.Le Doucen, C.Häusler, and J.C.Deroche, Can. J. Phys., 1984, 62, 322.
181 R.H.Kagann and A.G.Maki, J. Quant. Spectrosc. Radiat. Transfer 1984, 31, 173.
182 F.Bolduan, H.J.Jodi, and A.Loewenschuss, J. Chem. Phys., 1984, 80, 1739.
183 L.P.Giver, F.P.J.Valero, D.Goorvitch, and F.S.Bonomo, J. Opt. Soc. Amer.,B: Opt. Phys., 1984, 1, 715.
184 J.P.Devlin and K.Consani, J. Phys. Chem., 1984, 88, 3269.
185 M.R.Mohammed and W.F.Sherman, Spectrochim. Acta, Part A, 1984, 40, 763.
186 L.J.Schaad, B.A.Hess, P.Čársky, and R.Zahradník, Inorg. Chem., 1984, 23, 2428.
187 I.Silaghi-Dumitrescu and I.Haiduc, J.Mol. Struct., 1984, 206, 217.
188 P.B.Davies, P.A.Hamilton, W.Lewis-Bevan, and D.K.Russell, Proc. Roy. Soc. London,A, 1984, 392, 445.
189 P.B.Davies and P.A.Hamilton, Proc. Roy. Soc., London,A, 1984, 393, 397.
190 H.Hamaguchi, M.Tasumi, M.Yoshifuji and N.Inamoto, J. Am. Chem. Soc., 1984, 106, 508.
191 H.Hamaguchi, M.Tatsumi, T.Hoshida, and N.Inamoto, Chem. Lett., 1984, 463.
192 J.R.Anacona, P.B.Davies, and P.A.Hamilton, Chem. Phys. Lett., 1984, 104, 269.
193 M.Baudler and H.Tschäbunin, Z. Anorg. Allg. Chem., 1984, 511, 77.
194 M.A.Sarukhanov, I.A.Popova, and Yu.Ya.Kharitonov, Russ. J. Inorg. Chem., 1984, 29, 957.
195 L.Verdonck, P.H.De Ryck, F.F.Roelandt, and G.P. van der Kelen, Spectrochim. Acta, Part A, 1984, 40, 299.
196 V.P.Novkov, A.N.Yarkov, I.O.Umarova, O.A.Raevskii, A.N. Kurkin, and Z.S.Novkova, Izv. Akad. Nauk SSSR, Ser. Khim., 1983, 2252.
197 K.Hassler, Monatsh. Chem., 1984, 115, 713.
198 K.Hassler, J.Organomet. Chem., 1984, 266, Cl.
199 T.T.Bamgboye and O.A.Bamgboye, Spectrochim. Acta, Part A, 1984, 40, 329.
200 E.J.Baran, Z. Naturforsch., Teil A,1983, 38, 1152.
201 M.C.Apella and E.J.Baran, Z. Naturforsch.,Teil B,1984, 39, 449
202 A.Addou, P.Vast, and P.Legrand, Bull. Soc. Chim. France,I, 1984, 55.
203 B.C.Cornilsen, J. Mol. Struct., 1984, 117, 1.
204 M.Möbs and M.Jansen, Z. Anorg. Allg. Chem., 1984, 514, 39.
205 M.M.Ali, G.A.Bukhalova, and I.V.Mardirasova, Izv. Sev.-Kavk. Nauchn. Tsentra Vyssh. Shk., Estestv. Nauki, 1983, 71 (Chem. Abstr., 1984, 100, 13292).
206 J.Fuchs and R.Palm, Z. Naturforsch., Teil B, 1984, 39, 757.

207 G.H.Pieters, B.J. van der Veken, and M.A.Herman, J. Mol. Struct., 1983, 102, 221.
208 R.L.Odeurs, B.J. van der Veken, M.A.Herman, and J.R.Durig, J. Mol. Struct., 1984, 117, 235.
209 A.F.Shihada and I.A.M.Jasim, Z. Naturforsch., Teil B, 1984, 39, 783.
210 E.A.V.Ebsworth, C.M.Huntley, and D.W.H.Rankin, J. Chem. Soc., Dalton Trans., 1984, 2301.
211 H.Schnöckel and M.Lakenbrink, Z. Anorg. Allg. Chem., 1983, 507, 70.
212 U.Müller and A.T.Mohammad, Z. Anorg. Allg. Chem., 1984, 514, 164.
213 A.J.Blake and G.P.McQuillan, J.Chem. Soc., Dalton Trans., 1984 1849.
214 C.Wibbelmann, W.Brockner, B.Eisenmann, and H.Schäfer, Z. Naturforsch., Teil A, 1984, 39, 190.
215 C.Wibbelmann, W.Brockner, B.Eisenmann, and H.Schäfer, Z. Naturforsch., Teil B, 1983, 38, 1575.
216 R.Becker and W.Brockner, Z. Naturforsch., Teil A,1984, 39, 1120.
217 R.Becker, W.Brockner, and H.Schäfer, Z. Naturforsch., Teil A, 1984, 39, 357.
218 A.J.Blake, S.Cradock, E.A.V.Ebsworth, D.W.H.Rankin, and A.J. Welch, J. Chem. Soc., Dalton Trans., 1984, 2029.
219 S.Cradock, J. Mol. Struct., 1984, 117, 153.
220 M.Takami and H. Kuze, J. Chem. Phys., 1984, 80, 2314.
221 C.J.Marsden, J. Chem. Soc., Chem. Commun., 1984, 401.
222 G. Di Lonardo, L.Fusina, and J.W.C.Johns, J. Mol. Spectrosc., 1984, 104, 282.
223 R.K.Gupta, A.K.Rai, R.C.Mehrotra, and V.K.Jain, Polyhedron, 1984, 3, 721.
224 H.Schnöckel, M.Lakenbrink, and L. Zhengyan, J. Mol. Struct., 1983, 102, 243.
225 L.Bencivenni and K.A.Gingerich, Inorg. Chim. Acta,1984, 85, L11.
226 H.Ehrhardt and M.Jansen, Z. Anorg. Allg. Chem., 1983, 504, 128
227 L.Koudelka and M.Pisarčik, Mater. Chem. Phys., 1983, 9, 571.
228 L.Koudelka and M.Pisarčik, J. Non-Cryst. Solids, 1984, 64, 87.
229 F.Claus, M.Glaser, and R.Minkwitz, Z. Anorg. Allg. Chem., 1983, 506, 178.
230 R.Winter, H.Kruse, E.H.Fink, J.Wildt, and F.Zabel, Chem. Phys. Lett., 1984, 104, 383.
231 U.Ensinger and A.Schmidt, Z. Anorg. Allg. Chem., 1984, 514,137
232 G.Rajca, W.Schwarz, and J.Weidlein, Z. Naturforsch., Teil B, 1984, 39, 1219.
233 G.E.Binder, U.Ensinger, W.Dorsch, and A.Schmidt, Z. Anorg. Allg. Chem., 1983, 507, 163.
234 B.A.Arbuzov, Yu.M.Mareev, R.R.Shagidullin, V.S.Vinogradova, and I.Kh.Shakirov, Izv. Akad. Nauk SSSR, Ser. Khim., 1984,1846
235 H.J.Breunig and H.Jawad, J. Organomet. Chem., 1984, 277, 257.
236 M.M.de V.Steyn, A.M.Heyns, and R.B.English, J. Crystallogr. Spectrosc. Res., 1984, 14, 505.
237 M.G.B.Drew, J.M.Kisanyi, G.R.Willey, and S.O.Wandiya, J. Chem. Soc., Dalton Trans., 1984, 1717.
238 U.Müller and A.T.Mohammad, Z. Anorg. Allg. Chem., 1983, 506, 110.

239 S.W.Hedges and L.H.Bowen, Inorg. Chem., 1984, 23, 458.
240 T.Nanba, K.Hirota, S.Suto, and M.Ikezawa, J. Phys. Soc. Japan, 1984, 53, 449.
241 S.Hashimoto, G.Inoue, and H. Akimoto, Chem. Phys. Lett., 1984, 107, 198.
242 I.Benjamin and R.D.Levine, Chem. Phys. Lett., 1983, 101, 518.
243 H.A.Fry, L.H.Jones, and B.I.Swanson, Chem. Phys. Lett., 1984, 105, 547.
244 S.Krishnamurthy, R.Bansil, and J.Wiafe-Akenten, J. Chem. Phys. 1983, 79, 5863.
245 A.J.Barnes, R.M.Bentwood, and M.P.Wright, J. Mol. Struct., 1984, 118, 97.
246 Z.Kisiel, A.C.Legon, and D.J.Millen, J. Mol. Struct., 1984, 112, 1.
247 W.B.Monosmith and G.E.Walrafen, J. Chem. Phys., 1984, 81,669.
248 J.E.Bertie and J.P.Devlin, J.Phys. Chem., 1984, 88, 380.
249 D.D.Klug and E.Whalley, J. Chem. Phys., 1984, 81, 1220.
250 G.E.Stark, A.K.Garg, W.F.Sherman, and G.R.Wilkinson, J. Mol. Struct., 1984, 115, 161.
251 J.B.Bates, J.C.Wang, and W.J.Dudney, J. Chem. Phys., 1984, 80, 123.
252 A.L.Novozhilov and Yu.S.Zhoglik, Russ. J. Inorg. Chem., 1983, 28, 1376.
253 H.A.Fry, L.H.Jones, and J.E.Barefield, J. Mol. Spectrosc., 1984, 103, 41.
254 I.Kanesaka, M.Tsuchida, N.Kawai, and T.Takeuchi, J. Mol. Spectrosc., 1984, 104, 405.
255 K.O.Christe, P.Charpin, E.Soulie, R.Bougon, J.Fawcett, and D.R.Russell, Inorg. Chem., 1984, 23, 3756.
256 N.N.Haese and T.Oka, J. Chem. Phys., 1984, 80, 572.
257 P.A.Giguère and H.Chen, J. Raman Spectrosc.,1984, 15, 199.
258 S.M.Sonchik, L.Andrews and K.D.Carlson, J. Phys. Chem., 1984, 88, 5269.
259 M.Hawkins, L.Andrews, A.J.Downs, and D.L.Drury, J. Am. Chem. Soc., 1984, 106, 3076.
260 H.Willner, Z. Anorg. Allg. Chem., 1984, 514, 171.
261 R.Steudel, J.Steidel, and R.Reinhardt, Z. Naturforsch., Teil B 1983, 38, 1548.
262 H.Hattori and H.Kawamura, J. Non-Cryst. Solids., 1983, 59-60, 1063.
263 R.Steudel and E.M.Strauss, Angew. Chem., Int. Ed. Engl., 1984, 23, 362.
264 H.Schnöckel, Z.Anorg. Allg. Chem., 1984, 510, 72.
265 J.R.Magana and J.S.Lannin, J. Non-Cryst. Solids, 1984, 61-62, 373.
266 M.Hawkins and A.J.Downs, J. Phys. Chem., 1984, 88, 3042.
267 H.Henle, R.Hoppenheit, and R.Mews, Angew. Chem., Int. Ed. Engl. 1984, 23, 507.
268 M.Herberhold, W.Ehrenreich, and K.Guldner, Chem. Ber., 1984, 117, 1999.
269 C.S.Sass and B.S.Ault, J. Phys. Chem., 1984, 88, 432.
270 G.Kempe, D.Walther, and M.Bögel, Z.Chem., 1983, 23, 307.
271 J.Foropoulos and D.D.DesMarteau, Inorg. Chem., 1984, 23, 3720.
272 G.K.MacLean, J.Passmore, and P.S.White, J. Chem. Soc., Dalton Trans., 1984, 211.

273 U.Müller, E.Conradi, U.Demant, and K.Dehnicke, Angew. Chem., Int. Ed. Engl., 1984, 23, 237.
274 H.W.Roesky, R.Emmert, and T.Gries, Chem. Ber., 1984, 117, 404.
275 Y.Koga, H.Takeo, S.Kondo, C.Matsumura, and S.Saeki, Bull. Chem. Soc. Japan, 1984, 57, 2400.
276 M.H.Brooker, J. Mol. Struct., 1984, 112, 221.
277 N.F.Henfrey and B.A.Thrush, Chem. Phys. Lett., 1983, 102, 135.
278 H.M.Nagarathna, L.Bencivenni, and K.A.Gingerich, J. Chem. Phys. 1984, 81, 591.
279 A.M.Forster and A.J.Downs, J. Chem. Soc., Dalton Trans., 1984, 2827.
280 C.Chapadros and G.Birnbaum, J. Mol. Spectrosc., 1984, 105, 206
281 H.B.Levene and D.S.Perry, J. Chem. Phys., 1984, 80, 1772.
282 A.B.Altabef, E.L.Varetti, and P.J.Aymonino, Z. Anorg. Allg. Chem., 1983, 506, 161.
283 W.C.Lane, T.H.Edwards, J.R.Gillis, F.S.Bonomo, and F.J. Muroray, J. Mol. Spectrosc., 1984, 107, 306.
284 G.Andreev and L.Futekov, J. Mol. Struct., 1984, 114, 445.
285 S.Mohan and K.G.Ravikumar, Ind. J. Phys.,B, 1983, 57, 303.
286 G.Kempe, Z. Chem., 1984, 24, 137.
287 R.K.Chadha, J.E.Drake, M.A.Khan, and G.Singh, J. Organomet. Chem., 1984, 260, 73.
288 J.Milne and M.Mahadevan, Inorg. Chem., 1984, 23, 265.
289 L.J.Lawlor and J.Passmore, Can. J. Chem., 1984, 62, 1477.
290 C.A.Hogarth and A.A.Hosseini, J. Mater. Sci. Lett., 1984, 3, 359.
291 E.Gutierrez-Rios, Rev. R. Acad. Cienc. Exactas Fis. Nat., Madrid, 1983, 77, 621.
292 S.S.Abed-Ali and W.R.McWhinnie, J. Organomet. Chem., 1984, 277, 365.
293 R.J.H.Clark and M.J.Stead, Chem. Phys., 1984, 91, 113.
294 E.Schäfer and R.J.Saykally, J. Chem. Phys., 1984, 80, 2973.
295 C.J.Chunnilal, W.Fisherman, and G.R.Wilkinson, J. Mol. Struct., 1984, 115, 205.
296 R.L.Redington and D.F.Hamill, J. Chem. Phys., 1984, 80, 2446.
297 L.Andrews and G.L.Johnson, J.Phys. Chem., 1984, 88, 425.
298 A.Lautie, F.Romain, M.H.Limage, H.Novak, H.Prevost-Czeskleba, and J.Moret, J. Mol. Struct., 1984, 118, 227.
299 R.T.Arlinghaus and L.Andrews, J. Phys. Chem., 1984, 88, 4032.
300 K.Gholivand, H.Willner, D.Bielefeldt, and A.Haas, Z. Naturforsch., Teil B, 1984, 39, 1211.
301 S.C.Foster and J.W.C.Johns, J.Mol. Spectrosc., 1984, 103, 176.
302 C.Yamada and E.Hirota, J. Chem. Phys., 1984, 80, 4694.
303 F.Temps, H.G.Wagner, P.B.Davies, D.P.Stern and K.O.Christe, J. Phys. Chem., 1983, 87, 5068.
304 R.Wilbrandt, N.H.Jensen, A.H.Sillesen, and K.B.Hansen, Chem. Phys. Lett., 1984, 106, 503.
305 C.De Meyer, J.M.Levert, and A.Vanclef, Bull. Soc. Chim. France,I, 1983, 187.
306 M.O.Bulanin, V.N.Bukhmarina, E.G.Moisenko, and K.G.Tokhadze, Opt. Spektrosk., 1984, 56, 813.
307 D.W.T.Griffith, G.S.Tyndall, J.P.Burrows, and G.K.Moortgat, Chem. Phys. Lett., 1984, 107, 341.
308 A.M.Heyns and J.B.Clark, J. Raman Spectrosc., 1983, 14, 342.

309 L.Bencivenni, H.M.Nagarathna, K.A.Gingerich, and R.Teghil, J.Chem. Phys., 1984, 81, 3415.
310 C.I.Ratcliffe and D.E.Irish, Can. J. Chem., 1984, 62, 1134.
311 B.Háyek, O.Smrčkova, and P.Záruba, Coll. Czech. Chem. Commun., 1984, 49, 1756.
312 B.H.Loo and Y.G.Lee, J. Phys. Chem., 1984, 88, 706.
313 A.A.Bernussi and G.M.Gualberto, J. Raman Spectrosc., 1984, 15, 10.
314 S.Mohan and K.G.Ravikumar, Ind.J. Phys.,B, 1984, 58, 8.
315 H.Chang, S.G.Tien, and M.H.Hwang, Proc. Nat. Sci. Counc. Repub. China, B, 1983, 7, 420 (Chem. Abstr., 1984, 100, 182431).
316 M.R.Mohammad and W.F.Sherman, J. Mol. Struct., 1984, 115, 27.
317 J.W.C.Johns, J. Mol. Spectrosc., 1984, 106, 124.
318 R.Klein and P.Rosmus, Z. Naturforsch., Teil A, 1984, 39, 349.
319 G.A.Schumacher and G.J.Schrobilgen, Inorg. Chem., 1984, 23, 2923.

5

Vibrational Spectra of Transition-element Compounds

BY G. DAVIDSON

1 Detailed Studies

A complete assignment has been given for Raman spectra of oriented single crystals containing $M(OH_2)_6^{3+}$, where M = Ti, V, Cr or Fe. The totally symmetric stretch was at higher wavenumbers than had previously been thought: 517(Ti), 525(V), 540(Cr), and 523(Fe) cm^{-1}.[1] Assignments of i.r. and Raman spectra of Ln_2MgTiO_6, where Ln = La-Yb or Y, were helped by $^{24}Mg/^{26}Mg$ substitution for Ln = La, Sm or Yb. The Ti is six-, the Ln 10-12-coordinate.[2]

I.r. and Raman spectra for matrix-isolated Cs_2MO_4, where M = Cr, Mo or W, were assigned using the following isotopic shifts: $^{50}Cr/^{52}Cr/^{53}Cr/^{54}Cr$, $^{92}Mo/^{94}Mo/^{95}Mo/^{96}Mo/^{98}Mo/^{100}Mo$; and for Cs_2WO_4, $^{16}O/^{18}O$.[3] Among detailed assignments for $[Br(OC)MC]_2C_6X_4$, where M = Cr or W, X = H or D, were ν^{I}(M≡C) at 1372(Cr), 1352(W) cm^{-1}.[4] I.r. and Raman spectra, including $^{18}O_2$ and ^{13}CO substitution, of species formed by photolysis of $M(CO)_6$, where M = Mo or W, in O_2-doped Ar or CH_4 matrices at 10K show that they are $M(CO)_4O_2$, with a linear O=M=O group. $\nu_sMO_2(a_{1g})$ is at 820(Mo), 850(W)cm^{-1}, $\nu_{as}MO_2(b_{2u})$ at 758(Mo), 751(W)cm^{-1}. For $Mo(CO)_4O_2$, clear molybdenum isotopic fine structure was resolvable for $\nu_{as}MO_2$ in the i.r.[5]

$W_2O_3(CH_2CMe_3)_6$, and its ^{18}O analogue, gave the following bands: ν_{as}W=O 962(905)cm^{-1} (i.r.), ν_sW=O 942(894)cm^{-1} (Raman), ν_{as}W-O-W 693(668)cm^{-1} (i.r.), and ν_sW-O-W 205(205)cm^{-1} (Raman). The spectra were similar to those containing an O=Re-O-Re=O backbone.[6] The ν_3 band of WF_6 was studied at low rotational and vibrational temperatures in a pulsed supersonic free jet. Band origins were as follows: $^{182}WF_6$ 714.55767(12)cm^{-1}, $^{183}WF_6$ 714.23430(14)cm^{-1}, $^{184}WF_6$ 713.91520(10)cm^{-1}, $^{186}WF_6$ 713.28631(13)cm^{-1}.[7]

Two reports of i.r. and Raman spectra of $MReO_4$, where M = Li, Na, K, Cs, Rb or Tl, isolated in matrices showed that ReO_4^- in $TlReO_4$ has C_{2v} symmetry, but that the others have regular T_d ReO_4^-. There was some evidence for the formation of $M_2(ReO_4)_2$ dimers in N_2 and Ar matrices.[8,9]

The i.r. spectrum of Mg_2FeH_6 contains a band at 1720 cm^{-1} (1262 cm^{-1} in Mg_2FeD_6), while the Raman has a feature at 1870cm^{-1} (1340

cm^{-1} on deuteriation). These are consistent with O_h symmetry for FeH_6^{4-}.[10] $Fe_2O(phen)_4(H_2O)_2^{4+}$ has ν_sFe-O-Fe at 395cm^{-1} (shifts 5cm^{-1} on ^{18}O substitution) in the Raman, and ν_{as}Fe-O-Fe at 827cm^{-1} (40cm^{-1} isotopic shift) in both i.r. and Raman spectra. This is a good model for the spectroscopic properties of the binuclear iron proteins haemerythrin and ribonucleotide reductase.[11] A detailed set of vibrational assignments was proposed for $OsF_nCl_{6-n}^{2-}$, where n = 1-5.[12]

Cu-ligand modes were identified by $^{63}Cu/^{65}Cu$ isotopic shifts in the i.r. spectra of bis(L- and DL-phenylalaninato)copper(II) and their N-deuteriated analogues. νCu-O is greater than νCu-N for the DL- but reversed for the L-species.[13] Metal and H/D isotope shifts were used to assign metal-ligand modes in bis(L-serinato)-copper(II) and -zinc(II).[14]

I.r. and Raman spectra were analysed in $UO_2X_4^{2-}$, where X = Cl or Br, to give assignments for most fundamentals under D_{4h} symmetry.[15,16]

Resonance Raman Spectra.- Resonance and preresonance Raman spectra for Sc- and Mn-containing argon matrices were reported. For Sc, progressions due to both Sc_2 and Sc_3 were seen. For Sc_2, $\omega_e'' = 233.9$ cm^{-1}, $\omega_e''x_e'' = 0.93$cm^{-1}. For Sc_3 the data suggest an equilateral, or near-equilateral, triangle, with a symmetric stretch at 248cm^{-1} and a degenerate mode at 150cm^{-1}. The manganese systems gave progressions due to Mn_2 only ($\omega_e'' = 124.7$, $\omega_e''x_e'' = 0.24$cm^{-1}).[17]

The observed overtone sequence in νCr-N in the resonance Raman spectrum of $K_3[Cr(CN)_5NO]$ can be used to estimate distortions in the electronically excited state.[18] The resonance Raman excitation profile has been found for CrO_4^{2-} in aqueous solution throughout the $^1T_2 \leftarrow {}^1A_1$ transition at 372nm.[19] I.r. and resonance Raman spectra of $Mo_2(\mu_4\text{-}S_2)(\mu_2\text{-}SO_2)(CN)_8^{4-}$ gave the following features: νMo_2S_2 195, 347, 383, 428cm^{-1} (of which νMo-Mo is at 195cm^{-1}); νMo-CN 318, 461cm^{-1}; νMo-S 347, 383, 428cm^{-1}.[20] The resonance Raman spectra of $MMnO_4$, where M = Rb, Cs or Ag, $Sr(MnO_4)_2.3H_2O$ and $La(MnO_4)_3.nH_2O$ all give progressions in ν_1 of MnO_4^-.[21] Polycrystalline $AgMnO_4$ gave a resonance Raman spectrum which confirmed that there are important covalent contributions to the overall bonding.[22]

^{18}O substitution allows the assignment to ν_{as}Fe-O-Fe in the i.r. and resonance Raman spectra of $(HBpz_3)FeO(O_2CR)_2Fe(HBpz_3)$, where R = H, Me or Ph, of bands at 751cm^{-1} (^{16}O), 721cm^{-1}(^{18}O). ν_sFe-O-Fe is at 528cm^{-1} and δFe-O-Fe at 278cm^{-1}.[23]

As before, a large number of resonance Raman spectra have been recorded for biologically important iron-containing species. $^{12}CO/^{13}CO$ substitution in Fe(TpivPP)($C_6HF_4S^-$)(CO), where TpivPP = dianion of meso-α,α,α,α-tetrakis(o-(pivalamido)phenyl)porphyrin, shows that νFe-CO is at 479cm^{-1}.[24] Resonance Raman spectra of the carbon-bridged dimer (μ-carbido)[(5,10,15,20-tetraphenylporphinato)-iron] shows that ν_sFe-C-Fe is at 443cm^{-1}, compared to 424, 363cm^{-1} for the Fe-N-Fe and Fe-O-Fe modes in N- and O-analogues.[24] ^{54}Fe substitution in $(FeTPP)_2C$ and $(FeTPP)_2N^+$, where TPP = 5,10,15,20-tetraphenylporphinato, gives an assignment for ν_sFe-C-Fe which agrees with the above, but ν_sFe-N-Fe is said to be at 465cm^{-1}. Force constants for the Fe-X-Fe unit suggest that it involves substantial π-bonding.[26]

Soret-excited resonance Raman spectra gave information on Fe-C bonding in cyanomet- and carbonmonoxy-complexes of haemoglobin(III) from chironumus thummi thummi. Isotopic exchange ($^{13}CN^-$, $C^{15}N^-$, $^{13}C^{15}N^-$, ^{13}CO, $C^{18}O$, $^{13}C^{18}O$) shows that νFe-CN = 453cm^{-1}, δFe-C-N = 412cm^{-1}, νFe-CO = 500cm^{-1}, and δFe-C-O = 574cm^{-1}. These suggest that the Fe(III)-CN bond is weaker than the Fe(III)-CO bond.[27]

Other such systems to be studied by resonance Raman spectroscopy include: deoxymyo- and deoxyhaemoglobin;[28] ferric nitrosyl complexes of myoglobin, haemoglobin A and horseradish peroxidase;[29] OFe(TPP) (in O_2 matrices at 15K; νFe-^{16}O 852cm^{-1}; νFe-^{18}O 818 cm^{-1});[30] $(OEPFe)_2X$, where OEP = octaethylporphinato, X = O or N;[31] oxyhaemerythrin;[32] myoglobins with the haem isotopically substituted at the central iron atom or the pyrrole nitrogen atoms (showed that the iron-histidine stretch is ca.220cm^{-1});[33] high-spin haemoprotein of Chromatium;[34] azotobacter vinelandii ferredoxin I;[35] (2Fe-2S) and 2(4Fe-4S) ferredoxin from C. pasteuranium;[36,37] aa_3-type cytochrome oxidase of thermophilic bacterium PS3;[38] P. ovalis and M. smegmatis ferredoxins;[39] and ^{56}Fe- and ^{54}Fe-containing reduced cytochrome oxidase of T. thermophilus HB8.[40]

Resonance Raman spectra of $RuCl_6^{2-}$ and $RuBr_6^{2-}$ both contain progressions in the $\nu_1(a_{1g})$ mode, with subsidiary progressions in the $\nu_2(e_g)$ and $\nu_5(t_{2g})$ fundamentals.[41] The resonance Raman spectrum of $(NH_3)_4Ru^{II}$(riboflavin) contains a strong band at 275cm^{-1}, probably νRu-N(5).[42]

Molecular oxygen adducts of several Co(salen) derivatives, where salen = N,N'-ethylenebis(salicylideneiminato), have νCo-O 515-530cm^{-1}. The wavenumber is sensitive both to the geometry of the CoN_2O_2 core and to the degree of chelation over the chelate ring.[43] Long overtone progressions in the axial X-Pd(IV)-X symmetric stretch were seen in the resonance Raman spectra of $[Pd^{II}(L-L)_2][Pd^{IV}(L-L)_2X_2](ClO_4)_4$, where L-L = diaminoethane, 1,2- or 1,3-diaminopropane, and X = Cl (280cm^{-1}) or Br (150cm^{-1}).[44]

$^{63}Cu/^{65}Cu$ and H_2O/D_2O isotopic shifts were used to assign νCu-S(cys) and νCu-N(his) in the resonance Raman spectra of the blue copper proteins stellacyanin and laccase.[45]

2 Scandium, Yttrium, and the Lanthanoids

Reference has been made to detailed studies on Ln_2MgTiO_6, where Ln = La-Yb or Y,[2] and resonance Raman spectra of Sc_2 and Sc_3.[17]

$[Cp_2Y(\mu_3-H)]_2[(\mu_2-H)AlH_2NEt_3]_2$ has νY-H_{br} at 1275cm^{-1} and νAl-H_{br} at 1661cm^{-1}.[46] A normal coordinate analysis of Y_2O_3 gave good agreement with the experimental spectrum.[47]

I.r. and Raman spectra of $KLn(WO_4)_2$, where Ln = Y, Ce-Lu, were assigned in terms of monoclinic unit cells of C_{2h}^3 or C_{2h}^6 symmetry.[48] Skeletal modes were assigned for $MK[Fe(CN)_6].4H_2O$, where M = La, Nd, Sm, Eu, Gd, Tb, Dy, Ho or Yb.[49] A band ca. 250cm^{-1} was assigned as the Ln-cyclopentadienyl stretch in $[C_5H_4(CH_2)_3C_5H_4]$-LnCl.[50] Raman-active lattice modes have been assigned for LaOCl.[51]

The i.r. and Raman spectra of eleven lanthanoid acetylacetonate-α,β,γ,δ-tetraphenylporphyrin complexes, Ln(TPP)(acac), show that, because of bonding between Ln and N (of TPP), the Ln-O(acac) stretch is shifted to low wavenumbers.[52] Ln-O stretching wavenumbers were assigned for: $Ln(NO_3)_3.3L$, where L = di(2-ethylhexyl)isopropylphosphonate;[53] $K_3Ln(SO_4)_3$, where Ln = La, Pr, Nd, Sm, Ho, Er, Tm or Yb;[54] Ln_2MO_4, $LnSrMO_4$, where Ln = La, Pr, Nd, Sm, Gd, M = Fe, Al, Co or Cu;[55] Ln_2CuO_4, where Ln = La, Pr, Nd, Sm, Eu or Gd;[56] and M_2O_2Te, where M = La, Pr, Nd, Sm, Gd, Tb or Dy.[57]

Interaction of Eu or Yb atoms with Cl_2 in Ar matrices produces $EuCl_2$ (ν_3 283, ν_1 271cm^{-1}) or $YbCl_2$ (ν_3 294, ν_1 281cm^{-1}) of C_{2v} symmetry.[58] Gd reacts similarly but forms $GdCl_3$, while Ho gives a mixture of $HoCl_2$ and $HoCl_3$. $\nu_3(LnCl_3) > \nu_3(LnCl_2)$ wherever such a comparison is possible.[59]

$CeCl_6^{2-}$ vibrations were assigned from the i.r. and Raman spectra of benzyltriethylammonium hexachlorocerate.[60]

3 Titanium, Zirconium and Hafnium

Detailed studies were reported for $Ti(OH_2)_6^{3+}$,[1] and for Ln_2MgTiO_6, where Ln = La-Yb or Y.[2]

The i.r. spectrum of TiO_2 shows it to be non-linear, with νTi-O at 944 and 965cm^{-1}. $TiOCl_2$ has νTi-O at 818cm^{-1}, νTiCl ca. 535cm^{-1}.[61] $Ti_2O_2(cat)_4^{8-}$, where cat = 3,5-di-tert-butylcatechol, has an i.r. band at 605cm^{-1} due to the Ti_2O_2 unit - possibly a superposition of the two allowed bands, b_{2u} and b_{3u}, expected in this region.[62] The complexes (1), where R = Me, Et, Ph or p-ClC_6H_4, have

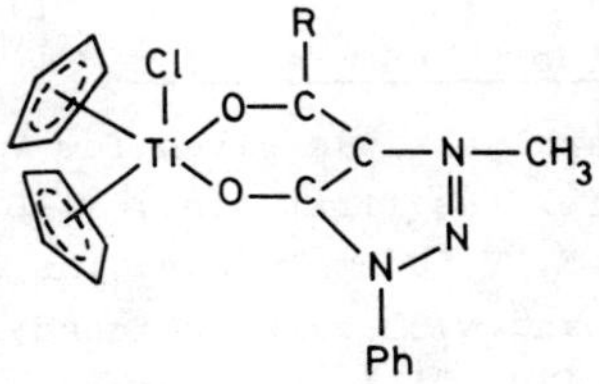

(1)

νTi-O at 610 ± 10cm^{-1} and 550cm^{-1}, and νTi-Cp 350 ± 10cm^{-1}.[63] There are two νTi-O bands in the i.r. spectra of $Ti(L)SO_4$, where L = octaethyl- or tetraphenylporphinato, at 572cm^{-1} (ν_s) and 610cm^{-1} (ν_{as}), showing that the SO_4 is bidentate.[64] Quite a detailed assignment has been given for the i.r. and Raman spectra of Ba_2TiS_4 (and Na_2WS_4).[65]

The Raman spectra of $(N_2H_5)_2TiF_6$, $(N_2H_6)TiF_6$ and $(N_2H_6)_2(TiF_6)F_2$ contain bands due to the previously unobserved ν_2 (455cm^{-1}) and the formally inactive ν_6 (ca. 210-220cm^{-1}).[66] A Ti-Ti stretching mode was identified at 281cm^{-1} in the vibrational spectrum of β-$TiCl_3$.[67]

The i.r. spectrum of $(NMe_4)(ZrF_5).H_2O$ shows that the H_2O is coordinated to the zirconium, that the anion is dimeric, with fluorine-bridging, and that each zirconium has pentagonal-bipyramidal geometry.[68] The i.r., Raman and resonance Raman spectra of HfS_3, together with data on TiS_3 and ZrS_3, were used to give a complete assignment of normal modes.[69] I.r. spectra of $HfCl_4$, $SnCl_4$ and $PbCl_4$ gave the expected isotopic pattern for tetrahedral molecules.[70]

Other vibrational assignments for titanium and zirconium compounds are summarised in Table 1.[71-75]

Table 1 Vibrational assignments in titanium and zirconium complexes

Species		Ref.
$(LiLn)_{\frac{1}{2}}TiO_3$	(Ln = La-Dy, Y)	71
$B_{12}TiO_{20}$	(single crystal)	72
$CpTi(S_2COR)_nCl_{3-n}$	(R = Me, Et, ^nPr, ^nBu, C_5H_{11}, n = 1,2,3)	73
$ZrOX_2L$	(L = 1,10-phenanthroline-N,N'-dioxide, X = Cl, Br, NO_3^-, NCS^-)	74
BaF_2/ZrF_4 glasses		75

4 Vanadium, Niobium and Tantalum

A detailed vibrational analysis was given for $V(OH_2)_6^{3+}$.[1] The anion VOF_3^- has been prepared for the first time; νV-O is at $970cm^{-1}$, νV-O..V $930cm^{-1}$, and νV-F..V $500cm^{-1}$, i.e. the solid salts are polymeric.[76] νV=O, νVO_2 and νVCl were assigned in $VO(O_2)_2Cl^{2-}$ and $V(O_2)_3Cl^{2-}$. νVO_2 is *ca.* $610cm^{-1}$ in each case, and νV-Cl *ca.* $420cm^{-1}$.[77] Similar assignments were given for $VO(O_2)F(H_2O)_2^+$.[78] νV=O is at $908cm^{-1}$ in oxovanadium(V) *meso*-tetraphenylporphine.[79] There is Raman evidence for a dimer of aquadioxovanadium(V) in concentrated $HClO_4$ and H_2SO_4 solutions, with a central unit O=V-O-O-V=O, and a non-linear V-O-V bridge.[80] $MVO_2SO_4.3H_2O$, where M = Rb^+ or Tl^+, have ν_sVO_2 *ca.* $930cm^{-1}$, $\nu_{as}VO_2$ *ca.* $900cm^{-1}$ and $\nu V\text{-}OH_2$ *ca.* $875cm^{-1}$.[81] Force constant calculations have been performed on $NaVO_3$ and other systems with chains of condensed V/O tetrahedra, of the type $(V_2O_6)_\infty$.[82] General assignments were proposed for $M(VO_3)_2$, where M = Mg, Ca, Sr or Ba, i.e. ν_s, $\nu_{as}VO_2$, ν_s, ν_{as}V-O-V.[83] Characteristic i.r. bands revealed the presence of VO_5 and VO_4 units in Fe_2O_3-V_2O_5 glasses.[84] Two new types of complex were isolated from MVO_3-(MF,HF)-H_2O_2-H_2O systems, where M = NH_4 or K, i.e. $M_3[HV_2O_2(O_2)_3F_4].2H_2O$ and $(NH_4)_3[V_2O_2(O_2)_4F].2H_2O$. Assignments were made to V-(μ_2-O_2)-V and V-F-V bridge unit vibrations.[85] A normal coordinate analysis was carried out on $PM_{12}O_{40}^{m-}$, where M = V, m = 15, or M = Mo or W, m = 3.[86] Other vanadium-oxy-species for which V-O ligand modes were assigned are listed in Table 2.[87-100]

νV-S in $V(ACDA)_4$ is at $460cm^{-1}$ (ACDA = 2-amino-1-cyclopentene-dithiocarboxylate).[101] The analogous mode in $S{=}V(SC_2H_4S)^{2-}$ is at $515cm^{-1}$,[102] and in VS(acen), where acen = Schiff base of acetyl-acetone and ethylenediamine, at $598cm^{-1}$.[103] Photolysis of VF_5

Table 2 Some vanadium-oxy-species in which V-O ligand modes are assigned

Species		Ref.
$\alpha VOXO_4$	(X = S, P, Mo)	87
$VO(SO_4)L_n$	(n = 4, L = thiourea; n = 1, L = phenylthiourea, etc.)	88
$VO(SO_4)py_2L'_{0.5}$	(L = thiourea, etc.)	88
VOL_2	(HL = hydrazides of propionic, butyric, isovaleric and substituted benzoic acids)	89
$VO(HL)(SO_4)$ VOL_2	(HL = 2-furoylacetone hydrazone)	90
$MVO(SO_4)_2$	(M = K, Rb, Cs)	91
$VO(LH)L$	(H_2L = dipicolinic acid, picolinic acid, nicotinic acid, etc.)	92
$VO(OR)_{3-n}NR'(SiMe_3)_n$	(n = 1,2; R = ^iPr, ^tBu; R' = ^tBu, $SiMe_3$, ^iPr)	93
$VO(MeSO_3)_xCl_{3-x}$ $VO(MeSO_3)_2$	(x = 1,2,3)	94
NH_4VO_3		95
MV_3O_8 $M_2V_6O_{16}$	(M = Li, Na) (M = Rb, Cs)	96
$xV_2O_5\cdot(1-x)B_2O_3$	glasses	97
$Ba(PO_3)_2-V_2O_5$	glasses	98
$Na_2O-V_2O_5-P_2O_5$	glasses	99
$V_2O_5-P_2O_5-TeO_2$	glasses	100

in inert-gas matrices forms VF_2, VF_3 and VF_4 - all identified by their i.r. spectra.[104]

νM-H was assigned in $MCl_2H_2(dmpe)_2$, where M = Nb ($1650cm^{-1}$) or Ta ($1610cm^{-1}$) - they showed the expected shifts on deuteriation. In $TaCl_2X_2(PMe_3)_4$, where X = H or D, νTa-H was at $1690cm^{-1}$, νTa-D at $1212cm^{-1}$.[105] A broad i.r. band near $1600cm^{-1}$ in $NbH(CO)_4(Ph_2P-CH_2CH_2PPh_2)$ and $NbH(CO)_3L$, where L = $PhP(CH_2CH_2PPh_2)_2$ or $P(CH_2CH_2PPh_2)_3$, is due to νNb-H, and is consistent with n.m.r. evidence for restricted fluxionality.[106] νNb=N was tentatively assigned to $1065cm^{-1}$ in $NbCl_2(SMe_2)(=NPh)_2(\mu-Cl)_2$.[107]

νNb-O modes were assigned from the i.r. and Raman spectra of $MNbO_3$, where M = Na or Ag(I).[108] Band assignments were made using i.r. and Raman data for α-$NbPO_5$, α-$MoPO_5$,[109] and β-MPO_5, where M = Nb or Ta.[110] Phase transitions in $Cd_2Nb_2O_7$ were followed by studying the Raman spectrum.[111] $\nu M-O_t$ ($900-750cm^{-1}$) and $\nu M-O_{br}$ ($750-450cm^{-1}$) were assigned for $nPbO.M_2O_5$, where n = 1.5 to 2.5, M = Nb or Ta.[112]

I.r. emission spectra of K_2NbF_7, at up to 900^oC in air, gave evidence for the formation of K_2NbOF_5 and $KNbOF_4$.[113] Raman spectra of $NbCl_5/AlCl_3$ melts contained bands due to $NbAlCl_8$. Some of these were also seen in the vapour-phase spectra.[114] νM-Cl modes were assigned in Cp_2MX_2Cl and $(Cp_2MXCl)^+$, where M = Cl, Br or I, and M = Nb or Ta.[115] νNb-Cl is in the range 250-260cm^{-1} in $(Cp_2NbClL)^+$, where L = PMe_3, MeCN or $OPPh_3$, and in $\{[Cp_2NbCl]_2(\mu\text{-}CH_2{=}CHCN)\}^{2+}$.[116]

νTa-O modes were assigned in MTa_7O_{19}, $MTaO_4$ and M_3TaO_7, where M = La, Nd or Eu.[117] Force-field calculations were published for $Cd_2Ta_2O_7$, $Tl_2Ta_2O_6$ and related species.[118] $TaCl(N^tBu)[N(SiMe_3)_2]_2$ has νTa-N at 410cm^{-1}, νTa-Cl at 353cm^{-1} and νTa=N^{t}Bu at 1150cm^{-1}.[119]

5 Chromium, Molybdenum and Tungsten

Detailed studies have been made of: $Cr(OH_2)_6^{3+}$;[1] Cs_2MO_4, where M = Cr, Mo or W;[3] $[Br(OC)MC]_2C_6X_4$, where M = Cr or W, X = H or D;[4] $M(CO)_4O_2$, where M = Mo or W;[5] $W_2O_3(CH_2CMe_3)_6$;[6] and WF_6.[7] Reference has already been made to resonance Raman results on $K_3[Cr(CN)_5NO]$,[18] CrO_4^{2-},[19] and $[Mo_2(\mu_4\text{-}S_2)(\mu_2\text{-}SO_2)(CN)_8]^{4-}$.[20]

νPt-M modes were assigned in trans-$Pt[M(CO)_3Cp]_2(PhCN)_2$, M = Cr (173cm^{-1}), Mo (148cm^{-1}) or W (134cm^{-1}), and in trans-$Pt[Mn(CO)_5]_2(PhCN)_2$ (157cm^{-1}). This was the first recorded observation of a νPt-Cr band.[120] νCr-NCS is at 355cm^{-1} in both $Cr(NCS)_4(dap)^-$ and $Cr(NCS)_2(dap)_2^+$.[121] νCr-N is at 470cm^{-1} and νCr-O at 420cm^{-1} in $Cr(pic)_3.H_2O$, where pic = picolinate.[122]

The anion in $Cs^+CrOF_4^-$ gives i.r. bands at 1005cm^{-1} (νCr=O), 650cm^{-1} (νCr-F_t) and 505cm^{-1} (νCr-F_{br}-Cr). These are very similar to the modes in VOF_4^-, which contains C_{4v} anions weakly linked by fluorine bridges to form infinite chains.[123] The i.r. spectrum of $(Ph_3SnO)_2CrO_2$ shows that the polymeric structure is built up from tetrahedral Ph_3SnO groups oxygen-bridged to C_{2v} CrO_4 units.[124] I.r. spectra have been obtained for CrO_2Cl_2 molecules in highly excited vibrational states.[125] The complex $[\{Cr(NCS)(TPyEA)\}_2O]^+$, where TPyEA = tetradentate ligand tris-(2-pyrazol-1-yl)amine, has ν_{as} Cr-O-Cr at 830cm^{-1} (790cm^{-1} in the ^{18}O analogue).[126] CrO_4^{2-} anchored on Al_2O_3 gives Raman bands which can be interpreted in terms of the species (2).[127] νCr-O and νAl-O were measured in the Raman spectra of the series $Mg(Cr_xAl_{2-x})O_4$.[128]

The complexes (3), where M, M' = Cr or Mo, give i.r. bands due to the μ_3-S-M stretch at 338cm^{-1} (M = M' = Mo), 462, 370, 340cm^{-1}

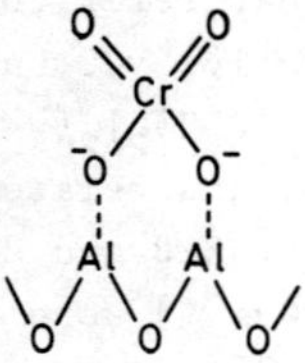

(2)

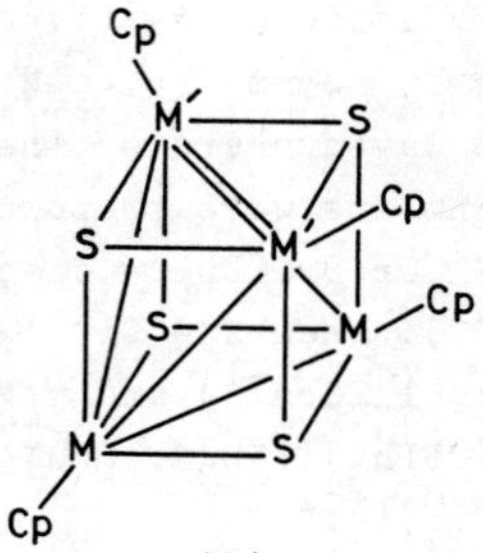

(3)

(M = Mo, M' = Cr), or 410, 325cm^{-1} (M = M' = Cr).[129] νCr-S-Cr modes were assigned for $(C_5Me_5)_2Cr_2S_4$ (447, 400cm^{-1}) and $[(C_5Me_5)_2Cr_2S_4Me]^+$ (463cm^{-1}).[130] CrF_6, from CrO_3 and F_2 at high temperature and pressure, has been characterised in Ar and N_2 matrices at 12K. A strong i.r. band at 758.9cm^{-1} was assigned as the t_{1u} Cr-F stretch - this showed the expected pattern due to chromium isotopes.[131] νCr-X modes were assigned in $CrLX_3$, where X = F, Cl, Br or I, L = terdentate P, As or S ligands such as $PPh(CH_2CH_2PPh_2)_2$, etc. It was possible to distinguish fac- and mer-isomers in many cases.[132] I.r. spectra of the vapours over $CrCl_2$ and $CrBr_3$ show that the following species are present: $(CrCl_2)_n$, where n = 1-3, $CrBr_4$, $CrBr_3$ and $(CrBr_2)_m$, where m = 1 or 2.[133]

The Mo-Mo stretch in $Mo_2(CNMe)_4(OAc)_2$ may be at 410cm^{-1}.[134] The first complexes of molecular hydrogen have been isolated: $M(CO)_3(PR_3)_2(H_2)$, where M = Mo or W, R = C_6H_{11} or ^{i}Pr. For M = W, R = C_6H_{11}, $\nu_{as}WH_2$ was at 1570cm^{-1}, ν_sWH_2 at 953cm^{-1}, νH-H at 2690 cm^{-1} and δWH_2 ca. 450cm^{-1}. These all showed the shifts expected for η^2-H_2 coordination on deuteriation.[135] The complex (4) has νMoH at

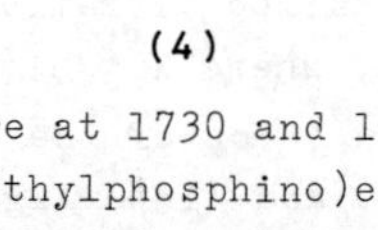

(4)

1800cm^{-1}.[136] νMoH bands are at 1730 and 1695cm^{-1} in $[Mo(dmpe)_3H_2]^{2+}$, where dmpe = bis-1,2-(dimethylphosphino)ethane.[137] Bands associated with the MoH_2 unit were assigned at 1945, 1920, 1670 and 1605cm^{-1} in $MoH_2(PMe_3)_5$.[138]

Solid-state i.r. and Raman spectra were reported for $Mo(CO)_5X$,

where X = pyridine derivatives. νMo-C, νMo-N and νC-O were assigned in terms of C_{4v} symmetry, and there was a linear relationship between the wavenumbers of the lowest Raman-active νCO and νMoN.[139] νMo-N assignments were proposed for $\{[MoX(CO)_2(DDH)_2]_2(\mu-CO)_2\}X_2$, where X = Cl or Br, DDH = diacetyldihydrazone, and related species ($213-248cm^{-1}$). When X = Cl, νMo-Cl were seen at 328 and $290cm^{-1}$.[140] $\nu_{as}MoNP$ (ca. $1120cm^{-1}$) and $\nu_s MoNP$ (ca. $569-590cm^{-1}$) were assigned for $Mo^VCl_4(NPPh_3)(PPh_3)$, $[Mo^{IV}Cl_4(NPPh_3)(PPh_3)]^-$ and $[Mo^VCl_5(NPCl_3)]^-$.[141]

$\nu Mo-^{16}O$ and $\nu Mo-^{18}O$ were assignable in terms of approximate C_{2v} symmetry for the MoO_2 groups in the following new species: $MoO_2(O^tBu)_2$, $MoO_2(O^tBu)_2(bipy)$, $MoO_2(OCH_2{}^tBu)(bipy)$ and $Mo_4O_8(O^iPr)_4py_4$.[142] The Raman spectrum of MoO_3 showed fundamental bands at 814 and $282cm^{-1}$ (a_1), 993 and $335cm^{-1}$ (e).[143, 144] Assignment of the i.r. and Raman spectra of $MoO_3.H_2O$, however, was made in terms of C_i symmetry.[145] All 13 Raman-active and 7 out of 8 i.r.-active bands of $PbMoO_4$ were assigned from the i.r. and Raman spectra of a single crystal.[146] A vibrational analysis was performed for MoO_4^{2-} in Pb_2MoO_5.[147] MoO_2L_2, where L = RR'NCR$''_2$CR$'''_2$S, R, R', R'', R''' = H, Me, Ph, etc., all have νMoO bands in the range $850-950cm^{-1}$ due to the MoO_2^{2+} unit; some had lower values than the others because they adopted 'skew-trapezoidal-bipyramidal' rather than octahedral coordination.[148] νMo=O (970, $920cm^{-1}$), νMo-O-Mo ($705cm^{-1}$), νMo-N ($400cm^{-1}$) and νMo-Cl ($320cm^{-1}$) for $Mo_2O_3Cl_4(glyH)_4$, where glyH = glycine, were consistent with the structure (5).[149]

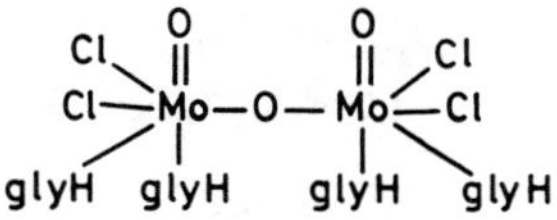

(5)

Cis-$[MoO_2\{(S)\text{-pen-OMe}\}_2]$, where (S)-pen-OMe = O-methyl(S)penicillaminato, shows $\nu Mo=O_t$ as two doublets: $918/913cm^{-1}$ and $878/875cm^{-1}$. This is presumably due to slight differences between the Λ and Δ isomers.[150] Quite detailed vibrational assignments were given for α- and $\beta-XM_{12}O_{40}^{n-}$, where X = Si(IV), Ge(IV), P(V) or As(V) and M = Mo(VI) or W(VI), in terms of the trimetallic 'ligand' vibrations ($\nu M-O_d$, $\nu M-O_c-M$) and 'interligand' vibrations ($\nu M-O_a$,

νM-O$_b$-M).[151] The reduction of $PMo_{12}O_{40}^{3-}$ was followed by changes in νMo-O-Mo and νMo-O bands in the i.r.[152] The resonance Raman spectrum of $Mo_3S_4(SCH_2CH_2S)_3^{2-}$ contains Mo-S stretching bands which show a striking similarity to those in Fe_3S_4 proteins.[153] *Fac*-$Mo(CO)_3L^1$, *fac*-$Mo(CO)_3L^2$, $[W(CO)_4]_2$-μ-L^1 and *cis*-$W(CO)_4L^2$, where L^1 = 1,4,8,11-tetrathiacyclotetradecane, L^2 = 3,6,10,13-tetrathiabicyclo[13.4.0]nona-1,16,18-triene, all have νM-S at 465-470 cm^{-1}.[154] The i.r. spectrum of $Mo_2Cl_{10}^{2-}$ has two $\nu MoCl_t$ (346, 310cm^{-1}) and two $\nu MoCl_2Mo$ (236,228cm^{-1}). Compared to neutral Mo_2Cl_{10} these show that the terminal bonds are weaker, but the bridge bonds are of comparable strength in the anion.[155]

Other molybdenum and tungsten oxygen and sulphur complexes for which vibrational assignments have been proposed are listed in Table 3.[48,65,86,87,109,156-174]

Table 3 Molybdenum and tungsten oxygen and sulphur species for which vibrational assignments have been proposed

Species	Ref.
MoO_3/Al_2O_3 catalysts	156
$MoOX_2(acda)_2$ (X = F,Cl,Br,I; Hacda=(6;R=H))	157
$MoOX(R_2dtc)_2$ (X = F,Cl,Br,I; R = Et, R_2 = piperidine)	158
MoO_2L_2 (HL = derivatives of 2-aminocyclopent-1-ene-carbodithioic acid)	159
$MoOCl_3(en)$, $Mo_2O_4Cl_2(en)_2$	160
$MoBr_2(NO)_2(OSMe_2)_2$	161
$Mo_2O_3(HR)_4$, $Mo_2O_4(HR)_4$ (H_2R= diphenylcarbazone)	162
$Mo_2O_4(OH)_3L$ (HL = anthranilic or nicotinic acid)	163
$Mo_2O_2L_2$ (HL = (6; R= H,Et,^{n}Bu or C_6H_{11})	164
$[(RNHCS_2)_2MoO]_2O$ (R= *o*-, *m*- or *p*-ClC_6H_4)	165
$Mo_2O_2(\mu\text{-}S)_2(S_2CR)_2$, $Mo_2O_2(\mu\text{-}O)(\mu\text{-}S)(S_2CR)_2$ (R= NMe_2, NEtPh, etc.)	166
$(CN)CuS_2MOS^{2-}$ (M = Mo,W)	167
$Mo_2O(S_2CNEt_2)_6^+$	168
$PM_{12}O_{40}^{3-}$ (M = Mo,W)	86
α-$VOMoO_4$	87
α-$MoPO_5$	109
$Bi_3FeMo_2O_{12}$, $Bi_2Fe_2Mo_2O_{12}$	169
$Mo(CO)_5L$ (L = substituted thiourea)	170
$Mo(CO)_3(SPPh_2)_2$, $M(CO)_2(PPh_3(SPPh_2)_2$ (M= Mo,W)	171
$KLn(WO_4)_2$ (Ln = Y, Ce-Lu)	48
$\{W(SEt)O(PMe_2Ph)\}_2(\mu\text{-}S)(\mu\text{-}SEt)_2$	172
$\{WS(S_2CNR_2)\}_2S_2$	173
Na_2WS_4	65
$WSCl_3$, $WSCl_4^-$, $WSCl_5^-$	174

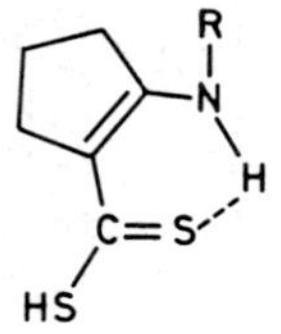

(6)

νWH_2 bands were observed at 2280 and 2260cm^{-1} in $WH_2S_2(PMe_2Ph)_3^{2+}$.[175] Fundamental vibrations of surface tungstate on WO_3/Al_2O_3 catalysts were assigned using ^{18}O substitution.[176] The complex $[WCl_4(Me_3SiC{\equiv}CSiMe_3)]_2$ has νWCl_t at 375, 361 and 340cm^{-1}, νWCl_2W at 313 and 280cm^{-1}.[177] In $[WCl_4(NSCl)]_2$, νWCl_t are at 378 and 368cm^{-1}, νWCl_{br} at 338 and 328cm^{-1}, while $WCl_5(NSCl)^-$ has νWCl at 320cm^{-1}. The former complex gives $\nu_{as}WNS$ at 1058cm^{-1}, the latter at 958cm^{-1}.[178]

6 Manganese, Technetium and Rhenium

Earlier reference has been made to detailed studies of Mn_2,[17] $MMnO_4$ (M = Rb, Cs, Ag), $Sr(MnO_4)_2.3H_2O$, $La(MnO_4)_3.nH_2O$,[21] $AgMnO_4$,[22] and $MReO_4$ (M = Li, Na, K, Rb, Cs or Tl),[8,9] as well as to more fragmentary reports on trans-$Pt[Mn(CO)_5]_2(PhCN)_2$,[120] $ReHS_3(PMe_2Ph)_3^{2+}$ and $ReHS_4(PPh_3)_2^{2+}$.[175]

The i.r. spectrum of MnH_2 in a xenon matrix at 10-12K provides evidence for the existence of two different species, one with νMnH at 1591cm^{-1}, δMnH_2 at 375cm^{-1}, the other with νMnH at 1565cm^{-1}, δMnH_2 at 366cm^{-1}.[179] $\nu Sn\text{-}Mn$ modes (150-167cm^{-1}) were assigned for $X_{4-n}Sn[Mn(CO)_5(PPh_3)]_n$, where X = Cl, Br or I, n = 2 or 3.[180] The complexes $Mn(NCS)_2L_n$, where L = $OPPh_3$, n = 4, L = $SC(NH_2)_2$, n = 2 or L = THF, n = 3, have $\nu Mn\text{-}N$ in the range 237-264cm^{-1}, with $\nu Mn\text{-}O$ for the O-ligands 300-420cm^{-1}.[181] Values for $\nu M\text{-}N$, of M-NCS and $M\text{-}N_2H_4$, for $M(N_2H_4)(NCS)_2$, where M = Mn, Fe, Co, Ni, Zn or Cd, follow the Irving-Williams series: $Cd^{2+} > Zn^{2+} > Ni^{2+} > Co^{2+} > Fe^{2+} > Mn^{2+}$.[182] $\nu M\text{-}N$(amide) and $\nu M\text{-}N$(CN) were in the ranges 270-330cm^{-1} and 200-260cm^{-1} respectively for $M[N(CN)_2]_2$, where M = Mn, Fe, Co, Ni, Cu or Zn.[183] In $M[C(CN)_3]_2$, where M = Mn, Fe, Co, Ni or Zn, $\nu M\text{-}N$(CN) is in the range 210-303cm^{-1}.[184] $\nu M\text{-}N$ was also assigned for $M(bipy)_3^{2+}$, where M = Mn, Co or Ni.[185] The i.r. spectra of M_2MnF_5, where M = NH_4, Na or K, are said to show that MnF_5^{2-} has D_{4h} symmetry.[186]

A band at 538cm^{-1} is tentatively assigned as $\nu Tc\text{-}P$ in $Tc[P(OMe)_3]_6^+\ BPh_4^-$.[187] $\nu Tc{=}O$ bands were reported for $TcO(L)(H_2O)^-$, $TcO_2L_2^-$, where L = various Schiff base ligands,[188] $TcOCl_3L^-$,

TcOCl(L) and $[TcO(L)]_2O$, where L = N,N'-propane-1,3-diylbis(salicylideneiminate) or N,N'-butane-1,4-diylbis(salicylideneiminate).[189] Assignments of νMO_4 bands were made in $Th_2O(MO_4)_6.6H_2O$, where M = Tc or Re,[190] and of νTc-O bands in $M(TcO_4)_3.4H_2O$, where M = Tb, Dy, Ho, Er, Tm, Yb, Lu or Y.[191] *Trans*-$O_2(TBP)_4Tc^+$, where TBP = 4-*tert*-butylpyridine, has $\nu_{as}TcO_2$ at $821cm^{-1}$.[192] νTc=O bands in the i.r. spectra of $TcO(H_2O)(C_{12}H_{18}N_2O_2)^+$ and related species were as expected for monoxo-technetium(V) species ($900-980cm^{-1}$).[193]

The values of νRe-H in the adducts $Cp_2ReHAlH_nX_{3-n}$, where X = Cl or Br, n = 0-3, are consistent with the existence of a direct Re-Al bond ($AlBr_3$, $AlCl_3$, $AlHCl_2$) or a Re-H-Al bridge (AlH_3, AlH_2Cl).[194] νRe-C modes were assigned for $Re(CNR)_6^+$, where R = Me, Ph or $4\text{-}MeC_6H_4$.[195]

νRe≡N in $Re(NBCl_3)Cl_2(PMe_2Ph)_3$ is higher than in $ReNCl_2(PMe_2Ph)_3$, due to the formation of a linear Re-N-B bridge.[196] $ReCl_4(N_2S_2)^-$ has ν_s, $\nu_{as}ReN_2$ at $750,760cm^{-1}$ respectively, and $\nu ReCl_4$ bands at 330, 310, 298 and $285cm^{-1}$.[197] Skeletal mode assignments were made similarly for $[ReCl_3(NSCl)_2]_2$, with the structure (7).[198] ReL_2X_2, where HL = 8-quinolinol, X = Cl or Br, have νRe-N

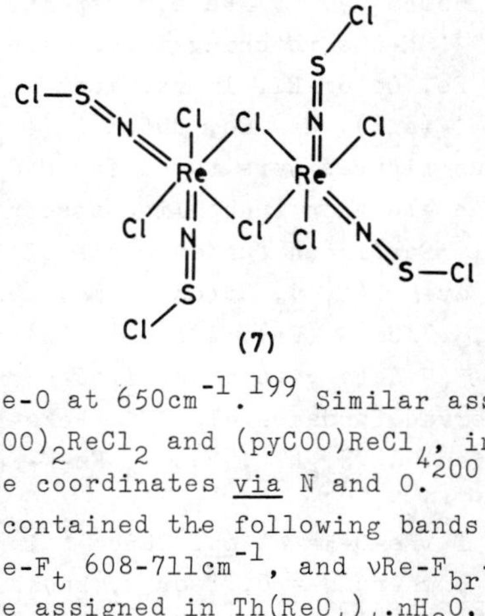

(7)

at $540cm^{-1}$, and νRe-O at $650cm^{-1}$.[199] Similar assignments were suggested for $(pyCOO)_2ReCl_2$ and $(pyCOO)ReCl_4$, in which the pyridinecarboxylate coordinates *via* N and O.[200] The i.r. spectrum of$ReNF_4.ReF_5(NCl)$ contained the following bands: νReNCl $1221cm^{-1}$, νRe-N $1102cm^{-1}$, $\nu Re\text{-}F_t$ $608-711cm^{-1}$, and $\nu Re\text{-}F_{br}\text{-}Re$ $525cm^{-1}$.[201]

νReO_4 modes were assigned in $Th(ReO_4)_4.nH_2O$, where n = 4, 3.5, 2 or 0.[202] νRe=O is *ca*. $965cm^{-1}$ and νReCl $270-335cm^{-1}$ in $ReOCl_3L^-$ and $[ReOCl_4(HL)]^-$, where HL = N-methylsalicylideneiminate and related species.[203] Skeletal mode assignments were proposed for $ReOCl_5^{2-}$, $ReO(NCS)_4(SCN)^{2-}$ and $ReOCl(NCS)_4^{2-}$.[204] $ReOF_4^+$ has νRe-O at $1060cm^{-1}$, ν_sReF_4 at $765cm^{-1}$ and $\nu_{as}ReF_4$ at $696cm^{-1}$. Assignments for

ReF_6^+ were: $\nu_1(a_{1g})$ 801cm^{-1}, $\nu_2(e_g)$ 737/730cm^{-1}, $\nu_3(t_{1u})$ 772cm^{-1} and $\nu_5(t_{1g})$ 385cm^{-1}.[205] Vibrational assignments were also given for $M_4Re_2OF_{10}$, where M = K or Cs,[206] and $ReCl_5(OPCl_3)$.[207] I.r. and Raman spectra of $(Re_3Se_2Cl_5)_2$ are consistent with an octahedral Re_6 cluster, with four μ_3-S, four μ_3-Cl ligands, and six chlorines terminally coordinated.[208] Raman spectra of aqueous alkaline solutions of $Re_3Se_2Cl_5$ and $Re_3Te_2Br_5$ were also assigned.[209]

7 Iron, Ruthenium and Osmium

Earlier reference has been made to the following studies: Mg_2FeH_6;[10] MK[Fe(CN)$_6$].4H$_2$O, where M = lanthanoid;[49] Fe[C(CN)$_3$]$_2$;[184] Fe(N$_2$H$_4$)$_2$(NCS)$_2$;[182] Fe[N(CN)$_2$]$_2$;[183] $Fe(OH_2)_6^{3+}$;[1] [Fe$_2$O(phen)$_4$-(H$_2$O)$_2$]$^{4+}$;[11] Ln_2FeO_4, $LnSrFeO_4$;[55] $Bi_{4-n}Fe_nMo_2O_{12}$;[170] many iron-containing bio-molecules;[23-40] $(NH_3)_4Ru^{II}$(riboflavin);[42] $RuCl_6^{2-}$, $RuBr_6^{2-}$;[41] and $OsF_nCl_{6-n}^{2-}$, where n = 1-5.[12]

HX and Fe atoms react in inert matrices to give the formerly unknown molecules HFeX, where X = Cl, Br or I. νFe-H is at 1755cm^{-1} (X = Cl, Ar matrix), 1739cm^{-1} (Cl, Kr matrix), 1738cm^{-1} (Br, Kr matrix), 1754cm^{-1} (I, Ar matrix). The Fe-H bond is stronger than in diatomic FeH (ν = 1660cm^{-1}).[210] The i.r. spectrum of FeD gives ω' at *ca*. 1070cm^{-1}.[211] M-C stretching modes were assigned for $M(C_6H_6)_2$, where M = Fe, Co or Ni, in matrices at -190°C: M = Fe, 395cm^{-1}(e_{1u}), 420cm^{-1} (a_{2u}); M = Co, 395cm^{-1} (e_{1u}); M = Ni, 405 cm^{-1} (e_{1u}). Analogous figures were found for $M(C_6H_5CH_3)_2$ systems.[212] Inelastic electron tunnelling spectroscopy was used to observe two optically forbidden fundamentals of $M(CN)_6^{4-}$, where M = Fe, Ru or Os. These are: $\nu_5(t_{1g})$, 354cm^{-1} (Fe), 344cm^{-1} (Ru), 362 cm^{-1} (Os); $\nu_{13}(t_{2u})$, 474cm^{-1} (Fe), 491cm^{-1} (Ru), 513cm^{-1} (Os). For M = Fe, ^{13}C and ^{15}N data gave a detailed normal coordinate analysis for 10 observed fundamentals.[213] PcFe-N-FePc, where Pc = phthalocyaninato anion, $C_{32}H_{16}N_8^{2-}$, has ν_{as}Fe-N-Fe at 915cm^{-1}, close to the value in (TPPFe)$_2$N.[214]

The observation of νFe-O-Fe at *ca*. 885cm^{-1} in all of [(*p*-Et$_2$NTPP)$_2$Fe$_2$]O^{n-}, where n = 0, 1 or 2, shows that all have similar structures.[215] $\nu Fe^{III}_2M^{II}O$ in $Fe_2M^{II}O(OAc)_6L_3$, where L = H$_2$O or py, M^{II} = Mn, Co or Ni, give two distinct bands, compared to the singlet in the $Fe_3^{III}O$ system. If M^{II} = Fe, there are still two bands, but they are not so clearly separated as in the other M^{II} compounds.[216]

I.r. spectra of Li_xMPS_3, where M = Fe or Ni, 1.5 > x > 0, contain characteristic bands at $336cm^{-1}$ (Fe) or $310cm^{-1}$ (Ni).[217] The Fe-histidine stretch in deoxyHb and photodissociated carbonyl-Hb derivatives varies with changes in the tilt of the proximal histidine with respect to the porphyrin plane.[218] νFe-X modes were assigned in the i.r. spectra of $Fe_2X_2(X_5)_2^{2-}$, where X = S or Se. These are: 414, 338, $315cm^{-1}$ (S) or 312, $258cm^{-1}$ (Se).[219] Similar assignments were made for $Fe_6S_9(SR)_2^{4-}$, where R = Me, Et or CH_2Ph.[220] I.r. and resonance Raman spectra were obtained for: $Fe_2X_2(SR)_4^{2-}$, where X = S or Se, $(SR)_2$ = $(SCH_2)_2C_6H_4$ or R = Ph, p-MeC_6H_4 or p-ClC_6H_4; $Fe_2X_2(SePh)_4^{2-}$; $Fe_2X_2Cl_4^{2-}$, where X = S or Se; and $Fe_2S_2Br_4^{2-}$. The νFeS_{br} wavenumbers are little affected by changes in the terminal ligands.[221] Skeletal assignments were proposed for $Fe_2S_2(NO)_4^{2-}$.[222]

νRu-H is at ca. $2000cm^{-1}$ in trans-RuH(Cl){$Ph_2P(CH_2)_3PR_2$}$_2$, where R_2 = Ph_2, MePh or Me_2.[223] In $Ru(C_6Me_6)(PHCy_2)HX$, νRu-H are as follows: $1945cm^{-1}$ (X = H), $1980cm^{-1}$ (Cl), $1930cm^{-1}$ (Ph).[224] Other νRu-H assignments were given for $RuH_2(dppm)(H_2O)^+$, where dppm = bis(diphenylphosphino)methane,[225] and RuHCl(π-C_6Me_6)L, where L = PPh_3, $AsPh_3$, $SbPh_3$, P^tBuPh_2,etc.[226]

νRu-C bands are at $460cm^{-1}$ (a_1) and 422, $410cm^{-1}$ (e), with νRu-N at $585cm^{-1}$ in $Na_2Ru(CN)_5NO\cdot 2H_2O$.[227] The νRu-N and νRu-O wavenumbers follow the sequence M = Rb > NH_4 > K in $M_3[Ru_2NCl_8$-$(H_2O)_2]$. These are related to the decrease in bond distances as the cation becomes smaller.[228] $Ru(TMP)O_2$, where TMP = 5,10,15,20-tetramesitylporphinato, has $\nu_{as}RuO_2$ in the i.r. spectrum at $821cm^{-1}$, shifted to lower wavenumber by $35cm^{-1}$ on ^{18}O substitution.[229] ν_1 - ν_5 for $Ru(H_2O)_6^{2+}$, $Ru(H_2O)_6^{3+}$ and their ^{18}O analogues were made from i.r. and Raman data on $Rb_2Ru(SO_4)_2\cdot 6H_2O$ and $CsRu(SO_4)_2\cdot$-$12H_2O$.[230]

Resonance Raman spectra of $(NH_3)_6Ru_2X_3^{2+}$, where X = Cl, Br or I, show enhancement of ν_2(Ru-X stretch) and ν_3(RuX_2 deformation) - both of a_1 symmetry. ν_2 is at $310cm^{-1}$ (Cl), $247cm^{-1}$ (Br) or $187cm^{-1}$ (I); ν_3 at $145cm^{-1}$ (Cl), $106cm^{-1}$ (Br) or $79cm^{-1}$ (I). The ground state has a localised, the excited state a delocalised odd electron.[231,232] νRu-Cl modes were assigned in $[(R_3P)Ru(CO)_2Cl_2]_2$, where R = Me or Et, and $(R_3P)_2Ru(CO)_2Cl_2$, where R = Me, ^nPr or ^nBu.[233]

νOs-H modes lie in the range 2005-$2060cm^{-1}$ for $Os(bipy)_2(CO)H^+$,

$Os(phen)_2(CO)H^+$, $Os(phen)(dppb)(PPh_3)H^+$ and trans-$Os(bipy)(PPh_3)_2$-$(CO)H^+$, where dppb = 1,2-bis(diphenylphosphino)benzene.[234] νOs-N is at 592cm^{-1} in $Os(NO)Cl_5^-$, while one νOs-Cl mode is much lower than the others: 354, 328, 317, 270cm^{-1}. This last is assigned to the bond trans to NO (which has a large trans-effect).[235]

Analysis of the vibrational fine-structure of the visible emission spectrum of OsO yields a ground-state vibrational wavenumber of 885cm^{-1}.[236] The ν_3 region of $^{192}OsO_4$ has been studied by high-resolution i.r. spectroscopy.[237] νOs-O bands at 852 and 834 cm^{-1} show that the Os-O bonds in OsO_4^- are significantly weaker than those in neutral OsO_4.[238] K_2 $Os(NO)F_5$ $.H_2O$ has νOs-F bands at 527 and 590cm^{-1} in its i.r. spectrum.[239] νOs-X modes were assigned in $Os(CNR)_4X_2$, where R = CMe_3, $CHMe_2$ or C_6H_{11}, X = Cl or Br, and in $Os(CNR)_4X_2^+$. νOs^{II}-Cl is ca. 300cm^{-1}, νOs^{III}-Cl ca. 330cm^{-1}, with νOs^{II}-Br ca. 218cm^{-1}, and νOs^{III}-Br ca. 238cm^{-1}. The single band in each case shows that the complexes have trans geometry.[240]

8 Cobalt, Rhodium and Iridium

νCo-CH_3 and νCo-CD_3 were assigned to 506, 478cm^{-1} respectively in methyl- and methyl-d_3-(4-tert-butylpyridine)cobaloxime complexes.[241] $Co_6(\mu_6$-$C_2)(\mu$-$CO)_6(CO)_8(\mu_4$-S) has i.r. bands predominantly due to νCo-C_2 at 698 and 610cm^{-1}. A νCo-S mode is seen at 230cm^{-1}.[242] Tentative assignments were proposed for νCo-CO and νCo-CN (360-470 cm^{-1}), νCo-ring (321-335cm^{-1}) and νCo-NCS (277-285cm^{-1}) in $RCo(C_2H_4)(NCS)_2$, where R = C_5Me_5, L = CO or C_2H_4.[243] The νM-C modes in $TiM(CN)_6$, where M = Co, Rh or Ir, are higher than in $K_3M(CN)_6$, due to Ti-NC-M bridge formation.[244] H/D substitution was used in assigning skeletal modes in the i.r. spectra of $[Co(NH_3)_5X]X_2$, where X = Cl or Br.[245] Fine structure in the band near 500cm^{-1} in the Raman spectra of $Co(NH_3)_6X_3$ and $Co(ND_3)_6X_3$, where X = Cl, Br or I, can be explained in terms of inequivalent sites, and different isotopic shifts for the a_{1g} and t_{1u}^b stretches.[246] $(Cp^*CoX_2)_2$, where Cp* = π-C_5Me_5, have νCo-Cl at 342 (t) and 309 (Br)cm^{-1} for X = Cl, νCo-Br at 265 (t) and 235(br) cm^{-1} for X = Br. $(Cp^*Co)_2HX_3$ complexes give νCo-H-Co bands ca. 1100cm^{-1} and ca. 830cm^{-1} for X = Cl, Br or I.[247]

Some skeletal vibrational mode assignments were proposed for (8) and related species, e.g. νRh-Rh 157cm^{-1}, νRh-C 460-490cm^{-1}, νRh-Cl 312, 325cm^{-1}.[248] νRh-O is ca. 490cm^{-1} in (9), where R^1, R^2,

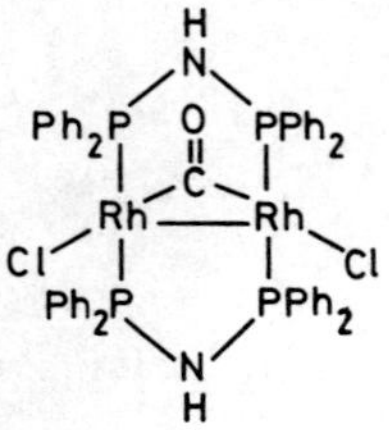

(8)

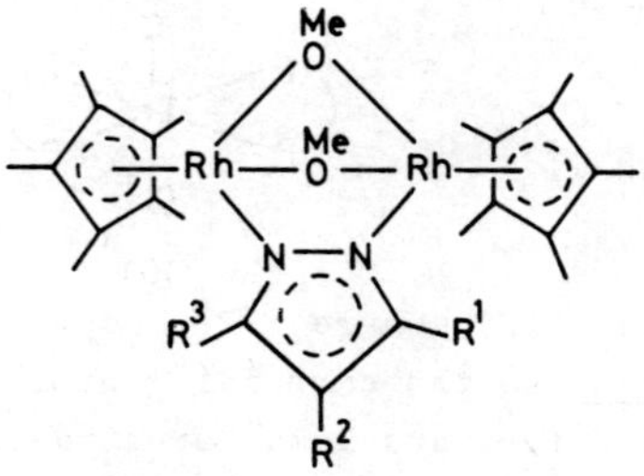

(9)

R^3 = H or Me.[249] Solution Raman spectra for $Rh_2(O_2CMe)_4$ showed that νRh-O bands were at 345cm^{-1} (a_{1g}), 323cm^{-1} (b_{1g} or e_g).[250] νRh-N and νRh-S were assigned for the isomers $Rh(NCS)_n(SCN)_{6-n}^{3+}$, where n = 0-3, and for $Rh_2(SCN)_{10}^{4-}$, with two Rh-SCN-Rh bridges, and the rest Rh-SCN terminal units.[251]

Other cobalt and rhodium complexes for which vibrational assignments have been given are summarised in Table 4.[43,55,182-5,212,252-261]

Table 4 Cobalt and rhodium complexes for which vibrational assignments have been proposed

Species	Ref.
$Co(C_6H_6)_2$	212
$CoX_2(HL)_2$ (X_2 = Cl_2, ClBr, Br_2, HL = 4,6-dimethyl-pyrimidine-2(1H)-one)	252
Co(salen) and related derivatives	43
$CoL[(EtS)_2PO_2]_2$ (L = py, α-aminopyridine, imidazole, quinoline, etc.)	
$Copy_2[(EtS)PO_2]_2$	253
$CoL'[(EtS)_2PO_2]_2$ (L' = bipy, phen)	
$CoCl_2L$ (L = 1-bis(dimethylamino)phosphino-2-diphenyl-phosphino-o-carbaborane)	254
$CoCl_2L'_2$ (L' = 1-diphenylphosphino-o-carbaborane)	
$Co(N_2H_4)(NCS)_2$	182
$Co[N(CN)_2]_2$	183
$Co[C(CN)_3]_2$	184
$Co(bipy)_2$	185
Co_2RhO_4, $CoRh_2O_4$	255
Ln_2CoO_4, $LnSrCoO_4$	55
$Co(SO_4)Pz_2$ (Pz = pyrazole)	256
$CoCl_4^{2-}$ (acetylpyridinium salts)	257
$Rh(PPh_3)_2H(\eta^3-B_4H_8)$	258
$Rh_2Cl_2(\mu-H)(\mu-CO)(DPM)_2$	
$Rh_2Cl_2(\mu-H)(\mu-CO)(DPM)_2L$ (L = FBF_3, THF)	259
$Rh_2(OAc)_2(SO_4)_2L_2^{2-}$ (L = py, H_2O)	260
$Rh_2(SO_4)_4L_2^{4-}$	
$M(P-i-Pr)_3(SO)Cl$ (M = Rh or Ir)	261

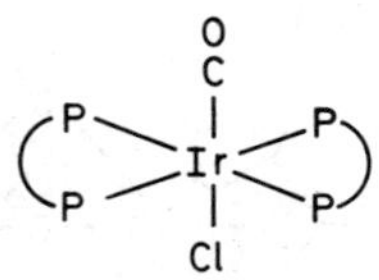

(10)

The complex (10), where P⌒P = dppe, has νIrCl at 310cm^{-1}. It is therefore <u>trans</u> to the carbonyl group.[262] A detailed assignment was given from the i.r. and Raman spectra of $IrF_nCl_{6-n}^{3-}$, where n = 1-5. For IrF_5Cl^{3-}, νIr-Cl gives two well resolved components due to ^{35}Cl and ^{37}Cl (361, 355cm^{-1}).[263] Assignments of νIr-H modes were made for the complexes listed in Table 5.[175,264-269]

Table 5 Iridium hydride complexes for which νIr-H assignments have been made

Species		Ref.
$IrCl(H)(SH)(CO)(PPh_4)_2$		264
$IrH_2S_2(PPh_3)_2$		175
$Ir_2H_4X_2(CO)_2(dppp)_2$, $IrH_2X(CO)(dppp)$, $IrH_2X(CO)(dppe)$	($Ph_2P(CH_2)_nPPh_2$: n = 2, dppe; n = 3, dppp)	265
$[IrH_2(COD)(dt)]_2^{2+}$	(Hdt = $(MeS)_2(CH_2)_3$)	266
$Ir(CH_2PPh_2C_6H_4)H(I)(CO)(PPh_3)$		267
<u>Trans</u>-$(C_5Me_5)IrH_2(Cl)(SiEt_3)$		268
$(\eta^5-C_5Me_5)Ir(CO)H(Me)$		269

<u>9 Nickel, Palladium and Platinum</u>

$[(np_3)NiH]BPh_4$, where np_3 = tris(2-(diphenylphosphino)ethyl)-amine, has νNi-H at 1928cm^{-1}.[270] νM-C bands were seen in the range 300-365cm^{-1} for the polymeric butadiene π-complexes $(MC_4H_6)_n$, where M = Ni or Pd.[271]

νM-O modes were assigned from F.T.i.r. spectra of $M(acac)_2$, where M = Pd, Pt or Co, and related β-diketonate complexes. For M = Pt, the νPt-O wavenumber correlates with ^{195}Pt n.m.r chemical shifts.[272] The <u>cis</u>-configuration of $Pd(SNNMe_2)(AsPh_3)Cl_2$ was shown by the presence of two νPd-Cl bands in the i.r. spectrum, at 323 and 282cm^{-1}. (11) has νPd-S at 390cm^{-1}, νPd-Cl_t at 340cm^{-1} and Pd-Cl_{br} at 295and 275cm^{-1}.[273]

X-ray diffraction shows that $(MeNC)_4Pd_2I_2$ has a non-bridged structure, with linear I-Pd-Pd-I. It does, however, have low νPd-I wavenumbers, which had previously been used to suggest Pd-I-Pd bridging. These low values are now attributed to the high

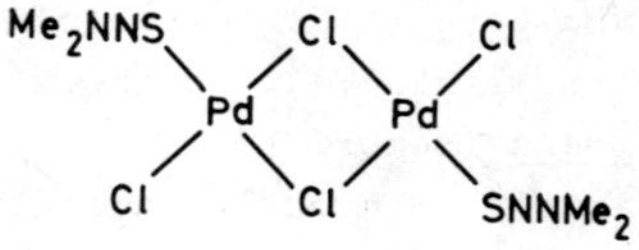

(11)

trans-effect of the Pd-Pd bond. The single i.r. band observed is assigned to ν_{as}Pd-I (skeletal symmetry D_{2h}).[274]

The high trans-effect of R in $PtH(R)(PR'_3)_2$ (where R = Ph, R'_3 = $^tBu_2Me$, $^tBuMe_2$, Ph_3, $^iPr_3$, Et_3, Me_3 or Me_2Ph; R = C_6H_4-4-Br, R'_3 = Et_3; R = Me, R'_3 = Ph_3, $^iPr_3$) gives rise to lower νPt-H wavenumbers than in trans-$PtHClL_2$ species, 1890-1990cm^{-1} compared to ca. 2100cm^{-1}.[275] νPt-H wavenumbers in $PtH(C_6H_4$-m- or-p-Y)$(PEt_3)_2$ increase as Y becomes more electron-withdrawing.[276] High-resolution electron enrgy loss spectroscopy (HREELS) of CO adsorbed on Pt(110) at 300K showed a band due to νPt-CO of a chemisorbed CO molecule, bonded end-on to a single platinum atom.[277] I.r. spectra of trans-$PtBr_2$(L)(R-an), where L = C_2H_4 or CO, R-an = variously substituted anilines, show that νPtC_2 or νPt-C and νPt-N are increased by electron-releasing substituents on R-an, i.e. there is effective transmission of electronic effects from the R groups.[278] Skeletal mode assignments for $(MeCN)_2PtCl_2$ show that it has cis-geometry; νPt-N are at 463, 408cm^{-1} (453, 401cm^{-1} in a ^{15}N-substituted sample), νPt-Cl at 362, 333cm^{-1} (although these are believed to couple strongly with δPt-N-C modes).[279]

$Pt(NO_2)_2(NH_3)_2$ and concentrated phosphoric acid produce a variety of platinum-phosphato complexes. These all have νPt-O (of PO_4) ca. 270cm^{-1} in their Raman spectra.[280] Skeletal mode assignments were given for trans-$Pt(NH_2OH)(NO_2)Cl_2$[281] and for $Pt(SO_3F)_{4+n}^{n-}$, where n = 0, 1 or 2.[282] Pt-Cl stretching modes gave unambiguous discrimination between cis and trans configurations for complexes related to dichlorobis(triphenylphosphine)platinum.[283] The presence of a trans-$PtCl_2$ unit in (12) is confirmed by a single νPt-Cl mode in its i.r. spectrum (355cm^{-1}).[284] Skeletal mode

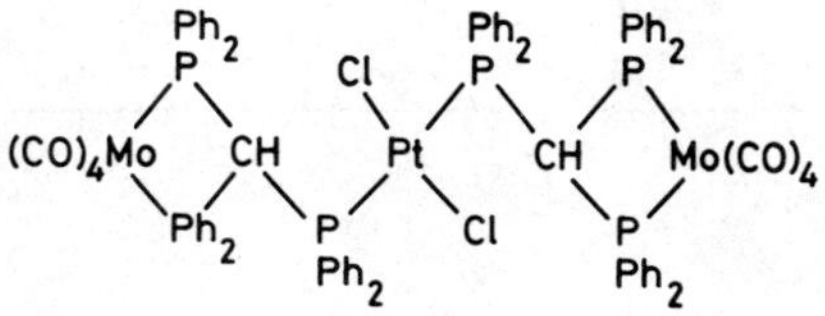

(12)

assignments for cis- and trans-$PtCl_2(oxazole)_2$ are consistent with Pt-N coordination for the ligand, C_{2v} symmetry for the cis- and D_{4h} symmetry for the trans-isomer.[285]

Other assignments involving nickel, palladium and platinum complexes are given in the references listed in Table 6.[44,120,182-5,212,256,257,286-300]

Table 6 Vibrational assignments in nickel, palladium and platinum complexes

Species		Ref.
$[Ni(CH_3CN)_6][ZnX_4]$	(X = Cl, Br)	286
$NiCl_2(DMP)_4$	(DMP = 3,5-dimethylpyrazole)	287
$Ni(N_2H_4)_2(NCS)_2$		182
$Ni[N(CN)_2]_2$		183
$Ni[C(CN)_3]_2$		184
$Ni(bipy)_2^{2+}$		185
$Ni(C_6H_6)_2$		212
NiX_2L	(L = 1-bis(dimethylamino)phosphino-2-phenylphosphino-o-carbaborane, X = Cl, Br, I, NCS)	288
$Ni(L-L)X_3$	(L-L = diphosphines, diarsines, X = Cl,Br)	289
$Ni(SO_4)Pz_2$	(Pz = pyrazole)	256
$NiCl_5^{3-}$	(as acetylpyridinium salt)	257
$[Pd(L-L)_2][Pd(L-L)_2X_2](ClO_4)_4$	(X = Cl, Br, L-L = 1,2-diaminoethane, etc.)	44
cis, trans-$PdCl_2(NH_2CH_2COX)_2$	(X = OH, Cl)	290
trans-$Pd_2Cl_4(NH_2CH_2COCl)_2$		290
PdL_2	(L = dipropyl-, dibutyl-, di-iso-butyl, ditrifluoroethyldithiocarbamates)	291
$(PdL_4)Cl_2$, trans-PdL_2Cl_2	(L = 2-aminopyridine)	292
PdL_2X_2	(X = Cl, Br, L = N-ethyl- or N-propyl-imidazole)	293
$[PdCl(plp)]_2$	(Hplp = 2-(1-pyrrolylpyridine))	294
$PdCl(plp)(PBu_3)$	(Hplp = 2-(1-pyrrolylpyridine))	294
Pd(plp)(acac)	(Hplp = 2-(1-pyrrolylpyridine))	294
trans-$Pt[M(CO)_3Cp](PhCN)_2$	(M= Cr,Mo,W)	120
trans-$Pt[Mn(CO)_5]_2(PhCN)_2$		120
PtL_2Cl_2, $Pt(L-H)_2$, $Pt(L-2H)_2$	(L = methyl ester of dithiocarbazic acid, $NH_2NHC(=S)SMe$)	295
$[Pt(DMSO)X_2]_2$	(DMSO, DESO = dimethyl-, diethylsulphoxides, X= Cl,Br)	296
$[Pt(DESO)(\mu-OH)X]_2$	(DMSO, DESO = dimethyl-, diethylsulphoxides, X= Cl,Br)	296
$Pt(PPh_3)(ETU)_n$	(n = 2,5, ETU = ethylenethiourea)	297
$Pt_2Ag_2Cl_4(C_6F_5)_4^{2-}$		298
$PtCl(C_6F_5)_3^-$		299
$PtCl_2(L)(py^{2,4,6})$	($py^{2,4,6}$=trimethyl-2,4,6-pyridine, L = OEt_2, $Me_2C=O$, MeCOOEt, etc.)	300

10 Copper, Silver and Gold

The i.r. and Raman spectra of $KCu(CN)_2$ show that it is centrosymmetric.[301] The complexity of the i.r. and Raman spectra of $Cu(ClO_4)_2$ at low wavenumbers rules out regular square-planar or octahedral coordination at Cu. The coordination of the Cu is probably greater than 4.[302] A skeletal symmetry of D_{2h} in the vapour phase for CuL_2, where HL = 2,4-pentanedione, 1,1,1-trifluoro-2,4-pentanedione or 1,1,1,5,5,5-hexafluoro-2,4-pentanedione, is shown by the observed number of νCu-O bands.[303] $(NEt_4)_2$ $Cu(SPh)_3$ gives νCu-S bands at $342cm^{-1}$ (i.r. and Raman) and $314cm^{-1}$ (Raman), consistent with D_{3h} symmetry for the CuS_3 unit.[304] ν_{as}S-Cu-S in the i.r. spectrum of $Cu_5(S^tBu)_6^-$ is at $270cm^{-1}$.[305] I.r. and Raman spectra were used to obtain metal-halogen vibrational assignments in $MX(PPh_3)_n$ (where M = Cl or Au, X = Cl or Br, n = 2; M = Au, X = Cl, n =3). The data show the expected dependence on coordination number, and confirm that $AuX(PPh_3)_2$ are indeed three-coordinate.[306] Approximately D_{4h} symmetry is indicated for $CuCl_5^{3-}$ as a $Co(NH_3)_6^{3+}$ salt, from its i.r. and Raman spectra.[307] CuX_3^{2-}, where X = Br or I, and AgI_3^{2-} have been prepared; they give $\nu_{as}MX_3$ at $190cm^{-1}$ ($CuBr_3^{2-}$), $163cm^{-1}$ (CuI_3^{2-}) and $133cm^{-1}$ (AgI_3^{2-}).[308]

Ag-Ag stretches and deformations were detected (150-200cm^{-1}, ca. 100cm^{-1} respectively) for Ag_3^{2+} and Ag_3 clusters. It was possible to probe the reversible cluster redox process:

$$Ag_3^{2+} \underset{O_2}{\overset{H_2}{\rightleftharpoons}} Ag_3^0$$

The data were consistent with linear structures for both clusters.[309] SERS for 0.5M NH_3 in 4.0M KCl at a silver electrode showed bands due to νAg-Cl ($216cm^{-1}$) and νAg-N ($325cm^{-1}$).[310] Anomalous behaviour was found for the linewidth and intensity of νAg-S modes in the Raman spectrum of Ag_3AsS_3.[311]

Heterotrimetallic Au(I) anionic complexes have been prepared: $(M\text{-}Au\text{-}M)^-$, where M = $Mn(CO)_5$, $Cr(CO)_4$, $Cr(CO)_3Cp$, $Mo(CO)_3Cp$, $W(CO)_3Cp$ or $Fe(CO)_2Cp$. All gave a strong i.r. band in the range 150-200cm^{-1}, due to ν_{as} of the linear M-Au-M system, e.g. Mn-Au-Mn $155cm^{-1}$, W-Au-W $152cm^{-1}$.[312] (13) has νAu-Cl at $350cm^{-1}$ (trans to N) and $300cm^{-1}$ (trans to phenyl).[313]

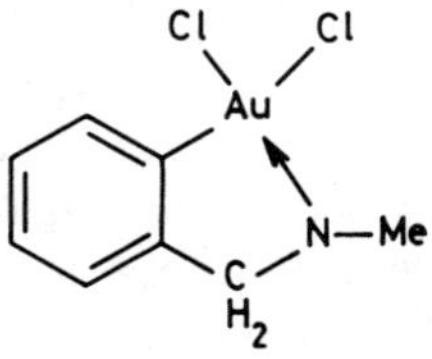

(13)

Other complexes of copper, silver and gold for which vibrational assignments have been made are listed in Table 7.[13,14,45,55,56,168,183,256,257,272,314-326]

Table 7 Vibrational assignments in copper, silver and gold complexes

Species	Ref.
Bis(L- and DL-phenylalaninato)copper(II)	13
Bis(L-serinato)copper(II)	14
Blue copper proteins	45
$Cu[N(CN)_2]_2$	183
$Cu(NCO)_2L_2$ (L = py, 4-Me-, 3,5-Me_2-, 4-Cl-, 4-Br-, 3-I-pyridine, NH_3, imidazole, etc.)	314
$Cu(NCS)_2(dmtp)_2$ (dmtp = 5,7-dimethyl 1,2,4-triazolo-1,5-a pyrimidine)	315
$Cu[C(CN)_3]L$ (L = 4-Me-, 2-Et-, 4-Et- or 3,4-Me_2-pyridine)	316
$Cu(NCO)_2(NH_3)_2$	317
$[Cu(dmphen)(CN)]_n$ (dmphen = 2,9-dimethyl-1,10-phenanthroline)	318
Ln_2CuO_4, $LnSrCuO_4$ (Ln = lanthanoid)	55, 56
CuL_4X_2 (X = ClO_4, BF_4, NO_3), CuL_2X_2 (X = Cl, Br), $CuLCl_3$ (L= 2-alkylamino-5-picoline-N-oxides)	319
$Cu(SO_4)Pz_2$ (Pz = pyrazole)	256
$Cu(acac)_2$	272
$(CN)CuS_2MOS^{2-}$ (M = Mo or W)	168
$Cu_2F_2(tmpz)_6^{2+}$, $Cu_2F_2(tmpz)_4(mpz)_2^{2+}$ (mpz = 5(3)-methyl-, dmpz = 3,5-dimethyl-, tmpz = 3,4,5-trimethylpyrazole)	320
CuCl	321
$Cu_2(PAP)(OH)Cl(SO_4).2H_2O$ (PAP = 1,4-bis(2-pyridylaminophthalazine))	322
$CuCl_3^-$, $CuCl_4^{2-}$ (acetylpyridinium salts)	257
M_2HgI_4 (M = Cu or Ag)	323
$[Ag(CNMe)_4]ClO_4$	324
AuS_9^-	325
trans-$AuBr_2(CN)(CH_2PPh_3)$	326

11 Zinc, Cadmium and Mercury

$ZnH_2.nEt_2O$, where n = 0.1-0.3, have νZnH_2 as very broad bands ($1250-1900cm^{-1}$, maximum at $1540cm^{-1}$), and δZnH_2 in the range

450-800cm^{-1}.[327] I.r. and Raman spectra of crystalline $ZnBr_2$ were assigned on the basis of a unit cell (space group $I4_1/acd$) built up of supertetrahedral Zn_4Br_{10} units.[328] $Cd_4O(OAc)_6$ has been reported for the first time. Skeletal modes gave i.r. bands at 245 and 160cm^{-1}.[329] Far-i.r. spectra of $CdHgI_4(PPr_3)_2$ contain bands due to $\nu M\text{-}I_t$ at 172, 147cm^{-1}, $\nu M\text{-}I_{br}$ at 118,87cm^{-1}.[330]

Polymeric $Hg(S^iPr)Cl$ has νHg-S at 231cm^{-1}, νHg-Cl at 218cm^{-1}; these low values reflect the interactions of ligand groups in the crystal.[331] The following νHg-X assignments have been proposed in HgX_2L_m, where X = Cl, Br or I, L = $P(CH_2CH_2CN)_3$, m = 1 or 2: $\nu Hg\text{-}Cl_t$ 323cm^{-1}, $\nu Hg\text{-}Cl_{br}$ 195cm^{-1}, $\nu Hg\text{-}Br_t$ 205cm^{-1}, $\nu Hg\text{-}I_t$ 177cm^{-1}.[332] νHg-Cl in $(bipy)_2Mo_2(CO)_6.HgCl_2$ and related species are at low wavenumbers (250-270cm^{-1}). The weakening of the Hg-Cl bond is thought to be due to the formation of the Hg-Mo bond.[333]

Other complexes of zinc, cadmium and mercury for which vibrational assignments have been made are listed in Table 8.[14,182-4,256,323,334-340]

Table 8 Vibrational assignments in zinc, cadmium and mercury complexes

Species	Ref.
Bis(L-serinato)zinc	14
$M(N_2H_4)_2(NCS)_2$ (M = Zn, Cd)	182
$Zn[N(CN)_2]_2$	183
$Zn[C(CN)_3]_2$	184
$(L\text{-}L)_nZn(NCS)_2Hg(SCN)_2$ (L-L = phen, bipy, etc., n = 1,2)	334
$BH^+(ZnXYB)^-$ (B = 2-, 3-picoline, quinoline, X,Y = Cl, Br, I)	335
$M(SO_3F)_2$ (M = Zn, Cd, Hg)	336
$Zn(SO_4)Pz_2$ (Pz = pyrazole)	256
ZnX_2Pz_2 (X = Cl, Br)	337
$[Hg(phosphine)_n](ClO_4)_2$ (n = 2,3,4, phosphine = P^tBu_3, PCy_3, etc.)	338
$(PPh_3HgX(NO_3)$ (X = Cl, Br, I)	339
$(L\text{-}L)(OC)_3MHgX$ (L-L = bipy, phen, M = Mo,W, X = Cl, SCN)	340
M_2HgI_4 (M = Cu, Ag)	323

12 The Actinoids

Earlier reference has been made to work on $ThCl_4$, UCl_4,[70] and $Th(ReO_4)_4.nH_2O$.[202] $ThCl_4(NMe_3)_3$ has νTh-N as a strong i.r. band at 491cm^{-1}, with Th-Cl at 256cm^{-1}.[341]

νU=O in $UF_4O.2SbF_5$ suggests that there are contributions to the bonding from ionic structures involving UF_2O^{2+}.[342] The association of UO_2^{2+} with Th(IV) hydrous polymer was followed by Raman spectroscopy (pH range 1.5 to 4.0). Attachment of the UO_2^{2+} is via hydroxyl bridges, giving a shift in $\nu_s UO_2$ to 665cm^{-1}.[343] The following vibrational assignments were proposed for the adduct $UO_2F_2.AsF_5$: $\nu_{as}UO_2$ 1000cm^{-1}; νU...F(bridging) and νAs...F(bridging) 625, 575, 544cm^{-1}; δUO_2 282, 270cm^{-1}.[344] Other UO_2^{2+} species for which assignments were made are listed in Table 9.[15,16,345-353]

Table 9 Species containing UO_2^{2+} for which vibrational assignments have been given

Species		Ref.
$UO_2L_3^-$	(HL = phenylacetohydroxamic acid)	345
$Na^+UO_2(OAc)_3^-$		346
$NH_4[UO_2(O_2CEt)_3]$ $[UO_2(O_2CEt)(urea)_3]UO_2(O_2CEt)_3$		347
$UO_2L.EtOH$	(H_2L = RC(O)$CH_2C(CH_3)$=NR', R = Me,Ph, R' = hydroxy- or thiol-containing groups)	348
UO_2X_2D	(X = acetylacetonate, benzoylacetonate, thenoyltrifluoroacetonate, hexafluoroacetylacetonate, D = DMSO, DMF, HMPA, etc.)	349
$Ca(UO_2)_2(PO_4)_2.6\text{-}8H_2O$ $Cu(UO_2)_2(PO_4)_2.8H_2O$		350
$UO_2F_5^{3-}$, UO_2X_2 (X = Cl, F) $UO_2(NO_3)_2$, $UO_2(NO_3)_3^-$ UO_2SO_4, etc.		351
$(pyH)_2UO_2Cl_4$		352
$UO_2X_4^{2-}$		15,16,353

I.r. and Raman spectra were reported for adducts of UF_5 with SF_4, AsF_5, TaF_5, etc. $SF_4.UF_5$ gives bands assignable to SF_3^+ UF_6^-, but the adducts with pentafluorides are covalent, fluoro-bridged species.[354,355] MUF_6, where M = K, Rb or Cs, were vapourised at ca. 400°C and the vapours condensed into Ar or N_2 matrices. I.r. spectra of these were consistent with the presence of species with C_{3v} (Ar matrices) or C_{2v} symmetry (N_2 matrices).[356] ν_3 of UF_6 was studied at low rotational and vibrational temperatures using a tunable diode laser. The $^{235}U/^{238}U$ shift was found to be 0.6040cm^{-1}.[357]

Raman spectroscopy was used to study interactions between VO^{2+}

and NpO_2^+ ions, forming the complex $NpVO_3^{3+}$. This contains a linear O=Np=O unit (ν_s at 775cm^{-1}). No band due to Np-O-V was observed; in $V_2O_3^{3+}$ νV-O-V was seen, hence it was suggested that the complex ions have different structures.[358] Raman spectra of PuO_2^{2+} and NpO_2^{2+} ions were studied over a range of pH values in complexing aqueous solutions. Symmetric stretches were found corresponding to the following species: PuO_2^{2+} (833cm^{-1}), $(PuO_2)_2(OH)_2^{2+}$ (817cm^{-1}), $(PuO_2)_4(OH)_7^+$ (826, 805cm^{-1}), NpO_2^{2+} (854cm^{-1}) and $(NpO_2)_2(OH)_2^{2+}$ (834cm^{-1}). At high pH a further Pu(VI) band was seen (794cm^{-1}), but not assigned to a specific species. ν_1 of PuO_2^+ was seen in Pu(V) solutions, at 748cm^{-1}.[359]

References

1 S.P.Best, J.K.Beattie and R.S.Armstrong, J. Chem. Soc., Dalton Trans., 1984, 2611.
2 N.V.Porotnikov, M.German, and L.M.Kovba, Russ. J. Inorg. Chem., 1984, 29, 825.
3 H.M.Nagarathna, L.Bencivenni and K.A.Gingerich, J. Chem. Phys., 1984, 81, 591.
4 N.Q.Dao, H.Fevrier, M.Jouan, E.O.Fischer and W.Röll, J. Organomet. Chem., 1984, 275, 191.
5 J.A.Crayston, M.J.Almond, A.J.Downs, M.Poliakoff and J.J.Turner, Inorg. Chem., 1984, 23, 3051.
6 I.Feinstein-Jaffe, D.Gibson, S.J.Lippard, R.R.Schrock, and A.Spool, J. Am. Chem. Soc., 1984, 106, 6305.
7 M.Takami and H.Kuze, J. Chem. Phys., 1984, 80, 5994.
8 L.Bencivenni, H.M.Nagarathna, D.W.Wilhire, and K.A.Gingerich, Inorg. Chem., 1984, 23, 1279.
9 L.Bencivenni, H.M.Nagarathna, K.A.Gingerich, and R.Teghil, J. Chem. Phys., 1984, 81, 3415.
10 J.J.Dicksheim, P.Zolliker, K.Yvon, P.Fischer, J.Schefer, M.Gubelmann, and A.F.Williams, Inorg. Chem., 1984, 23, 1953.
11 J.E.Plowman, T.M.Loehr, C.K.Schauer, and O.P.Anderson, Inorg. Chem., 1984, 23, 3553.
12 W.Preetz, D.Ruf, and D.Tensfeldt, Z. Naturforsch., Teil B, 1984, 39, 1100.
13 J.Odo, Y.Saito, and Y.Tanaka, Chem. Pharm. Bull. 1983, 31,3796.
14 J.Odo, M.Nishio, Y.Saito, Y.Tanaka, and K.Machida, Chem. Pharm. Bull., 1983, 31, 3802.
15 C.D.Flint and P.A.Tanner, J. Chem. Soc., Farad. Trans.,2, 1984, 80, 219.
16 C.D.Flint and P.A.Tanner, Mol. Phys., 1983, 50, 625.
17 M.Moskovits, D.P.DiLella and W.Limm, J. Chem. Phys., 1984, 80, 626.
18 Y.-Y.Yang and J.I.Zink, J. Am. Chem. Soc., 1984, 106, 1500.
19 E.Campani, F.Ferri, G.Gorini, E.Polacco, and G.Massetti, Chem. Phys. Lett., 1984, 107, 91.
20 J.M.Manoli, J.M.Bregeault, C.Potvin, and G.Chottard, Inorg. Chim. Acta, 1984, 88, 75.
21 A.H.Jubert, S.G.Manca, E.L.Varetti, and E.J.Baran, J.Mol. Struct. 1984, 115, 15.
22 A.H.Jubert, E.L.Varetti, and E.J.Baran, J. Raman Spectrosc., 1984, 15, 139.

23 W.H.Armstrong, A.Spool, G.C.Papaefthymiou, R.B.Frankel, and S.J.Lippard, J. Am. Chem. Soc., 1984, 106, 3653.
24 G.Chottard, M.Schappacher, L.Ricard, and R.Weiss, Inorg. Chem., 1984, 23, 4557.
25 J.A.Hofmann and D.F.Bocian, Inorg. Chem., 1984, 23, 1177.
26 M.A.Crisanti, T.G.Spiro, D.R.English, D.N.Hendricksen, and K.S.Suslick, Inorg. Chem., 1984, 23, 3897.
27 N.T.Yu, B.Bojan, E.A.Kerr, and K.Gersonde, Proc. Natl. Acad. Sci.,U.S.A. 1984, 81, 5106.
28 O.Bangcharoenpaurpong, K.T.Schomacker, and P.M.Champion, J. Am. Chem. Soc., 1984, 106, 5688.
29 B.Benko and N.T.Yu, Proc. Natl. Acad. Sci.,U.S.A., 1983, 80, 7042.
30 K.Bajdor and K.Nakamoto, J. Am. Chem. Soc., 1984, 106, 3045.
31 J.A.Hoffmann and D.F.Bocian, J. Phys. Chem., 1984, 88, 1472.
32 A.K.Shiemke, T.M.Loehr, and J.Sanders-Loehr, J. Am. Chem. Soc., 1984, 106, 4951.
33 P.V.Argade, M.Sassardi, D.L.Rousseau, T.Innbushi, M.Ikeda-Saito, and A.Lapidot, J. Am. Chem. Soc., 1984, 106, 6593.
34 M.R.Ondrias, E.W.Findsen, D.F.Gaul, and D.B.Kneff, Biochim. Biophys. Acta, 1984, 788, 81.
35 M.Lutz, J.M.Moulis, and J.Meyer, FEBS Lett., 1983, 163, 212.
36 J.Meyer, J.M.Moulis, and M.Lutz, Biochem. Biophys. Res. Commun., 1984, 119, 828.
37 J.M.Moulis, J.Meyer, and M.Lutz, Biochem. J., 1984, 219, 829.
38 T.Ogura, N.Sone, K.Tagawa, and T.Kitagawa, Biochemistry, 1984, 23, 2826.
39 T.Imai, H.Saito, J.Tobari, D.Ohmori, and K.Suzuki, FEBS Lett., 1984, 165, 227.
40 T.Ogura, K.Honnami, T.Oshima, S.Yoshikawa, and T.Kitagawa, J. Am. Chem. Soc., 1984, 106, 7781.
41 R.J.H.Clark and T.J.Dines, Mol. Phys., 1984, 52, 859.
42 M.J.Benecky, M.G.Dowling, M.J.Clarke, and T.G.Spiro, Inorg. Chem., 1984, 23, 865.
43 K.Bajdor, K.Nakamoto, H.Kanatonu, and I.Murase, Inorg. Chim. Acta, 1984, 82, 207.
44 R.J.H.Clark, V.B.Croud, and M.Kurmoo, Inorg. Chem., 1984, 23, 2499.
45 L.Nestor, J.A.Larrabee, G.Woolery, B.Reinhammer, and T.G.Spiro, Biochemistry, 1984, 23, 1084.
46 V.K.Belsky, A.B.Erofeev, B.M.Bulychev, and G.L.Soloveichik, J. Organomet. Chem., 1984, 265, 123.
47 N.V.Porotnikov, O.I.Kondratov, L.L.Kochergina, and K.I.Petrov, Russ. J. Inorg. Chem., 1984, 29, 954.
48 J.Hanuza, J. Mol. Struct., 1984, 114, 471.
49 O.Zakharieva-Pencheva, V.A.Dement'ev, and H.Förster, J. Mol. Struct., 1984, 112, 273.
50 C.Qian, C.Ye, H.Lu, Y.Li, and Y.Huang, J. Organomet. Chem., 1984, 263, 333.
51 H.Haeuseler, J. Raman Spectrosc., 1984, 15, 120.
52 G.Liu, Y.Zhang, Y.Zhao, G.Wang, and Y.Wang, Gadeng Xuexiao Huaxue Xuebao, 1984, 5, 713 (Chem. Abstr., 1984, 101,237537).
53 C.Yuan, E.Ma, and X.Yan, Zhongguo Xitu Xuebao, 1983, 1, 17 (Chem. Abstr.,1984, 101, 16224).
54 A.K.Vazhnov, F.A.Degtyarev, A.N.Pokrovskii, and V.V.Fomichev, Russ. J. Inorg. Chem., 1983, 28, 1710.
55 K.K.Singh and P.Ganguly, Spectrochim. Acta, Part A, 1984, 40, 539.

56 I.S.Shaplygin and N.V.Porotnikov, Russ. J. Inorg. Chem., 1984, 29, 491.
57 A.L.Voloshina, L.A.Ivanchenko, V.A.Obolonchik, and V.A. Serdyuk, Poroshk. Metall. (Kiev),1984, 90 (Chem. Abstr., 1984, 100, 164529).
58 N.S.Loktyushina, S.B.Osin, and A.A.Mal'tsev, Russ. J. Inorg. Chem., 1983, 28, 1384.
59 N.S.Loktyushina, S.B.Osin, and A.A.Mal'tsev, Russ. J. Inorg. Chem., 1984, 29, 986.
60 A.Brandt, Yu.M.Kiselev, and L.I.Martynenko, Russ. J. Inorg. Chem., 1983, 28, 1593.
61 T.C.DeVore and T.N.Gallaher, High. Temp. Sci., 1983, 16, 269.
62 B.A.Borgias, S.R.Cooper, Y.B.Koh, and K.N.Raymond, Inorg. Chem. 1984, 23, 1009.
63 S.Saxena, Y.P.Singh, and A.K.Rai, J. Organomet. Chem., 1984, 270 , 301.
64 A.R.Miksztal and J.S.Valentine, Inorg. Chem., 1984, 23, 3548.
65 M.A.Ulibarri-Cormenzana, F.Gonzalez-Garcia, P.S.Valerga-Jimenez, and F.Gonzales-Vilches, An. Quim.,B, 1984, 80, 23.
66 S.Milićev and J.Maček, J. Chem. Soc., Dalton Trans., 1984, 297.
67 I.Kanesaka, K.Kawai, T.Miyatake, and M.Kakugo, Spectrochim. Acta, Part A, 1984, 40, 705.
68 R.L.Davidovich, M.A.Medkov, V.B.Timchenko, and B.V.Burvetskii, Izv. Akad. Nauk SSSR, Ser. Khim., 1983, 2427.
69 S.P.Gwet, Y.Matthey, and C.Sourisseau, Phys. Stat. Solidi,B, 1984, 123, 503.
70 S.A.Arthers and I.R.Beattie, J.Chem. Soc., Dalton Trans., 1984, 819.
71 L.L.Kochergina, N.B.Khanin, N.V.Porotnikov, and K.I.Petrov, Russ. J. Inorg. Chem., 1984, 29, 506.
72 Yu.G.Zaretskii, M.V.Krasin'kova, G.A.Kurbatov, Yu.I.Ukhanov, and Yu.V.Shmartsev, Fiz. Tverd. Tela (Leningrad),1984, 26,2233.
73 O.P.Pandey, S.K.Sengupta, and S.C.Tripathi, Polyhedron, 1984, 3, 695.
74 R.K.Agarwal, G.Singh, M.Srivastava, and A.K.Srivastava, Thermochim. Acta, 1984, 73, 243.
75 Y.Kawamoto, Phys. Chem. Glasses, 1984, 25, 88.
76 M.K.Chaudhuri, S.K.Ghosh, and J.Subramanian, Inorg. Chem., 1984, 23, 4439.
77 M.K.Chaudhuri and S.K.Ghosh, Inorg. Chem., 1984, 23, 534.
78 M.K.Chaudhuri and S.K.Ghosh, J. Chem. Soc., Dalton Trans., 1984, 507.
79 M.Hoch, D.Rehder, and C.Weedemann, Inorg. Chim. Acta, 1984, 92, L5.
80 C.Madic, G.M.Begun, R.L.Hahn, J.P.Launay, and W.E.Thiessen, Inorg. Chem., 1984, 23, 469.
81 V.N.Krasil'nikov, Russ. J. Inorg. Chem., 1984, 29, 678.
82 A.N.Lazarev, A.E.Chisler, B.F.Shchegolev, and A.P.Mirgorodskii, Khim. Fiz., 1982, 1170.
83 L.V.Kristallov, M.P.Tsvetkova, and A.A.Fotiev, Russ. J. Inorg. Chem., 1984, 29, 990.
84 V.Dimitrov, Y.Dimitriev, and V.Mihailova, Monatsh., 1983, 114, 669.
85 P.Schwendt and D.Joniaková, Polyhedron, 1984, 3, 287.
86 E.N.Yurchenko and V.I.Bugaev, J.Mol. Struct., 1984, 115, 71.
87 G.T.Stanford and R.A.Condrate, Spectrosc. Lett., 1984, 17, 85.

88 M.B.Adi and A.S.R.Murty, J. Ind. Chem. Soc., 1983, 60, 729.
89 Z.M.Musaev, Ya.S.Usmankhodzhaeva, and O.K.Khodzhaev, Russ. J. Inorg. Chem., 1984, 29, 993.
90 B.Singh, R.N.Singh, and R.C.Aggarwal, Synth. React. Inorg. Met.-Org. Chem., 1984, 14, 413.
91 V.N.Krasil'nikov, M.P.Glazyrin, and A.A.Ivakin, Russ. J. Inorg. Chem., 1984, 29, 52.
92 S.K.Sengupta, S.K.Sahni, and R.N.Kapoor, J. Ind. Chem. Soc., 1983, 60, 821.
93 F.Preuss, E.Fuchslocher, and W.Towae, Z.Naturorsch., Teil B, 1984, 39, 61.
94 S.Kumar, K.K.Bhasin, V.P.Kapila, and R.K.Mahajan, Ind. J. Chem. Sect. A, 1984, 24, 200.
95 S.Mohan and S.Krishnamurthi, Ind. J. Pure Appl. Phys., 1984, 22, 519.
96 Y.Kera, J.Solid State Chem., 1984, 51, 205.
97 E.Culea, A.Nicula, and I.Bratu, Phys. Stat. Solidi, A, 1984, 83, K15.
98 J.Wasylak, J.Koprowski, E.Czerwosz and E.Sewicka, Glastech. Ber. 1984, 57, 12.
99 O.Ya.Miroshnichenko and V.V.Mombelli, Fiz. Khim. Stekla, 1983, 9, 521.
100 C.A.Hogarth and A.A.Hosseini, J. Mater. Sci. Lett.,1984, 3, 359
101 R.D.Bereman and J.R.Dorfman, Polyhedron, 1983, 2, 1013.
102 D.Szeymies, B.Krebs, and G.Heukel, Angew. Chem., Int. Ed. Engl. 1984, 23, 804.
103 R.Seangprasertkij and T.L.Riecher, Inorg. Chem.,1984, 23, 991.
104 O.V.Blinova, D.D.Moldavskii, Yu.B.Predtechenskii, and L.D. Shcherba, Khim. Fiz., 1983, 1363.
105 M.L.Luetkens, W.L.Elcesser, J.C.Huffman, and A.P.Sattelberger, Inorg. Chem., 1984, 23, 1718.
106 K.Bachmann and D.Rehder, J. Organomet. Chem., 1984, 276, 177.
107 F.A.Cotton, S.A.Duraj, and W.J.Roth, J. Am. Chem. Soc.,1984, 106, 4749.
108 E.Husson and Y.Repelin, Spectrochim. Acta, Part A, 1984, 40,315
109 G.T.Stanford and R.A.Condrate, J. Solid State Chem., 1984, 52, 248.
110 G.T.Stanford and R.A.Condrate, J. Mater. Sci. Lett., 1984, 3, 303.
111 G.A.Smolenskii, N.N.Kolpakova, V.V.Tikhonov, and F.R.Muratova, Fiz. Tverd. Tela, 1984, 26, 1398.
112 M.T.Vandenborre, E.Husson, and M.Chubb, Spectrochim. Acta, Part A, 1984, 40, 361.
113 A.I.Agulyanskii and V.A.Bessonova, Russ. J. Inorg. Chem., 1984, 29, 44.
114 B.Krebs, H.Janssen, N.J.Bjerrum, R.W.Berg, and G.N.Papatheodorou, Inorg. Chem., 1984, 23, 164.
115 A.Antiñolo, F.Fajardo, A.Otero, and P.Royo, J. Organomet. Chem. 1984, 265, 35.
116 F.A.Urbanos, M.Mena, P.Royo, and A.Antiñolo, J. Organomet. Chem., 1984, 276, 185.
117 M.G.Zuev, A.A.Fotiev, and M.P.Tsvetkova, J. Mol. Struct., 1984, 115, 79.
118 M.T.Vandenborre and E.Husson, J. Solid State Chem., 1984, 53, 253.
119 D.C.Bradley, M.B.Hursthouse, K.M.A.Malik, A.J.Nielsen, and G.B.C.Vuru, J.Chem. Soc., Dalton Trans.,1984, 1069.
120 R.Bender, P.Braunstein, J.-M.Jud, and Y.Dusausoy, Inorg. Chem., 1984, 23, 4489.

121 E.Blasius and E.Mernke, Z. Anorg. Allg. Chem., 1984, 509, 167.
122 V.I.Spitsyn, S.V.Mozgin, N,A.Subbotina, and M.G.Felin, Russ. J. Inorg. Chem., 1984, 29, 680.
123 E.G.Hope, P.J.Jones, W.Levason, J.S.Ogden, M.Tajik, and J.W. Turff, J. Chem. Soc., Dalton Trans., 1984, 2445.
124 M.Handlíř, J.Holeček, M.Nádvarník, A.Lyčka and M.Hucl, Coll. Czech. Chem. Commun., 1984, 49, 1497.
125 A.V.Evseev, V.M.Krivtsun, Yu.A.Kuritsyn, A.A.Makarov, A.A. Puretskii, E.A.Ryabov, E.P.Snegirev, and V.V.Tyakht, Zh. Eksp. Teos. Fiz., 1984, 87, 111.
126 M.Di Vaira and F.Mani, Inorg. Chem., 1984, 23, 409.
127 A.Iannibello, S.Marengo, P.Tittarelli, G.Morelli, and A.Zecchina, J.Chem. Soc., Farad. Trans,1, 1984, 80, 2209.
128 J.M.Malezieux, J.Barbillat, B.Cervelle, J.P.Coutures, M.Couzi, and B.Piriou,Tschermak's Mineral. Petrogr., 1983, 32, 171.
129 H.Brunner, H.Kauermann, and J.Wachter, J. Organomet. Chem., 1984, 265, 189.
130 H.Brunner, H.Kauermann, W.Meier, and J.Wachter, J. Organomet. Chem., 1984, 263, 183.
131 E.G.Hope, P.J.Jones, W.Levason, J.S.Ogden, and M.Tajik, J. Chem. Soc., Chem. Comm., 1984, 1355.
132 L.R.Gray, A.L.Hale, W.Levason, F.P.McCullough, and M.Webster, J. Chem. Soc., Dalton Trans., 1984, 47.
133 V.M.Kovba, Russ. J Inorg. Chem., 1983, 28, 1525.
134 J.Telser and R.S.Drago, Inorg. Chem., 1984, 23, 1798.
135 G.J.Kubas, R.R.Ryan, B.I.Swanson, P.J.Vergamini, and H.J. Wasserman, J. Am. Chem. Soc., 1984, 106, 451.
136 M.Canestrari, M.L.H.Green, and A.Izquierda, J. Chem. Soc., Dalton Trans., 1984, 2795.
137 F.G.N.Cloke, P.J.Fyne, V.C.Gibson, M.L.H.Green, M.J.Ledoux, R.N.Perutz, A.Dix, A.Gourdon, and K.Prout, J. Organomet. Chem., 1984, 277, 61.
138 D.Lyons, G.Wilkinson, M.Thornton-Pett, and M.B.Hursthouse, J. Chem. Soc., Dalton Trans., 1984, 695.
139 H.O.Desseyn, B.J.van der Veken, J.R.Moss, B.J.Smith, P.Verhoev-en and D.A.Thornton, Spectrochim. Acta, Part A, 1984, 40, 467.
140 M.Iglesias, A.Llorente, C.del Pino, and A.Santos, J. Organomet. Chem., 1984, 263, 193.
141 I.Schmidt, U.Kynast, J.Hanich, and K.Dehnicke, Z. Naturforsch., Teil B, 1984, 39, 1248.
142 M.H.Chisholm, K.Folting, J.C.Huffman, and C.C.Fitzpatrick, Inorg. Chem., 1984, 23, 1021.
143 S.Mohan and K.G.Ravikumar, Bull. Soc. Chim. France,I,1983, 267.
144 S.Mohan and K.G.Ravikumar, Curr. Sci., 1984, 53, 471.
145 S.S.Saleem and G.Aruldhas, Pramana, 1983, 21, 283 (Chem. Abstr. 1984, 100, 42110).
146 Sh.M.Efendiev, N.G.Darvishov, V.M.Nagiev, N.M.Gasanly, and V.T. Gabrielyan, Elektron Yavleniya Tverd. Telakh Gazakh., 1983, 82 (Chem. Abstr., 1984, 100, 58953).
147 L.Liao and P.Zhang, Beijing Gongye Daxue Xuebao,1983, 9, 45 (Chem. Abstr., 1984, 101, 62836).
148 J.L.Corbin, K.F.Miller, N.Pariyadath, J.Heinecke, A.E.Bruce, S.Wherland, and E.I.Stiefel, Inorg. Chem., 1984, 23, 3404.
149 V.I.Spitsyn, M.G.Felin, N.A.Subbotina, M.I.Aizenberg, and S.V.Mozgin, Russ. J. Inorg. Chem., 1984, 29, 407.
150 I.Buchanan, C.D.Garner, and W.Clegg, J. Chem. Soc., Dalton Trans., 1984, 1333.

151 R.Thouvenot, M.Fournier, R.Franck, and C.Rocchiccioli-Deltcheff Inorg. Chem., 1984, 23, 598.
152 N.Yamazoe and K.Eguchi, Kenkyu Hokoku-Asahi Garasu Kogyo Gijutsu, 1983, 42, 247 (Chem. Abstr., 1984, 100, 60917).
153 T.R.Halbert, K.McGauley, W.H.Pan, R.S.Czernuszewicz, and E.I. Stiefel, J. Am. Chem. Soc., 1984, 106, 1849.
154 C.G.Young, J.A.Broomhead, and C.J.Boreham, J. Organomet. Chem., 1984, 260, 91.
155 E.Hey, F.Weller, and K.Dehnicke, Z. Anorg. Allg. Chem., 1984, 508, 86.
156 J.M.Stencel, L.E.Makovsky, T.A.Sarkus, J.de Vries, R.Thomas, and J.A.Moulijn, J. Catal., 1984, 90, 314.
157 M.Chaudhury, Inorg. Chem., 1984, 23, 4434.
158 S.Vasanthi and K.S.Nagaraja, Ind. J. Chem.,A, 1984, 23, 70.
159 M.Chaudhury, J. Chem. Soc., Dalton Trans., 1984, 115.
160 R.Lozano Fernandez, A.L.Doadrio Villarejo, and E.Parrondo Iglesias, An. Quim.,Ser. B, 1983, 79, 33.
161 C.Schumacher, F.Weller, and K.Dehnicke, Z. Anorg. Allg. Chem., 1984, 508, 79.
162 A.T.Pilipenko, L.L.Shevchenko, N.V.Nemazanyan, and V.V. Trachevskii, Russ. J. Inorg. Chem., 1983, 28, 1437.
163 V.I.Spitsyn, S.V.Mogin, M.G.Felin, N.A.Subbotina, and M.I. Aizenberg, Russ. J. Inorg. Chem., 1984, 29, 58.
164 M.Chaudhury, J. Chem. Soc., Dalton Trans., 1984, 115.
165 B.Bhushan, A.K.Sharma, and N.K.Kaushik, Bull. Soc. Chim. France,I, 1983, 297.
166 A.L.Doadrio Villarejo, M.P.Alonso, A.Doadrio, and R.Lozano, An. R. Acad. Farm., 1983, 49, 501.
167 S.F.Gheller, T.W.Hambley, J.R.Rodgers, R.T.C.Brownlee, M.J. O'Connor, M.R.Shaw, and A.G.Wedd, Inorg. Chem., 1984, 23, 2519.
168 J.A.Broomhead, M.Sterns, and C.G.Young, Inorg. Chem.,1984, 23, 729.
169 B.Grzybowska, E.Payen, L.Gengembre, and J.P.Bonnelle, Bull. Pol. Acad. Sci. Chem., 1983, 31, 245.
170 A.O.Baghlaf, M.Ishaq, and A.S.Daifiliah, Polyhedron,1984, 3,235
171 H.P.M.M.Ambrosius, A.W.van Hemert, W.P.Bosman, J.H.Noordijk, and G.J.A.Ariaans, Inorg. Chem., 1984, 23, 2678.
172 D.C.Povey and R.L.Richards, J. Chem. Soc., Dalton Trans., 1984, 2585.
173 R.Lozano, E.Alarcon, J.Roman, and A.Doadrio, Rev. Chim. Min., 1984, 21, 177.
174 U.Müller and P.Klingelhöfer, Z. Anorg. Allg. Chem.,1984, 510, 109.
175 R.H.Crabtree, G.G.Hlatky, C.P.Parnell, B.E.Segmüller, and R.J. Uriarte, Inorg. Chem., 1984, 23, 354.
176 J.M.Stencel, L.E.Makovsky, J.R.Diehl, and T.A.Sarkus, J. Raman Spectrosc.,1984, 15, 282.
177 E.Hey, F.Weller, and K. Dehnicke, Z. Anorg. Allg. Chem., 1984, 514, 18.
178 U.Kynast, U.Müller, and K.Dehnicke, Z. Anorg. Allg. Chem., 1984, 508, 26.
179 G.A.Ozin and J.G.McCaffrey, J. Am. Chem. Soc., 1984, 106, 807.
180 H.-J.Haupt, P.Balsaa, B.Schwab, U.Flörke, and A.Preut, Z. Anorg. Allg. Chem., 1984, 513, 22.
181 C.G.Benson,M.G.Little, and C.A.McAuliffe, Inorg. Chim. Acta, 1984, 87, 169.

182 K.C.Patil, J.P.Vittal, and C.C.Patel, Proc. Ind. Acad. Sci., Chem. Sci., 1983, 92, 83.
183 A.Yu.Tsivadze and H.Köhler, Z. Anorg. Allg. Chem., 1984, 510,25
184 A.Yu.Tsivadze and H.Köhler, Z. Anorg. Allg. Chem.,1984, 510, 31
185 I.Lupu, Bul. Univ. Brasov.,Ser. C, 1981, 23, 215 (Chem. Abstr., 1984, 100, 14652).
186 M.N.Bhattacharjee and M.K.Chaudhuri, Ind. J. Chem., Sect. A, 1984, 23, 424.
187 D.W.Wester, D.H.White, F.W.Miller, and R.T.Dean, Inorg. Chem., 1984, 23, 1501.
188 J.G.H.Du Preez, T.I.A.Gerber, P.J.Fourie, and A.J. van Wyk, Inorg. Chim. Acta, 1984, 82, 201.
189 G.Bandoli, M.Nicolini, U.Mazzi, and F.Refosco, J. Chem. Soc., Dalton Trans., 1984, 2505.
190 L.L.Zaitseva and A.Yu.Vakhrushin, Russ. J. Inorg. Chem., 1984, 29, 99.
191 E.G.Teterin and L.L.Zaitseva, Russ. J. Inorg. Chem., 1983, 28, 1717.
192 M.E.Kastner, P.H.Fackler, M.J.Clarke, and E.Deutsch, Inorg. Chem., 1984, 23, 4683.
193 S.Jurisson, L.F.Lindoy, K.P.Dancey, M.McPartlin, P.A.Tasker, D.K.Uppal, and E.Deutsch, Inorg. Chem., 1984, 23, 227.
194 V.M.Ishchenko, G.L.Soloveichik, B.M.Bulychev, and T.A.Sokolova, Russ. J. Inorg. Chem., 1984, 29, 66.
195 C.J.Cameron, R.A.Walton, and D.A.Edwards, J. Organomet. Chem., 1984, 262, 335.
196 R.Dantona, E.Schweda, and J.Strähle, Z. Naturforsch., Teil B, 1984, 39, 733.
197 W.Hiller, J.Mohyla, J.Strähle, H.G.Hauck, and K.Dehnicke, Z. Anorg. Allg. Chem., 1984, 514, 72.
198 H.G.Hauck, P.Klingelhöfer, U.Müller, and K.Dehnicke, Z. Anorg. Allg. Chem., 1984, 510, 180.
199 V.I.Spisyn, P.E.Kazin, A.I.Zhirov, and N.A.Subbotina, Russ. J. Inorg. Chem., 1983, 28, 1445.
200 V.I.Spitsyn, P.E.Kazin, A.I.Zhirov, and N.A.Subbotina, Russ. J. Inorg. Chem., 1983, 28, 1606.
201 W.Kafitz, K.Dehnicke, E.Schweda, and J.Strähle, Z. Naturforsch., Teil B, 1984, 39, 1114.
202 L.L.Zaitseva, A.Yu.Vakhrushin, and S.V.Shepel'kov, Russ. J. Inorg. Chem., 1984, 29, 443.
203 A.Marchi, A.Duatti, R.Rossi, L.Magon, U.Mazzi, and A.Pasquetto, Inorg. Chim. Acta, 1984, 81, 15.
204 V.Yatirajam and M.LKantam, Ind. J. Chem. Sect. A, 1984, 23, 771
205 G.J.Schrobilgen, J.H.Holloway, and D.R.Russell, J. Chem. Soc., Dalton Trans., 1984, 1411.
206 T.K.Jana, S.Sen, B.Pandyopadhyay, and B.K.Sen, Ind. J. Chem., Sect. A, 1984, 23, 689.
207 P.Klingelhöfer, U.Müller, H.G.Hauck, and K.Dehnicke, Z. Naturforsch., Teil B, 1984, 39, 135.
208 V.E.Fedorov, A.V.Mishchenko, B.A.Kolesov, S.P.Gubin, Yu.L. Slovkhotov, and Yu.T.Struchkov, Izv. Akad. Nauk SSSR, Ser. Khim 1984, 2159.
209 B.A.Kolesov, V.E.Fedorov, A.V.Mishchenko, Koord. Khim., 1984, 10, 907.
210 S.F.Parker, C.H.F.Peden, P.H.Barrett, and P.G.Pearson, J. Am. Chem. Soc., 1984, 106, 1304.
211 W.J.Balfour, B.Lindgren, and S.O'Connor, Phys. Scr., 1983, 28, 551.

212 A.L.Shobert, I.C.Hisatsune, and P.S.Skell, Spectrochim. Acta, Part A, 1984, 40, 609.
213 K.W.Hipps, S.D.Williams, and U.Mazur, Inorg. Chem., 1984, 23, 3500.
214 V.L.Goedken and C.Ercolani, J.Chem. Soc., Chem. Commun., 1984, 378.
215 D.Chang, P.Cocolias, Y.T.Wu, and K.MKadish, Inorg. Chem., 1984, 23, 1629.
216 R.D.Cannon, L.Montri, D.B.Brown, K.M.Marshall, and C.M.Elliott, J. Am. Chem. Soc., 1984, 106, 2591.
217 M.Barj, C.Sourisseau, G.Ouvrard, and R.Brec, Solid State Ionics, 1983, 11, 179.
218 J.M.Friedman, Time-Resolved Vib. Spectrosc. (Proc. Int. Conf. TRVS), 1982(publ. 1983), 307.
219 H.Strasdeit, B.Krebs, and G.Henkel, Inorg. Chim. Acta, 1984, 89, L11.
220 H.Strasdeit, B.Krebs, and G.Henkel, Inorg. Chem., 1984, 23,1816
221 P.Beardwood and J.F.Gibson, J.Chem. Soc., Dalton Trans., 1984, 1507.
222 L.Zhang, Z.Liu, L.He, Y.Zheng, and D.Huang, Jiegou Huaxue, 1983, 2, 11 (Chem. Abstr., 1983, 99, 221373).
223 J.C.Briggs, C.A.McAuliffe, and G.Dyer, J. Chem. Soc., Dalton Trans., 1984, 423.
224 R.H.Morris and M.Shiralian, J. Organomet. Chem., 1984, 260, C47
225 B.Chaudret, G.Commenges, and R.Poilblanc, J. Chem. Soc. Dalton Trans., 1984, 1635.
226 M.A.Bennett, T.-N.Huang, and J.L.Latten, J. Organomet. Chem., 1984, 272, 189.
227 J.A.Olabe, L.A.Gentil, G.Rigotti, and A.Navaza, Inorg. Chem., 1984, 23, 4297.
228 M.L.Good, M.D.Patil, L.M.Trefonas, J.Dodge, C.J.Alexander, R.J.Majeste, and M.A.Cavanaugh, J. Phys. Chem., 1984, 88, 483.
229 J.T.Groves and R.Quinn, Inorg. Chem., 1984, 23, 3844.
230 P.Bernhard and A.Ludi, Inorg. Chem., 1984, 23, 870.
231 R.S.Armstrong, J.K.Beattie, P.Del Favero, V.M.Ellis, and N.S. Hush, Inorg. Chim. Acta, 1984, 89, L33.
232 N.S.Hush, J.K.Beattie, and V.M.Ellis, Inorg. Chem.,1984, 23, 3339.
233 J.Lippold and H.Singer, J. Organomet. Chem., 1984, 266,C5.
234 B.P.Sullivan, J.V.Caspar, and T.J.Meyer, Organometallics, 1984, 3, 1241.
235 R.Weber and K.Dehnicke, Z.Naturforsch., Teil B,1984, 39, 262.
236 W.J.Balfour and R.S.Ram, J. Mol. Spectrosc., 1984, 105, 360.
237 E.N.Bazarov, G.A.Gerasomov, V.P.Gubin, V.L.Derbov, A.D.Novikov, S.Yu.Otrokhov, A.I.Sazonov, and V.V.Fomin, Opt. Spektrosk., 1984, 57, 434.
238 E.Bilger, J.Pebler, R.Weber, and K.Dehnicke, Z.Naturforsch., Teil B, 1984, 39, 259.
239 N.M.Sinitsyn, N.M.Septsova, and A.A.Svetlov, Russ. J. Inorg. Chem., 1983, 28, 1464.
240 S.M.Tetrick and R.A.Walton, Inorg. Chem., 1984, 23, 379.
241 A.M.Ervin, S.I.Shupack, and D.M.Byler, Spectrosc. Lett., 1984, 17, 603.
242 G.Gervasio, R.Rosetti, P.L.Stanghellini, and G.Bor, Inorg. Chem 1984, 23, 2013.
243 S.Delgado, M.J.Macazaga, and J.R.Masaguer, J. Organomet. Chem., 1984, 268, 79.
244 H.Inoue and E.Fluck, Z. Naturforsch., Teil B, 1984, 39, 185.

245 C.Tellez and G.Diaz, Acta Sud. Am. Quim.,1983, 3, 85.
246 L.Dubicki and B.Williamson,Inorg. Chem., 1984, 23, 3779.
247 M.J.Macazaga, S.Delgado, R.M.Medina, and J.R.Masaguer, J. Organomet. Chem., 1984, 277, 423.
248 J.Ellermann, G.Szucsányi, K.Geibel, and E.Wilhelm, J. Organomet. Chem., 1984, 263, 297.
249 L.A.Oro, D.Carmona, M.P.Lamata, M.C.Apreda, C.Foces-Foces, F.H.Cano, and P.M.Maitlis, J. Chem. Soc., Dalton Trans., 1984, 1823.
250 V.M.Miskowski, W.P.Schaefer, B.S.Sadeghi, B.D.Santarsiero, and H.B.Gray, Inorg. Chem., 1984, 23, 1154.
251 H.-H.Fricke and W.Preetz, Z. Anorg. Allg. Chem., 1984, 507, 12.
252 R.Battistuzzi and G.Peyronel, Z. Anorg. Allg. Chem., 1984, 509, 192.
253 I.M.Oglezneva and V.N.Kirichenko, Zh. Strukt. Khim., 1984, 25, 164.
254 J.G.Contreras, J.P.Pena, and L.M.Silva-Trevino, Bol. Soc. Quim. Chil,, 1983, 28, 3.
255 C.Cascales and I.Rasines, Mater. Chem. Phys.,1984, 10, 199.
256 M.S.Barvinok and L.G.Lukina, Russ. J. Inorg. Chem., 1983, 28, 1129.
257 D.X.West, T.J.Parsons, and R.K.Bonting, Inorg. Chim. Acta, 1984, 84, 7.
258 S.K.Boocock, M.A.Toft, K.E.Inkrott, L.-Y.Hsu, J.C.Huffman, K.Folting, and S.G.Shore, Inorg. Chem., 1984, 23, 3084.
259 B.R.Sutherland and M.Cowie, Inorg. Chem., 1984, 23, 1290.
260 I.B.Baranovskii and A.N.Zhilyaev, Russ. J. Inorg. Chem., 1984, 29, 607.
261 W.A.Schenk, J.Leissner, and C.Burschka, Angew. Chem., Int. Ed. Engl., 1984, 23, 806.
262 M.A.Lilga and J.A.Ibers, Inorg. Chem., 1984, 23, 3538.
263 D.Tensfeldt and W.Preetz, Z. Naturforsch., Teil B, 1984, 39, 1185.
264 A.M.Mueting, P.Boyle, and L.H.Pignolet, Inorg. Chem., 1984, 23, 44.
265 B.J.Fisher and R.Eisenberg, Inorg. Chem., 1984, 23, 3216.
266 J.C.Rodriguez, C.Clower, and A.Ruiz, J. Organomet. Chem., 1984, 272, C67.
267 G.R.Clark, W.R.Roper, and A.H.Wright, J. Organomet. Chem., 1984, 273, C17.
268 M.J.Fernandez and P.M.Maitlis, J. Chem. Soc., Dalton Trans., 1984, 2063.
269 A.J.Rest, I.Whitwell, W.A.G.Graham, J.K.Hoyano, and A.D. McMaster, J. Chem. Soc., Chem. Commun., 1984, 624.
270 F.Cecconi,C.A.Ghilardi, P.Innocenti, C.Mealli, S.Midollini, and A.Orlandini, Inorg. Chem., 1984, 23, 922.
271 I.C.Hisatsune, Spectrochim. Acta, Part A, 1984, 40, 391.
272 I.M.Oglezneva and I.K.Igumenov, Koord. Khim., 1984, 10, 313.
273 G.Tresoldi, G.Bruno, P.Piraino, G.Faraone, and G.Bombieri, J. Organomet. Chem., 1984, 265, 311.
274 N.M.Rutherford, M.M.Olmstead, and A.L.Balch, Inorg. Chem., 1984, 23, 2833.
275 D.P.Arnold and M.A.Bennett, Inorg. Chem., 1984, 23, 2110.
276 D.P.Arnold and M.A.Bennett, Inorg. Chem., 1984, 23, 2117.
277 S.R.Bare, P.Hofmann, and D.A.King, Surf. Sci., 1984, 144, 347.
278 G.A.Foulds, P.S.Hall, and D.A.Thornton, J. Mol. Struct., 1984, 117, 95.

279 F.D.Rochon, R.Melanson, H.E.Howard-Lock, C.J.L.Lock, and G. Turner, Can. J. Chem., 1984, 62, 860.
280 V.E.Abashkin, G.S.Muraveiskaya, I.F.Golvaneva, A.N.Smirnov, A.Yu.Tsivadze, and R.N.Shchelokov, Russ. J. Inorg. Chem., 1983, 28, 1626.
281 M.A.Sarukhanov, M.Sh.Mridka, Yu.Ya.Kharitonov, N.D.Mitkinova, and T.G.Abzaeva, Russ. J. Inorg. Chem., 1983, 28, 1158.
282 K.C.Lee and F.Aubke, Inorg. Chem., 1984, 23, 2124.
283 H.A.Brune, M.Falck, R.Hemmer, G.Schmidtberg, and H.G.Alt, Chem. Ber., 1984, 117, 2791.
284 S.Al-Jibori and B.L.Shaw, J. Organomet. Chem.,1984, 273, 213.
285 M.Mehicic, F.A.Pesa, and J.G.Grasselli, J. Phys. Chem., 1984, 88, 581.
286 T.Nortia, J.Arpalahti, and M.Kappinen, Spectrochim. Acta, Part A, 1984, 40, 257.
287 I.M.Oglezneva, L.G.Lavrenova, and S.V.Larionov, Russ. J. Inorg. Chem., 1984, 29, 849.
288 J.G.Contreras, J.Pena, and L.M.Silva-Trevino, Inorg. Chim. Acta, 1984, 87, 181.
289 L.R.Gray, S.J.Higgins, W.Levason, and M.Webster, J. Chem. Soc., Dalton Trans., 1984, 459.
290 M.Castillo, A.Romero, and E.Ramirez, Inorg. Chem., 1984, 23, 2668.
291 M.L.Riekkola, T.Pakkanen, and L.Niinistö, Acta Chem. Scand.,A, 1983, 37, 807.
292 C.Gomez-Vaamonde, A.Alvarez-Valdes, M.C.Navarro-Ranninger, and J.R.Masaguer, Transition Met. Chem., 1984, 9, 52.
293 M.Biddau, M.Massacesi, G.Ponticelli, G.Devoto, and I.A.Zakharovo, Polyhedron, 1983, 2, 1261.
294 M.Nonoyama, J. Organomet. Chem., 1984, 262, 407.
295 C.Battistoni, A.M.Giuliarni, E.Paparazzo, and F.Tarli, J. Chem. Soc., Dalton Trans., 1984, 1293.
296 I.V.Pakhomova, Yu.N.Kukushkin, L.V.Konovalov, and V.V.Strukov, Russ. J. Inorg. Chem., 1984, 29, 575.
297 B.Singh, R.J.Sinha, R.N.Pandey, H.N.Tiwari, and U.Bhanu, J. Ind. Chem. Soc., 1983, 60, 415.
298 R.Uson, J.Forniés, M.Tomás, F.A.Cotton, and L.R.Favello, J. Am. Chem. Soc., 1984, 106, 2482.
299 R.Uson, J.Forniés, M.Tomás, and R.Fandos, J. Organomet. Chem., 1984, 263, 253.
300 J.Auffret, P.Courtot, R.Pichon, R.Rumin, and J.Y.Salaun, J. Organomet. Chem., 1984, 262, C19.
301 D.A.Long, B.M.Chadwick, and S.U.Qureshi, J. Chem. Soc. Pak., 1983, 5, 163.
302 J.-L.Pascal, J.Potier, D.J.Jones, J.Rozière, and A.Michalowicz, Inorg. Chem., 1984, 23, 2068.
303 S.V.Volkov, L.I.Zheleznova, and E.A.Mazurenko, Koord. Khim., 1983, 9, 1338.
304 C.D.Garner, J.R.Nicholson, and W.Clegg, Inorg. Chem., 1984, 23, 2148.
305 G.A.Bowmaker, G.R.Clark, J.K.Seadon, and I.G.Dance, Polyhedron, 1984, 3, 535.
306 G.A.Bowmaker and D.A.Rogers, J. Chem. Soc., Dalton Trans., 1984, 1249.
307 D.Reinen and C.Friebel, Inorg. Chem., 1984, 23, 791.
308 G.A.Bowmaker, G.R.Clark, D.A.Rogers, A.Camus, and N.Marsich, J. Chem. Soc., Dalton Trans., 1984, 37.
309 M.D.Baker, G.A.Ozin, and J.Godber, J. Phys. Chem.,1984, 88,4902

310 L.A.Sanchez, J.R.Lombardi, and R.L.Birke, Chem. Phys. Lett., 1984, 108, 45.
311 W.Taylor, P.J.S.Ewen, and T.Han, Ferroelectrics, 1984, 55, 751.
312 P.Braunstein, U.Schubert, and M.Burgard, Inorg. Chem., 1984, 23, 4057.
313 J.Vicente, M.T.Chicote, and M.D.Bermúdez, J. Organomet. Chem., 1984, 268, 191.
314 J.Kohout, M.Hvastijová, and J.Mroziński, J. Mol. Struct., 1984, 116, 211.
315 J.G.Haasnoot, W.L.Driessen, and J.Reedijk, Inorg. Chem., 1984, 23, 2803.
316 M.Hvastijová, J.Kohout, H.Wusterhausen, and H.Köhler, Z. Anorg. Allg. Chem., 1984, 510, 37.
317 M.Hvastijová and J.Kohout, Z. Chem., 1984, 24, 105.
318 G.O.Morpurgo, G.Dessy, and V.Fares, J. Chem. Soc., Dalton Trans., 1984, 785.
319 D.X.West, Polyhedron, 1983, 2, 999.
320 F.J.Rietmeijer, R.A.G. de Graaff, and J.Reedijk, Inorg. Chem., 1984, 23, 151.
321 W.J.Balfour and R.S.Ram, J.Phys.,B, 1984, 17, L19.
322 L.K.Thompson, A.W.Hanson, and B.S.Ramaswamy, Inorg. Chem., 1984, 23, 2459.
323 R.Sudharsanan, T.K.K.Srinivasan, and S.Radhakrishna, Phys. Stat. Solidi,B, 1984, 124, 481.
324 A.Bell and D.A.Edwards, J. Chem. Soc., Dalton Trans., 1984,1317
325 G.Marbach and J.Strähle, Angew. Chem. Int. Ed. Engl., 1984, 23, 246.
326 G.A.Bowmaker and H.Schmidbaur, J. Chem. Soc., Dalton Trans., 1984, 2859.
327 N.T.Kuznetsov, N.S.Kedrova, N.N.Mal'tseva, S.N.Ryabchenko, and G.A.Terekhova, Russ. J. Inorg. Chem., 1983, 28, 1682.
328 M.A.White, C.Chieh, A.Anderson, and L.A.K.Staveley, J. Chem. Phys., 1984, 80, 1254.
329 A.I.Grigor'ev and N.V.Donchenko, Russ. J. Inorg. Chem., 1984, 29, 376.
330 N.A.Bell, M.Goldstein, L.A.March, and I.W.Nowell, Inorg. Chim. Acta, 1984, 83, 75.
331 P.Biscarini, E.Foresti, and G.Pradella, J. Chem. Soc., Dalton Trans., 1984, 953.
332 N.A.Bell,. M.Goldstein, L.A.March, and I.W.Nowell, J. Chem. Soc. Dalton Trans., 1984, 1621.
333 M.Panozo and M.Cano, J. Organomet. Chem., 1984, 266, 247.
334 A.Santos and P.Tigeras, Inorg. Chim. Acta , 1984, 81, 175.
335 M.Kanellaki, Th.F.Zafiropolus, S.P.Perlepes, and A.G.Galinos, Izv. Khim., 1982, 15, 387.
336 S.P.Mallela and F.Aubke, Can. J. Chem.,1984, 62, 382.
337 M.S.Barvinok and L.G.Lukina, Russ. J. Inorg. Chem., 1984, 29, 562.
338 T.Allman and R.G.Goel, Can. J. Chem., 1984, 62, 621.
339 G.A.Bowmaker and D.A.Rogers, J. Chem. Res.(S), 1984, 158.
340 M.P.Pardo and M.Cano, J. Organomet. Chem., 1984, 270, 311.
341 M.G.B.Drew and G.R.Willey, J. Chem. Soc., Dalton Trans., 1984, 727.
342 J .Holloway, D.Laycock, and R.Bougon, J. Fluorine Chem.,1984, 26, 281.
343 L.M.Toth, H.A.Friedman, G.M.Begun, and S.E.Dorris, J. Phys. Chem., 1984, 88, 5574.
344 D.Gantar, B.Frlec, and B.Volavšek, J. Chem. Soc., Dalton Trans., 1984, 93.

345 P.C.Kundu and C.R.Das, Ind. J. Chem.,Sect. A,1983, 22, 901.
346 J.Szöke, J. Mol. Struct., 1984, 115, 51.
347 V.I.Spitsyn, V.V.Kolesnik, V.E.Mistryukov, Yu.N.Mikhailov, and K.M.Dunaeva, Dokl. Akad. Nauk SSSR, 1983, 273, 355.
348 A.Syamal and O.P.Singhal, Rev. Roum. Chim., 1983, 28, 981.
349 V.E.Karasev, I.V.Shchukina, and R.N.Shchelokov, Russ. J. Inorg. Chem., 1983, 28, 1316.
350 J.Cejka, A.Muck, and B.Hayek, Sb. Vys. Chem.-Technol. Praze Anorg. Chem., 1984, B29, 175 (Chem. Abstr.,1984, 101, 80922).
351 M.Gal, P.L.Goggin, and J.Mink, J. Mol. Struct., 1984, 114,459.
352 V.I.Berezin, V.V.Ganin, V.V.Sikorskii, and D.S.Umreiko, Koord. Khim., 1984, 10, 826.
353 E.M.Nour and A.N. Abd-el-Rahman, Acta Chim. Hung., 1984, 116, 377.
354 J.H.Holloway, G.M.Staunton, K.Rediess, R.Bougon, and D.Brown, J. Chem. Soc., Dalton Trans., 1984, 2163.
355 D.Gantar and B.Frlec, J. Chem. Soc., Dalton Trans.,1984, 2345.
356 S.A.Arthers, I.R.Beattie, and P.J.Jones, J. Chem. Soc., Dalton Trans., 1984, 711.
357 M.Takami, T.Oyama, T.Watanabe, S.Namba, and R.Nakane, Jpn. J. Appl. Phys., Part 2, 1984, 23, 88.
358 C.Madic, G.M.Begun, D.E.Hobart, and R.L.Hahn, Radiochim. Acta, 1983, 34, 195.
359 C.Madic, G.M.Begun, D.E.Hobart, and R.L.Hahn, Inorg. Chem., 1984, 23, 1914.

6

Vibrational Spectra of Some Co-ordinated Ligands

BY G. DAVIDSON

1 Carbon, Silicon, and Tin Donors

The complexes $CpTi(CH_2Ph)_2$ and $CpV(CH_2Ph)_2$ give i.r. bands characteristic of η^1-benzyl ligands at 1590cm^{-1} (Ti) or 1588cm^{-1} (V).[1] Cp_2ZrLCl and Cp_2ZrL_2, where LH = carboxylic acid, have i.r. spectra showing the presence of one h^5- and one h^1-C_5H_5 ring ligand.[2] The complex (1) has νC=O at 1668cm^{-1} - higher than would have been seen for η^2-acyl ligands (confirmed by ^{13}C substitution).[3]

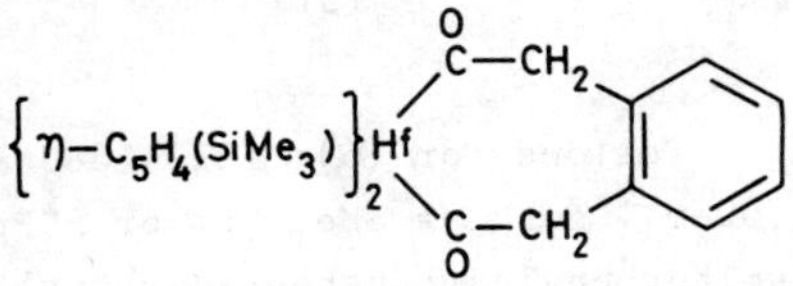

(1)

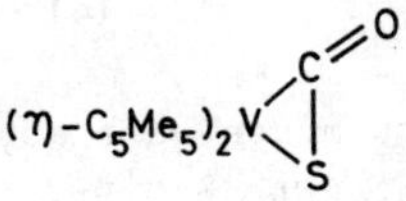

(2)

νC=O in (2) is at 1625cm^{-1}.[4] $Cp_2V(C{\equiv}CCMe_3)_2$ has an i.r. band at 2075cm^{-1}, as expected for a σ-alkynide.[5]

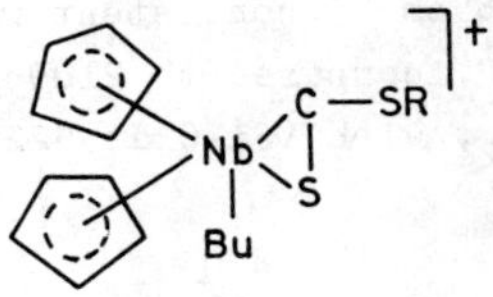

(3)

The species (3), where R = Me, Et, ^{i}Pr or CH_2COOEt, give νCS in the range 1110-1145 cm^{-1}, in agreement with the cationic formulation, and bidentate dithioalkylester (unidentate coordination via C would have given νCS ca. 1000cm^{-1}).[6] Assignments for νC≡C (1640-1695cm^{-1}) and νC-C (1170-1270cm^{-1}) modes were proposed in the following diphenylacetylene (= L) complexes: $[NbCl_3L]_4$, $[MX_4L]_2$, where X = Cl, Br or I, M = Mo or W, $ReCl_5L$, $[ReCl_4L]_2$ and $[MCl_5L]^-$, where M = W or Re.[7]

Detailed vibrational assignments have been given for phenyl-(trans-halotetracarbonyl)carbyne metal complexes and their deuteriated analogues, $[Br(OC)_4MC]_2C_6X_4$, where X= H or D, M = Cr or W.

These show that the carbyne fragment influences the π-electron distribution of the ring by delocalising the π-electrons on the bonds parallel to the axis of the complex.[8]

The complexes (4) have νC=N at 1655cm^{-1} (M = Cr) or 1660cm^{-1} (W).[9] The carbomethoxy compound $(\eta^7-C_7H_7)Mo(CO)_2(CO_2Me)$ has νC=O at 1631cm^{-1}.[10] The acyl complexes $MoX(\eta^2-COR)(CO)(PMe_3)_3$ have νC=O in the range 1480-1530cm^{-1} (X = Cl, Br or NCO, R = CH_2SiMe_3 or CH_2CMe_2Ph) - consistent with the formulation (5).[11]

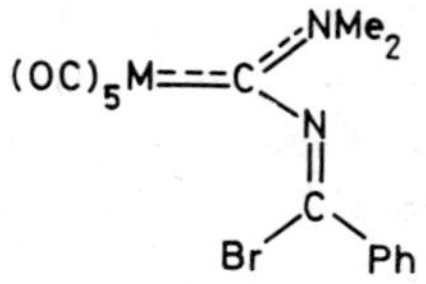

(4)

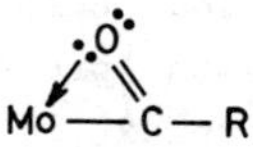

(5)

νC=N modes were assigned as follows for (6): 1640-1680cm^{-1} for L = OPh, OEt or O^{i}Pr, 1625-1760cm^{-1} for L = SMe, SEt or S^{i}Pr.[12] νCO is at 2060cm^{-1} in methylaminophenylcarbenepentacarbonyl-tungsten(0), implying analogous bonding to other zero-valent carbonyl carbene complexes, including those lacking a hetero-atom bonded to the carbene carbon.[13] $[L(CO)(CN)_2W(O-C-C-R)]^-$, where L = o-phen or bipy, R = Me or Ph, have νCN 2092cm^{-1}, νC≡O 1870-1850cm^{-1} and νC=O(ketene) 1665-1685cm^{-1}, consistent with the formulation (7).[14] νCC is at 1621cm^{-1}, compared to 2109cm^{-1} in the free ligand, in $[WCl_4(Me_3SiC{\equiv}CSiMe_3)]_2$, with νSi-C at 922cm^{-1}.[15]

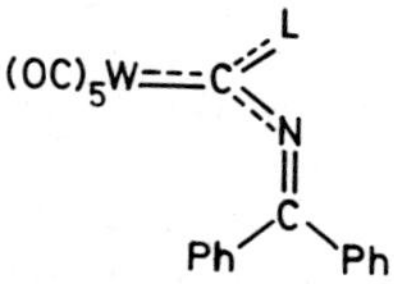

(6)

A detailed assignment has been given for the low-wavenumber i.r and Raman spectra of $(\eta^5-C_5H_5)M(CO)_3$, where M = Mn or Re, and of $(\eta^5-C_5D_5)Mn(CO)_3$.[16] The i.r. spectra of various isotopically substituted (H/D) derivatives of $MeM(CO)_5$, where M = Mn or Re, gave values for the "average" "isolated" νCH in CHD_2 derivatives of

2955.0cm^{-1} (Mn) or 2934.6cm^{-1} (Re). These correspond to r_0^{CH} values of 1.095(9)Å (Mn) or 1.098(8)Å (Re).[17]

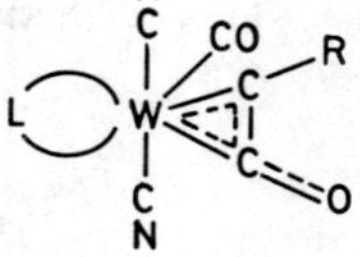

(7)

The complexes (8) have νC=C at 1587cm^{-1} (R = H), 1585cm^{-1} (R = Me). νC=O for the η^2-ketene ligand is in the range 1712-1785cm^{-1} for (9), where R = H or Me.[18] νCO(carbamoyl) is between 1675 and 1685cm^{-1} in $Re(CO)_4(COL)$ (10), where R = Me, Ph or H, R' = Ph or p-MeC_6H_4, R" = H or Me. ν_{as}NCN (ca. 1570cm^{-1}) is about 50cm^{-1} lower than in the parent amidine.[19]

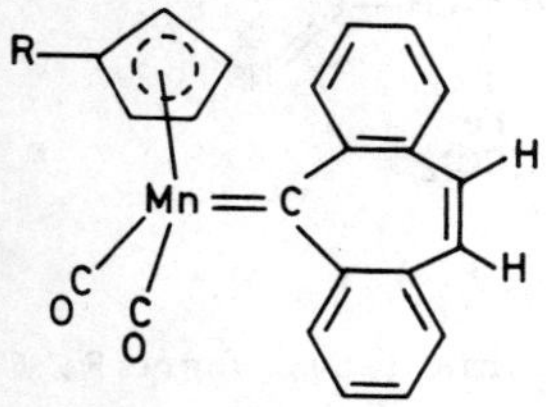

(8)

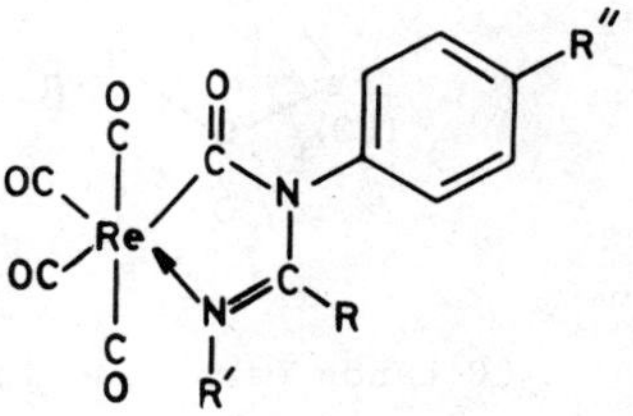

(10)

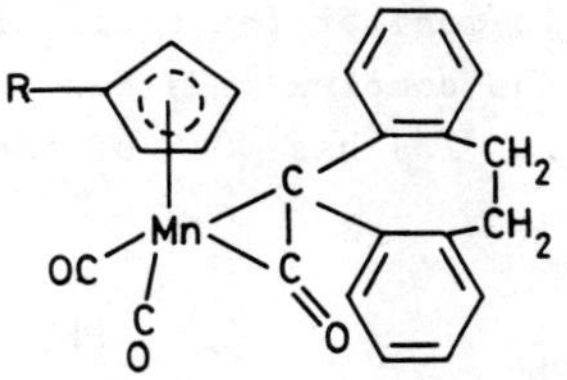

(9)

νC=S of the thioalkoxy carbonyl group gives an intense i.r. band at 1200cm^{-1} in (11).[20] (12) has νC=C at 1551cm^{-1}, while the corresponding complex with one PMe_3 replaced by CO has a similar band at 1580cm^{-1}. There is thus a high electron density at the terminal β-carbon atom.[21]

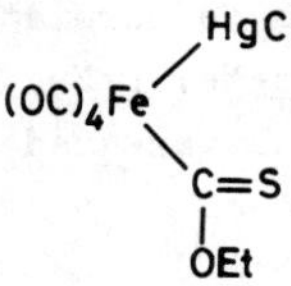

(11)

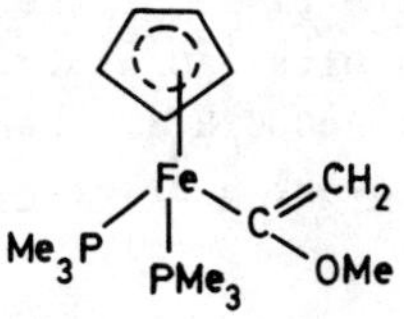

(12)

Bonding between the CN carbon and the sulphur atom of the dithiocarbamate ligand gives a decrease in νCN to $1760cm^{-1}$ in (13).[22] νCN in (14) is at $1540cm^{-1}$, and in (15) at $1565cm^{-1}$, i.e. an increase of $25cm^{-1}$ in the six-, compared to the five-membered ring.[23]

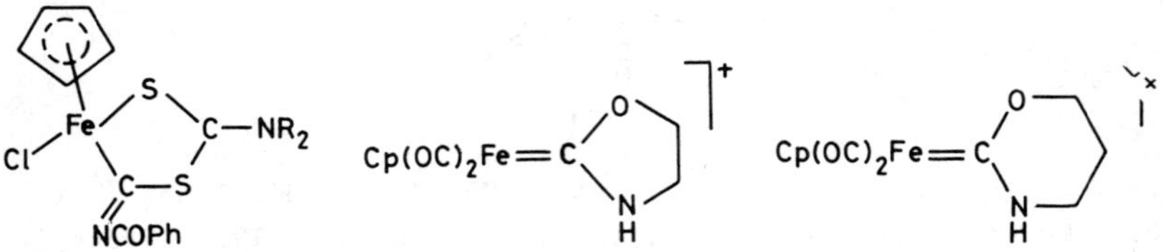

(13) (14) (15)

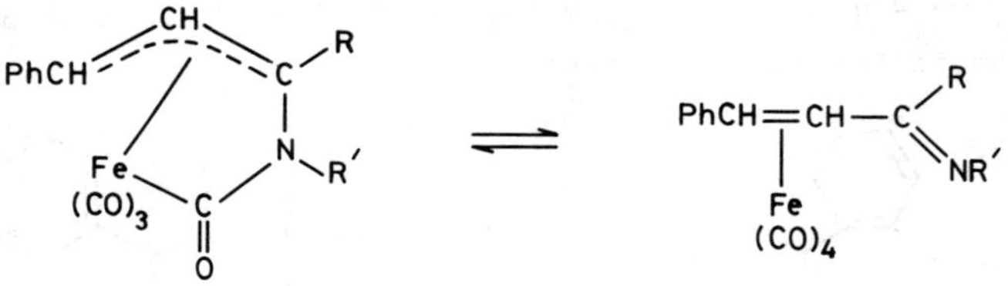

(16)

νCO bands were used to monitor the equilibria (16), where R, R' = H or various alkyls.[24] The complex (17) has νC=C of the furan group at $1595cm^{-1}$,[25] while (18) has νC=O of the cyclobutanone group at $1765cm^{-1}$.[26]

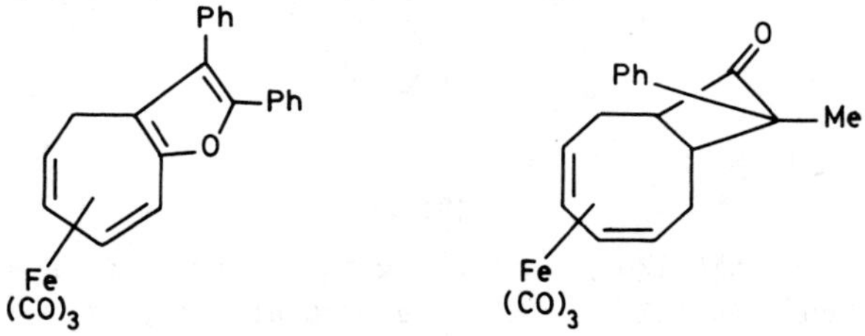

(17) (18)

νC=O in (19), $1674cm^{-1}$, is higher than that in (20), $1666cm^{-1}$, due to tilting of the axis of the CO group in the former, with respect to the plane of the cyclopentadienyl ring.[27] νC=O correlates roughly with ^{13}C n.m.r. chemical shifts for aromatic ferrocenyl ketones, $FcCOC_6H_4X$, where X = H, o-Me, m-Me, p-Me, p-OMe, o-Cl, m-Cl, p-Cl, o-Br, m-Br or p-CN.[28] Assignments were made for νN=X in

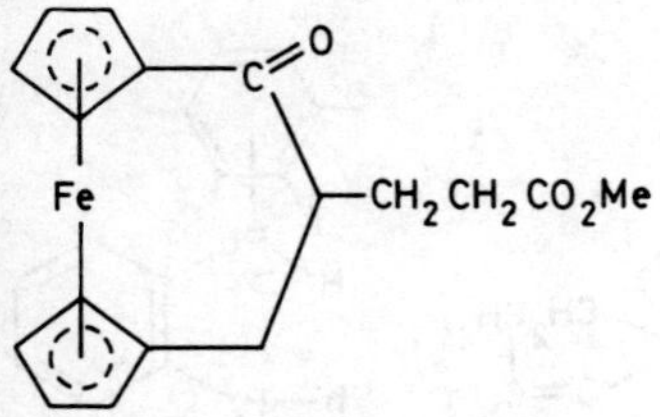
(19)

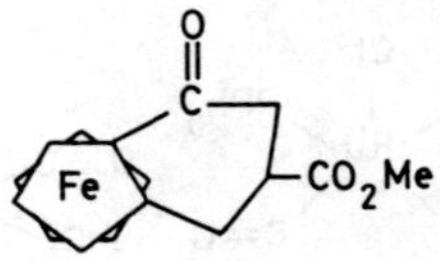
(20)

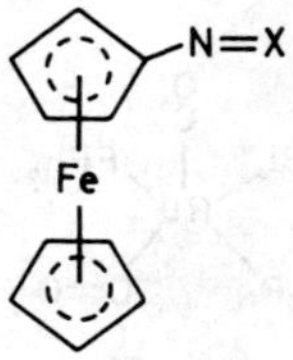
(21)

(21), as follows: X = O, 1465cm^{-1}; $C(CN)_2$, 1520cm^{-1}; PEt_3, 1290cm^{-1}.[29]

I.r. bands due to both 1:1 and 2:1 arene/metal complexes are seen in matrices at -190°C containing benzene or toluene and Fe, Co or Ni. The 2:1 complexes gave spectra analogous to that for the known $(C_6H_6)_2Cr$.[30] Complex (22) has νSiH at 2069cm^{-1}.[31]

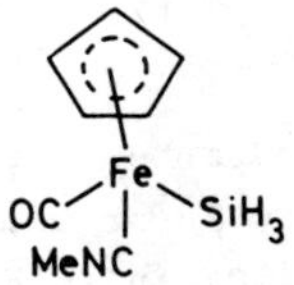
(22)

The νC=O mode of $RuCl[C(O)Et](CO)(PPh_3)_2$ is at 1645cm^{-1}, consistent with the monohapto-formulation. The presence of two νC≡O bands in solution suggests that two isomers are present, possibly (23) and (24).[32] The complexes (25), where R = Me, Et or Bu, contain i.r. bands due to νC=O and νC-O at 1640, 1230cm^{-1}, respectively, compared with values of 1720, 1200cm^{-1} for the free ester. These are consistent with the bonding shown.[33] In (26), the bands due to the ortho-metallated phenyl group are at 1555, 1413, and 728cm^{-1}, with νRu-H at 1955 cm^{-1}.[34]

I.r. bands due to νSO are at 1120, 1100, and 788cm^{-1} in $Os(\overline{CH_2S[O]O})Cl(NO)(PPh_3)_2$. These are explicable by the coordination (27), which is confirmed by X-ray diffraction.[35] νCF (1125cm^{-1}) and

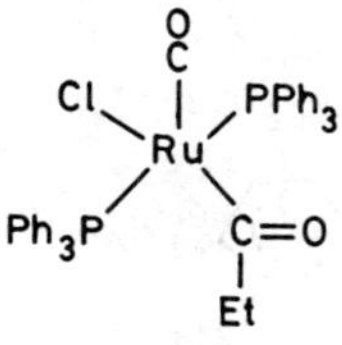
(23)

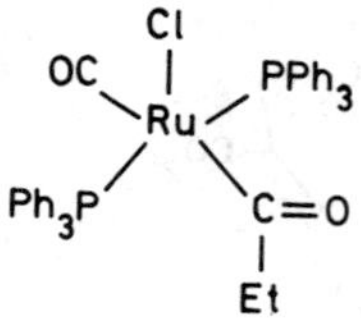
(24)

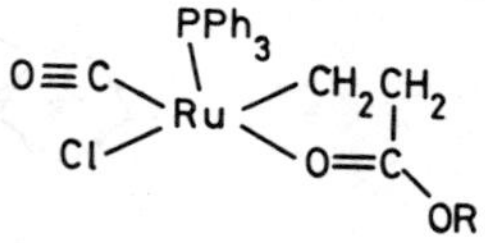
(25)

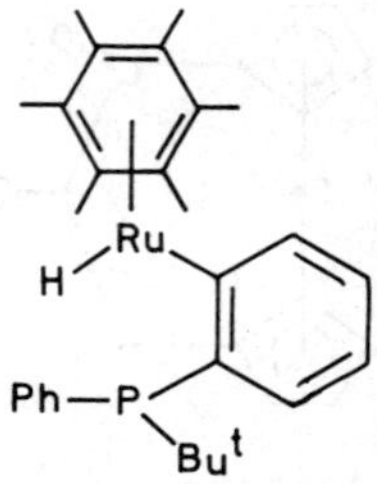
(26)

νCCl (885-909cm^{-1}) modes were assigned in $OsCl_2(CFCl)(CO)L_2$ and $OsCl(CFCl)(MeCN)(CO)L_2^+$, where L = PPh_3.[36] The complex (28) has νNH at 3356cm^{-1}, and a feature due to νC=O + νC-N at 1356cm^{-1}.[37]

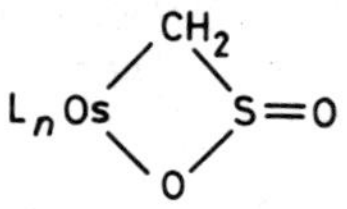
(27)

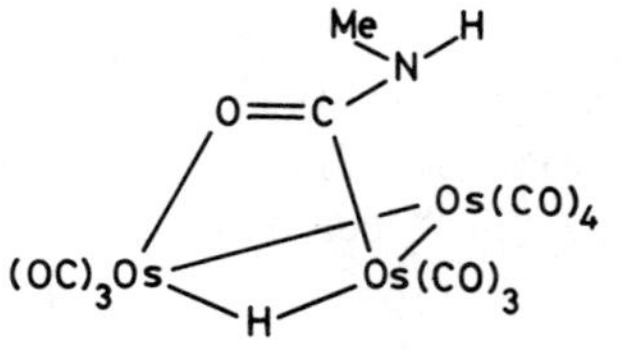
(28)

Vibrational spectra were reported for $Os_3(CO)_{11}(\eta^2-C_2H_4)$, $M_3(CO)_9(\mu-H)(\mu_3-CMe)$, where M = Ru or Os, $Os_3(CO)_{10}(\mu_3-\eta^2-HCCH)$, $Os_3(CO)_9(\mu-H)(\mu_3-\eta^2-CCH)$, $Ru_2(CO)_3(\eta^5-C_5H_5)_2(\mu-CHMe)$, $Ru_2(CO)_3(\eta^5-C_5H_5)_2(\mu-CMe)^+$ and $Os_3(CO)_9(\mu-H)(\mu-S^nPr)L$, where L = CO, C_2H_2 or C_2H_4. Vibrational assignments were proposed for C_2H_n, where n = 1-4, and on this basis it is suggested that acetylene rearranges to a $\mu_3-\eta^2$-vinylidene unit on Ni(111) or Fe(110) surfaces. Assignments for μ_3-CMe are summarised in Table 1.[38]

Table 1 Vibrational assignments for the μ_3-CMe ligand in $M_3(CO)_9(\mu-H)_3(\mu_3-CMe)$ (/cm^{-1})

M =	Ru	Os	Assignment
	2915	2916	$\nu_{as}CH_3$
	2889	2881	$\nu^s CH_3$
	1438	1432	$\delta^s CH_3$
	1361	1356	$\delta^{as} CH_3$
	1147	1125	$\nu^s CC$
	1037	1028	$\rho\ CH_3$

CH_3/CD_3 substitution in methyl(4-tert-butylpyridine)cobaloxime gave definitive assignments of νCH/νCD and $\delta CH_3/\delta CD_3$ modes.[39] νCC of the dicarbide ligand in $Co_6(\mu\text{-}C_2)(\mu\text{-}CO)_6(CO)_8(\mu_4\text{-}S)$ gave a weak i.r. band at $1444cm^{-1}$. In the νCO region, 6 terminal νCO modes were seen, with a very peculiar pattern due to νCO_{br}. It appears from this that there is one regular plus 5 'semi-bridging' CO groups.[40] CS_2O and CS_3 ligand mode assignments were made in $Co(\eta\text{-}C_5H_5)L(CS_2X)$ where X = O or S, L = phosphines or arsines. For X = O, νC-S is ca. $840cm^{-1}$, νC=O $1585\text{-}1710cm^{-1}$; for X = S, νC-S is $855\text{-}870cm^{-1}$, νC=S $1030\text{-}1050cm^{-1}$.[41] $(C_5Me_5)Co(C_2H_4)L(NCS)$, where L = CN or NCS, have νC=C of the coordinated ethene between 1480 and $1490cm^{-1}$.[42]

^{13}CO substitution gave the assignments shown in (29), all $/cm^{-1}$.[43]

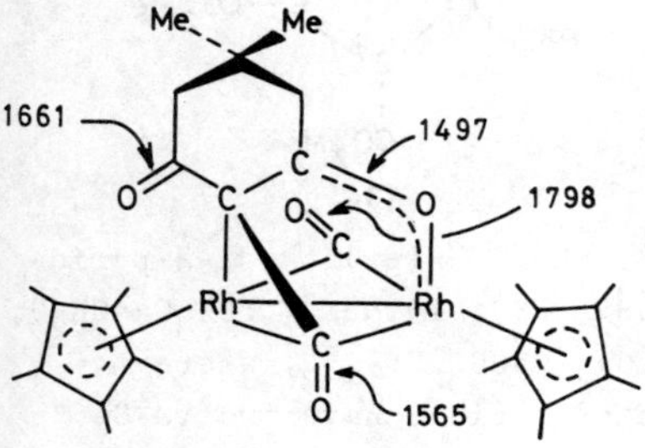

(29)

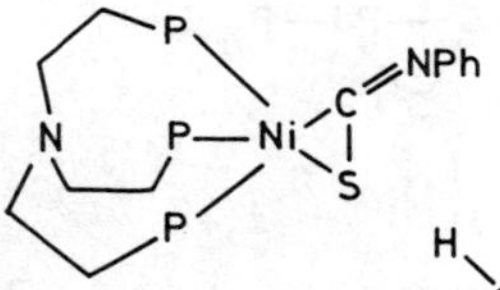

(30)

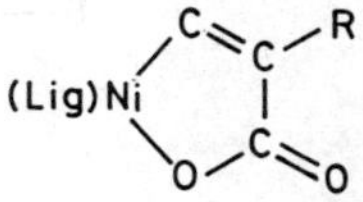

(31)

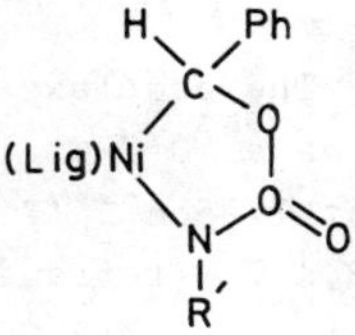

(32)

$Ir(SnCl_3)(TFB)L_2$, where TFB = tetrafluorobenzobarrelene, and L = various N, P, As, O or S ligands, have νSnCl at higher wavenumber than in free $SnCl_3^-$, confirming the Ir-Sn coordination.[44]

(30) has νC=N at $1640cm^{-1}$, and $\nu\overline{Ni\text{-}S\text{-}C}$ at $650cm^{-1}$.[45] νC=O bands lie in the ranges $1610\text{-}1630cm^{-1}$ in (31), and $1640\text{-}1646cm^{-1}$ in (32), where Lig = bipy or tetramethylethylenediamine, R = H, Me or Ph, R' = Me or p-MeC_6H_4.[46,47]

$L_2Ni(RC{\equiv}CR)$, where L = PPh_3, $P(O\text{-}o\text{-}tol)_3$ or $P(C_6H_3\text{-}2\text{-}{}^iC_3H_7\text{-}5\text{-}CH_3)_3$, R = CH_2OH, CH_2OMe, CH_2OEt, $CH_2\text{-}N(CH_2)_4O$ or $CH_2\text{-}N(CH_2)_5$, all have νC≡C in the range $1800\text{-}1900cm^{-1}$. The nature of L has a significant effect on the wavenumber of νC≡C for a given R.[48] I.r. spectra were reported for low-temperature (-190°C) matrices containing 1,3-butadiene and Ni, Pd or Fe atoms. Nickel forms a π-complex $Ni(C_4H_6)_2$, which in turn produces a red, polymeric π-complex $(NiC_4H_6)_n$. The assignments for C_4H_6 modes in both systems were consistent with those for $(C_4H_6)_2Fe(CO)$.[49]

Inelastic electron tunnelling spectra of species at $Al/Al_2O_3/Pb$ junctions showed both i.r. and Raman-active modes. Interaction of HC≡CH with Pd at such a junction gave bands due to several different Pd(HC≡CH) surface species.[50]

νCS_2 modes were assigned for $Pd_2(CS_2)_2L_2$, where L = PCy_3, P^tBuPh_2, P^tBu_2Ph or P^tBu_3, containing the bridging unit (33).[51] (34) has νC=O at 1725cm^{-1}, νC=C at 1655cm^{-1}, νC-O at 1210cm^{-1} and νN-O at 900cm^{-1}.[52] $Pt(C_6F_5)_3(C_2Ph_2)^-$ has bands due to the diphenylethyne ligand at 2005 (νC≡C), 1595, 1500, 755, 688, 567 and 505cm^{-1}.[53]

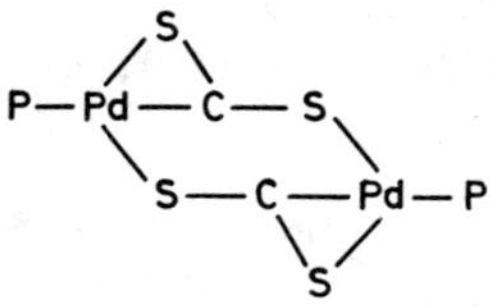

(33)

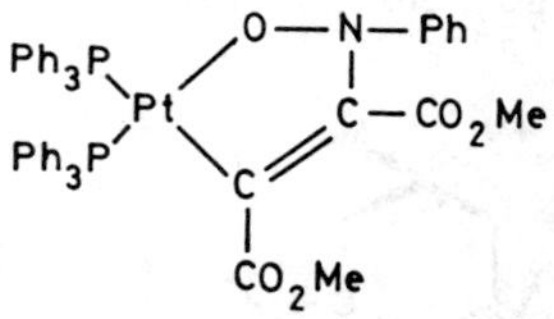

(34)

The complexes $Cu[NH(py)_2]X^+$, where $NH(py)_2$ = di-2-pyridylamine, have νC=C at 1505cm^{-1} (X = CH_2=CH_2), 1515cm^{-1} (CH_3CH=CH_2), or νC≡C at 1795cm^{-1} (X = HC≡CH).[54] $[LCu_2(C≡CCO_2Et)_2]_2$, where L = phen or 3,4,7,8-tetramethyl-1,10-phenanthroline, have two νC≡C bands in the i.r. (2100, 1930cm^{-1}). The higher wavenumber is due to a C≡C not involved in coordination to another copper atom. The lower band is due to C≡C interacting rather strongly with a metal centre.[55] $(CuX)_4(DMSO)_2$(alkyne), where alkyne = RR'C(OH)C≡C-Y, R, R' = Me or Et, X, Y = CN, Cl or Br, have i.r. spectra suggesting that the alkyne is π-bonded to Cu(II) (νCC ca.1650cm^{-1}).[56] $Cu(O_3SCF_3)$ forms π-complexes with the terminal alkynes 1-hexyne, 3-methyl-1-pentyne etc. These have νCH 68-107cm^{-1} lower than in the free alkynes, as well as νC≡C decreased by 81-173cm^{-1}. For diynes (1,5-hexa-, 1,7-octadiyne), two νC≡C bands due to coordinated alkyne were seen, consistent with the presence of more than one complex.[57] There is i.r. evidence that adsorption of C_3H_6 or C_3D_6 on CuO/Al_2O_3 or Cu(II) YA-zeolites is accompanied by reduction of Cu(II) and formation of Cu(I) π-complexes of C_3H_6 or C_3D_6.[58]

The Raman spectrum of ethene adsorbed on silver films contains bands at 983, 1326, 1582, 2970 and 3000cm^{-1}.[59] C_6F_5 modes in $Au(C_6F_5)_2^-$ (780, 950cm^{-1}) are shifted to higher energies on formation of binuclear complexes such as $[AuAg(C_6F_5)_2py]_n$ (789, 965cm^{-1}).[60] The i.r. and Raman spectra of solutions of R_2Zn or RZnCl, where R = allyl or substituted allyl, contain νC=C characteristic of η^1-allyl coordination (1655-1606cm^{-1}). Crystalline $(C_3H_5)_2Zn$ or $(CH_2{=}CMe{=}CH_2)_2Zn$, however, have νC=C in the range 1480-1550cm^{-1}, so there may be delocalised bonding involving the allyl groups.[61] I.r. intensity measurements of νC≡C in $Hg(C{\equiv}CR)_2$ and $R'Hg(C{\equiv}CR)_3$, where R = H, Me, ^{n}Bu, ^{t}Bu etc., R' = Me, Et, Ph, $CH{=}CH_2$ etc., are consistent with the presence of weak d_π-p_π interactions between Hg and C.[62]

$Cp_3U(\eta^2$-COR), where R = CMe_3, $CHMe_2$, ^{n}Bu, Et or Me, have νCO of the <u>dihapto</u>-acyl group in the range 1490-1505cm^{-1}, compared to the <u>monohapto</u>-acyl range of 1630-1680cm^{-1}.[63] Several ligand modes were assigned from the resonance Raman spectrum of $U(TMCOT)_2$, where TMCOT = tetramethylcyclo-octatetraene: 879cm^{-1}, in-phase C-Me stretch; 750cm^{-1}, ring breathing mode; 580, 513cm^{-1}, in-plane and out-of-plane ring deformations.[64]

Only one νC=C band was seen in $(CH_2{=}CH)_3SnX$, where X = Cl, Br or I, i.e. there is very little coupling between the vinyl groups.[65]

<u>2 Carbonyl and Thiocarbonyl Complexes</u>

I.r. spectra were recorded for CO adsorbed on evaporated films of CaX_2, where X = F, Cl, Br or I. For CaF_2, νCO is at 2156cm^{-1}, for molecules adsorbed on cations of the (111) face. For the other surfaces, the main bands were in the range 2170-2190cm^{-1}, depending on coverage.[66] $[LM(CO)_3H]^+BF_4^-$, where L = 1,4,7-triazacyclononane, M = Cr, Mo or W, have νCO 100-150cm^{-1} higher than in $LM(CO)_3$, as expected for formally M(II) species.[67] kCO values for $M(CO)_4(R_2PCH_2CH_2PR_2)$, where M = Cr or Mo, R = F, Cl, C_6F_5, CH_3O, Ph, Me or C_6H_{11}, suggest the following order of π-acceptor strength: R = F > Cl > C_6F_5 > CH_3O > CH_3 > Ph > C_6H_{11}.[68] Values for νCO show that in THF solution Na^+ or Li^+ form contact ion-pairs with <u>cis</u>-$HM(CO)_4PR_3^-$, where M = Cr or W, R = Me, Ph or OMe. The interaction is specifically at the hydride ligand or metal-hydride bond site. This should be compared with $HM(CO)_5^-$ (M = Cr or W), $HFe(CO)_4^-$, and $CpV(CO)_3H^-$, where contact interaction occurs at the CO oxygen site.[69] νCN and νCO for $(CF_3NC)M(CO)_5$, where M = Cr or W, show that the new CF_3NC ligand is a very strong π-acceptor.[70]

Linear correlations were found between νCO(a_1, eq.), the n.m.r. chemical shift $\delta(^{13}C(CO))$ and the XPS C1s(CO) and O1s(CO) binding energies for $(Ph_{3-n}PX_n)Cr(CO)_5$, where X = Cl, Br, I or H, n = 0-3. Hence all of these parameters can be used as a measure of

($d_{Cr} \rightarrow \pi^*_{CO}$) π-back-bonding.[71,72] High-resolution F.T.i.r. spectra were reported for $Cr(^{12}CO)_6$ and $Cr(^{12}CO)_5(^{13}CO)$ in argon matrices at 10K. The t_{1u} νCO mode gave three closely spaced bands. It is possible to deduce a quantitative model for the matrix trapping site.[73] νCO bands for $Fe_3M(CO)_{14}{}^{2+}$, where M = Cr, Mo or W, are all very similar. Hence the anions are isosructural. νCO_{br}, ca.1750-1775cm^{-1}, is characteristic of an edge-sharing carbonyl. There was no evidence for a band due to an η^2-CO ligand.[74]

$Mo(acac)Cl(CO)(PMe_3)_3$ has νCO at a very low value (1755cm^{-1}) for a terminal Mo-CO unit.[75] Very low νCO values were also found for solutions of $Mo(CO)_2(L\text{-}L)_2$, where L-L = bipy or phen (1720-1795cm^{-1}). Together with data for related ligands, these gave the following π-acidity series: PF_3 > dppe > dmpe > phen > bipy.[76] Photoreactions of $M(\eta^5\text{-}C_5H_5)(CO)_3R$, where M = Mo or W, R = Me or Et, in PVC-film matrices (12-298K) were followed by νCO i.r. It was possible to obtain quite detailed information on the species produced.[77] Similar results were also reported for $M(CO)_3R$-(η-$C_5R'_5$), where M = Mo or W, R = Et, ^{n}Pr, ^{i}Pr, ^{n}Bu, Ph or CH_2Ph, R' = H or Me, in solution and in low-temperature matrices,[78] and for $MH(CO)_3(\eta^5\text{-}C_5H_5)$, where M = Mo or W, and trans-$WH(CO)_2(C_2H_4)$-(η^5-C_5H_5). One product in N_2 matrices was $MH(CO)_2(N_2)(\eta^5\text{-}C_5H_5)$ for which νNN = 2192cm^{-1} (Mo) or 2163.5cm^{-1} (W).[79] νCO, together with ^{13}CO substitution, helped to identify the photoproducts of reaction of $M(CO)_6$, where M = Mo or W, and O_2 in Ar or CH_4 matrices as $M(O)_2(CO)_4$.[80] νNO and νCO in $Mo(NO)(CO)_4(ClAlCl_3)$ are at 1713 and 2065cm^{-1} respectively. Formation of $Mo(NO)(CO)_4(ClAlCl_3)(AlEtCl_2)$ shifts these to 1758, 2085cm^{-1} respectively.[81]

The series of complexes cis-$W(CO)_2(bipy)_2^{n+}$, where n = 0, 1 or 2, show the expected trends in νCO with increased formal oxidation state - Table 2.[82]

Table 2 νCO wavenumbers in cis-$W(CO)_2(bipy)_2^{n+}$ (/cm^{-1})

n = 0	1778, 1708
1	1895, 1815
2	1982, 1895

The presence of νNO at 1708cm^{-1} and only one νCO band in the i.r. (2058cm^{-1}) for $W(NO)(CO)_4^+\ AlCl_4^-$ shows that the four CO groups are in square-planar arrangement.[83] Carbonyl force constants and

^{13}C n.m.r data for 23 anionic and neutral $W(CO)_5L$ species show good correlation between $\delta(^{13}CO)$ and k(CO) even when the L's include ligands with widely different donor/acceptor properties ($L = H^-$, Me^-, Ph_2C, CO, CN^-, py, SCN^-, Ph_3P, Cl^- etc.).[84] For $W(CO)_5L$, where $L = R_2EER'_2$, $R_2EE'R$, R,R' = CH_3 or CF_3, E = P or As, E' = S, Se or Te, νCO increased by 20-30cm^{-1} on substituting CH_3 by CF_3 on the coordinated donor atom, but only by 5-10cm^{-1} for the non-coordinated donor atom.[85] $W(CO)_5$ $As(OTeF_5)_3$ has νCO at 2107 cm^{-1} (a_1), 2004cm^{-1} (a_1), and 1975cm^{-1} (e), suggesting that PF_3 and $As(OTeF_5)_3$ have approximately equal π-acidities.[86] $(\mu_2\text{-H})Os_3S\text{-}W(\eta^5\text{-}C_5H_5)(CO)_{11}$ has five terminal νCO bands in its i.r. spectrum, and one νCO due to 'semibridging' CO, at 1846cm^{-1}.[87]

νCO bands are seen in $Cp(OC)_2Mn{\equiv}C\text{-}CH{=}CR_2^+$, where R = ^tBu or Ph, ca. 80-100cm^{-1} higher than in the neutral species $Cp(OC)_2Mn{=}C{=}CR_2$, for the same R.[88] The complex (35) gives only two νCO bands in the i.r. spectrum, corresponding to the centrosymmetric structure shown.[89]

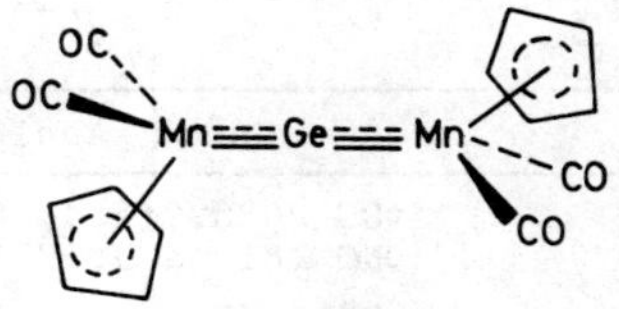

(35)

Solid $[(Me_5C_5)_2Yb(\mu\text{-}OC)_2Mn(CO)_3]_\infty$ has νCO bands in the i.r. at 1965, 1937, 1928, 1882, 1840 and 1775cm^{-1}. In solutions, bands typical of the contact ion-pair $[(C_5Me_5)_2Yb]^+[Mn(CO)_5]^-$ are seen.[90] The complex $[C_5(CO_2Me)_5]Mn(CO)_3$ has νCO bands at 2020 and 1940cm^{-1}, rather low for a system containing so many electron-withdrawing groups.[91] The i.r. spectra of $[(OC)_4M]_3M'.nTHF$, where M = Mn, Re or Co, M' = Sm, Tm or Yb, n = 1-4, show the presence of both terminal and bridging CO groups.[92]

νCO i.r. spectra were measured and Cotton-Kraihanzel force constants calculated for $Fe(Ar)(CO)L(\eta^5\text{-}C_5H_5)$ (where L = CO, Ar = Ph, C_6F_5, $2\text{-}XC_6H_4$ (X = F, Cl, Br, I, CH_3 or CH_2Ph); L = PPh_3, Ar = Ph or $2\text{-}FC_6H_4$). In the dicarbonyls, the CO force constants decrease in the sequence: Ar = C_6F_5 > $2\text{-}FC_6H_4$ ∿ $2\text{-}BrC_6H_4$ > $2\text{-}IC_6H_4$ > C_6H_5 >

2-$CH_3C_6H_4$ ∿ 2-$PhCH_2C_6H_4$.[93] I.r. spectra (νCO) in PVC-film matrices (12-200K) show that photo-induced decarbonylation of Fe(η^5-C_5H_5)-$(CO)_2$(COMe) is thermally reversible, and that photolysis of Fe(η^5-C_5H_5)$(CO)_2$Me leads to the reversible formation of Fe(CO)(η^5-C_5H_5)Me.[94]

Observation of a deuterium isotopic shift in νCO of ferrous cytochrome c peroxidase carbonyl suggests that there is hydrogen-bonding between a distal amino acid side chain and the haem-coordinated CO.[95] A similar shift in νCO on replacement of H_2O by D_2O is seen for the carbonyl co-enzyme C_2 of horseradish.[96] I.r. bands due to νCO were measured for CO bound to mitochondria from bovine and porcine hearts, bovine brain, rat kidney and blowfly flight muscle. A band (ca. 1963cm^{-1}) is found similar to that in carbonyl cytochrome c oxidase purified from bovine heart. A second band (ca. 1959cm^{-1}) is ascribed to a less stable conformer of the purified oxidase carbonyl, detected in mitochondria. The CO binding sites in all of the oxidases examined were all very similar.[97]

Table 3 Vibrational assignments for $Fe(CO)_{2-n}(N_2)_n(NO)_2$ (/cm^{-1}) in liquid krypton

Species	Assignment
$Fe(CO)_2(NO)_2$	νCO 2086.9(a_1), 2038.5(b_1) νNO 1817(a_1), 1775.8(b_2)
$Fe(CO)(N_2)(NO)_2$	νNN 2257.5 νCO 2046.0 νNO 1805(a'), 1761.5(a")
$Fe(N_2)_2(NO)_2$	νNN 2250.5(a_1), 2228.0(b_1) νNO 1796.3(a_1), 1749.4(b_2)

νCO, νNO and νNN assignments were given for $Fe(CO)_{2-n}(N_2)_n(NO)_2$ (n = 0, 1 or 2) in liquid krypton - Table 3.[98] I.r. evidence (νCO) was found for the formation of $Fe(CO)_4$ and $Fe(CO)_3$ by photolysis of $Fe(CO)_5$ in low-density polyethylene or PTFE films.[99] I.r. spectra of gas-phase mixtures produced by u.v. photolysis of $Fe(CO)_5$ also showed the presence of these two species, also $Fe(CO)_2$.[100]

νCO, νCS and νCN bands for $Cp_2Fe_2(CO)_2$(CS)(CNR), where R = Me, Et, ^{i}Pr or ^{t}Bu, in solution show that four isomers are present, (36) - (39), except for R = ^{t}Bu, for which no bridging CNR groups are present.[101] I.r. evidence was found for the formation of a triply bridged carbonyl species on photolysis of trans-$[Cp_2Fe(CO)_2]_2$ and related species.[102]

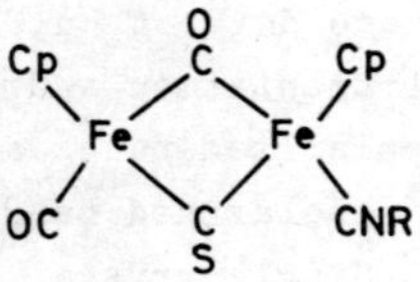

(36)

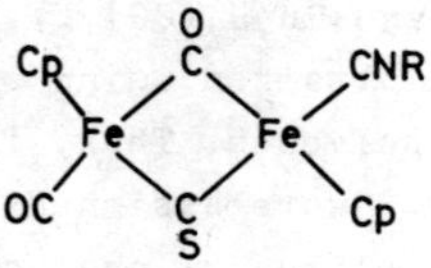

(37)

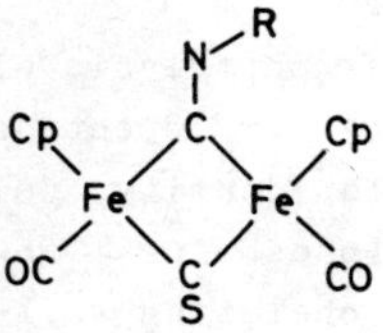

(38)

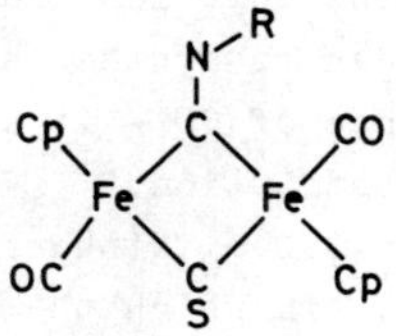

(39)

I.r. spectra (νCO, νNN) were used to identify $CpFeCo(CO)_5$, $[CpFe(CO)_3]^+[Co(CO)_4]^-$, and (in N_2-doped matrices) $CpFeCo(CO)_3$-$(N_2)(\mu\text{-}CO)_2$ as photoproducts of $CpFe(CO)_2Co(CO)_4$.[103] I.r. bands at 1940 and 1960cm^{-1} were assigned as νCO of "$Fe(CO)_4(SiO_2)$", i.e. a complex between $Fe(CO)_4$ and a surface hydroxyl or siloxane bridging oxo-group - produced by laser photolysis (λ = 337nm) of $Fe(CO)_5$ adsorbed on SiO_2.[104] Ion-pairing in salts of $[M_3(\mu\text{-}H,\mu\text{-}CO)(CO)_{10}]^-$, where M = Fe or Ru, was followed by using νCO, in the bridging region. The anions were found to be only weakly paired with unipositive cations.[105]

CO adsorption on Ru/SiO_2 gives three νCO bands: 2030, 2080 and 2140cm^{-1}. They were assigned to $Ru(CO)_n$, where n >1, $Ru(CO)X_m$ and $Ru(CO)_nX_m$ respectively, where X = H_2O, OH, O or Cl.[106] Reactions between $Ru_3(CO)_{12}$ and SiO_2, TiO_2 or Al_2O_3 were followed by i.r. spectroscopy in the νCO region (with ^{13}CO substitution). On SiO_2, $Ru_3(\mu\text{-}H)(CO)_{10}(\mu\text{-}O\text{-}Si\equiv)$ is generated initially, decaying to $Ru^{II}(CO)_{2\,n}$, where n = 2 or 3, in air.[107] There is i.r. evidence for the formation of $H_4Ru_4(CO)_{11}$ by u.v. irradiation of $H_4Ru_4(CO)_{12}$ in an alkane matrix at 77K.[108]

Changes in the wavenumbers of the a_1' νCO modes of $Ru_3(CO)_{12}$ and $Os_3(CO)_{12}$ in mixed crystals suggest that these originate in minor changes in the shape and size of the molecules, to adapt to the sites available in the crystal.[109] More detailed νCO Raman solution data than ever before have been obtained for $Ru_3(CO)_{12}$,

$Os_3(CO)_{12}$ and $Ru_4H_4(CO)_{12}$, together with some data on $Ru_5C(CO)_{15}$. These were obtained using specialised instrumentation and careful choice of solvents. Thus, for $Ru_3(CO)_{12}$, polarised bands at 2039 and 2119cm^{-1} were assigned to a_1' modes, a depolarised band at 2011 cm^{-1} to an e' mode (local symmetry b_1).[110] Single crystal i.r. and Raman spectra gave a complete assignment of νCO modes for solid $Ru_3(CO)_{12}$. These show that D_{3h} symmetry is retained in the crystal and confirm the solution-phase assignments.[111]

$Os(CO)_5$ can be synthesised by i.r. laser photolysis of $H_2Os(CO)_4$ in the presence of CO - it was identified by i.r. spectroscopy, with ^{13}CO substitution.[112] It is possible to identify, using νCO, isomers of $Os_3(CO)_{10}$(diphosphine), where diphosphine = $Ph_2P(CH_2)_n$-PPh_2, n = 2-4, in which the diphosphine is chelating (1,1-isomer, 40) or bridging (1,2-isomer, 41).[113]

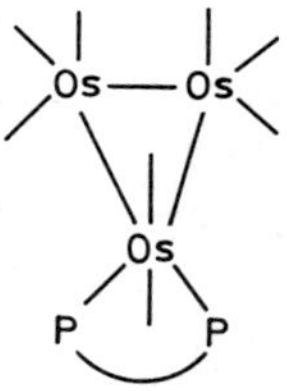

(40)

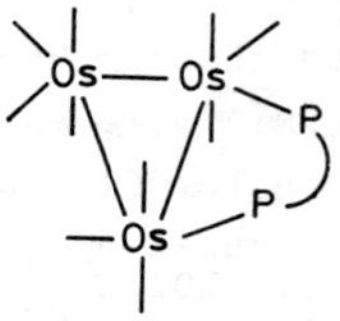

(41)

$Co(CN)_5CO^{2-}$ has $\nu^{12}CO$ at 2108cm^{-1}, $\nu^{13}CO$ at 2061cm^{-1} - high values as expected for Co(III). Three νCN bands were seen, at 2152, 2142 and 2132cm^{-1}.[114] CO on silica-supported cobalt catalysts gives bands at ca. 2000, ca. 2180cm^{-1}, due to adsorption at Co^0, Co^{2+} sites respectively.[115] A new Raman band at 1797cm^{-1} for $Co_2(CO)_6(PPh_3)_2$ on Al_2O_3 is assigned to a carbonyl species hydrogen-bonded to OH groups on the surface of γ-Al_2O_3.[116] $XCCo_3(CO)_7(NO)^-$, where X = Me, Ph, COOH or $CpFeC_5H_4$, give 5 νCO bands. The four highest are from the terminal $Co(CO)_3$ groups (42), the lowest (1857-1870cm^{-1}) from the terminal Co(CO)(NO) group. The νNO values, 1715-1740cm^{-1}, show that the NO is always linear and terminal.[117] $^{12}C/^{13}C$ isotopic shifts were used in νCO i.r. studies of the formation of $Co_4(CO)_{11}L$, where L is a phosphorus ligand, from $Co_4(CO)_{12}$. No cluster fragmentation occurs.[118]

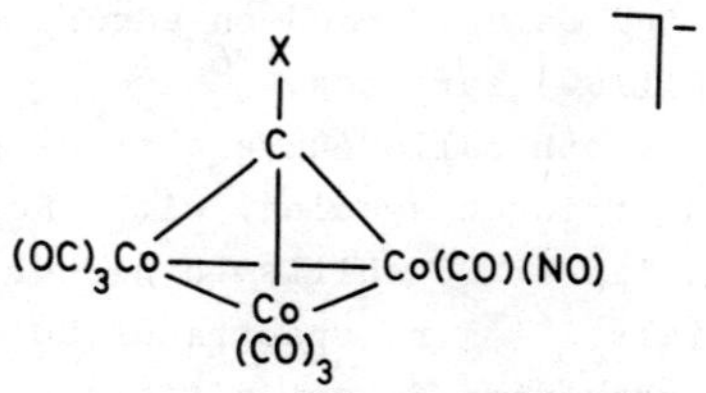

(42)

CO_2 undergoes dissociative adsorption on rhodium dispersed on Al_2O_3 to form an adsorbed carbonyl species, with νCO at 2015 cm^{-1}.[119] νCO bands were used to characterise a variety of intra-zeolite rhodium carbonyl complexes prepared by CO reduction of Rh(III)-exchanged zeolites. $Rh(CO)_2^+$ is the best characterised species, and this reacts with various phosphines to give e.g. $Rh(CO)(PMe_2Ph)_x$, where x = 1, 2 or 3.[120] CO adsorbed on Rh(I) sites on Al_2O_3 forms the gem-dicarbonyl $Rh^I(CO)_2$. It is also possible to observe $Rh^I(CO)_3$; this gives three νCO bands (2120, 2078, 2026 cm^{-1}) and thus has low symmetry.[121] A band at 1650cm^{-1} in the solid state i.r. spectrum of $(\eta^5-C_5H_5)_2Zr(CH_2PPh_2)_2Rh(CO)Cl$ was attributed to a strong interaction between the oxophilic zirconium atom and the carbonyl oxygen of the rhodium in one possible rotamer, (43).[122] Interaction of $Rh_4(CO)_{12}$ or $Rh_6(CO)_{16}$ with Al_2O_3, TiO_2 or

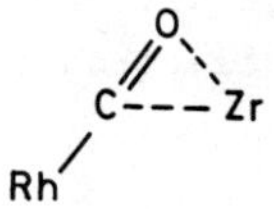

(43)

SiO_2 was monitored by νCO i.r. spectroscopy, including ^{13}C substitution. The conversion of $Rh_4(CO)_{12}$ to $Rh_6(CO)_{16}$ is seen on Al_2O_3 and TiO_2. Oxidative fragmentation gives $Rh^I(CO)_2$ sites in all cases.[123]

νNiC and νCO were assigned for 'on-top' and 'bridged' species for CO adsorbed on Ni(111).[124] I.r. emission spectra of CO adsorbed on Ni(100) at 310K show that no bridge-bonded carbonyl is present at less than half coverage.[125] There is, however, evidence for bridged carbonyl species being formed on adsorption of CO on to sulphur-modified Ni(100) surfaces.[126]

CO adsorbed on a smooth palladium electrode gives i.r. bands due to both linear and bridged species, with the latter predominating. The νCO of the linear CO shifts to higher wavenumbers at more positive potentials.[127] I.r. spectra of CO adsorbed on Pd/ZrO_2 show that the ZrO_2 causes significant differences compared to the Pd/SiO_2 system. Both electronic and steric effects were thought to be important.[128]

There is evidence for both terminal (ca. 2060cm^{-1} for νCO) and bridged (ca. 1860cm^{-1}) species from the adsorption of CO on Pt/ZnO.[129] I.r. spectra of CO adsorbed on Pt(111) at 85K and 300K shed light on coupling between adjacent carbon monoxide molecules.[130]

CO adsorbed on oxidised Cu(111) surfaces gives bands at 2100 and 2117cm^{-1}, on oxidised Cu(110) surfaces at 2140cm^{-1}.[131] The high wavenumber band seen when CO is adsorbed on Cu(110) is due to adsorption at step or defect sites.[132] CO adducts of a number of binuclear Cu(I) complexes have νCO values which correlate with the electronic properties of the other ligands present.[133]

The interaction of CO and NO with matrix-isolated silver clusters gives shifts of 40cm^{-1} (CO) or 45cm^{-1} (NO) to lower wavenumber compared to the free molecules. This is the first direct measurement of such shifts.[134] The Raman spectra of CO on a silver electrode exhibit variations in wavenumber and intensity of νCO on changing the potential at the electrode, i.e. there is changing Ag→CO electron transfer.[135]

3 Boron-containing Donors

νBH modes in $M(BH_4)_3$.DME, where M = Sc or Y, DME = 1,2-dimethoxyethane, show that the BH_4^- ligands are tridentate.[136] The νBH band in $CoH(BH_3CN)(PPh_3)_3$ is at 2340cm^{-1} - in the range previously suggested for unidentate N-coordination of the ligand (shifts to 1755cm^{-1} on deuteriation). νCN is only shifted by 11cm^{-1} compared to free BH_3CN^-, although the ligand is shown by X-ray diffraction to be strongly bonded. It is difficult to rationalise this observation.[137]

$ZnCl_2.B_2H_4.2PMe_3$ has νBH_t at 2346cm^{-1} and νB-H-Zn at 2010cm^{-1}.

The CuI analogue gives similar features at 2292 and 1925cm^{-1}, and that with CuCl at 2280 and 1970cm^{-1}.[138] $U(BH_4)_3(Ph_2Ppy)_2$ has i.r. bands showing tridentate BH_4^- coordination (2420, 2300, 2200 and 1120cm^{-1}).[139]

4 Nitrogen Donors

4.1 Molecular nitrogen, azido- and related complexes.- The transient i.r. spectrum of $Cr(CO)_5N_2$ has been obtained at room temperature in cyclohexane. νNN is seen at 2240cm^{-1} - very close to the low-temperature values in liquid xenon solution or CH_4/N_2 matrices. νCO bands were seen at 1976cm^{-1} (e) and 2086cm^{-1} (a_1).[140] $\nu_{as}N_3$ in $M(CO)_5N_3^-$, where M = Cr or W, is ca. 2060-2080cm^{-1}. In $[(\mu\text{-}N_3)\{Cr(CO)_5\}_2]^-$ and $[(\mu\text{-}N_3)\{M(CO)_3bipy\}_2]^-$, where M = Cr, Mo or W, two such bands are seen (2020-2085cm^{-1}), and in $[\mu\text{-}(N_3)_3\{M(CO)_3\}_2]^-$ there are 2-4 in the same region.[141]

The complex (44) has ν$N\equiv N$ at 1980cm^{-1}, characteristic of terminally bound dinitrogen.[142] Trans-$Mo(N_2)_2(PMePh_2)_2(PPh_2CH_2CH_2$-SMe) has two νN_2 bands in the i.r., at 2014 and 1942cm^{-1}.[143] In (45) νN_2 is at 1517cm^{-1}, with νCO at 2052 and 1990cm^{-1}.[144] νN=N is at 1570cm^{-1} in trans-$WI(NNSiMe_3)(PMe_2Ph)_4$, while in mer-$WI_2(NNHSiMe_3)(PMe_2Ph)_3$ νNN is at 1355cm^{-1}, νNH at 3450cm^{-1}.[145]

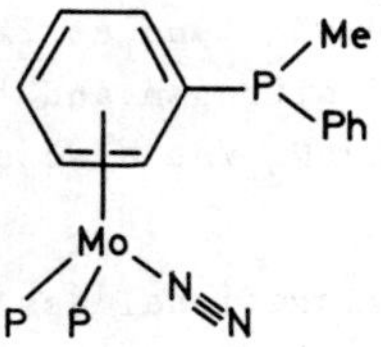

(44)

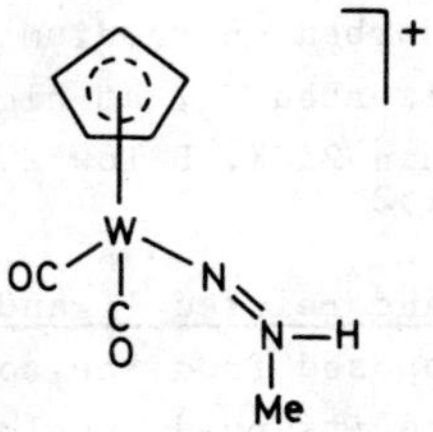

(45)

Co-condensed Fe atoms and N_2 in Kr or N_2 matrices form $Fe(N_2)_x$, where x = 1-5. νNN assignments are summarised in Table 4. It appears that the structures are analogous to those for $Fe(CO)_x$.[146] ν$N\equiv N$ in $Fe(EDTA)N_2^{2-}$ is at 2040cm^{-1}.[147]

The complexes (46) have νN=N at 1575cm^{-1} (E = P) or 1576cm^{-1} (As).[148] νNN is at 2139cm^{-1} in $OsCl_2(N_2)$ N(OH)$CHCO_2Et$ $(PPh_3)_2$.[149]

Table 4 νNN assignments in $Fe(N_2)_x$ ($/cm^{-1}$)

Species	Symmetry	νNN
$Fe(N_2)$	$C_{\infty v}$	2002
$Fe(N_2)_5$	D_{3h}	2010
$Fe(N_2)_2$	$D_{\infty h}$	2018
$Fe(N_2)_3$	C_{3v}	2036
$Fe(N_2)_4$	C_{2v}	2086, 2098, 2108
$Fe(N_2)_5$	D_{3h}	2108

νN=N in the η^1-species $HOs_3(CO)_{12}(N_2Ar)$ (47) is ca. $1500cm^{-1}$, while in the η^2-isomers (48) it is ca. $1480cm^{-1}$ (Ar = Ph, p-FC_6H_4, p-MeC_6H_4).[150]

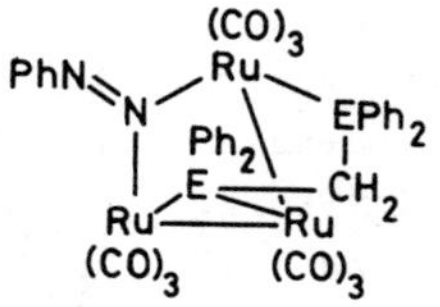

(46)

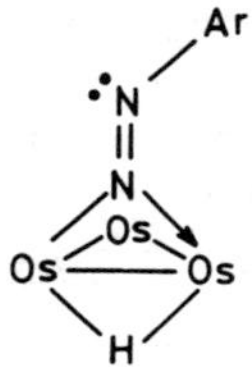

(47)

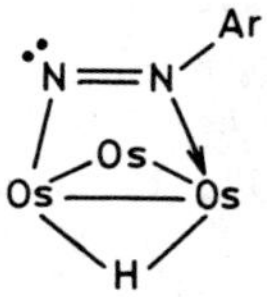

(48)

The i.r. band of the N_2 species adsorbed on a Co/Al_2O_3 catalyst is at $2214cm^{-1}$.[151]

$^{14}N_2$ adsorbed on rhodium gives a single νN_2 band at $2257cm^{-1}$, due to chemisorbed N_2, at high pressures of nitrogen and temperatures less than 240K. Below 200K, physisorbed N_2 was also detected ($2331cm^{-1}$).[152]

4.2 Amines and related ligands. A complete vibrational assignment has been proposed from the solid-state i.r. and Raman spectra of $Mo(CO)_5py$ and its py-d_5 analogue.[153]

I.r. and Raman spectra of $MC_2O_4.N_2H_4.xH_2O$, where M = Mn, Co, Ni or Zn, show that for M = Co, Ni or Zn only bridging hydrazine is present. For Mn(II) the hydrazine may be unsymmetrically coordinated.[154] Adducts of Mn(II), Fe(II), Co(II), Ni(II) and Cu(II) β-diketonates and hydrazine give i.r. bands characteristic of bidentate binding of N_2H_4.[155]

Assignments of ligand mode vibrations in the resonance Raman spectra of $M(bipy\text{-}d_8)_3^{2+}$, where M = Fe(II) or Ru(II), and of $Ru(bipy\text{-}d_8)_2Cl_2$ were made by correlation with the hydrogen-containing analogues.[156]

Resonance Raman spectra were also reported for specifically deuteriated 2,2'-bipyridine complexes of Fe(II), Ru(II) and Os(II), including the Fe(II) and Ru(II) perdeuteriated species, as well as $Ru(bipy\text{-}3,3'\text{-}\underline{d}_2)_3^{2+}$ and $Ru(bipy\text{-}6,6'\text{-}\underline{d}_2)_3^{2+}$. Again, the assignments were made by comparison with the all-hydrogen species.[157] Surface-enhanced Raman spectra were seen for $Ru(NH_3)_5^{3+}$ and $Os(NH_3)_5^{3+}$ systems containing pyrazine or bipy ligands, adsorbed at a silver-aqueous interface.[158] The resonance Raman spectra of $Ru(azpy)_2L_2^{n+}$ (where azpy = 2-(phenylazo)pyridine, L = Cl, Br, CN, NO_2, or N_3, n = 0; L_2 = en, n = 2) show that they have a relatively localised MLCT excited state, with large distortions of the azo N=N bond. νN=N was in the range 1337-1392cm^{-1}.[159]

The i.r. spectrum of $(4'\text{-MeLH})RuCl_3$, where 4'-MeLH = (49), contains two strong bands above 1600cm^{-1}, characteristic of a coordinated non-deprotonated isoindoline ligand.[160]

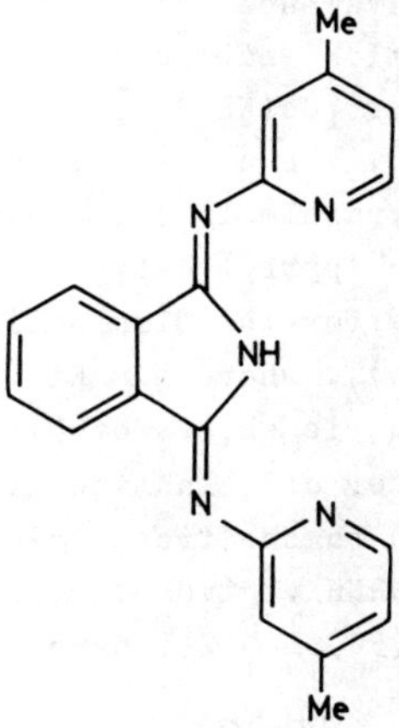

(49)

I.r. and resonance Raman spectra of $[(NH_3)_5Ru(pyz)Ru(NH_3)_5]^{n+}$, where n = 4, 5 or 6, pyz = pyrazine, gave some data apparently consistent with D_{2h} symmetry for the (II, III) species. However, some pyrazine modes showed i.r./Raman coincidences, hence a non-centrosymmetric structure (C_{2v}) is more likely.[161] I.r., Raman and resonance Raman spectra of $Ru(bipy)_3Cl_2$ ($\underline{h}_{24}$- and $\underline{d}_{24}$-species) and of $[Ru(bipyO_3)]_{0.25}Mn_{0.5}PS_3Na_{0.5}(H_2O)_2$ gave a complete assignment of $Ru(bipy)_3^{2+}$ modes.[162] S.E.R.S. for $Ru(bipy)_3^{2+}$ and <u>in situ</u>-reduced $Ru(bipy)_3^{+}$ at a silver electrode show that reduction leads to shifts , to lower wavenumbers, in ligand modes of up to $25cm^{-1}$. There was no obvious similarity between the spectra of $Ru(bipy)_3^+$ and the photoexcited state of $Ru(bipy)_3^{2+}$.[163] The resonance Raman spectra of $Ru(bipy)_3^n$, where n = +2, +1, 0 or -1, show that for n = +1 the extra electron is definitely localised on a bipyridyl anion. For n = 0 or -1, however, the electrons seem to be much more delocalised.[164]

Resonance Raman and time-resolved resonance Raman spectra were used to probe ground and metal-to-ligand (MLCT) excited states of $Os(bipy)_n(P_2)_{3-n}^{2+}$, where P_2 = <u>cis</u>-$Ph_2PCH=CHPPh_2$, n = 1-3. The ligand modes suggest that the CT electron density is localised in the lowest π* orbital of one bipy ligand only.[165] Electrochemical oxidation of $Os(bipy)_3^{2+}$ gives 3+ and 4+ species. The resonance Raman spectra of the Os(II) and Os(III) compounds are as expected, but that for the Os(IV) species is anomalously complex, due to considerable ligand inequivalence.[166]

H/D isotopic substitution enabled a complete assignment of the ligand modes to be made for $[Co(NH_3)_5X]X_2$, where X = Cl or Br.[167] I.r. spectra of $M(SO_4)Pz_2$, where M = Co, Ni, Cu or Zn, Pz = pyrazole, show that the pyrazole is coordinated <u>via</u> the 'pyridine'-type nitrogen, and not the 'pyrrole'-type nitrogen.[168]

I.r. and Raman spectra for the diquinolinemetal(II) tetracyanonickelates, $M(C_9H_7N)_2Ni(CN)_4$, where M = Ni or Cd, show that they have two-dimensional polymeric sheets of $Ni(CN)_4^{2-}$, bridged by $M(C_9H_7N)_2^{2+}$ cations. A number of quinoline modes are shifted to higher wavenumber compared to the free ligand.[169] $PdLCl_2$, where LH = carbohydrazide, $H_2NNHCONHNH_2$, gives an i.r. spectrum suggestive of the coordination mode (50).[170] 2(1-pyrrolyl)pyridine (Hplp) has

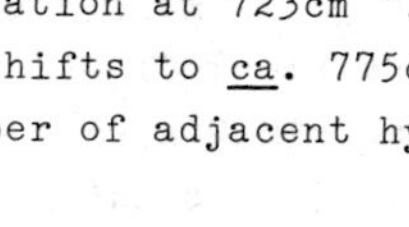

(50)

a CH out-of-plane deformation at $723cm^{-1}$. In $[PdCl(plp)]_2$, Pd(plp)acac etc., this shifts to <u>ca</u>. $775cm^{-1}$. This was related to the decrease in the number of adjacent hydrogen atoms on aromatic rings from 4 to 3.[171]

νNH_2 bands in $Pd(C_6F_5)_2(R-C_6H_4-NH_2)_2$, where R = <u>o</u>-MeO, <u>p</u>-MeO, <u>p</u>NH_2, <u>o</u>-HO, or <u>p</u>-HO, suggest that there is partially hindered rotation of the coordinated amine about the Pd-N bond. The chelate ligand in $Pd(C_6F_5)_2(o\text{-}NH_2C_6H_4NH_2)$ has νNH_2 at 3340 and $3295cm^{-1}$ only.[172] Some ligand mode assignments were proposed for cis- and trans-$PdCl_2(NH_2CH_2COOH)_2$ and for cis- and trans--$PdCl_2(NH_2CH_2CO\text{-}Cl)_2$.[173] Internal modes of NH_2OH and NO_2 ligands were assigned in trans-$Pt(NH_2OH)_2(NO_2)Cl$.[174] Ligand modes were also assigned for N-bonded 1-aminocamphene in trans-$PtCl_2L_2$, where L = 1-aminocamphene.[175]

I.r spectra of copper(II) complexes of N^3-phenylbenzamidrazone (BA) and N^3-phenylpicolinamidrazone (PA) show that BA coordinates via NH and NH_2 groups, PA via pyridine nitrogen and NH_2.[176]

Bands due to $\nu_{as}NC_3(1234cm^{-1})$, ρCH_3 $(987cm^{-1})$, and ν_sNC_3(811 cm^{-1}) of the coordinated NMe_3 were seen in $ThCl_4(NMe_3)_3$.[177]

4.3 Ligands containing >C=N< groups.- A good correlation is found between the wavenumber of the core-size marker mode, ν_{10}, of pyrrole-substituted metallo-porphyrins and the wavelength λ_α (the α-band maximum) for a wide variety of main-group and transition metals.[178]

Resonance Raman spectra were recorded for Mg, Zn, and Cu tetraphenylporphyrins (MTPP) and their electrochemical reduction products. Possible electronic, structural and geometrical changes onreduction were discussed.[179] Some ligand mode assignments were made from the Raman spectra of thin solid films of Mg, Cu and Zn phthalocyanine complexes. The most prominent bands were due to pyrrole-group stretches (e.g. 1340, 1506, amd $1141cm^{-1}$) and vibrations of the inner macrocycle (e.g. 682, 744, and $827cm^{-1}$ - all for MgPc).[180]

SERS were recorded for meso-tetraphenylporphyrin (TPP) and its Mg, Ag(II) and Zn complexes . It was possible to make a number of assignments for TPP modes for all these M(TPP) complexes.[181]

Unusual resonance Raman effects were found for excitation within the 'hyper'-type absorption of oxothiocyanatomolybdenum(V) tetraphenylporphyrin. Enhancement of a_{1g} and a_{2g} modes was found for excitation within the CT absorption in the near-u.v., but of b_{1g} and b_{2g} modes for excitation within the strongly red-shifted Soret band.[182]

In matrices at 10K, the primary step in the photochemical reactions of $Fe(CO)_3$(α-di-imine) complexes is from the σ,σ-N,N'-ligand (51) to the π,π-ligand (52). The latter has ν_sCN and νCC at 1365 and $1349cm^{-1}$ respectively.[183]

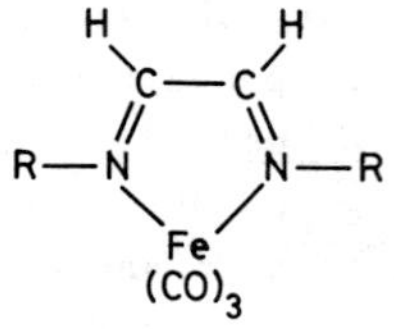

(51)

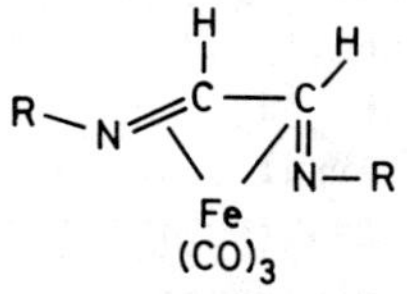

(52)

Resonance Raman spectra of $(FeTPP)_2C$ and $(FeTPP)_2N^+$ show that the core-size marker bands are at 1568, 1564cm^{-1} respectively. This shows that there is core expansion in the cation, due to increased non-bonded interactions.[184] Resonance Raman excitation profiles were determined for two totally symmetric modes (690, 1362cm^{-1}) in ferrocytochrome c.[185] Resonance Raman spectra of $(OEPFe)_2X$, where X = O or N, OEP = octaethylporphinato, gave no evidence for dimer vibrational splitting of OEP modes. The core-size markers (ν_3 and ν_{10}) are 1495, 1627cm^{-1} (X = O), 1513, 1647cm^{-1} (X = N), i.e. they contain high- and low-spin iron respectively. The oxidation state marker (ν_4) is at 1377cm^{-1} in each case, i.e. they both contain high-valent iron.[186]

Bis-pyridine adducts of the metallo-octaethylporphyrins $(OEP)Mpy_2$, where M = Fe(II), Ru(II) or Os(II), give resonance Raman spectra in which wavenumber variations in porphyrin modes with changing M reflect changes in both σ- and π-bonding between M and porphyrin, changes in porphyrin core size and formal metal charge. Resonance enhancement was also found for pyridine modes.[187] I.r. and resonance Raman spectra for various liganded haemoglobins show that there are no energetically significant structural changes at the haem upon the quaternary structure transition in six-coordinate haemoglobins.[188] Resonance Raman spectra of ferrous forms of horse myoglobin, soybean leghaemoglobin (isoenzyme a), monomeric haemoglobin from Glycera etc. showed that oxygen-carriers and peroxidases have distinct haem structures which produce significant changes in a number of vibrational bands.[189]

Resonance Raman scattering (1355cm^{-1} oxidation-marker band) was used to probe pH-induced symmetry distortions in deoxyhaemoglobin.[190] Raman scattering tensors were determined for the haem group in myoglobin from the depolarisation ratios and excitation profiles of bands at 1355, 1564cm^{-1} (deoxymyoglobin) or 1373, 1580cm^{-1} (cyanometmyoglobin).[191] The resonance Raman behaviour of some vibrational bands of oxyhaemoglobin (1375, 1583, 1683cm^{-1}) relates to distortions of the haem group from idealised D_{4h} symmetry.[192] A transient Raman study of carbon monoxide photolysis of cytochrome $\underline{a}_3$ reveals significant changes in both ligand and Fe-N bonds which are very useful in characterising transient haem species generated by ligand photolysis.[193] In the resonance Raman spectrum of the $\underline{aa}_3$-type cytochrome oxidase of thermophilic bacterium PS3, the 1667cm^{-1} band (peripheral C=O stretch) altered in intensity with temperature and pH in the same way as the oxidase activity.[194]

Resonance Raman spectra of 2-methylimidazole complexes of ferroprotoporphyrin IX, ferrohaematoporphyrin IX, ferrodeuterioporphyrin IX and ferroetioporphyrin I showed that it was possible to identify modes having major contributions from pyrrole peripheral group deformations.[195] Resonance Raman spectra could also be used to characterise the local haem environment of a high-spin ligand-binding haem protein from Chromatium vinosum.[196] The resonance Raman spectra of alkaline forms of horseradish peroxidase isoenzymes are indicative of six-coordinate low-spin haems. The ν_4 porphyrin wavenumbers of Fe(IV) derivatives were found not to be significantly higher than those of Fe(III) haems of the alkaline forms.[197]

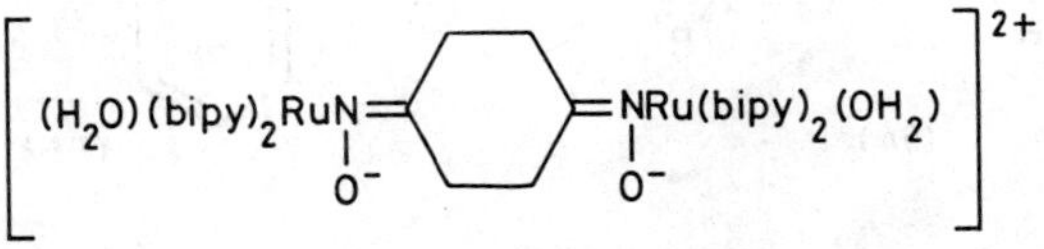

(53)

The complex (53) has νC=N of the oxime group at 1629cm^{-1}, compared to the free oxime value of 1679cm^{-1}. The assignment was confirmed by ^{15}N substitution (shift to 1611cm^{-1}), and its position was consistent with coordination of N to the Ru atom.[198] General ligand mode assignments were given for $(Me_2M)_2[M'(Me_2C_2O_2N_2)_2]$ (where M = Al, M' = Ni; M = Ga, M' = Ni, Pd, Pt or Cu; M = In, M' = Ni), i.e. bis(dimethylmetal(III)glyoximato)metallates(II).[199]

Resonance Raman spectra of Ni(DP), Ni(2-acetylDP), Ni(4-acetylDP), Ni(2,4-diacetylDP) and Ni(2,4-diformylDP), where DP =

deuterioporphyrin IX methyl ester. Nearly all of the in-plane porphyrin skeletal modes were assigned on the basis of deuteriation shifts. Several specific effects of substituent groups were detected. The effects of 2- and 4-substitution were virtually identical.[200] Resonance Raman data on CHO, CHNR, $CHNHR^+$ and pyrrolidine adducts of the nickel porphyrin chlorin demonstrate the isolation of the peripheral C=O and $C=N^-$ groups from the macrocyclic π-system, and selective enhancement of νC=N for those species with a split Soret band.[201] Resonance CARS spectra (1200-1650cm^{-1}) of Ni and Cu octaethylporphyrins contain bands corresponding to vibrations in both ground and excited electronic states.[202] Resonance Raman spectra of <u>meso</u>-formyl or <u>meso</u>-nitro derivatives of octaethylporphyrinatonickel(II) gave no evidence for direct conjugation between the <u>meso</u>-substituent and the π-electron system of the porphyrin ring.[203] The resonance Raman spectrum of the nickel tetrapyrrole, cofactor F_{430} of methanogenic bacteria, contained strong bands at 1530, 1562 and 1628cm^{-1}. The spectrum was consistent with a low degree of unsaturation in the complex.[204]

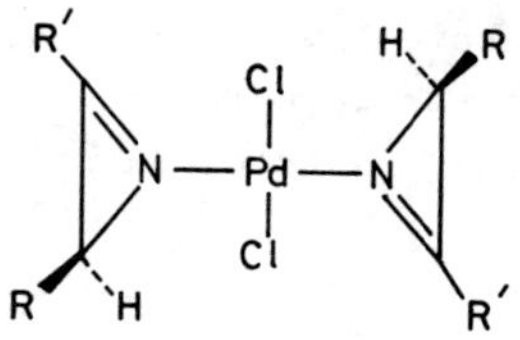

(54)

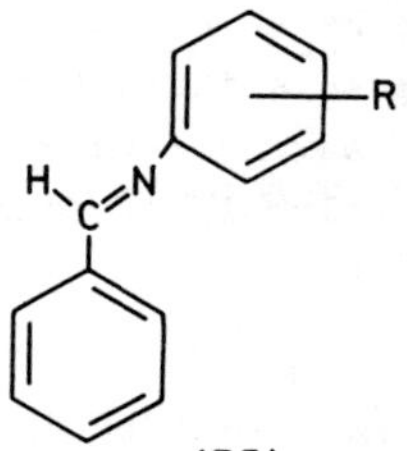

(55)

The complexes (54) (where R = Ph, R' = CH=CHPh; R = CH=CHPh, R' = Ph) give two νC=N bands of the azirine ligands, at lower wavenumbers than for the free ligand.[205]

The pH-dependent resonance Raman spectra of copper cytochrome <u>c</u> show that the species in neutral media have a significantly larger core size than those in either acidic or basic media.[206]

The i.r. spectra of SnX_4L_2, where L = (55, R = H, 2-Me, 3-Me or 4-Me), all have νC=N 20-40cm^{-1} higher than in the free ligand. Hence, coordination occurs <u>via</u> the azomethine nitrogen.[207]

4.4 Cyanides, Isocyanides and Related Ligands.- νCN wavenumbers were assigned in $CD_3CN...M^{n+}$ as follows: M = Al^{3+} (2330cm^{-1}), Mg^{2+} (2302cm^{-1}), Li^+ (2276cm^{-1}). There was also evidence for $CD_3CN...H$-O(H)...Al^{3+} (νCN at 2283cm^{-1}).[208] νCN in $TiM(CN)_6$, where M = Co, Rh or Ir, are ca. 30cm^{-1} higher than in $K_3M(CN)_6$ - due to Ti-NC-M bridging.[209] νCN and νCO modes were assigned in $E[NCCr(CO)_5]_6^{2-}$, where E = Si, Ge or Sn, with the help of ^{13}CN substitution of the silicon derivative. The νCN bands indicated octahedral coordination of E by NC groups. The νCO bands were as expected for local C_{4v} symmetry at the $Cr(CO)_5$ groups.[210]

The Raman spectra of the tricyanomethanides, $M[C(CN)_3]_2$, where M = Mn, Fe, Co, Ni or Zn, contained the following: ν_sCN 2250-2270cm^{-1}, ν_{as}CN 2150-2210cm^{-1}, ν_eCC 1230-1270cm^{-1}, ν_sCC 680-720 cm^{-1}.[211] Raman polarisation measurements showed that ν_sCN bands were at 2215-2290cm^{-1} and ν_{as}CN 2150-2210cm^{-1} in $M[N(CN)_2]_2$, where M = Mn, Fe, Co, Ni, Cu or Zn, and $Mpy_2[N(CN)_2]$, where M = Mn, Co, Ni, Cu or Zn.[212]

Bands due to νCN in $Re_2(\mu\text{-}H)_3H_2(PPh_3)_4(CN^tBu)_2^+$ were at 2090, 2040cm^{-1}, i.e. there are only terminal t-butylisocyanide ligands.[213] Raman and i.r. spectra were assigned for $Re(CNR)_6^+$, where R = Me, Ph, or 4-MeC_6H_4, in terms of local O_h symmetry for the $Re(CNC)_6$ skeleton. Internal modes of MeNC and PhNC were assigned under C_{3v}, C_{2v} symmetry respectively. νN≡C (e_g) was more intense in the Raman spectra than the a_{1g} mode - as often found for carbonyls.[214]

I.r. spectra (νCN) of $Fe_3(\mu\text{-}CNCF_3)(CO)_{11}$ suggest that the trifluoromethylisocyanate ligand is bridging both in the solid state and in solution.[215] I.r. data for $Cp_2Fe_2(CO)_3L$, where L = CO, CNPh or $CNCH_2Ph$, and $Cp_2Fe_2(CO)_3(CNHPh)^+$ show that μ-CN^+HPh is a very effective σ-electron withdrawing ligand, and a more powerful π-acceptor than either CO or μ-CNR.[216] The i.r. spectra of species formed from $Fe(CN)_6^{3-}$, $Ru(CN)_6^{3-}$ etc. at nickel electrode surfaces show that they are present as cyano-bridged bimetallic complexes.[217]

The complex (56) has νC≡N ca. 300cm^{-1} higher than in the free ligand. νSi-O is ca. 840cm^{-1}.[218]

Ligand mode assignments for bis(acetonitrile)dichloroplatinum (II) are consistent with *cis*-geometry. The νCN mode assignments were confirmed by ^{15}N substitution, e.g. ν_1CN (a_1) 2332.8cm^{-1} (^{14}N), 2312.5cm^{-1} (^{15}N); ν_{12}CN (b_1) 2304.7cm^{-1} (^{14}N), 2282.1cm^{-1}(^{15}N).[219]

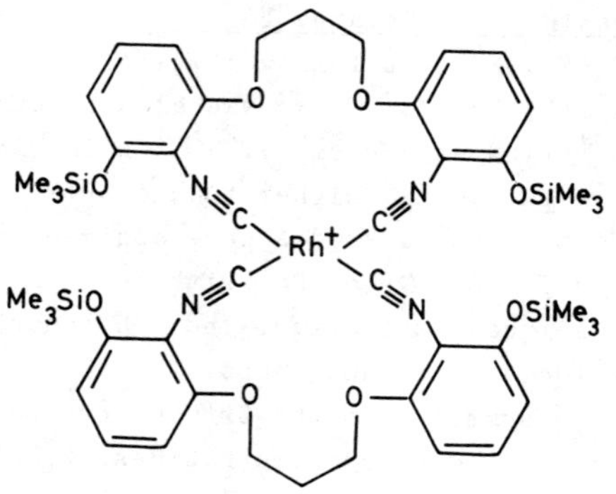

(56)

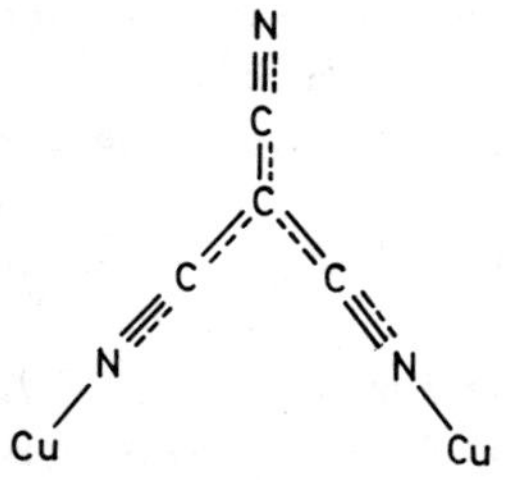

(57)

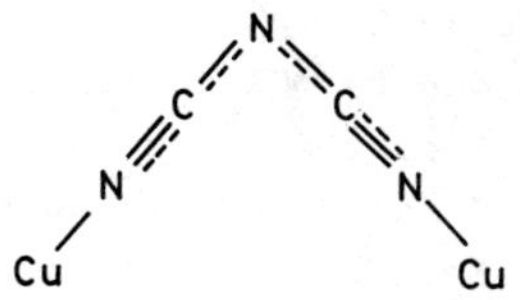

(58)

νCN values in $Cu[C(CN)_3]L$ and $Cu[N(CN)_2]L$, where L = 4-Me-, 2-Et-, 4-Et- or 3,4-Me_2-pyridine, are consistent with (57) and (58) respectively.[220] $(Ph_2MeP)_3Cu(NC)_2BH_2$ has νBH at 2410, 2360cm^{-1}, νC≡N at 2210cm^{-1} (bridging, i.e. Cu-NC-B) and 2192cm^{-1} (terminal B-CN).[221] Quite a detailed assignment was given for the ligand modes in $[Ag(CNMe)_4]^+ClO_4^-$. C_{3v} local symmetry can be used for the CH_3 modes, T_d symmetry for the skeletal features.[222]

Four different types of acetonitrile molecules were detected from Raman spectra of the $AlCl_3$-H_2O-MeCN system. Only one was thought to be directly bonded to the Al^{3+}, $CH_3CN...Al^{3+}$, with νCN at 2342cm^{-1} (2335cm^{-1} in CD_3CN).[223]

<u>4.5 Nitrosyl and Thionitrosyl Complexes</u>.- I.r. spectra of NO adsorbed on selectively photoreduced V/SiO_2 catalysts show that at low coverages the NO forms strongly bonded mononitrosyl species V^{3+}...NO. At higher coverages, oxidation to V^{4+} takes place, with only weak adsorption of NO on V^{4+}.[224] $V(NO)_2(CN)_4^{2-}$ has two νNO

bands - corresponding to <u>cis</u>-geometry (1650, 1520cm^{-1}). Formation of $V(NO)_2(CN)_2$(L-L), where L-L = bipy or phen, shifts these to 1680, 1540cm^{-1}.[225]

νNX modes were assigned in $CpCr(CO)_2NX$ and $HB(3,5-Me_2pz)_3M(CO)_2NX$, where M = Mo or W, X = O or S.[226] A broad i.r. band <u>ca</u>. 840cm^{-1} in $Mo(NO)_2(PF_6)_2$ suggests weak coordination of PF_6^-, with a possible structure (59).[227]

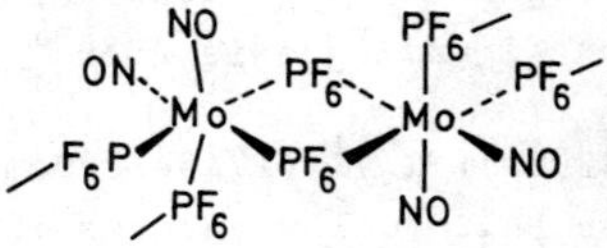

(59)

Several new octahedral $W(NO)_2Cl_2L_2$ complexes were reported, where L = variety of Lewis bases. Geometries were deduced from the νNO bands, e.g. when L_2 = dppe, νNO were at 1756, 1651cm^{-1}, i.e. the complex is <u>cis</u>.[228] νNS bands were at 1127, 1140cm^{-1} for the terminal NS in $HB(Me_2pz)_3W(CO)_2NS$, where $HB(Me_2pz)_3$ = hydridotris-(3,5-dimethylpyrazolyl)borato.[229]

The complexes $MnX_2(PR_3)(NO)$, where R_3 = $^nPr_3$, $^nBu_3$, $PhMe_2$ or $PhEt_2$, X = Cl, Br or I, have νNO in the range 1595-1607cm^{-1}.[230]

The νNO (and νCO) wavenumbers in $Fe(\eta^2-CH_2CHCH_2PR_3)(CO)_2(NO)$, where R = Me, Et, Bu or C_6H_{11}, are all <u>ca</u>. 100cm^{-1} lower than in σ-allyl analogues, showing an increased effective negative charge on the iron.[231] νNO in $Fe(L-L)(CN)_3(NO)^-$, where L-L = bipy or phen, is <u>ca</u>. 1635cm^{-1}, i.e. higher than in $Fe(CN)_5(NO)^{3-}$, due to the different σ/π donor/acceptor properties of L-L compared to CN^-.[232] Raman bands due to a new metastable state of $Na_2[Fe(CN)_5NO].2H_2O$, below 160K, show that νNO is shifted by 118cm^{-1}, νCN <u>ca</u>. 8cm^{-1}, compared to the normal form.[233] The fine structure of νNO bands in the Sr^{2+} and Ba^{2+} salts of $Fe(CN)_5NO^{2-}$, using $N^{16}O$/$N^{18}O$ shifts was analysed in terms of the vibrational dipole/dipole coupling between nitrosyl groups.[234] Fe(Por)(R)NO, where Por = tetraphenyl- or octaethylporphyrin, R = Me, Ph, <u>p</u>-MeC_6H_4 or <u>p</u>-$MeOC_6H_4$, all have νNO 1780-1805cm^{-1}, consistent with a <u>trans</u> configuration.[235]

$(\eta-C_5R_5)Ru(NO)X_2$, where R = H or Me, X = Cl, Br or I, have νNO between 1764 and 1818cm^{-1}, as expected for terminal nitrosyls. For both R = H and Me, the sequence is Br > Cl > I.[236] νNO in

$Na_2[Ru(CN)_5NO].2H_2O$ ($1926cm^{-1}$) is lower than in the Fe(II) analogue as Ru(II) is showing a higher π-donor ability than Fe(II).[237]

Phosphonium salts of $Os(NO)Cl_5^-$ have νNO at $1823cm^{-1}$, and of $Os(NS)Cl_5^-$ νNS at $1340cm^{-1}$.[238] The bridging NO group in $Os_{10}C(CO)_{24}(\mu_2\text{-}NO)^-$ has νNO at $1603cm^{-1}$. The terminal NO in $Os_{10}C(CO)_{23}NO^-$ has νNO at $1760cm^{-1}$.[239] The νNS wavenumber in $Os(NS)Cl_3(PPh_3)_2$ ($1310cm^{-1}$) is $106cm^{-1}$ higher than in free NS. This is consistent with NS being a better σ-donor and worse π-acceptor than NO (since νNO in the NO analogue is $40cm^{-1}$ lower than in free NO).[240]

Changes in νNO for $Ni(NO)(PR_3)_3^+$ ($1745\text{-}1790cm^{-1}$), where R_3 = Me_3, Et_3, Me_2Ph, $MePh_2$ or Ph_3, follow changes in the basicity of PR_3.[241]

5 Phosphorus Donors

νPH modes were assigned in $(C_6H_{11})_2NPH_2$ $M(CO)_5$, where M = Cr (2360, $2250cm^{-1}$) or W ($2320cm^{-1}$), and in $(C_6H_{11})_2NPh_2$ $Fe(CO)_4$ (2330, $2250cm^{-1}$).[242] C≡C bands in $M(CO)_3L$, where M = Cr or W, L = $P(C_2Ph)_nPh_{3-n}$, n = 1-3, cis-$M(CO)_4L_2$ and fac-$Cr(CO)_3L_3$ are hardly shifted from the free-ion positions.[243] The complexes (60, M = Cr, Mo or W) all have νP=O ca. $1200cm^{-1}$.[244]

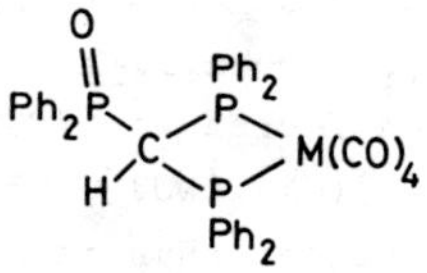

(60)

The coordinated PMe_3 group in $MoO(acac)_2(PMe_3)$ gives a characteristic i.r. band at $940cm^{-1}$, with acac modes at 1580 and 1500 cm^{-1}.[245]

Cis-$Re(dppe)Cl_4$, where dppe= 1,2-bis(diphenylphosphino)ethane, has characteristic i.r. bands at 880 and $895cm^{-1}$, due to the coordinated dppe ligand in the gauche conformation.[246] νP-H bands were assigned for $Ru(C_6Me_6)(PHCy_2)XY$, where XY = H_2, Cl_2, HCl, HPh or $D(C_6D_5)$ (all ca. $2230cm^{-1}$).[247] Some ligand mode assignments were proposed for $Os(PXPh)Cl(CO)_2(PPh_3)_2$, where X = H (νP-H 2264 cm^{-1}, δPH $905cm^{-1}$) or I (νP-I $394cm^{-1}$), and for OsCl[PX(OMe)Ph]-

$(CO)_2(PPh_3)_2^+$, where X = I (νP-I 378cm^{-1}) or X = H (νP-OMe 1020 cm^{-1}).[248] $\{RhL_2[P(OR)_2(OH)]_2\}ClO_4$, where L = PPh_3, PPh_2Me or $PPhMe_2$, all have νP-OH ca. 890cm^{-1}.[249] A large number of approximate ligand assignments were made for (61) and related species; e.g. νNH, νNH...O, 3310, 3000cm^{-1} respectively; $\nu_{as}NP_2$ ca. 790cm^{-1}; $\nu_s NP_2$ ca. 590cm^{-1}; νCO ca. 1780cm^{-1}.[250]

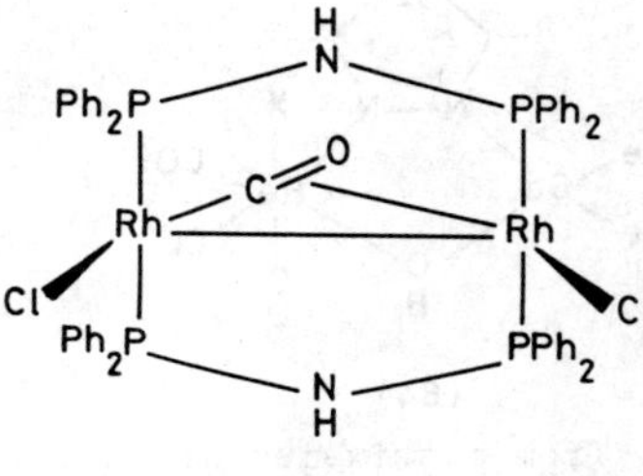

(61)

6 Oxygen Donors

6.1 Molecular oxygen, peroxo-, aquo- and related complexes.- The presence of bridging -OH-rather than -O- in (62) is shown by the presence of νOH at 3450cm^{-1}.[251] νOO in $VO(O_2)F(H_2O)_2^+$, as NH_4^+, K^+, Rb^+ or Cs^+ salts, is in the range 880-895cm^{-1}, associated with the unit (63).[252] νOO bands were assigned in $VO(O_2)_2Cl^{2-}$ (870, 885cm^{-1}) and $V(O_2)_3Cl^{2-}$ (855cm^{-1}), showing that units (63) were present here also, and that the O-O bond order decreases with an increasing number of peroxo-ligands coordinated to the metal centre.[253] νOO modes (856-900cm^{-1}) were also assigned in the new complexes $V_2O_2(O_2)_4F^{3-}$ and $HV_2O_2(O_2)_3F_4^{3-}$, containing V-(μ_2-O_2)-V units.[254]

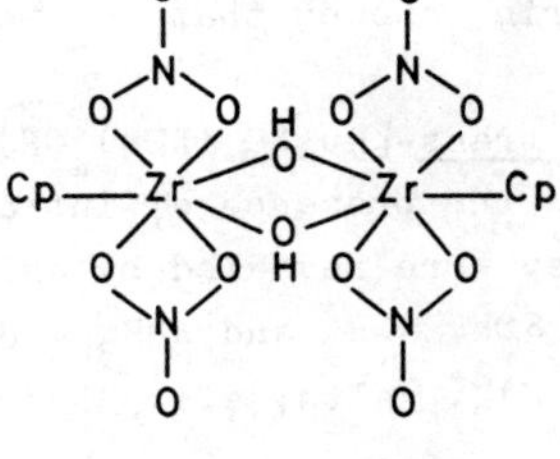

(62)

(63)

The reaction of O_2 with $MnBr_2(PR_3)$, where R_3 = Me_3, Et_3 or $PhMe_2$, in films on i.r. windows leads to coordination of the O_2. The extent of reversibility of this reaction was markedly dependent on the nature of PR_3.[255] νOH for the bridging hydroxyl group in (64, R = H or Me, X = CO or PPh_3) lie in the range 3570-3600 cm^{-1}.[256]

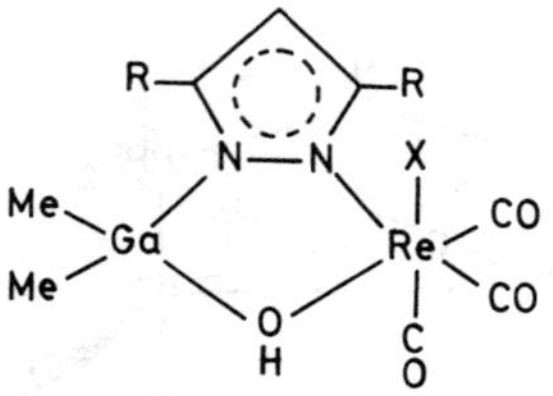

(64)

Assignments for νO_2 from matrix-isolation i.r. spectra of O_2 adducts of iron(II) porphyrins and related compounds are summarised in Table 5, where TPP = tetraphenylporphyrinato, OEP = octaethylporphyrinato, Pc = phthalocyaninato, salen = N,N-bis(salicylideneiminato). The first two are present as two isomers, and it was

Table 5 νO_2 for O_2 adducts of iron(II) porphyrins and related compounds ($/cm^{-1}$)

Species	O_2
$Fe(TPP)O_2$	1196, 1105
$Fe(OEP)O_2$	1190, 1104
$Fe(Pc)O_2$	1207
$Fe(salen)O_2$	1106

suggested that one isomer had an open, bent Fe-O-O unit (angle less than 130°, ca. 1190cm^{-1}) and the other a cyclic FeO_2 unit (ca. 1100cm^{-1}).[257] Oxyhaemerythrin, an Fe-O-Fe system, gave νOO of coordinated O_2 at 844cm^{-1}.[258] $^{18}O_2$ substitution in Fe(TpivPP)-$(C_6F_4HS^-)(O_2)$, where TpivPP = dianion of meso-α,α,α,α-tetrakis-(o-(pivalamido)phenyl)porphyrin, showed that νO_2 was at 1140 cm^{-1}.[259]

I.r and Raman spectra of trans-$[Ru(NO)(NH_3)_4OH]X$, where X = Cl, Br, I, NCS, NO_3 or ClO_4, show the presence of intra- and intermolecular hydrogen bonds. They were revealed by shifts in the wavenumbers of the νOH, νOD, δOH, δNH_3 and ρNH_3 modes.[260] νOH modes were assigned in $OsO_4(OH)_2^{2-}$ salts, e.g. 3533cm^{-1} for the Ca^{2+} salt.[261]

Resonance Raman spectra of O_2 adducts of six Co(salen) derivatives gave νO-O 1143-1146cm^{-1}, insensitive to substitution in the ligand.[262]

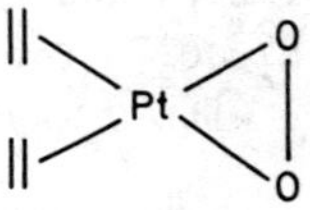

(65)

νO-O in $Pt(C_2H_4)_2O_2$ was at 874cm^{-1}, with νC=C at 1481 and 1498 cm^{-1}, consistent with (65).[263] The first unsymmetrical η^2-O_2 complex of platinum, $(P^tBu_3)(PPh_3)Pt(O_2)$, has νO-O at 838cm^{-1}.[264]

The i.r. band at 1015cm^{-1} in Cu[HB{3,5-Me_2(pz)}$_3$]($^{18}O_2$) was assigned as νO-O (the analogous band in the $^{16}O_2$ complex was obscured by a ligand mode), in the region for an unsymmetrically bound superoxide ion (HB[3,5-Me_2(pz)]$_3$ = hydridotris(3,5-dimethyl-1-pyrazolyl)borate).[265] $Cu(OH)_4^{2-}$ and $Cu(OH)_6^{4-}$ both have νOH 3660-3545cm^{-1} due to almost free OH groups, and 2600-3500cm^{-1} (very broad) due to strongly hydrogen-bonded hydroxyl groups.[266] Bands due to νOH(bridging) were assigned in $Cu_2(PAP4Me)(OH)(H_2O)_2(NO_3)_2^+$ (3580cm^{-1}) and $Cu_2(PAP)(OH)Cl(SO_4)$ (3620cm^{-1}), where PAP4Me = 1,4-bis((4-methyl-2-pyridyl)amino)phthalazine, PAP = 1,4-bis-(2-pyridylamino)phthalazine.[267] A series of hydroxo-bridged

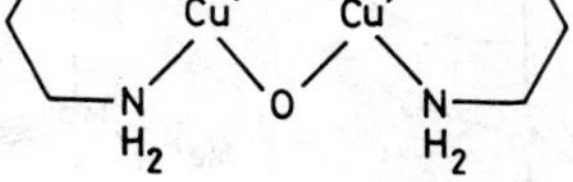

(66)

copper(II) complexes of the type (66) all have νOH as a broad band at ca. 3540cm^{-1} in the i.r.[268]

νOH was assigned (ca. 3630cm^{-1}) in $M(OH)_6^{3-}$, where M = Yb or Lu.[269]

6.2 Carboxylato- and related complexes.- Reactions of Na, K and Cs with CO_2 have been followed in Ar, N_2 and CO_2 matrices. The i.r. spectra show that $M^+CO_2^-$ have rhombic ring structures (C_{2v} symmetry) where M^+ interacts equally with the two O atoms. For M = Cs, bands of a_1 symmetry were seen at 1345.5 and 1608.3cm^{-1} in an N_2 matrix. There was some evidence that $M_2^{2+}CO_2^{2-}$ (for M = K or Cs) can exist as two isomers, one with C_{2v}, one with C_s symmetry.[270]

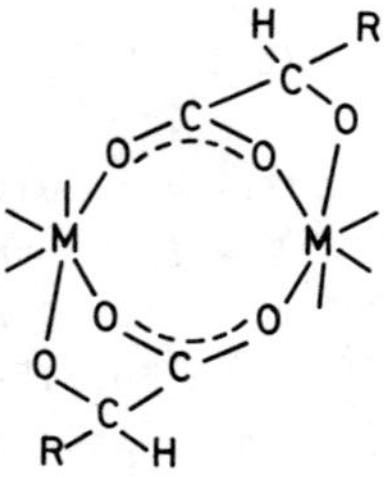

(67)

νCO_2 modes in the mandelate complexes $Ti_2(PhCHOCOO)_2Cl_2$ and $Cu_2(PhCHOCOO)_2$ are consistent with the presence of bridging mandelate ligands, probably forming the unit (67), where M = Ti or Cu.[271] The i.r. spectrum of $Cp_2Zr(bzac)^+$ shows that the benzoyl-acetonate (= bzac) ligand is bidentate.[272] I.r. spectra of $M_3O(O_2CCF_3)_7$, where M = Cr, Mn or Fe, show that the trifluoroacetate groups are also bidentate.[273] Complexes of malonic, ethyl- and benzylmalonic acids with Cr(III), Fe(III), Co(III), Co(II), Ni(II), Cu(II), Zn(II), Cd(II) and Al(III) have been prepared. νCO_2 bands were correlated with the ionic radii and structure. Ethylmalonates have stronger M-O bonding than the other ligands.[274]

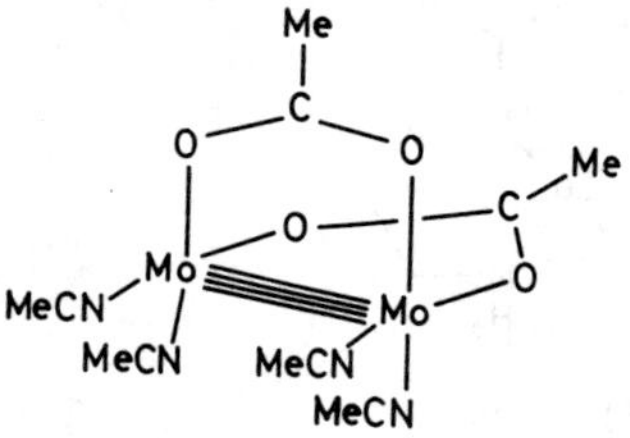

(68)

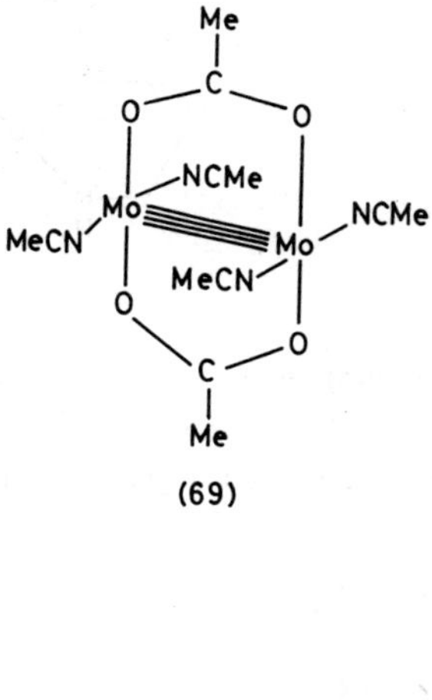

(69)

The presence of δCO_2 at 685cm^{-1} is indicative of bridging acetate in $Mo_2(MeCN)_4(OAc)_2$, and (68) rather than (69). The νCN bands were as expected for end-on nitrile coordination, with little π-back-bond stabilisation.[275]

I.r. and Raman spectra of $M(C_2O_4).2N_2H_4$, where M = Mn, Fe, Co, Ni or Zn, show that the oxalato group is bidentate, except for M = Mn, where it is tetradentate.[276]

The values of $(\nu_{as}-\nu_s)$ CO_2 for $Fe_2Cl_2(O_2CCF_3)_4$, $Fe_2Cl(O_2CCF_3)_5$ and $Fe(O_2CCF_3)_3$ suggest the presence of bridging,bidentate trifluoroacetate groups.[277] The bridging acetato groups in $Ru_2(\mu\text{-}O_2CMe)_4$-$(THF)_2$ gave $\nu_{as}CO_2$ at 1560cm^{-1}, ν_sCO_2 at 1440cm^{-1}.[278]

Bands due to the acetato ligands in $Rh_2(OAc)_2(SO_4)_2L_2^{2-}$, where L = H_2O or py, show that both are bridging.[279] Dirhodium tetrasalicylate, $Rh_2(OOCC_6H_4OH)_4$, gives νOH at 3255cm^{-1}, corresponding well with the value for the hydrogen-bonded OH in salicylic acid.[280] Solution-phase Raman data for $M_2(O_2CMe)_4$ gave the following νC-C (a_{1g}) assignments: M = Rh, 981cm^{-1} (in MeCN), 953cm^{-1} (in H_2O); M = Cu, 945cm^{-1} (in MeCN).[281]

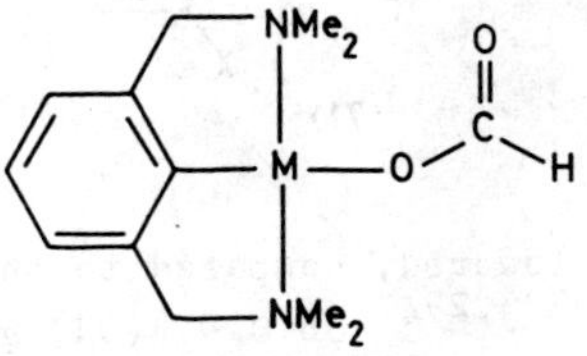

(70)

The complexes (70), where M = Ni or Pt, both gave νCO_2 bands characteristic of unidentate formato groups, with ν_{as} ca. 1610, ca. 1595cm^{-1}, ν_s ca. 1320cm^{-1}.[282] $\{(C_7H_{11}O_4)_2Pd\}_2(C_2O_4)^{2-}$, where $C_7H_{11}O_4$ = diethylmalonato, has νCO (b_{2u}) of the quadridentate bridging oxalato group at 1618cm^{-1}. The molecular symmetry appears to approximate to D_{2h}.[283]

$\nu_{as}CO_2$ of formate adsorbed on Cu(110) is seen at 1635cm^{-1}, with ν_s 1348-1358cm^{-1}.[284] ν_s, $\nu_{as}CO_2$, ν_s, $\nu_{as}SO_2$ and νSN were assigned for Cu(Tsgly)(bipy)(EtOH), Cu(Tsgly)(bipy)(H_2O) and Cu(Tsgly)(bipy), where Tsgly = N-tosylglycinate dianion.[285] νCO_2 modes were assigned for complexes of copper with L-isoleucine and L-tryptophan, to obtain evidence on the relative covalence of the different Cu-O bonds.[286]

Ligand mode assignments revealed bidentate formate groups in $Y_2(O_2CH)_3.2H_2O$ and $Y(O_2CD)_3.2H_2O$.[287] ν_s and $\nu_{as}CO_2$ show bidentate coordination of the carboxylate in MR_3, where M = La, Ce, Pr, Nd or Sm, R = 3,4-dimethoxybenzoate.[288] A general assignment of ligand modes was given for $Ln(BA)_2(acid).H_2O$, where Ln = Nd, Eu, Gd or Yb, BA = benzoylacetonato, acid = acrylate, methacrylate, fumarate or maleate. These showed bidentate coordination of the carboxylate residues, with only one carboxyl group of fumarate and maleate coordinated.[289] $UO_2(OAc)_2$ and $UO_2(OAc)_3^-$ both give νCO_2 bands from both uni- and bidentate acetato groups.[290]

$M(C_2O_4)_2F_2^-$, where M = Al or Ga, give oxalate bands showing that the ligands are chelating.[291] The i.r. spectra of mixed dialkyltin(IV) trifluoroacetates, $RR'Sn(O_2CCF_3)_2$ (where R = R' = Et; R = Et, R' = ^{n}Pr; R = ^{n}Pr, R' = ^{n}Bu), all show the presence of bidentate trifluroacetates.[292] The complexes (71), where X = NR_2 or OR, R =

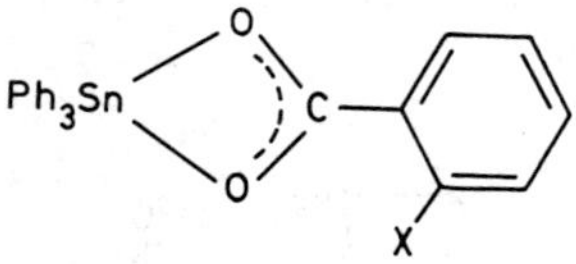

(71)

H or Me, all have $\nu C{=}O$ lowered, compared to the free acids, showing C=O group coordination.[293,294] $Sb(C_2O_4)(OH)$ gave characteristic i.r. bands at 1720, 1680, 1300, 805, 560 and 400cm^{-1}.[295]

<u>6.3 Keto-, Alkoxy, Ether and related complexes</u>.- The i.r. spectra of DMF complexes $LiCl.DMF.0.5H_2O$, $NaClO_4.2DMF$, $CaCl_2.2DMF.2H_2O$ and $Mg(ClO_4)_2.6DMF$ all show $\nu C{=}O$ at lower (by 10-17cm^{-1}) wavenumbers than in the free amide.[296] There is i.r. evidence that D-glucoranate binds to Ca^{2+} *via* 3 sites: carbonyl oxygens combined with O(5) and O(4), and one involving the O(1)O(2) pair of OH groups.[297]

The two $\nu C{-}O$ bands (1515-1582cm^{-1}) on $RTiCl_2(acac)$, where R Cp, MeCp or Ph_2CHCp, show *O,O*-coordination of acac.[298] $MO(NCS)_2$-(DMF), where M = Zr or Hf, give $\nu C{=}O$ of DMF at *ca*. 1650cm^{-1}, i.e. there is M-O=C coordination.[299]

I.r. (especially $\nu C{=}O$) and n.m.r. data on $Hf[\eta\text{-}C_5H_4(SiMe_3)]_2$-$[o\text{-}(CO)_2(CH_2)_2C_6H_4]$ shows that it exists as two conformers of

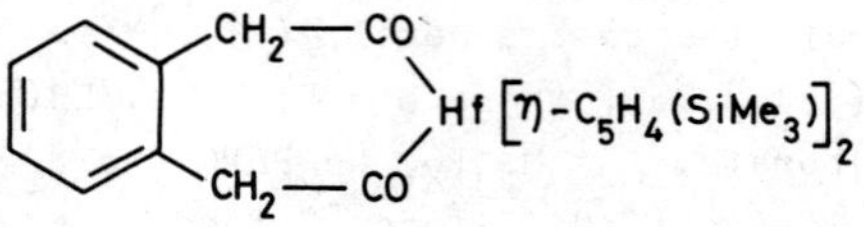

(72)

(72).[300] $Ru_2(\eta\text{-}C_6H_6)_2(^iPrO)_3^+$ has νC-O at $1050cm^{-1}$.[301]

The resonance Raman spectrum of copper pheophytin shows that νC_9=O from chlorophyll a is absent here. The strongest band in the present case is at $1497cm^{-1}$, which correlates with the chlorophyll a band at $1495cm^{-1}$.[302] νC-O assignments were proposed for tetraalkoxy derivatives of Cu(II), $Cu[Au(OR)_4]_2$, where R = Me, Et, ^iPr or ^nBu.[303]

I.r. spectra of $M(PhCONHO)_3$, where M = La, Ce, Pr, Nd, Sm, Eu, Gd, Tb, Dy, Ho, Er, Tm, Yb, Lu or Y, show that the benzohydroxamate coordinates *via* two oxygen atoms.[304] Some vibrations of internal ligand modes were assigned from Raman spectra of crystalline β-diketonates of Al(III), Ga(III), and In(III), $M(DK)_3$, where DK = acetylacetonate, dipivaloylmethanate, trifluoroacetylacetonate or hexafluoroacetylacetonate.[305]

Ligand mode assignments for $R_3Sn(dbzm)$, where R = Ph, Me or C_6H_{11}, dbzm = 1,3-diphenyl-1,3-propanedionato, and $R_3Sn(trop)$, where R = Ph or C_6H_{11}, trop = tropolonato, all show that the bidentate ligand is chelating, to give monomeric five-coordinate species.[306]

6.4 Ligands containing O-N, O-P or O-As bonds.- νP=O is at $1150cm^{-1}$ ($140cm^{-1}$ lower than in free $POCl_3$) in $ReCl_5(OPCl_3)$, i.e. coordination occurs *via* oxygen.[307] Internal mode assignments were proposed for $H_2AsO_4^-$, $HAsO_4^{2-}$ and AsO_4^{3-} ligands in $Ni(H_2AsO_4)_2$, $Ni(HAsO_4)$ and $Ni_3(AsO_4)_2$.[308]

νP=O in CuL_2^+, where L = $Ph_2P(=O)CH_2P(=O)Ph_2$, is at $1155cm^{-1}$,

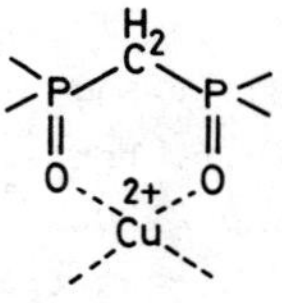

(73)

compared to 1203, $1190cm^{-1}$ in the free ligand. Hence both P=O groups are coordinated to the metal, (73).[309]

$M(NCS)_2[HB(pz)_3]_2.nL$, where M = U or Th, $HB(pz)_3$ = hydridotris(pyrazolyl)borate, L = Me_3PO, Me_2PhPO, $MePh_2PO$ or Ph_3PO, n = 1, 1.5 or 2, all have νP=O 45-155cm^{-1} lower than in the free phosphine oxide ligands, i.e. O-bonded.[310] I.r spectra show the presence of unidentate, covalent nitrato groups in $Ph_2Te(NO_3)_2L$, where L = py-N-O, Ph_3PO, DMSO, py, etc.[311]

6.5 Ligands containing O-S or O-Te bonds.- $Ti(L)SO_4$, where L = octaethyl- or tetraphenylporphinato, have νSO_4 bands indicative of a bidentate, chelating sulphato group.[312] νSO_4 bands in $CsVO_2(SO_4)$-$2H_2O$, on the other hand, suggest unidentate coordination of the sulphato group.[313]

$Mn(CO)_5(OTeF_5)$ has νTe-O at 841cm^{-1}. The νCO data show that $OTeF_5$ is a poorer σ-donor than Cl^-.[314] FeL_2SO_4, where L = cyclopentanone- or cyclohexanone-semicarbazone, have i.r. spectra showing unidentate SO_4 coordinated to Fe.[315] There is Raman evidence for the formation of contact ion-pairs, $Zn^{2+}SO_4^{2-}$, in concentrated aqueous $ZnSO_4$ solutions.[316] νSO_4 modes were assigned for the double sulphates $Cs_2SO_4.Ce(SO_4)_2$ and $Cs_2SO_4.Ce(SO_4)_2.H_2O$. The spectra were discussed in terms of factor-group splitting.[317]

6.6 Ligands containing O-Cl bonds.- I.r. bands due to ClO_4 were assigned in $Hf(ClO_4)_n^{(4-n)+}$, where n = 5,6 or 7.[318] mer-$[Fe(OClO_3)$-$(diphos)(CNR)_3]^+$ ClO_4^-, where R = Ph or 4-MeC_6H_4, diphos = dppm or dppe, all give νCl-O bands due to both C_{3v} and T_d perchlorato groups.[319] Ligand mode assignments in the i.r. and Raman spectra of $Cu(ClO_4)_2$ show that more than one type of perchlorato ligand is present.[320] $NO_2^+[UO_2(ClO_4)_3]^-$ shows perchlorato bands typical of bidentate, C_{2v}, coordination.[321]

7 Sulphur and Selenium Donors

νSO bands were assigned in the S,S-bonded complexes $Cp_2Nb(S_2O)Cl$ (1020cm^{-1}), $Ir(S_2O)(dppe)_2^+$ (1049cm^{-1}) and $Ir(S_2O_2)(dppe)_2^+$ (1045, 1019cm^{-1}).[322] $(C_5Me_5)_2Mo_2(\mu\text{-}S)_2(\mu\text{-}SH)_2$ has νSH at 2419cm^{-1}.[323] $Mo_2(\mu_4\text{-}S_2)(\mu_2\text{-}SO_2)(CN)_8^{4-}$ has i.r. bands at 520cm^{-1} (νSS), 1167, 1140cm^{-1} (νSO_2), 2135, 2127 and 2120cm^{-1} (νCN).[324] Values of νCN (1500cm^{-1}) and νCS (960cm^{-1}) suggest that there are bidentate dithiocarbamate ligands in $MoO(NCS)(R_2NCS_2)_2$, where R = Et or $PhCH_2$, R_2 = piperidinyl.[325] The complexes (74) (R = H, Et, ^nBu or

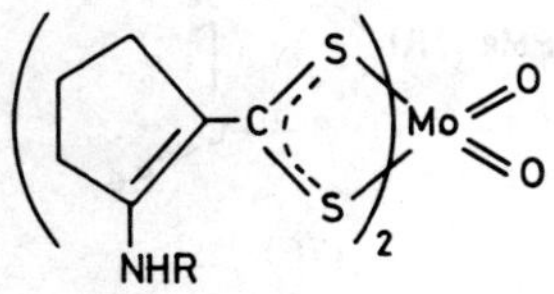

(74)

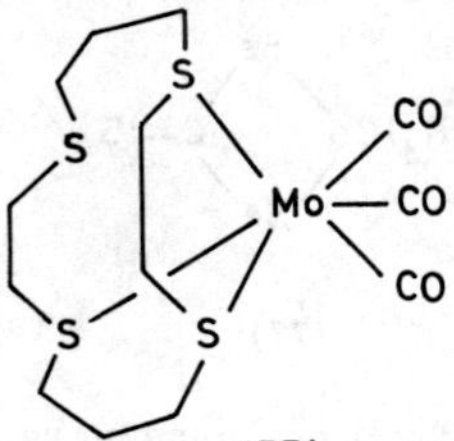

(75)

C_6H_{11}) have $\nu_{as}CS_2$ as a single band (805-820cm^{-1}). This is consistent with symmetric bidentate chelation.[326] (75) has νCO as expected for a fac-isomer. The νCSC bands of the free ligands have disappeared, due to Mo-S coordination.[327]

$Fe_2X_2(X_5)_2^{2-}$, where X = S or Se, have νX-X at 474cm^{-1} (S), 264 or 247cm^{-1} (Se).[328] I.r. bands due to the dithiocarboxylate group are seen at 1091, 917 and 751cm^{-1} in (76).[329]

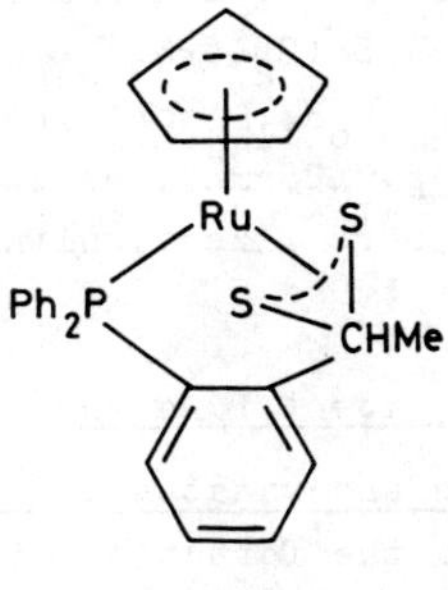

(76)

I.r. spectra of ML_n (where L = $Me(EtO)PS_2^-$, $(EtO)_2PS_2^-$ or $Et_2PS_2^-$; M = Co, Rh or Bi, n = 3; M = Ni, Pd, Zn, Cd or Pb, n = 2; M = Cu or Ag, n = 1) all show a shift in νPS to lower wavenumber.[330] (77), where R_3 = Me_3 or Me_2Ph, have νC=S ca. 1010-1020cm^{-1}; (78) has νC=S at 1016cm^{-1}, νC=Se at 879cm^{-1}.[331] $M(P^iPr_3)_3(SO)Cl$, where M = Rh or Ir, have νSO of coordinated sulphur monoxide at ca. 1075cm^{-1}.[332]

Normal coordinate analyses for $M(Et_2dtc)_2$, where M = Ni, Cu, Zn or Cd, reveal extensive mixing of ligand modes.[333] νPS modes were assigned (605-645cm^{-1}) in $(PPh_3)Pt[Ph_2P(S)C(S)R]$, where R =

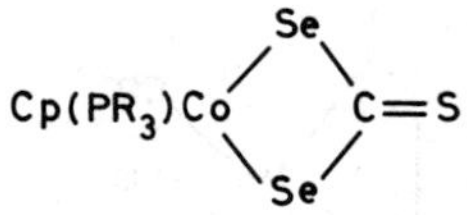

(77)

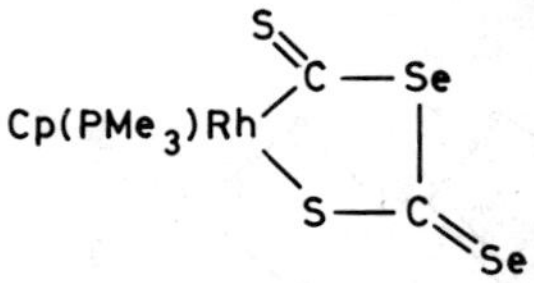

(78)

SMe, CH_2Ph etc. These were almost unchanged from the free ligand values, hence there is no coordination involving the PS group. It is presumably the CS unit which is responsible for the coordination, since νCS has disappeared (1050cm^{-1} in the free ligand).[334]

νS-S bands have been assigned in the novel complexes $(S_6)Cu(S_8)Cu(S_6)^{4-}$ (468, 455, 437, 425cm^{-1}) and $Cu_3(S_4)_3^{3-}$ (484, 456, 406cm^{-1}).[335] $(S_6)Ag(S_8)Ag(S_6)^{4-}$ has νS-S bands at 490, 454, 442, 411 and 394cm^{-1}.[336] $Hg(S_6)_2^{2-}$, with seven-membered chelate rings, has a symmetric ring-breathing mode at 460cm^{-1} in the Raman spectrum, with the symmetric ring deformation at 196cm^{-1}.[337] The diphenyl dithioarsinates, $R_nSn(S_2AsPh_2)_{4-n}$ (where n = 2, R = Me, Bu or Ph; n = 3, R = Me, Ph or C_6H_{11}) contain unidentate ligands, with νAs-S 410-420cm^{-1}, νAs=S ca. 485cm^{-1}, except for R = Ph, where bidentate coordination is more likely, with $\nu_s AsS_2$ ca. 420cm^{-1}, $\nu_{as} AsS_2$ ca. 460cm^{-1}.[338]

8 Potentially Ambident Ligands

8.1 Cyanates, Thio- and Selenocyanates and their Iso-analogues.-

There is i.r. evidence for the formation of CpW(CO)(NCO) and CpW(NCO) on photolysis of $CpW(CO)_2NO$ in inert matrices, by intramolecular attack of CO on NO. Use of $^{12}C/^{13}CO$ and $^{16}O/^{18}O$ isotopic shifts confirmed the proposed assignments, e.g $\nu_{as}NCO$ at 2265.7cm^{-1} in CpW(CO)(NCO), 2233.7cm^{-1} in CpW(NCO).[339]

Values of νCN, νCSe and δNCSe were all consistent with Mn-N bonding in $Mn(NCS)_2.nL$ (where L = PPh_3O, n = 4; L = $SC(NH_2)_2$, n = 2; L = THF, n = 3).[340] $M_3(LH)_6(NCS)_6(H_2O)_2$, where M = Mn, Co or Ni, LH = 3,5-dimethyl- or 3,5-diethyl-1,2,4-triazole, all gave an i.r. band ca. 2000cm^{-1}, assignable to a bridging unit (79). Other νCN

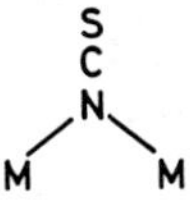

(79)

modes (2070-2110cm^{-1}) were due to terminal M-NCS groups.[341]

Terminal NCO or NCS bands were seen in $(C_5Me_5)Co(CO)(X)PX_2$, where X = NCO or NCS, but it was not possible to differentiate between the two types of group.[342] The i.r. spectrum of trans-$[Co(NCS)(dmgH)_2L].H_2O$, where $dmgH_2$ = dimethylglyoxime, L = tetramethylguanidine, shows that a mixture of Co-N and Co-S bonded isomers is present.[343] I.r. studies of the NO reduction by CO over Rh/SiO_2 gave a clear indication that a Rh-NCO species is produced (a band ca. 2170-2200cm^{-1}). There was also some evidence for Si-NCO.[344] νCN(N), νCN(S), νCS(N), νCS(S), δNCS and δSCN were assigned for the isomers $Rh(NCS)_n(SCN)_{6-n}^{3-}$, where n = 0-3, and also for $Rh_2(SCN)_{10}^{4-}$. The latter contains two Rh-SCN-Rh bridges and eight terminal Rh-SCN groups.[345]

$Cu(NCO)_2(NH_3)_2$ has ν_{as}NCO at 2225cm^{-1}, ν_sNCO at 1234cm^{-1}, and δNCO at 628 and 605cm^{-1}. The coordinated NH_3 gives bands at 3350, 3268cm^{-1} (νNH), 1616cm^{-1} ($\delta_d NH_3$), 1231cm^{-1} ($\delta_s NH_3$), 719 and 650cm^{-1} ($\rho_r NH_3$).[346] νCN and νCS were seen for the terminal and -NCS-bridging NCS groups in $[Cu(NCS)_2(dmtp)_2]_2$, where dmtp = 5,7-dimethyl[1,2,4]triazolo[1,5-a]pyrimidine.[347] Surface-enhanced Raman scattering was observed for the νCN band of SCN^- in acetonitrile at a silver electrode. Ag-SCN coordination is most likely.[348]

νCN_{br}, νCN_t and δSCN were assigned in $(L\text{-}L)_nZn(NCS)_2Hg(SCN)_2$, where L-L = phen, bipy, dimethylphenanthroline, n = 1 or 2.[349] νCN modes in 1- and 2-$(NCS)B_{10}H_9^{2-}$ and 1,10-$(NCS)_2B_{10}H_8^{2-}$ show B-S coordination of the NCS groups.[350] The i.r. spectrum of $ISn(NCS)_3$ shows that it has a polymeric structure, with thiocyanate bridges (νCN 2058, 1985cm^{-1}, νCS 794, 760cm^{-1}, δNCS 470cm^{-1}).[351] ν_s and ν_{as} NCO were assigned in F_5E-NCO, where E = S, Se or Te. For S and Te the data were consistent with isocyanate formation. For Se the data were ambiguous, but the isocyanate form was shown by electron diffraction.[351]

8.2 Ligands containing N and O or P and O atoms.- The i.r. spectra of $VOSO_4.2L.nH_2O$, where L = MeC(Me)=NNHC(O)R, R = Me, Et, Pr, Bu etc., show that L are bidentate, via O and azomethine N.[352] The i.r. spectra of $Mo_2O_4(OH)_3L$, where LH = anthranilic or nicotinic acid, and $Mo_2O_5(OH)L'$, where L'H = α- or β-alanine, valine or isonicotinic acid, contain νCO_2 bands for bidentate coordination of the carboxylate group and bands showing bridging N, i.e. the bonding is as shown in (80).[353]

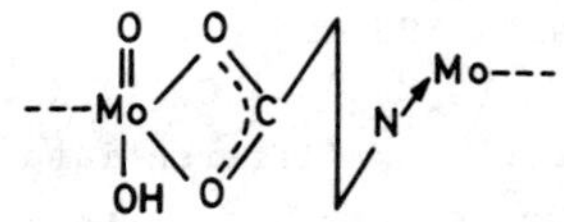

(80)

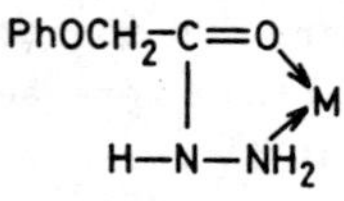

(81)

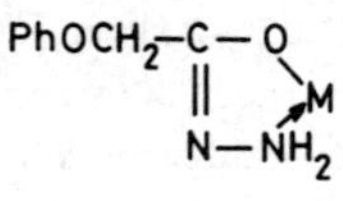

(82)

I.r. spectra of phenoxyacetylhydrazine (POAH) complexes show that for $Cu(POAH)_2X_2$ (X = Cl, $\frac{1}{2}SO_4$), $Cd(POAH)_2Cl_2$, $M(POAH)_3X_2$-nH_2O (M = Mn, Co, Ni, X = Cl, NO_3, $\frac{1}{2}SO_4$) and $M(POAH)_2(NCS)_2$ (M = Co, Ni Cd) the coordination is (81). For $M(POAH)_n$ (M = Co, Cu, Zn, n = 2 or 3) it is thought to be (82).[354] The i.r. spectra of $ML_2(H_2O)$, where HL = uracil, M = Mn, Fe, Co, Ni or Cu, show coordination of uracil through C(2)=O and N(3).[355] The complex (83) has $\nu_{as}CO_2$ at 1591cm^{-1}, ν_sCO_2 at 1409cm^{-1} - close to values in manganacarboxylates.[356]

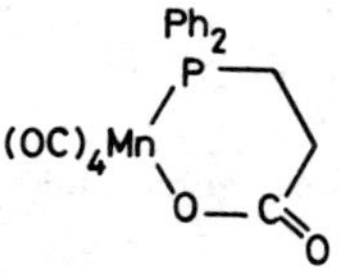

(83)

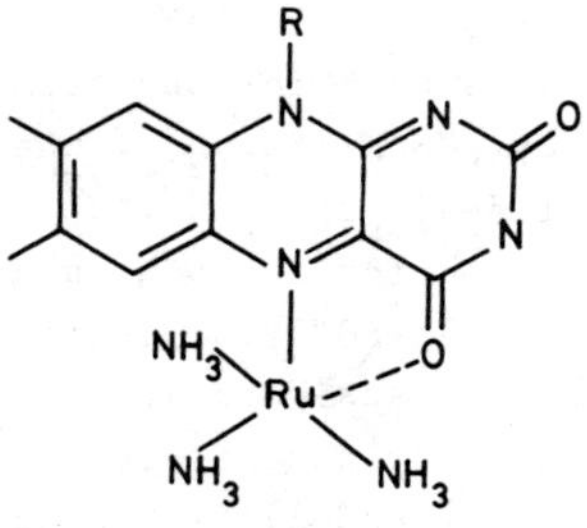

(84)

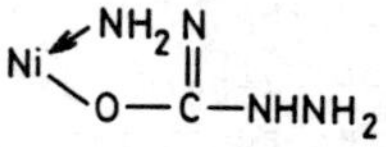

(85)

Resonance Raman spectra of $(NH_3)_4Ru^{II}$(riboflavin) and $(NH_3)_4Ru^{II}$(10-methylisoalloxazine) (84) show strong enhancements in the range 1150-1350cm^{-1}, associated with pyrazine and uracil modes, plausibly due to molecular distortion of the charge-transfer excited state.[357]

$Ni(L - H)_3^-$, where L = $H_2NNHCONHNH_2$, gives i.r. bands due to νC=N and νC-O, suggesting that coordination involves the imido-alcohol form (85).[358]

Characteristic 1-methyluracil (1-MeU) bands were seen in cis- $(NH_3)_2(ONC_2)Pt$(i-MeU)Ag NO_3 at 801cm^{-1} (vs R, vw i.r.), 647 cm^{-1} (s R, w i.r.), 601cm^{-1} (s R, s i.r.) and 495cm^{-1} (s R, s i.r.), indicating N(3)-Pt binding. There was no strong evidence for silver(I) coordination to exocyclic oxygens of 1-MeU, but some of the above band positions are not inconsistent with it.[359] ν_{as}, $\nu_s NO_2$ and νCO(of phenoxo-ligands) were assigned in $(PPh_3)_2Cu$-$(2\text{-}NO_2C_6H_4O)$ and related species. The complex shown involves chelation <u>via</u> unidentate NO_2 and CO, with $\nu_{as}NO_2$ 1545cm^{-1}, $\nu_s NO_2$ 1326 cm^{-1}, νCO 1246cm^{-1}. Related species had bridging <u>via</u> the CO group and uncoordinated NO_2, i.e. $\nu_{as}NO_2$ 1586, 1564cm^{-1}, $\nu_s NO_2$ 1337cm^{-1} and νCO 1291cm^{-1}.[360]

Resonance Raman spectra were obtained for phenolato-bridged binuclear copper complexes (models for haemocyanin and tyrosinase), e.g. $Cu_2L(OH)^{2+}$, where L = (86) (R' = H, Me, R = C_5H_4N,etc.). These show that CO is always above 1300cm^{-1}. This is significantly higher than in mononuclear analogues.[361] Fairly detailed assignments were proposed for ligand modes from i.r. and resonance Raman spectra of <u>N,N'</u>-ethylenebis(salicylideneiminato)copper(II), Cu(Salen).[362] Ligand mode assignments for Ag(Suc) and M Ag$(Suc)_2$, where M = Na, Li, HSuc = succinimide, reveal shifts due to Ag-N coordination. Ag(Suc) involves both N- and O-coordination, but no definite effects due to the latter were seen in the i.r. spectra.[363]

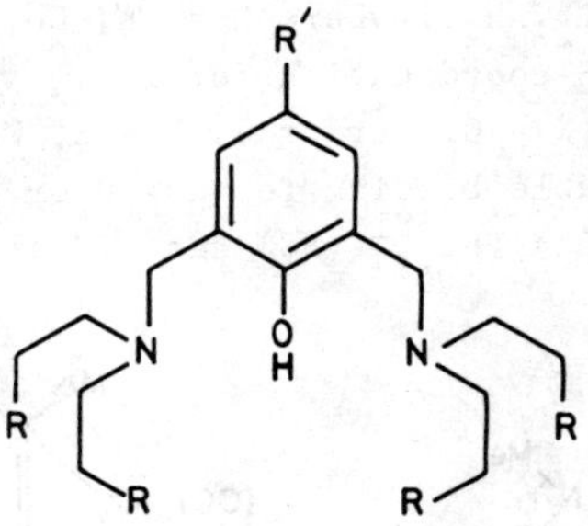

(86)

I.r. spectra show that riboflavin coordinates to Gd <u>via</u> N(1) and O(2), to Cu(II) <u>via</u> O(4), and to Ca(II) <u>via</u> O(2) and the ribitol hydoxy groups.[364]

<u>8.3 Ligands containing N and S or P and S atoms</u>.- Bidentate complexes of S_2N_2 have been reported, e.g. $MCl_4.S_2N_2$, where M = Ti or Sn, and $BeCl_2.S_2N_2$. The b_{3u} νS-N mode is shifted 20-110cm^{-1} to higher wavenumbers. The S_2N_2 is thought to be bridging, to give

infinite polymeric chain structures.[365] VL_4, where L = 2-amino-1-cyclopentenedithiocarboxylate, has νCS at 950, 900cm^{-1}, due to the delocalised CS_2^- unit.[366]

ν_{as}S-N in $M(CO)_5$ $S(NR_2)_2$ complexes [where M = Cr, NR_2 = NMe_2, $N(CH_2)_5$, $N(CH_2)_2O(CH_2)_2$ or $N(CH_2Ph)_2$; M = Mo or W, NR_2 = $N(CH_2)O(CH_2)_2$] are all lower than in free thiobisamines - consistent with M-S bonding. The νCO modes are largely in agreement with local C_{4v} symmetry.[367] Ligand mode assignments in $CpM(CO)_2[Ph_2P(S)$-$C(S)R]$, where M = Mo or W, R = Me or Ph, gave evidence for the formation of two isomers in each case. One is S,S-bonded (higher νCN, lower νP=S), the other S,N-bonded (lower νCN, higher νP=S).[368] νSCC bands in the chlorine-bridged dimer $[WCl_4(NSCl)]_2$ are at 508, 498cm^{-1}. In the monomeric anion $WCl_5(NSCl)^-$ the analogous mode is at 493cm^{-1}.[369]

Ligand modes in $ML_2.nH_2O$, where M = Mn, Co, Ni, Cu or Zn, HL = $H_2NNHC(S)OEt$, show N,S-coordination for L.[370] ν_1(NCS) and ν_2(NCS) were assigned in (87) (X = Cl, Br or I, R = H, Me or Ph). ν_1(NCS) corresponds to C=N double bonding for R = H or Me, but it is at lower wavenumber for R = Ph.[371] (88) has PC at 707cm^{-1}.[372]

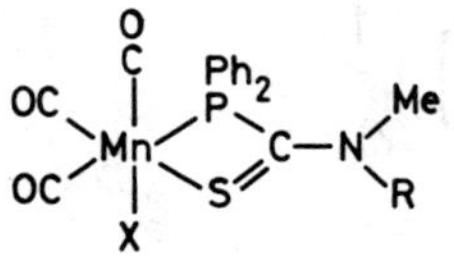

(87)

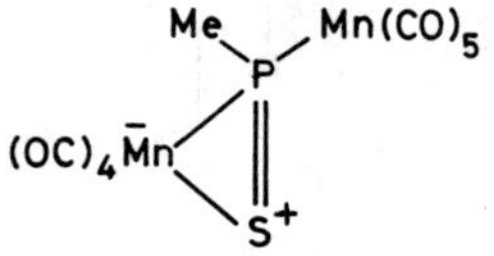

(88)

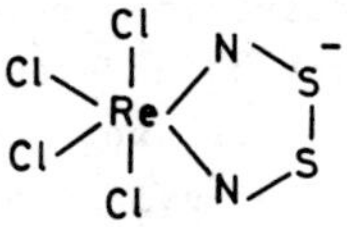

(89)

The dimer $[ReCl_3(NSCl)_2]_2$ has ν_{as}S-Cl at 475cm^{-1}, ν_sS-Cl at 452cm^{-1} (shifted to 440cm^{-1} on ^{34}S enrichment).[373] The complex (89) gives ν_{as}NS at 1070cm^{-1}, ν_sNS at 1040cm^{-1}.[374]

In (90), νNH is at $3365cm^{-1}$, νNS at $760cm^{-1}$, and the deformation of the H with respect to the Fe_2NS skeleton at $984cm^{-1}$.[375] In trans-Rh(CO)(NSO)$(PPh_3)_2$, ν_{as}NSO is at $1240cm^{-1}$, ν_sNSO at 1050 cm^{-1} and δNSO at $580cm^{-1}$.[376] The i.r. spectrum of cis-Pd($SNNMe_2$)-$(AsPh_3)Cl_2$ contains νN-S at $785cm^{-1}$ and νN-N at $1135cm^{-1}$ (compared to 910, $1105cm^{-1}$ respectively in the free ligand). These show that the N-thionitrosodimethylamine is S-bonded to the metal.[377]

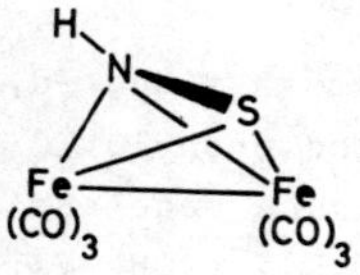

(90)

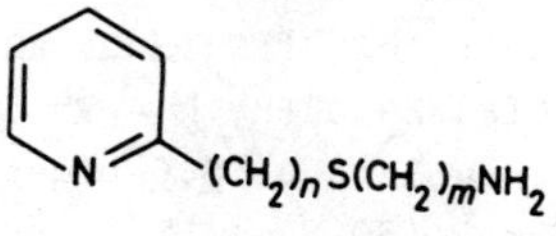

(91)

Copper(II) complexes with thio-aminopyridines [(91) (n = 1 or 2, m = 2 or 3) = L], e.g. CuL_2^{2+} or $CuLCl_2$, have νNH lower and the pyridyl ring stretch higher than in the free ligand. Thus both aliphatic and aromatic nitrogen atoms are coordinated. This also implies coordination by the sulphur, otherwise a very large chelate ring would result.[378] $M(BCTA)_2X_2$, where M = Zn or Hg, X = Cl, Br or I, BCTA = benzenocarbothioamide, and $Cd(BCTA)X_2$ have νCN at higher, νCS at lower wavenumber than in the free ligand. Thus, coordination to the metal is via S. C_{2v} monomers were thought to exist for M = Zn and Hg, while the Cd species are dimeric, symmetry C_{2h}.[379] I.r. and Raman spectra for $HSCH_2CH_2CONHCH_2CH_2COOH(COO^-)$ complexes with Hg(II) or CH_3Hg show that the mercury is covalently bonded to the cysteinic sulphur atom.[380]

8.4 Ligands containing S and O atoms.- Internal ligand mode assignments for the methylsulphates $V(CH_3SO_3)_3$, $V(CH_3SO_3)_3$(bipy), $Cr(CH_3SO_3)_3$(bipy), $Mn(CH_3SO_3)_2$ and $Fe(CH_3SO_3)_2$ show that they only contain bidentate CH_3SO_3. $V(CH_3SO_3)_3py_3$, $Cr(CH_3SO_3)_3py_3$, $Mn(CH_3SO_3)_2py_4$ and $Fe(CH_3SO_3)_2(bipy)_2$, on the other hand, contain unidentate groups.[381] $MoBr_2(NO)_2(OSMe_2)_2$ has ν_{as}SO at $955cm^{-1}$, ν_sSO at $938cm^{-1}$, confirming O-coordination. These bands and those due to νNO show that both NO and $OSMe_2$ groups are cis.[382] The following νSO modes were assigned for mer, trans-$W(CO)_2(dppe)(PR_3)(\eta^1-SO_2)$: $R_3 = Me_3$, 1178, $1050cm^{-1}$; $R_3 = {}^iPr_3$, 1190, $1048cm^{-1}$; $R_3 = Ph_2Me$, 1190, $1054cm^{-1}$.[383] Raman spectra of WO_2^{2+} complexes with thiomalic acid show that the ligand is bidentate *via* S and an adjacent carboxylate oxygen.[384]

νSO in $RuBr_2(Me_2SO)_4$ gives a single i.r. band at $1082cm^{-1}$, consistent with the presence of all S-bonded DMSO. This was confirmed by X-ray diffraction.[385] Low-wavenumber νSO modes, at 1160, $1040cm^{-1}$, were seen, due to the bridging SO_2 group in $Rh_2(\eta\text{-}C_5H_5)_2(\mu\text{-}SO_2)(\mu\text{-dppm})$, where dppm = bis(diphenylphosphino)-methane.[386] νSO bands in $Rh_2(O_2CCF_3)_4(Me_2SO)_2$ are at 948 and 931 cm^{-1}, showing the presence of O-bonded DMSO.[387]

I.r. spectra of $NiLQ_2(H_2O)$ and MLQ_2, where M = Co, Pd, Pt or Cu, L = RNHC(S)NHC(O)Ph, R = Ph or substituted phenyl, HQ = phthalimide, all show L to be S,O-bidentate.[388] Ligand mode assignments were made for $Pt(SO_3F)_4$, $Pt(SO_3F)_5^-$ and $Pt(SO_3F)_6^{2-}$. The first gave poor spectra, but these were consistent with a polymeric structure and bidentate bridging fluorosulphato groups. $Pt(SO_3F)_5^-$ contains both uni- and bidentate SO_3F groups, while these are all unidentate in $Pt(SO_3F)_6^{2-}$.[389]

The Raman spectra of $M(SO_3F)_2$, where M = Zn, Cd or Hg, show the bands expected for C_{3v}, tridentate fluorosulphato groups. These coordinate so as give polymeric structures.[390] $\nu S{=}O$ modes in $UO_2X_2.D$, where X = β-diketonate, D = dimethyl sulphoxide or dihexyl sulphoxide, show that the D is coordinated *via* an oxygen atom.[391]

νSO (1360, $1180cm^{-1}$) and $\nu S\text{-}F$ ($812cm^{-1}$) show the presence of bidentate O-coordination in $Al(SO_3F)_3$. The adduct $Al(SO_3F)_3py_3$ has νSO at 1290, $1075cm^{-1}$, i.e. unidentate SO_3F^-, but in $Al(SO_3F)_3$-(bipy) there seems to be both uni- and bidentate coordination.[392] SO_2 and Tl_2O react in a matrix to form $Tl_2^{2+}SO_3^{2-}$, with SO_3^{2-} bands at 1067, 974, 588, 548 and $442cm^{-1}$ (the symmetry is less than C_{3v} due to ion-pairing). Further bands are found at higher SO_2 concentrations, due to $S_2O_5^{2-}$ (in a symmetrical, oxygen-bridged form).[393]

I.r. and Raman spectra of $GeF_2(SO_3F)_2$ and $SnF_2(SO_3F)_2$ show that the fluorosulphato groups are bidentate, giving six-coordination at the metal. Unidentate SO_3F^- groups are found in $M(SO_3F)_6^{2-}$, M = Ge or Sn.[394] Fairly detailed ligand mode assignments were made for $Me_3Sn(OSPRR')$, where R, R' = Me or Ph. These were consistent with bidentate S,O-coordination to Sn. X-ray diffraction shows that this is bridging, not chelating.[395]

Evidence from a Raman band at $1636cm^{-1}$ suggests the formation of $C_{14}^+\ AsF_5(SO_3F)^-$ as a graphite intercalation compound.[396] *Bis*-

xanthates of phenylarsenic(III), $PhAs(S_2COR)_2$, where R = Me, Et, ^nPr, ^iPr, allyl or ^nBu, give only one characteristic i.r. band, at ca. $1040cm^{-1}$, i.e. there is bidentate, S,S-coordination of the xanthate ligands.[397] $xSbF_3.yMe_2SO$ (where x = y = 1; x = 1, y = 2; x = 2, y = 1) all have νSO due to O-bonded DMSO.[398] The i.r. spectra of $M(O_3SCF_3)_3$, where M = Sb or Bi, show the presence of both uni- and bidentate trifluoromethanesulphonate groups.[399]

REFERENCES

1 J.Nieman, J.W.Pattiasina, and J.H.Teuben, J. Organomet. Chem., 1984, 262, 157.
2 S.K.Sengupta and B.P.Baranwal, Rev. Roum. Chim., 1983, 28, 487.
3 G.S.Bristow, M.F.Lappert, T.R.Martin, J.L.Atwood, and W.F. Hunter, J. Chem. Soc., Dalton Trans., 1984, 399.
4 S.Gambarotta, M.L.Fiallo, C.Floriani, A.Chiesi-Villa, and C.Guastini, Inorg. Chem., 1984, 23, 3532.
5 W.J.Evans, I.Bloom, and R.J.Doedens, J.Organomet. Chem., 1984, 265, 249.
6 J.Amaudrut, A.Kadmiri, J.Sala-Pala, and J.E.Guerchais, J. Organomet. Chem., 1984, 266, 53.
7 E.Hey, F.Weller, and K.Dehnicke, Z. Anorg. Allg. Chem., 1984, 514, 25.
8 N.Q.Dao, H.Fevrier, M.Jouan, E.O.Fischer, and W.Röll, J.Organomet. Chem., 1984, 275, 191.
9 R.Märkl and H.Fischer, J. Organomet. Chem., 1984, 267, 277.
10 D.A.Brown, N.J.Fitzpatrick, W.K.Glass, and T.H.Taylor, J.Organomet. Chem., 1984, 275, C9.
11 E.Carmona, L.Sanchez, J.M.Marin, M.L.Poveda, J.L.Atwood, R.D.Priester, and R.D.Rogers, J. Am. Chem. Soc., 1984, 106,3214
12 H.Fischer, S.Zeuner, K.Ackermann, and U.Schubert, J.Organomet. Chem., 1984, 263, 201.
13 J.T.Guy and D.W.Bennett, Transition Met. Chem., 1984, 9, 43.
14 E.O.Fischer, A.C.Filippou, and H.G.Alt, J. Organomet. Chem., 1984, 276, 377.
15 E.Hey, F.Weller, and K.Dehnicke, Z. Anorg. Allg. Chem., 1984, 514, 18.
16 K.Chhor and G.Lucazeau, Inorg. Chem., 1984, 23, 462.
17 C.Long, A.R.Morrisson, D.C.McKean and G.P.McQuillan, J. Am. Chem. Soc., 1984, 106, 7418.
18 W.A.Herrmann, J.Plank, G.W.Kriechbaum, M.L.Ziegler, H.Pfisterer J.L.Atwood, and R.D.Rogers, J. Organomet. Chem., 1984, 264, 327
19 J.A.Clark and M.Kilner, J. Chem. Soc., Dalton Trans.,1984, 389.
20 W.Petz, J.Organomet. Chem., 1984, 270, 81.
21 G.Grötsch and W.Malisch, J. Organomet. Chem., 1984, 262, C38.
22 D.Catheline, E.Roman E., and D.Astruc, Inorg. Chem., 1984, 23, 4508.
23 M.M.Singh and R.J.Angelici, Inorg. Chem., 1984, 23, 2691.
24 L.V.Rybin, N.A.Stelzer, I.A.Garbusova, B.V.Lokshin, and M.I. Rybinskaya, J. Organomet. Chem., 1984, 265, 295.
25 Z.Goldschmidt, S.Antebi, and I.Goldberg, J. Organomet. Chem., 1984, 260, 105.
26 Z.Goldscmidt, S.Antebi, D.Cohen, and I. Goldberg, J.Organomet. Chem.,1984, 273, 347.

27 V.Rapić and L.Koratoš, J. Organomet. Chem., 1984, 260, 219.
28 R.J.Ranson and R.M.G.Roberts, J. Organomet. Chem.,1984, 260, 307.
29 M.Herberhold and L.Haumaier, Angew. Chem., Int. Ed. Engl., 1984, 23, 521.
30 A.L.Shobert, I.C.Hisatsune, and P.S.Skell, Spectrochim. Acta, Part A, 1984, 40, 609.
31 H.-U.Wekel and W.Malisch, J.Organomet. Chem., 1984, 264, C10.
32 K.R.Grundy and J.Jenkins, J. Organomet. Chem.,1984, 265, 77.
33 S.Gopinathan, K.Joseph, and C.Gopinathan, J. Organomet. Chem., 1984, 269, 273.
34 M.A.Bennett, T.-N.Huang, and J.L.Latten, J.Organomet. Chem., 1984, 272, 189.
35 W.R.Roper, J.M.Watersm and A.H.Wright, J. Organomet. Chem., 1984, 275, C13.
36 S.V.Hoskins, R.A.Pauptit, W.R.Roper, and J.M.Waters, J. Organomet. Chem., 1984, 269, C55.
37 Y.-C,Lin, A.Mayr, C.B.Knobler, and H.D.Kaesz, J. Organomet. Chem., 1984, 272, 207.
38 J.Evans and G.S.McNulty, J. Chem. Soc., Dalton Trans.,1984, 79.
39 A.M.Ervin, S.I.Shupack, and D.M.Byler, Spectrosc. Lett., 1984, 17, 603.
40 G.Gervasio, R.Rosetti, P.L.Stanghellini, and G.Bor, Inorg. Chem., 1984, 23, 2073.
41 J.Doherty, J.Fortune, A.R.Manning, and F.S.Stephens, J. Chem. Soc., Dalton Trans., 1984, 1111.
42 S.Delgado, M.J.Macazaga, and J.R.Masaguer, J. Organomet. Chem., 1984, 268, 79.
43 W.A.Herrmann, G.W.Kriechbaum, C.Bauer, B.Koumbouris, H. Pfisterer, E.Guggolz, and M.L.Ziegler, J. Organomet. Chem., 1984, 262, 89.
44 R.Usón, L.A.Oro, D.Carmona, and M.A.Esteruelas, J. Organomet. Chem., 1984, 263, 109.
45 C.Bianchini, C.A.Ghilardi, A.Meli, and A.Orlandini, J. Organomet. Chem., 1984, 270, 251.
46 H.Hoberg, D.Schaefer, G.Burkhart, C.Krüger, and M.J.Romão, J. Organomet. Chem., 1984, 266, 203.
47 H.Hoberg and K.Sümmermann, J. Organomet. Chem.,1984, 264, 379.
48 U.Rosenthal, Z. Anorg. Allg. Chem., 1984, 513, 151.
49 I.C.Hisatsune, Spectrochim. Acta, Part A, 1984, 40, 391.
50 W.C.Kaska, P.K.Hansma, A.Bayman, and R.Kroeker, A.C.S. Symp. Ser., 1984, 248, 427.
51 D.H.Farrar, R.R.Gukathasan, and K. Won, J. Organomet. Chem., 1984, 275, 263.
52 S.Cenini, F.Porta, M.Pizzotti, and G.La Monica, J. Chem. Soc., Dalton Trans., 1984, 355.
53 R.Usón, J.Forniés, M.Tomas, and R.Fandos, J. Organomet. Chem., 1984, 263, 253.
54 J.S.Thompson and J.F.Whitney, Inorg. Chem., 1984, 23, 2813.
55 G.La Monica, G.Ardizzoia, S.Cenini, and F.Porta, J.Organomet. Chem., 1984, 273, 263.
56 A.C.Ukwueze and S.R.Lander, Spectrochim. Acta, Part A, 1984, 40, 433.
57 J.G.Hefner, P.M.Zizelman, L.D.Durfee, and G.S.Lewandos, J. Organomet. Chem., 1984, 260, 369.
58 A.A.Efremov and A.A.Davydov, Kinet. Katal., 1983, 24, 1180.
59 A.V.Bobrov, A.N.Gass, O.I.Kopsta, and N.M.Omel'yanovskaya, Fiz. Tverd. Tela, 1984, 26, 341.

60 R.Usón, A.Laguna, M.Laguna, B.R.Manzano, P.G.Jones, and G.M. Sheldrick, J. Chem. Soc., Dalton Trans., 1984, 285.

61 E.G.Hoffmann, H.Nehl, H.Lehmkuhl, K.Seevogel, and W.Stempfle, Chem. Ber., 1984, 117, 1364.

62 C.Cauletti, C.Furlani, M.N.Piancastelli, A.Sebald, and B.Wrackmeyer, Inorg. Chem., 1984, 23, 1113.

63 G.Paolucci, G.Rossetto, P.Zanella, K.Yünlü, and R.D.Fischer, J. Organomet. Chem., 1984, 272, 363.

64 S.Yamaguchi and T.G.Spiro, Chem. Phys. Lett., 1984, 110, 209.

65 E.Vincent, L.Verdonck, L.Naessens, and G.P.van der Kelen, J.Organomet. Chem., 1984, 277, 235.

66 J.Schmidt, O.Marcovitch, A.Lubezky, and Y.Kozirovski, J. Chem. Soc., Farad. Trans., 1, 1984, 80, 1.

67 P.Chaudhuri, K.Wieghardt, Y.-H.Tsai, and C.Krüger, Inorg. Chem., 1984, 23, 427.

68 R.L.Cook and J.G.Morse, Inorg. Chem., 1984, 23, 2332.

69 S.C.Kao, M.Y.Darensbourg, and W.Schenk, Organometallics, 1984, 3, 871.

70 D.Lentz, Chem. Ber., 1984, 117, 415.

71 E.Vincent, L.Verdonck, D.F.van de Vondel, and G.P.van der Kelen, J. Mol. Struct., 1984, 112, 119.

72 L.Verdonck, E.Vincent, and G.P.van der Kelen, J. Mol. Struct., 1984, 117, 257.

73 A.Horton-Mastin and M.Poliakoff, Chem. Phys. Lett., 1984, 109, 587.

74 C.P.Horwitz, E.M.Holt, and D.F.Shriver, Inorg. Chem., 1984, 23, 2491.

75 E.Carmona, A.Galindo, L.Sanchez, A.J.Nielson, and G.Wilkinson, Polyhedron, 1984, 3, 347.

76 M.H.Chisholm, J.A.Connor, J.C.Huffman, E.M.Kober, and C.Overton Inorg. Chem., 1984, 23, 2298.

77 R.H.Hooker and A.J.Rest, J. Chem. Soc., Dalton Trans., 1984, 761

78 K.A.Mahmoud, A.J.Rest, H.G.Alt, M.E.Eichner, and B.M.Jansen, J. Chem. Soc., Dalton Trans., 1984, 175.

79 K.A.Mahmoud, A.J.Rest, and H.G.Alt, J. Chem. Soc., Dalton Trans., 1984, 187.

80 J.A.Crayston, M.J.Almond, A.J.Downs, M.Poliakoff, and J.J. Turner, Inorg. Chem., 1984, 23, 3051.

81 K.Seyferth and R.Taube, J.Organomet. Chem., 1984, 262, 179.

82 J.A.Connor, E.J.James, C.Overton, and N.El Murr, J. Chem. Soc., Dalton Trans., 1984, 255.

83 K.Seyferth, K.Rosenthal, G.Kühn, and R.Taube, Z. Anorg. Allg. Chem., 1984, 513, 57.

84 W.Buchner and W.A.Schenk, Inorg. Chem., 1984, 23, 132.

85 J.Grobe and D.LeVan, J. Fluorine Chem., 1984, 24, 25.

86 S.H.Strauss and K.D.Abney, Inorg. Chem., 1984, 23, 515.

87 G.Süss-Fink, U.Thewalt, and H.-P.Klein, J. Organomet. Chem., 1984, 262, 315.

88 N.E.Kolobova, L.L.Ivanov, O.S.Zhvanko, O.M.Khitrova, A.S. Batsanov, and Yu.T.Struchkov, J. Organomet. Chem., 1984, 262, 39.

89 D.Melzer and E.Weiss, J. Organomet. Chem., 1984, 263, 67.

90 J.M.Boncella and R.A.Andersen, Inorg. Chem., 1984, 23, 432.

91 C.Arsenault, P.Bougeard, B.G.Sayer, S.Yeroushalmi, and M.J.McGlinchey, J.Organomet. Chem., 1984, 265, 283.

92 G.Z.Suleimanov, L.F.Rybakova, L.T.Abdullaeva, A.A.Pasynskii, and I.P.Beletskaya, Dokl. Akad. Nauk SSSR, 1983, 272, 885.

93 I.R.Butler, W.E.Lindsell, and M.J.K.Thomas, J. Organomet. Chem., 1984, 262, 59.
94 R.H.Hooker, A.J.Rest, and I.Whitwell, J. Organomet. Chem., 1984, 266, C27.
95 J.D.Satterlee and J.E.Erman, J. Am. Chem. Soc., 1984, 106,1139.
96 M.L.Smith, P.I.Ohlsson, and K.G.Paul, FEBS Lett., 1983, 163, 303.
97 L.J.Young, O.Einarsdottir, C.R.Vossbrink, and W.S.Caughey, Biochem. Biophys. Res. Commun., 1984, 123, 247.
98 G.E.Gadd, M.Poliakoff, and J.J.Turner, Inorg. Chem., 1984, 23, 630.
99 M.A.De Paoli, S.M.De Oliveira, E.Baggio-Saitovitch, and D.Guenzburger, J. Chem. Phys., 1984, 80, 730.
100 A.J.Ouderkerk, T.A.Seder, and E.Weitz, Proc. SPIE-Int. Soc. Opt. Eng., 1984, 458, 148.
101 J.A.S.Howell, P.Mathur, R.Kumar, A.R.Manning, and F.S.Stephens, J. Organomet. Chem., 1984, 262, 227.
102 A.F.Hepp, J.P.Blaha, C.Lewis, and M.S.Wrighton, Organometallics, 1984, 3, 174.
103 S.C.Fletcher, M.Poliakoff, and J.J.Turner, J. Organomet. Chem., 1984, 268, 259.
104 R.L.Jackson, Proc. SPIE-Int. Soc. Opt. Eng., 1984, 459, 33.
105 K.P.Schick, N.L.Jones, P.Sekula, N.M.Boag, J.A.Labinger, and H.D.Kaesz, Inorg. Chem., 1984, 23, 2204.
106 H.-W.Chen, Z.Zhong, and J.M.White, J.Catal., 1984, 90, 119.
107 J.Evans and G.S.McNulty, J. Chem. Soc., Dalton Trans., 1984, 1123.
108 J.G.Bentsen and M.S.Wrighton, J. Am. Chem. Soc., 1984, 103, 4041.
109 S.F.A.Kettle and P.-L.Stanghellini, Inorg. Chem., 1984, 23, 4469.
110 T.R.Gilson, J. Chem. Soc., Dalton Trans., 1984, 149.
111 T.R.Gilson and J.Evans, J. Chem. Soc., Dalton Trans., 1984, 155
112 N.J.Bristow, B.D.Moore, M.Poliakoff, G.Ryott, and J.J.Turner, J.Organomet. Chem., 1984, 260, 181.
113 A.J.Deeming, S.Donovan-Mtunzi, and S.E.Kabir, J. Organomet. Chem., 1984, 276, C65.
114 R.Palmans, L.Viaene, and J.D'Olieslager, Inorg. Chem., 1984, 23, 1792.
115 K.Sato, Y.Inoue, I.Kojima, E.Miyazaki, and I.Yasumori, J. Chem. Soc., Farad. Trans.,1, 1984, 80, 841.
116 S.I.Woo and C.G.Hill, J. Mol. Catal., 1984, 24, 165.
117 S.B.Colbran, B.H.Robinson, and J.Simpson, J.Organomet. Chem., 1984, 265, 199.
118 D.J.Darensbourg and D.J.Zalewski, Inorg. Chem., 1984, 23,4382.
119 E.A.Hyde, R.Rudham, and C.H.Rochester, J. Chem. Soc., Farad. Trans., 1, 1984, 80, 531.
120 B.E.Hanson, M.E.Davis, D.Taylor, and E.Rode, Inorg. Chem., 1984, 23, 52.
121 H.P.Wang and J.T.Yates, J. Catal., 1984, 89, 79.
122 R.Choukroun and D.Gervais, J.Organomet. Chem., 1984, 266, C37.
123 J.Evans and G.S.McNulty, J.Chem. Soc., Dalton Trans., 1984,587.
124 K.A.Arunkumar, H.A.Marzouk, and E.B.Bradley, Proc. SPIE-Int. Soc. Opt. Eng.,1984, 482, 130.
125 R.G.Tobin, S.Chiang, P.A.Thiel, and P.L.Richards, Surf. Sci., 1984, 140, 393.
126 J.L.Gland, R.J.Madix, R.W.McCabe, and C.DiMaggio, Surf. Sci., 1984, 143, 46.
127 K.Kunimatsu, J. Phys. Chem., 1984, 88, 2195.

128 B.Y.Lee, Y.Inoue, and I.Yasumori, Bull. Chem. Soc. Japan, 1984, 57, 1283.
129 J.Hu, Y.Song, Z.Hong, and Z.Liu, Guangpuxue Yu Guangpu Fenxi, 1984, 4, 22 (Chem. Abstr., 1984, 101, 62887).
130 M.W.Severson, W.J.Tornquist, and J.Overend, J. Phys. Chem., 1984, 88, 469.
131 P.Hollins and J.Pritchard, Surf. Sci., 1983, 134, 91.
132 P.Hollins, K.J.Davies, and J.Pritchard, Surf. Sci., 1984, 138, 75.
133 S.Kitagawa, M.Munukata, and N.Miyaji, Bull. Chem. Soc. Japan, 1983, 56, 2258.
134 F.Frank, W.Schulze, B.Tesche, and F.W.Froben, Chem. Phys. Lett. 1984, 103, 336.
135 M.R.Mahoney, H.W.Martin, and R.P.Cooney, J.Electroanal. Chem., Interfacial Electrochem., 1984, 161, 163.
136 V.D.Makhaev, A.P.Borisov, and E.B.Lobkovskii, Russ. J. Inorg. Chem., 1984, 29, 666.
137 R.J.Barton, D.G.Holah, H.Shengzhi, A.N.Hughes, S.I.Khan, and B.E.Robertson, Inorg. Chem., 1984, 23, 2391.
138 S.A.Snow, M.Shimoi, C.D.Ostler, B.K.Thompson, G.Kodama, and R.W.Parry, Inorg. Chem., 1984, 23, 511.
139 H.J.Wasserman, D.C.Moody, R.T.Paine, R R.Ryan, and K.V.Salazar, J. Chem. Soc., Dalton Trans., 1984, 533.
140 S.P.Church, F.-W.Grevels, H.Hermann, and K.Schaffner, Inorg. Chem., 1984, 23, 3830.
141 D.Sellmann, W.Weber, G.Liehr, and H.P.Beck, J. Organomet. Chem. 1984, 269, 155.
142 R.Luck and R.H.Morris, Inorg. Chem., 1984, 23, 1489.
143 R.H.Morris, J.M.Ressner, J.F.Sawyer, and M.Shiralian, J. Am. Chem. Soc., 1984, 106, 3683.
144 W.A.Herrmann, G.W.Kriechbaum, and I.Steffl, Z. Anorg. Allg. Chem., 1984, 508, 100.
145 M.Hidai, K.Komori, T.Kodama, D.-M.Jin, T.Takahashi, S.Sugiura, Y.Uchida, and Y.Mizabe, J.Organomet. Chem., 1984, 272, 155.
146 M.M.Doeff, S.F.Parker, P.H.Barrett, and R.G.Pearson, Inorg. Chem., 1984, 23, 4108.
147 J.M.Lopez-Alcala, M.C.Puerta, and F.Gonzalez-Vilchez, Polyhedron, 1984, 3, 623.
148 M.I.Bruce, E.Horn, M.R.Snow, and M.L.Williams, J.Organomet. Chem., 1984, 276, C53.
149 M.A.Gallop, C.E.F.Rickard, and W.R.Roper, J. Organomet. Chem., 1984, 269, C21.
150 D.E.Samkoff, J.R.Shapley, M.R.Churchill, and A.J.Wasserman, Inorg. Chem., 1984, 23, 397.
151 N.Kinoshita, K.Domen, K.Aika, and T.Onishi, Appl. Surf. Sci., 1984, 18, 342.
152 H.P.Wang and J.T.Yates, J.Phys. Chem., 1984, 88, 852.
153 H.O.Desseyn, B.J.van der Veken, J.R.Moss, B.J.Smith, P. Verhoeven, and D.A.Thornton, Spectrochim. Acta, Part A,1984, 40 467.
154 E.A.Nikonenko, I.N.Marenkova, I.I.Olikov, and L.N.Margolin, Russ. J. Inorg. Chem.,1984, 29, 590.
155 Zh.N.Bublik, S.V.Volkov, and E.A.Mazurenko, Russ. J. Inorg. Chem., 1984, 29, 73.
156 T.C.Strekas and S.K.Mandal, J. Raman Spectrosc.,1984, 15, 109.
157 S.McClanahan and J.Kincaid, J. Raman Spectrosc.,1984, 15, 173.
158 S.Farquharson, K.L.Guyer, P.A.Lay, R.H.Magnuson, and M.J.Weaver J. Am. Chem. Soc., 1984, 106, 5123.

159 S.Wolfgang, T.C.Strekas, H.D.Gafney, R.A.Krause, and K.Krause, Inorg. Chem., 1984, 23, 2650.
160 R.R.Gagne and D.N.Marks, Inorg. Chem., 1984, 23, 65.
161 U.Fürholz, H.B.Bürgi, F.E.Wagner, A.Stebler, J.H.Ammeter, E.Krausz, R.J.H.Clark, M.J.Stead, and A.Ludi, J. Am. Chem. Soc. 1984, 106, 121.
162 O.Poizat and C.Sourisseau, J. Phys. Chem.,1984, 88, 3007.
163 H.R.Virdee and R.E.Hester, J.Phys. Chem., 1984, 88, 451.
164 S.M.Angel, M.K.DeArmond, R.J.Donohoe, K.W.Hanck, and D.W. Wertz, J. Am. Chem. Soc., 1984, 106, 3688.
165 J.V.Caspar, T.D.Westmoreland, G.H.Allen, P.G.Bradley, T.J. Meyer, and W.H.Woodruff, J. Am. Chem. Soc., 1984, 106, 3492.
166 J.G.Gaudiello, P.G.Bradley, K.A.Norton, W.H.Woodruff, and A.J.Bard, Inorg. Chem., 1984, 23, 3.
167 C.Tellez and G.Diaz, Acta Sud. Am. Quim., 1983, 3, 85.
168 M.S.Barvinok and L.G.Lukina, Russ. J. Inorg. Chem.,1983, 28, 1129.
169 S.Agustoslu, S.Akyüz, and J.E.D.Davies, J. Mol. Struct., 1984, 114, 437.
170 M.G.Ivanov and I.I.Kalinichenko, Russ. J.Inorg. Chem., 1984, 29, 708.
171 M.Nonyama, J.Organomet. Chem., 1984, 262, 407.
172 G.Lopez, G.Garcia, C.de Haro, G.Sanchez, and M.C.Vallero, J. Organomet. Chem., 1984, 263, 247.
173 M.Castillo, A.Romero, and E.Ramirez, Inorg. Chem.,1984, 23, 2668.
174 M.A.Sarukhanov, M.Sh.Mridkha, Yu.Ya.Khritonov, N.D.Mitkinova, and T.G.Abzaeva, Russ. J. Inorg. Chem., 1983, 28, 1158.
175 E.M.Poteshnykh, L.A.Leites, B.I.Ignat'eva, G.N.Kartsev, and I.I.Kritskaya, Izv. Akad. Nauk SSSR, Ser. Khim.,1984, 888.
176 A.D.Toleva, N.M.Samus', and A.N.Shishkov, Russ. J. Inorg. Chem. 1983, 28, 1617.
177 M.G.B.Drew and G.R.Wliiey, J. Chem. Soc., Dalton Trans., 1984, 727.
178 J.A.Selnutt and M.R.Ondrias, Inorg. Chem., 1984, 23, 1175.
179 H.Yamaguchi, A.Soeta, H.Toeda, and K.Itoh, J. Electroanal. Chem. Interfacial Electrochem., 1983, 159, 347.
180 C.Jennings, R.Aroca, A.M.Hor, and R.O.Loutfy, J. Raman Spectrosc., 1984, 15, 34.
181 K.Shoji, Y.Kobayashi, and K.Itoh, Chem. Phys. Lett.,1983, 102, 179.
182 J.Terner and J.Topich, Chem. Phys. Lett., 1984, 106, 508.
183 M.W.Kokkes, D.J.Stufkens, and A.Oskam, J. Chem. Soc., Dalton Trans., 1984, 1005.
184 M.A.Crisanti, T.G.Spiro, D.R.English, D.N.Hendrickson, and K.S.Suslick, Inorg. Chem., 1984, 23, 3897.
185 B.R.Stallard, P.R.Callis, P.M.Champion, and A.C.Albrecht, J. Chem. Phys., 1984, 80, 70.
186 J.A.Hofmann and D.F.Bocian, J. Phys. Chem., 1984, 88, 1472.
187 G.A.Schick and D.F.Bocian, J. Am. Chem. Soc., 1984, 106, 1682.
188 D.L.Rousseau, S.L.Tan, M.R.Ondrias, and R.W.Noble, Biochem., 1984, 23, 2857.
189 A.Desbois, G.Mazza, F.Stetzkowski, amf M.Luz, Biochim. Biophys. Acta, 1984, 785, 161.
190 R.Schweitzer-Stenner, W.Dreybrodt, and S.El-Naggar, Biophys. Struct. Mech., 1984, 10, 241.
191 S.El-Naggar, R.Schweitzer-Stenner, W.Dreybrodt, and A.Mayer, Biophys. Struct. Mech., 1984, 10, 257.

192 R.Schweitzer-Stenner, W.Dreybrodt, D.Wedekind, and S.El-Naggar, Eur. Biophys. J., 1984, 11, 61.
193 E.W.Findsen and M.R.Ondrias, J. Am. Chem. Soc., 1984, 106, 5736.
194 T.Ogura, N.Sone, K.Tagawa, and T.Kitagawa, Biochem., 1984, 23, 2826.
195 A.Desbois, Y.Henry, and M.Lutz, Biochim. Biophys. Acta, 1984, 785, 148.
196 M.R.Ondrias, E.W.Findsen, D.F.Gaul, and D.B.Knaff, Biochim. Biophys. Acta, 1984, 788, 87.
197 J.Terner and D.E.Reed, Biochim. Biophys. Acta, 1984, 789, 80.
198 M.J.K.Geno and J.H.Dawson, Inorg. Chem., 1984, 23, 1182.
199 U.Kohler, H.-D.Hausen, and J.Weidlein, J. Organomet. Chem., 1984, 272, 337.
200 D.L.Willems and D.F.Bocian, J. Am. Chem. Soc., 1984, 106, 880.
201 L.K.Hanson, C.K.Chang, B.Ward, P.M.Callahan, G.T.Babcock, and J.D.Head, J. Am. Chem. Soc., 1984, 106, 3950.
202 P.A.Apanasevich, J. Mol. Struct., 1984, 115, 233.
203 T.M.Ivanova and V.V.Berdyugin, J. Mol. Struct., 1984, 117, 287.
204 A.K.Shiemke, L.D.Eirich and T.M.Loehr, Biochim. Biophys. Acta, 1983, 748, 143.
205 P.F.dos Santos Filho and U.Schuchardt, J. Organomet. Chem., 1984, 263, 385.
206 J.A.Shelnutt, K.D.Staub, P.M.Rentzepis, M.Gouterman, and E.R.Davidson, Biochem., 1984, 23, 3946.
207 J.A.Garcia-Vazquez, M.Lopez-Becerra, and J.R.Masaguer, Z. Anorg. Allg. Chem., 1984, 548, 187.
208 H.H.Emons, K.Pollmer, and G.Kabisch, Z. Chem., 1984, 24, 28.
209 H.Inoue and E.Fluck, Z.Naturforsch., Teil B,1984, 39, 185.
210 R.Bär, W.P.Fehlhammer, D.K.Breitinger, and J.Mink, Inorg. Chim. Acta, 1984, 82, L17.
211 A.Yu.Tsivadze and H.Köhler, Z. Anorg. Allg. Chem., 1984, 510,31
212 A.Yu.Tsivadze and H.Köhler, Z.Anorg. Allg. Chem., 1984, 510, 25
213 J.D.Allison, F.A.Cotton, G.L.Powell, and R.A.Walton, Inorg. Chem., 1984, 23, 159.
214 C.J.Cameron, R.A.Walton, and D.A.Edwards, J. Organomet. Chem., 1984, 262, 335.
215 I.Brüdgam, H.Hartl, and D.Lentz, Z. Naturforsch., Teil B, 1984, 39, 721.
216 H.H.Wei and K.-T.Dam, J. Chin. Chem. Soc.,Taipei,1983, 30, 131.
217 S.Sinha, B.D.Humphrey, and A.B.Bocarsly, Inorg. Chem., 1984, 23, 203.
218 D.T.Plummer, B.A.Karcher, R.A.Jacobson, and R.J.Angelici, J.Organomet. Chem., 1984, 260, 347.
219 F.D.Rochon, R.Melanson, H.E.Howard-Lock, C.J.L.Lock, and G.Turner, Can. J. Chem., 1984, 62, 860.
220 M.Hvastijova, J.Kohout, H.Wusterhausen, and H. Köhler, Z. Anorg. Allg. Chem., 1984, 510, 37.
221 P.G.Egan, D.G.Holan, and K.W.Morse, Inorg. Chem., 1984, 23,2203
222 A.Bell and D.A.Edwards, J. Chem. Soc. Dalton Trans., 1984, 1317
223 H.H.Emons, E.Janneck, G.Kabisch, and K.Pollmer, Z. Anorg. Allg. Chem., 1984, 511, 148.
224 S.F.Gerasomov and V.N.Filimonov, React. Kinet. Catal. Lett., 1984, 25, 273.
225 R.Bhattacharyya, P.S.Roy, and A.K.Dasmahapatra, J. Organomet. Chem., 1984, 267, 293.

226 M.Minelli, J.L.Hubbard, D.L.Lichtenberger, and J.H.Enemark, Inorg. Chem., 1984, 23, 2721.

227 P.Legzdins and J.C.Oxley, Inorg. Chem., 1984, 23, 1053.
228 A.D.Hunter and P.Legzdins, Inorg. Chem., 1984, 23, 4198.
229 D.L.Lichtenberger and J.L.Hubbard, Inorg. Chem.,1984, 23, 2718.
230 D.S.Barratt and C.A.McAuliffe, J. Chem. Soc., Chem. Comm., 1984, 594.
231 G.Cardaci, J. Chem. Soc., Dalton Trans., 1984, 815.
232 J.Fiedler and J.Masek, Inorg. Chim. Acta, 1984, 81, 117.
233 W.Krasser, T.Woike, P.S.Bechtold, and S.Haussühl, J. Mol. Struct., 1984, 114, 57.
234 S.R.Gonzalez, P.J.Aymonino, and O.E.Piro, J. Chem. Phys., 1984, 81, 625.
235 G.Lagrange, P.Cocolios, and R.Guilard, J. Organomet. Chem., 1984, 260, C16.
236 A.Efraty and G.Elbaze, J. Organomet. Chem., 1984, 260,331.
237 J.A.Olabe, L.A.Gentil, G.Rigotti, and A.Navaza, Inorg. Chem., 1984, 23, 4297.
238 R.Weber and K.Dehnicke, Z. Naturforsch., Teil B, 1984, 39, 262.
239 B.F.G.Johnson, J.Lewis, W.J.H.Nelson, J.Puga, D.Braga, K.Henrick, and M.McPartlin, J. Organomet. Chem.,1984, 266, 173.
240 H.W.Roesky, K.K.Pandey, W.Clegg, M.Noltemeyer, and G.M. Sheldrick, J. Chem. Soc., Dalton Trans., 1984, 719.
241 G.Elbaze, F.Dahan, M.Dartiguenave, and Y.Dartiguenave, Inorg. Chim. Acta, 1984, 87, 91.
242 R.B.King and N.D.Sadanani, J. Chem. Soc., Chem. Comm.,1984, 955
243 A.Hengefeld and R.Nast, J. Organomet. Chem., 1984, 267, 151.
244 S.Al-Jibori and B.L.Shaw, J. Organomet. Chem.,1984, 273, 213.
245 R.D.Rogers, E.Carmona, A.Galindo, J.L.Atwood, and L.G.Canada, J. Organomet. Chem., 1984, 277, 403.
246 K.V.Kotegov, Z.A.Lovchikova, G.B.Avetikyan, Russ. J. Inorg. Chem., 1983, 28, 1799.
247 R.H.Morris and M.Shiralian, J. Organomet. Chem., 1984, 260, C47
248 D.S.Bohle and W.R.Roper, J. Organomet. Chem., 1984, 273, C4.
249 J.A.S.Duncan, T.A.Stephenson, M.D.Walkinshaw, D.Hedden, and D.M.Roundhill, J. Chem. Soc., Dalton Trans., 1984, 801.
250 J.Ellermann, G.Szucsanyi, K.Geibel, and E.Wilhelm, J.Organomet. Chem., 1984, 263, 297.
251 W.Lasser and U.Thewalt, J. Organomet. Chem., 1984, 275, 63.
252 M.K.Chaudhuri and S.K.Ghosh, J. Chem. Soc., Dalton Trans., 1984, 507.
253 M.K.Chaudhuri and S.K.Ghosh, Inorg. Chem., 1984, 23, 534.
254 P.Schwendt and D.Joniakova, Polyhedron, 1984, 3, 287.
255 V.F.Newberry, H.D.Burkett, S.D.Worley, and W.E.Hill Inorg. Chem., 1984, 23, 3911.
256 B.M.Louie and A.Storr, Can. J. Chem., 1984, 62, 1344.
257 T.Watanabe, T.Ama, and K.Nakamoto, J. Phys. Chem., 1984, 88, 440.
258 A.K.Shiemke, T.M.Loehr, and J.Sanders-Loehr, J. Am. Chem. Soc., 1984, 106, 495.
259 G.Chottard. M.Schappacher, L.Ricard, and R.Weiss, Inorg. Chem., 1984, 23, 4557.
260 I.A.Khartonik, G.G.Novitskii, and N.M.Sinitsyn, Russ. J. Inorg. Chem., 1984, 29, 1075.
261 B.N.Ivanov-Emin, N.A.Nevskaya, B.E.Zaitsev, N.N.Nevskii, and A.S.Izmailovich, Russ. J. Inorg. Chem., 1984, 29, 710.
262 K.Bajdor, K.Nakamoto, H.Kanatomi, and I.Murase, Inorg. Chim. Acta, 1984, 82, 207.
263 W.-S.Hwang, Polyhedron, 1984, 3, 513.
264 A.B.Goel and S.Goel, Angew. Chem., Int. Ed. Engl., 1984, 23, 375.

265 J.S.Thompson, J. Am. Chem. Soc., 1984, 106, 4057.
266 B.N.Ivanov-Emin, L.P.Petrishcheva, B.E.Zaitsev, and A.S. Izmailovich, Russ. J. Inorg. Chem., 1984, 29, 860.
267 L.K.Thompson, A.W.Hanson, and B.S.Ramaswamy, Inorg. Chem., 1984, 23, 2459.
268 S.K.Mandal and K.Nag, J. Chem. Soc., Dalton Trans., 1984, 2141.
269 B.N.Ivanov-Emin, N.A.Nevskaya, B.E.Zaitsev, and A.S. Izmailovich, Russ. J. Inorg. Chem., 1984, 29, 777.
270 Z.H.Kafafi, R.H.Hauge, W.E.Billups, and J.L.Margrave, Inorg. Chem., 1984, 23, 177.
271 A.M.Glebov, Yu.I.Sal'nikov, and V.V.Ustyak, Russ. J. Inorg. Chem., 1983, 28, 1526.
272 A.K.Sharma and V.K.Kaushik, Acta Chim. Hung., 1983, 114, 41.
273 M.Puri and R.D.Verma, Monatsh. Chem., 1984, 115, 533.
274 M.E.Farago and S.Amirhaeri, Inorg. Chim. Acta, 1984, 81, 205.
275 J.Telser and R.S.Drago, Inorg. Chem., 1984, 23, 1798.
276 E.A.Nikonenko, L.N.Margolin, I.N.Marenkova, and I.I.Olikov, Koord. Khim., 1984, 10, 970.
277 M.Puri and R.D.Verma, Ind. J. Chem., Sect. A, 1984, 23, 612.
278 A.J.Lindsay, R.P.Tooze, M.Motevalli, M.B.Hursthouse, and G.Wilkinson, J. Chem. Soc., Chem. Comm., 1984, 1383.
279 I.B.Baranovskii and A.N.Zhilyaev, Russ. J. Inorg. Chem., 1984, 29, 607.
280 D.P.Bancroft, F.A.Cotton, and S.Han, Inorg. Chem.,1984, 23, 2408.
281 V.M.Miskowski, W.P.Schaefer, B.Sadeghi, B.D.Santarsiero, and H.B.Gray, Inorg. Chem., 1984, 23, 1154.
282 D.M.Grove, G.van Koten, H.J.C.Ubbels, R.Zoet, and A.L.Spek, J.Organomet. Chem., 1984, 263, C10.
283 T.Kawato, T.Uechi, H.Koyama, H.Kanatomi, and Y.Kawanami, Inorg. Chem., 1984, 23, 764.
284 B.E.Heyden, K.Prince, D.P.Woodruff, and A.M.Bradshaw, Surf. Sci., 1983, 133, 589.
285 L.Antolini and L.Menabue, Inorg. Chem., 1984, 23, 1418.
286 B.N.Misra and R.Kripal, Ind. J. Pure Appl. Phys.,1984, 22, 428.
287 A.S.Yasyukevich, Yu.I.Atrashevskii, M.V.Nikanovich, G.G. Novitskii, Yu.I.Krasilov, and S.M.Portnova, Koord. Khim., 1984, 10, 812.
288 G.N.Makushova and S.B.Pirkes, Russ. J. Inorg. Chem.,1984, 29, 531.
289 V.T.Panyushkin, A.A.Mastrakov, and N.N.Bukov, Russ. J. Inorg. Chem., 1983, 28, 1578.
290 M.Gal, P.L.Goggin and J.Mink, J. Mol. Struct., 1984, 114, 459.
291 A.K.Sengupta and K.Sinha, J. Fluorine Chem., 1984, 24, 153.
292 A.Midha, R.D.Verma, K.Brown, and R.V.Parish, Monatsh. Chem., 1983, 114, 661.
293 R.G.Swisher, J.F.Vollano, V.Chandrasekhar, R.O.Day, and R.R. Holmes, Inorg. Chem., 1984, 23, 3147.
294 J.F.Vollano, R.O.Day, D.N.Rau, V.Chandrasekhar, and R.R.Holmes, Inorg. Chem., 1984, 23, 3153.
295 V.P.Karlov, G.N.Butuzov, and T.F.Dobrokhotova, Russ. J. Inorg. Chem., 1983, 28, 1218.
296 C.P.Rao, A.M.Rao, and C.N.R.Rao, Inorg. Chem., 1984, 23, 2080.
297 H.A.Tajmir-Riahi, Carbohydr. Res., 1983, 122, 241.
298 R.Martin-Benito, R.Jimenez-Aparicio, and M.C.Barral, J.Organomet. Chem., 1984, 269, 267.
299 Yu.Ya.Kharitonov, M.M.Bekhit, and S.F.Belevskii, Russ. J. Inorg. Chem., 1983, 28, 1803.

300 G.S.Bristow, M.F.Lappert, T.R.Martin, J.L.Atwood, and W.F. Hunter, J. Chem. Soc., Dalton Trans., 1984, 399.
301 R.O.Gould, T.A.Stephenson, and D.A.Tocher, J. Organomet. Chem., 1984, 263, 375.
302 J.S.Keller, J.J.McMahon, R.Chang, and R.L.Carter, Spectrosc. Lett., 1984, 17, 145.
303 R.C.Mehrotra and J.V.Singh, Z.Anorg. Allg. Chem.,1984, 512, 221
304 R.N.Shchelokov, G.T.Bolotova, and V.N.Perov, Russ. J. Inorg. Chem., 1984, 29, 41.
305 B.A.Kolesov and I.K.Igumenov, Spectrochim. Acta, Part A, 1984, 40, 233.
306 S.K.Brahma, W.H.Nelson, and N.F.Howard, J. Organomet. Chem., 1984, 273, 47.
307 P.Klingelhöfer, U.Müller, H.G.Hauck, and K.Dehnicke, Z. Naturforsch., Teil B, 1984, 39, 135.
308 A.Z.Beilina, K.M.Zhumanova, and Z.M.Mudlakhmetov, Russ. J. Inorg. Chem., 1983, 28, 1733.
309 E.I.Sinyavskaya, Russ. J. Inorg. Chem.,1983, 28, 1148.
310 I.Ahmed, K.W.Bagnall, L.Xing-Fu, and P.Po-jung, J. Chem. Soc., Dalton Trans., 1984, 19.
311 T.N.Srivastava and P.Kumar, Natl. Acad. Sci. Lett. (India), 1982, 5, 399.
312 A.R.Miksztal and J.S.Valentine, Inorg. Chem., 1984, 23, 3548.
313 A.A.Ivakin and I.G.Chufarova, Russ. J. Inorg. Chem., 1984, 29, 748.
314 S.H.Strauss, K.D.Abney, K.M.Long, and O.P.Anderson, Inorg. Chem., 1984, 23, 1994.
315 S.Chandra and K.K.Sharma, Acta Chim. Hung., 1984, 115, 9.
316 A.C.Hayes, P.Kruus, and W.A.Adams, J. Solution Chem., 1984, 13, 61.
317 S.D.Nikitina and S.A.Bondar', Russ. J. Inorg. Chem., 1983, 28, 1275.
318 L.S.Skogareva, V.P.Babaeva, and V.Ya.Rosolovskii, Russ. J. Inorg. Chem., 1983, 28, 1432.
319 V.Riera, J.Ruiz, and F.Mayor, J. Organomet. Chem., 1984, 276,47
320 J.-L.Pascal, J.Potier, D.J.Jones, J.Rozière, and A.Michalowicz, Inorg. Chem., 1984, 23, 2068.
321 V.P.Babaeva, V.G.Sevast'yanov, and V.Ya.Rosolovskii, Russ. J. Inorg. Chem., 1983, 28, 1634.
322 J.E.Hoots, D.A.Lesch, and T.B.Rauchfuss, Inorg. Chem., 1984, 23, 3130.
323 H.Brunner, H.Kauermann, W.Meier, and J.Wachter, J. Organomet. Chem., 1984, 263, 183.
324 J.M.Manoli, J.M.Bregeault, C.Potvin, and G.Chottard, Inorg. Chim. Acta, 1984, 88, 75.
325 K.S.Nagaraja and M.R.Udupa,Transition Met. Chem., 1984, 9, 290.
326 M.Chaudhury, J. Chem. Soc., Dalton Trans., 1984, 115.
327 C.G.Young, J.A.Broomhead, and C.J.Boreham, J. Organomet. Chem., 1984, 260, 91.
328 H.Strasdeit, B.Krebs, and G.Henkel, Inorg. Chim. Acta, 1984, 89, L11.
329 M.I.Bruce, T.W.Hambley, M.R.Snow, and A.G.Swincer, J. Organomet. Chem., 1984, 273, 361.
330 S.V.Usova, E.V.Rakitina, and O.N.Grishina, Russ. J. Inorg. Chem., 1984, 29, 80.
331 O.Kolb and H.Werner, J. Organomet. Chem., 1984, 268, 49.
332 W.A.Schenk, J.Leissner, and C.Burschka, Angew. Chem., Int. Ed. Engl., 1984, 23, 806.

333 N.Trendafilova, R.Kellner, and G.Nikolov, J. Mol. Struct., 1984, 115, 439.
334 S.W.Carr, R.Cotton, and D.Dakternieks, Inorg. Chem.,1984, 23, 720.
335 A.Müller, F.W.Baumann, H.Bögge, M.Römer, E.Krickemeyer, and K.Schmitz, Angew. Chem., Int. Ed. Engl., 1984, 23, 632.
336 A.Müller, M.Römer, H.Bögge, E.Krickemeyer, F.W.Baumann, and K.Schmitz, Inorg. Chim. Acta, 1984, 89, L7.
337 A.Müller, J.Schimanski, and U.Schimanski, Angew. Chem., Int. Ed. Engl., 1984, 23, 159.
338 L.S.Dumitrescu, I.Haidiuc, and J.Weiss, J. Organomet. Chem., 1984, 263, 159.
339 R.B.Hitam, A.J.Rest, M.Herberhold, and W.Kremnitz, J. Chem. Soc., Chem. Comm., 1984, 471.
340 C.G.Benson, M.G.Little, and C.C.McAuliffe, Inorg. Chim. Acta, 1984, 87, 169.
341 G.A.van Albado, R.A.G.de Graaff, J.G.Haasnoot, and J.Reedijk, Inorg. Chem., 1984, 23, 1404.
342 M.S.Delgado, M.J.Macazaga, and J.R.Masaguer, Z. Naturforsch., Teil B, 1984, 39, 142.
343 Yu.Ya.Kharitonov, V.N.Shafranskii, I.V.Dranka, and A.A. Stratulat, Koord. Khim., 1984, 10, 359.
344 W.C.Hecker and A.T.Bell, J. Catal., 1984, 85, 389.
345 H.-H.Fricke and W.Preetz, Z.Anorg. Allg. Chem.,1984, 507, 12.
346 M.Hvastijova and J.Kohout, Z. Chem., 1984, 24, 105.
347 J.G.Haasnoot, W.L.Driessen, and J.Reedijk, Inorg. Chem., 1984, 23, 2803.
348 D.A.Guzonas, G.F.Atkinson, and D.E.Irish, Chem. Phys. Lett., 1984, 107, 193.
349 A.Santos and A.Tigeras, Inorg. Chim. Acta, 1984, 81, 175.
350 H.-G.Srebny and W.Preetz, Z. Anorg. Allg. Chem.,1984, 513, 7.
351 D.Tudela, V.Fernandez, and J.D.Tornero, Z. Naturforsch., Teil B, 1984, 39, 791.
351a P.Huppmann, G.Klöter, J.S.Thrasher, K.Seppelt, and D.D. DesMarteau, Inorg. Chem., 1984, 23, 2317.
352 Ya.S.Usmankhozhaeva, O.F.Khozhaev, Z.M.Musaev, and N.A.Parpiev, Uzb. Khim. Zh., 1984, 6.
353 V.I.Spitsyn, S.V.Mozgin, M.G.Felin, N.A.Subbotina, and M.I. Aizenberg, Russ. J. Inorg. Chem., 1984, 29, 58.
354 R.I.Machkhoshvili, D.P.Metreveli, G.Sh.Mitaishvili, and R.N. Shchelokov, Russ. J. Inorg. Chem., 1984, 29, 586.
355 P.Ghosh, T.K.Mukhopadhyay, and A.R.Sakkar, Transition Met. Chem. 1984, 9, 46.
356 A.Behr, U.Kanne, and G.Thelen, J. Organomet. Chem., 1984, 269, C1.
357 M.J.Benecky, M.G.Dowling, M.J.Clarke, and T.G.Spiro, Inorg. Chem., 1984, 23, 865.
358 M.G.Ivanov and I.I.Kalinichenko, Russ. J. Inorg. Chem., 1984, 29, 688.
359 U.Thewalt, D.Neugebauer, and B.Lippert, Inorg. Chem.,1984, 23, 1713.
360 L.Naldini, A.Panzanelli, G.Rassu, F.Cariati, F.Demartin, M. Manassero, and N.Masciocchi, Inorg. Chim. Acta, 1984, 83, L71.
361 J.W.Pyrz, K.D.Karlin, T.N.Sorrell, G.C.Vogel, and L.Que, Inorg. Chem., 1984, 23, 4581.
362 A.Bigotto, E.Reisenhofer, and R.Giordano, Spectrochim. Acta, Part A, 1984, 40, 203.
363 J.Perron and A.L.Beauchamp, Inorg. Chem., 1984, 23, 2853.

364 L.N.Lugina, Russ. J. Inorg. Chem., 1984, 29, 390.
365 H.W.Roesky, J.Anhaus, and W.S.Sheldrick, Inorg. Chem., 1984, 23, 75.
366 R.D.Bereman and J.R.Dorfman, Polyhedron, 1983, 2, 1013.
367 C.Diaz and G.Gonzalez, Inorg. Chim. Acta, 1984, 85, 61.
368 H.P.M.M.Ambrosius, A.H.van Hemert, W.P.Bosman, J.H.Noordijk, and G.J.A.Ariaans, Inorg. Chem., 1984, 23, 2678.
369 U.Kynast, U.Müller, and K.Dehnicke, Z. Anorg. Allg. Chem., 1984, 508, 26.
370 N.K.Singh, S.C.Srivastava, and R.C.Aggarwal, Ind. J. Chem., Sect. A, 1983, 22, 704.
371 U.Kunze, A.Bruns, and D.Rehder, J. Organomet. Chem., 1984, 268, 213.
372 E.Lindner, K.Auch, W.Hiller, and R.Fawzi, Angew. Chem., Int. Ed. Engl., 1984, 23, 320.
373 H.G.Hauck, P.Klingelhöfer, U.Müller, and K.Dehnicke, Z. Anorg. Allg. Chem., 1984, 510, 180.
374 W.Hiller, J.Mohyla, J.Strähle, H.G.Hauck, and K.Dehnicke, Z. Anorg. Allg. Chem., 1984, 514, 72.
375 M.Herberhold and W.Böhlemeyer, Angew. Chem., Int. Ed. Engl., 1984, 23, 80.
376 H.W.Roesky, K.K.Panday, B.Krebs, and M.Dartmann, J. Chem. Soc., Dalton Trans., 1984, 2271.
377 G.Tresoldi, G.Bruno, P.Piraino, G.Faraone, and G.Bombieri, J.Organomet. Chem., 1984, 265, 311.
378 C.T.Huys and A.M.Goeminne, Z. Anorg. Allg. Chem.,1984, 509, 201
379 A.Castineiras, A.Arquero, and J.R.Masaguer, Transition Met. Chem., 1984, 9, 273.
380 C.Perchard, M.H.Baron, and C.De Loze, J. Mol. Struct., 1984, 112, 247.
381 S.K.Sharma, R.K.Mahajan, B.Kapila, and V.P.Kapila, Polyhedron, 1983, 2, 973.
382 C.Schumacher, F.Weller, and K.Dehnicke, Z. Anorg. Allg. Chem., 1984, 508, 79.
383 W.A.Schenk and F.-E.Baumann, J. Organomet. Chem.,1984, 260, C6.
384 M.M.Caldeira and J.J.C.Teixeira-Dias, Spectrochim. Acta, Part A, 1984, 40, 265.
385 J.D.Oliver and D.P.Riley, Inorg. Chem., 1984, 23, 156.
386 F.Faraone, G.Bruno, S.Lo Schiaro, and G.Bombieri, J. Chem. Soc., Dalton Trans., 1984, 533.
387 J.Telser and R.S.Drago, Inorg. Chem., 1984, 23, 2599.
388 C.L.Sharma and V.P.Mishra, Acta Chim. Hung., 1983, 114, 3.
389 K.C.Lee and F.Aubke, Inorg. Chem., 1984, 23, 2124.
390 S.P.Mallela and F.Aubke, Can. J. Chem., 1984, 62, 382.
391 V.E.Karasev, I.V.Shchukina, and R.N.Shchelokov, Russ. J. Inorg. Chem., 1983, 28, 1316.
392 S.Singh and R.D.Verma, Polyhedron, 1983, 2, 1209.
393 S.J.David and B.S.Ault, Inorg. Chem., 1984, 23, 1211.
394 S.P.Mallela, K.C.Lee, and F.Aubke, Inorg. Chem., 1984, 23, 653.
395 A.-F.Shihada, I.A.-A.Jassim, and F.Weller, J. Organomet. Chem., 1984, 268, 125.
396 S.Karunaithy and F.Aubke, J. Fluorine Chem., 1984, 25, 339.
397 R.K.Gupta, A.K.Rai, R.C.Mehrotra, and V.K.Jain, Polyhedron, 1984, 3, 721.
398 L.A.Zemnukhova, R.L.Davidovich, and T.L.Semenova, Koord. Khim., 1983, 9, 1369.
399 S.Singh, Amita, and R.D.Verma, Ind. J. Chem. Sect. A, 1983 22, 814.

7
Mössbauer Spectroscopy

BY S. J. CLARK, J. D. DONALDSON and S. M. GRIMES

1 Introduction

Twenty five years have passed since the first demonstration that hyperfine interactions in chemical compounds could be measured by γ-resonance spectroscopy. In the twenty five year period the technique has developed from the status of an interesting novel observation to a widely used method for studying solid-state samples containing appropriate Moessbauer nuclei. The spectra obtained provide valuable information to chemists, physicists, biologists, zoologists, metallurgists, and geologists on many aspects of the structural, electronic and magnetic properties of their samples. To the chemist in particular the development of a sophisticated Moessbauer-atom-specific technique for determining the ways in which the atom uses and distributes its outer electrons has proved to be of immense value. In the area of main-group element chemistry particularly Moessbauer spectroscopy has provided evidence on the use of bonding electrons that could not easily have been obtained using any other method of study. The seventh volumes of the literature services 'Moessbauer Effect Data and Reference Journal'[1] and 'Moessbauer Spectroscopy Abstracts'[2] were published in 1983 and continue to provide a valuable title list and abstract service respectively.

This introductory section provides details of books and review articles published during the review year and is followed by sections that deal with advances in instrumentation and methodology. Sections 4, 5 and 6 contain detailed reviews on the iron-57, tin-119 and other isotope Moessbauer effects respectively while section 7 reviews papers published on conversion electron and back-scattering γ-resonance spectroscopy. The isotopes (energies in keV in parentheses) that have been mentioned, some of them in review articles, during the year include: Fe-57(14.412), Ni-61(67.4), Zn-67(93.26), Ge-73(13.3), Ru-99(89.36),

Sn-119(23.875), Sb-121(37.15), Te-125(35.46), I-127(57.6), I-129(27.72), La-139(165.9), Pr-141(145.4), Nd-145(67.3), Sm-149(35.8), Eu-151(21.64), Gd-155(86.54), Dy-161(80.7), Er-166(80.56), Tm-169(8.4), Yb-170(84.26), Ta-181(6.24), Ir-191(82.4), Ir-193(73.0), Au-197(77.35), and Np-237(59.54).

Books and Reviews. In a volume[3] dedicated to Rudolf Moessbauer to mark the first twenty-five years of the use of nuclear γ-resonance spectroscopy, R.H. Herber has edited the proceedings of a symposium on Moessbauer spectroscopy held during the 187th. meeting of the American Chemical Society. The volume contains review papers on most of the important developments in the technique over the last quarter century. The papers are by B.A. Sosinsky on aspects of organoiron Moessbauer spectroscopy, by P. Guetlich on spin transitions in iron compounds, W.M. Reiff on zero and high field Moessbauer spectroscopy studies of the magnetic ordering behaviour of one, two and three dimensional systems, R.H. Herber and H. Eckert on Moessbauer spectroscopy of intercalation compounds, H. Sano on hot-atom chemistry and trapped species, R.H. Herber on structure, bonding, and the Moessbauer lattice temperature, L.H. Bowen and S.B. Weed on Moessbauer spectroscopy of soils and sediments, D.L. Ulrich on Moessbauer studies of liquid crystals, J.J. Zuckerman on organotin-119m Moessbauer spectroscopy: the first quarter century, H. de Waard on the Moessbauer spectroscopy of Iodine, J.F. Stevens on Moessbauer spectroscopy of Antimony compounds, G.K. Shenoy on rare-earth Moessbauer studies of chemical problems, and D.L. Karaker on chemical aspects of Np-237 Mossbauer spectroscopy.

A reference text[4] on Moessbauer spectroscopy applied to inorganic chemistry has also been produced. This text contains an Introduction and a chapter on basic concepts by the editor G.J. Long. These are followed by chapters on Moessbauer spectrometers and their calibration by T.E. Cranshaw; on spectral data reduction and refinement by G. Longworth; on Moessbauer effect isomer shifts by G.K. Shenoy; on the electric field gradient and the quadrupole interaction by H. Spiering; on Moessbauer spectroscopy and the coordination chemistry of iron by W.M. Reiff and G.J. Long; on spin transitions in iron complexes by P. Guetlich; on applications to biological systems by D.P.E. Dickson; on Moessbauer spectroscopy in heterogeneous catalysis by F.J. Berry; on Moessbauer spectroscopy of silicate minerals by J.M.D. Coey; on studies of ceramics and archaeological materials by G. Longworth, on structure and bonding in tin compounds, by R.V. Parish; on Gold-197 Moessbauer spectroscopy in the characterization of gold compounds, by R.V. Parish; and on one-dimensional magnetism by C.E. Johnson.

An article on Moessbauer spectroscopy has been published in Analytical chemistry as part of its series of biennial reviews,[5] and the developments in the use of γ-resonance spectroscopy by Chinese scientists have been surveyed.[6]

A number of publications that have appeared since the last report have been concerned with applications of the Moessbauer effect to studies of a number of important systems including magnetic materials[7], micro-organisms[8], proteins[9], coal[10] and low-rank coals[11]. The uses of the technique in providing information in a number of fields of study have been reviewed including applications in environmental and geochemical studies,[12] archaeology[13], energy studies[14], and nutritional research.[15] The use of γ-resonance data to provide chemical state analyses of iron compounds has been considered[16]. Moessbauer data have also been included in general review articles on a number of elements and types of compounds including iron[17], tin[18], tellurium[19], lanthanides[20], gold[21], antimony[22], light actinides[23], ferrocenes[24], iron hydroxyoxime complexes[25] and tin dialkyldithio phosphates[26]. Other topics reviewed include the uses of Moessbauer spectroscopic data in supplementing neutron spectroscopic data in studies of the diffusion and correlation of lattice defects[27], in relaxation studies of domain wall dynamics in pure and impure Ising-type magnetic chains,[28] in static and dynamic structure studies[29], in investigations of surface and catalytic phenomena[30], and in studies of metallic glasses[31]. The uses of Moessbauer emission spectroscopy in hot atom chemistry have also been surveyed.[32]

During the review year of this chapter, the proceedings of the 1983 Second Seeheim Workshop on trends in Moessbauer spectroscopy became available.[33] This workshop was one of the symposia designed to celebrate the twenty-fifth anniversary of the discovery of the Moessbauer effect. A list of the papers presented at a symposium on industrial applications of Moessbauer spectroscopy was printed in the Moessbauer Effect Reference and Data Journal.[34] Applications described in the papers presented included studies on catalysis, corrosion, alloys, ceramics, glasses, minerals, battery materials, magnetic recording materials, thin films, and superconductors.

2 THEORETICAL

Spin-lattice relaxation rates for paramagnetic $Fe^{3+}(^6S)$ have been calculated[35] explicitly using a dynamic spin Hamiltonian and have been compared with observed Moessbauer data. The relative importance of the contributions from a number of relaxation mechanisms including one-phonon, direct two-phonon, and indirect 2-phonon pathways at temperatures below 300 K was considered in the light of the calculated and observed data. Iron-57 Moessbauer spectra for amorphous $Fe_{24}Zr_{76}$ alloys were shown[36] to have almost exclusive quadrupole splittings. This observation is in conflict with the predictions of the dense random packing models for amorphous alloys and does indicate the presence of strong short-range order. It has been claimed[37] that the isomer shift and quadrupole splitting data can be calculated for a series of isostructural compounds provided the correct sign of the e.f.g is assigned to the splitting data. The correlation was used to predict the sign of the e.f.g for some monosubstituted pentacyanoferrate(II) complexes but there are dangers in the use of such correlations because they assume a direct relationship between changes in s-electron density (shift) and the imbalance in *p*- or *d*-electron densities (splitting) not the *p*- and *d*-electron densities themselves.

A mathematical model has been devised to explain the Moessbauer parameters obtained from liquids and solutions trapped in the pores of porous silicate glasses[38]. Three glass-liquid models were considered, *viz* (1) situations in which only solvent molecules are in contact with the capillary walls and in which all Moessbauer nuclei have similar environments because they are only surrounded by solvent molecules, (2) situations in which the Moessbauer nuclei are in molecules large enough to assume that they are all in direct contact with the capillary walls, (3) situations in which the Moessbauer nuclei are part of small molecules only some of which are in contact with the capilllary walls.

Theoretical analysis of the Debye-Waller factor and the dynamics of the presence of substitutional Moessbauer impurity atoms in harmonic solids has been presented.[39,40] The double-time Green's function method was used in the analysis. The presence of the impurity is found to broaden the resonance line with an increase in broadening as the level of the impurity increases. The results have been discussed in terms of the dependences of the mean-square vibrations on the various masses and force-constants in the lattice. The Moessbauer f-values were estimated from the theoretical model making allowances for mass disordering and force constant changes and good agreement was found between the calculated f-values and previously published experimental data. The effects of the presence of a surrounding macromolecule on the spectrum of a Moessbauer label atom have also been considered.[41] The resonance spectra for the label atom tightly bound to a macromolecule and therefore sensitive to its low frequency thermal vibrations has been calculated. The deformation inner mobility of atoms or groups of atoms within the macromolecule was described by collective excitations of the surface with small deviations from the equilibrium spherical form. The γ-resonance f-factors were determined by the coefficient of surface elongation and the internal viscosity of the macromolecule. The model explains the observed temperature dependence of the Moessbauer probability in terms of its strong dependence on the viscosity of the host macromolecule. Gupta[42] has proposed a model to describe the dynamics of a Moessbauer atom in some biological systems in which the observed spectra show two components - a narrow line and a broad line with an energy range several orders of magnitude larger than the natural line width. The model suggests that the broad resonance arises from quasi-elastic damped harmonic motion rather than inelastic processes. The predictions of the model agree well with the observed temperature dependence of the width and intensity of the broad-line component. The shapes and temperature dependencies of the γ-resonance spectra of iron-containing proteins have also been analysed[43] in terms of a model involving N-coupled Brownian oscillators for the internal dynamics of the proteins. The changes in isomer shift and hyperfine field in the first coordination shell of impurity atoms in iron-based 3d-metal alloys have been discussed[44] in terms of different charge screening models. Although the isomer shift changes can be described in terms of the asymptotic Friedel model, interpretation of the hyperfine field data is more difficult.

Two papers available since the last report are concerned with theoretical treatments of cluster atoms. For trinuclear clusters of transition metals the appearance of structural features of the exchange clusters in their Moessbauer spectra were considered[45] in terms of a simplified model which shows that non-Heisenberg interactions of the Dzyaloshinskii type have substantial effect on the quadrupole interactions of the clusters. Geometric distortions of the cluster are not reflected in the spectra because of the significant dependence of the quadrupole parameters on the ratio between the distortion parameters and the Dzyaloshinskii interaction. The antisymmetric Dzyaloshinskii exchange interaction tends to eliminate the differences in quadrupole splittings brought about by structural inequivalences in the cluster atoms. The presence of a strong magnetic field, however, depresses the Dzyaloshinskii interaction and unfreezes static deformations. These effects are reflected in the γ-resonance spectra. The model thus explains the observed spectra in terms of interactions between the Moessbauer nuclei and the exchange subspectrum. It was shown[46] that Moessbauer spectra of exchange-coupled mixed valence clusters can be used to determine the parameters of the inter-atom interaction responsible for electron transfer. The presence of double exchange leads to splitting of the hyperfine structures of the spectra.

An interpretation of the Moessbauer specta of di-octahedral Fe^{3+}-containing layer silicates has been presented[47] including calculation of e.f.g. values and the relative intensities of the quadrupole-split lines.

A theoretical description of the unusual features found in secondary emission spectra under conditions of total reflectance of the Moessbauer radiation has been derived.[48] The calculated spectra consist of substantially broadened asymmetric double peaked lines. Under the total reflectance conditions the inhomogeneous secondary emission wave is shown to be concentrated near the surface of the sample with its depth of penetration varying with the sliding angle and along the resonance curve.

The possibility of the application of the invariance principle to solve the problem of γ-radiation transfer in crystals containing Moessbauer nuclei was considered for the case where nuclear resonance and Rayleigh scattering take place.[49] An analytical expression was derived to describe the radiation intensity after resonance scattering with spherical indicatrix and complete frequency redistribution in crystals of finite thickness. The model shows that the width of the γ-radiation spectrum increases with increasing thickness of the scatterer. The sideband spectra of the Fe-57 Moessbauer radiation measured as a function of frequency and of the voltage of a quartz crystal on which the Co-57 source was deposited have been analysed by considering the distribution of the modulation index.[50] The width of the distribution was found to be frequency dependent and the quantum beat measurements found to be in agreement with previous optical theory.

The effects of source thickness on the apparent recoil-free fractions and linewidths of γ-resonance spectra have been determined[51] taking into account both resonant self-absorption and non-resonant absorption. An analytical series expansion in powers of the effective source thickness has been obtained to calculate the source line width which can be used in cases of small thickness sources. It is known that both the inhomogeneity of the sample thickness and the dead-time effect in collecting γ-resonance spectra have an effect upon the determination of Debye-Waller factors from Moessbauer spectroscopic data. This arises in the case of the inhomogeniety because of lowering of the spectral intensity and deformation of the line shape. Kansey <u>et al.</u>[52] showed that it is necessary to take into consideration the electronic absorption coefficient and the term proportional to the second derivative of the transmission integral to obtain the correct theoretical Moessbauer spectrum. Using these methods, the room temperature Debye-Waller factors obtained for sodium nitroprusside, iron(II) sulphide, and Fe-57 in a chromium matrix were found to be 0.376 ± 0.002, 0.855 ± 0.026, and 0.763 ± 0.005 respectively. Two different methods were devised for solving the transmission integral in Moessbauer spectroscopy,[53] viz. (1) approximating the transmission integral by a single channel calculation taking account of the nearby cross-section contributing to the absorbtion by comparing the cross-section at the energy of the channel used with the approximate solution in the thin absorber approximation and (2) a more elaborate approximation in which the inner part of the integral is convoluted numerically and the outer part calculated by an asymptotic approximation.

Two papers published during the last year were concerned with the inverse problem of Moessbauer spectroscopy in radio-frequency-vibrating crystals.[54] In one, a method is described for the evaluation of statistical errors in the solution of the problem of recognition of a periodic perturbation acting on a Moessbauer atom in a crystal by use of its frequency-modulated γ-ray resonance spectrum. The periodicity of the method was confirmed by model calculations. In the other paper an approach to incoherent ultrasonic vibrations is presented. The method for the recognition of periodical perturbations at ultrasonic frequencies acting on a Moessbauer atom in a crystal has now been generalised to include the case when not all nuclei in the crystal vibrate at the same amplitude. Again results from numerical calculations for several model situations are shown to be satisfactory.

Studies have been carried out on the use of coincidence Moessbauer spectroscopy in the presence of a hyperfine field and electronic spin relaxation.[55] An expression was derived for the transmission through a resonant absorber as a function of time using joint solutions of the Maxwell and density matrix equations. The radiation that reaches the detector behind the absorber is shown to be modulated by the hyperfine field. A number of polarisation schemes for the observation of coincidence Moessbauer spectra have been analysed and numerical calculations made for the Fe-57 nucleus.

A dynamical theory of Moessbauer diffraction in the presence of hyperfine interactions for total external reflection Bragg diffraction geometry was described by Andreeva and Kuzmin[56]. They showed that the oversurface diffractive wave intensity had a maximum near the critical angles and that it also had an unusual angular dependence. Anomalously high reflectivity has been observed in nuclear resonance diffraction in the Bragg geometry in nearly perfect $Fe(57)BO_3$ single crystals. These observations can be explained in terms of the dynamical theory for resonance diffraction with the high reflectivities resulting as consequence of the suppression effect.[57] For γ-resonance crystal diffraction at low temperature a number of new phenomena, including rotation of the plane of polarisation of the γ-quanta and double refraction in the polarised media, have been considered and the possibility of studying the distribution of the degree of nuclear polarisation and temperature gradients within crystals discussed.[58] A general scattering theory of photons by field targets periodically dependent upon time has been described for the scattering of γ-resonance radiation by a crystal.[59] Included in the theory was scattering of γ-quanta by a crystal that performs periodically enforced vibrations under the influence of alternating fields. The cross-sections for scattering and absorption of γ-rays by the crystal were also calculated.

3 INSTRUMENTATION AND METHODOLOGY

The design of a multidimensional parametric γ-resonance spectrometer has been described.[60] The spectrometer is equipped with several Doppler modulation systems and an adaptive method of automatic control. The sweep rate reproducibility is better than 5% for several vibrators working simultaneously. A spectrometer with an electron recording capability has also been described[61]. This system consists of a Doppler modulation unit, the source, a device with an inlet window for separating electrons on an energy basis, an ellipsoidal electrostatic mirror with the analyte sample at the first focus and an electron detector at the second focus, a protective screen and a recording device. A series of additional electron detectors are also placed in parallel to minimise any measurement errors. Although most Moessbauer spectrometers operate in the time mode, pulse-height modulation provides an alternative, and a system using the alternative mode has been designed.[62] A simple multiscaler interface has been constructed to permit the use of the Apple II microcomputer in γ-resonance data collection[63] and a microcomputer-base system using a data acquisition interface circuit described.[64]

A new electronic technique for the rapid determination of the zero velocity channel in a velocity scanning spectrometer was developed.[65] The applications of the technique are that it permits study of time stability of the system, of drive and signal quality and of the presence of any off-set voltage superimposed on the drive channel.

The accuracies of the modulation techniques used in Moessbauer spectroscopy to measure small energy shifts between source and absorber line positions have been compared.[66] The techniques considered were sinusoidal and constant velocity Doppler modulation at low drive frequency and sinusoidal and square-wave-shaped phase modulation at high drive frequency. The use of the transient frequency region was also discussed. A strong dependence of the result on the thickness of the absorber was found and in many cases high frequency and transient methods were found to give smaller statistical errors than conventional methods. The Moessbauer effect in Zn-67 was used to illustrate the measurement of small shifts.

A simple, inexpensive, nitrogen flow cryostat for Moessbauer measurements in the temperature range 77-300K has been described[67] and the design of a high pressure chamber with polycrystalline cubic boron nitride Bridgeman anvils for use in high-pressure γ-resonance studies on tin-containing materials reported.[68] The high-pressure chambers can produce static pressures up to 200 kbar on 1mm^2 samples. A Spanish patent describes the synthesis of a (Co-57)MgO source.[69]

A Co-57 source and a stainless steel absorber have been used in a study of the time dependence of resonant γ-radiation modulated by acoustic excitations.[70] In the experiment the absorber was oscillated at 3400Hz and spectra were obtained for different voltages at the vibrosystem transducer. The time dependent system permits (1) measurement of isomer shifts in a suitable manner, (2) the determination of the initial phase of oscillations, and (3) the study of the form of the complex oscillations of the vibrosystem.

The shift measured for the Co-57 source-stainless steel absorber pairing was 0.029mms^{-1}. A new method of nanosecond modulation of Fe-57 Moessbauer radiation has been described by Smirnov et al.[71] In the method, the time of arrival of the γ-radiation at the nuclear system under study is fixed by rapid opening of the beam of Moessbauer quanta using a magnetic resonance chopper. This was achieved by exploiting the low inertia of $FeBO_3$ crystals on remagnetisation. Modulation of the Fe-57 radiation on rotation of the magnetisation of the crystal occurs because of the strong polarisation dependence of the resonance absorption. The radiation pulses obtained had a rectangular shape and a front deviation of less than 10ns.

The total electronic absorption coefficient for 14.4keV iron radiation in Al has been measured for the first time using the Moessbauer effect as a monochromator.[72] The data obtained are important for the determination of the background in Moessbauer recoilless fraction measurements especially if the energy of γ-radiation from the source host lattice is close to 14.4keV. Improved precision in hyperfine field measurements was said to be achieved by carrying out measurements on absorbers containing both the sample under study and an internal standard.[73] Using the method, the standard deviations of the hyperfine fields of the antiferromagnetic phases in a series of alumino-haematites was reduced from 1.6 to less than 0.4 kOe and this improved precision allowed the determination of the degree of Al substitution to about ± 1%.

A new method for studying Fe-containing surfaces that involves visible light emitted following an Fe-57 Moessbauer event has been developed.[74] The sample to be studied by the method is placed in a chamber in an elliptical reflector with the γ-resonance source and a photoelectric detector positioned at right angles to each other.

A unique algorithm applicable to any Moessbauer nucleus was developed for fitting γ-resonance spectra in the general case of combined hyperfine interaction of the nucleus with the magnetic and electric field surrounding the nucleus[75]. A new fast method of analysing spectra based on least-squares fitting has also been described[76]. The principles of an algorithm for automated preliminary data processing [77] and interactive computer program written in FORTRAN for a VAX11/780 computer for fitting Moessbauer spectra of either mirror image or sawtooth type to Lorentzian peaks[78] and a spectrum stripping technique[79] are among the topics described in papers on the computer fitting of Moessbauer spectra. The least squares and utmost correlation methods of fitting the resonance spectra have been compared.[80] The utmost correlation method is based on the Schwartz inequality and consists of searching for the function that provides the maximum correlation with the experimental spectrum. A user-orientated versatile Mossbauer spectra analysis program to simulate transmission spectra has also been devised[81]. The calculation involved considers a number of experimental situations and comparisons are made with simulated spectra either by least-squares sums or by plotting the simulated and measured spectra. A fitting routine that minimises the least squares sum is used to obtain the Moessbauer parameters. A precise measurement method called the residual area analysis method has been described for determining small parameter changes of single symmetrical or nearly symmetrical Moessbauer resonance lines.[82] The advantage of the method is that it does not require either a prior knowledge of line-shape or a curve fitting procedure. Close correspondence between experimental results and theoretical conclusions from the residual area analysis method is claimed. Another aspect of the use of computers in Moessbauer spectroscopy is dealt with in a paper by Stevens on the operation of a small data centre for Moessbauer data.[83]

Three papers published during the year described the use of the Moessbauer effect in chemical analysis applications. A rapid Moessbauer method for the non-destructive determination of iron-bearing phases has been described.[84] A theoretical analysis is used to stimulate spectra that are based on the differences in the gamma resonance given with the active source stationary and in vibration. The method can be used to determine pyrites in coal and ores and iron in steel and slag samples. The concentrations of Mössbauer elements or the probabilities of their resonance absorption in multicomponent samples have been calculated by accounting for the photoeffect on the atoms of all the elements.[85] Mossbauer spectral lines that have been broadened by environmental effects can be fitted to pseudo-Lorentzian line shapes. In the case of mineral mixtures these lines can be used for analytical purposes and the example of analysis of bauxite containing haematite and goethite is described. The major haematite component of the spectrum has an influence on the goethite component but this can be overcome and analysis achieved by constraining the line width and the intensity of the goethite spectral lines during the curve-fitting procedure.[86]

Previous techniques for analysing the six-line Fe-57 Moessbauer spectra obtained from iron-metalloid ferromagnetic glasses have been based on extracting the detailed shape of the hyperfine field that dominated the nature of the outer lines. More subtle aspects of the spectra that provide correlations between the hyperfine field, the chemical isomer shift and the Zeeman quadrupole energy are usually neglected or assumed to be of simple linear form. It has been shown[87] that, if some aspects of the shape of the hyperfine field distribution, such as its mean and its variance, are considered, then the Moessbauer spectra can be analysed in a consistent manner that retains all relevant parameter correlations even if no analytic relationships exist between the parameters. Room temperature Moessbauer Zeeman spectra for six different boron

TABLE 1 Mean values and root mean square deviations and correlations between Moessbauer parameters shift (δ), hyperfine field (H), and a first order function (u) describing the e.f.g. - nuclear quadrupole moment

	$Fe_{76}B_{24}$	$Fe_{82}B_{18}$	$Fe_{83}P$
δ/mms^{-1}	0.092(1)	0.064(2)	0.143(5)
$\sigma(\delta)/mms^{-1}$	0.105(12)	0.093(9)	0.12(3)
u/mms^{-1}	-0.009(5)	-0.004(5)	-0.012(12)
$\sigma(u)/mms^{-1}$	0.120(10)	0.132(8)	0.19(2)
H/kOe	241.2(3)	248.8(3)	250.7(7)
$\sigma(H)/kOe$	34.0(3)	33.2(3)	35.7(6)
$\mu_N\langle\Delta H\,\Delta\delta\rangle/(mms^{-1})^2$	+0.035(5)	+0.046(4)	-0.08(1)
$\mu_N\langle H\,\Delta u\rangle/(mms^{-1})^2$	-0.046(5)	-0.042(4)	-0.02(1)
$\mu_N\langle(\Delta H)^2\Delta\delta\rangle/(mms^{-1})^3$	-0.04(1)	-0.05(1)	+0.10(3)
$\mu_N\langle(\Delta u)^2\Delta\mu\rangle/(mms^{-1})^3$	+0.00(1)	+0.01(1)	-0.10(3)

and phosphorous metalloid glasses have been analysed in this way and the mean values and variances for hyperfine fields, isomer shifts and pure quadrupole distributions obtained are in Table 1. Variances obtained for the isomer shifts and quadrupole splittings are the first reported for a ferromagnetic phase. Analysis of the parameter correlations for the amorphous phase with the corresponding correlations for the tetragonal crystalline Fe_3B and Fe_3P phases shows that, although the phosphorous glasses are closely related to Fe_3P, no similar relationship exists between the boron glasses and Fe_3B. In another paper Billard et al.[88] describe a very simple model for analysing the Moessbauer spectra of metallic glasses. The model is based on a soft-sphere relaxed description of the glass and involves analysis of the structure from the point of view of the electric-field tensor to provide a distribution law for the quadrupole splittings and the dipolar terms to the internal field. Fe-57 and Sn-119 γ-resonance measurements have also been made on the amorphous alloys $Fe_{79}SnB_{20}$, $Fe_{69}SnCr_{10}B_{20}$, $Fe_{40}Ni_{39}SnB_{12}$ and $Fe_{35}Ni_{34}SnCr_{10}B_{20}$.[89] Window's Fourier expansion was used to obtain hyperfine field distribution data from the spectra. The hyperfine field distribution for Fe-57 is a single well defined peak for alloys containing no Cr and a broadened bimodal curve for alloys containing Cr. Fe-57 Moessbauer spectra and soft X-ray appearance potential spectroscopy were combined in a study of the iron-nickel-boron metallic glasses $(Fe_{0.5}Ni_{0.5})_{1-x}B_x$, where x= 0.16-0.26. The combination of both sets of data for iron permits[90] the derivation of the charge of sub-band occupation as a function of B concentration and shows that charge transfer occurs not only from B to empty minority iron states but from majority to minority iron states. A special cell was constructed to permit the study by Moessbauer spectroscopy of the chemisorbtion of nitrogen and hydrogen onto highly dispersed iron particles supported on magnesium oxide.[91] The intensities of the resonance spectra for dilute substitutional Fe-57 in $NbH_{0.84}$ and $VD_{0.78}$ show an anomalous decrease near 250 and 200K respectively. This was attributed[92] to Arrhenius behaviour of the hydrogen jump rates with excitation energies of about 0.20 and 0.14 eV respectively.

A number of papers published during the year were concerned with the methodology of the Moessbauer effect in specific applic-

-ations. Included in these applications are the uses of gamma-resonance spectroscopy in studies of oxygen diffusion electrodes[93], of interfaces[94], of photochromic centres in Fe-$SrTiO_3$[95], of the corrosion of boiler tubes[96], of archaeological artefacts[97], of magnetic anisotropic constants in iron[98] and of aqueous phases in perfluororocarboxylate membranes[99].

A number of papers published in the review year refer to the use of Moessbauer resonance for scattering experiments.[100-102] The high energy resolution of the Moessbauer effect was used to separate elastically and inelastically scattered radiation from a silicon single crystal. The measured ratios of the elastically scattered radiation at different Bragg angles were then used to determine the temperature factor for silicon.[100] A method has been described[101] that makes it possible to determine the moment of arrival of gamma-resonance radiation in a scattering system of nuclei[102] using radiation from Fe-57 in $FeBO_3$ crystals. Moessbauer spectra reflected from and transmitted through a thin nearly perfect $^{57}FeBO_3$ single-crystal platelet have been measured at different angular positions in the vicinity of a Bragg angle.[102] An angular-dependent shift of the nuclear resonance dip was found in the transmission spectra in common with a shift of the resonance peaks in the reflection spectra. A pronounced asymmetry of the transmitted intensity was also observed above and below the resonances. The Borrmann effect induced by nuclear resonance diffraction was observed for the first time in this study and the angular dependence of the transmission spectra showed that, under Bragg reflection conditions, diffraction becomes as important as internal conversion and incoherent scattering.

4 Iron-57

__General Topics.__- __Nuclear Parameters and Metallic Iron.__ An MgO:^{57}Co Moessbauer source has been patented.[103] It has δ = -1.05 mms^{-1} (relative to $K_4Fe(CN)_6.3H_2O$), Δ = 0.26 mms^{-1}, and intensity 8.9. Substitution induced spin and charge density changes in <u>bcc</u>-Fe have been studied on substitution of its atoms by Al, Si, Ge, V, W, and Cr.[104] Changes in spin and charge density were calculated from the linear correlations between the hyperfine fields and chemical shift and between the average hyperfine field and the average number of solutes within the first two neighbour shells. The surface and interface magnetism of a very thin ^{57}Fe layer deposited onto a thick ^{56}Fe film and the magnetic properties of iron interfaces in contact with various nonmagnetic materials have been studied.[105] Moessbauer spectroscopy and high energy ion beam scattering have been used to study amorphous, sputtered, iron stabilised by carbon.[106] A possible relationship between s-electron densities and and magnetic hyperfine fields at the iron nuclei was discussed. Nitrogen and hydrogen chemisorption on ultrafine iron particles has been studied.[107]

Iron-57 Impurity Studies, Polymers, and Ion Exchange. The e.f.g. at ^{57}Fe probe nuclei in Se has been measured over the temperature range 120-453K and shown to follow a $T^{3/2}$ law.[108] The Moessbauer spectrum of Ga_2Se_3 containing 2 at.% of ^{57}Fe has been measured at 80K.[109] A doublet was obtained with δ = 0.65 mms^{-1} (rel. to ^{57}Co in Cr) and Δ=2.4 mms^{-1}. This is consistent with about 1 at.% Fe actually entering the matrix and bonding with matrix atoms.

There have been several Moessbauer studies on iron doped carbonates. The temperature dependence of the quadrupole splitting of Fe-57 in $MgCO_3$ has been explained by including the effects of orbit-lattice interactions in a crystal field calculation.[110] It was suggested that the orbit-lattice interaction could also affect the temperature dependence of other Moessbauer parameters. The spin-lattice relaxation times of Fe^{2+} in $ZnCO_3$ and $MgCO_3$ have been calculated in terms of the normal coordinates of the complex composed of the paramagnetic ion plus its nearest neighbours and compared with experimental results obtained from Moessbauer data.[111] A reinterpretation of Moessbauer data for the $MnCO_3$:Fe system has been published.[112] A phase diagram was established for phases containing up to 0.04 at.% iron. Despite broadening by internal random magnetic fields the boundaries of the mixed phase were clearly observed in the Moessbauer data. The Moessbauer spectrum of a single crystal of $(Fe_{0.002}Co_{0.998})Cl_2$ at 4.2K has been analysed in terms of the effective-field approximation to the hyperfine interactions.[113] Normal state and optically pumped Moessbauer spectra have been studied to identify photochromic centres in $SrTiO_3$ doped with up to 3 at.% iron.[114] The type and fraction of the centres was found to depend on the impurity concentration. Site geometry and charge compensation effects have been considered in an interpretation of the Moessbauer spectra of $LiNbO_3:Fe^{3+}$.[115]

A Moessbauer study of polymers prepared from poly(vinylpyridine) and $FeCl_3$ or $Fe(NO_3)_3$ has been described.[116] Three types of iron centre were identified corresponding to amorphous hydrated ferric oxide particles, a small number of high-spin Fe^{2+} sites, and to oxygen bridged ferric dimers. The last type of site was only found in the chloride preparations and was assigned to $[Cl_3Fe\text{-}O\text{-}FeCl_3]^{2-}$. Studies involving Moessbauer spectroscopy have been made on [Fe(phthalocyaninato)-(μ-pyrazine)]$_n$[117] and on complexes of iron(III) with polyvinylhalogen carbazoles.[118]

A perfluorocarboxylate membrane has been characterised by a number of techniques including Moessbauer spectroscopy and shown to be similar to sulphonated fluoropolymer membranes.[119] No Moessbauer spectrum could be obtained above 260K, but at lower temperatures two ferric doublets were observed and assigned to hydrated and anhydrous oxy-bridged Fe^{2+}. Moessbauer spectroscopy of Nafion membranes in their ferrous and ferric forms have indicated that the ions are entirely located in a liquid environment at and below room temperature.[120] Both Moessbauer and electron microprobe data showed that the iron was precipitated as small particles during the reexchange by another cation. The distribution and structural configurations of iron in commercial Grace 4A Zeolite exchanged with Fe^{3+}, pumice samples and in hydrothermally derived zeolites have been studied.[121] Several different iron sites were revealed by Moessbauer spectroscopy.

<u>Frozen Solutions and Matrix Isolation.</u> Frozen solutions of acetone and acetonitrile containing mixed valence iron complexed with triflouroacetic and trichloroacetic acids have been studied by Moessbauer spectroscopy in the temperature range 80-190K.[122] Solvation was found to shift the decolourisation of electrons to higher temperatures and to decrease the Moessbauer line width due

to the removal of crystal lattice defects.

Moessbauer evidence has been presented for the formation of a bridging $H_3O_2^-$ group via the hydrolysis of the mononuclear iron complex Fe(EHGS), where H_3(EHGS) = o-$HOC_6H_4CH:NCH_2NHCH$-(o-HOC_6H_4)COOH.[123] The spectra were strongly pH dependent, and at values of pH near the pK_a contributions from a fast relaxing component were observed. These results were interpreted in terms of the formation of a strong hydrogen bond between the hydroxide group of one complex and an aquo ligand of another to complete the bridge. Frozen solutions of $FeCl_3$ and $Fe(SCN)_3$ in pyridine have been studied.[124] The Moessbauer parameters of the thiocyanate complex were very similar to those for solid $Fe(SCN)_3.py_3$ but the chloride had an unusually large quadrupole splitting of 1.45 mms^{-1} attributed to the formation of polymeric species.

The reactions of iron atoms and small iron clusters with the small molecules H_2, CH_4, CO, HI, N_2O, and CH_2N_2 have been investigated.[125] The state of dispersion and photochemical behaviour of tris(β-dikenato)iron(III) complexes isolated at low temperatures have been characterised by Moessbauer spectroscopy.[126] Iron hydride molecules have been characterized from the Moessbauer emission spectra of ^{57}Co implanted in solid hydrogen.[127] The compound, which was stable up to at least 70K, had δ = 0.59 mms^{-1} (relative to α-Fe) and a temperature independant quadrupole splitting, ΔE_Q, of 2.4 mms^{-1}. Photofragmentation processes of pentacarbonyl iron dispersed in polyethene and polytetrafluoroethene have been studied.[128] The reaction products, together with information from i.r. and Moessbauer spectroscopy, suggested the formation of $Fe(CO)_4$ and $Fe(CO)_3$ as intermediate species. It was found that the polymer matrix allows the isolation of otherwise unstable species, and the high permeability of low density polyethene to gaseous CO prevents the reformation of $Fe(CO)_5$ thus excluding the "cage effect" observed in frozen gas matrices. Self-consistent M.O. calculations were also performed and compared with the experimental results.[129]

A Moessbauer study of 1,1'-diacetylferrocene in the liquid crystal 4-pentylphenyl-4'-heptyloxythiobenzoate ($\bar{7}$S5) has examined the crystalline phase at 110-300K and the smectic-C phases in the temperature range 110-240K.[130] From the deviation of the ln f vs. T data from linear behaviour the smectic-glass to supercooled smectic transition was found to be 215K and information was obtained on the ordering of the glass phase.

Emission Studies. Moessbauer emission spectroscopy has been used to study the segregation behaviour of cobalt in silicon.[131] Emission spectra of Co-57 doped La_2NiO_4 obtained at 4.2 and 77K have revealed magnetic order increasing with paramagnetism in a sample which contained a small concentration of Co-57 impurities and $La_{2+n}Ni_{1+n}O_{4+3n}$ intergrowth.[132] The Moessbauer spectra of Co-57 doped lithium niobate and lithium tantalate have been studied at 4.2K in an applied magnetic field of 2 - 7T.[133] The electronic field and magnetic field dependence of the line intensities were related to irreversible processes such as the decoupling of electrons from the lattice or a field-dependent slow relaxation.

Emission Moessbauer studies have been made on the iron ions produced from ^{57}Co adsorbed onto α-Fe_2O_3, Cr_2O_3, and α-Al_2O_3.[134] Clustering of high-spin Co(III) was identified in the surface layer of a $Co_{3-x}O_4$ film by using Moessbauer and XPS data.[135] $^{57}Co_{1-x}O$ sources, prepared at 1000°C in various oxygen pressures, have been studied from room temperature up to 1000°C.[136] It was found that, above 500°C, electronic relaxations occurred between the ^{57m}Fe impurity and the valence band. A theoretical model which describes this behaviour was presented.

Compounds of Iron.- High-spin Iron (II) Compounds. The magnetostatic properties of $K_2Co_xFe_{1-x}F_4$ have been examined by measuring the magnetic susceptibility and Moessbauer effect at different concentrations and temperatures.[137] Three magnetically ordered phases were identified and described. A different group of authors have examined the corresponding rubidium compound and identified two magnetic phases.[138] In both sets of results an antiferromagnetic phase was described in which the spins were aligned at an oblique angle to the c-axis. Another competing anisotropy system, $Fe_{1-x}Co_xCl_2$, has been studied at 4.2K as single crystals and over a range of temperatures (for x = 0.32, 0.35, and 0.44).[139] The observed hyperfine fields and quadrupole splittings were compared with parameters calculated using the rigid-crystal model. Non-linear domain wall dynamics in the quasi-one-dimensional ferromagnet $FeCl_2(pyridine)_2$ have been studied by Moessbauer and susceptibility experiments with particular reference to the effects of doping with nonmagnetic impurities.[140]

An investigation of the states of iron in the $NaCl-FeCl_2$ system has revealed the formation of a Suzuki phase in crystals containing <0.1 mol.% Fe.[141] This phase was identified by a singlet line with $\delta = 1.38$ mms^{-1}. The Moessbauer spectra of $xKCl.(90-x)ZnCl_2.10FeCl_2$, where x < 27, have suggested that the ferrous ions are present, as a network modifier, at interstitial sites of the 3-dimensional network of $ZnCl_2$ tetrahedra.[142] A continuous decrease in the isomer shift observed with increasing KCl content was ascribed to the presence of non-bridging chlorine atoms at neighbouring iron sites. Moessbauer data for a series of mixed metal(II)-iron(II) chloride phases $CsM_xFe_{1-x}Cl_3.2H_2O$ (M = Mn, Co or Ni, x = 0, 0.5) have been reported.[143] Slight differences in the quadrupole splittings of these materials were explained in terms of changes in the site symmetry resulting in a change of orbital occupancy.

Two studies on $FeSiF_6.6H_2O$ have been described.[144,145] An examination of the quadrupole splitting over the temperature range 4.2 to 300K was used in a detailed analysis of its electronic eigen states and spin relaxation,[144] and the phase transition at 230K was identified as an order-disorder transition.[145] Powdered samples of $FeSO_4.7H_2O$ and $(NH_4)_2Fe(SO_4)_2.6H_2O$ were studied in an external magnetic field, and an induced Stark effect was observed in ferrous sulphate.[146]

High-spin Iron (III) Compounds. The 4.2K Moessbauer spectrum of amorphous FeF_3 has been reported.[147] Six broad lines were observed from which a statistical distribution of iron environments was obtained. New hexagonal tungsten bronze phases of FeF_3 and $(H_2O)_{0.33}FeF_3$ have been characterised by Moessbauer spectroscopy.[148,149] They had saturation hyperfine fields of 577 and 560kOe and magnetic ordering temperatures of 97 and 128.7K respectively. Both parameters were sensitive to the presence of zeolitic water in the system. At 100K the hexagonal tungsten bronze FeF_3 has $\delta = 0.57$ mms^{-1}, $\Delta = 0$, and H = 370kOe. Two structural phase transitions have been found in $RbFeF_4$, a 1st order transition at 381K and a 2nd order transition at 417K.[150] The tilting of the FeF_6 octahedra with respect to the c-axis by both transitions was reflected in the area ratios of the single crystal quadrupole lines. The trigonal phases α- and β-$LiMnFeF_6$ have been studied by neutron diffraction, magnetization and Moessbauer methods.[151] Below 113K the Fe(1a) in the ferrimagnetic β-phase was weakly coupled to the net magnetisation and behaved as an idle spin.

A Moessbauer investigation of both polycrystalline and single crystal samples of ammonium ferric alum has shown that the anomalous variations in linewidth with temperature should not be directly associated with the ferroelectric transformation at 88K.[152] These variations were, instead, interpreted as arising from competing spin-spin and spin-lattice relaxation processes. The form of the results confirmed the presence of a critical spin-lattice process in the magnetically concentrated salt. The Moessbauer spectrum of an Fe_3BO_6 single crystal has been found to be strongly affected by interference from transitions in iron nuclei in non-equivalent positions.[153] Moessbauer studies on alkali metal iron citrates have been described.[154,155] The doublets obtained at room temperature for $M_3[FeL_3].xH_2O$ (M = Li, Na, Rb, Cs, NH_4, L = citrate) indicated that these compounds all had octahedral symmetry. However, for the potassium salt only a broad singlet was found. For all compounds the isomer shifts decreased with increasing cation electronegativity. The photolysis of these compound was also investigated. Moessbauer spectroscopy has been used to identify the thermal decomposition products of iron(III) benzoate[156] and characterise the iron benzoxazole (bo) complexes $[Fe(bo)_6]Cl_3.H_2O$ and $[Fe(bo)_3]Cl_3.2H_2O$.[157] A series of μ-oxo dimers of Fe(III) with tetradentate ligands based on S-alkylisothiosemicarbazides have been examined by Moessbauer spectroscopy at 80 and 300K.[158] The Moessbauer parameters corresponded to high-spin Fe(III) and were independent of the character of the alkyl substituents.

Intercalation Compounds Containing Iron. Moessbauer spectra of first stage intercalated $FeCl_3$-graphite and of the inverse ferrocene-FeOCl intercalation compounds have been measured between room temperature and 4K.[159] The room temperature spectra consisted of resonance lines characteristic of Fe^{3+} only with high-spin Fe(II) sites appearing at low temperatures. This result was explained by an incomplete redox reaction caused by partial transfer of the π-electrons of the graphite (host layer) or of the ferricinium ions (guest layers) to the adjacent layers. A Moessbauer study of FeOCl intercalated with 0.25 mol.% α-picoline has been described.[160] The spectra of paramagnetic samples (T> 64K) showed charge transfer effects reflected in the presence of Fe(II) (14% at 65K), in a time dependent population ratio Fe(III)/Fe(II), and in the appearance of an intermediate valency iron site. In a related study, eleven intercalates of nitrogen Lewis acids with FeOCl have been prepared and characterised.[161] Detailed temperature dependent Moessbauer studies showed that these materials were mixed valence compounds in which electron exchange occurred, on the Moessbauer time scale, at temperatures >100K.

Several techniques including Moessbauer spectroscopy have been used to study samples of commercial iron-graphimet (reduced iron chloride intercalated graphite).[162] From the 13K Moessbauer spectra this intercalate was shown to contain metallic iron, ultrafine non-stoichiometric magnetite, and a broad asymmetric quadrupole doublet which had parameters close to those of anhydrous $FeCl_2$ and $FeCl_2$-intercalated graphite. A Moessbauer and e.p.r. spectroscopic study of polyacetylene intercalated with $FeCl_3$ and $FeBr_3$ has been described.[163] The Moessbauer Debye temperature of the chloride doped system was found to be 87K.

Mixed-Valence Compounds And Unusual Electronic States. A series of fluoride glasses of the type A-B-FeF_3 (A = the fluoride of Pb, Zn, Al or Zr, B = CaF_2 or BaF_2) have been examined.[164,165] At room temperature the lead fluoride based glass contained only Fe(III) whereas the other glasses also contained Fe(II). The fraction of Fe(II) was found to increase with the basicity of the glass. This suggests that Fe(II) behaves as a glass modifier and Fe(III) as a glass former. The shifted hexagonal tungsten bronze $Fe^{II}Fe_2^{III}F_8.2H_2O$ has been studied.[166] "Idle spin" behaviour was found, and while the ferric spins ordered at T_N the Fe^{2+} remained paramagnetic down to 35K. At lower temperatures the field at the ferrous nuclei was very weak, only 47kOe at 4.2K.

A series of bidentate bridging complexes $[Fe(acac)_2L.I_x]_n$ (where L = pyrazine, 1,2-bis(4-pyridyl)ethane, trans-1,2-bis(4-pyridyl)ethylene, 4,4'-bipyridine, and 4,4'-trimethylenebipyridine) containing iron(II) and iron(III) have been prepared and their Fe(III)/Fe(II) ratios determined by Moessbauer spectroscopy.[167] Mixed-valence tri-μ-oxo iron acetates, $[Fe^{III}{}_2Fe^{II}]O(OOCCH_3)_6L_3$, L = aquo or pyridine, have been studied from the standpoint of their intramolecular electron transfer rates.[168] Data from Moessbauer and i.r. spectra and solution redox chemistry indicated that these compounds are in the Robin and Day class II, but with appreciable electron delocalization. Variable temperature Moessbauer data have been presented for solvates of the acetate where L = 4-ethylpyridine.[169] Changing the solvate molecule was found to dramatically change the electron transfer rate in the solid state, and those Fe_3O complexes showing a change from two doublets at low temperature to one at higher temperature show no line broadening effects. A dynamic disorder of certain ligands and/or solvate molecules which was faster than the Moessbauer time scale was suggested.

Spin-crossover Systems and Unusual Spin States. Detailed investigations of the Moessbauer spectra of the spin-crossover iron(II) complexes $Fe(phy)_2(ClO_4)_2$ and $Fe(bts)_2(NCS)_2$ as a function of temperature (phy = 1,10-phenanthroline-2-carbaldehydephenylhydrazone (I) and bts = 2,2'-bis(5-methyl-2-thiazoline) (II)) were made.[170] The high-spin to low-spin ($^5T_2$ - $^1A_1$) transition in the phy complex was discontinuous (T_c = 247-252K), while that of the bts complex was of the continuous type (T_c = 219.5K). For both compounds the high-spin quadrupole splitting showed a discontinuity near T_c which arose because the values of ΔE_Q for the residual paramagnetic molecules were different from

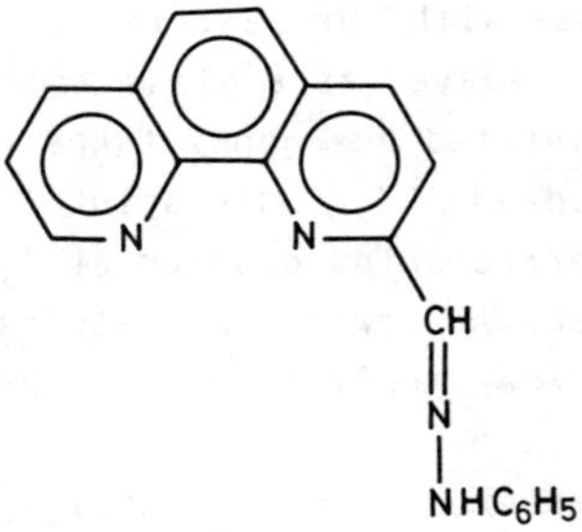

I

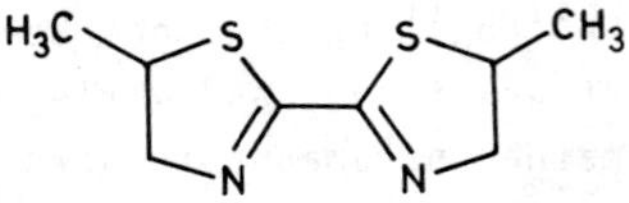

II

those of the bulk paramagnetic phase above T_c. The compound $Fe(phy)_2(BF_4)_2$ also showed a discontinuous spin-crossover, with T_c= 286-277K, which was associated with a crystallographic phase change.[171] In this case the area of the Moessbauer absorption showed pretransitional effects and a pronounced discontinuity at the transition temperature. Another ferrous complex, $[Fe(4,7\text{-}Me_2phen)_2(NCS)_2]$.α-picoline, was found to have a similar spin transition associated with a phase change.[172] For a powder sample the hysteresis in T_c was 56K (T_c = 202-146K), butthis was very dependent on the crystallinity of the material and was affected by the mobility of the α-picoline in the lattice. The same workers have also studied the gradual high-spin to low-spin (S=2 to S=0) transition found in the complex $[Fe(dppen)_2Cl_2]$.2(acetone) (dppen = cis-1,2-bis(diphenylphosphino)ethylene).[173] Evidence was found for different Debye-Waller factors with the characteristic temperatures 138K for the high-spin and 173K for the low-spin states respectively. Two slightly different iron sites observed below 240K were attributed to two different orientations of the acetone molecule.

A high-spin to low-spin transition has been observed in the central iron atom of $[Fe_3(Ettrz)_6](H_2O)_6(CF_3SO_3)_6$ (ettrz = 4-ethyl-1,2,4-triazole) in which the two pairs of metal centres are linked by three triazole ligands.[174] The effects of particle size on the high-spin to low-spin equilibrium in supported bis(N-phenyl-2-pyridinaldimine)iron(II) isothiocyanate were studied by γ-resonance spectroscopy.[175] Variable temperature

Moessbauer, magnetic susceptibility, and i.r. studies of $Fe(phen)_2(NCBH_3)_2$ have revealed a slow, reversible, singlet - quintet intersystem crossing at 359K.[176] A novel fast intersystem crossing was observed in the transition region.

The iron(III) 2-pyridylmethylamine complexes $Fe(vapa)_2Y$, $Fe(sapa)_2Y.3/2H_2O$, and $Fe(bzpa)_2Y.2H_2O$ ($Y = PF_6$, BPh_4, and NO_3) have been investigated by temperature dependent magnetic and Moessbauer measurements.[177] All showed a fast or slow S = 1/2 to S = 3/2, low-spin to high-spin, spin relaxation. Another fast crossover iron(III) complex to be investigated was $[Fe(SalAPA)_2]ClO_4$ (where SalAPA is the monoionic Schiff base derived from salicaldehyde and N-(3-aminopropyl)aziridine).[178] The dynamics of spin-state interconversion of various perturbations of this complex, which changes spin faster than the ^{57}Fe Moessbauer time scale, were studied. Tris(N-methyl-N-n-butyldithiocarbamato) iron(III) has been prepared and its crystal structure and magnetic behaviour studied.[179] This complex exhibits strong high-spin/ low-spin thermal equilibrium with a predominant contribution from the S = 5/2 state at room temperature and a drastic shift towards the S = 1/2 state at liquid nitrogen temperatures. The Fe-S bond distance and room temperature magnetic moment were similar to those for other homologous compounds. The normally rapid spin-spin relaxation in the Fe(III) adduct $Fe(phen)Cl_3.DMF$ was slowed down by the application of a longitudinal magnetic field.[180]

A Moessbauer and e.s.r. study of iron(III) phthalocyanines containing carboxyl groups has been described.[181] These compounds were mixtures of high- and low-spin Fe(III) complexes, and evidence was found for the presence of intermediate spin complexes in 2,9,16,23-tetracarboxyphthalocyaninatoiron. Moessbauer and magnetic susceptibility measurements on iron(II) phthalocyanine have been used to determine that its electronic ground state has major ^3A character.[182] A crystallographic and Moessbauer study found that the complex Fe(tsalen)Cl contains three crystal forms of which one was predominant (96%).[183] At 4.2K the Moessbauer spectrum was a symmetrical doublet with $\delta = 0.38$, $\Delta = 3.13$ mms^{-1} consistent with the presence of a single, S = 3/2, intermediate spin, site. Chlorophthalocyanineiron(III) has been shown to be a spin-admixed (S = 3/2/S = 5/2) system.[184] Its Moessbauer parameters at 4.2K show a single doublet with δ = 0.28 mms^{-1} (relative to α-Fe) and ΔE = 2.94 mms^{-1}, and are similar to those reported for other chloroiron(III) chelates with $^6A_1$ or mixed $^6A_1/^4A_2$ ground states.

Low-spin and Covalent Compounds. The Moessbauer spectra and magnetic susceptibilities have been obtained for a series of Prussian Blue analogues in which iron was partly replaced by Mn, Cu, Co, Cr, or Ru.[185] In each case it was possible to assign a site (A or B), oxidation state and spin state to each transition metal ion. All the Moessbauer linewidths were consistent with the Ludi model of Prussian Blue. The effects of γ-irradiation on the Moessbauer spectra of $K_4Fe(CN)_6$ and $K_3Fe(CN)_6$ have been reported.[186] Moessbauer spectra of $K_4Fe(CN)_6$ complexes with the cationic polyelectrolytes poly(N,N-diallyl-N,N-dimethylammonium chloride) and poly(N,N,N-trimethyl-N,N-(4-vinylbenzyl)ammonium chloride) showed that their interaction was accompanied by the oxidation of Fe(II) to Fe(III).[187] The effects of outersphere cations on the Moessbauer and i.r. spectra of $MK(Fe(CN)_6).4H_2O$, where M = La, Nd, Sm, Eu, Gd, Tb, Dy, Ho, or Yb, have been investigated.[188,189] A reduction in isomer shift as Z increased was related to an increase in the polarising power of the outer-sphere rare earth cation.

A series of twenty-two $Na_3[Fe(CN)_5PX_3].2H_2O$ (PX_3 = phosphine or phosphite) complexes have been prepared.[190] The phosphine complexes were found to have smaller isomer shifts than the phosphite complexes, and the complexes containing bulky or aryl phosphines had larger quadrupole splittings than the phosphite complexes. These trends were discussed in terms of the π-acceptor capabilities of the ligands. Complexes between bis-(2-furanthiocarboxyhydrazido)iron(II) and pyridine-2- and -4-carboxaldehydes have been prepared which have Moessbauer parameters consistent with low-spin Fe(II) in an octahedral geometry.[191] The Moessbauer spectra of the adduct $Li_3[Fe(CN)_6]2C_6H_{12}N_4.5H_2O$ have been recorded.[192] At all temperatures in the range 290-77K two doublets were observed with similar isomer shifts but different quadrupole splittings, thus indicating two types of hexacyanoferrate(III) ion in slightly different iron environments.

An investigation of the complex hexakis(2-ethylimidazole) iron(II) hydridotetracarbonylferrate(II) has discovered anomalies in the Moessbauer spectrum of the $[HFe(CO)_6]^-$ anion around 200K.[193] The crystal structure showed three of the carbonyls on normal tetrahedral sites while the fourth was found to occupy three disordered sites. The unusual spectra were ascribed to jump behaviour, similar to trapped diffusion, of this carbonyl. Moessbauer investigations have been carried out on the aryliron derivatives Fe(Ar)(CO)L(η^5Cp) (where Ar = an aryl group, L = Co or PH_3, Cp = cyclopentadienyl).[194] The data were discussed in relation to variations of the aryl ligand and the Fe-C bonds. Moessbauer data have been reported for (η^5Cp)$(CO)_2$FeHgX complexes (X = Cl, Br, I, S_2COEt, S_2CNEt_2, $S_2P(OEt)_2$, and $Fe(CO)_2$(η^5Cp)).[195]

A linear relationship has been found between oxidation potentials and Fe-57 Moessbauer quadrupole splittings for mono- and disubstituted ferrocenes.[196] MO energy levels in ferrocenes were related to the variation of both parameters with Hammett substituent parameters. The Moessbauer spectra of 1,1',2,2'-tetrachloroferrocene[197] and the plastic phase of formylferrocene[198] have been reported. A series of new ferrocenyl alcohols FcCH(OH)C_6H_4R (Fc = ferrocenyl, R= Cl, Br, Me) and FcCH(OH)R' (R' = 1-adamantyl, 9-anthracyl, mesityl) have been prepared and their Moessbauer spectra obtained in frozen trifluoroacetic acid.[199] Moessbauer parameters from 31 alkylene-bridged and oxoalkylene-bridged ferrocenes have been used to study the effect of the carbonyl group.[200] The perturbing influence of ring tilt and bond lengths on electron withdrawal by the carbonyl group was shown by some trimethylene-bridged ferrocenes. Interpretation of the results was based on changes of electron population of ring-based orbitals and by the non-coplanarity of the carbonyl with the cyclopentadienyl ring. Moessbauer spectroscopy has been used in the conformational analysis of FcC_6H_4X compounds (X = H, o-, m-, and p-Me, NO_2, and OMe).[201] A small mesomeric effect was found on the quadrupole splittings of the p-substituted derivatives.

The smaller quadrupole splittings in [1]ferrocenophanes than in ferrocene have been accounted for by assuming a smaller e.f.g. at the iron nucleus rather than by electron distribution among the iron 3d orbitals.[202] Polyiodides of [2]ferrocenophanes have been examined and a quadrupole splitting of 3.70 mms^{-1} was found, suggesting a strong direct interaction between iron and iodine.[203] Moessbauer parameters have been determined for 5 bis(pentadienyl) iron compounds and compared with the corresponding data for ferrocene and azaferrocene.[204] The decrease in isomer shift in going from the cyclic to the open structure was attributed to a change in the orbital population around the metal atom resulting in increased electron density at the iron nucleus.

Stabilisation modes in Fc_2CH^+ and protonated Fc_2CO have been studied.[205] From the quadrupole splitting data it was suggested that $\underline{e}$ rather than $\underline{e}_2$ orbitals were involved in Fc_2CH^+. Moessbauer spectroscopy has been used to study the oxidation states of oxidized biferrocenophanes and biferrocenylethylenes.[206] Mercuric chloride adducts of biferrocene and biferrocenylene have been investigated.[207] The former showed a large quadrupole splitting suggesting the presence of an Fe-Hg bonded species, whereas the latter showed a smaller quadrupole splitting which was ascribed to a time-averaged mixture of Fe(II) and Fe(III). Moessbauer spectroscopy was used to investigate the electronic structure of salts of various mixed-valence biferrocenium and mononuclear ferrocenium cations in high magnetic fields.[208] Spin-relaxation effects were observed in the PF_6^- salts of four substituted ferrocenium cations and confirmed by the Moessbauer spectra of a series of tri-iodide salts in applied magnetic fields of up to 60 kG.

The γ-resonance spectra of phosphaferrocenes have indicated that the presence of phosphorus in the cyclopentadiene moiety causes a reduction of about 0.3 mms^{-1} in the quadrupole splitting compared with that of ferrocene.[209] Both this effect and the effects of ring substituents were found to be additive. Reorientational motion of the tricarbonyl(1-5-hapto-cyclohexadienyl)iron cations was found to cause a reversible temperature dependence of their Moessbauer spectra.[210] The paramagnetic iron(I) compound (η^5-Cp)Fe($\eta^6 C_6Me_6$), its oxidised monocationic diamagnetic form, and the related diamagnetic (η^5-Cp)Fe(cyclohexadienyl) complexes were studied by Moessbauer spectroscopy on powders and aligned crystals.[211]

Biological Systems and Related Compounds. A series of protoporphyrin IX compounds have been prepared containing ligands that allow the iron to be in a 5-coordinate high-spin iron(II) electronic environment.[212] These compounds are characterised by large quadrupole splittings (4.0 - 4.4 mms^{-1}) and have characteristic electronic spectra. A number of sulphur containing ligands in solution with protoporphyrin IX iron(II) have been found to give Moessbauer data similar to those recorded for cytochromes P450 in their deoxy-high-spin ferrous state.[213] A series of low-spin iron(II) complexes, $Fe(Q)L_2$, where Q = meso-tetraphenylporphine (TPP) dianion or phthalocyanine and L = trialkylphosphine or trialkylphospite, have been prepared and characterised.[214] The Moessbauer and electronic spectra gave strong evidence for the existence of an electron sink or electron buffer capability of macrocyclic ligands.

The intermediate spin complex [TPP(2,4,6-OMe)]$FeClO_4$,$CHCl_3$ has been prepared.[215] Its Moessbauer parameters, δ = 0.41 mms^{-1} and Δ = 3.55 mms^{-1}, and its magnetic properties were consistent with its description as an S = 3/2 species. Its bis(tetrahydrofuran) adduct, however, had the parameters δ = 0.39, Δ = 3.28 mms^{-1}, which reflect some S = 5/2 admixture in this complex. Frozen solution Moessbauer spectra were reported for tetra(p-sulphophenyl)porphine iron(III).[216] Above pH4 the complex exists only as the μ-oxo-oligomer, while at or below pH2 only the monomer is found. An unusual adduct between O_2 and $((TPP)Fe)_2N$ in frozen polar glasses was identified using a number of techniques including Moessbauer spectroscopy.[217] The γ-resonance spectra of a series of mononuclear complexes of iron(III) with meso-$\alpha,\alpha,\alpha,\alpha$-tetrakis(nicotinamidophenyl)porphyrin have been reported.[218]

Nonequilibrium hemoglobin (Hb) states formed at 77K by reduction of its derivatives MetHb and HbO_2 by thermolysed electrons have been studied by Moessbauer spectroscopy.[219] Correlations between the relaxation temperatures and the changes in the protein dynamic structure were investigated. Another study has investigated the dynamics of ^{57}Fe-enriched horse Hb and sperm whale myoglobin (Mb) in the temperature range 80-260K.[220] Analysis of the temperature dependence of the recoil-free fraction showed that it is sensitive to conformational fluctuations which affect the mean square displacement of the iron atom. The protein conformation had a dramatic effect on these measurements. The temperature evolution of the Moessbauer spectrum of oxyHb has been explained in terms of an excited triplet state at about 150K above the ground state.[221] However, another group were unable to find any evidence for any such triplet state.[222] This group examined the Moessbauer spectra of human oxyHb, its isolated β-chains, and oxyhemoglobin from horse and sperm whale measured in fields of 4 and 6 T at 4.2-200K.

Moessbauer experiments on the ^{57}Fe enriched oxy derivatives of human hemoglobin and the products of its digestion with carboxypeptidases were made.[223] The enzymic digestions which remove residues at least 16 Å from the iron of the nearest heme appeared to modify the electronic environment of the metal. Comparative analyses of the Moessbauer spectra of oxyHb in red blood cells from healthy people and leukaemia patients have been made.[224] In the leukaemia spectra the quadrupole splittings were increased, suggesting impaired oxygen transport by the Hb. The dynamic properties of iron protein in crystals of Mb and ferritin have been compared by Moessbauer spectroscopy.[225] Phase transitions associated with coordinated water were seen in both proteins. In ferritin the transition was sharp and correlated with the melting of interior water, while Mb showed a gradual transition associated with thawing of adsorbed water. From the temperature dependences, a greater rigidity in the bonding between iron and ferritin compared with iron and Mb was proposed. Ferritin and hemosiderin isolated from iron-overloaded human spleens were investigated at 1.3 - 200K and in applied magnetic fields.[226] The differences between the two materials lie in the temperature dependence of the superparamagnetic Moessbauer spectra between 10 and 100K. Combined Moessbauer and electron microscopy data showed that the magnetic anisotropy constant of hemosiderin was much greater than that of ferritin, and that hemosiderin could be derived from ferritin. Moessbauer spectra of the heme complexes of poly(1-vinyl-2-methylimidazole) (PMI) and poly(1-vinylimidazole) and of the heme derivatives with covalently bound imidazole and 2-methylimidazole embedded in poly(1-vinyl-2-pyrrolidine) film have been measured.[227] The quadruplole splitting data for the CO adduct of PMI-heme indicated a large electronic field gradient at the iron nuclei, which was attributed to steric hindrance of the polymer chain.

The electronic and magnetic properties of the Se-substituted $2[4Fe\text{-}4Se]^{2+/+}$ ferredoxin from Clostridium pasteurianum have been investigated by e.p.r. and Moessbauer spectroscopy.[228] The $[4Fe\text{-}4Se]^{2+}$ clusters of oxidised ferredoxin were diamagnetic and had Moessbauer spectra nearly identical to those of oxidised $[4Fe\text{-}4S]^{2+}$ ferredoxin. While the 77K spectra of $2[4Fe\text{-}4Se]^{+}$ ferredoxin were similar to those of the sulphur compound, two extra antiferromagnetic sites appeared below 20K. These new sites were attributed to S = 3/2 and S = 7/2 spin states accompanying the normal S = 1/2 state. A highly active nickel-free Periplasmic hydrogenase from Desulfovibrio vulgaris has been purified and studied by e.p.r and Moessbauer spectroscopy.[229] Of the eleven non-heme iron atoms per molecule, the majority were found in two ferredoxin-type [4Fe-4S] clusters with normal parameters; however no trace of a 3Fe cluster was found, and anomalous spectra indicated the presence of atypical non-heme iron centres. The bidirectional hydrogenase from Clostridium pasteurianum W5 has been studied.[230] Moessbauer evidence suggested that at least eight of the 12 iron atoms were organised in two ferredoxin-type [4Fe-4S] clusters, while combined Moessbauer and ENDOR data for the oxidised enzyme identified an iron centre with 2 - 4 iron atoms which was formed by oxidation from a structure with a [4Fe-4S] core. A protocatechuate 3,4-dioxygenase with exceptionally sharp features and a new subunit composition has been isolated from Brevibacterium fuscum.[231,232] Moessbauer and other data suggested that the enzyme was homogeneous in the vicinity of the iron site, while the anaerobic compound gave complex spectra with several iron environments present.

Moessbauer and e.p.r measurements of horseradish peroxidase in its native state and its reaction intermediates with peroxide and chlorite have been reported.[233] The evidence suggested that the heme iron of some compounds was in a ferryl, OFe^{IV}, state. Similar experiments were carried out on the green primary compound of chloroperoxidase.[234] Subtraction of the Moessbauer spectrum of the native enzyme from the samples gave a well resolved six-line spectrum at 4.2K with the parameters (δ = 0.15 mms^{-1} and Δ = 1.02 mms^{-1}) consistent with the model of an exchange-coupled spin, S = 1, ferryl iron and a spin, S' = 1/2, porphyrin radical. Moessbauer and e.p.r. studies of the hemoprotein subunit of Escherichia coli sulphite reductase have shown that the siroheme and the [4Fe-4S] cluster were exchange coupled in several states.[235] A spin-coupling model was developed in which the reduced Fe-S cluster, with cluster spin S = 1/2, was exchange coupled to the siroheme iron. The Moessbauer spectra of the sulphide complex of the oxidised protein (Figure 1) showed the presence of both a low-spin iron(III) heme iron and a $[4Fe-4S]^{2+}$ cluster. Moessbauer, e.p.r and optical studies of the P-460 centre of hydroxylamine oxidoreductase from Nitrosomonas europae were described.[236] The P-460 iron was identified as high-spin Fe(II) with δ = 0.96 mms^{-1} (at 4.2K) and a very large quadrupole splitting of 4.21 mms^{-1}. Oxidation converts iron to a high-spin Fe(III) species with parameters very similar to those of metMb. From this evidence it was suggested that the P-460 iron resides in a heme-like macrocycle with some unusual features.

^{57}Fe-enriched rat erythrocytes, infected by Plasmodium berghei malaria parasites, have been examined by Moessbauer spectroscopy in order to characterise the malarial pigment iron and test the effects of the drug chloroquine on these parameters.[237] The iron in the malarial pigment was found to be high-spin iron(III) with Moessbauer parameters different from any known iron-porphyrin containing compound. No differences were found between the parameters obtained for erythrocytes infected by drug-sensitive and drug-resistant strains of the parasite, both before and after treatment with chloroquine. A Moessbauer study of F-series nitrogen fixation model compounds of the type $(Et_4N)_2[Mo_xFe_{4-x}S_4(SR)_4]$ has been described and the isomorphous substitution of molybdenum and iron demonstrated.[238] The iron-gallocyanine system, a potential model for the anti-cancer drug Bleomycin, has been studied,[239] and a series of microbial iron transport compounds, Fe(III) complexes with dihydroxamate ligands, were prepared and characterised.[240,241] A study of the effect of sugars (D-fructose, D-sorbose, D-mannose, and D-glucose) on hydrogen evolution from sugar solutions in the absence of oxygen through the Moessbauer spectra of the disproportionation products of $Fe(OH)_2$ has been described.[242]

Oxide and Chalcogenide Compounds Containing Iron.- General and Hydroxides. Iron microcrystals with a slowly oxidised surface have been studied by Moessbauer spectroscopy.[243] A six-line spectrum was obtained at low temperatures which changes into a very broad singlet at room temperature. This behaviour was explained using a model in which an asymmetric uniaxial anisotropy was assumed for each oxide crystallite in the surface layer. In a similar study, very thin oxide layers on iron metal produced spectra similar to γ-Fe_2O_3 which were superparamagnetic at 300K and had a collinear spin arrangement at 4.2K.[244]

A Moessbauer study of aged and separated iron(III) hydrolysis products has been reported.[245] The high MW fraction aged in the presence of nitrate precipitated as α-FeOOH while that from chloride solution yielded β-FeOOH and the low MW fractions precipitated Fe(III) hydroxide and ferrihydrite. The Néel temperatures of a number of synthetic β-FeOOH (akaganite) samples have been studied.[246] Moessbauer spectra in external fields of up to 6T revealed the transition to be a true magnetic order-disorder transition with no trace of superparamagnetic relaxation. The value of T_N and the strength of the magnetic interactions were highly variable due to the variable amounts of interstitial water present. A more detailed examination, over the temperature range 4.2-56K, was made to examine the low temperature antiferromagnetic state.[247] Two of the three observed hyperfine fields were identified with Fe^{3+} in $O_2(OH)_4$ and $O_3(OH)_3$ octahedra respectively, while the third contribution was shown to be a thermally induced effect.

A Moessbauer and X-ray diffraction study of δ-FeOOH has been reported,[248] and the products of the aging of magnetite prepared in urea solution have been identified as γ-FeOOH, γ-Fe_2O_3 or

$Fe(OH)_3$.[249] Iron(III) hydroxide prepared by reaction of $FeSO_4$ with $[Co^{III}(en)(dien)]_2O_2[ClO_4]_4$ has been found to have properties markedly different from $Fe(OH)_3$ gels.[250] Two, approximately equally occupied, magnetically ordered sites were observed at 4.2K with the Moessbauer parameters $\delta = 0.366$, $\Delta = 0.509$ mms^{-1}, H = 486kOe, and $\delta = 0.365$, $\Delta = 0.949$ mms^{-1}, H = 437kOe respectively.

Wustite, Hematite, and Related oxides. Fe_yO samples have been studied by Moessbauer spectroscopy at 4.2 and 78K and by polarized neutron diffraction.[251] When y = 0.926 or 0.943 antiferromagnetic order was maintained around the defect clusters. When y = 0.918, however, an extremely complex spectrum was obtained. This was fitted to one Fe(III) and two Fe(II) sextets and one Fe(III) quadrupole doublet with a splitting of 3.8 mms^{-1}. The same workers obtained similar results for the $(Mn_xFe_{1-x})_yO$ solid solutions.[252] For samples containing almost no Fe(III), the spectra were fitted to a model with three Fe(II) sextets, corresponding to sites with differing numbers of Mn neighbours.

The effect of particle size on the Moessbauer parameters of α-Fe_2O_3 have been studied for particle sizes of 0.1, 0.47, and 1.35 μm.[253] It was found that the internal magnetic field falls with decreasing particle size while the isomer shift and quadrupole splitting increase slightly with increasing particle size. The electronic states of Fe_2O_3 have been studied at pressures up to 60 GPa.[254] Two components were identified in the high pressure phase, a low-spin Fe(III) doublet ($\delta = 0.07$ mms^{-1}, $\Delta = 0.89$ mms^{-1}) and an Fe(II) sextet ($\delta = 0.4$ mms^{-1}, H = 824kOe). Moessbauer spectroscopy has been used to identify the products of the hydrothermal reaction between hematite and carbonate solutions.[255] The effects of substitution of diamagnetic Ga^{3+} for Fe^{3+} in $DyFeO_3$ have been investigated using magnetization and Moessbauer methods[256] and the Moessbauer spectra of a BaO-5.7Fe_2O_3 ceramic described.[257] An electric field induced texture effect has been observed in the Moessbauer spectra of $NiTiO_3$-Fe_2O_3 pellets.[258] The quadrupole doublet became asymetric when the γ-rays were at 45° to the surface and an electric field of at least 1000V/mm was applied.

Magnetite and Spinel-type Oxides. The oxidation kinetics of non-stoichiometric magnetite to γ-Fe_2O_3 in air[259] and the magnetic behavior of small magnetite particles in a magnetic field[260] have been studied by Moessbauer spectroscopy.

The magnetic structures of the small particle ferrites $CoFe_2O_4$ and $MnFe_2O_4$[261] and of the ordered phase of $ZnFe_2O_4$[262] have been studied. Moessbauer spectra of RFe_2O_4 (R = Y, Ho, Er, Tm, Yb) spinels have been measured at 200-500K and all but the Y and Er compounds analysed using a stochastic theory.[263] The effects on the Moessbauer spectra of doping nickel ferrites with Al and Cr have been studied.[264] Magnetic properties of a non-stoichiometric Zn-Mn-Fe spinel have been examined.[265] Several studies of nickel-zinc ferrites have been described. Substitution of Fe^{2+} into $Zn_xNi_{1-x}Fe_2O_4$ produced a large increase in the Néel temperature, which was explained by a 4-lattice model.[266] The effects of doping with Ti[267] and Sn[268] were also investigated. Anomalous lineshapes in the Moessbauer spectra of zinc-nickel ferrites (with x = 0.45) in an external magnetic field were investigated and the anomalies ascribed to either superparamagnetic or electronic relaxation effects.[269] The magnetic properties of manganese-cadmium[270], copper-cadmium[271,272], and germanium-copper ferrites[273], and spin structures in nickel-cadmium ferrite[274] have all been investigated. A Moessbauer study on Ti substituted lithium ferrites has been described[275] and the magnetic properties of chromite substituted calcium ferrite were studied over the temperature range 4.2-1000K using applied magnetic fields of up to 7T.[276] Magnesium ferrite-aluminate[277] and fluoride substituted nickel and cobalt ferro-spinels[278] were also studied.

Next-nearest neighbour effects in the Moessbauer spectra of chromium-aluminium spinel[279] and the interacting Jahn-Teller and magnetic transitions in $Fe_{1-x}Cr_{2-x}O_4$[280] have been investigated and the Moessbauer spectra of $CoGaFeO_4$[281] and the (Fe,Zn)V_2O_4 system[282] studied. The iron-europium exchange interaction in $SrEuFeO_4$ has been studied by ^{57}Fe and ^{151}Eu Moessbauer spectroscopy.[285] Cation distributions in series of spinel antimonides, $M_{12}Fe_3{}^{III}Sb_3{}^{V}O_{24}$, where M = Zn, Mg, Co, or Ni, and of

$Ni_{10}Fe_6Sb_2O_{24}$ have been studied at room temperature and 4.2K. It was found that, when M = Ni, Co, or Mg, iron was displaced from octahedral sites into tetrahedral sites.

Other Oxides. Magnetic exchange and supertransferred hyperfine field effects in the perovskite solid solutions $EuFe_{0.8}M_{0.2}O_3$ (where M= Al, Ga, Sc, Mn, Ni, Cr, or Co) have been studied using ^{57}Fe, ^{119}Sn and ^{151}Eu Moessbauer resonances.[285] The Fe^{3+}-O-Cr^{3+} superexchange coupling in orthoferrite and orthochromite solid solutions was shown to be antiferromagnetic. Moessbauer and magnetic susceptibility measurements were used to characterise highly frustrated antiferromagnetic behaviour in the oxygen defect perovskite $Ca_3Mn_{1.35}Fe_{1.65}O_{8.02}$.[286] E.f.g calculations for the rare earth iron garnets $M_3Fe_5O_{12}$ (M = Sm, Eu, Gd, Tb, Dy, Y, Ho, Er, Tm, Yb, Lu) have been published[287] and the cation distributions[288] and the sign of the e.f.g. at the tetrahedral iron sites of $Eu_3Sc_2Fe_5O_{12}$ have been determined.[289] Magnetic behaviour in Eu^{4+} doped yttrium iron garnet has been studied.[290] Two magnetic phases were identified above 80K and only one at lower temperatures. A series of polycrystalline iron garnets, $Y_{3-x}Ca_xFe_{5-x}Ti_xO_{12}$, have been studied at room temperature by Moessbauer spectroscopy.[291] The hyperfine fields at both the octahedral and tetrahedral sites were found to decrease with increasing titanium content.

Cation distributions and phase compositions in mixed barium-calcium hexaferrites have been studied[292] and samples with low calcium concentrations were found to contain pentacoordinate Fe(III). The site occupancy in the bipyramidal sublattice of the paramagnetic phase $BaFe_{12-x}Mn_xO_{19}$ has been deduced.[293] The absence of Mn^{2+} cations in the five-fold lattice site was cited as proof of the pseudotetrahedral character of this site. Local field parameters at the iron nuclei and the electronic states of the iron in $MFe_{12}O_{19}$, M-type hexaferrites (M = Ba, Pb, $Sr_{0.5}Ca_{0.5}$, $La_{0.5}Na_{0.5}$)[294] and the effects of the Zn^{2+} distribution on the magnetic properties of $Ba_2Zn_2Fe_{12}O_{22}$(Zn_2-Y) hexaferrite[295] have been described.

The room temperature Moessbauer spectrum of $FeAlO_3$ was

reexamined.[296] From the observed area ratios it was shown that all but 10% of the tetrahedral sites were occupied by Fe, with the remainder being occupied by Al. Magnetic order and charge-density wave interactions in La_2NiO_4 have been studied using ^{57}Fe and ^{57}Co doped samples.[297] Other oxides which were studied are Fe_2TiO_5[298], Fe_3BO_6[299], $Fe_{3-x}Ga_xBO_6$[300] and γ-Fe_2SiO_4[301].

Inorganic Oxide Glasses Containing Iron. Room temperature Moessbauer measurements have been made on ^{57}Fe in Na_2O-Fe_2O_3-SiO_2 and Na_2O-Fe_2O_3-SiO_2-B_2O_3 glasses and a model for the iron in these glasses proposed from the data.[302] The effects of nucleating agents (P_2O_5 or TiO_2) on the crystallisation of YIG in Na_2O-SiO_2-Y_2O_3-Fe_2O_3 glass ceramics have been investigated.[303] The coordination of Fe, Ga and Ge in high pressure sodium silicate glasses has been studied using Moessbauer, Raman, and X-ray spectroscopy.[304] No coordination changes were observed at pressures up to 30kbar. Moessbauer spectroscopy has been used to relate the redox equilibria of iron to the melt structure of MO-SiO_2 glasses (M= Ba, Sr, Ca, Mg)[305] and in examining the role of iron in calcium silicate glasses.[306]

Magnetism and Moessbauer measurements have been reported for iron(II) aluminosilicate glasses.[307] A linear correlation between isomer shift and quadrupole splitting distributions was established and attributed to changing covalency of the bonding of 6-coordinate iron(II). The effect of heat treatment at 750-830K on the behaviour of ferric oxide doped potassium aluminium borate glass has been studied.[308] It was found that the size of magnetically ordered particles was overestimated by Moessbauer spectroscopy. Ionic ratios in xFe_2O_3-$(1-x)[B_2O_3.PbO]$ glasses have been determined by Moessbauer spectroscopy and used to analyse the thermal dependence of the Curie constants.[309]

Minerals. The accuracy of the Moessbauer spectroscopy in mineral analysis has been studied using two mineral standards at seven laboratories.[310] Errors on different aliquots of the same sample were estimated to be ±0.02 mms^{-1} for isomer shift and quadrupole splittings and $\pm1.5\%$ on area data for well resolved peaks.

Moessbauer spectral studies made on selected banded iron formations have shown that the youngest rocks are the most oxidised.[311] An examination of the Moessbauer spectra of type 3 Allende carbonaceous chondrite meteorite has found no evidence for the presence of any appreciable amounts of ferromagnetic materials.[312] The iron oxides, metallic iron and troilite present all produced six-line patterns.

Moessbauer data for the alkali amphiboles riebeckite and arfvedsonite have been presented.[313] A series of papers on the interpretation of the Moessbauer spectra of dioctahedral iron(III) containing layered silicates have been published.[314-316] A model was presented which accounted for a spatial distribution of cations in 2:1 layers of the mineral and for variations of the octahedral shape and of OH group position. This model was used to interpret the spectra of nontronite and celadonite. Two papers on orthopyroxenes have been published[317,318] and the Moessbauer spectra of helvite group rocks studied.[319] Papers on the Moessbauer spectra of andradite[320], cordierite[321], and ludwigite and vosenite[322] have been published.

Moessbauer spectroscopy was used to study the effectiveness of dithionite/citrate/bicarbonate treatment in removing iron coatings from clay particles.[323] The effects of crystallinity on the Moessbauer spectra of lepidocrocite have been determined.[324] These effects increased the quadrupole splittings of poorly crystallised paramagnetic lepidocrocites and produced distributions of magnetic hyperfine fields at 4.2K. Moessbauer spectra of five samples of ferrihydrite have been recorded at 4.2K, and it was concluded that even impure ferrihydrite could be distinguished from lepidocrocite.[325] The influence of aluminium substitution and crystallinity on the Moessbauer spectra of goethite[326] and iron substitution in ferruginous aluminosilicates[327] has been studied. Two studies on ilvaites have been described.[328,329] Ferripyrophyllite[330], hydroxycarbonate green rusts[331], chloritoids[332] and the interaction between smectite and flavomononucleotide[333] have all received attention.

Cation distributions in synthetic and natural iron-manganese silicate olivines[334] and the location of the ferric ions in sapphirine[335] have been studied. Iron localisation patterns and structural similarities in columbite-tantalite and ixiolite have been determined.[336] A non-Lorentzian spectral line broadening in bauxite has been attributed to environmental effects.[337] This effect was ascribed to the influence of the dominant haematite subspectrum on the minor goethite subspectrum.

Chalcogenides. Moessbauer spectra of natural pyrite were measured at 290-430K and the data used in a lattice dynamical study.[338] The room temperature mean-square amplitudes $\langle z^2 \rangle = 0.14 \times 10^{-22}$ m^2 and $\langle x^2 \rangle = \langle y^2 \rangle = 0.30 \times 10^{-22}$ m^2 were obtained. A similar investigation of natural and synthetic marcasite obtained a value of $\langle r^2 \rangle = 1.3 \times 10^{-22}$ m^2 at room temperature.[339] Electronic-structure calculations have been performed on the pyrites MS_2 (M = Fe, Co, Ni, Cu, Zn), using a self-consistent-charge LCAO method, and their Moessbauer spectra calculated.[340]

Structural changes in mechanically and thermally activated chalcopyrite have been studied by Moessbauer spectroscopy.[341,342] The effects of vanadium doping on FeS[343] and iron doping on MnS[344] have been examined. (Fe,Mn)S and (Fe,Mg)S solid solutions have been studied to examine the properties of Fe^{2+} in two different B1-structured hosts,[345] and the quenched niningerite, (Mg,Fe)S,[346] system has been investigated. The layer compound $FePS_3$, a 2-dimensional Ising antiferromagnet, has been investigated using Moessbauer and magnetic susceptibility measurements.[347] A first order phase and magnetic transition was found at 120K. A similar study of the Chevrel-phase compound $FeMo_6S_8$ has been described.[348]

Preliminary results from a room temperature Moessbauer study of ternary compositions in the system $(Cu,Fe)Se_{2-x}$ have been published.[349] The Moessbauer spectra of FeV_2Se_4,[350] $FeCr_2Se_{0.75}S_{3.25}$[351] and $Fe_{1+x}Nb_{3-x}S_{10}$[352] have been reported. The valence state, local environment and coordination of iron impurity atoms in chalcogenide glasses have been studied.[353]

Applications of Iron-57 Moessbauer Spectroscopy.- Catalysts. Particle sizes of catalytic magnetite on Grafoil have been determined by Moessbauer and magnetic susceptibility methods.[354] Good agreement was obtained between the two techniques. Spherical iron catalysts for ammonia oxidation have been examined.[355] The influence of carbon on iron catalysts used in the hydrogenolysis of isopentane has been studied.[356] Used catalysts contained α-Fe at high hydrogen to hydrocarbon feed ratios, and the formation of Fe_3C at low ratios was associated with changes in selectivity. An in situ gradientless Moessbauer catalytic reactor has been constructed for the study of iron molybdate conversions during partial oxidation of propene.[357]

The hydrodesulfurisation of organosulphur compounds by Fe-Mo/Al_2O_3 has been studied by Moessbauer spectroscopy and the sulphided catalysts characterised.[358] One octahedral and one tetrahedral cobalt site have been identified in the Moessbauer emission spectra of calcined Co-Mo/Al_2O_3 catalysts.[359] Two papers have been published on cobalt molybdenum sulphide catalysts.[360,361] The direct participation of cobalt in the hydrodesulfurization was found by both groups.

Silica supported Rh-Fe Fischer-Tropsch catalysts were characterized.[362-364] The Fe^{3+}/Fe^0 ratio and Moessbauer parameters depended on the Fe/Rh atomic ratio and were related to the catalytic properties.[362] Catalyst changes in various gaseous environments were also studied.[364] The CO/H_2 reaction over Fe/TiO_2 after reduction at different temperatures[365] and the use of Moessbauer spectroscopy as a probe of FeO and Ir-Fe on titania[366] have been described. Decomposition of Fe_3O_{12} on a series of mordenites has been studied by magnetic and Moessbauer measurements.[367] The changes in isomer shift of these Fischer-Tropsch catalysts was used to deduce the nature of the metal-support interactions. Magnetic, Moessbauer and i.r. studies on iron-zeolite catalysts have been described.[368,369]

Coal and Related Topics. A series of in situ Moessbauer measurements have been carried to study the high temperature transformations of four bituminous coals.[370] Iron bearing minerals in some coals from British Columbia have been characterised.[371] Pyrite, marcasite, siderite, jarosite, rozenite or melanterite, and Fe(II) containing clays were all found in the coals. A series of four fly ashes from various sources, a bottom ash sample from the combustion of lignite-natural gas mixtures[372], and brown coal ash and fireside deposits[373] have been studied.

<u>Ores, Soils, and Sediments.</u> The states of iron-containing impurities in North Onega bauxite ores were investigated by γ-resonance spectroscopy.[374] Hydrated goethite was identified as the main impurity and hydrochlorination was examined as a method for the removal of iron. Characteristic Moessbauer spectra and parameters have been published of standard iron oxides and iron-bearing minerals found in soils.[375] Chemical and Moessbauer methods for the analysis of soils have been compared for six French[376] and several Bulgarian[377] soils. Moessbauer spectroscopy was used to determine iron forms and valence states in various soils from the Alazani valley (U.S.S.R.) which could not be subjected to destructive treatment.[378] Fe(II)/Fe(III) ratios in sediments from a seasonally anoxic lake were determined by Moessbauer and wet chemical methods.[379] The chemical methods were found to overestimate the ferrous component in the presence of acid-labile sulphide ligands.

<u>Cements and Ceramics.</u> The states of iron in cement raw mixes up to the clinkering temperature[380] and high iron concentrations in cement clinker[381] have been studied by Moessbauer spectroscopy. Moessbauer and X-ray diffraction studies of dry and hydrated samples of four portland cements have been described.[382] Iron occupying distorted sites in the dry cement formed ferrite monosulphate on hydration without producing $Fe(OH)_3$ or its gel. Pottery manufactured during the Chinese Henan Longshan period[383] and the firing behaviour of six Peruvian clays[384] were studied.

<u>Other Applications.</u> Rust formed on carbon steel during simulated atmospheric corrosion[385] and the corrosion products of boiler tubes[96] have been characterized by Moessbauer spectroscopy. The boiler tubes first formed Fe_3O_4 which then further oxidized to α-Fe_2O_3. Moessbauer spectra have been used in the phase analysis

of materials used in the manufacture of iron(III) containing pigments and magnetic tapes[386] and of ferrovanadium production waste.[387] In situ Moessbauer spectroscopy has been used to examine an operating fuel cell.[388] The high surface-area oxygen-fed cathode was preadsorbed with iron phthalocyanine in alkaline solution and had a hydrogen-fed electrode as anode.

5 TIN

General Topics. A new source for gamma-resonance studies has been prepared by melting the Sn-119m isotope with neptunium to form Np_3Sn(119m). When this source is annealed at 1000-1200°C in a nitrogen atmosphere for about an hour the probability of recoiless emmission is increased.[389]

A quantum chemical interpretation of the isomer shift data for tin compounds was carried out on a series of tin halides.[390] Not suprisingly a dependence of the shift on covalent bonding effects was confirmed. A value for $\Delta R/R$ was obtained from the calculation. The pressure dependence of the electron density at the nucleus in metallic tin has also been calculated.[391] Bulk metal and thin-film samples of α_2-tin with very high Moessbauer chemical isomer shift values have been obtained in studies of the reduction of aqueous solutions of tin(II) chloride with Mg in the presence of ferromagnetic impurities such as Fe, Co or Ni or in an external magnetic field.[392] The temperature dependence of the shifts of the α_2-tin samples was also studied.

A number of papers published during the past year were concerned with the use of Moessbauer spectroscopic data to characterise tin-containing supported heterogeneous catalysts.[393-396] The role of Sn in Ir-Sn/Al_2O_3 systems was investigated[393] but most of the work on the characterisation of the catalysts was carried out on Pt-Sn/Al_2O_3 systems. The valence states of tin in Pt-Sn/Al_2O_3 reforming catalysts with different tin contents (0.3-1.5 wt%) was studied[394] at various pretreatment stages by Moessbauer spectroscopy. The spectra of all of the catalysts showed the presence of Sn(IV), Sn(II) and Sn(0) but the amount of tin present as Sn(IV) in catalysts prepared by co-precipitation was found to be significantly greater than the amount present in the other valence states. The authors claim that the tin in the reduced catalyst system is present mainly as tin(IV) oxide phases with small amounts of Sn(II) and Sn(0). Any Pt-Sn alloy present was said to result from reaction between the small amount tin metal present and Pt. As a result of a comparison[395] of γ-resonance data for the Pt-Sn/Al_2O_3 and Sn/Al_2O_3 systems, it has been suggested that the reduction of the tin present is promoted by Pt with the formation of Pt-Sn clusters. Heating of the sample to 300, 400 and 500°C gave materials that contained Sn(IV), Sn(II) and Pt-Sn alloy phases with the shifts for the alloy phases being 1.54, 1.65 and 1.83 mms^{-1} respectively. The authors state that identification of the alloy phases is important because these are likely to be the major components in deciding the catalytic activity of the Pt-Sn/Al_2O_3 materials. The recoil-free factor(f) has also been used to characterise Pt-Sn/Al_2O_3 catalyst materials prepared in different ways[396]. A simple model based on the experimental data was proposed and was shown to be in agreement with changes in f-values with different methods of preparation. The heat treatment process in the catalyst manufacture was also shown to result in an increase in f presumably because of stronger bonding between the Sn moieties present and the oxygen cations of the alumina support.

The Sn-119 nucleus has been used as a probe impurity atom in various systems. Both Moessbauer spectroscopy and lifetime positron measurements were used in investigations of the $BaPb_{0.99-x}Sn_{0.01}Bi_xO_3$ system[397] with x= 0-0.4. A correlation between both the dependence on the room temperature Moessbauer parameters on x and the dependence on the positron spectra on x and the transition temperature to the superconducting state of the samples was obtained. The data were interpreted in terms of bielectron transfer over Bi centres in the lattices. Moessbauer data have also been used by other workers[398] to study dielectric and superconductive transitions in the $BaPbO_3$-$BaBiO_3$ system. Samples containing 80-100% $BiBO_3$ were shown to have concentration-dependent dielectric anomalies that are accompanied by structural changes that can be detected with a Sn-119 probe. The spectra of samples doped with Sn-119 have concentration dependent chemical isomer shifts with a maximum value at 78%

$BiBO_3$. This maximum was interpreted as being associated with an 0.04 electron increase in the 5s-electron density at the Sn(IV) atom in the 78% $BiBiO_3$ composition. A Moessbauer resonance study of tin in CrO_2 prepared by hydrothermal oxidation of $^{119}Sn(IV)$ doped Cr_2O_3 has shown[399] that some of the Sn probe atoms are contained in the resulting CrO_2 lattice and that a positive field is transferred to the probe atom. The data were interpreted in terms of a transfer mechanism. Tin metal has been introduced into the alpha-Fe_2O_3 lattice by diffusion at 1000°C and the Moessbauer spectra of the tin impurities obtained.[400] The spectra consist of well resolved six-line patterns arising from a supertransferred magnetic field.

The first successful measurement of the spin-flop transition phenomenon in metallic Cr was achieved with the use of Moessbauer effect measurements on doping Sn-119 nuclei. The phenomenon showed hysteresis with the change occuring at 123K when the sample was being heated and at 110K when the sample was being cooled.[401] The hyperfine interaction at Sn-119 impurity atoms in metallic gadolinium was investigated by gamma-resonance emission spectroscopy.[402] The studies were carried out on Gd doped with about 0.5 at.% Sn and in the temperature range 4.5-320K. At 4.5K the measured quadrupole interaction was close to zero, the magnetic hyperfine field had a value of -32.3T and the shift was 1.91 mms^{-1} from $CaSnO_3$. At temperatures below 170K the temperature dependence of the hyperfine field was correlated with changes in the magnetisation matrix for the depolarizing effect of the non-magnetic impurity, and the data suggest that the mean of the Gd ion spins does not depend upon temperature. In the temperature range 200K to 291K (the Curie temperature) the data showed an appreciable negative anomaly and for temperatures above 230K the changes in the Sn-119 spectra in external applied magnetic fields indicate a non-collinear ordering of the Gd spin moments. The alloy phases studied by Moessbauer spectroscopy on tin impurity atoms include $SnAl_2$,[403], $Cr_{27}Fe_{25}$[404], and Heusler alloys.[405,406] A strong anisotropy was observed in the hyperfine interaction of the tin atoms doped into $SnAl_3$ and was interpreted in terms of a non-spherical distribution of 4f-electron density around the tin; samples of $Cr_{27}Fe_{25}$ doped with 0.6 at.% Sn-119 were shown to be chemically heterogeneous but magnetically heterogeneous over the temperature range 4.2-331K. The alloy was described in terms of the presence of a magnetic phase and a non-magnetic phase that persists down to temperatures of about 9K. The Heusler alloys studied by tin doping were $Co_2MnAl_{1-x}Si_x$[405] with x = 0.02 - 1 and $Co_2YZ_{1-x}Zn_x$ with Y = Ti, V, or Cr, Z= Al or Ga, and x = 0.02.[406] The measurements on $Co_2MnAl_{1-x}Si_x$ showed an approximately linear decrease in the field from +40 kOe with x = 0.02 to -16KOe for the Al-free alloy. These data and those of the other Heusler alloys[406] were discussed in terms of current theoretical predictions of hyperfine fields at non-magnetic sites in ferromagnetic alloys.

A range of amorphous alloys with the compositions listed in Table 2 have been prepared by quenching from appropriate melts using the single roller method. Moessbauer effect spectra were then obtained at 77K using both Fe-57 and Sn-119 sources.[89] The Sn-119 data showed: (1) that addition of 10 at.%Cr reduces the most probable Sn field by 6kOe in Fe-Sn-B phases and 10kOe in Fe-Ni-Sn-B phases, (2) that the Sn field in Fe-Ni and Sn-B alloys is 8kOe more positive than the field in Fe-Sn-B alloys, and (3) that additions of Cr to either Fe-Sn-B or Fe-Ni-Sn-B alloys does not increase ΔH for the tin field distribution. The tin hyperfine field distribution does not show the same trends as the iron field distribution. Table 2 contains data for the most probable fields ($\bar{H}$) and the field distributions (ΔH). Unlike the data for the Fe-57 effect, there is no broadening of the Sn-119 hyperfine field distribution

Table 2 Hyperfine field parameters in amorphous alloys

Alloy Compositions	Fe $\bar{H}$	Fe ΔH	Sn $\bar{H}$	Sn ΔH
$Fe_{79}SnB_{20}$	285	108	30	113
$Fe_{69}SnCr_{10}B_{20}$	185	163	24	63
$Fe_{40}Ni_{39}SnB_{20}$	253	95	38	146
$Fe_{35}Ni_{34}SnCr_{10}B_{20}$	188	181	28	115

for Cr-containing alloys, and a narrower distribution is actually suggested. These observations were discussed in terms of the local environment of iron and tin atoms in the amorphous alloys. The author suggests that, for $Fe_{69}SnCr_{10}B_{20}$, iron atoms tend to cluster near the tin probe atom so that the near-neighbour environment of the probe is hardly changed when chromium is added. The decrease in the value of H from 30 to 24kOe at the tin atom would then be due essentially to a fairly large average moment disturbance on the Fe atoms caused by the addition of Cr. Dunlap and Stroink[417] have also reported on the crystallisation and magnetic properties of amorphous Fe-Sn-B-Si alloys. Moessbauer studies have been made[418] on tin and tin plus zirconium carbide coatings obtained under non-steady-state electrolysis conditions, and preliminary data on tin-substituted Fe_4N show a preferential substitution of Fe(I) sites.[409]

Moessbauer studies have been carried out on glassy and crystalline phases in the Sn-Se-As system. The Sn-119 chemical isomer shift was found to be independent of composition in all of the vitreous phases and to have a value of $1.6 mms^{-1}$. This shift is consistent with tin being tetrahedrally surrounded by Se atoms in all of the glasses. The data for crystalline phases obtained from materials of composition $Se_{60}Sn_3As_{34}$ and $Se_{47}Sn_3As_{50}$ showed that both SnSe and $SnSe_2$ were precipitated during the crystallisation process.[410] Moessbauer data have also been discussed in two papers dealing with spin glasses of the type M_2SnTe_4 (M = Cr, Mn, Fe or Co).[411,412] Solid samples of $(In_{15}Te_{85})_xSn_{1-x}$ have been obtained by splat-cooling from molten mixtures at different initial temperatures and analysed by Moessbauer spectroscopy and electron diffraction.[413] Evidence was found for the presence of two phases, viz. (1) InTe with some tin in solid solution and (2) an SnTe phase.

A review[414] has been published on the effects of radiation damage in thin solid films, in Nb_3Sn and in SnS_2. The advantages in using the Moessbauer effect for these studies was stressed. Other gamma-resonance studies of tin-containing alloys described recently have included investigations into the conditions for obtaining a single phase Nb_3Sn superconducting layer[415]. The measurement of the hyperfine field systematics in the Rh_2MnSn Heusler alloy[416] and an investigation into the effects of the treatment of Al-Mg-Si-Sn alloys on the formation of the Mg_2Sn

intermetallic phase[417] were described.

Tin(II) Compounds. A major review of the relationship between Moessbauer parameters and the known crystallographic environments of tin in tin(II) compounds has been published.[18] The review deals with the background to Sn-119 Moessbauer spectroscopy applied to tin(II) compounds and in particular with the difficulties associated with the interpretation of the data for the low symmetry lone-pair distorted compounds. It also considers the important problems in tin(II) chemistry that have been solved by Sn-119 Moessbauer spectroscopy over the past twenty years including the high *s*-character of Sn(II)-F bonds, the weak Sn-Sn interactions in $Sn[CH(SiMe_3)_2]_2$, the formation of anionic tin(II)-fluoride networks in ternary tin(II) halides and the solid state effects associated with delocalisation of the non-bonding electron pair into the three-dimensional band structure of solids. The data for tin(II) compounds that are of known crystal structure are tabulated under the headings: trigonal pyramidal sites; tin(II) compounds with tin in trigonal pyramidal sites; tin(II) compounds with tin in square pyramidal sites; and tin(II) compounds with tin in high symmetry sites. For compoumds with the tin in trigonal pyramidal SnL_3 sites with Sn-L bonds to the same ligand, a close relationship between the chemical isomer shift and the Sn-L bond distance is reported. The theories that can be applied to descriptions of the bonding in tin(II) complexes are discussed critically in terms of the structural and Moessbauer spectroscopic data available. Theoretical models including the VSEPR theory, the concept of average bond valence, the geometric model assuming that the non-bonding electrons have volume, the crystal field approach and the molecular orbital approach do not satisfactorily explain all of the data, and a new model based on orbital energy matching was proposed.

The Moessbauer effect has been used to determine the concentrations of both Sn(II) and Sn(IV) in tin-containing sodium silicate glasses.[418] The amount of Sn(II) was found to increase with increasing SnO_2 content of the initial glass melt. The rate constants for oxidation of the tin(II) oxide component in air at 843-1173K were determined and the oxidation activation energies for short- and long-term heating determined to be about 115.5 and 93.8

kJ mol^{-1} respectively.

A Sn-119 Moessbauer study of single crystals of $Pb_{0.2}Sn_{0.8}S$ irradiated with X-radiation gave a spectrum containing a new doublet with shift and splitting parameters of 3.35 and 0.88 mms^{-1} respectively.[419] Irradiation of the crystals along the c-axis affects mainly the sulphide sublattice, while at high irradiation doses the crystals undergo a gradual conversion to a new electronic state. Nasredinov et al[420,421] showed by Moessbauer spectroscopy that the change in state of the Sn atoms in $Pb_{1-x}Sn_xS$ and $Pb_{1-x}Sn_xSe$ was dependent upon the position of the Fermi levels in the solids. Both neutral (Sn^{2+}) and ionized (Sn^{4+}) states of the Sn centres were shown to be present, and evidence for a two-electron exchange between these neutral and ionized states was found for the partially compensated solid solutions of compositions $Pb_{1-x}Sn_xSe$ doped with Na and Tl impurities. Two distinct peaks found in the 80K Moessbauer spectra of $Pb_{96.3}Sn_{3.7}Se$ doped with Na and Tl were taken as evidence of a slow exchange. Evidence for the precipitation of SnSe in crystalline phases obtained from glasses having the compositions $Se_{60}Sn_3As_{34}$ and $Se_{47}Sn_3As_{50}$ was been obtained from their Moessbauer spectral data.[410] Moessbauer evidence has also been found for the presence of a phase based on the SnTe lattice in splat-cooled indium-tellurium-tin phases.[413] A gamma-resonance study of the $(PbTe)_{1-x}(SnSe)_x$ system for x = 0.2, 0.5, 0.6, 0.7, and 1.0 provided evidence for the presence of a two-phase (cubic and orthorhombic) region at x = 0.7 with about 54% of Sn in the cubic phase. Shift and line width data change abruptly during the phase transition but remain constant in the cubic phase with changes in x.[422]

Edwards and Khallow[423] have reported that the Sn-119 isomer shift for the single resonance line of $Sn(SbF_6)_2(AsF_3)_2$ is the highest so far obtained for a tin(II) species. They also report on the crystal structure determination of the compound which shows a nine-coordinated arrangement of fluorine atoms around the tin in a distorted geometry that is consistent with the presence of a non-bonding electron pair. Their suggestion that the tin(II) species could be a bare Sn^{2+} ion is, however, not consistent with the observed lone-pair distortion which would involve mixing of the

$5s^2$ electrons with 5p in order to obtain directional lone-pair character. Previous Moessbauer spectroscopic studies on the species present in frozen solutions of tin(II) fluoride have been extended[424] to the determination of equilibrium constants. The stability constants obtained showed that hydrogen-bonding was important for frozen perchloric acid solutions. The shifts and splittings for frozen solutions of $Sn(ClO_4)_2$, $KSnF_3$, and NH_4SnF_3 were 3.55 and 1.50 mms^{-1}, 3.00 and 1.91 mms^{-1}, and 3.25 and 1.85 mms^{-1} respectively. Birchall and Denes[425] have carried out a detailed ^{19}F and ^{119}Sn n.m.r. study of the SnF_2-MF-H_2O systems for M = Li, Na, K, Rb, Cs, and NH_4 and have complemented the data with Sn-119 Moessbauer parameters. When SnF_2 is dissolved in water and frozen,the spectrum shows a small increase in shift and a considerable increase in splitting compared to solid SnF_2. This suggests that the cyclic tetrameric SnF_2 structure has been disrupted possibly with the formation of $SnF_2.H_2O$. Addition of MF (M= Na, K, Rb, Cs, NH_4) to these solutions results in a decrease in isomer shift and an increase in quadrupole splitting. These changes indicate a greater involvement of the two 5s-electrons in bonding between the metal and the ligands surrounding it. If the species in frozen solutions of SnF_2 is assumed to be $SnF_2.H_2O$ then the water molecules would be replaced by F^- added as MF. The changes in parameters are progressive with the amount of MF added until more or less constant values are obtained for MF:SnF_2 ratios greater than 1:1 - that is when there is an excess of F^- over the composition $MSnF_3$. The Moessbauer parameters obtained from frozen solutions containing an excess of F^- are very similar to the data reported previously for solid $MSnF_3$ compounds. The degree of cation interaction with the SnF_3^- species in the frozen solutions does, however, appear to be less that in the solid $MSnF_3$ phases. The n.m.r. data do not suggest the presence of appreciable amounts of $Sn_2F_5^-$ species in the solutions.

The Moessbauer spectrum of $BaSnF_4$ has been recorded[426] at 80K. It consists of a quadrupole doublet characteristic of tin(II) in a non-cubic fluoride environment with shift and splitting values of 3.45 and 1.95 mms^{-1} respectively. The Moessbauer data for trifluoroditinhexafluoroarsenate and trifluoroditinhexafluoro-antimonate have also been reported.[427]

A Moessbauer study of tin(II) chloride hydroxide and tin(II) oxide hydroxide has been reported.[428] A series of papers by Chinese workers on the gamma-resonance spectra of some halogenostannate solutions has been published.[429-431] The Moessbauer parameters of frozen solutions of tin(II) in perchloric acid solution were obtained relative to the amount of added NaCl. Addition of NaCl does not change the shift data but significant changes are found in the measured splittings. Stability constants that are in good agreement with published values were calculated from the splitting data.[429] Moessbauer effect studies also suggested[430] that there was a correlation between quadrupole splitting data and the apparent degree of dissociation of frozen solutions containing $SnCl_3^-$ and $SnBr_3^-$ ions. Solutions of $KSnCl_3$, NH_4SnCl_3 and $KSnBr_3$ were studied and for all systems a decrease in the splitting was found with increasing halogenostannate(II) concentrations. The stability constants of the $SnBr_3^-$ complexes in frozen aqueous solutions at 78K were estimated[431] from the quadrupole splitting against solution concentration graphs. A value of $\delta = 0.25$ mms^{-1} was obtained for the pure $SnBr_3^-$ species by extrapolation, and this value was also assumed for solid $KSnBr_3.2H_2O$. As part of a study of the crystal structure of $(NH_4)_2SnCl_4.H_2O$, the Moessbauer spectra for $(NH_4)_2SnCl_4.H_2O$, $SnCl_2$.tu, $SnCl_2.H_2O$ and $SnCl_2$ were recorded.[432] In agreement with previous data on the K derivative $(NH_4)_2SnCl_4.H_2O$ is better formulated as $(NH_4)_2(SnCl_3)(Cl).H_2O$.

The Moessbauer parameters for monoclinic $Sn_3(PO_4)_2$ have been obtained.[433] The data are said to be in agreement with the crystal structure but differ from those reported previously. Both i.r. and γ-resonance data have also been used to characterise the phosphates Sn_2PO_4Cl, $SnPO_4H$, $Sn_2PO_4(OH)$ and $Sn_3(PO_4)_2$ in terms of the known structural data.[434] The Moessbauer parameters of $[Sn_2(H_2O)L].3H_2O$, where L= hedp, are found to be similar to those of $Sn_2P_2O_7$.[435]

Three papers by the reviewers' group have been concerned with various aspects of the non-bonding electron pair in tin(II) chemistry.[436-438] The Moessbauer parameters of a series of oxalatostannates(II), $M_2Sn(C_2O_4)_2.xH_2O$, have been rationalised in terms of distorted square-pyramidal tin environments. The chemical isomer shifts are smaller and the quadrupole splittings are larger than those for the parent $Sn(C_2O_4)$ as would be expected for complex formation and the known crystal structures.[436] A relationship between isomer shift and bond length was found in the oxalate series. Moessbauer data have also been reported for the malonate series $M_2Sn_2[CH_2(COO)_2]_3.xH_2O$ (M = NH_4, K, Na, Rb, or Cs, x = 0, 1 or 3) and the data have been shown to be consistent with the presence of non-chelating, bridged, malonate groups and trigonal pyramidal Sn environments similar to that found in potassium malonatostannate(II).[436]

The crystal structures of $K_3Sn_2(SO_4)_3X$ (X=Br or Cl) have been determined. These structures consist of three-dimensional networks of tin atoms and bridging sulphate groups with discrete potassium and halide ions in holes within the networks. The arrangement of the tin atoms, taken along with the Moessbauer data and the Sn-O bond lengths, is consistent with cluster formation in which lone-pair electron density on the tin atoms is delocalised in cluster molecules.[437] The shift data for the two compounds are compared with those of the parent material tin(II) sulphate in Table 3. The considerable increase in the average Sn-O bond length

Table 3 Bond lengths and Moessbauer isomer shifts for tin(II) complexes

Compound	Average Sn-O bond length / Å	δ/mms^{-1}
$K_3Sn_2(SO_4)_3Br$	2.55	3.86
$K_3Sn_2(SO_4)_3Cl$	2.52	3.76
$SnSO_4$	2.26	3.95
$SnHPO_3$	2.17	3.15
$KSn(OOCH)_3$	2.16	3.08
$Ca[Sn(OOCMe)_3]_2$	2.14	2.90

in the halide sulphates is not reflected in an increase in chemical shift, but instead a significant decrease is observed. The longer than ionic Sn-X bond is not expected to have much effect on the shift and so this loss of s-electron density must be explained in terms of solid-state effects rather than tin-ligand interactions. The cluster arrangement of the Sn and K atoms around the halide atoms in the structures, however, provides the correct geometry for Sn-Sn cluster interactions. The non-bonding electron pairs on the tin atoms must point towards the centre of the cluster and must overlap empty cluster orbitals to give the Sn-Sn interactions that result in the low chemical isomer shifts observed.

The Sn-119 Moessbauer parameters for thiourea complexes of tin(II) compounds have been interpreted in terms of new X-ray crystal structures.[438] The study illustrates the importance of using all available information to interpret bonding and structural details because predictions based on Moessbauer data alone would have led to wrong assumptions. The data show that the chemical shift of $SnSO_4.2tu$ is lower than that for $SnSO_4$ because of the formation of a short Sn-S bond to thiourea but that the lower shift for $Sn_2Br_4.5tu.2H_2O$ must be due to shortening of Sn-Br bonds on complex formation. The increased shift observed when $Sn(CH_3COO).2tu$ is formed from the parent acetate can also be rationalised in terms of the fact that no shortening of the Sn-O bonds or formation of short Sn-S bonds occurs as a result of the production of the complex.

Dong and Zuckerman[439] have studied the synthesis and characterisation by Sn-119 Moessbauer spectroscopy of ring-bridged di-L^5-cyclopentadienyl tin(II) derivatives of α,α'-dicyclopentadienyl-ortho-, meta- and para-xylene. The spectra show that the derivatives are tin(II) compounds and that the splittings for the ortho and para derivatives at 1.84 and 1.94 mms^{-1} respectively are much larger than that for the parent stannocene. The C-trimethylsilyl substituted stannacarboranes III have been

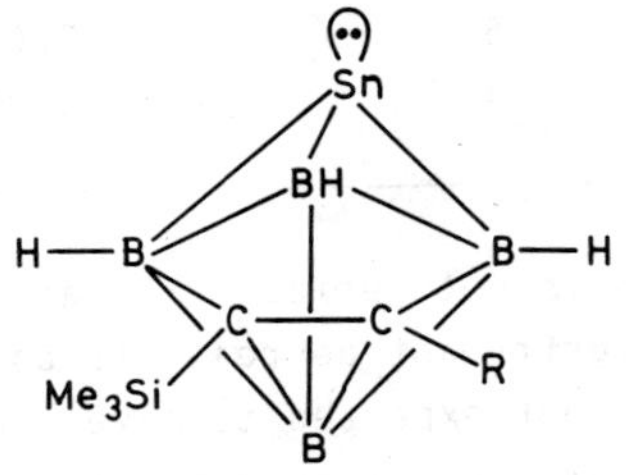

III

prepared and characterised with R = $SiMe_3$, Me or H. The shifts for the compounds are in the tin(II) region and are consistent with the structure **III**.[440]

Inorganic Tin(IV) Compounds. The Moessbauer recoil-free fraction data have been obtained for samples of tin(IV) oxide containing Fe impurities.[441] The graph of the concentration dependence of f shows a minimum for a concentration of 0.5 mol.% Fe_2O_3 in SnO_2. The results were explained in terms of the effects of the Fe impurity atoms on the phonon spectrum of the tin dioxide matrix. Moessbauer effect studies were also carried out on heavily Sb-doped $Pb_xSn_{1-x}O_2$ sputtered thin films. The changes in the shift and splitting parameters observed in the films were interpreted in terms of two possible models and the results used to obtain (1) the homogeneity of the Sb atom distribution, (2) the effective charges on the cations present, and (3) the Sternheimer antishielding factor for Sn in SnO_2.[442] In another study of thin films, data were presented for thin films of SnO_2 having the calcium fluoride type structure.[443] The concentration of Sn(IV) and Sn(II) in silicate glasses has been determined by Moessbauer spectroscopy and the kinetics of the oxidation state changes in the glasses studied.[418] Both X-ray diffraction and gamma-resonance methods were used[444] to study the cation distribution and quadrupole interaction in a series of Sn-substituted zinc ferrites of the type $Zn[Zn_xSn_xFe_{2-2x}]O_4$. The Sn-119 spectra of these ferrites showed that the Sn atoms were present in octahedral states only and that Sn substitution does not appear to affect Fe at. distributions. The oxidation products of oxidised epitaxial layers on SnSe have been studied by Moessbauer spectroscopy before and after hydrogenation.[445] The data were found to be consistent with the presence of both $SnTe_3O_8$ and SnO_2 in the oxidised layer and with the formation of β-Sn on hydrogenation of the sample.

The intermediate products in the thermal decomposition of SnS_2 have been identified in a Moessbauer study.[446] Samples were heat treated and quenched and the Moessbauer spectra of the quenched samples recorded. The decomposition in an atmosphere of nitrogen was found to give Sn_2S_3 which subsequently was reduced to SnS. A new peak in the Moessbauer spectra of decomposed samples having a composition of approximately $SnS_{1.15}$ was assigned to a phase

$Sn_{1-x}S$ co-existing with Sn_2S_3. Herber and Kalish[447] have discussed the use of the Moessbauer effect in studies of radiation damage in heavy ion implanted thin layer crystals of SnS_2. The samples used in the study were single-crystal platelets mounted with the easy cleavage plane normal to the incident ion beam and then implanted at 100K with 150 and 350KeV O^+ ions at a total dose of about 2.5 $\times 10^{17}$ cm^{-2}. Data are given for samples mounted with a thin layer of thermal transfer grease (thermally clamped samples) and in the absence of a thermal transfer grease (thermally unclamped samples). With the exception of the thermally unclamped sample the Sn-119 Moessbauer spectra of all of the SnS_2 samples consisted of a single well defined resonance maximum having a linewidth ca. 0.76 to 0.90 mms^{-1}. The observed isomer shifts at 78K are in good agreement with previously recorded values. The isomer shift is only weakly temperature dependent. For SnS_2 samples in good thermal contact with the low temperature heat sink during the heavy ion implantation, the Moessbauer spectra showed only a singlet peak, at an isomer shift indistinguishable from that of the unirradiated sample. No evidence was found to indicate irradiation induced changes in the tin valency and there was no significant broadening of the resonance lines. In contrast to the spectra of the thermally clamped implanted samples, those of unclamped SnS_2 samples showed very clear evidence of decomposition, presumably arising from thermal effects during implantation. In addition to the major resonance peak at 1.1 mms^{-1}, there is a further doublet observable in the spectra of the unclamped samples. The isomer shift (3.32 mms^{-1}) and quadrupole splitting (0.97 mms^{-1}) leave little doubt that this pattern arises from SnS in the irradiated matrix. It is well known that Sn(IV) sulphide decomposes thermally and irreversibly to Sn(II) sulphide, starting at about 600K, and that in a vacuum this decomposition is rapid above about 870K. The present results show that the formation of SnS in SnS_2 by thermal decomposition during heavy atom implantation is a major consequence of the energy deposition by the heavy ion in the matrix of the thermally isolated sample. Moreover, it is clear from the present results that Moessbauer spectroscopy constitutes a sensitive probe of such decomposition in the irradiated samples. To ensure that the thermal decomposition is unrelated to the chemical identity of the implanted ion, the authors carried out a replicate experiment with a thermally lightly clamped sample of SnS_2, using neon as the projectile. The results for this experiment were indistinguishable from those in which oxygen ions were used. The authors also considered the possibility of formation of tin oxide in oxygen implanted SnS_2 and showed that there was no evidence for the formation of a material with properties similar to bulk SnO.

Sn-119 Moessbauer spectroscopy was used[448] in a study of the oxygen defects in the Cheverel phase $SnMo_5S_8$. A detailed analysis of the data provided confirmation that the Sn sites are associated with an oxygen containing point defect. The Moessbauer parameters for the two sites (δ = 3.64, Δ = 0.92 mms^{-1} for the Sn site associated with the point defect, and δ = 3.83, Δ = 0.71 mms^{-1} for the oxygen-free site) were considered to be sufficiently different to be distinguishable. It was also shown that the reduction in transition temperature by oxygen substitution for sulphur in the Cheverel phase was not due to charge transfer from the metal ion. The gamma-resonance spectra of tin-containing spinel phases, $CdCr_2(Se_xS_{1-x})_4$, consist of single lines with shifts of 1.15-1.44 mms^{-1}. The data suggest that the maximum distortion of the octahedral distribution of chalcogenide atoms around the tin atoms present occurs for x= 0.5.[449] The ternary alloy glasses $Ge_{1-x}Sn_xSe_x$ have been prepared for x= 0.002 to 0.01 and then crystallised by heating them to the crystallisation temperature. Moessbauer spectroscopy was then used to follow any changes in bonding at the tin probe atoms. The crystallisation process was found to lead to formation of tetrahedral Sn-Se sites and to octahedral sites such as those found in SnSe but not those found in $SnSe_2$. The $SnSe_2$ phase is, however, found among the products of the crystallisation of sputtered $SnSe_2$. The tetrahedral sites found in the crystallised products must be due to Sn substition of Ge and cannot arise from a tin-rich amorphous phase.[450] Both SnSe and $SnSe_2$ have been found among the precipitated products from the crystallisation of Sn-Se-As glasses.[410]

The reaction of KSn with Te at high temperatures gives a mixture of the congruently melting phase K_4SnTe_4 and tin.[451] The Sn-119 Mossbauer spectrum of K_4SnTe_4 is a singlet with a shift of 1.71 mms^{-1} and is in the range expected for tetrahedrally co-ordinated tin in the Zintl phase. A variety of products can be obtained from solutions of K_4SnTe_4 in the presence of various cations. Treatment of methanolic solutions of K_4SnTe_4 with Me_4NBr, for example, gives $(Me_4)_4Sn_2Te_6$ which contains the $Sn_2Te_5{}^{4-}$ ion, a possible mechanism for the formation of the ion being:

$$SnTe_4{}^{4-} + Br^- \rightleftharpoons SnTe_3Br^{3+} + Te^{2-}$$
$$SnTe_3Br^{4+} + SnTe_4{}^- \rightleftharpoons Sn_2Te_7{}^{6-} + Br^-$$
$$Sn_2Te_7{}^{6-} \rightarrow Te^{2-} + Sn_2Te_6Br^{4-}$$

An X-ray crystal structure determination on $(Me_4N)_4Sn_2Te_6$ confirmed the presence of an $Sn_2Te_6{}^{4-}$ ion with bridging and terminal Sn-Te bonds. The Moessbauer spectrum of the compound, unlike that of K_4SnTe_4, is a quadrupole doublet because the tin sites are no longer ideally tetrahedral.

Two papers published during the review year dealt with graphite inter-calation compounds of $SnCl_4$.[452,453] The layered $C_xSnCl_{6.0}$ (x= 54-66) was, for example, studied at 77 and 300K. The Moessbauer spectra at 77K are concentration independent and have shifts of about 0.36mms^{-1} and linewidths close to 1 mms^{-1}. The spectra of samples heated to about 300K lose the features due to the layered graphite compound and show only the presence of $SnCl_4$. Vertes[454] has continued his work on the Debye Waller factors of tin halides trapped in the porous glasses.

The isomer shifts obtained from frozen mixed solutions of $SnCl_4.5H_2O$ and K_3SnF_6 were found to be linearly related to the amounts of each component present.[455] The concentration dependence of chemical isomer shifts of K_2SnCl_4 and $SnCl_4$ in frozen aqueous solution has also been studied and again correlation found with the average number of ligands.[456] Samples of polyacetylene film containing varying amounts of the trans isomer have been oxidised with $SnCl_4$ and the tin containing product studied by Moessbauer spectroscopy.[457] The data (δ = 0.504, Δ = 0.605 mms^{-1}) were taken to suggest the existence of $SnCl_5{}^-$ as the inserted species. The low value (95K) for the Moessbauer lattice temperature obtained from recoil-free fraction data suggests that the $SnCl_5{}^-$ species has considerable mobility at room temperature.

X-ray diffraction, Moessbauer and other spectroscopic data have been used in studies on some tin(IV) chloride complexes with various N-, O- and S-donor ligands.[458] The changes in Moessbauer parameters were found to be related to the nature of the ligand donor atoms. A similar study[459] of $SnX_n(NCS)_{4-n}L$ complexes (where X = Cl, Br or I, n= 2, 3 or 4, and L = 2,2'-bipyridine, phenanthroline and substituted phenanthrolines) also showed a linear relationship between isomer shift and the average Mulliken electronegativities of the halide-isothiocyanate ligands around the tin.[459] The reactions between $SnCl_2$ and Br_2 and I_2 in weak donor solvents have been studied by Moessbauer spectroscopy and other methods and have been shown to result in the formation of tin(II) halide complexes of the type $SnCl_2Br_2.2L$ and $SnCl_2Br_2.4L$, where L = THF, OEt_2 or CH_3CN, in the bromide system and in $SnCl_4$ and SnI_3Cl in the iodide system.[460]

The new flourosulphates $M[Sn(SO_3)F_5]$ (M = K or Cs) have been prepared and characterised by Moessbauer and vibrational spectroscopies. A polynuclear anionic structure with hexa-co-ordination around the tin is suggested by the data.[461] The Moessbauer parameters of the tin(IV) complex with h.e.d.p. have also been reported.[437]

Two papers have described Sn(IV) complexes with crown ethers (L) with formulae $SnCl_4.L.2H_2O$ and $SnCl_4.L.4H_2O.nCHCl_3$, where n= 1 or 2. The complexes were precipitated and characterised by a number of techniques including Sn-119 Moessbauer spectroscopy.[462] The spectra for the crown ether complexes of $SnCl_4$ are singlets or doublets with small quadrupole splittings (less than 1 mms^{-1}) and chemical shifts in the range 0.26 to 0.62 mms^{-1}. The crystal structure data for $SnCl_4.2H_2O.18\text{-crown-}6.CHCl_3$ shows tin surrounded octahedrally by four Sn-Cl bonds and two $Sn\text{-}OH_2$ bonds in cis positions, and the Moessbauer parameters for the complex are consistent with this structure. Russo et al[463] have described the use of the Sn-119 Moessbauer effect and other spectroscopic techniques to characterise dihydrated 1:1 and 2:1 $SnCl_4$:crown ether complexes. Again the data suggest octahedral tin environments with four bonds to chlorine and two to water oxygen atoms.

Organotin(IV) Compounds. Barbieri and Silvestri[464] have correlated Sn-119 Moessbauer chemical isomer shift data with partial atomic charges on tin calculated by a valence state electronegativity equalisation procedure for 1112 compounds in a number of homologous series of five-coordinated Sn(IV) and $Sn(IV)R_n$ (n = 1-3) derivatives. They suggest that results indicate that the shift/atomic charge functions are generally suitable for structural assignments.

Calogero, Valle and Russo[465] have described a Moessbauer and single crystal X-ray study of hexa- and penta-co-ordinated organotin compounds containing urea type derivatives. The bonding in triorganotin pyridines has also been studied by the Moessbauer effect. In the study 4-bromopyridine was lithiated and treated with R_3SnCl compounds (R= Me or Ph) to give R_3SnR' derivatives with R' = 4-pyridyl. These compounds could then be oxidised to give the 4-pyridyl-N-oxide (R") derivatives R_3SnR''. Reaction of Me_3SnR' with an excess of Me_3SnCl gave $Me_3Sn(\mu\text{-}R')SnMe_3Cl$. The Moessbauer parameters[466] were consistent with all of these pyridine derivatives being covalent monomers in the solid state and with the presence of both four- and five-co-ordinate tin in $Me_3Sn(\mu\text{-}R')SnMe_3Cl$. The triorgano-tin(IV) derivatives of saccharin R_3Sn(N-saccharyl) have been prepared for R = Me, Et, Bu, Ph, Pr and cyclohexyl.[467] The Moessbauer data suggest that the tin co-ordination in these compounds is greater than four and that they contain approximately planar equatorial R_3Sn groups. The Moessbauer parameters for the compounds R_3SnOSO_2R' (where for R= Bu then R'= Me, Et, or C_6H_4(4-Me), and for R= Ph, R'= Et or Ph) have values in the range 1.36-1.51 mms^{-1} (shift) and 3.90-4.46 mms^{-1} (splitting) that are consistent with infinite chain polymeric structures in which the sulphonate groups bridge planar R_3Sn moieties. The Sn-119 Moessbauer data for $(Bu_3SnO)_3PO$, $(Bu_3SnO)P(O)Ph_2$ and $(Bu_3SnO)_2P(O)Ph$ have been obtained.[468] The shifts and splittings for the phosphorus oxyacid derivatives are consistent with the presence of both four- and five-coordinated tin environments in $(Bu_3SnO)_3PO$ and $(Bu_3SnO)_2P(O)Ph$. The shifts for both sites are similar and are in the range 1.36-1.40 mms^{-1} but the splittings for the two sites differ, being in the ranges 2.57-2.76 and 3.74-3.82 mms^{-1}

respectively. The spectra for $Bu_3SnOP(O)Ph_2$ show as the presence of one tin site with parameters of $\delta = 1.43$ mms^{-1} and $\Delta = 3.83$ mms^{-1}. Some five and six-co-ordinated di- and tri-n-butyl tin(IV) semicarbazones were prepared and characterised by a number of techniques including Sn-119 Moessbauer spectroscopy.[469] The data for the compounds Bu_2SnL and Bu_3SnL are consistent with trigonal tin co-ordination and those for the compound Bu_2SnL with distorted cis-octahedral geometry. The parameters for Bu_2SnL_2 are $\delta = 1.08$, $\Delta = 2.36$ mms^{-1} and $\delta = 0.98$, $\Delta = 2.23$ mms^{-1} for L= $C_8H_8N_3O_2$ and $C_9H_{10}N_3O_2$ respectively.

Both i.r. and Sn-119 Moessbauer spectroscopy have been used to identify the chemical reactions between the hydroxyl groups of silica gel and the $Bu_3SnOSnBu_3$ molecule.[470] The reaction is said to involve a cleavage of the Sn-O-Sn bond to form Si-O-Sn bonds on the surface of the silica gel. Complexes of the type $(SnPh_3X)_2.L.2H_2O$, where X is halogen atom and L is a crown ether have also been prepared and characterised[462] by Moessbauer spectroscopy along with $(SnR_2X_2)_n.L.2H_2O$ type complexes, where n= 1 or 2. These crown ether complexes have shifts in the range 1.1-1.43 mms^{-1} and splittings in the range 3.13-4.03 mms^{-1}. The structure of triphenyltin acetate has been shown[471] to contain tin environments that can be described as distorted six-co-ordinateed mer-Ph_3SnO_3 sites in a lattice in which carboxylate bridges link subunits together to form a flattened helical polymer. Variable temperature Moessbauer data suggest that the vibrational freedom of the tin atoms in the lattice is similar to that normally associated with non-interacting lattices.

Both Sn-119 n.m.r. and Sn-119 Moessbauer studies have been carried out on the series of heteroaryl tin compounds SnR_4X_{4-n} (R = 3-furyl, X = Cl or Br, and n = 2 or 3).[472] The γ-resonance data for the $SnR_3X(Ph_3PO)$ compounds are consistent with the presence of five-co-ordinated trigonal pyramidal tin sites with equatorial 3-furyl groups. The data for $SnR_2X_2(bpy)$ compounds are consistent with octahedral Sn environments with cis-3-furyl groups. Moessbauer evidence for the presence of non-equivalent four- and five-co-ordinated tin sites has been found in bis(trialkyltin) carbonates.[473] The temperature dependence of the gamma-resonance spectra of cyclic dibutyltin stannoxanes has been measured for solid and frozen solution samples.[474] The data suggest that the compounds consist of dimeric species with five-co-ordinated tin sites. Davies and Price[475] found evidence for five-co-ordinated tin sites in some complexes of 2,2-dibutyl-1,3,2-dioxastannolanes with pyridine, DMF, THF and DMSO. The compounds Bu_2SnL and $[Me_2Sn(OH)_2]L$, where L = 3-thiouracil, were also characterised by a number of techniques including Moessbauer spectroscopy.[476] Twelve complexes of R_2Sn moieties with nucleoside ligands were found to have Moessbauer shift and splitting data in the ranges 1.14-1.29 mms^{-1} and 2.97-3.10 mms^{-1} that were interpreted in terms of dimeric units.[477] Moessbauer and other spectroscopic evidence has been used to show the anisobidentate character of dithiocarbamate ligands in its complexes with the R_2SnCl_2 compounds with R = Bu and n-hexyl.[478]

The crystal structure of bis[2-(aminocarbonyl)anilinium] dimethyltetrachlorostannate(IV) dihydrate at 138K shows the presence of trans-Me_2SnCl_4 units held in a three-dimensional associated lattice by hydrogen bonding. The Moessbauer parameters (δ = 1.19 mms^{-1}, Δ = 4.09 mms^{-1} at 77K) are consistent with this structure.[479] The Sn species formed in the reaction between long chain monoalkyltin compounds and cellulose fibres has been studied by Moessbauer spectroscopy.[480] For a fibre treated with sodium butanestannoate the data (shift = 0.71 mms^{-1}, splitting = 1.83 mms^{-1}) are consistent with the presence of butanestannanic acid in the fibre. For fibres treated with $RSnCl_3$ (R = Bu or Oct), the gamma-resonance data are consistent with the presence of $RSn(OH)_2Cl$ on the fibre.

6 OTHER ELEMENTS

This section reviews the data for elements other than iron or tin. In each of the three main subsections (main-group elements, transition-metal elements, and lanthanide and actinide elements) the isotopes are treated in order of increasing atomic number.

Main-Group Elements.- Antimony (^{121}Sb). A review[481] of Moessbauer spectroscopy of antimony compounds has been published in the past year, which concentrates on the developments that have taken place since 1972. Antimony-121 Moessbauer data have also been included in a general review article by Berry[22]. Pebler and Dehnicke have reported[482] the Sb-121 Moessbauer effect of octahedral antimony(V) compounds. Transmission Moessbauer spectra at 4.5<T<150K have been recorded[483] for polycrystalline antimony. From the temperature dependence of the relative area of the single absorption line, the Debye temperature was determined to be 163(3)K. The isomer shift obtained, interpolated to 100K was -11.42(7) mms^{-1} (relative to a $BaSnO_3$ source cooled to 100K), and within the experimental accuracy no quadrupole broadening of the single line was detected. The electronic and structural properties of antimony in stage 1 antimony pentafluoride-graphite have been investigated[484] using Sb-121 Moessbauer spectroscopy. The valence configuration of Sb is essentially (>94%)Sb^{5+}, and the intercalated molecular state has been assigned to SbF_5 in a configuration similar to that of solid SbF_5 in a distorted octahedral coordination. Large lattice vibration anisotropy was demonstrated and analysed in terms of an anisotropic Debye model. It has been suggested that the small (6%)Sb^{3+} component is not actually intercalated. Friedt et al.[485] have extended this study and presented a detailed analysis of ^{121}Sb Moessbauer spectroscopic data obtained from two $SbCl_5$.HOPG samples. All spectra showed resonances due to Sb^{5+} and Sb^{3+} (present as $SbCl_3$) with a 2:1 intensity ratio. Analysis of the vibrational anisotropy showed both species were intercalated and the results were compared with previously reported data for SbF_5.HOPG[484].

A series of substituted pyridinium bromoantimonate(III) compounds with anions $SbBr_4^-$, $SbBr_5^{2-}$ and $Sb_2Br_9^{3-}$ have been prepared and studied by Raman and ^{121}Sb Moessbauer spectroscopies.[486] Moessbauer spectra were obtained at both liquid-helium and liquid-nitrogen temperatures. Some of the data recorded at 4K are presented in Table 4. The general trend observed in the spectra from $SbBr_3$ to $SbBr_5^{2-}$ is a decrease in both the isomer shift and quadrupole coupling constant as the number of bromines surrounding the antimony increases. The isomer shift (relative to InSb) ranges from -7.64 to -8.20 mms^{-1} for the $SbBr_4^-$ anion, from -8.90 to -9.01 mms^{-1} for $Sb_2Br_9^{3-}$ and from -8.20 to -9.33mms^{-1} for $SbBr_5^{2-}$. The quadrupole coupling constant ranges from +5.9 to +8.0 mms^{-1} for

$SbBr_4^-$ and from 0 to +5.2mms^{-1} for the $Sb_2Br_9^{3-}$ and $SbBr_5^{2-}$ anions. Substitution on the pyridinium ring has a significant effect on the Moessbauer parameters of a given anion which, the authors consider, is probably due to distortion of the anion geometry.

Table 4 Moessbauer Parameters for $R_xSb_yBr_2$ at 4K

Compound	(+0.05) δ/mms^{-1}	(+0.05) Δ/mms^{-1}
$SbBr_3$	-5.9	11.6
	$RSbBr_4$	
$pyHSbBr_4$	-8.01	5.9
2-$MepyHSbBr_4$	-8.2	7.2
3-$MepyHSbBr_4$	-8.0	7.6
4-$MepyHSbBr_4$	-8.08	8.0
2,4,6-$Me_3pyHSbBr_4$	-7.64	8.1
2-Br(py)$HSbBr_4$	-7.82	6.4
	R_2SbBr_5	
(3-MepyH)$_2SbBr_5$	-8.35	4.7
(4-MepyH)$_2SbBr_5$	-9.33	0.0
(2-Br(py)H)$_2SbBr_5$	-8.20	3.5
	$R_3Sb_2Br_9$	
(2,4,6-Me_3pyH)$_2Sb_2Br_9$	-9.01	0.0

The Fe-57 and Sb-121 Moessbauer-effect spectra of six iron-containing antimonate spinels of the general formula $M^{II}{}_{12}Fe^{III}{}_3Sb^V{}_3O_{24}$, where M^{II} is divalent Zn, Mg, Co and Ni, and the spinel $Ni_{10}Fe_6Sb_2O_{24}$ have been measured[284] and the 4.2K ^{57}Fe results used to determine the cation distribution between tetrahedral (A) and octahedral (B) sites. The Sb-121 Moessbauer effect spectra were measured at 4.2K and room temperature, and at 4.2K both $Zn_{12}Fe_3Sb_3O_{24}$ and $Mg_{12}Fe_3Sb_3O_{24}$ show a single absorption envelope with a slightly resolved quadrupole interaction. The isomer shift values are characteristic of Sb(V), and there is no evidence in these spectra of a hyperfine field at the antimony nucleus. However, the compounds containing magnetic M^{II} dications exhibit ^{121}Sb Moessbauer spectra that are considerably broadened at 4.2K by the existence of small supertransferred hyperfine fields at the antimony nucleus, the magnitudes of which are considered to be related to the nature and distribution of the dications. Pyrochlore oxides of the type $Mn_2Sb_2O_7$ and $(Mn_{1-x}Cd_x)_2Sb_2O_7$ have been synthesized by high temperature solid state reactions.[487] The ^{121}Sb Moessbauer spectra have been recorded at 4.2K on some representative samples in the $(Mn_{1-x}Cd_x)_2$ Sb_2O_7 solid solution, i.e. x = 0.0, 0.2, and 0.8, in order to ascertain the valence states of Sb in the pyrochlores. All materials gave a single line spectrum, showing only one type of Sb, with an isomer shift of 0.0mms^{-1} (relative to $^{121}SnO_2$), indicating antimony in the 5+ state. A single Lorentzian fit to the data showed Sb to be in a symmetric environment, the linewidth of the spectrum being about 4 mms^{-1}. The magnetic data, taken along with Moessbauer data, indicated that the compound $Mn_2Sb_2O_7$ should be nearly stoichiometric with all the Sb present as Sb^{5+} in octahedral coordination sites (site B) and all the Mn^{2+} occupying the A site. Progressive substitution of Cd in $Mn_2Sb_2O_7$ is found to decrease the rhombohedral distortion giving a solid solution which is cubic at $x \geq 0.6$. The bonding in Sb(III) complexes of the type $(Pr^iO)SbL$ and Sb_2L_3 (where L is the dianion of semicarbazone or thiosemicarbazone) has been discussed in terms of the Sb-121 Moessbauer spectra.[488] The data recorded at 4.2K are characteristic of lone-pair distorted Sb(III) in that the shifts have values of about -4 mms^{-1} (from a $BaSnO_3$ source) and large coupling constants of 12-18 mms^{-1}. The thiosemicarbazone complexes have more negative isomer shifts than their oxo analogues. This fact suggests that the X group (X = O, S) is bound to the antimony. It has also been noted that the S-containing complexes have smaller quadrupole coupling constants than the corresponding semicarbazones. The authors consider that this, taken along with the trend in isomer shifts, indicates a less active participation of the antimony 5<u>s</u> electrons in the bonding scheme of the thio complexes than in the oxo species.

The synthesis, structure and antimony-121 Moessbauer spectra of covalent and ionic tetraorganoantimony(V) compounds have been reported.[489] The ^{121}Sb Moessbauer spectra of the stibonium complexes prepared from R_5Sb (R = Me, Ph) and 8-hydroxyquinoline or anthranilic acid are consistent with covalent structures for the materials. However, the spectra of the hydrogen-dicarboxylates, e.g. $[Me_4Sb][O_2CC_6H_4CO_2H]$, suggest ionic structures with, only weak, if any, donor interactions between the anions and the stibonium cations.

Tellurium (^{125}Te). Moessbauer data have been included in a general review article on tellurium[19].

Moessbauer emission spectra of ^{125}I-containing organic carboxylate salts have shown[490] that in aqueous solution K-capture in ^{125}I leads to Te compounds with quadrupole splittings of 13-14 mms^{-1}. A previous suggestion that 5s electrons participate in the chemical bonding of Te(VI) has been confirmed by a comparative ESCA study of TeO_2 and $AgTeO_6$.[491] Also the chemical forms of the Te(IV) and Te(VI) sources, $[^{129}TeCl_6]^{2-}$ and $[^{129}TeO_4]^{2-}$, were confirmed by emission Moessbauer spectroscopy. The 4.2K ^{125}Te Moessbauer parameters of eleven square planar and six octahedral Te(IV) complexes have been recorded and the results discussed in terms of the s and p character of the bonds involved.[492] A paper has appeared[493] reporting the preparation of some donor-acceptor complexes of organotellurium compounds, the acceptor being 7,7,8,8-tetracyano-p-quinodimethane (tnq). Generally, the stoichiometry is 1:1, but two ditellurides provided exceptions giving R_2Te_2.2tnq (R = Ph or C_6H_4OEt-p). Tellurium-125 Moessbauer spectroscopy has been used to assess the degree of charge transfer for the donor-acceptor complexes of organotellurium compounds. The Moessbauer data were recorded at 4K using a ^{125}Sb-Cu source, and the chemical shifts (δ(I-Cu)[δ(Sb-Cu)+0.15]mms^{-1}) for the tcnq complexes lie in the range 0.19-0.46, while the quadrupole splitting data lie in the range 8.72-10.09 mms^{-1}. The preparations of di-tert-butylditelluride and di-tert-butyltelluride have been reported and the Te-125 n.m.r. and Moessbauer spectra of some dialkyl tellurides and ditellurides discussed.[494]

Iodine (^{127}I, ^{129}I). Moessbauer spectroscopy of iodine has been reviewed for the period since 1972 by De Waard.[495]

Iodine Moessbauer-effect sources in the form of Mg_3TeO_6 and Ca_3TeO_6 have been developed.[496] Due to the unusual solid state properties of these crystals and the high co-ordination symmetry of Te in their matrices it was possible to obtain single, narrow, emission lines with a recoiless fraction at room temperature of 0.55, ten times larger than that of conventional ZnTe and SnTe sources. The compounds can be prepared and repeatedly irradiated without any perceivable radiation damage and with negligible residual activity from the other constituent elements. This allows for the first time the use of ^{129}I Moessbauer spectroscopy at room temperature and significantly improves the performance of ^{127}I at low temperatures. It is considered that Mg_3TeO_6 should be adopted as a standard iodine Moessbauer source. Pasternak, Van der Heyden and Langouche[497] have also reported a high efficiency, single-line absorber for iodine Moessbauer spectroscopy. The authors have described the properties and preparation procedures of a new single-line absorber in the form of Na_5IO_6 with a recoiless fraction of 0.25 at room temperature. This f value is ten times larger than the conventional CuI or alkali-iodide single-line absorbers. Therefore it is possible for the first time to employ a ^{129}I absorber at 300K with sources at cryogenic temperatures. The linewidth of Na_5IO_6 measured for zero effective thickness is 0.98 mms^{-1}, also somewhat larger than that of CuI (0.78 mms^{-1}).

Graphite intercalation compounds with IF_5 have been studied at 4.2K by Moessbauer spectroscopy with the 57.5 keV γ-resonance of ^{127}I.[498] The identical hyperfine parameters observed for three differently prepared samples revealed the presence of square-pyramidal IF_5 intercalated molecules as in the pure molecular solid. Apart from absorber thickness effects, no changes in the intrinsic spectral shapes of the resonance spectra were found for different orientations of the γ-ray direction relative to the graphite c-axis. The observed anisotropy of the recoil-free fraction shows that the IF_5 molecules are much more strongly bound in the c direction compared to the direction parallel to the graphite planes.

Iodine-127 Moessbauer spectra at 4.2K have been obtained[499] for several compounds that contain the phenyl-iodine moiety including the alkane-activating Mn porphyrin complexes containing iodosylbenzene ligands $[XMn^{IV}TPP(OIPh)_2]O$ (TPP = tetraphenylporphinato dianion, X= Cl, Br (**1-Cl** and **1-Br** respectively)) that were recently purified and characterised. Sufficient iodine Moessbauer data now exist on compounds of known structure that the authors have been able to use this method to investigate compounds whose structures have not yet been determined. The relation of the ^{127}I Moessbauer and structural parameters of several iodine(III) compounds strongly implies that the observed positive quadrupole constants (2410-2507MHz) and large asymmetry parameters (0.47-0.79) of these compounds reflect a disruption of the classical T-shape symmetry around I(III) by the proximity of a fourth ligand. The I-127 Moessbauer data for **1-Cl** and **1-Br** are similar to those of other I(III) compounds, but quite distinct from those of I(I) or I(V) compounds. The iodine(III) in each of these complexes is coordinated by carbon, oxygen and halogen atoms to give distorted T-shaped arrangements.

Transition Metals. The assumption that the measured isomer shifts in Moessbauer spectra of transition metals caused by absorption of hydrogen are a consequence of an H-induced decrease in the local electronic charge distribution at the cores of the transition metal atoms has been proved[500] directly by scattered wave (SW)-X_α calculations of the electronic hydrogen density in octahedrally symmetric clusters $M_{14}H$ (where M = transition metal).

Nickel (^{61}Ni, ^{63}Ni).- A scattering experiment on observing the Moessbauer effect in the 87keV transition ($t_{1/2}$ = 1.7 μs) in radioactive ^{63}Ni ($t_{1/2}$ = 100 yrs) has been reported[501]. Because of the long lifetime of this state, its natural linewidth has the highest relative energy resolution of all Moessbauer resonances in use, which means the use of Ni-63 should enable detailed Moessbauer studies in Ni to be performed. The size and the sign of the magnetic hyperfine field in the nickel spinels $NiFe_2O_4$, $GeNi_2O_4$, $NiFe_{0.3}Cr_{1.7}O_4$, $NiCr_2O_4$ and $NiRh_2O_4$ have been determined by Ni-61 Moessbauer studies in applied magnetic fields[502]. The hyperfine fields were found to be negative, of the order of -10T, in octahedral sites and positive, in the range +2.5T to 63T, in tetrahedral sites. The signs and magnitudes of the hyperfine fields have been explained in terms of ligand field theory including the effects of 3*d*-4*p* orbital mixing.

Zinc (^{67}Zn). Different modulation methods for the determination of small isomer shifts have been compared[503]. The methods examined were sinusoidal and constant velocity Doppler modulation at low drive frequencies. The use of the transient frequency region has also been discussed and corrections due to inhomogeneous line broadening and electronic absorption included. The results are found to strongly depend on absorber thickness and the experimental data for ^{67}ZnO found to be in good agreement with theoretical predictions, and a relative energy resolution of about 3 $\times 10^{-18}$ was obtained. Both Potzel *et al.*[504] and Pound *et al.*[505] have reported on the anisotropy of the recoil-free fraction in Moessbauer resonance of Zn-67 in single crystals of zinc metal. In the first paper the anisotropy of the mean square atomic displacement was determined for Zn metal between 4.2 and 47K. The corresponding f factors perpendicular and parallel to the *c* axis vary by a factor of 2100 at the highest temperature investigated. This very large anisotropy and second-order Doppler shift were well described by an extended Debye model characterised by two Debye temperatures $\Theta_{\perp} = (242 \pm 10)$K and $\Theta_{\parallel} = (149 \pm 20)$K. The authors concluded from their results that the 67-Zn Moessbauer resonance is a most sensitive tool for studying lattice dynamical effects. Pound and his co-workers[505], however, have shown that cyclotron activation of ^{67}Ga in single crystals of metallic zinc produces sources capable of giving a sharp recoil-free resonance of ^{67}Zn in oriented single crystals of natural zinc metal. A linewidth of 0.53 μms^{-1} was observed and the measured anisotropy of the f-fraction was 20.4 x $10^{-24} m^2$ perpendicular to the *c*-axis and 30 x 10^{-24} m^2 parallel to it.

Ruthenium (^{99}Ru). The Moessbauer absorption of the 89.4keV line of ^{99}Ru has been measured[506] at 4.2K for samples of Ru in TiO_2 and in TiO_2 doped with Ta. Ruthenium replaces Ti substitutionally in TiO_2 as Ru(IV) and without delocalisation of the four Ru(IV) 4*d* electrons. If the donor Ta is added to the system, the ion Ru(II) is formed, demonstrating that Ru behaves as an acceptor in TiO_2. As part of a study of nitrido-bridged complexes of Ru, single-crystal analyses, Moessbauer isomer shifts, and infrared stretching frequency studies have been carried out.[507] Moessbauer spectra of the complexes of the type $K_3[Ru_2NX_8(H_2O)_2]$ measured at 4.2K showed isomer shifts (-0.42 to -0.21 mms^{-1}) which parallel the position of the X ligand in the spectrochemical series viz. Br<Cl<NCS<NO_2. The higher shift (-0.025 mms^{-1}) observed for $K_2[Ru_2N(CN)_{12}]$ was explained in terms of the cyanide ligand capability of accepting *d*-electron density from the metal. In general, the anionic complexes showed small or unresolved quadrupole splittings while the cationic species showed a well resolved splitting indicating a greater electronic charge asymmetry around the Ru in the cationic complexes. Ru-99 Moessbauer spectra and X-ray single crystal data for three binuclear complexes $[(NH_3)_5Ru(pyz)Ru(NH_3)_5]^{n+}$, where n = 4, 5, 6, and pyz = pyrazine, have been reported.[508] The combined studies have shown that the Creutz-Taube ion cannot be described as the combination of its mononuclear precursors or as the average of the [II,II] and [II,III] ions. The slight structural differences between $[II,III]Cl_5.5H_2O$ and $[II,III](tos)_5.4H_2O$ (tos = *p*-toluenesulphonate) do not show up in the Ru-99 Moessbauer spectra, both consisting of a slightly broadened quadrupole-split doublet, δ around 0.53 mms^{-1} and Δ of about 0.51 mms^{-1}. The spectrum is similar to that of a [III,III] complex; consistent with this suggestion is the fact that the spectrum of $[Ru^{II},Ru^{III}]$, where the valences are assumed to be trapped, is also shifted significantly towards that of [III,III] and away from that of [II,II]. Even with the sum of the structural and spectroscopic evidence reported, there is still doubt as to whether the odd electron in the Creutz-Taube ion is trapped on one ruthenium or delocalised equally over both.

Tantalum (^{181}Ta). Static and dynamic influences on the Ta-181 Moessbauer resonance have been reviewed[509] with particular reference to applications in metallurgy and diffusion. The potential of the application of the resonance to the Ta compounds has been summarised for the 2-dimensional type Ta dichalcogenides and other chalcogenide compounds. Potzel et al.[510] have used Moessbauer spectroscopy along with neutron scattering spectroscopy to study the effect of atmospheric hydrogen in Ta at high hydrogen concentrations. Pfeiffer et al. have reported the Ta-181 Moessbauer spectra for the charge-density waves in both 1T and 2H polytypes of $TaSe_2$ and suggested a method for calibration of their results. The isomer shift of the 6.2keV nuclear transition of Ta-181 in Cu_3TaTe_4 has recently been determined by Moessbauer spectroscopy.[512] The value obtained was +55.95 ±0.06 mms^{-1} (relative to Ta metal). The isomer shifts of the three isotypic compounds Cu_3TaX_4 (X= S, Se, Te) were found to be related linearly to the Mulliken electronegativities of the chalcogens (for Cu_3TaS_4 δ = 15.22, and for Cu_3TaSe_4 δ = 32.64mms^{-1}). A correlation between isomer shifts and the unit cell volumes was deduced for these compounds.

Gold (^{197}Au). One chapter in the reference text[4] on Moessbauer spectroscopy applied to Inorganic Chemistry has dealt with gold-197 Moessbauer spectroscopy in the characterisation of gold compounds. Silver[21] has included Au-197 Moessbauer data in a general review on gold. Trautwein and co-authors[513] have shown how electronic structure calculations can be used to quantitatively evaluate Moessbauer parameters to gain insight in the atomic and electronic structure of a system. Some of the applications deal with gold complexes in terms of the quadrupole splitting - isomer shift relationship for Au(I) and Au(III) complexes. The value of ^{197}Au Moessbauer spectroscopy in distinguishing the two major oxidation states, the identity of the ligands, and the coordination numbers of gold compounds has been reviewed.[514] Data for the compounds $Cat[Au(Ar)_2]$ (Cat = Bu_4N, Ag, Ag(thiophene), Ar = C_6F_5, $C_6F_3H_2$) have been interpreted in terms of cation-anion interaction. When Cat = $[Au(SbPh_3)_4]$ or $[Au(pdma)_2]$ (pdma = o-$C_6H_4(AsMe_2)_2$), the parameters of the cations are consistent with the known four-coordination. In these compounds and in $(Bu_4N)_2[(pdma)Au(RR)Au(RR)Au(RR)Au(pdma)]$ (RR = $C_6F_4.C_6F_4$) the two different gold atoms give very different intensities.

A large number of Au(I) thiolates [Au(SR)] have been prepared in which the R group is an alkyl, aryl, carboxylic acid, amino acid, or heterocycle.[515] The gold(I) thiolates have been characterised by 197-Au Moessbauer, ^1H and ^{13}C n.m.r. spectroscopy and these data compared with those of the drugs 'Myocrisin', 'Solganol', and 'Auranofin'. The Moessbauer spectra, recorded at liquid helium temperature, are consistent with linear coordination of the gold by two sulphur ligands and, in agreement with the n.m.r. data, show broadening indicative of the presence of non-equivalent gold atoms. It has been demonstrated[516] for the first time that ^{197}Au Moessbauer spectroscopy and XANES/EXAFS can be usefully applied to Au-protein complexes and, in particular, that the co-ordination geometry and oxidation state of gold in the albumin-gold thiomalate complex can be assigned unambiguously by the combination of these two techniques. The complexes $BSA(AuSTm)_n$ (where n = 0.44, 1.30 and 2.65, BSA = bovine serum albumin, and (AuSTm) is sodium gold(I) thiolate) were prepared and the 197-Au Moessbauer spectrum of $BSA(AuSTm)_{1.3}$ recorded at 4.2K. The spectrum of this complex was resolved into two quadrupole doublets(δ_1= 1.88, Δ_1= 6.68 mms^{-1}, and δ_2= 1.70, Δ_2= 6.50 mms^{-1}) consistent with two slightly different Au^IS_2 environments.

A number of papers have appeared in the literature[517-520] reporting the use of Au-197 Moessbauer spectroscopy in the study of Au complexes with tertiary phosphines. As a follow-up to previous work, Parish <u>et al.</u>[517] have reported the ^{197}Au Moessbauer and $^{31}P-\{^1H\}$ n.m.r. spectroscopic studies on gold(I) complexes with tertiary phosphines. The Moessbauer data reported for complexes of the types [AuXL] and $[AuL_2]Y$ (X= Cl or I, Y= BPh_4 or ClO_4, L=$PBu^n{}_2Ph$, $PPr^i{}_2Ph$, $PPr^n{}_2Ph$, PEt_2Ph, $PMePh_2$ or PMe_2Ph) are consistent with values for related compounds. The first type of complex shows an unusual trend that the isomer shift and quadrupole splitting are lower for X = I than for X = Cl. The four-coordinate complexes $[AuL_4]BPh_4$ (L=PMe_2Ph or $PMePh_2$) are observed to follow previous trends, with zero quadrupole splittings unambiguously demonstrating tetrahedral geometry and the isomer shifts increasing in the expected order, L = PPh_3<$PMePh_2$ <PMe_2Ph. The lower symmetry complex $[AuCl(PPh_3)_3]$ shows a low shift consistent with the known four-coordinate configuration of the cation, and the small quadrupole splitting is considered to arise from the difference in electron density between the bonds to the three phosphine ligands and that to the fourth ligand.

The remaining three papers reporting Moessbauer studies of Au tertiary phosphine clusters were published by Van der Welden and co-workers[518-520]. Extended Hueckel molecular orbital calculations have been shown[518] to be useful in obtaining insight into the bonding of the gold skeletons of gold cluster compounds and also in the understanding of the quadrupole splittings observed in the Au-197 Moessbauer spectra. An empirical correlation has been established between the Au(6<u>p</u>) electron population and the Au-197 Moessbauer quadrupole splitting providing a better understanding of the ^{197}Au Moessbauer spectra of gold clusters. The fact that no separate signal was observed for the central Au atoms in the two clusters <u>viz.</u> $[Au_9(PPh_3)_8]^{3+}$ and $[Au_8(PPh_3)_7]^{2+}$ has been explained in terms of the asymmetry parameter for the central Au atom in the complexes being rather large and comparable to those of the peripheral gold atoms. Attempts to isolate[517] a small intermediate containing 2-4 gold atoms, from the reaction of gaseous gold atoms with L (where L=PPh_3) in the synthesis of gold clusters, failed, but led to the discovery of two novel Au cluster compounds, the $[Au_8L_6I]^+$ and $[Au_7L_7]^+$ clusters. The Au-197 Moessbauer spectrum of $[Au_8L_6I]^+$ has been fitted reasonably well with two quadrupole pairs attributable to a gold-phosphine site and a gold-iodine site without a separately observable central gold atom (Table 5).

<u>Table 5</u> The Au-197 Moessbauer parameters for $[Au_7L_7]^+$ and $[Au_8L_6I]PF_6$

Compound	Au site	IS*/mms^{-1}	Q.S./mms^{-1}
$[Au_7L_7]^+$	axial	1.48	4.12
	equatorial	1.95	6.95
$[Au_8L_6I]PF_6$	Au-P	1.58	6.65
	Au-I	1.03	3.48

*relative to ^{197}Pt source

The complex $[Au_9(PPh_3)_8]PF_6$ has been prepared[520] by cathodic deposition during the coulometric reduction of $[Au_9(PPh_3)](PF_6)_3$ in MeCN. The Au-197 Moessbauer data were compared with those of $[Au_9(PPh_3)_8]^{3+}$, and an X-ray crystal structure determination has shown that the nine gold atoms in the cluster are in a centred cubic arrangement.

Calis and Hadjiliadis[521] have used ^{197}Au Moessbauer spectroscopy to study the structure and bonding of Au(I) and Au(III) nucleosides. The spectra for the series of complexes of $Au(nucl)_2Cl$, $Au(nucl)Cl_3$, $Au(nucl-H^+)Cl_2$ and $Au(nucl)_2Cl_3$ (where nucl = nucleoside) measured at 4.2K showed that the Au(I) nucleosides have linear Cl-Au-N coordination with one coordinated nucleoside molecule per Au(I) ion bound via the N(7) atom. The significantly higher shift and splitting values for the complexes Au(inosine)Cl_3 and Au(guanosine)Cl_3 than the corresponding complexes with triacetylnucleosides were explained in terms of the weak O(6) interaction with Au(III) in a nearly trigonal-bipyramidal configuration in the former complexes and by the presence of the strongly electron-withdrawing acetyl groups in the latter, which reduces the donor strength of their N(7) atoms. The complexes of the $Au(nucl-H^+)Cl_2$ series all appear to have a polymeric stucture with Au(III) bound to the N(7) atom and the O(6) or the N(1) atom of the nucleosides. The Moessbauer spectra of the series $Au(nucl)_2Cl_3$ have been explained by assuming approximately octahedral AuN_2Cl_4 structures with bridging Cl atoms.

Lanthanide and Actinide Elements. Shenoy[522] in his article on 'Rare Earth Moessbauer Studies of Chemical Problems' has presented, in tabular form, the characteristics of useful rare earth Moessbauer transitions. The list includes these details for La-139, Pr-141, Nd-145, Sm-149, Eu-151, Gd-155, Dy-161, Er-166, Tm-169, and Yb-170. During the past year papers have been published on Moessbauer effect studies using isotopes of europium, gadolinium, dysprosium, thulium, ytterbium, and neptunium.

Europium (^{151}Eu). Eu-151 Moessbauer and optical reflectivity measurements have been carried out[523] on intercalation compounds formed either by reaction of graphite with $EuCl_3$ or with $EuCl_3.AlCl_3$ mixtures in the presence of chlorine. The results on graphite flakes and highly oriented pyrolytic graphite (HOPG) indicate the presence of a single Moessbauer line characteristic of Eu^{3+}, which is identified with intercalated $EuCl_3$. No evidence was found for Eu^{2+} in the Moessbauer spectra. Schaefer-Stahl and Von Eynatten[524] have reported on the europium-ammonia-hydrogen-graphite system. The Moessbauer spectra of the quaternary graphite co-intercalates show a mixture of Eu^{2+} and Eu^{3+} at room temperature. The intensity of the Eu^{2+} line is observed to decrease in the higher stage compounds. The Eu-151 Moessbauer data of the ternary hydride Eu_2IrH_5 in the temperature range 4.2-300K have been reported.[525] The exchange mechanism causing ferromagnetism in Eu_2IrH_5 is considered to be the same as that in EuH_2.

A study of the quadrupole interactions in the orthorhombic perovskite series $EuMO_3$ (M = Co, Cr, Mn, Fe, and Sc) has been made using the ^{151}Eu Moessbauer effect.[526] The quadrupole coupling constant at room and liquid nitrogen temperatures was found to be negative, lying in the range -9.3 to -5.8 mms^{-1}, and has a weak temperature dependence caused by the presence of low lying excited states in Eu^{3+}. It was shown that for Eu^{2+} the magnetic hyperfine field from 4*f* electrons induced by an external magnetic field is of opposite sign and almost equal intensity to the external field. A major investigation of magnetic exchange and supertransferred hyperfine field effects in perovskite solid solutions has been carried out using ^{57}Fe, ^{119}Sn, and ^{151}Eu Moessbauer resonances. Results on these investigations have been published in four parts, the fourth part having appeared in the literature during the review year.[285] In this work the ^{57}Fe and ^{151}Eu data have been presented for the solid solutions $EuFe_{1-x}M_xO_3$ (0<x<1), where M= Al, Ga, Sc, Mn, or Ni, and a detailed and critical discussion of the superexchange interactions has been developed in the light of the data obtained. The data for the ^{57}Fe, ^{119}Sn and ^{151}Eu resonances in solid solutions of rare earth orthochromites and orthoferrites have been found to be completely consistent with an antiferromagnetic Fe^{3+}-O-Cr^{3+} superexchange.

High pressure-high field Moessbauer effect studies on magnetic interactions in pure and magnetically diluted europium monochalcogenides have been reported.[527] By the simultaneous application of pressures $\leq$19.6kbar and magnetic fields $\leq$13T the various contributions to the magnetic hyperfine fields of EuX (X=S, Se, Te) and $Eu_xSr_{1-x}S$ were studied using the ^{151}Eu Moessbauer effect. A relationship between the magnetic exchange interaction and the transferred hyperfine fields was found. The concentration dependence of the saturation hyperfine field, as deduced from ^{151}Eu Moessbauer spectroscopy in EuS_xSe_{1-x}, has shown[528] that there is a continuous change of the local magnetic structure along the series. It was found that the spin glass phases consist of nearly two-dimensional ferromagnetic clusters whereas the ferromagnetic states close to the ferromagnetic-spin-glass boundary behave as 'frustrated ferromagnets'. Relaxation phenomena were observed in both the spin glass and 'frustrated ferromagnetic' phase. Takabatake and another group of coauthors have investigated the low temperature magnetic properties of $EuMo_6S_8$ and $EuMo_6Se_8$ by measuring the Eu-151 Moessbauer effect at 0.006-300K.[529] In another paper, the superconducting transition temperature of $Pb_{1-x}Eu_xMo_6S_8$ was studied as a function of x; both the temperature and composition dependences of the electrical resistivity and the Moessbauer spectra of the compound were measured.[530] The data indicated a stronger interaction between the Eu moments and the conduction electrons than in $Sn_{1.2(1-x)}Eu_xMo_6S_8$, a material investigated in the mid-seventies by another group of authors.

X-ray, magnetisation and Moessbauer (^{151} Eu, ^{155}Gd, ^{161}Dy, and dilute ^{57}Fe) studies of RM_2Si_2 (R= rare earth, M= Rh, Ru) have revealed[531] that when R is a magnetic rare earth ion the compounds order antiferromagnetically. For M = Rh a second antiferromagnetic phase transition has been observed corresponding to Rh itinerant electron magnetic ordering. The Moessbauer studies of the Eu compounds were performed using the 21.6 keV transition of ^{151}Eu with a $^{151}SmF_3$ source. In $EuRh_2Si_2$, the Eu ion is found to be predominantly divalent with a mixed valency component, while in $EuRu_2Si_2$ it is predominantly trivalent. The mixed valent intermetallic compound $EuCu_2Si_2$ has been studied[532] at $4.2 \leq T \leq 320K$ and in external magnetic fields up to B_0= 12.8T by using the Eu-151 Moessbauer effect. In a limited temperature range around T= 180K the application of B = 12.8T was found to induce a valence shift towards Eu^{2+} which is reflected by a change in δ of 0.2 mms^{-1}. This observation, in another paper by Scherzberg, was explained in terms of the ionic interconfigurational fluctuation model.[533] The author also reports, in the same paper, the results of the Moessbauer effect studies on $EuPd_2Si_2$. In this system an anomaly of the quadrupole interaction was found in the temperature region $130 \leq T \leq 180K$ where the valency changed rapidly. This effect is considered to be due to an admixture of 4_f_-contributions to the conduction band states. Besides the main resonance line, an additional satellite line was observed in all samples. Two papers report studies of the valence behaviour of Eu in some new boron containing rare earth intermetallic compounds.[534,535] The results show that the valence of europium, which is trivalent in $EuPd_3$ ($\delta = 4.5 \pm 0.1$ mms^{-1} at 300K), changes with boron content. The absorption in $EuPd_3B_{0.25}$, which occurs at the same position as in $EuPd_3$, was found to be temperature independent down to 1.4K but was found to become temperature dependent as the boron level is increased still further. In $EuPd_3B$, a charge ordering of the Eu ions is considered to occur at low temperature. The authors concluded that this is the first metallic rare-earth system known to exhibit such behaviour. A Moessbauer spectroscopic study[536] has confirmed that Eu in the europium-mercury system exhibits a divalent core and the isomer shift is found to depend on the number density of Eu atoms in the unit cell.

The ^{151}Eu isomer shift has been investigated[537] in several intermetallic compounds containing Eu(III), comprising $EuIr_2$, $EuRh_2$, $EuPd_3$, $EuPt_5$, $EuNi_5$ and Eu_3Ni^7. Analyses of the isomer shift data in terms of the model of Mieddema and Van der Woude, which makes the assumption that the isomer shift can be described in terms of electonegativity and charge density at the atomic cell boundary, showed that in the compounds with trivalent Eu the isomer shift behaves anomalously. This anomalous behaviour was ascribed to valence instabilities of europium in most of these materials. ^{151}Eu Mössbauer spectroscopy[538] taken along with time-resolved luminescence, EXAFS and e.p.r. spectroscopy have been used to study the relations between the zeolite framework and electron-transfer properties of Eu^{2+} and Eu^{3+}.

Gadolinium (^{155}Gd). Moessbauer studies of ^{155}Gd on $GdCoB_2$ were made[539] using a $SmPd_3$ source at 4.1 and 77K. The electric field gradient for Gd^{3+} was observed to be temperature independent and the value of eqQ = 580 $MHzs^{-1}$ at 4.1K reported to be the largest e.f.g. ever observed in a metallic Gd-containing compound.

In their study of RRh_2Si_2 and RRh_2Si_2 (R= rare earth element), Felner and Nowik[531] have presented the Gd-155 Moessbauer data for the Gd compounds of the series. The spectra were recorded with a $SmPd_3$ source. The orientations of the Gd magnetic moment with respect to the crystallographic axes of $GdRh_4B_4$ and $RE_xGd_{1-x}Rh_4B_4$ (RE= Ho and Er) materials have been studied with the ^{155}Gd Moessbauer effect.[540] The complex nature of the spin ordering (even well below the magnetic ordering temperature) in $GdRh_4B_4$ and the influence of anisotropic rare-earth ions on the orientation of the Gd spins have been demonstrated. The results have shown that the ^{155}Gd Moessbauer effect can be used as a probe of magnetic ordering directions in rare-earth materials.

Dysprosium (^{160}Dy, ^{161}Dy). Sources of ^{161}DyGd, prepared by thermal neutron irradiation of ^{160}Gd-enriched Gd metal, were used[541] to study Moessbauer line broadening in the temperature range $3x10^{-4} \leqslant t \leqslant 3x10^{-2}$ above T_c. Using the assumption that the spin fluctuations are isotropic and described by dynamical scaling theory, the authors found an anomalously low value for the dynamical exponent of 1.3-1.52. The authors consider that this anomalous value may originate from incorrect assumptions made in interpreting the Moessbauer line broadening. They also consider it may arise from the fact that spin fluctuations may become anisotropic as $T \rightarrow T_c$ with different temperature dependences for the longitudinal and transverse correlation times. A possible origin of the excess magnetic hyperfine field at the ^{161}Dy nuclei in DyFe compounds has been discussed.[542] Contributions to the excess were estimated on the basis of ^{57}Fe and ^{161}Dy Moessbauer effects and magnetisation on data. The hyperfine fields and sub-lattice magnetisations were calculated versus temperature using the molecular field approximation. It was found that the predominant contribution was the transferred hyperfine field from the Fe ions. Dy-161 Moessbauer data were presented for Dy, $DyFe_2$, $DyFe_3$, and Dy_2Fe_{17} at 4.2K. Dy-161 Moessbauer spectroscopy and static and dynamic magnetic properties have been investigated[543] in a series of amorphous Dy alloys ($Dy_{80}Au_{20}$, $Dy_{80}Ga_{20}$, $Dy_{65}Al_{35}$) presenting different structural short-range orders. The Moessbauer spectra of these alloys were recorded at room temperature. The temperature dependences of the Moessbauer spectra revealed three distinct regimes in all the alloys. Below T_f, static magnetic hyperfine spectra were observed. Above T_f, in an intermediate temperature range ($T_f < T < 2T_f$), smeared magnetic hyperfine spectra appeared, whilst at still higher temperatures ($T > 2T_f$) the spectra collapsed into a broad single resonance. The temperature dependence of the hyperfine parameters and of their distribution below T_f was interpreted in the molecular-field approximation using a non-axial quadratic crystal field model, which was found to be necessary and sufficient in order to represent experiments in the considered alloys.

Felner and Nowik[531] have included in their paper reporting studies on the RRh_2Si_2 and RRu_2Si_2 systems (R= rare earth element) the Moessbauer spectra for the Dy compounds, which were recorded using a GdF_3 source. X-ray, magnetisation and Moessbauer studies of ^{161}Dy in DyM_2Si_2 (M= Mn, Fe, Co, Ni, Cu, Ru, Rh, Pd, Pt, Au) and DyM_2Ge_2 (M= Cr, Mn, Fe, Co, Ni, Cu) have been performed.[544] Almost all the compounds order antiferromagnetically at low temperatures, and some of them exhibit time-dependent magnetisation phenomena. Correlations were observed between the ordering temperature and interatomic distances and between the ^{161}Dy isomer shift and magnetic hyperfine field. Moessbauer studies of dysprosium-iron-aluminium ($Dy_2Fe_{17-y}Al_7$) hydrides using the ^{57}Fe and ^{161}Dy effects have been performed.[545] Only a small variation of the hyperfine field at ^{151}Dy nuclei has been observed after hydrogenation.

Thulium (^{169}Tm). Moessbauer spectroscopy has been used[546] to measure the magnetic and electrical hyperfine interaction of Tm-169 nuclei in silver at 4.2K as a function of an external magnetic field of up to 3.2T. From the measured hyperfine field and quadrupole interaction the authors derived the parameters of the cubic crystalline electric field at the Tm site: $A_4\langle r^4\rangle = -6(5)K$ and $A_6\langle r^6\rangle = 6.5(4)K$. These values disagree strongly with previous susceptibility measurements on this system, but are close to values found from Moessbauer spectroscopy on $^{161}DyAg$.

Investigations of the temperature dependence of the ^{169}Tm Moessbauer spectra in $TmNi_2$, $TmRh_2$ and $TmIr_2$ have shown that magnetic ordering of the 4*f* moments does not occur in any of these compounds at 2K or higher temperatures.[548] From the analysis of the quadrupole splitting it can be concluded that none of the three compounds can be regarded as a van Vleck paramagnet at low temperatures. In all three compounds a strongly temperature dependent quadrupole splitting was observed. In view of the absence of magnetic or quadrupolar ordering, the authors were tempted to attribute this quadrupole splitting to the occurrence of a rapidly relaxing distortion as expected in the case of a dynamical Jahn-Teller effect. However, it was considered that further investigations were necessary to corroborate this view. The crystallographic and magnetic properties of Tm_6Fe_{23} before and after hydrogen absorption were studied[548] using X-ray diffraction, bulk magnetic measurements, and ^{169}Tm and ^{57}Fe Moessbauer spectroscopies. By means of ^{169}Tm Moessbauer spectroscopy the values of the thulium moments at 4.2K in Tm_6Fe_{23} and its hydride, $Tm_6Fe_{23}H_{14}$, were shown to be only slightly smaller than the free-ion value. The ^{169}Tm and ^{57}Fe spectra of the hydride were shown to exhibit no substantial line broadening with respect to the uncharged intermetallic compounds. An intepretation of the sharp Moessbauer spectrum considered by the authors was the partial occupation of more than one hole site and, associated with it, a jump frequency which is high compared with the Moessbauer frequency. Based on the ^{169}Tm spectra this jump frequency was considered to be of the order of $10^{12}s^{-1}$ or higher.

^{169}Tm Moessbauer measurements have been reported[549] for the distorted perovskite compounds $TmAlO_3$, $TmFeO_3$, $TmCrO_3$, and $TmVO_3$ and analysed to provide information concerning the crystal field and magnetic properties of the Tm^{3+}. Tm^{3+} has no magnetic moment in $TmAlO_3$ and low temperature saturated moments of 0.09, 0.8, and 2.1 μB in $TmFeO_3$, $TmCrO_3$, and $TmVO_3$ respectively. Bingham *et al.*,[550] using Moessbauer spectroscopy, have made measurements of the electric gradient at Tm^{3+} in $TmAsO_4$ and have been able to report a set of crystal field parameters that are consistent with known data.

The Moessbauer effect has been reported[551] for ^{169}Tm in $TmPdSi_2$. For Tm, the data have been interpreted in terms of a Tm^{3+} charge state and the crystal field parameters, magnetic ordering temperature, and paramagnetic relaxation rate obtained.

The measurement of the Moessbauer hyperfine spectra of $TmFe_2Si_2$ has been recorded at different temperatures using the 8.4keV transition in ^{169}Tm.[552] The separation and intensities of the quadrupole split lines vary with temperature, with the weaker lines becoming stronger at higher temperatures. The detailed temperature dependence of the quadrupole interaction has enabled the authors to evaluate the crystalline electric fields acting on the Tm atom, and analysis indicated the ground state of Tm to be a singlet state. The hyperfine coupling of ^{169}Tm with the excited crystalline electric field states generates a second order magnetic hyperfine interaction which was observed in these experiments.

Ytterbium (^{170}Yb, ^{174}Yb). A ^{174}Yb γ-ray source for Moessbauer spectroscopy has been claimed and its fabrication patented.[553] The Moessbauer emission spectra of $^{170}Yb^{3+}$ were recorded[554] in $Au^{170}Tm$ sources at $0.09<T<4.2K$ and interpreted using a relaxation line-shape theory which allowed for the existence of populations out of thermal equilibrium in the hyperfine levels. In another paper, by Bonville et al.[555], the very low temperature Moessbauer emission spectra were reported for ^{170}Yb, in a study of the electronic properties of the Yb^{3+} ion diluted in Pd.

The atomic scale electronic structures of the amorphous alloys $Yb_{80}X_{20}$ (X= Cu, Ag, Au) were studied by ^{174}Yb Moessbauer spectroscopy.[556] For all materials investigated, the distributions of the quadrupole splitting and asymmetry parameters corresponded to the distribution of charges for dense random packing of ions. Hodges and Jehanno[551] have presented the Moessbauer effect measurements for ^{170}Yb diluted into both $TmPd_2Si_2$ and $YbPd_2Si_2$. In both cases there was no evidence of any magnetic character, but intermediate valence behaviour was observed on Yb in both the diluted and concentrated samples.

Neptunium (^{237}Np). A review of the 'Chemical aspects of ^{237}Np Moessbauer spectroscopy' has been published during the past year.[557]

The Np-237 Moessbauer spectrum for $CoNpF_6.2H_2O$ was recorded at 4.2K and shows magnetic hyperfine structure associated with the quadrupole splitting.[558] The spectrum was fitted assuming that the principal axes of the magnetic field and the e.f.g. were collinear and that the asymmetry parameter was zero. Jove and another group of co-authors[559] have published the results of measuring the ^{237}Np Moessbauer effect and magnetic susceptibility of the $U_{1-x}Np_xO_2$, fluorite, solid solution for x = 0.15-0.75. All the magnetic ordering temperatures are less than for NpO_2, and in all cases the Np ions are in the 4^+ state.

Two papers have appeared[560,561] reporting crystal structure and Moessbauer studies of Np(IV) complexes with polycarboxylic acids. NpO_2^+ complexes with pyromellitic (benzene tetracarboxylic) and mellitic (benzene hexacarboxylic) acids were prepared for these studies. The Moessbauer spectrum of the $Na_4(NpO_2)_2$ mellitite has the parameters (recorded at 4.2K) of H_{eff}= 5490kOe, e^2qQ= 97mms^{-1}, and δ = -19.2mms^{-1}, which correspond to Np(IV).

^{237}Np Moessbauer spectroscopy measurements have been recorded at 4.2 and 77K for some Np polychalcogenides: NpS_3, β-NpS_2, $NpTe_3$, α-Np_2S_3, and γ-Np_2Se_3.[562] From the isomer shift values the oxidation state of the Np ion could be assigned, viz. Np(IV) in NpS_3 and β-NpS_2, Np(III) in $NpTe_3$, α-Np_2S_3, and γ-Np_2Se_3, and a mixture of Np(III) and Np(IV) ions in Np_3Se_5. Comparisons of the isomer shift values showed the influence of the electronegativity of the chalcogen atom and of the Np-X bond length. In a second paper, the results of a study of the magnetic properties and hyperfine interactions of neptunium sulphides and selenides have been reported.[563] Thermal variations of magnetic susceptibility and Moessbauer resonance showed magnetic order at low temperature in NpS_3 (45K), $NpSe_3$ (55K), Np_2Se_5 (15K), Np_3S_5 (13-17K) and Np_3Se_3 (26K). The charge on Np, given by isomer shift values, was found to be 4+ for NpS_3, $NpSe_3$, and Np_2Se_3 and a mixture of 3+ and 4+, in the ratio 2:1, in Np_3S_5 and Np_3Se_5.

A high-pressure low-temperature Moessbauer spectrometer[564] for the ^{237}Np resonance and a contactless conductivity susceptometer have been constructed. Magnetic properties of some novel neptunium intermetallics were studied using Moessbauer spectroscopy. In addition, hydrogen absorption in actinide intermetallics was studied. Finally, Moessbauer experiments were carried out[565] on $NpAl_2$, $NpOs_2$, $NpCo_2Si_2$ and NpAs using the 60keV transition in Np-237. The magnetic hyperfine field and isomer shift, as well as the magnetic ordering temperatures, were determined at pressures up to 70kbar and in the temperature range 1.6-77K. The spectra were discussed with reference to the Np-Np separation and the delocalisation of 5*f*-electrons.

7 Back Scatter and Conversion Electron Moessbauer Spectroscopy

The noticeable increase in papers describing the developments and application of the CEMS technique cited during the period of 1983 has not been matched over the past year. Considerably fewer papers have been published, details of which will now be given in terms of the theoretical developments in instrumentation and methodology followed by descriptions of the recent applications of the technique in studies of iron-, tin-, and europium-containing materials.

The reference text on Moessbauer spectroscopy applied to inorganic chemistry[4] includes a chapter on X-ray and conversion electron Moessbauer scattering for chemists. In their review of Moessbauer spectroscopy Stevens and Bowen[5] refer to the articles dealing with the technique of conversion electron Moessbauer spectroscopy. Meisel has summarised[566] the main aspects of the Moessbauer technique, both in the transmission and conversion electron mode, and has illustrated these by means of examples of its application in corrosion research.

A theory of conversion electron Moessbauer spectroscopy including second order effects, i.e. secondary electron emission and detection coincidence corrections, has been derived[567]. The theory is applicable to surface films containing any number of distinct layers. Partial spectra are given in terms of analytical functions of the Moessbauer parameters and the physical characteristics of each layer in the sample, and numerical results have been compared with available experimental data. A theory[568] of the secondary processes at total reflection of Moessbauer irradiation from a flat surface containing resonant nuclei has been developed. The CEM spectrum was shown to have a sufficiently broadened 2-humped shape rather than a simple Lorentzian lineshape. Direct calculations were presented for resonant ^{57}Fe films of different thicknesses and substrates.

Two types of gas flow proportional counters have been built.[569] One was a hexagonal pillar type detector by which angular dependent (90° and 45° incident γ-ray to sample surface) conversion electron Mössbauer spectra can be measured. The other was a thin plate type, which was designed to reduce the background. The variation of energy distribution of electrons reemitted from ^{57}Fe-enriched stainless steel was examined as a function of the distance between the anode wire of the detector and the sample surface. With an increase in the distance between the γ-ray source and the detector, the full widths at half maximum height of the peaks are independent of the Doppler velocity and the background is flatter. The incident angle of the γ-ray was found to change the electron energy distribution and the peak shape of the CEM spectrum. When high energy electrons were detected with the flow gas, asymmetric peaks appeared in the CEM spectrum due to the count loss of electrons. The phenomenon was explained in terms of the space-charge effect coupled with the scanning speed of time scaler of the multichannel pulse height analyser. A CEM spectrometer has been constructed[570] in which a mini-orange magnetic filter focuses the conversion electrons emitted by a moving absorber onto a twin window Si surface barrier detector, which serves as an electron spectrometer. A special multiscaler was used to record a number of spectra corresponding to different conversion electron energy groups simultaneously. A large reduction in counting time, compared to other methods of energy differential CEMS, was achieved by the relatively high transmission of the mini-orange (ca. 2%), the high efficiency of the detector, and the feature of simultaneous recording. The system was tested using a sandwich absorber of Fe-57 evaporated onto stainless steel enriched in Fe-57 irradiated by a ^{57}Co-Rh source.

Kordesch *et al.*[571] have presented a technique which allows CEMS detection on a continuously emersed iron electrode. The method involves the electrochemical formation of the passive film on a substrate that can be continuously emersed from the electrolyte, withdrawn into the conversion electron detector, and subsequently returned to solution. The iron electrode is a disk with its lower half under potential control in the electrolyte and its upper half in the conversion electron detector surrounded by the counting gas. The disk is mounted on a motor that rotates a freshly passivated surface continuously into the beam of the γ-ray source returning the previously measured area into the electrolyte. The results obtained have shown that analogous spectral behaviour is observed for a rotating, continuously emersed electrode and an electrode fully immersed in the solution using CEMS and transmission Moessbauer measurements respectively. This provides evidence that the electrical state of the iron/solution interface appears to remain unaltered upon emersion of the electrode under potential control. This newly developed technique allows an increase in surface sensitivity reducing at the same time the acquisition time to a fraction of that required in transmission Moessbauer spectroscopy. A new retarding-grid electron spectrometer for CEMS has been described.[572] The structure and performance of the spectrometer were described and the response functions for various settings of resolution determined by using the energy spectrum of a very thin ^{57}Co source. The best energy resolution obtained was 0.3% at 7.3 keV with a detection efficiency of 1.2% 4π.

<u>Iron.</u> The favourable properties of the ^{57}Fe Moessbauer γ-radiation and its associated conversion electrons result in most of the applications of the CEMS technique over the past year being concerned with the iron-57 effect. These applications are discussed under the headings: films and implantations studies, steels and corrosion products, and chemical reactions.

<u>Films and implantation studies</u>. Depth-selective CEMS has been used for the analysis of Fe-containing surface layers.[573] The results on $^{57}Fe/^{56}Fe$ thin-film sandwich layers have indicated strong interdiffusion near the interface or possibly island formation, whilst in another experiment, drastic changes with time of the ion-implanted ^{57}Fe concentration profile in a Cu foil were observed and the spectral changes related to the changes of depth in the ^{57}Fe local surroundings. The authors have also shown that Moessbauer measurements using 585 eV Auger electrons of iron were feasible, and enhanced surface sensitivity was expected. Godbole <u>et al.</u>[574,575] have published two papers reporting, for the first

time, the use of a novel interface-sensitive conversion electron Moessbauer spectroscopic study of ion beam induced reactions at the iron-silicon surface[574] and the iron-aluminium surface.[575] The samples were prepared by depositing a thin layer (<50Å) of enriched ^{57}Fe isotope onto the metal substrate, where the metal is Si(111) or Al, followed by deposition of a film of natural iron. The samples were then subjected to either Xe^+ or Ar ion bombardment at an incident ion energy of 100keV and a dose of 6 x 10^{15} ions cm^{-2}. CEMS was then used to characterise the formation and growth of different phases as a function of annealing treatment. The spectra for the as-deposited and bombarded surfaces showed considerable broadening which was attributed to the formation of a large concentration of defects at the interface due to ion beam induced collision cascades. Continued annealing of the samples was found to sharpen the spectral lines, showing recovery of the phases after the initial defective state. The phases of FeSi and Fe_3Si, and FeAl and Fe_3Al from the reactions at the Fe-Si interface and Fe-Al interface respectively were formed. The formation of a Ti-Fe alloy surface was observed upon YAG:Nd^{3+} laser irradiation[576] (λ= 1.06 μm, τ = 2.5 μs, E = 0.5 - 0.7 J) of solid titanium with a vacuum-deposited ^{57}Fe film ((8 - 10)x $10^{-5}gcm^{-2}$). The CEM spectra indicated that iron oxides were not formed under these conditions. Almost identical results were obtained using laser radiation flux densities varying over the range 4 x10^2 - 6 x10^3 Jcm^{-2}. A study involving another type of garnet has also been reported[577] in which Marest et al. have performed CEMS measurements on $Y_3Fe_5O_{12}$ (YIG) enriched by 25% ^{57}Fe. The samples were implanted with H^+ and Ne^+ respectively and the CEMS spectra measured at room temperature. A comparison of the H^+ and Ne^+ as implanted samples has confirmed the anomalous character of the H^+ implantation. The recorded spectra show that one of the implanted effects involves a reduction of the hyperfine field, which is smaller for H_2^+ than for Ne^+ implantation. The linewidths of all the components present in the spectra are larger for both implantations but the effect is more noticeable in the case of H_2^+ than for Ne^+. The broader linewidth is considered to be due to implantation defects, and the enhanced effect observed with H_2^+ implants may be explained in terms of a possible variation of iron valence state. The authors have also studied the influence of the SiO_2 layer on the annealing behaviour of hydrogen implantations. Changes in spectral linewidths and hyperfine field have been observed and it is considered that further studies must be carried out to fully understand the effects of the presence of the silica layer.

CEMS has been used[578] to study the effect of ion implantation and subsequent annealing on the oxidation kinetics of iron foils. 80keV N_2^+ ions were implanted in iron foils at a dose of 2×10^{16} ions/cm^2 and a number of such samples were subjected to thermal oxidation treatment. The Moessbauer spectroscopic and gravimetric studies have demonstrated that nitrogen ion implantation has a significant effect on the oxidation kinetics of iron foils. When implanted with 80keV nitrogen ions at a dose of 2×10^{16} ions/cm^2, the iron foils exhibit an enhancement in the oxidation rate which is primarily due to the oxygen indiffusion assisted by point defects produced by the implantation process. On subjecting the implanted foils to a vacuum annealing treatment at a temperature of 300°C for half an hour, the resistance of the films to oxidation is seen to improve considerably and is higher even as compared to that of the virgin foils. The authors have attributed this behaviour to the composite action of (i) the blocking of short circuit diffusion paths due to migration of nitrogen atoms towards dislocations during the annealing treatment and nitride phase formation, and (ii) stabilisation of compressive stress which tends to inhibit oxidation. Confirmatory support for this analysis has been obtained by carrying out oxidation studies on Ar^+-implanted iron foils and iron films in which stresses were purposely produced by suitably choosing the substrate material and its temperature during the film deposition.

Steels and Corrosion Products. A study[579] of the retained austenite phase transformation in low alloy steel using CEMS and X-ray diffraction techniques has been reported. Fujinami and Ujihara have published two papers[580,581] on the use of conversion electron Moessbauer spectroscopy in the study of chemical reactions on steel surfaces. In one paper,[580] the CEMS technique was used to study the chemical state analysis of corrosion products on a steel by exposing it in a silicon dioxide (0.16 vol%) environment. In their second paper,[581] the chemical state analysis of corrosion products on a steel in hydrogen sulphide-nitrogen and hydrogen sulphide-oxygen-nitrogen environments was investigated. In another paper, Ujihara and his co-author, Nomura, have used conversion electron Moessbauer spectroscopy to analyse the black layers produced on the surfaces of iron and steel after treatment in alkaline solutions at boiling point.[582] The broad doublet and magnetically split peaks in the CEM spectra observed at room temperature were attributed to the magnetically relaxed magnetite caused by particle size effects. The black coating on steel formed by alkaline treatment was found to consist of nonstoichiometric magnetite, the thickness of which was less than about 400nm and the particle size of which lay in the range 7-10nm. The effects of (1) an oxidising agent in alkaline solution on the formation of surface products, (2) immersion time on coating, and (3) chloride ions in alkaline solution on the formation of the coating were investigated. The presence of an oxidising agent or chloride ions in alkaline solution was found to accelerate the deposition rate of the surface product and the formation of a surface coating respectively. Although the thicknesses of the black coatings became saturated after several minutes immersion, the effect of longer immersions (of up to 25 minutes) increased the surface roughness.

Two papers by Meisel[566,583] report on the applications of conversion electron Moessbauer spectroscopy in corrosion research. The author considers the technique of Moessbauer spectroscopy as a very useful tool to perform phase analyses. Some examples of the use of the backscattering technique reported in this work include: a study of the corrosion of the inner surface of hot water pipes, the effect of corrosion inhibitors, the influence of aggressive components, and the nondestructive analysis of sublayer rusting beneath a coating. CEMS spectra were recorded for steel samples, one of which had been exposed for 16 hours in a pure test solution of 18° hardness and 355ppm chloride concentration, and the other steel sample had been exposed for 24 days to a test solution containing a corrosion inhibitor (800ppm of hydroxyethanediphosphonic acid). In the former case a thick layer had formed on the surface of the steel, giving rise to the doublet observed in the CEMS spectrum, whilst in the latter case the spectrum consists of a sextet due to the metallic substrate. Depth-resolved conversion electron Moessbauer spectroscopy has been attempted[584] with an electron spectrometer for 6.6-7.3keV ^{57}Fe electrons scattered from solid surfaces, and depth profiling of such surfaces has been achieved by this technique with much better depth-selectivity than using a proportional counter. Depth-resolved conversion electron Moessbauer spectra of corrosion products in the surface layer of an iron foil were measured and the thickness of the product layer estimated. One of the products identified at low temperature was β-FeOOH. The technique was also used to characterise the products of the photolytic reactions in the top surface layer in solid potassium trisoxalatoferrate(III).

Chemical Reactions. Tests have been performed[585] on an electron detector built for CEMS to study the effects on the detector resolution of high voltage, gas flow rate, and counting rate. CEMS spectra were obtained using standard ^{57}Fe enriched and natural Fe samples and also for iron-doped indium phosphide samples. There appears to be evidence for the presence of an FeP_2 component and a region of pure iron. Conversion electron Moessbauer spectroscopy has confirmed the results achieved[586] in a non-destructive analytical study of sputter-etching on the magnetic surface properties of manganese zinc ferrite (100) surfaces using ellipsometry and the polar magneto-optical Kerr effect. The characterisation of the iron oxide phase in desert varnish from Antartica using back-scattered conversion electron and X-ray Moessbauer spectroscopy has been reported.[587] Johnston and Cardile have published further work on the desert varnish in Antartica.[588] In this paper, together with several co-authors, they have reported the results of CEMS studies on samples of dolecite and sandstone. The spectra show that the surface is comprised of Fe^{2+} and Fe^{3+}. Iron in fine grain dolecite, at a depth of ca. 10 μm from the surface is primarly Fe^{2+} whilst in the coarse grain dolecite the iron at 10 μm is predominantly Fe^{3+}. The Moessbauer spectra of the sandstone were poorly resolved due to the low iron concentration. A peak due to Fe^{3+} was discernible, indicating that the varnish was oxidised at a greater depth than in the dolecites.

Other Elements. In principle, the CEMS technique can be used for any of the Moessbauer isotopes. However, in practice, in addition to those papers reporting applications of the ^{57}Fe Moessbauer effect, only two papers appear to have been published in the past year describing studies involving the use of ^{119}Sn isotope and one referring to the use of the ^{151}Eu isotope.

The first paper[589] reports the determination of the change of nuclear charge radius of the tin-119 Moessbauer transition by internal conversion electron Moessbauer spectroscopy, and in the second paper Peneva and Djuneva have described[590] some reflection high energy electron diffraction and depth selective Moessbauer spectroscopic investigations on the structures in evaporated thin tin films before and after heating in vacuum of ca. 5 x 10^{-6} torr at 260-800°C.

Shenoy in his review[591] has presented the conversion electron spectra for three europium compounds Eu_2O_3, EuF_3 and Eu_2SiO_4 along with their absorption spectra measured in the transmission mode for comparison.

References

1. 'Mössbauer Effect References and Data Journal', ed. J.G. Stevens, V.E. Stevens, and W.L. Gettys, Mössbauer Effect Data Center, University of North Carolina, Asheville, N.C. 28814, 1984.
2. 'Mössbauer Spectroscopy Abstracts', PRM Science and Technology, 261A Finchley Road, London, 1984, Vol. 7.
3. 'Chemical Mössbauer Spectroscopy', ed. R.H. Herber, Plenum Press, New York, 1984.
4. 'Mössbauer Spectroscopy Applied to Inorganic Chemistry', ed. G.J. Long, University of Missouri at Rolla, Rolla, MO. 65401, 1984.
5. J.G. Stevens and L.H. Bowen, Anal. Chem., 1984, 56(5), 199R.
6. J.G. Stevens and L. Zhe, Magn. Reson. Rev., 1984, 9(1-3), 1.
7. S.J. Campbell, Aust. J. Phys., 1984, 37, 29.
8. V.F. Babanin, N.V. Verkhovtseva, I.N. Glebova, L.O. Karpachevskii, and V.V. Morozov, Biol. Nauki, 1983, 11, 101.
9. H. Appel and H. Haffner, GIT Fachz. Lab., 1984, 28(11), 989, 993.
10. P.A. Montano, NATO ASI Ser., Ser. C, 1984, 124(Magn. Reson.), 177.
11. G.P. Huffman and F.E. Huggins in 'The Chemistry of Low Rank Coals', ed. H.M. Schobert, Am. Chem. Soc. Symp. Ser., No. 264, The American Chemical Society, Washington D.C., 1984, 159.
12. T. Tomihaga and Y. Minai, Nucl. Sci. Appl. (N.Y.), 1984, 1(9), 749.
13. G. Jin, X. Pan, and Z. Sun, Hejishu, 1984, 1, 71.
14. A.M. Van der Kraan, J.W. Niemantsverdriet, and E. Gerkema, Energiespektrum, 1984, 8(4), 104.
15. L. May, Am. Chem. Soc. Symp. Ser., 1984, 258(Stable Isotope Nutrition), 53.
16. Y. Ujihira, Tracer, 1984, 9, 2.
17. B.W. Fitzsimmons, Annu. Rep. Prog. Chem., Sect. A, 1984, 80, 211.
18. J.D. Donaldson and S.M. Grimes, Rev. Silicon, Germanium, Tin and Lead, 1984, VIII(1), 1.
19. F.J. Berry, Annu. Rep. Prog. Chem., Sect. A, 1984, 80, 117.
20. S.J. Lyle, Annu. Rep. Prog. Chem., Sect. A, 1984, 80, 305.
21. K.P. Hall and D.M.P. Mingos, Prog. Inorg. Chem., 1984, 32, 237.
22. L.D. Freedman and G.O. Doak, J. Organomet. Chem., 1984, 261, 31.
23. G.M. Kalvius and J. Gal, Proc. Int. Conf. Magn. Rare Earths Actinides, 1983, 1, 40.

24. G. Marr and B.W. Kockett, J. Organomet. Chem., 1984, 257, 275.
25. K.A. Woode, Coord. Chem. Rev., 1984, 59, 141.
26. R.C. Mehrotra, G. Srivastava, and B.P.S. Chauhan, Coord. Chem. Rev., 1984, 207.
27. R. Raab, H. Wipf, J. Voelkl, G. Roth, and U. Potzel, Spez. Ber. Kernforschungsanlage Juelich, 1984, Juel.-Spez. 250, 510.
28. H.J.M. De Groot, R.C. Theil, and L.J. De Jongh, Springer Ser. Solid-State Sci., 1984, 54, 21.
29. F. Parak and A.X. Trautwein, Brussels Hemoglobin Symp. 1983, 1984, 299.
30. B.J. Tatarchuk and J.A. Dumesit, Springer Ser. Chem. Phys., 1984, 35, 65.
31. A.K. Bhatnagar and R. Jagannathan in 'Mössbauer Studies in 'Metallic Glasses' Production, Properties and Applications', Trans. Tech. Publications Ltd., Switzerland, 1984, 89.
32. H. Sano and P. Gutlich, Stud. Phys. Theor. Chem., 1984, 31, 265.
33. 'Trends in Mössbauer Spectroscopy', Proc. Second Seeheim Workshop, ed. P. Gutlich and G.M. Kalvius, 1983.
34. See ref. 1, p. 94.
35. K.K.P. Srivastava and S.N. Mishra, Proc. Nucl. Phys. Solid State Symp., 1983, 23C, 770.
36. G. Le Caer, J.M. Cadogan, R.A. Brand, J.M. Dubois, and H.J. Guntherodt, J. Phys. F, 1984, 14, L73.
37. D.V. Parwaterand and A.N. Garg, J. Radioanal. Nucl. Chem. Lett., 1984, 87, 379.
38. A. Vertes, K. Burger, L. Takacs, and I. Horvath, J. Radioanal. Nucl. Chem. Lett., 1984, 86, 195.
39. S.K. Roy and D.L. Bhattacharga, Indian J. Phys., 1984, 589, 94.
40. S.K. Roy and D.L. Bhattacharga, Phys. Status Solidi B, 1984, 123, 45.
41. A.V. Zatovskii and V.N. Lisy, Biofizika, 1984, 29, 35.
42. A. Gupta, Phys. Lett. A, 1984, 106, 332.
43. K.V. Shaitin and A.B. Rubin, Biofizika, 1984, 29, 735.
44. L. Cser, I. Gladkih, and G. Hargitai, Phys. Status Solidi B, 1984, 124, 271
45. A.I. Sames, V.E. Fainzilberg, M.I. Belinskii, and B.S. Tsukerblat, Khim. Fiz., 1983, 723.
46. S.A. Borshch, Fiz. Tverd. Tela, 1984, 26, 1883.
47. L.G. Dainyak, B.A. Dainyak, A.S. Bukin, and V.A. Drits, Kristallografiya, 1984, 29, 94.
48. M.A. Andreeva, S.F. Borisova, and R.N. Kuz'min, Opt. Spektrosk., 1984, 56(3), 546.
49. A.R. Mkrtchyan and R.G. Gabrielyan, Astrofizika, 1984, 20(3), 607.
50. E.H. DuMarchie van Voorthuysen, G.L. Zhang, and H. De Waard, Phys. Rev. A, 1984, 30(5), 2356.
51. P.R. Sarma and K.C. Tripathi, Nucl. Instrum. Methods Phys. Res., Sect., A, 1984, 226(2-3), 463.
52. J. Kansy, T.J. Panek, and M. Szuszkiewicz, Phys. Status Solidi A, 1983, 80(2), 643.
53. P. Jernberg, Nucl. Instrum. Methods Phys. Res., Sect. B, 1984, 232(3), 412.
54. L. Tsankor, J. Phys. D, 1984, 17(7), 1511.
55. A.V. Mitin and N.V. Polyakov, Hyperfine Interact., 1984, 20(3), 135.
56. M.A. Andreeva and R.N. Kuz'min, Solid State Commun., 1984, 49(8), 743.
57. H.J. Maurus, U. Van Buerck, G.V. Smirnov, and R.L. Mössbauer, J. Phys. C, 1984, 17(11), 1991.
58. V.A. Baryshevsky, J. Phys. C, 1984, 17(18), 3255.

59. A. Ya.Dzyublik, Phys. Status Solidi B, 1984, 123(1), 53.
60. S.M. Irkaev, M.M. Sokolov, and V.A. Semenkia, Geofiz. Appar., 1983, 77, 86.
61. O.A. Gordeev, S.M. Irkaev, V.V. Kupriyanov, V.A. Semenkin, and M.M. Sokolov, Otkritiya Izobret. Prom. Obraztsy. Touarnye Znaki, 1984, 6, 160.
62. L. Xue, Z. Liu, W. Zhou, L. Liu, and M. Hu, Hejishu, 1984, (1), 23.
63. J. Linares and T. Sundqvist, J. Phys. E, 1984, 17(5), 350.
64. M. Wang, Jilin Daxue Ziran Kexue Xuebao, 1984, (4), 58.
65. N. Rajan and A.K. Nigam, J. Phys. E, 1984, 17(11), 942.
66. P. Helistoe, E. Ikonen, T. Katila, W. Potzel, and K. Riski, Gov. Rep. Announce. Index (U.S.), 1984, 84(20), 74.
67. S. Das, M. Bhattacharya, and R. Bhattacharya, Indian J. Cryogen., 1984, 9(1), 61.
68. E.V. Kapitanov and E.N. Yakovlev, Mater. Res., Soc. Symp. Proc., 1984, 22, 305.
69. O.D. Ramon and T.P. Javier, Span. Pat. ES II J14, 836.
70. A.R. Mkrtchyan, A.R. Arakelyan, R.G. Gabrielyan, L.A. Kocharyan, G.R. Grigoryan, and M.M. Slavinskii, Phys. Status Solidi B, 1984, 125(1), K67.
71. G.V. Smirnov, Yu.V. Shvyd'ko, O.S. Kolotov, V.A. Pogzhev, M. Kotrbova, S. Kadeckova, and J. Novak, Zh. Eksp. Teor. Fiz., 1984, 86(4), 1495.
72. N. Rajan and A.K. Nigam, Nucl. Instrum. Methods Phys. Res., Sect. A, 1984, 219(3), 565.
73. Y.S. Lia, Phys. Rev. B: Condens. Matter, 1984 29(1), 79.
74. G.N. Belozerskii, Poverkhnost, 1983, (12), 41.
75. B.E. Vintaikin and R.N. Kuzmin, Deposited Doc., 1983, VINITI 5522.
76. T. Kobayashi, Nucl. Instrum. Methods Phys. Res., Sect. B, 1984, 232, 202.
77. V.N. Gittsovich and A.A. Golubev, Zavod. Lab., 1984, 50(11), 46.
78. L.P. Aldridge, Rep. N.Z. Dep. Sci. Ind. Res. Chem. Div., 1984, 2336.
79. Y. Zheng, Z. Yu, and J. Lu, Nucl.-Tech., 1983, (5), 19, 26.
80. A. Delunas, V. Maxia, and G. Bussetti, Nucl. Instrum. Methods Phys. Res., Sect. A, 1984, 221(2), 443.
81. P. Jernberg and T. Sundqvist, Uppsala Univ., Inst. Phys. (Tech. Rep.), 1983, UUIP-1090.
82. E. Ratner, Nucl. Instrum. Methods Phys. Res., Sect. B, 1984, 233(3), 518.
83. J.G. Stevens, Comput. Phys. Commun., 1984, 33, 105.
84. F.E. Huggins and G.P. Huffman, J. Appl. Phys., 1984, 55(5), 1404.
85. A.B. Bateev, O.M. Vakar, P.L.G. Gruzin, and Yu. V. Petrikin, Isv. Vyssh. Uchebn. Zaved., Fiz., 1983, 26(11), 42.
86. S.A. Fysh and P.E. Clark, Phys. Status Solidi A, 1984, 84(1), 31.
87. M. Eibschutz, M.E. Lines, H.S. Chen, and M. Touyoshi, J. Phys. F: Met. Phys., 1984, 14, 505.
88. L. Billard, F. Lancon, B. Rodmacq, and A. Chamberod, J. Phys. F: Met. Phys., 1984, 14, 555.
89. R.A. Dunlap, J. Phys. F: Met. Phys., 1984, 14, 549.
90. V. Dose, A. Härtl, H. Kraus, H. Langhoff, and J. Rogozik, J. Phys. F: Met. Phys., 1984, 14, 1541.
91. G. Kiss, Petrochemia, 1983, 23(1), 17.
92. F.E. Wagner, R. Wordel, and M. Zelger, J. Phys. F: Met. Phys., 1984, 14, 535.
93. F. Barz, R. Holze, and J. Willsam, Glas. Hem. Drus., Beograd, 1983, 48, 339.

94. T. Shinjo, Hyomen Kagaku, 1984, 5, 47.
95. M. Muljani, Mater. Res. Bull., 1984, 19, 27.
96. R.-Y. Hong and Z.-H. Lin, Xuexiao Huexue Xuebao, 1984, 5, 407.
97. G.-Q. Jin, X.-J. Pan, and Z.-T. Sun, Hejishu, 1984, 71.
98. K. Mahesh, Indian J. Pure Appl. Phys., 1984, 22(7), 419.
99. N.G. Boyle, J.M.D. Coey, A. Meaguer, V.S. McBrierty, Y. Nakano, and W.J. McKnight, Macromolecules, 1984, 17, 1331.
100. K. Krek and W. Steiner, Acta Crystallogr., Sect. A, 1984, 40(4), 459.
101. G.V. Smirnov, Yu.V. Shvyd'ko, and E. Realo, Pis'ma Zh. Eksp. Teor. Fiz., 1984, 39(1), 33.
102. U. Van Buerck, H.J. Maurus, G.V. Smirnov, and R.L. Mössbauer, J. Phys. C, 1984, 17(11), 2003.
103. R.O. Dominguez and J.T. Placios, Span. Pat. ES 514, 836.
104. S.M. Dubiel and W. Zinn, J. Magn. Magn. Mater., 1984, 45, 298.
105. T. Shinjo and T. Takada, Bull. Instrum. Chem. Res., Kyoto Univ., 1983, 61, 397.
106. S. Bjarman and R. Waeppling, J. Magn. Magn. Mater., 1983, 40, 219.
107. G. Kiss, Petrochimia, 1983, 23, 197.
108. K. Venugopalan, J. Varma, and D.R.S. Somayajulu, Phys. Lett. A, 1984, 103, 84.
109. B.G. Tagiev, G.D. Sultanov, G.M. Niftiev, I.M. Askerov, and G.K. Aslanov, Phys. Status Solidi A, 1984, 84, K59.
110. K.K.P. Sivastava, Phys. Rev. B, 1984, 29, 4890.
111. L. Cianchi, M. Mancini, and G. Spina, J. Phys. C, 1984, 17, 4493.
112. M.C.K. Wiltshire and D.C. Price, Solid State Commun., 1984, 51, 531.
113. D.C. Price, B.D. Howes, and M.C.K. Wiltshire, J. Phys. C, 1984, 17, 3645.
114. M. Multani, Mater. Res. Bull., 1984, 19, 25.
115. A. Putzka and H.D. Pfannes, Ferroelectrics, 1984, 54, 1137.
116. J.M.D. Coey, A. Meagher, J.M. Kelly, and J.G. Vos, J. Polym. Sci., Polym. Chem. Ed., 1984, 22, 303.
117. B.N. Diel, T. Inabe, N.K. Jaggi, J.W. Lyding, O. Schneider, M. Hanack, C.R. Kanewurf, T.J. Marks, and L.H. Schwarz, J. Am. Chem. Soc., 1984, 106, 3207.
118. J. Pielschowski and A. Romaniseyn, J. Polm Sci., Polym. Lett. Ed., 1984, 22, 395.
119. N.G. Boyle, J.M.D. Coey, A. Meagher, V.J. McBrierty, Y. Nakano, and W.J. MacKnight, Macromolecules, 1984, 17, 1332.
120. A. Meagher, B. Rodmacq, J.M.D. Coey, and M. Pineri, React. Polym. Ion Exchange, Sorbents, 1984, 2, 51.
121. N. Burriesci, N. Giordano, M. Petrera, and J.C.J. Bart, J. Chem. Technol. Biotechnol., 1984, 34A, 317.
122. K.I. Turte, S.A. Bolokova, and R.A. Stukan, Zh. Neorg. Khim., 1983, 29, 731.
123. C.J. Carrano and K. Spartallan, Inorg. Chem., 1984, 23, 1993.
124. N. Marques and A. Pires de Matos, Inorg. Chim. Acta, 1984, 84, L29.
125. C.H.F. Heyden, Diss. Abstr. Int. B, 1984, 44, 3421.
126. Y. Yamada, Y. Minai, H. Sato, and T. Tominaga, J. Radioanal. Nucl. Chem., 1984, 87, 359.
127. M. Pasternak, M. van der Heyden, and G. Langouche, Chem. Phys. Lett., 1984, 104, 398.
128. M.A. de Paoli, S.M. Oliveira, E. Baggio-Saitovitch, and D. Guenzburger, J. Chem. Phys., 1984, 80, 730.
129. D. Guenzburger, E. Baggio-Saitovitch, M.A. de Paoli, and H. Manela, J. Chem. Phys., 1984, 80, 735.

130. D.L. Uhrich and R.P. Marande, Mol. Cryst. Liq. Cryst., 1984, 111, 171.
131. M. Pasemann, W. Bergholz, and W. Schroeter, Phys. Status Solidi A, 1984, 81, 273.
132. J. Fontcuberta, G. Longworth, and J.B. Goodenough, Phys. Rev. B, 1984, 30, 6320.
133. R. Doerfler, D.L. Nagy, H.D. Pfannes, A. Putzka, G. Ritter, and N. Zeman, Phys. Status Solidi B, 1984, 124, 767.
134. F. Ambe, T. Okada, S. Ambe, and H. Sekizawa, J. Phys. Chem., 1984, 88, 3015.
135. F.Kh. Chibirova, I.D. Belova, D.I. Zakharin, S.I. Reiman, Yu.E. Roginskaya, and Yu.N. Venevtsev, Fiz. Tverd. Tela, 1984, 26, 890.
136. T. Harami, J. Loock, E.H. Venges, J. Fontcuberta, J. Tejada, and F. Parak, J. Phys. Chem. Solids, 1984, 45, 181.
137. K. Fendler and G. von Eynatten, Z. Phys. B, 1984, 54, 313.
138. Y.S. Hiraoka and A. Ito, J. Phys. Soc. Jpn., 1984, 53, 3959.
139. B.D. Howes, D.C. Price, and D.C.K. Wiltshire, J. Phys. C, 1984, 17, 3669.
140. H.J.M. de Groot, L.J. de Jongh, R.C. Thiel, and J. Reedijk, Phys. Rev. B, 1984, 30, 4041.
141. K.F. Tsitskishvili, Fiz. Tverd. Tela, 1984, 26, 2858.
142. T. Nishida, A. Nakamura, and Y. Takashima, Chem. Lett., 1984, 1683.
143. B.Y. Enwiya, J. Silver, and I.E.G. Morrison, Inorg. Chim. Acta, 1984, 86, 113.
144. K.K.P. Srivastava and S.N. Mishra, Phys. Status Solidi B, 1984, 121, K83.
145. P. Orlandi and A. Rigamonti, Phys. Status Solidi A, 1984, 82, K129.
146. N.V. Nair and D.C. Khan, Proc. Nucl. Phys. Solid State Phys. Symp. 1980, 1983, 23(C), 815.
147. M. Eibshuetz, M.E. Lines, L.G. Van Ultert, H.J. Guggenheim, and G.J. Zydzik, Phys. Rev. B, 1984, 29, 3843.
148. Y. Calage, M. Leblanc, G. Ferey, and F. Varret, J. Magn. Magn. Mater., 1984, 43, 195.
149. M.E. Lopez-Herrera, F. Varret, Y. Calage, and G. Ferey, J. Magn. Magn. Mater., 1984, 44, 304.
150. H. Baumeler, H. Keller, W. Kuendig, I.M. Savic, and B.M. Wanklyn, Helv. Phys. Acta, 1984, 57, 101.
151. G. Courbion, R. De Pape, J. Teillet, F. Varret, and J. Pannetier, J. Magn. Magn. Mater., 1984, 42, 217.
152. J.M. Roberts and J.R. Sambles, J. Phys. Chem. Solids, 1984, 45, 937.
153. P.P. Kovalenko, V.G. Labushkin, S.R. Ovsepyan, E.R. Sarkisov, and E.V. Smirnov, Pis'ma Zh. Eksp. Teor. Fiz., 1984, 39, 471.
154. A.S. Brar and B.S. Randhawa, J. Phys. (Les Ulis, Fr.), 1983, 44, 1345.
155. A.S. Brar and B.S. Randhawa, Polyhedron, 1984, 3, 169.
156. A. Abras, M.M. Braga, and J.C. Machado, J. Radioanal. Nucl. Chem., 1984, 86, 111.
157. A. Delunas, V. Maxia, S. Serci, R. Pinna, and G. Ponticelli, Lett. Nuova Cimento Soc. Ital. Fis., 1983, 37, 413.
158. N.V. Gerbeleu, S.G. Shova, B.Ya. Kuyavskaya, K.I. Turte, M.A. Yampol'skaya, and M.Kh. Veksel'man, Zh. Neorg. Khim., 1984, 29, 157.
159. H. Schaefer-Stahl, Synth. Met., 1983, 8, 61.
160. G.A. Fatseas, P. Palvadeau, and J.P. Venien, J. Solid State Chem., 1984, 51, 17.
161. H. Eckert and R.H. Herber, J. Chem. Phys., 1984, 80, 4526.
162. M.C. Lin and R.G. Barnes, J. Appl. Phys., 1984, 55, 2994.

163. Z. Kucharski, A. Pron, J. Suwalski, I. Kulszewica, D. Billaud, and P. Bernier, Solid State Commun., 1984, 50, 397.
164. Y. Kawamoto, I. Nohara, K. Hirao, and N. Soga, Solid State Commun., 1984, 51, 769.
165. Y. Kawamoto, I. Nohara, K. Hirao, and N. Soga, Chem. Lett., 1984, 1441.
166. M. Leblanc, G. Ferey, Y. Calage, and R. De Pepe, J. Solid State Chem., 1984, 53, 360.
167. H.H. Wei and H.L. Sharpe, Trans. Met. Chem. (Weinheim, Ger.), 1984, 9, 211.
168. R.D. Cannon, L. Montri, D.B. Brown, K.M. Marshall, and C.M. Elliot, J. Am. Chem. Soc., 1984, 106, 2591.
169. S.M. Oh, D.N. Hendrickson, K.L. Hassett, and R.E. Davis, J. Am. Chem. Soc., 1984, 106, 7984.
170. E. Koenig, G. Ritter, and S.K. Kulshreshtha, Inorg. Chem., 1984, 23, 1144.
171. E. Koenig, G. Ritter, S.K. Kulshreshtha, J. Waigel, and H.A. Goodwin, Inorg. Chem., 1984, 23, 1896.
172. E. Koenig, G. Ritter, S.K. Kulshreshtha, and N. Csatary, Inorg. Chem., 1984, 23, 1903.
173. E. Koenig, G. Ritter, S.K. Kulshreshtha, J. Waingel, and I. Sarranni, Inorg. Chem., 1984, 23, 1241.
174. G. Vos, R.A.G. De Graff, J.G. Haasnoot, A.M. van der Kraan, P. De Vaal, and J. Reedjik, Inorg. Chem., 1984, 23, 2905.
175. H.H. Wei and J.C. Jean, Chem. Phys. Lett., 1984, 106, 523.
176. M.P. Edwards, C.D. Hoff, B. Curnette, J.S. Eck, and K.F. Purcell, Inorg. Chem., 1984, 23, 2613.
177. Y. Maeda, N. Tsutsumi, and Y. Takashima, Inorg. Chem., 1984, 23, 2440.
178. W.D. Federer and D.N. Hendrickson, Inorg. Chem., 1984, 23, 3870.
179. A. Terzis, S. Filippakis, D. Mentzalos, M.V. Petrouleas, and A. Malliaris, Inorg. Chem., 1984, 23, 334.
180. W.M. Reiff and E.H. Witten, Polyhedron, 1984, 3, 443.
181. N. Kobayashi, H. Shirai, and N. Hojo, J. Chem. Soc., Dalton Trans., 1984, 2107.
182. A. Labarta, E. Molins, X. Vinas, J. Tejada, A. Caubet, and S. Alvarez, J. Chem. Phys., 1984, 80, 444.
183. G.D. Fallon, B.M. Gatehouse, P.J. Marini, K.S. Murray, and B.O. West, J. Chem. Soc., Dalton Trans., 1984, 2733.
184. B.J. Kennedy, G. Bain, and K.S. Murray, Inorg. Chim. Acta, 1984, 81, L29.
185. P.G. Rasmussen and E.A. Meyers, Polyhedron, 1984, 3, 183.
186. K.S. Sengupta and M.N. Ray, Proc. Nucl. Chem. Radiochem. Symp. 1981, 1983, 653.
187. E.A. Bekturov, S.E. Kudaiberger, S.S. Salybaeva, A.S. Khlystov, and V.I. Yaskevich, Makromol. Chem., Rapid Commun., 1984, 5, 763.
188. O. Zakharineva-Pencheva and V.A. Dement'ev, J. Mol. Struct., 1984, 115, 75.
189. O. Zakharineva-Pencheva, L. Tsankov, and T.S. Bonchev, Koord. Khim., 1984, 10, 776.
190. H. Inoue, M. Sasagawa, E. Fluck, and T. Shirai, Bull. Chem. Soc. Jpn., 1983, 56, 3434.
191. N.K. Singh, S. Agrawal, and A.C. Aggarwal, Polyhedron, 1984, 3, 1271.
192. J. Pickardt, J. Kaehler, N. Rautenberg, and E. Rediel, Z. Naturforsch., Teil B, 1984, 39, 1162.

193. E. Bill, N. Blaes, K.F. Fischer, U. Gonser, K.H. Pauli, R. Preston, F. Seel, R. Straab, and A.X. Trautwein, Z. Naturforsch., Teil B, 1984, 391, 333.
194. I.R. Butler, W.E. Lindsell, and M.J.K. Thomas, J. Organomet. Chem., 1984, 262, 59.
195. M.M. Kubicki, R. Kergoat, J.Y. De Vanssay, J.L. Le Gall, and J.E. Guerchais, Polyhedron, 1984, 3, 365.
196. R.M.G. Roberts and J. Silver, J. Organomet. Chem., 1984, 263, 235.
197. K. Sato, M. Konno, and H. Sano, Chem. Lett., 1984, 17.
198. K. Sato, M. Iwai. H. Sano, and M. Konno, Bull. Chem. Soc. Jpn., 1984, 57, 634.
199. G. Neshvad, R.M.G. Roberts, and J. Silver, J. Organomet. Chem., 1984, 260, 319.
200. A.G. Nagy, J. Organomet. Chem., 1984, 270, 327.
201. R.M.G. Roberts, J. Silver, and B. Yamin, J. Organomet. Chem., 1984, 270, 221.
202. S. Ikuta, I. Motoyama, and H. Sano, Bull. Chem. Soc. Jpn., 1984, 57, 299.
203. M. Watanabe, K. Sato, I. Motoyama, and H. Sano, Chem. Lett., 1984, 1775.
204. R.D. Ernst, D.R. Wilson, and R.H. Herber, J. Am. Chem. Soc., 1984, 106, 1646
205. D. Bickar, B. Lukas, G. Neshvad, R.M.G. Roberts, and J. Silver, J. Organomet. Chem., 1984, 263, 225.
206. H.H. Wei and S.J. Wang, J. Chin. Chem. Soc. (Taipei), 1984, 31, 9.
207. M. Watanabe, K. Suto, I. Motoyama, and H. Sano, Chem. Lett., 1984, 1317.
208. M.J. Cohn, M.D. Timken, and D.N. Hendrickson, J. Am. Chem. Soc., 1984, 106, 6683.
209. B. Lukas, R.M.G. Roberts, J. Silver, and A.S. Wells, J. Organomet. Chem., 1984, 256, 103.
210. M.J. Buckingham, B.W. Fitzsimmons, and I. Sayer, J. Chem. Soc., Chem. Commun., 1984, 339.
211. J.P. Mariot. P. Michaud, S. Lauer, D. Astruc, A.X. Trautwein, and F. Varret, J. Phys. (Les Ulis, Fr.), 1983, 44, 1377.
212. J. Silver, B. Lukas, and G. Al-Jaff, Inorg. Chem. Acta, 1984, 91, 125.
213. J. Silver and B. Lukas, Inorg. Chem. Acta, 1984, 91, 279.
214. T. Ohya, H. Morohoshi, and M. Sato, Inorg. Chem., 1984, 23, 1303.
215. G.E. Toey, L.E. ter Haar, J.E. Savfin, A. Gold, W.E. Hatfield, and R. Sangaiah, Inorg. Chem., 1984, 23, 2561.
216. J. Silver and B. Lukas, Inorg. Chem. Acta, 1984, 91, 259.
217. D.F. Bocian, E.W. Findsen, J.A. Hofmann, C.A. Schick, D.R. English, D.N. Hendrickson, and K.S. Suslich, Inorg. Chem., 1984, 23, 800.
218. M.J. Gunter, G.M. McLaughlin, K.J. Bery, K.S. Murray, M. Irving, and P.E. Clark, Inorg. Chem., 1984, 23, 283.
219. V.E. Prusakov, R.A. Stukan, and R.M. Davydov, Biofizika, 1984, 29, 359.
220. A. Levy, K. Alston, and J.M. Rifkind, J. Biomol. Struct. Dyn., 1984, 1, 1299.
221. I. Cianchi, F. Pieralli, F. Del Giallo, M. Mancini, G. Spina, and L. Fiesoli, Phys. Lett. A, 1984, 100, 57.
222. B. Boso, P.G. Debrunner, G.C. Wagner, and T. Inubushi, Biochim. Biophys. Acta, 1984, 791, 244.
223. B. Balko, E. Bucci, R.L. Berger, L.J. Swarzendruber, and J.X. Montemarano, J. Biochem. Biophys. Methods, 1984, 10, 55.

224. R.I. Mints, M.I. Oshtrakh, and V.A. Semenkin, Biofizika, 1984, 29, 310.
225. S. Ofer, E.R. Bauminger, S.G. Cohen, and I. Nowik, Struct. Funct. Iron Storage Transp. Proteins, Proc. Int. Conf., 6th., 1983, 1983, 59.
226. S.H. Bell, M.P. Weir, D.P.E. Dickson, J.F. Gibson, G.A. Sharp, and T.J. Peters, Biochim. Biophys. Acta, 1984, 787, 227.
227. E. Tsuchida, H. Nishide, H. Yokoyama, H. Inoue, and T. Shirai, Polym. J. (Tokyo), 1984, 16, 525.
228. J.M. Moulin, P. Auric, J. Gaillard, and J. Meyer, J. Biol. Chem., 1984, 259, 11 396.
229. B.H. Huynh, M.H. Czechowski, H.J. Kruger, D.V. DerVartanian, H.D. Peck, jun., and J. LeGall, Proc. Natl. Acad. Sci. U.S.A., 1984, 81, 3728.
230. G. Wong, M.J. Benecky, H.H. Boi, J.F. Cline, M.W.W. Adams, L.E. Mortenson, B.M. Hoffman, and E. Muenck, J. Biol. Chem., 1984, 259, 14 328.
231. J.W. Whittaker, J.D. Lipscomb, T.A. Kent, and E. Muenck, J. Biol. Chem., 1984, 259, 4466.
232. J.W. Whittaker, Diss. Abstr. Int. B, 1984, 44, 3392.
233. C.E. Schulz, R. Rutter, J.T. Sage, P.G. Debruner, and L.P. Hagar, Biochemistry, 1984, 23, 4743.
234. R. Rutter, L.P. Hagar, H. Donau, M. Hendrich, M. Valentine, and P. Debrunner, Biochemistry, 1984, 23, 6809.
235. J.A. Christner, E. Muenck, T. . Kent, P.A. Janick, J. . Salerno, and L.M. Siegel, J. Am. Chem. Soc., 1984, 106, 6786.
236. K.K. Andersson, T.A. Kent, J.D. Lipscomb, A.B. Hooper, and E. Muenck, J. Biol. Chem., 1984, 259, 6833.
237. A. Yayon, E.R. Bauminger, and S. Ofer, J. Biol. Chem., 1984, 259, 8163.
238. W. Youtong, Heijishu, 1984, 37.
239. K.T. Douglas, B. Howlin, and J. Silver, Inorg. Chim. Acta, 1984, 92, 135.
240. S.J. Barclay, R.E. Riley, and K.N. Raymond, Inorg. Chem., 1984, 23, 2005.
241. S.J. Barclay, B.H. Huynh, and K.N. Raymond, Inorg. Chem., 1984, 23, 2011.
242. J. Zhang, R. Liu, S. Wang, Y. Xia, and S. Qi, Kexue Tongbao (For. Lang. Ed.), 1984, 29, 1091.
243. I. Tamura and M. Hayashi, Surf. Sci., 1984, 146, 501.
244. T. Shinjo, T. Iwasaki, T. Shigematsu, and T. Takada, Jpn. J. Appl. Phys., Part 1, 1984, 23, 283.
245. D. Hantze and H. Bilinski, Mater. Chem. Phys., 1984, 10, 277.
246. D. Chambaere and E. De Grave, J. Magn. Mater., 1984, 42, 263.
247. D. Chambaere and E. De Grave, J. Magn. Mater., 1984, 44, 245.
248. A Vertes, T.C. Huang, J.W. Vanderhoff, K. Lazar, M. Bennetch, and W. Reiff, Radiochem. Radioanal. Lett., 1984, 59, 307.
249. K. Matsuda, Proc. Fac. Sci. Tokyo Univ., 1984, 19, 69.
250. S.C.F. Au-Yeung, G. Denes, J.E. Greedan, D.R. Eaton, and T. Birchall, Inorg. Chem., 1984, 23, 1513.
251. C. Wilkinson, A.K. Cheetham, G.J. Long, P.D. Battle, and D.A.O. Hope, Inorg. Chem., 1984, 23, 3136.
252. G.J. Long, D.A.O. Hope, and A.K. Cheetham, Inorg. Chem., 1984, 23, 3141.
253. T. Tsuji, K. Naito, and K. Ishigure, Phys. Status Solidi A, 1984, 82, K57.
254. Y. Syono, S. Morimoto, T. Suzuki, T. Yagi, and S. Akimoto, Solid State Commun., 1984, 50, 97.

255. V.B. Koval, V.P. Ivanitskii, I.V. Matyash, V.A. Samsonov, and P.O. Voznyuk, Geol. Zh., 1984, 44, 110.
256. O. Nikolov, T. Ruskov, T. Tomov, A.M. Kadomtseva, I.B. Krimetskii, and M.M. Lukina, J. Magn. Magn. Mater., 1984, 44, 181.
257. P.P. Kirichok, V.A. Pashhchenko, T.O. Dem'yaniv, G.N. Ryabova, and A.M. Lisovskii, Izv. Akad. Nauk SSSR, Neorg. Mater., 1984, 20, 659.
258. J.K. Srivastava, S.M. Rao, and R. Vijayaraghavan, Proc. Nucl. Phys. Solid State Phys. Symp., 1980, 1983, 23, 689.
259. A.I. Mateev, Yu.V. Maksimov, I.P. Suzdalev, A.V. Golub'ev, and V.M. Erkin, Poverkhnost, 1984, 120.
260. W. Winkler, Phys. Status Solidi A, 1984, 84, 193.
261. R.E. Vandeberghe, R. Vanleerberghe, and D. Gryffroy, Phys. Lett. A, 1984, 101, 297.
262. O.M. Srivatsava, S.N. Shringi, and M. Vijaybabu, Bull. Mater. Sci., 1984, 6, 27.
263. M. Tanaka, K. Siratori, and N. Kimizuka, J. Phys. Soc. Jpn., 1984, 53, 760.
264. E.A. Zhurakovskii, T.D. Bakuma, V.Ya. Garmash, and G.S. Podval'nykh, Poroshk. Metall. (Kiev), 1984, 76.
265. Z.M. Staknik and Y. Kawai, J. Phys. Soc. Jpn., 1984, 53, 2761.
266. C.M. Srivastava, S.N. Shringi, M.J. Patni, and S.M. Jogekar, Bull. Mater. Sci., 1984, 6, 7.
267. C. Prakash and J.S. Baijal, Solid State Commun., 1984, 50, 557.
268. R.K. Puri, U. Varshney, and R.G. Mendiratta, Indian J. Pure Appl. Phys., 1983, 21, 686.
269. J.K. Srivastava, K. Muraleedharan, and R. Vijayaraghavan, Phys. Lett. A, 1984, 104, 482.
270. R.V. Upadhayay and R.G. Kulkarni, Mater. Res. Bull., 1984, 19, 655.
271. R.G. Kulkarni and R.V. Upadhayay, J. Mater. Sci., 1984, 19, 1622.
272. R.V. Upadhayay, S.N. Rao, and R.G. Kulkarni, J. Mater. Sci. Lett., 1984, 3, 636.
273. G.J. Baldha and R.G. Kulkarni, Solid State Commun., 1984, 49, 169.
274. V.G. Panicker, R.V. Upadhayay, S.N. Rao, and R.G. Kulkarni, J. Mater. Sci. Lett., 1984, 3, 385.
275. P. Kishan, C. Prakash, J.S. Baijal, and K.K. Laroia, Phys. Status Solidi A, 1984, 84, 535.
276. A. Apostlov, M. Michov, Z. Charizanova, K. Zaveta, Kh. Khristov, and Sp. Cholakov, Czech. J. Phys., 1984, B34, 46.
277. M.D. Sundararajan, A. Narayanasamy, T. Nagarajan, C.S. Swamy, and K.V. Ramanujachary, J. Phys. C, 1984, 17, 2953.
278. Sh.Sh. Bashkirov, A.B. Liberman, and A.M. Khasanov, Fiz. Tverd. Tela, 1984, 26, 1086.
279. M.D. Osborne, M.E. Fleet, and G.M. Bancroft, J. Solid State Chem. 1984, 53, 174.
280. K. Kose and S. Iida, J. Appl. Phys., 1984, 55, 2321.
281. G.D. Sultanov, S.G. Ibragimov, I.G. Kerimov, and N.G. Aliev, Dokl. Akad. Nauk Az. SSR, 1984, 40, 30.
282. E. Reidel and H. Azak, Z. Naturforsch., Teil B, 1983, 38, 1630.
283. F. Sevsek and D. Hanzel, J. Phys. C, 1984, 16, 6047.
284. F. Grandjean, G.J. Long, G. Longworth, and B.J. Laundy, Inorg. Chem., 1984, 23, 1886.
285. T.C. Gibb, J. Chem. Soc., Dalton Trans., 1984, 667.

286. N. Nguyen, Y. Calage, F. Varret, G. Ferey, V. Caignaert, M. Hervieu, and B. Raveau, J. Solid State Chem., 1984, 53, 398.
287. Z.M. Stadnik, J. Phys. Chem. Solids, 1984, 45, 311.
288. Z.M. Stadnik and E. De Boer, J. Phys. Chem. Solids, 1984, 45, 113.
289. Z.M. Stadnik, Solid State Commun., 1984, 51, 79.
290. G. Balestrino, F. Scarinci, and A. Tucciarone, J. Magn. Magn. Mater., 1984, 44, 249.
291. C.M. Srivastava, S.N. Shringi, B.M. Rao, D. Bahadur, and O.M. Parakash, Bull. Mater. Sci., 1984, 6, 17.
292. M.M. Shipko, Ya.G. Rozin. V.I. Bondar, and L.A. Bashkirov, Vestni Akad. Navuk BSSR, Ser. Fiz.-Mat. Nayuk, 1984, 70.
293. X. Obradors, J. Tejada, A. Isalgue, and J.C. Joubert, Solid State Commun., 1984, 50, 821.
294. Yu.A. Mamalui and L.P. Ol'khovik, Fiz. Tekh. Vys, Davlenii, 1983, 12, 17.
295. G. Antonioli, E. Calabrese, A. Deriu, and U. Emiliani, Nuovo Cimento Soc. Ital. Fis. D, 1984, 40, 255.
296. K.J.D. MacKenzie and I.W.M. Brown, J. Mater. Sci. Lett., 1984, 3, 159.
297. J. Fontcuberta, G. Longworth, and J.B. Goodenough, Phys. Rev. B, 1984, 30, 6320.
298. D. Hanzel and F. Sevseh, Mater. Res. Bull., 1984, 19, 35.
299. P.P. Kovalenko, V.G. Labushkin, A.K. Ovsepyan, E.R. Sarkisov, and E.V. Smirnov, Pis'ma Zh. Eksp. Teor. Fiz., 1984, 39, 471.
300. M. Vithal, R. Jagannathan, and N. Ravi, Bull. Mater. Sci., 1984, 6, 33.
301. K. Suito, Y. Tsutsui, S. Nasu, A. Onodera, and F.E. Fujita, Mater. Res. Soc. Symp. Proc., 1984, 22, 295.
302. Y.F. Xia, R.C. Liu, S.K. Wang, C. Yu, S.Y. Pan, and Y.B. Cheng, Wuli Xuebao, 1984, 33, 132.
303. A. Prasad, R.M. Singru, D. Bahadur, and D. Chakravorty, J. Mater. Sci., 1984, 19, 3021.
304. M.E. Fleet, C.T. Herzberg, G.S. Henderson, E.D. Crozier, M.D. Osborne, and C.M. Scarfe, Geochim. Cosmochim. Acta, 1984, 48, 1455.
305. B.O. Mysen, D. Virgo, and F.A. Seifert, Am. Mineral., 1984, 69, 834.
306. M.M. Nasrallah, A.K. Arif, and A.M. Bishay, Mater. Res. Bull., 1984, 19, 269.
307. J.P. Sanchez, S.W. Friedt, R. Muras, and A.J. Van Duyneveldt, J. Phys. C, 1984, 17, 127.
308. G.N. Belozerskii, A.V. Kalyamin, E.E. Kornilova, G.T. Petrovskii, and S.A. Stephanov, Fiz. Khim. Stekla, 1984, 10, 289.
309. E. Buezo, D. Ungur, and I. Ardelean, Solid State Commun., 1984, 50, 1083.
310. M.D. Dyar, Am. Mineral., 1984, 69, 1127.
311. V.W. Viera, H.G. Jensen, J.M. Knudsen, and M. Olsen, Phys. Scr., 1984, 30, 284.
312. F.W. Oliver, E.E. Isuk, and C. Wynter, Meteorites, 1984, 19, 25.
313. T.A. Valetov and I.L. Lapides, Dokl. Akad. Nauk SSSR, 1984, 276, 116.
314. L.G. Dainyuk, B.A. Dainyuk, A.S. Bukin, and V.A. Drits, Kristallografiya, 1984, 29, 94.
315. L.G. Dainyuk, A.S. Bukin, and V.A. Drits, Kristallografiya, 1984, 29, 304.
316. L.G. Dainyuk, A.S. Bukin, and V.A. Drits, Kristallografiya, 1984, 29, 312.
317. S. Friedrich, Neues Jahrb. Mineral., Abh., 1984, 148, 141.

318. V.M. Khomenko, A.N. Platonov, E.V. Pol'shin, R.I. Shiroshtan, and N.I. Polovko, Probl. Kristallokhim. Genezisa Miner., 1983, 60.
319. K. Ishida, Y. Maeda, and Y. Aoki, Kobutsugaku Zasshi, 1984, 16, 317.
320. E. Murad, Am. Mineral., 1984, 69, 722.
321. E.R. Vance and D.C. Price, Phys. Chem. Miner., 1984, 10, 200.
322. X. Kan, H. Xie, Y. Li, and E. Zhang, Kexue Tongbao (For. Lang. Ed.), 1984, 29, 928.
323. T. Ericsson, L. Linares, and E. Lotse, Clay. Miner., 1984, 19, 85.
324. E. Murad and U. Schwertmann, Mineral. Mag., 1984, 48, 507.
325. C.W. Childs, D.P.E. Dickson, B.A. Goodman, and D.G. Lewis, Aust. J. Soil Res., 1984, 22, 149.
326. E. Murad and U. Schwertmann, Clay Miner., 1984, 18, 301.
327. M.B. McBride, V.C. Farmer, J.D. Russell, J.M. Tait, and B.A. Goodwin, Clay Miner., 1984, 19, 1.
328. J.M.D. Coey, J. Allan, K. Xuemin, N. Van Dang, and S. Ghose, J. Appl. Phys., 1984, 55, 1963.
329. F.J. Litterst and G. Amthauer, Phys. Chem. Miner., 1984, 10, 250.
330. J.M.D. Coey, F.V. Churkhrov, and B.B. Zvyagin, Clays Clay Miner., 1984, 32, 198.
331. E. Murad and R.M. Taylor, Clay Miner., 1984, 19, 77.
332. E. De Grave, R. Vanleerberghe, L. Verdonck, and G. De Geyter, Phys. Chem. Miner., 1984, 11, 85.
333. M.M. Mortland, J.G. Lawless, H. Hartman, and R. Frankel, Clays Clay Miner., 1984, 32, 279.
334. H. Annersten, J. Adetunji, and A. Filippidis, Am. Mineral., 1984, 69, 1110.
335. G. Steffen, F. Seifert, and G. Amthuer, Am. Mineral., 1984, 69, 339.
336. V.V. Koruvushkin, V.L. Laverova, and N.V. Petrova, Mineral. Rudn. Mestorozhd., 1983, 82.
337. S.A. Fysh and P.E. Clark, Phys. Status Solidi A, 1984, 84, 31.
338. J. Kansy, T.J. Panek, and M. Szuszkiewicz, J. Phys. C, 1984, 17, 1585.
339. R. Garg, E. Galvano da Silva, Y.S. Liu, and V.K. Garg, Solid State Commun., 1984, 50, 1.
340. S. Lauer, A.X. Trautwein, and F.E. Harris, Phys. Rev. B, 1984, 29, 6774.
341. K. Tkachova, V.V. Boldyrev, Yu.T. Pavlyuchin, E.G. Avvakumov, R.Sh. Sadykov, and P. Balaz, Izv. Sib. Otd. Akad. Nauk SSSR, 1984, 9.
342. R.Sh. Sadykov, P. Balaz, V.V. Boldyrev, Yu.T. Pavlyuchin, and E.G. Avvakumov, Dokl. Akad. Nauk SSSR, Phys. Chem., 1983, 273, 643.
343. K.S. Baek, Y.S. Park, and N.H. Ok, Phys. Rev. B, 1984, 30, 404.
344. G.V. Loseva, L.I.R. Mabinkina, S.A. Ovchinnikov, and O.A. Baunkov, Fiz. Tverd. Tela, 1983, 25, 3717.
345. C.A. McCammon, I. Jackson, A.E. Ringwood, and J.D. Cashion, Phys. Chem. Miner., 1984, 11, 182.
346. M.D. Osborne and M.E. Fleet, Phys. Chem. Miner., 1984, 10, 245.
347. P. Jernberg, S. Bjarman, and R. Waeppling, Uppsala Univ., Inst. Phys. (Tech. Rep.), 1984, UUIP-1106.
348. J.M. Friedt, C.W. Kimball, A.T. Aldred, R.A. Dunlap. F.Y. Fradin, and G.K. Shenoy, Phys. Rev. B, 1984, 31, 3863.
349. G.P. Bernardi, G. Mazzetti, and D.J. Vaughan, Rend. Soc. Ital. Mineral. Petrol., 1982, 1983, 38, 1407.
350. H.N. Ok and C.S. Lee, Phys. Rev. B, 1984, 29, 5168.
351. C.S. Lee, Pusan Suasan Taehak Yongu Pogo, Chayan Kwahak, 1983, 23, 22.
352. J.W. Brill, P. Boochand, and G.H. Lemon, Solid State Commun., 1984, 51, 9.

353. E.A. Bychkov, Z.U. Borisova, Yu.G. Vlasov, G.N. Belozerskii, and V.N. Gottsovich, Fiz. Khim. Stekla, 1984, 10, 278.
354. J. Phillips, Y. Chen, and J.A. Dumesic, Prepr. Pap. - Am. Chem. Soc., Div. Pet. Chem., 1984, 29, 944.
355. L. Liu and W. Wang, Gaodeng Xuexiao Huaxue Xuebao, 1983, 4, 763.
356. J. Monnier, G. Denes, and R.B. Anderson, Can. J. Chem. Eng., 1984, 62, 419.
357. Yu. V. Maksimov, M.Sh. Zurmukhtashvili, I.P. Suzdalev, and L. Ya. Margolis, Kinet. Katal., 1984, 25, 948.
358. H. Qi, X. Yang, Y. Gu, C. Zhu, J. Ren, and N. Zhou, Cuihua Xuebao, 1984, 5, 190.
359. C. Wivel, B.S. Clausen, R. Candia, S. Moerup, and H. Topsoee, J. Catal., 1984, 87, 497.
360. M. Breyesse, I. Fretz, B. Benaichouba, and P. Bussiere, Radiochem. Radioanal. Lett., 1983, 59, 265.
361. B.S. Clausen, H. Topsoee, R. Canda, and B. Lengeler, Am. Chem. Soc. Symp. Ser., 1984, 248, 71.
362. Y. Minai. T. Fukushima, M. Ichikawa, and T. Tominaga, J. Radioanal. Nucl. Chem., 1984, 87, 189.
363. J.W. Niemantsverdriet, A.M. Van Der Kraan, and W.N. Delgass, J. Catal., 1984, 89, 138.
364. J.W. Niemantsverdriet, D.P. Aschenbeck, F.A. Fortunato, and W.N. Delgass, J. Mol. Catal., 1984, 25, 285.
365. L.M. Tau and C.O. Bennett, J. Catal., 1984, 89, 285.
366. L.L. Murrell and R.L. Garten, Appl. Surf. Sci., 1984, 19, 218.
367. M. Oskooie-Tabrizi, C. Lo, and L.N. Mulay, IEEE Trans. Magn., 1983, MAG-19, 2001.
368. T. Pannaparayil, M. Oskooie-Tabrizi, C. Lo, L.N. Mulay, G.A. Melson, and V.U.S. Rao, J. Appl. Phys., 1984, 55, 2601.
369. T. Pannaparayil, M. Oskooie-Tabrizi, C. Lo, and L.N. Mulay, Mater. Res. Bull., 1984, 19, 1411.
370. P.A. Monatano and A.S. Bommannavar, Prepr. Pap. - Am. Chem. Soc. Div. Pet. Chem., 1984, 29, 66.
371. S.P. Taneja and C.H.W. Jones, Fuel, 1984, 63, 695.
372. M.D. Patil, H.C. Eaton, and M.E. Tittlebaum, Fuel, 1984, 63, 788.
373. T.L.H. Aikin, J.D. Cashion, and A.L. Ottrey, Fuel, 1984, 63, 1269.
374. R.G. Chernova, S.A. Shcherban, Kh.N. Nurmagambetov, and A.A. Kozlovskii, Kompleksn. Ispol'z Miner. Syr'ya, 1984, 67.
375. C.W. Childs and J.G. Baker-Sherman, N.Z., Soil Bur., Sci. Rep., 1984, 66.
376. E. Jeanroy, B. Guillet, P. Delacroix, and C. Janot, Bull. Assoc. Fr. Etude Sol., 1983, 185.
377. I. Georgieva and S. Mirchev, Pochvozn. Agrokhim., 1984, 19, 72.
378. V.V. Morozov, V.F. Babanin, L.V. Ivanov, E.M. Samoilova, and A.A. Solov'ev, Pochvovodenie, 1984, 135.
379. W. Davison and D.P.E. Dickson, Geol. Chem., 1984, 42, 177.
380. N.A. Eissa, M.Y. Hassaan, H.A. Sallam, S.H. Salah, and S.A. Abo-El-Enein, J. Mater. Sci. Lett., 1984, 3, 88.
381. N.A. Eissa, M.Y. Hassaan, H.A. Sallam, S.H. Salah, J. Mater. Sci. Lett., 1984, 3, 262.
382. K.S. Harchand, R. Kumar, K. Chandra, and Vishwamittar, Cem. Concr. Res., 1984, 14, 170.
383. X. Pan, Z. Sun, and G. Jin, Nucl. Tech., 1983, 49.
384. R. Salazar, U. Wagner, F.E. Wagner, and E. Murad, Radiochem. Radioanal. Lett., 1983, 59, 299.
385. H. Leidheiser and I. Czako-Nagy, Corros. Sci., 1984, 24, 569.

386. J. Lipka and M. Hucl, Chem. Prum., 1984, 34, 411.
387. A.M. Tseitlin, Yu.B. Voitkovskii, and A.G. Rusakova, Izv. Vyssh. Uchebn. Zaved., Chern. Metall., 1984, 129.
388. D.A. Scherson, C. Fierro, E.B. Yeager, M.E. Kordesch, J. Eldrige, R.W. Hoffman, and A. Barnes, J. Electroanal. Chem. Interfacial Electrochem., 1984, 149, 287.
389. Y.F. Barbikova, P.L. Gruzin, I.A. Evstyukhina, and A.G. Sokolov, Okrytiya, Izobret., Prom. Obtaztsy, Tovarmye, Znakis, 1984, 148.
390. W. Winkler and H. Mehner, J. Radioanal. Nucl. Chem., 1984, 82, 151.
391. J.N. Farrell, Diss. Abstr. Int. B, 1984, 45, 1227.
392. V. Rusanov, T. Bontchev, S.K. Peneva, K.H. Chakarova, L.A. Spasov, and S.L. Petrov, J. Solid State Chem., 1984, 51, 336.
393. R. Frety, B. Benaichouba, P. Bussiere, D.S. Cunha, and Y.L. Lam, J. Mol. Catal., 1984, 25, 173.
394. Y.-X. Li, Y.-F. Zhang, and Y.-F. Shia, Cuihua Xuebao, 1984, 5, 311.
395. P.-Q. Zhang, H.-R. Shao, X.-Y. Yang, and L. Pang, Cuihua Xuebao, 1984, 5, 101.
396. S. Zhang, B. Xie, L. Pan, and J. Zhang, Cuihua Xuebao, 1984, 5, 77.
397. I.M. Groznov, S.P. Ionov, J.B. Kevolina, V.S. Lyubimov, E.F. Makerov, O.V. Marchenko, N.N. Mikhailov, V.A. Onishchuk, and V.P. Shantarovich, Phys. Status Solidi B, 1984, 123, 183.
398. Y.N. Venevtsves, V.V. Bogatko, M.V. Plotnikova, and S.I. Reiman, Ferroelectrics, 1984, 56, 1065.
399. G. Demazeau, M. Pouchard, P. Hagenmuller, P.B. Fabrichnyi, L.P. Fefilatev, and A.M. Babechkin, Mater. Res. Bull., 1984, 19, 1337.
400. S. Iahiba and T. Yamaguchi, Chem. Lett., 1984, 1681.
401. S.M. Dubiel, Phys. Rev. B, 1984, 29, 2816.
402. S.I. Reiman, N.I. Rokholov, V.S. Shpinel, and E.P. Kaminskaya, Zh. Eksp. Teor. Fiz., 1984, 86, 330.
403. N.N. Delyagin, V.I. Krylov, N.I. Moreva, G.T. Mudzhiri, V.I. Nesterov, and S.I. Reiman, Pis'ma Zh. Eksp. Teor. Fiz., 1984, 39, 430.
404. S.M. Dubiel, C. Sauer, and W. Zinn, Phys. Rev. B, 1984, 30, 6285.
405. R.A. Dunlap and S.P. Ritcey, Hyperfine Interact., 1983, 14, 195.
406. S.P. Ritcey and R.A. Dunlap, J. Appl. Phys., 1984, 55, 2051.
407. K.A. Dunlap and G. Stroink, J. Appl. Phys., 1984, 55, 1068.
408. I.M. Valeev, I.A. Abdullin, A.B. Liberman, and S.S. Tsakevskii, Elektrokhimiya, 1984, 20, 1269.
409. D. Andriamandroso, L. Fefilatiev, G. Demeseay, L. Fouines, and M. Pouchard, Mater. Res. Bull., 1984, 19, 1187.
410. H. Cremers, B.D. Basel, W. Muller-Warmuth, and G.H. Frischat, J. Non-Cryst. Solids, 1984, 63, 329.
411. R.C. Haushalter, C.M. O'Connor, A.M. Umarji, G.K. Shenoy, and G.K. Saw, Solid State Commun., 1984, 49, 929.
412. R.C. Hanshaeter, C.M. O'Connor, J.P. Haushalter, A.M. Umariji, and G.K. Slanoy, Angew. Chem., 1984, 96, 147.
413. B. Arcondo, G. Ceuintat, H. Sirkin, and F. Cernnschi, J. Non-Cryst. Solids, 1984, 65, 435.
414. R.H. Herber, Mater. Sci. Forum, 1984 (Mod. Nucl. Methods Mater. Sci., 1984, 2, 49).
415. B. Glowacki and J. Chojcan, Phys. Status Solidi A, 1983, 80, K93.
416. R.A. Dunlap, S. Jha, and G.M. Julian, Can. J. Phys., 1984, 62, 396.
417. A. Vertes, T. Turmezey, A. Griger, M.Z. Awad, and S. Nagy, J. Radioanal. Nucl. Chem., 1984, 85, 123.
418. G.T. Petrovskii, E.A. Vasilev, N.D. Soloveva, T.M. Tkachenko, A.V. Shablovskii, and D.M. Yudin, Fiz. Khim. Stekla, 1984, 10, 385.

419. S.S. Bashkirov, A.B. Liberman, and S.S. Tsarevoskii, Zh. Telkh. Fiz., 1984, 544, 202.
420. F.S. Nasredinov, L.V. Prokofeva, and P.P. Seregin, Zh. Eksp. Teor. Fiz., 1984, 87, 951.
421. F.S. Nastredinov, L.V. Prokofeva, A.N. Kurmantaev, and P.P. Seregin, Fiz., Tverd. Tela, 1984, 26, 862.
422. D. Baltrunas, S. Motejunas, P.M. Starik, and V.I. Mikityuk, Izv. Akad. Nauk SSSR, Neorg. Mater., 1984, 20, 1312.
423. A.J. Edward and I.K. Khallow, J. Chem. Soc., Chem. Commun., 1984, 50.
424. H.S. Cheng, C.-M. Hsu, and H.H. Wei, J. Chin. Chem. Soc. (Taipei), 1984, 31, 107.
425. T. Birchall and G. Denes, Can. J. Chem., 1984, 62, 591.
426. G. Denes, T. Birchall, M. Sayer, and M.F. Bell, Solid State Ionics, 1984, 13, 213.
427. T. Birchall, J.E. Vkris, B. Frlec, D. Ganter, and D. Hanzel, J. Fluorine Chem., 1984, 27, 61.
428. S. Ichiba and M. Takeshito, Bull. Chem. Soc. Jpn., 1984, 57, 1087.
429. H.S. Cheng, C.M. Hsu, and L.L. Shih, J. Chin. Chem. Soc. (Taipei), 1984, 31, 117.
430. C.N. Hsu and H.-S. Cheng, J. Chin. Chem. Soc. (Taipei), 1984, 31, 217.
431. C.M. Hsu, H.-S. Cheng, H. Luh, and M.-Y. Chen, J. Chin. Chem. Soc. (Taipei), 1984, 31, 211.
432. P.G. Harrison, B.J. Haylett, and T.J. King, Inorg. Chim. Acta, 1983, 265.
433. R.C. Mercader, E.J. Baran, and A.R. Lopez-Garcia, J. Radioanal. Nucl. Chem., 1984, 85, 13.
434. M.C. Apella, E.J. Baran, S.B. Etcheverry, and R.C. Mercader, Monatsh. Chem., 1983, 114, 1149.
435. R. Zhuang, W. Xu, A. Tai, and J. Zhang, Nanjing Daxue Xuet. Ziran Kexue, 1983, 437.
436. Z. Arifin, E.J. Filmore, J.D. Donaldson, and S.M. Grimes, J. Chem. Soc., Dalton Trans., 1984, 1965.
437. J.D. Donaldson and S.M. Grimes, J. Chem. Soc., Dalton Trans., 1984, 1301.
438. J.D. Donaldson, S.M. Grimes, S. Calogero, U. Russo, G. Valle, and P.J. Smith, Inorg. Chim. Acta, 1984, 84, 173.
439. T.S. Davy and J.J. Zuckerman, J. Organomet. Chem., 1984, 264, 295.
440. N.S. Hosmane, N.N. Sirmokadam, and R.H. Herber, Organometallics, 1984, 3, 1665.
441. D.A. Khramov, V.S. Ursov, and O.G. Ellert, Pis'ma Zh. Eksp. Teor. Fiz., 1984, 39, 174.
442. E. Kusior, S.M. Dubiel, M. Jachimowski, and M.K. Drop, Thin Solid Films, 1984, 112, 237.
443. S.K. Peneva, P.K. Rudarska, D.D. Nihtianova, and J. Amarov, Thin Solid Films, 1984, 112, 247.
444. J.S. Baijal, C. Prakash, P. Kishan, and K.K. Laroia, J. Phys. C, 1984, 17, 5993.
445. D. Baltrunas, Phys. Status Solidi A, 1984, 85, K97.
446. S. Ichiba and K. Kunita, Bull. Chem. Soc. Jpn., 1984, 57, 51.
447. R.H. Herber and R. Kalish, J. Phys. Chem. Solids, 1984, 45, 89.
448. G.K. Shenoy, D.G. Hinks, A.M. Umarji, and C.W. Kimball, Solid State Commun., 1984, 51, 621.
449. O.A. Bayukov, S.A. Gavrichkov, A.F. Savitskii, and T.G. Aminov, Fiz. Svoistava Magn. Poluprovodn., 1983, 149.
450. P. Boolchand and M. Stevens, Phys. Rev. B, 1984, 29, 1.
451. S.C. Huffman, J.P. Haushalter, A.M. Umarji, G.K. Shenoy, and R.C. Haushalter, Inorg. Chem., 1984, 23, 2312.

452. R. Schlogl and H.P. Boehm, Z. Naturforsch., Teil B, 1984, 39, 112.
453. V.L. Solozhenko, P.B. Fabrichnyi, M.E. Leonovg, and Y.A. Kalashnikov, Khim. Fiz., 1983, 711.
454. A. Vertes, K. Burger, L. Takacs, and I. Horvath, J. Radioanal. Nucl. Chem., 1984, 86, 195.
455. H.-S. Cheng and C.-M. Hsu, J. Chin. Chem. Soc., 1984, 31, 101.
456. H.-S. Chong, C.-M. Hsu, T.-M. Uen, and P.K. Txeng, J. Chin. Chem. Soc., 1984. 31, 221.
457. Z. Kucharski, A. Pron, M. Jozefowick. J. Suwalski, S. Lefrant, and S. Krichene, Solid State Commun., 1984, 51, 853.
458. A. Soxene, J.P. Tandon, and A.J. Crowe, Inorg. Chem. Acta, 1984, 84, 195.
459. P.A. Cussack and B.N. Patel, Inorg. Chem. Acta, 1984, 86, 1.
460. D. Tudela, V. Fernandez, and J. Tornero, Z. Anorg. Allg. Chem., 1984 509, 174.
461. S.P. Mallela, K.C. Lee, and F. Aubke, Inorg. Chem., 1984, 23, 653.
462. P.A. Cusack, B.N. Patel, P.J. Smith, D.W. Allen, and J.W. Nowell, J. Chem. Soc., Dalton Trans., 1984, 1239.
463. U. Russo, A. Cassol, and A. Silvestri, J. Organomet. Chem., 1984, 260, 69.
464. R. Barbieri and A. Silvestri, J. Chem. Soc., Dalton Trans., 1984, 1019.
465. S. Calogero, G. Valle, and U. Russo, Organometallics, 1984, 3, 1205.
466. J.E. Phillips and R.H. Herber, J. Organomet. Chem., 1984, 272, 411.
467. S.J. Blunden and R. Hill, Inorg. Chim. Acta, 1984, 87, 83.
468. S.J. Blunden, R. Hill, and D.G. Gillies, J. Organomet. Chem., 1984, 270, 39.
469. A. Sanexa and J.P. Tandon, Polyhedron, 1984, 3, 681.
470. N.A. Stepanova, A.A. Maligin, G.N. Kuzenetsova, T.K. Egoiov, A.V. Evdokimov, and T.M. Katkova, Zh. Obshch. Khim., 1984, 54, 111.
471. K.C. Molloy, T.G. Purcell, K. Quill, and I.W. Nowell, J. Organomet. Chem., 1984, 267, 237.
472. D.W. Allen, D.J. Derbyshire, J.S. Brooks, S.J. Blunden, and P.J. Smith, J. Chem. Soc., Dalton Trans., 1984, 1889.
473. S.J. Blunden, R. Hill, and J.N.R. Ruddich, J. Organomet. Chem., 1984, 267, C5.
474. R.H. Herber, A. Shanzer, and J. Libman, Organometallics, 1984, 3, 586.
475. A.G. Davies and A.J. Price, J. Organomet. Chem., 1983, 258, 7.
476. G.C. Stocco, L. Pellerito, M.S. Girasolo, and A.G. Osborne, Inorg. Chim. Acta, 1984, 83, 79.
477. G. Ruisi, M.T. LoGiudice, and L. Pellerito, Inorg. Chim. Acta, 1984, 93, 161.
478. C.P. Sharma, N. Kumar, S. Chandra, and B.S. Garg, J. Indian Chem. Soc., 1983, 60, 434.
479. F.A.K. Nassu, H.B. Hossain, D. Van der Helm, and J.J. Zuckerman, Inorg. Chem., 1984, 23, 606.
480. M.J. Smith, B.N. Patel, and J.S. Brooks, Chem. Ind., 1984, 804.
481. J.G. Stevens in ref. 3, p. 319.
482. J. Pebler and K. Dehnicke, Z. Anorg. Allg. Chem., 1984, 517, 103.
483. P. Jernberg, Uppsala Univ., Inst. Phys. (Tech. Rep.), 1984, UUIP-1114.
484. J.M. Friedt, L. Soderholm, R. Poinsot, and R. Vangelisti, Synth. Met., 1983, 8(1-2), 99.
485. J.M. Friedt, R. Poinsot, and L. Soderholm, Solid State Commun., 1984, 49(3), 223.

486. S.W. Hedges and L.H. Bowen, Inorg. Chem., 1984, 23(10), 1458.
487. M.A. Subramanian, A. Clearfield, A.M. Umarji, G.K. Shenoy, and G.V. Subba Rao, J. Solid State Chem., 1984, 52, 124.
488. A. Saxena, J.P. Tandon, T. Birchall, B. Ducourant, and G. Mascherpa, Polyhedron, 1984, 3(6), 661.
489. H. Schmidbaur. B. Milewski-Mahrla, and F.E. Hagner, Z. Naturforsch., Teil B, 1983, 38(11), 1477.
490. L.G. Shtyrkov, V. Khrapov. T.A. Babushkina, A.F. Volkov, and V.I. Stanko, Khim. Fiz., 1983, (3), 362.
491. S.J. Bondarevskii, Yu.P. Kostikov, and S.A. Timofeev, Teor. Eksp. Khim., 1984, 20(1), 89.
492. S. Calogero, U. Russo, F.E. Wagner, and R. Wordell, Inorg. Chim. Acta, 1984, 85(2), L51.
493. H.B. Singh, W.R. McWhinnie, R.F. Ziolo, and C.H.W. Jones, J. Chem. Soc., Dalton Trans., 1984, (7), 1267.
494. C.H.W. Jones and R.D. Sharma, J. Organomet. Chem., 1983, 255, 61.
495. H. De Waard in ref. 3, p. 295.
496. M. Pasternak, M. Van der Heyden, and G. Langouche, Nucl. Instrum. Methods Phys. Res., Sect. B, 1984, 232(1), 152.
497. M. Pasternak, M. Van der Heyden, and G. Langouche, J. Chem. Phys., 1984, 80(1), 36.
498. G. Wortmann, I. Nowik, G. Kaindl. H. Selig, and I. Palchan, Synth. Met., 1984, 10(2), 141.
499. T. Birchall, J.A. Smegal, and C.L. Hill, Inorg. Chem., 1984, 23(13), 1910.
500. H.G. Fritsche and F. Duebler, J. Less-Common Met., 1984, 97, 277.
501. H. Muthig, E. Huenges, and H. Morinaga, Nature (London), 1984, 312(5994), 532.
502. P. Gutlich, K.M. Hasselbach, H. Rummel, and H. Spiering, J. Chem. Phys., 1984, 89(3), 1396.
503. P. Helisto, E. Ikonen, T. Katila, W. Potzel, and K. Riski, Phys. Rev. B: Condens. Matter, 1984, 30(5), 2345.
504. W. Potzel, W. Adlassnig, U. Naerger, T. Obenhuber, K. Riski, and G.M. Kalvius, Phys. Rev. B: Condens. Matter, 1984, 30(9), 4980.
505. R.V. Pound, L. Niesen, H. De Waard, and G.L. Zhang, Phys. Rev. B: Condens. Matter, 1984, 29(11), 6086.
506. P. Triggs, F. Levy, and F.E. Wagner, Mater. Res. Bull., 1984, 19(2), 197.
507. M.L. Good, M.D. Patil, L.M. Trefonas, J. Dodge, C.J. Alexander, R.J. Majeste, and M.A. Cavanaugh, J. Phys. Chem., 1984, 88(3), 483.
508. U. Furholz, H.B. Burgi, F.E. Wagner, A. Stebler, J.H. Ammeter, E. Krausz, R.J.H. Clark, M.J. Stead, and A. Ludi, J. Am. Chem. Soc., 1984, 106, 121.
509. D. Salomon in ref. 33, p. 205.
510. U. Potzel, R. Raab, J. Voelkl, H. Wipf, A. Magerl, D. Salomon, and G. Wortmann, J. Less-Common Met., 1984, 101(1), 343.
511. L. Pfeiffer, T. Kovacs, and F.J.D. Salvo, Phys. Rev. Lett., 1984, 52(8), 687.
512. K. Zitter, J. Schmand, and R. Schoellhorn, Mater. Res. Bull, 1984, 19(6), 801.
513. A.X. Trautwein, R. Bales, S. Lauer, M. Grodzicki, and K.M. Hasselbach in ref. 33, p. 68.
514. R.V. Parish in ref. 33, p. 56.
515. A.K.H. Al-Sa'ady, K. Moss, C.A. McAuliffe, and R.V. Parish, J. Chem. Soc., Dalton Trans., 1984, 1609.
516. F.C. Shaw, N.A. Schaeffer, R.C. Elder, M.K. Eidsness, J.M. Trooster, and G.H.M. Calis, J. Am. Chem. Soc., 1984, 106(12), 3511.

517. A.K.H. Al-Sa'ady, C.A. McAuliffe, K. Moss, R.V. Parish, and R. Fields, J. Chem. Soc., Dalton Trans., 1984, 491.
518. J.W.A. Van der Welden and Z.M. Stadnik, Inorg. Chem., 1984, 23(17), 2640.
519. J.W.A. Van der Welden, P.T. Beurskens, J.J. Bour, W.P. Bosman, J.H. Noordik, M. Kolenbrander, and J.A.K.M. Buskes, Inorg. Chem., 1984, 23(2), 146.
520. J.W.A. Van der Welden, J.J. Bour, W.P. Bosman, J.H. Noordik, and P.T. Beurskens, Recl. J. Neth. Chem. Soc., 1984, 103(1), 203.
521. G.H.M. Calis and N. Hadjiliadis, Inorg. Chim. Acta, 1984, 91(3), 203.
522. G.K. Shenoy in ref. 3, p. 343.
523. P. Boolchand, G. Lemon, W.J. Bresser, D.H. McDaniel, P.C. Eklund, R.E. Heinz, E. Stumpp, and G. Nielfeld, Solid State Commun., 1984, 52(7), 675.
524. H. Schaefer-Stahl and G. Von Eynatten, Synth. Met., 1983, 7(1-2), 73.
525. Z.M. Stadnik and R.O. Moyer, J. Less-Common Met., 1984, 98(1), 159.
526. Z.M. Stadnik and E. De Boer, Solid State Commun., 1984, 50(4), 335.
527. C. Sauer, A.M. Zaker, and W. Zinn in ref. 33, p. 147.
528. T. Bauermann, M.M. Abd-Elmeguid, T. Takabatake, and H. Micklitz, J. Phys. C, 1983, 16(33), 6435.
529. T. Takabatake, R.W. McCallum, M. Kubota, and F. Pobell, J. Low Temp. Phys., 1984, 55(1-2), 111.
530. M.P. Janawadkar, T.S. Rahhakrishnan, S.H. Devare, H.G. Devare, R.G. Pillay, R. Janaki, A.M. Umarji, and G.V. Stubba Rao, Proc. Nucl. Phys. Solid State Phys. Symp., 1980, 1983, 23, 15.
531. I. Felner and I. Nowik, J. Phys. Chem. Solids, 1984, 45(4), 419.
532. A. Scherzberg, C. Sauer, U. Koebler, W. Zinn, and J. Roehler, Solid State Commun., 1984, 49(11), 1027.
533. A. Scherzberg, Ber. Kernforschungsanlage Juelich, 1984, Juel. 1895, 71 pp.
534. S.K. Malik, S.K. Dhar, and R. Vijayaraghavan, Pramana, 1984, 22(3-4), 329.
535. S.K. Dhar, R. Nagarajan, S.K. Malik, R. Vijayaraghavan, M.M. Abd-Elmeguid, and H. Micklitz, Phys. Rev. B: Condens. Matter, 1984, 29(10), 5953.
536. S.J. Lyle and W.A. Westall, J. Less-Common Met., 1984, 99(2), 265.
537. J.W.C. De Vries, R.C. Thiel, and K.H.J. Buschow, Physica B + C (Amsterdam), 1984, 124(3), 291.
538. G.D. Stucky, L. Iton, T. Morrison, G. Shenoy, S. Suib, and R.P. Zerger, J. Mol. Catal., 1984, 27(1-2), 71.
539. I. Felner, Solid State Commun., 1984, 52(2), 191.
540. D.R. Noakes, A.M. Umarji, G.K. Shenoy, and D.G. Hinks, J. Appl. Phys., 1984, 55(6), 1847.
541. A.R. Chowdhury, G.S. Collins, and C. Hohenemser, Phys. Rev. B, 1984, 30(11), 6277.
542. J. Pszczola, J. Zuckrowski, K. Krop, and J. Suwalski, J. Magn. Magn. Mater., 1984, 44(1 + 2), 223.
543. M. Maurer and J.M. Friedt, Phys. Rev. B: Condens. Matter, 1984, 30(5), 2369.
544. I. Nowik, I. Felner, and M. Seh, Proc. Int. Conf. Magn. Rare Earths, Actinides, 1983, 1, 112.
545. J. Zubrowski, A. Barnasik, K. Krop, R. Radwanski, J. Pszczola, J. Suwalski, Z. Kucharski, and M. Lukasiak, Hyperfine Interact., 1983, 16(1-4), 801.
546. K. Post and L. Niesen, J. Phys. F, 1984, 14(9), 2203.
547. P.C.M. Gubbens, A.M. Van der Kraan, and K.H.J. Buschow, J. Phys. F, 1984, 14(9), 2195.

548. P.C.M. Gubbens, A.M. Van der Kraan, and K.H.J. Buschow, J. Phys. F, 1984, 14, 235.
549. J.A. Hodges, P. Imbert, and A. Schuhl, J. Magn. Magn. Mater., 1984, 43, 101.
550. D. Bingham, B. Bleaney, R.K. Day, and J.B. Dunlop, J. Phys. C, 1984, 17, 1511.
551. J.A. Hodges and G. Jehanno, J. Phys. (Les Ulis, Fr.), 1984, 45, 1663.
552. A.M. Umarji, G.K. Shenoy, D.R. Noakes, and S. Dattagupta, J. Appl. Phys., 1984, 55, 2297.
553. D. Weschenfelder, V. Oestreich, H. Schmidt, and G. Czjzek, Ger. Offen. DE 3, 318, 611.
554. P. Bonville, P. Imert, G. Jehanno, F. Gonzalez-Jimenez, and F. Hartmann-Boutron, Phys. Rev. B, 1984, 30, 3672.
555. P. Bonville, F. Gonzalez-Jimenez, P. Imbert, G. Jehanno, L.C. Lopes, A.K. Bhattacharjee, and B. Coqblin, J. Phys. (Les Ulis, Fr.), 1984, 45(3), 467.
556. G. Czjzek, D. Weschenfelder, V. Oestreich, H. Schmidt, A. Vaures, and M. Maurer, J. Non-Cryst. Solids, 1984, 61-62, 433.
557. D. Karaker in ref. 3, p. 355.
558. H. Abazli, A. Cousson, J. Jove, M. Pages, and M. Gasperin, J. Less-Common Met., 1984, 96, 23.
559. A. Tabuteau, J. Jove, M. Pages, C.H. De Novion, and J. Gal, Solid State Commun., 1984, 50, 357.
560. F. Nectoux, H. Abazli, J. Jove, A. Cousson, M. Pages, M. Gasperin, and G. Choppin, J. Less-Common Met., 1984, 97, 1.
561. A. Cousson, S. Dabos, H. Abazli, F. Nectoux, J. Jove, M. Gasperin, and M. Pages, Inorg. Chim. Acta, 1984, 94, 1.
562. T. Thevenin, J. Jove, and M. Pages, Hyperfine Interact., 1984, 20, 173.
563. T. Thevenin, INIS Atomindex, 1984, 15, Abstr. No. 15, 037 079.
564. J. Gal, Spez. Ber. Kernforschungsanlage Juelich, 1984, Juel.-Spez. 250, 601.
565. W. Potzel, J. Moser, M. Wunsch, J. Zankert, Th. Obenhuber, L. Asch, G.M. Kalvius, J.C. Spidet, J. Gal, and M. Boge, Inorg. Chim. Acta, 1984, 94, 21.
566. W. Meisel in ref. 33, p. 15.
567. F. Salvat and J. Parellada, Nucl. Instrum. Methods Phys. Res., Sect. B, 1984, 229, 70.
568. M.A. Andreeva and R.N. Kuz'min, Phys. Status Solidi B, 1984, 125, 461.
569. K. Nomura and Y. Ujihira, Bunseki Kagaku, 1984, 33, T81.
570. S.C. Pancholi, H. De Waard, J.L.W. Petersen, A. Van der Wijk, and J. Van Klinken, Nucl. Instrum. Methods Phys. Res., Sect. A, 1984, 221, 577.
571. M.E. Kordesch, J. Eldridge, D. Scherson, and R.W. Hoffman, J. Electroanal. Chem. Interfacial Electrochem., 1984, 164, 393.
572. T. Toriyama, K. Asano, K. Saneyoshi, and K. Hisatake, Nucl. Instrum. Methods Phys. Res., Sect. B, 1984, 232, 170.
573. W. Keune, T. Shigematsu, S. Staniek, H.D. Pfannes, R.H. Mussbaum, R.A. Brand, K. Saneyoshi, and D. Liljequist in ref. 33, p. 27.
574. V.P. Godbole, V.G. Bhide, S.V. Ghaisas, S.M. Kanetkar, R.S. Joshee, and S.B. Ogale, Mater. Res. Soc. Symp. Proc., 1984, 25, 241.
575. V.P. Godbole, V.G. Bhide, S.V. Ghaisas, S.M. Kanetkar, S.M. Chaudhari, and S.B. Ogale, Mater. Res. Soc. Symp. Proc., 1984, 25, 247.

576. I. Ya Dekhtyar, V.V. Nemoshkalenko, M.M. Nishchenko, O.N. Razumov, and N.A. Tomashevskii, Ukr. Fiz. Zh. (Russ. Ed.), 1984, 29(1), 140.
577. G. Marest, A. Perez, J.L. Ponthenier, P. Gerard, and J.M. Robertson, J. Appl. Phys., 1984, 55(6), 2560.
578. Y.S. Dorik, V.G. Bhide, S.M. Kanetkar, S.V. Ghaisas, S.M. Chaudhari, and S.B. Ogale, J. Appl. Phys., 1984, 56(9), 2566.
579. S. Skrzypek, E. Kolawa, J.A. Sawicki, and T. Tyliszczak, Mater. Sci. Eng., 1984, 66(2), 145.
580. M. Fujinami and Y. Ujihira, Appl. Surf. Sci., 1984, 17, 265.
581. M. Fujinami and Y. Ujihira, Appl. Surf. Sci., 1984, 17(3), 276.
582. K. Nomura and Y. Ujihira, J. Mater. Sci., 1984, 19(8), 2664.
583. W. Meisel, Spectrochim. Acta, Part B, 1984, 39(12), 1505.
584. M. Matsuo. H. Sato, and T. Tominaga, Radiochim. Acta, 1984, 35(4), 227.
585. G. Hoy, D.C. Cook, and E. Agyekum, Gov. Rep. Announce. Index (U.S.), 1984, 84(20), 64.
586. J.W.D. Martens. M. Erman, and H.M. Van Noort, J. Magn. Magn. Mater., 1984, 42(3), 279.
587. J.H. Johnston and C.M. Cardile, Chem. Geol., 1984, 45(1-2), 73.
588. J.H. Johnston, C.M. Cardile, G.E. Coote, R.J. Sparks, G. Wallace, J.W. McMillan, F.C. Pummery, and G. Longworth, Chem. Geol., 1984, 42(1/4), 189.
589. H. Muramatsu, T. Miura, H. Nakahara, M. Fujioka, E. Tanaka, and A. Hashizume, Hyperfine Interact., 1984, 20(4), 305.
590. S.K. Peneva and K.D. Djuneva, Thin Solid Films, 1984, 113(4), 297.
591. G.K. Shenoy in ref. 3, p. 343.

8

Gas-phase Molecular Structures Determined by Electron Diffraction

By D. W. H. RANKIN and H. E. ROBERTSON

1 Introduction

In this chapter we report the results of structural studies of gaseous inorganic compounds published during 1984. The new method of producing this review reduces the amount of information we can present, and so we limit ourselves to discussion of structures, and must omit references to new techniques, however important they may be. We also omit new work on old molecules, where the new results are not significantly different from the old. And the index has also had to go - so if you want to find a particular molecule, you may have to read the whole chapter!

2 Compounds of Elements in Main Groups II and III

The most interesting molecule in this section is undoubtedly $Be_4O(NO_3)_6$,[1] which has been studied in the gas phase at 430 K. The structure is very similar to that of the well known 'basic beryllium acetate', in that it has a central oxygen atom surrounded by four beryllium atoms, with r_g Be-O 1.665(21) Å. Each edge of the Be_4 tetrahedron is then bridged by a nitrate group, with two Be-O bonds of 1.620(8) Å, and two O-N bonds of 1.298(4) Å. The terminal N=O bonds are 1.185(5) Å long, and the O-N-O angles are 117.0(9)°. Each planar NO_3 group is twisted about an axis passing through the nitrogen atom and the oxygen atom at the centre of the molecule, by 25.2(18)°, so that the overall symmetry is reduced to $\underline{T}$. The force field for this unusual molecule was refined, using both the electron-diffraction data and spectroscopic data.

Potassium metaborate, KBO_2, has been studied at 1280 K.[2] The metal atom is bound to one oxygen atom, with r_a(K-O) 2.402(36) Å, and the BO_2 ion is linear, with a mean B-O distance of 1.260(8) Å. The angle KOB in the r_g structure is 107(6)°,

1. N.I. Tuseev, V.A. Sipachev, R.F. Galimzyanov, A.V. Golubinskii, E.Z. Zasorin, and V.P. Spiridonov, J.Mol.Struct., 1984, 125, 277.
2. Yu.S. Ezhov and S.M. Tolmachev, Zh. Strukt. Khim., 1984, 25(3), 169.

compared with 109(5), 113(5), and 127(12)° in the analogous metaborates of sodium, rubidium, and caesium, respectively. Of course, there may be large shrinkages, and it is possible that angles in the equilibrium configurations may be much larger than these, and it is even possible that the structures are linear.

Molecules of boric anhydride, B_2O_3, have two almost linear OBO units sharing a common oxygen atom, giving an overall structure which is V-shaped.[3] The data were interpreted in terms of the harmonic equilibrium distances, r_e^h, and harmonic force constants, and yielded BOB and OBO angles of 134.2(50) and 173.4(44)°, respectively. The outer B-O bonds, which are formally double bonds, are 1.195(6) Å long, while the lengths of the inner, single bonds are 1.329(10) Å. The harmonic force constants were used to derive vibrational frequencies, most of which are in good agreement with experimental values. However, two modes observed at over 500 cm^{-1} were calculated to lie near 280 cm^{-1}. The observed structural parameters are also in good agreement with those calculated ab initio.

Structures of trimethylamine adducts of borane[4] and gallane[5] have been reported. Data for the boron compound were collected at 346 K, and rotational constants were also used in the analysis. In the r_g structure the BH_3 group had r(B-H) 1.261(6) Å and <HBH 112.8(4)°, while in the NMe_3 group r(C-N) was 1.485(1) Å and <CNC was 109.2(2)°. The connecting B-N distance was 1.656(2) Å, and it was estimated that the barrier to rotation about this bond was more than 20 kJ mol^{-1}. The NMe_3 group in the gallane adduct was similar to that in the borane adduct, with r(C-N) 1.482(5) Å and <CNC 109.0(5)°. The GaH_3 group had r(Ga-H) 1.497(15) Å, and the angle HGaH was approximately 116°, while the Ga-N bond length was 2.124(7) Å. It is noteworthy that this bond is some 0.06 Å longer than the Al-N bond in the analogous AlH_3 adduct, but that the Al-H bonds in the latter compound are longer than Ga-H bonds in the GaH_3 adduct. This observation is consistent with reported vibrational frequencies and force constants for hydrides of aluminium and gallium.

3. V.P. Spiridonov, A.G. Gershikov, E.Z. Zasorin, A.A. Ivanov and L.I. Ermolayeva, High Temp.Sci., 1983, 16, 325.
4. K. Iijima, N. Adachi and S. Shibata, Bull.Chem.Soc.Jpn., 1984, 57, 3269.
5. P.L. Baxter, A.J. Downs and D.W.H. Rankin, J.Chem.Soc., Dalton Trans., 1984, 1755.

3 Methyl-substituted silyl compounds

There is a mixture of two conformers in the gas phase of $CH_2Br-SiClMe_2$.[6] The *gauche* form is slightly more predominant (55 ± 9%) and has a BrCSiCl dihedral angle of 58(15)°, while the equivalent angle in the *anti* form is 157(9)°. Other geometrical parameters were assumed to be the same for the two conformers, and refined bond lengths (r_g) are r(Si-C)(mean) 1.857(3), r(C-Br) 1.995(7), and r(Si-Cl) 2.065(4) Å. The angles were all close to the tetrahedral angle. Although the parameters are obviously strongly correlated, there can be no doubt as to the conclusions about the conformations, as the Br...Cl peaks are clearly visible in the radial distribution curve. In the same paper structural results are also reported for $CMe_2Br-CClH_2$.

In octamethyltrisilane[7] there must be considerable steric strain, and this is minimised by the trimethylsilyl groups taking up staggered positions with respect to the groups on the central atom, so that the molecule has (very nearly) C_{2v} symmetry overall. But with this conformation there must be strong 1,3 interactions, and these are relieved partly by slight twisting away from the C_{2v} structure, and partly by widening of the central SiSiSi angle, to 118.0(25)°. All other angles are close to the tetrahedral angle, and bond lengths are only marginally greater than those in unstrained molecules. The Si-Si distance (r_a) is 2.325(12) Å, and the mean Si-C distance is 1.887(3) Å, with the bonds to the central atom being 0.02 Å longer than those to the outer atoms.

A number of compounds containing two or three $SiMe_3$ groups bound to a common carbon atom have been studied, and they all show effects of crowding of these bulky groups. In bis(trimethylsilyl)methyl lithium,[8] which is monomeric in the gas phase at 413 K, the effects of crowding are shown in the wide SiCSi and SiCLi angles, which are 119.0(15) and 115(2)°, respectively, and in lengthening of the inner Si-C bonds to 1.892(2) Å, compared with 1.877(2) Å for the Si-C(methyl) bonds. The position of the lithium atom could not be determined reliably: r(Li-C) is given as 2.03(6) Å. In this case the $SiMe_3$ groups are said to adopt configurations which are twisted only 9°

6. Q. Shen, J.Mol.Struct., 1984, 112, 169.
7. A. Almenningen, T. Fjeldberg and E. Hengge, J.Mol.Struct., 1984, 112, 239.
8. J.L. Atwood, T. Fjeldberg, M.F. Lappert, N.T. Luong-Thi, R. Shakir and A.J. Thorne, J.Chem.Soc., Chem.Commun., 1984, 1163.

from the eclipsed positions, and the methyl groups are slightly compressed, with C(central)SiC(methyl) angles of 111.0(12)°. In the same paper data are presented for the crystalline phase, in which the compound is polymeric.

Preliminary structural data have been reported[9] for $Ge[CH(SiMe_3)_2]_2$, 430 K. The Ge-C and mean Si-C bond lengths are 2.04(2) and 1.89(1) Å respectively, and GeCSi and SiCSi angles are 111(1) and 113(1)°. Most of these parameters show little effect from steric crowding, but the Ge-C distance is some 0.08 Å longer than in tetraalkylgermanes. The angle CGeC is 107(2)°, which is 10° greater than the corresponding angle in the analogous tin compound, so it may be that the two bis(trimethylsilyl)methyl groups are pushing each other apart. The overall symmetry was assumed to be C_2, so we cannot tell what the effects of conformation on the apparent parameters may be.

Tris(trimethylsilyl)methyl phosphine[10] is one of a number of $-C(SiMe_3)_3$ compounds whose structures have been investigated recently as solids or as gases. In this compound the SiCSi angles are actually less than the tetrahedral angle, at 106.9(13)° (r_a structure). This is a consequence of two-bond Si...P contacts, which also contribute to lengthening of the inner Si-C bonds to 1.941(5) Å, compared to 1.883(2) Å for the Si-C(methyl) bonds. The strain within the $-C(SiMe_3)_3$ group is minimised by compression of the methyl groups within each $SiMe_3$ group, so that the C(methyl)SiC(methyl) angles are only 104.3(4)°, by twisting each of the $SiMe_3$ groups by 21.2(4)° away from the fully staggered conformation, and by tilting the local C_3 axes of the $SiMe_3$ groups by 7.3(15)° away from each other. The effect of this last distortion is that the individual C(central)SiC(methyl) angles range from 108 to 121°. The P-C bond is 1.808(9) Å long, which is somewhat shorter than usual for an alkyl phosphine.

There has been a lot of structural work done on methyl-substituted silyl amines and amido-compounds, many of which would have been regarded, a few years ago, as too difficult subjects for gas-phase electron-diffraction studies. The SiH_3, SiH_2Me, and $SiHMe_2$ derivatives of dimethylamine[11] all have nitrogen atoms

9. T. Fjeldberg, A. Haaland, B.E.R. Schilling, H.V. Volden, M.F. Lappert and A.J. Thorne, J.Organometal.Chem., 1984, 276, C1.
10. A.H. Cowley, J.E. Kilduff, E.A.V. Ebsworth and D.W.H. Rankin, J.Chem.Soc., Dalton Trans., 1984, 689.
11. G. Gundersen, R.A. Mayo and D.W.H. Rankin, Acta Chem.Scand., Ser.A., 1984, 38, 579.

with a shallow pyramidal configuration, with the sum of the three angles being between 352 and 356° in the r_a structures. It may be that the equilibrium structures are planar at nitrogen, and the observed non-planarity may be a shrinkage effect, but whether this is so or not, it is clear that these amines are intermediate between the truly planar trisilylamines and alkyldi(silyl)-amines, and the pyramidal trialkylamines. Important parameters for $SiH_3N(CH_3)_2$ are: *r*(Si-N) 1.713(5), *r*(N-C) 1.457(6) Å, <SiNC 120.9(3) and <CNC 112.0(6)°. Parameters for $CH_3SiH_2N(CH_3)_2$ include: *r*(Si-C) 1.867(6), *r*(Si-N) 1.715(6), *r*(N-C) 1.455(3) Å, <SiNC 121.5(8), <CNC 112.7(8), and <NSiC 113.3(23)°. The preferred conformation of this compound has one $Si-CH_3$ bond *gauche* to the lone pair of electrons on the nitrogen atom, with the smallest CNSiC dihedral angle 72(3)°. Principal parameters for $(CH_3)_2SiHN(CH_3)$ are: *r*(Si-C) 1.869(3), *r*(Si-N) 1.719(5), *r*(N-C) 1.460(4) Å, <SiNC 119.3(8), <CNC 113.7(15), <NSiC 109.9(18), and <CSiC 107.5(38)°. In this case also the preferred conformation had one $Si-CH_3$ bond *gauche* to the nitrogen lone pair, and the smallest CNSiC dihedral angles were 68(6) and 50(5)°.

The structures of two closely related silyl amines, $NH(SiHMe_2)_2$ [12] and $NH(SiMe_3)_2$,[13] have been studied by two different groups, and the results are in excellent agreement. Important parameters of the r_a structures, with that for the $SiHMe_2$ compound given first in each case, are: *r*(Si-C) 1.867(3) and 1.876(1) Å, *r*(Si-N) 1.727(3) and 1.738(5) Å, <SiNSi 130.4(15) and 131.3(15)°, mean <NSiC 110.2(3) and 110.7(5)°. The CSiC angle in $NH(SiHMe_2)_2$ was 112.8(11)°. In both cases the silyl groups are twisted in such a way that the methyl groups are staggered, when viewed along the non-bonded Si...Si axis. This suggests that non-bonded 1,3 interactions between methyl groups are important, and this is supported by the observation of exceptionally wide SiNSi angles in both compounds. The rather longer Si-C and Si-N bonds in $NH(SiMe_3)_2$ are also consistent with this idea.

Several compounds in which two bis(trimethylsilyl)amido groups are linked by metal atoms have been studied by electron diffraction. Both $Mg[N(SiMe_3)_2]_2$ [14] and $Zn[N(SiMe_3)_2]_2$ [15] at ca. 400 K have linear N-M-N arrangements, staggered Si_2NMNSi_2

12. G. Gundersen and D.W.H. Rankin, Acta Chem.Scand., Ser.A., 1984, 38, 647.
13. T. Fjeldberg, J.Mol.Struct., 1984, 112, 159.
14. T. Fjeldberg and R.A. Andersen, J.Mol.Struct., 1984, 125, 287.
15. A. Haaland, K. Hedberg and P.P. Power, Inorg.Chem., 1984, 23, 1972.

conformations with D_{2d} symmetry at equilibrium, and large-amplitude torsional vibrations about the axis, with barriers to internal rotation of 13(4) and 10(6) kJ mol^{-1}, respectively. Some important parameters (r_a) are as follows, with those for the magnesium compound preceding those for the zinc analogue: *r*(M-N) 1.91(3) and 1.824(14), *r*(Si-N) 1.703(6) and 1.728(7), *r*(Si-C) 1.886(5) and 1.889(6) Å, <SiNSi 128(2) and 132(2), and <NSiC 116.4(12) and 113.5(20)°. The short length of the Si-N bonds in the magnesium amide is remarkable as is the apparent size of the NSiC angles. The latter correspond to CSiC angles of 101°, which seems unreasonably small, but it may be that these angles are correlated with tilts of the $SiMe_3$ groups, which do not seem to have been considered.

The bis(trimethylsilyl)amide of lithium[16] exists solely as dimers in the gas phase at 400 K, with the two lithium atoms bridging between two nitrogen atoms. The overall molecular symmetry of the dimer is D_{2h}. The central, planar Li_2N_2 ring has r_a(Li-N) 1.99(3) Å and <NLiN 100(3)°. In the $N(SiMe_3)_2$ groups *r*(Si-N) is 1.712(7) Å, and *r*(Si-C) is 1.889(4) Å, while the SiNSi and NSiC angles are 129.8(17) and 112.2(9)°, respectively. This group therefore has a structure more or less the same as is found in the singly bridged magnesium and zinc amides, and in the free amine, with the major difference again being in the lengths of the Si-N bonds. As in most of the amides, the $SiMe_3$ groups adopt a conformation in which the methyl groups are almost perfectly staggered, when viewed along the Si...Si axis.

4 Other Compounds of Elements in Main Group IV

In this section we include derivatives of organic compounds that have several halogen atoms. The three halomethanes CH_2BrCl, $CHBrCl_2$, and $CHBr_2Cl$ have been studied,[17] and the very small differences between their structures have been discussed in terms of valence-shell electron-pair repulsion theory and the electronegativities of the various substituents. In the three compounds, in the order given, the C-Cl distances (r_g) are 1.755(8), 1.758(4), and 1.749(10) Å, and the C-Br distances are 1.928(7), 1.929(9), and 1.930(4) Å. All the angles between bonds to halogen atoms are between 111.0 and 111.7°, with the exception

16. T. Fjeldberg, M.F. Lappert and A.J. Thorne, J.Mol.Struct., 1984, 125, 265.
17. S. Konaka, K. Kiyoto, K. Yamamoto, N. Suzuki and M. Kimura, J.Mol.Struct., 1984, 125, 143.

of the BrCCl angle in CH_2BrCl, which is 113.3(6)°.

A report[18] of the structure of 1,1,2-trichloro-3,3-difluoroprop-1-ene, $CCl_2=CClCF_2H$, is of unusual interest, as it shows clearly how local symmetry, which is often assumed in electron-diffraction work, may be significantly perturbed by substituents. In this molecule there are three different CCCl angles, involving chlorine atoms which are cis, trans, and geminal with respect to the CF_2H group. These angles ($<_\alpha$) are 123.3(6), 121.8(6), and 120.2(10)°, respectively. The fourth angle adjacent to the double bond, CCC, is even larger at 124.5(9)°. However, even in this case it was necessary to assume that all C-Cl bond lengths were equal. Refined distances (r_a) were: r(C=C) 1.365(18), r(C-C) 1.509(19), r(C-Cl) 1.713(4), and r(C-F) 1.353(11) Å. Only one conformer was found to be present. It had C_s symmetry, with the C-H bond eclipsing the C=C bond, but with an r.m.s. torsional amplitude of 10.5(32)°.

Perchloro-1,5-hexadiene-3-yne, $Cl_2C=CCl-C\equiv C-CCl=CCl_2$, and its hydrocarbon parent have been studied at 410 and 290 K, respectively.[19] Both molecules have the possibility of having extended conjugation, but only in the chloro-compound is there a significant proportion of molecules having a planar arrangement of the atoms. In this case 30% of the molecules have the anti conformation, while the rest have a twisted, gauche conformation, with the C=C....C=C dihedral angle being 38°. In the hydrocarbon there appears to be essentially free rotation. In C_6Cl_6 the important parameters (r_a) are: r(C≡C) 1.218(8), r(C=C) 1.353(4), r(C-C) 1.415(5), r(C-Cl) 1.715(2) Å, <C-C-Cl 115.9(4), <C=C-C 123.6(3) and <C=C-Cl (terminal) 122.8(1)°. The parameters for the hydrocarbon are generally similar, but the C=C-C angle is 125.4(2)°, which is significantly wider than in the chlorine compound.

An r_{av}^* structure for 1,1,1-trifluoro-4-silabut-2-yne, $CF_3C\equiv CSiH_3$, has been reported.[20] This is based on electron-diffraction data and one rotational constant, measured by microwave spectroscopy, and is supported by other spectroscopic observations. The geometry of the CF_3 group is unremarkable,

18. S.H. Schei and R. Seip, Acta Chem.Scand., Ser.A., 1984, 38, 345.
19. A. Almenningen, E. Gogstad, K. Hagen, H. Schei, R. Stølevik, O. Thingstad and M. Traetteberg, J.Mol.Struct., 1984, 116, 131.
20. D.W.W. Anderson, S. Cradock, E.A.V. Ebsworth, A.R. Green, D.W.H. Rankin and A.G. Robiette, J.Organometal.Chem., 1984, 271, 235.

with $\underline{r}$(C-F) 1.341(3) Å and <FCF 108.7(4)°. The C≡C distance, 1.218(8) Å, is only very slightly longer than those in $CF_3C≡CH$ and $SiH_3C≡CSiH_3$, but the C-C distance in CF_3-C≡C- compounds seems to vary widely, being 1.453(2) Å in $CF_3C≡CCl$, 1.500(5) Å in $CF_3C≡CH$, and 1.482(8) Å in the present case. The Si-C bond appears to be somewhat long [$\underline{r}_{av}^{*}$ is 1.827(6) Å and $\underline{r}_a$ is 1.847 Å] in comparison with those found in other silyl acetylenes (typically around 1.82 Å for $\underline{r}_a$), and is more like that in silyl cyanide, for which $\underline{r}_a$(Si-C) is 1.850 Å.

Structures of two trifluorosilyl compounds, $(SiF_3)_2CH_2$ and $(SiF_3)_2CCl_2$, have been reported this year,[21] and they show very clearly the steric and electronic effects of the fluorine and chlorine substituents. Thus the Si-C distance ($\underline{r}_a$ structure) in $(SiF_3)_2CH_2$ is 1.827(4) Å, which is very short, reflecting the electronegativity of the fluorine substituents, but the mutual repulsion of the SiF_3 groups leads to widening of the SiCSi angle to 117.7(4)°, compared with 114.4° in $(SiH_3)_2CH_2$. Replacement of the hydrogen atoms on the central carbon by chlorine increases the steric strain, and this leads to lengthening of the Si-C bonds, to 1.871(8) Å, and to further widening of the SiCSi angle to 119.4(6)°. The C-Cl bonds in this compound are also unusually long, at 1.798(7) Å. In both compounds the SiF_3 groups are twisted away from the perfectly staggered arrangement, by 10.3(25) and 8.9(6)° in $(SiF_3)_2CH_2$ and $(SiF_3)_2CCl_2$, respectively, thus reducing the effects of 1,3 fluorine-fluorine interactions while reducing the molecular symmetry from $\underline{C}_{2v}$ to $\underline{C}_2$. Other important parameters for $(SiF_3)_2CH_2$ are $\underline{r}$(Si-F) 1.567(3) Å and <FSiF 108.0(1)°, while in $(SiF_3)_2CCl_2$ $\underline{r}$(Si-F) is 1.560(3) Å, <FSiF is 109.4(1)°, and <ClCCl is 112.6(6)°.

The structure of 1,1-dichloro-1-sila-2-oxacyclohexa-3,5-diene is not an easy one to determine by electron diffraction,[22] and it is necessary to make several assumptions in order to reduce the problem to a manageable size. One curious assumption is that the C=C and C-O bonds are of equal length, but fortunately this should not have a great effect on other parameters. Refined bond lengths ($\underline{r}_g$) are: Si-O 1.653(9), Si-C 1.851(13), Si-Cl 2.032(6), C-O/C=C 1.317(9), and C-C 1.481(9) Å. In the ring the bond angles are OCC 119.7, CCC 123.7(7), CCSi

21. E. Vajda, M. Kolonits, G. Fritz, J. Thomas and E. Sattler, J.Mol.Struct., 1984, 117, 329.

22. M.E. Nipan, N.I. Sadova, A.V. Golubinskii, L.V. Vilkov, T.L. Krasnova, E.A. Chernyshev and M.O. Labartkava, Zh.Strukt.Khim., 1984, 25(5), 41.

121.4, CSiO 97.2, and SiOC 134.3(15)°, and there is a small twist about the C-C bond linking the two C=C double bonds, of 8(6)°. Thus the ring is distinctly non-planar, and is considerably distorted from regularity. The wide angle at oxygen is similar to that in CH_2=CHOSiMe$_3$, but in the ring compound it leads to a narrow angle at silicon. The ClSiCl angle is 108.5(6)°.

Some years ago a structure of Ge(C_5Me_5)$_2$ was reported, but it was subsequently shown that the data related to Ge(C_5Me_5)Cl. Now both structures have been determined, at ca.400 K.[23] In Ge(C_5Me_5)$_2$ the two rings are not parallel, at least in the thermal average, and the angle between them is 22(2)°, compared with 34(7)° in Ge(C_5H_4Me)$_2$. The two rings are staggered, so that the molecule has $\underline{C_s}$ symmetry, and the mean Ge-C distance is 2.519(7) Å. The lateral positions of the rings relative to the germanium atom are not precisely determined, but it seems probable that the central atom is not far displaced relative to the local $\underline{C_5}$ axes of the cyclopentadienyl groups. The structure is very similar to those of the analogous tin and lead compounds in the crystalline phase. In Ge(C_5Me_5)Cl the germanium atom is displaced 0.43(6) Å laterally from the $\underline{C_5}$ axis of the C_5Me_5 group, towards the mid-point of one C-C bond, and the distance between the germanium atom and the plane of the ring is 2.11(2) Å, about 0.1 Å less than in Ge(C_5Me_5)$_2$. The Ge-Cl bond lies at an angle of 110(2)° to the ring normal through germanium, and is 2.258(12) Å long. This structure is also closely related to that of its tin analogue, in the crystalline phase.

5 Compounds of Elements in Main Group V

Derivatives of organic compounds having -NO_2 substituents are traditionally included in this chapter, the N-O bonds being sufficient for us to regard the compounds as inorganic. This year there have been three reports of the structures of aromatic nitro compounds, nitrobenzene,[24] p-bromonitrobenzene,[25] and 2,6-dinitrochlorobenzene;[26] parameters given below ($\underline{r}_g$, $\underline{r}_a$, and $\underline{r}_g$,

23. L. Fernholt, A. Haaland, P. Jutzi, F.X. Kohl and R. Seip, Acta Chem.Scand., Ser.A., 1984, 38, 211.
24. I.F. Shishkov, N.I. Sadova, V.P. Novikov and L.V. Vilkov, Zh.Strukt.Khim., 1984, 25(2), 98.
25. A. Almenningen, J. Brunvoll, M.V. Popik, L.V. Vilkov and S. Samdal, J.Mol.Struct., 1984, 118, 37.
26. O.G. Batyukhnova, L.V. Vilkov, N.I. Sadova and Yu.A. Pankrushev, Zh.Strukt.Khim., 1984, 25(3), 1669.

respectively) apply to the three compounds in this order. In each case the nitro group is planar, with $\underline{r}$(N=O) 1.218(4), 1.239(2), and 1.229(2) Å and <ONO 123.4(16), 125.0(3), and 123.0(3)°, and with the group twisted out of the plane of the benzene ring. In two cases static models were used, and the twist angles are reported as 22.7(41) and 54.0(10)°, but for the bromo compound a dynamic model was used, giving a barrier to rotation of 18(3) kJ mol^{-1}. The electronegative substitutents distort the ring, and CC_NC angles are 125.1(14), 121.6(2), and 121.6(4)°, while the angles $CC_{Br}C$ and $CC_{Cl}C$ are 122.6(2) and 118.9(6)°, respectively. The C-Br distance is 1.896(2) Å, and the C-Cl distance is 1.712(5) Å. Two other nitrobenzene derivatives are described in section 6 of this chapter.

Methyl(vinyl)-$\underline{N}$-nitramine, $CH_3N(NO_2)CH=CH_2$, has also been studied by electron diffraction.[27] In this compound $\underline{r}_g$ (N=O) is 1.213(3) Å, and the angle ONO is 128.0(21)°, somewhat wider than in the aromatic compounds. The other major bond lengths are: C=C 1.343(15), N-N 1.385(8), N-C(methyl) 1.459(15), and N-C(vinyl) 1.432(15) Å. The bond angles are: NCC 119.3(15), NNC(methyl) 113.5(12), and CNC 124.1(10)°. It is not easy to determine the conformation of a molecule like this, but the final conclusion of the authors was that all the heavy atoms are probably coplanar, with the NO_2 group lying anti with respect to the C=C bond.

The remaining nitrogen compound in this section is trifluoromethyl isocyanide, which has been the subject of a combined electron-diffraction and microwave study.[28] In the $\underline{r}_{av}$ structure the C-F, C-N, and N=C distances are 1.324(1), 1.407(3), and 1.171(3) Å, respectively, and the FCF angle is 108.8(1)°. The CNC group is linear, when proper account is taken of vibrational corrections for distances, and so the molecule has $\underline{C}_{3v}$ symmetry.

In the $\underline{r}_\alpha$ structure of tricyanophosphine[29] the PCN groups are linear, the central CPC angle is 98.9(9)°, and the P-C and C≡N distances are 1.806(4) and 1.162(3) Å, respectively. Attempts to refine an $\underline{r}_a$ structure were not very successful, and the apparent PCN angles varied enormously, depending on what

27. O.G. Batyukhnova, N.I. Sadova, L.V. Vilkov and Yu.A. Pankrushev, Zh.Strukt.Khim., 1984, 25(6), 47.
28. D. Christen, K. Ramme, B. Haas, H. Oberhammer and D. Lentz, J.Chem.Phys., 1984, 80, 4020.
29. V.A. Naumov, V.Yu. Nesterov, V.P. Novikov and S.A. Katsyuba, Zh.Strukt. Khim., 1984, 25(6), 129.

assumption was made about overall symmetry. This example shows well how proper attention to vibrational effects is important in the study of easily deformed molecules.

Vinyldichlorophosphine consists of two conformers in the gas phase at 300 K.[30] The two forms, present in almost equal amounts, had the bisectors of the ClPCl angles lying *syn* or *anti* with respect to the C=C double bond, *i.e.* with the phosphorus lone pair of electrons *anti* or *syn* to the double bond.

In dimethyl(phenyl)phosphine[31] the P-C(methyl) and P-C(phenyl) bond lengths were indistinguishable, at 1.844(5) and 1.845(9) Å (r_g), respectively, but the angles at phosphorus are far from being equal. The two angles C(methyl)PC(phenyl) ($<_\alpha$) are 103.4(10)°, but the remaining angle, C(methyl)PC(methyl), is 96.9(25)°. The preferred orientation of the phenyl group is that in which the ring plane bisects the angle between the two P-C(methyl) bonds, and the barrier to rotation of the phenyl group was estimated to be 5 kJ mol^{-1}.

There are two ring orientations to consider for diphenylphosphine, and in this case a static model (r_g structure) was used.[32] Two structures fit the data equally well, but in each case one ring is twisted 14° and the other 66° (with estimated errors of 4 and 9°), the reference (zero twist) position being that in which a phenyl ring eclipses the further P-C bond. Other parameters are essentially the same for both structures, with r(P-C) 1.831(6) Å and <CPC 100.8(17)°.

The best fit to the data for $MeSPCl_2$ was achieved for a mixture of two conformers, 70% of the form with the *S*-methyl bond *gauche* to the phosphorus lone pair of electrons, and 30% of the *anti* form.[33] Important bond lengths and angles (r_a) are: r(P-Cl) 2.038(5), r(P-S) 2.082(12), r(S-C) 1.795(23) Å, <ClPCl 99.6(12), <ClPS 100.7(6), and <PSC 98.8(25)°. For the analogous bromine compound, $MeSPBr_2$, a similar mixture of *gauche* and *anti* conformers fitted the data well, although a marginally better fit was obtained with 100% *gauche*. Important parameters are: r(P-Br) 2.244, r(P-S) 2.116(16), r(S-C) 1.829(30) Å, <BrPBr 100.6(6),

30. V.A. Naumov, V.Yu. Nesterov and I.A. Aleksandrova, *Zh.Strukt.Khim.*, 1984, 25(4), 137.
31. V.P. Novikov, V.I. Kolomeets, Yu.N. Syshchikov, L.V. Vilkov, A.V. Yarkov, E.N. Tsvetkov and O.A. Raevskii, *Zh.Strukt.Khim.*, 1984, 25(5), 27.
32. V.A. Naumov and O.N. Kataeva, *Zh.Strukt.Khim.*, 1984, 25(4), 140.
33. V.A. Naumov, O.N. Kataeva and O.G. Sinyashin, *Zh.Strukt.Khim.*, 1984, 25(3), 79.

<BrPS 101.3(6), and <PSC 99.3(32)°.

We have only one structure of a four-coordinate phosphorus compound to report. Tris(trifluoromethyl)phosphine oxide[34] has C_3 symmetry, with the CF_3 groups twisted 16° away from the perfectly staggered arrangement. These groups are also said to be tilted 2° *away* from the P=O bond, *i.e.* *towards* each other. The P-C bond length (r_g) is 1.897(4) Å, almost 0.1 Å longer than in PMe_3O. The C-F and P=O distances are 1.336(2) and 1.476(8) Å, and the angles FCF and CPC are 108.5(2) and 104.3(4)°, respectively. The latter corresponds to an OPC angle of 114.2(4)°.

Diaminotrifluorophosphorane[35] has a trigonal-bipyramidal structure with equatorial amino groups, the planes of which are perpendicular to the equatorial plane. Overall symmetry is therefore C_{2v}, but the amino groups undergo torsional oscillations with an r.m.s. amplitude of 8.9(11)° at 320 K, corresponding to a barrier to rotation of *ca.* 55 kJ mol^{-1}. The only significant distortion of the angles at phosphorus from those of a regular trigonal bipyramid is in the equatorial plane. The angle NPN is 127.4(10)°, and the NPF(equatorial) angles are therefore 116.3(5)°.

In recent years several five-membered ring compounds containing phosphorus or arsenic have been studied, and this year there are three more structures to report. In the gas phase 2-chloro-1,3,2-dioxaphospholane[36] adopts an unsymmetrical structure. A group of ring atoms POCC are almost coplanar, while the remaining oxygen atom lies below this plane, and the chlorine atom is above it. Ring angles in the r_α structure are: OPO 96.4(12), POC 110.3(28) and 112.2(18), and OCC 103.4(30) and 110.1(35)°. The two OPCl angles are very close to 100°. The bond lengths are: P-Cl 2.092(4), P-O 1.608(4), C-O 1.440(8), and C-C 1.547(28) Å.

In contrast, both 1-chloro-arsolane and 1-methyl-1,3,2-dithiaarsolane have C_2 symmetry.[37] In the chloro-arsolane the angles at the ring carbon atoms are between 108 and 110° (r_a

34. C.J. Marsden, *Inorg.Chem.*, 1984, *23*, 1703.
35. C.J. Marsden, K. Hedberg, J.M. Shreeve and K.D. Gupta, *Inorg.Chem.*, 1984, *23*, 3659.
36. N.M. Zaripov and R.M. Galiakberov, *Zh.Strukt.Khim.*, 1984, *25*(4), 143.
37. N.M. Zaripov, R.M. Galiakberov, A.V. Golubinskii, L.V. Vilkov and N.A. Chadaeva, *Zh.Strukt.Khim.*, 1984, *25*(4), 25.

structure), and the remaining angle, CAsC, is very small, 88.6(13)°. In this compound the CAsCl angles are 97.5(7)°, and the As-C and As-Cl bond lengths are 1.974(8) and 2.213(4) Å, respectively. However, in the dithiaarsolane there are small ring angles at arsenic [93.8(7)°] and at sulphur [100.5(8)°], and so the angles at carbon are wider [111.2(22)°], and the whole ring is more twisted away from planarity. The As-C, As-S, and C-S bond lengths are 1.968(20), 2.229(2), and 1.805(9) Å, respectively.

6 Compounds of Elements in Main Group VI

The chemical properties of trifluoromethyl hypochlorite, CF_3OCl, are very different from those of CF_3OF, and so its structure has been investigated to see if it sheds any light on the problem.[38] The main difference in the structures is that the angle at oxygen ($<\alpha$) is 112.9(5)° in the chloro-compound, compared to 104.9° in the fluorine analogue. This difference is attributed to the different polarities of the halogen atoms, as chlorine carries a small positive charge, and so is repelled from the carbon atom, whereas the negatively charged fluorine is attracted. The bond lengths ($\underline{r}_g$) are: C-F 1.325(3), C-O 1.365(7), and O-Cl 1.679(3) Å. The FCF angle is 109.2(7)°, and the CF_3 group is tilted 3.9(8)° away from the O-Cl bond.

Dimethyl peroxide[39] has been found to have a non-planar arrangement of the heavy atoms, with the dihedral angle COOC 119(10)°, and a barrier to rotation via the anti conformation of 1 kJ mol^{-1}. This result is consistent with data obtained by infra-red, Raman, and microwave spectroscopy but is in conflict with an interpretation of the photo-electron spectrum. The angle COO in this molecule is 105.2(5)°, and the C-O and O-O bond lengths ($\underline{r}_a$) are 1.420(7) and 1.457(12) Å, respectively.

The compound 3,3-dimethyl-3-silathietane is of interest as having four-membered rings with two different hetero-atoms.[40] The bonds forming the ring are all rather long, Si-C being 1.894(30) Å ($\underline{r}_a$) and S-C 1.854(10) Å. The Si-C(methyl) distance is 1.879(30) Å. The ring is not planar, but the angle between

38. H. Oberhammer, T. Mahmood and J.M. Shreeve, J.Mol.Struct., 1984, 117, 311.
39. B. Haas and H. Oberhammer, J.Am.Chem.Soc., 1984, 106, 6146.
40. V.S. Mastryukov, S.A. Strelkov, A.V. Golubinskii, L.V. Vilkov, S.V. Kirpichenko, E.N. Suslova and M.G. Voronkov, Dokl.Akad.Nauk S.S.S.R., 1983, 271, 384.

the CSiC and CSC planes is only 13.2(32)°, compared with an equivalent angle of 34° in silacyclobutane and 26° in thietane. Angles within the ring are all close to 90°, the smallest being <CSiC, which is 87.0(8)°, and the angle C(methyl)SiC(methyl) is also rather narrow, at 105.1(30)°.

Independent studies of two closely related compounds, *o*-nitrophenylsulphenyl chloride[41] and *o*-nitrophenylselenyl bromide,[42] have been reported. In both cases all the atoms are essentially co-planar, there is a short interaction between one oxygen atom of the nitro group and the adjacent sulphur or selenium atom (O...S 2.41, O...Se 2.36 Å), and the O...S/Se-Cl/Br groups of atoms are colinear. Further evidence of attractive interactions is provided by the observations that the C-N and C-S or C-Se bonds are bent towards each other, in each case by about 1.5° from the position that is symmetrical with respect to the ring. In the -SCl compound important bond lengths (r_a) and angles are: *r*(S-Cl) 2.046(4), *r*(S-C) 1.776(5), *r*(N-C) 1.459(9), *r*(N=O) 1.229(3) Å and <ClSC 102.3(14). The two CNO angles (which are not forced to be equal) are 119.4(28) and 115.5(44)°. Main parameters for the -SeBr compound are: *r*(Se-Br) 2.354(3), *r*(Se-C) 1.917(11), *r*(N-C) 1.466(24), *r*(N=O) 1.229(8) Å, <BrSeC 98.5(7) and <CNO(mean) 117.6(6)°.

Methyl(phenyl)sulphone is the latest in a long series of sulphones to be studied by electron diffraction.[43] At 440 K it adopts a conformation in which the ring plane lies perpendicular to a projection of the S-C(methyl) bond, and the molecule has C_s symmetry. In the r_a structure the S-C(methyl) and S-C(phenyl) bond lengths are 1.784(8) and 1.770(8) Å, *r*(S=O) is 1.444(1) Å, and the angles at sulphur, OSC(methyl), OSC(phenyl), CSC, and OSO, are 108.6(4), 108.2(3), 103.9(5), and 118.4(4)°, respectively.

In the trigonal-bipyramidal molecule $S(CF_3)_2F_2O$ the CF_3 groups and the oxygen atom occupy the equatorial positions,[44] as predicted by electron-pair repulsion theory. The CF_3 groups and fluorine atoms are displaced from their positions in a regular trigonal-bipyramidal structure, away from the oxygen atom, so

41. G. Schultz, I. Hargittai, K. Kapovits and A. Kucsman, J.Chem.Soc., Faraday Trans.II, 1984, 80, 1273.
42. N.M. Zaripov, A.V. Golubinskii, C.V. Solokov, L.V. Vilkov and T.P. Mannapov, Dokl.Akad.Nauk S.S.S.R., 1984, 278, 664.
43. J. Brunvoll, O. Exner, I. Hargittai, M. Kolonits and P. Scharfenberg, J.Mol.Struct., 1984, 117, 317.
44. H. Oberhammer, J.M. Shreeve and G.L. Gard, Inorg.Chem., 1984, 23, 2820.

that <CSC is only 97.8(8)° and <FSF is 173.1(6)°. The FCF angle is 109.4(3)°, and the bond lengths (r_a) are: C-F 1.314(3), S-F 1.422(7), S-C 1.891(5), and S=O 1.422(7) Å.

Tetrafluorosulphur chloroimide is dimeric in the gas phase,[45] and consists of planar four-membered rings, in which six-coordinate sulphur atoms alternate with nitrogen atoms, each of which is bound to chlorine as well as to two sulphur atoms. The two chlorine atoms lie above and below the ring plane at angles of 34.8(14)° to the plane, so that the molecule has overall C_{2h} symmetry. Strain in the molecule is indicated by the SNS angle, 99.3(6)°, the NSN angle, 80.7(6)°, and the exceptionally large S-N bond length, 1.734(4) Å (r_a). The sulphur atoms show further distortion from regular octahedral coordination in that the bonds to axial fluorines (i.e. those not lying in the ring plane) are 1.590(6) Å long, whereas the other S-F bonds are only 1.545(5) Å long. The axial fluorines are bent away from the ring, with <FSF 176.4(8)°. Despite the narrow SNS angle the sum of angles at nitrogen is 344°, with <SNCl 122.1(7)°. The N-Cl distance is 1.638(10)°, which is very much less than the 1.759 Å reported for NCl_3.

7 Compounds of Transition Metals

Simple metal dihalides have been of interest for some time, as it is often difficult to tell whether they are linear or bent. Nickel difluoride has been studied at 1470 K.[46] The Ni-F bond length (r_g) is 1.729(4) Å, and the observed F...F distance is 3.400(34), corresponding to an effective FNiF angle of 159(7)°. The equilibrium structure of this molecule is probably linear, and the apparent shrinkage of the F...F distance is caused by the large-amplitude, low-frequency bending vibration. In contrast, samarium diodide is definitely bent.[47] The Sm-I distance (r_a) is 2.998(6) Å and the I...I distance is 5.371(11) Å, and these give an angle ISmI of 127(2)°. In this case the so-called harmonic equilibrium structure has been determined, and the angle in this structure is 138(10)°, with a bond length of 2.944(7) Å. In this

45. H. Oberhammer, A. Waterfeld and R. Mews, Inorg.Chem., 1984, 23, 415.
46. G.V. Girichev, N.Yu. Subbotina, K.S. Krasnov and V.V. Ostropikov, Zh.Strukt.Khim., 1984, 25(3), 1702.
47. E.V. Erokhin, V.P. Spiridonov, A.G. Gershikov, N.I. Raevskii and Yu. M. Kiselev, Zh.Strukt.Khim., 1984, 25(3), 75.

second form of the refinement force constants were determined and from these vibrational frequencies were calculated. The frequency of the bending mode was thus found to be 26 cm^{-1}.

The structure of iron dichloride has been the subject of debate. Electron-diffraction data have implied that the equilibrium structure is linear, whereas vibrational spectra of the compound isolated in inert matrices have been interpreted in terms of a bent structure, with $<$ClFeCl 160°. New calculations[48] show that the gas-phase structure is indeed linear, and it is concluded that the molecules are deformed by the matrix.

Tungsten tetrachloride has a distorted trigonal-bipyramidal structure with C_{2v} symmetry,[49] and in this respect it is similar to the tetrahalides of main group VI elements. The axial bond length, 2.295(6) Å (r_g), is greater than the equatorial bond length, 2.202(6) Å, and all the bonds are bent away from the vacant site. The angle Cl(ax)WCl(ax) is 170(10)°, and $<$Cl(eq)WCl(eq) is 90(4)°.

Chromium pentafluoride is formally derived from the d^1 configuration of Cr^{5+} and as such may undergo Jahn-Teller distortion, lowering the symmetry of the molecule from D_{3h}. Electron-diffraction data[50] recorded at 350 K show that a D_{3h} model does not fit the data so well as one with C_{2v} symmetry, with the axial fluorine atoms displaced away from one of the equatorial atoms, and the remaining equatorial atoms displaced towards the unique one. In this C_{2v} model the angles in the equatorial plane were 128.1(18) and 115.9(9)°, while the F(ax)CrF(ax) angle was 168.4(8)°. The axial and equatorial bond lengths (the latter assumed to be all equal) were 1.742(10) and 1.695(6) Å, respectively, (r_g). However, an even better fit was obtained with a model which does not correspond to any rigid structure, but just involves one Cr-F bond length [1.714(2) Å] and three F...F distances, of 2.411(6), 2.878(22), and 3.470(26) Å. Thus the odd electron in the valence shell of the chromium atom probably results in dynamic Jahn-Teller distortions, which on average resemble a static C_{2v} structure. It is interesting therefore that the data for molybdenum pentachloride at 390 K[51] have been interpreted in terms of a

48. M. Hargittai and I. Hargittai, J.Mol.Spectrosc., 1984, 108, 155.
49. Yu.S. Ezhov and S.A. Komarov, Zh.Strukt.Khim., 1984, 25(1), 82.
50. E.J. Jacob, L. Hedberg, K. Hedberg, H. Davis and G. Gard, J.Phys.Chem., 1984, 88, 1935.
51. J. Brunvoll, A.A. Ischenko, V.P. Spiridonov and T.G. Strand, Acta Chem.Scand., Ser.A, 1984, 38, 115.

mixture of square-pyramidal molecules [$\underline{C}_{4\underline{v}}$ symmetry, 51(4)%], trigonal-bipyramidal molecules [$\underline{D}_{3\underline{h}}$ symmetry, 26(5)%], and dimers with $\underline{D}_{4\underline{d}}$ symmetry [24(3)%]. However, a satisfactory fit to the data could also be obtained with 56% $\underline{C}_{4\underline{v}}$, 44% $\underline{D}_{3\underline{h}}$, and no dimer. The possibility of Jahn-Teller distortion was not considered in this case, and it may well be that $MoCl_5$ is also rapidly changing its conformation. In these circumstances the parameters are not very meaningful, and the original paper should be consulted for details.

In contrast, tetrafluorosulphidotungsten, WF_4S, is simple.[52] The molecules, studied at ca. 500 K, have $\underline{C}_{4\underline{v}}$ symmetry, with $\underline{r}_a$(W=S) 2.104(7), $\underline{r}$(W-F) 1.847(3) Å, and <SWF 104.5(11)°. The angle FWF is therefore 86.6(5)°. The structure is thus a square pyramid, with the tungsten atom well above the plane of the four fluorine atoms.

Copper(II) nitrate has been reinvestigated,[53] at ca. 420 K, to resolve previously conflicting results concerning the coordination of the nitrate groups around the copper atom. When dissociation (70%) to give NO_2 and O_2 was taken into account it was found that the copper has square-planar coordination, and all the atoms are coplanar. THe Cu-O distance ($\underline{r}_a$) is 1.946(3) Å, the distance from nitrogen to the coordinating oxygen atoms is 1.298(3) Å, and the terminal N-O distance is 1.205(5) Å. The great difference between Cu-O and N-O distances in the ring means that the angles are also very different, with <ONO 113.5(6)°, but <OCuO 67.8(2)°. The angles at oxygen are very close to 90°.

Finally we come to four zinc compounds, the only transition-metal organometallic compounds to be included in this year's review. In both bis-3-methoxypropyl zinc and bis-4-methoxybutylzinc, $Zn[(CH_2)_{\underline{n}}OCH_3]$ ($\underline{n}$=3 or 4), at ca. 350 K, the chains are coiled back to allow intramolecular Zn-O bonding.[54] These extra interactions are not very strong, as the Zn-O distances ($\underline{r}_a$) in the two compounds are 2.37(3) and 2.38(5) Å, compared with 1.974(4) and 1.984(6) Å for the Zn-C bond lengths. The Zn-C bonds are not very greatly distorted from the linear arrangement, the CZnC angles being 175(20) and 158(13)°, and the

52. D.A. Rice, K. Hagen, L. Hedberg, K. Hedberg, G.M. Staunton and J.H. Holloway, Inorg.Chem., 1984, 23, 1826.
53. S. Shibata and K. Iijima, J.Mol.Struct., 1984, 117, 45.
54. J. Boersma, L. Fernholt and A. Haaland, Acta Chem.Scand., Ser.A, 1984, 38, 523.

two interacting oxygen atoms are fairly close together, with angles OZnO 111(20)° and 92(10)°. With such complicated molecules it is impossible to make definitive statements about conformations, and the conformational parameters quoted are simply those for the best-fitting model in each case. Both structures have $\underline{C}_2$ symmetry. For n=3 the chelate rings are fairly flat, but for n=4 they are strongly buckled, presumably reflecting the extra degrees of freedom available in making the Zn-O contacts.

Similar interactions occur between zinc and C=C double bonds in at least 80% of the molecules of dibut-3-enyl zinc and dipent-4-enyl zinc, $Zn[(CH)_2)_{\underline{n}}\ CH{=}CH_2]_2$ ($\underline{n}$=2 or 3).[55] Again the chains are coiled back to allow the extra interactions, the energies of which are estimated to be 4.5 and 6 kJ mol^{-1} in the two cases. The shortest Zn-C($\underline{sp}^2$) distance is 3.15(6) Å for $\underline{n}$ = 3 and 3.00(8) Å for $\underline{n}$=4, the longer chain in the pentenyl compound allowing stronger interaction to occur. In both compounds the Zn-C distances ($\underline{r}_a$) are 196 pm, the CZnC angles are not significantly changed from 180°, and the ZnCC angles are 116°.

55. A. Haaland, H. Lehmkuhl and H. Nehl, Acta Chem.Scand., Ser.A, 1984, 38, 547.